Horst D. Schulz
Matthias Zabel
Marine Geochemistry
2nd revised, updated and extended edition

Horst D. Schulz
Matthias Zabel
Editors

Marine Geochemistry

2nd revised, updated and extended edition

With 303 Figures, 49 in color

 Springer

Professor Dr. Horst D. Schulz
Universität Bremen, FB 5
Geowissenschaften
Klagenfurter Straße
28359 Bremen
Germany

Dr. Matthias Zabel
Universität Bremen, FB 5
Geowissenschaften
Klagenfurter Straße
28359 Bremen
Germany

Library of Congress Control Number: 2006920339

ISBN-10   3-540-32143-8 Springer Berlin Heidelberg New York
ISBN-13   978-3-540-32143-9 Springer Berlin Heidelberg New York

ISBN   3-540-66453-X (1st edition) Springer Berlin Heidelberg New York

Springer is a part of Springer Science+Business Media
Springeronline.com
© Springer-Verlag Berlin Heidelberg 1999, 2006
Printed in Germany

Cover design: E. Kirchner, Heidelberg
Typesetting: camera-ready by editors
Production: Almas Schimmel
Printing and binding: Stürtz AG, Würzburg

Printed on acid-free paper   32/3141/as 5 4 3 2 1 0

PREFACE 2ND EDITION

Almost six years have passed since the launching of the first edition of this textbook. The friendly acceptance and brisk demand of the "Marine Geochemistry" have encouraged us to prepare the second edition. Six years mean a comparatively long period for a relatively young discipline such as marine geochemistry. Data and measurements from numerous expeditions, results of different new methods and techniques, as well as findings of various research programs justify a careful revision of the first edition.

The authors of the first edition have revised and substantially updated and enlarged their chapters. We had to give up the partial chapter on sedimentary magnetism of the first edition. In chapter 6 the authors now concentrate on benthic cycles, and the new co-author Heide N. Schulz especially reports on phosphorus cycles and the microbial parts. Chapter 8 was extended to the complete marine sulfur cycling in connection with anaerobic methane oxidation. Gerhard Bohrmann and Marta E. Torres contribute their completely new chapter 14 on methane hydrates in marine sediments, representing a well-rounded presentation of this exciting discipline, which will be of major interest in the future.

We also had to consider that other new textbooks (Boudreau and Jørgensen 2001, Fasham 2003, Zhu and Anderson 2002) are now available, whereas revised editions of books (Chester 2000, 2003, Hoefs 2004) were published.

Horst D. Schulz and Matthias Zabel   Bremen, November 2005

References

Anderson, R.N., 1986. Marine geology - A planet Earth perspective. Wiley & Sons, NY, 328 pp.
Berner, R.A., 1980. Early diagenesis: A theoretical approach. Princeton Univ. Press, Princeton, NY, 241 pp.
Boudreau, B.P., 1997. Diagenetic models and their impletation: modelling transport and reactions in aquatic sediments. Springer-Verlag, Berlin, Heidelberg, NY, 414 pp.
Boudreau, B.P. and Jørgensen, B.B., 2001. The Benthic Boundary Layer. Oxford University Press, 404 pp.
Broecker, W.S. and Peng, T.-H., 1982. Tracer in the Sea. Lamont-Doherty Geol. Observation Publ., 690 pp.
Chester, R., 1990. Marine Geochemistry. Chapman & Hall, London, 698 pp.
Chester, R., 2000, 2003. Marine Geochemistry. Blackwell Science Ltd, Oxford, 2nd ed., pp. 506.
Clark, I. and Fritz, P., 1997. Environmental isotopes in hydrogeology. Lewis Publ., NY, 328 pp.
Engel, M.H. and Macko, S.A., 1993. Organic Geochemistry. Plenum Press, 861 pp.
Faure, G., 1986. Principles of Isotope Geology. Wiley & Sons, NY, 589 pp.
Grasshoff, K., Kremling K. and Ehrhardt, M., 1999. Methods of Seawater Analysis. Wiley-VCH, Weinheim, NY, 3rd ed., 600 pp.
Hoefs, J., 1997. Stable Isotope Geochemistry. Springer, Berlin Heidelberg NY, 4th ed., 201 pp.
Hoefs, J., 2004. Stable Isotope Geochemistry. Springer, Berlin Heidelberg NY, 5th ed., 244 pp.
Kennett, J.P., 1982. Marine Geology. Prentice Hall, New Jersey, 813 pp.
Stumm, W. and Morgan, J.J., 1996. Aquatic Chemistry. Wiley & Sons, 1022 pp.
Zhu, C. and Anderson, G., 2002. Environmental Applications of Geochemical Modelling. Cambridge University Press, 284 pp.

PREFACE 1ST EDITION

Today, most branches of science have been extensively described. As to their objectives and interrelationships they are also well distinguished from the adjoining science disciplines. In this regard, marine geochemistry makes an exception, as does geochemistry in general, for - depending on the scientist's educational and professional career - this particular field of research can be understood more or less in terms of geology, chemistry, biology, even mineralogy or oceanography. Despite some occasional objection, we conceive our interdisciplinary approach to marine geochemistry rather as an opportunity - even if our own origins most certainly lie in the geosciences. R. Chester (1990) emphasized the chemistry of the water column and the relations to continental inputs in his book on "Marine Geochemistry". For us, however, the investigation of the marine surface sediments and the (bio)geochemical processes taking place therein will be of major concern. We therefore see our book as a continuation of what R.A. Berner (1980) initiated with his classical work "Early Diagenesis", with which he had a determining influence in pointing the way ahead. The concept and the contents we present here is addressed to graduated students of earth science who specialize in marine geochemistry.

Before the background of a continually expanding field of research, it appears impossible for a textbook on marine geochemistry to cope with the task of achieving completeness. Some parts of marine geochemistry have been described in more detail elsewhere and with an emphasis placed on a different context. These parts were permitted to be treated in brief. The classical subjects of "Marine Geology" are therefore to be found in J.P. Kennet (1982), or with a different perspective in R.N. Andersen (1986). No book on "Aquatic Chemistry" can hardly be better than the one written by W. Stumm and J.J. Morgan (1996). There is furthermore the textbook written by K. Grasshoff, K. Kremling, and M. Ehrhardt (1999) which is concerned with the analytical measurements in seawater. "Tracers in the Sea" by W.S. Broecker and T.-H. Peng (1982) still remains an essential work of standard, albeit a new edition is dearly awaited for. The important field of "Isotope Geochemistry" is exquisitely represented by the books written by G. Faure (1986), Clark and Fritz (1997) and J. Hoefs (1997), as much as "Organic Geochemistry" is represented by the book published by M.H. Engel and S.A. Macko (1993). "Diagenetic Models and Their Implementation" are described from the perspective of a mathematician by B.P. Boudreau (1997), owing to which we were able to confine ourselves to the geochemist's point of view.

Marine geochemistry is generally integrated into the broad conceptual framework of oceanography which encompasses the study of the oceanic currents, their interactions with the atmosphere, weather and climate; it leads from the substances dissolved in water, to the marine flora and fauna, the processes of plate tectonics, the sediments at the bottom of the oceans, and thus to marine geology. Our notion of marine geochemistry is that it is a part of marine geology, wherefore we began our book with a chapter on the solid phase of marine sediments concerning its composition, development and distribution. The first chapter written by Dieter K. Fütterer is therefore a brief summarizing introduction into marine geology which describes all biochemically relevant aspects related to the subject. Monika Breitzke and Ulrich Bleil are concerned in the following chapter with the physical properties of sediments and sedimentary magnetism in a marine-geophysical context, which we deem an important contribution to our understanding of geochemical processes in the sediment.

In the third chapter, Horst D. Schulz demonstrates that the method to quantify biogeochemical processes and material fluxes in recent sediments affords the analysis of the pore-water fraction. Jürgen Rullkötter subsequently gives an overview of organic material contained in sediments which ultimately provides the energy for powering almost all (bio)geochemical reactions in that compartment. The fifth chapter by Bo. B. Jørgensen surveys the world of microorganisms and their actions in marine sediments. The chapters six, seven, and eight are placed in the order of the oxidative agents that are involved in the oxidation of sedimentary organic matter: oxygen and nitrate (Christian Hensen and Matthias Zabel), iron (Ralf R. Haese) and sulfate (Sabine Kasten and Bo. B. Jørgensen). They close the circle of primary reactions that occur in the early diagenesis of oceanic sediments.

In the ninth chapter, marine carbonates are dealt with as a part of the global carbon cycle which essentially contributes to diagenetic processes (Ralf R. Schneider, Horst D. Schulz and Christian Hensen). Ratios of stable isotopes are repeatedly used as proxy-parametes for reconstructing the paleoclimate and paleoceanography. Hence, in the succeeding chapter, Torsten Bickert discusses the stable isotopes in the marine sediment as well as the processes which bear influence on them. Geoffrey P. Glasby has dedicated a considerable part of his scientific work to marine manganese. He visited Bremen several times as a guest scientist and therefore it goes without saying that we seized the opportunity to appoint him to be the author of the chapter dealing with nodules and crusts of manganese. Looking at the benthic fluxes of dissolved and solid/particulate substance across the sediment/water interface, the processes of early diagenesis contribute primarily to the material budgets in the world's oceans and thus to the global material cycles. In chapter 12, Matthias Zabel, Christian Hensen and Michael Schlüter have ventured to make initial methodological observations on global interactions and balances, a subject which is presently in a state of flux. A summarizing view on hot vents and cold seeps is represented by a chapter which is complete in itself . We are indebted to Peter M. Herzig and Mark D. Hannington for having taken responsibility in writing this chapter, without which a textbook on marine geochemistry would always have remained incomplete. In the final chapter, conceptual models and their realization into computer models are discussed. Here, Horst D. Schulz is more concerned with the biogeochemical processes, their proper comprehension, and with aspects of practical application rather than the demonstration of ultimate mathematical elegance.

This book could only be written because many contributors have given their support. First of all, we have to mention the Deutsche Forschungsgemeinschaft in this regard, which has generously funded our research work in the Southern Atlantic for over ten years. This special research project [Sonderforschungsbereich, SFB 261] entitled "*The South Atlantic in the Late Quaternary: Reconstruction of Material Budget and Current Systems*", covered the joint activities of the Department of Geosciences at the University Bremen, the Alfred-Wegener-Institute for Polar and Marine Research in Bremerhaven, and the Max-Planck-Institute for Marine Microbiology in Bremen.

The marine-geochemical studies are closely related to the scientific publications submitted by our colleagues from the various fields of biology, marine chemistry, geology, geophysics, mineralogy and paleontology. We would like to thank all our colleagues for the long talks and discussions we had together and for the patient understanding they have showed us as geochemists who were not at all times familiar with the numerous particulars of the neighboring sciences.

We were fortunate to launch numerous expeditions within the course of our studies in which several research vessels were employed, especially the RV METEOR. We owe gratitude to the captains and crews of these ships for their commitment and services even at times when duty at sea was rough. All chapters of this book were subject to an international process of review. Although all colleagues involved have been mentioned elsewhere, we would like to express our gratitude to them once more at this point. Owing to their great commitment they have made influential contributions as to contents and character of this book.

Last but not least, we wish to thank our wives Helga and Christine. Although they never had the opportunity to be on board with us on any of the numerous and long expeditions, they still have always understood our enthusiasm for marine geochemistry and have always given us their full support.

Horst D. Schulz and Matthias Zabel   Bremen, May 1999

Acknowledgements

Along with the authors we are deeply indebted to the reviewers of the first and the second edition. Their helpful comments have considerably improved the contents of this book:

David E. Archer, University of Chicago, USA; Wolfgang H. Berger, Scripps Institution of Oceanography, La Jolla, USA; Ray Binns, CSIRO Exploration & Mining, North Ryde, Australia; Walter S. Borowski, Exxon Exploration Company, Housten, USA; Timothy G. Ferdelman, Max-Planck-Institut für marine Mikrobiologie, Bremen, Germany; Henrik Fossing, National Environmental Research Institute, Silkeborg, Denmark; John M. Hayes, Woods Hole Oceanographic Institution, USA; Bo B. Jørgensen, Max-Planck-Institut für marine Mikrobiologie, Bremen, Germany; David E. Gunn, Southampton Oceanography Centre, United Kingdom; Richard A. Jahnke, Skidaway Institute of Oceanography, Savannah, USA; Karin Lochte, Institut für Ostseeforschung Warnemünde, Germany; Philip A. Meyers, University of Michigan, USA; G.M. McMurtry, University of Hawaii; Jack Middelburg, Netherland Institute of Ecology, Yerske, Netherlands; C.L. Morgan, Planning Solutions, Hawaii; Nikolai Petersen, Universität München, Germany; Christophe Rabouille, Unité Mixte de Recherche CNRS-CEA, Gif-sur-Yvette, France; Jürgen Rullkötter, Universität Oldenburg, Germany; Graham Shimmield, Dunstaffnage Marine Laboratory, Oban, United Kingdom; Ulrich von Stackelberg, Bundesanstalt für Geowissenschaften und Rohstoffe (BGR), Hannover, Germany; Doris Stüben, Universität Karlsruhe, Germany; Bo Thamdrup, Odense Universitet, Denmark; Cornelis H. van der Weijden, Utrecht University, Netherland; John K. Volkman, CSIRO, Hobart, Tasmania, Australia.

We would also like to express our gratitude to numerous unnamed support staff at the University of Bremen.

Table of Contents

1   The Solid Phase of Marine Sediments

DIETER K. FÜTTERER

2   Physical Properties of Marine Sediments

MONIKA BREITZKE

3 Quantification of Early Diagenesis: Dissolved Constituents in Marine Pore Water

HORST D. SCHULZ

4 Organic Matter: The Driving Force for Early Diagenesis

JÜRGEN RULLKÖTTER

5   Bacteria and Marine Biogeochemistry

Bo Barker Jørgensen

6 Benthic Cycling of Oxygen, Nitrogen and Phosphorus

CHRISTIAN HENSEN, MATTHIAS ZABEL AND HEIDE N. SCHULZ

7 The Reactivity of Iron

RALF R. HAESE

8 Sulfur Cycling and Methane Oxidation

Bo Barker Jørgensen and Sabine Kasten

9 Marine Carbonates: Their Formation and Destruction

Ralph R. Schneider, Horst D. Schulz and Christian Hensen

10 Influences of Geochemical Processes on Stable Isotope Distribution in Marine Sediments

TORSTEN BICKERT

11 Manganese: Predominant Role of Nodules and Crusts

GEOFFREY P. GLASBY

12 Quantification and Regionalization of Benthic Refluxe

MATTHIAS ZABEL AND CHRISTIAN HENSEN

13 Input from the Deep: Hot Vents and Cold Seeps

PETER M. HERZIG AND MARK D. HANNINGTON

14   Gas Hydrates in Marine Sediments

GERHARD BOHRMANN AND MARTA E. TORRES

15   Conceptual Models and Computer Models

HORST D. SCHULZ

Answers to Problems

Index

Authors

Torsten Bickert   Universität Bremen, Fachbereich Geowissenschaften, 28359 Bremen, Germany, bickert@allgeo.uni-bremen.de

Gerhard Bohrmann   Universität Bremen, Fachbereich Geowissenschaften, 28359 Bremen, Germany, gbohrmann@uni-bremen.de

Monika Breitzke   Alfred-Wegener-Institut für Polar- und Meeresforschung, 27515 Bremerhaven, Germany, mbreitzke@awi-bremerhaven.de

Dieter K. Fütterer   Alfred-Wegener-Institut für Polar- und Meeresforschung, 27515 Bremerhaven, Germany, dfuetterer@awi-bremerhaven.de

Geoffrey P. Glasby   Universität Bremen, Fachbereich Geowissenschaften, 28359 Bremen, Germany, g.p.glasby@talk21.com

Ralf R. Haese   Marine and Coastal Environment Group Geoscience Australia, Canberra ACT 2601 (Australia), Ralf.Haese@ga.gov.au

Mark. D. Hannington   Geological Survey of Canada, Ottawa, Canada, markh@gsc.NRCan.gc.ca

Christian Hensen   IFM – GEOMAR, Leibniz-Institut für Meereswissenschaften an der Universität Kiel, 24105 Kiel, Germany, chensen@ifm-geomar.de

Peter M. Herzig   IFM – GEOMAR, Leibniz-Institut für Meereswissenschaften an der Universität Kiel, 24105 Kiel, Germany, pherzig@ifm-geomar.de

Bo Barker Jørgensen   Max-Planck-Institut für marine Mikrobiologie, 28359 Bremen, Germany, bjoergen@mpi-bremen.de

Sabine Kasten   Alfred-Wegener-Institut für Polar- und Meeresforschung, 27515 Bremerhaven, Germany, skasten@awi-bremerhaven.de

Jürgen Rullkötter   Universität Oldenburg, Institut für Chemie und Biologie des Meeres, 26111 Oldenburg, Germany, j.rullkoetter@ogc.icbm.uni-oldenburg.de

Ralph R. Schneider   Institut für Geowissenschaften, Christian-Albrechts-Universität zu Kiel, 24118 Kiel, Germany, schneider@gpi.uni-kiel.de

Heide N. Schulz   Universität Hannover, Institut für Mikrobiologie, 30167 Hannover, Germany, schulz@ifmb.uni-hannover.de

Horst D. Schulz   Universität Bremen, Fachbereich Geowissenschaften, 28359 Bremen, Germany, hdschulz@uni-bremen.de

Marta E. Torres   College of Oceanic and Atmospheric Sciences, Oregon State Univ., Corvallis OR 97331-5503, USA, mtorres@coas.oregonstate.edu

Matthias Zabel   Universität Bremen, Fachbereich Geowissenschaften, 28359 Bremen, Germany, mzabel@uni-bremen.de

1 The Solid Phase of Marine Sediments

Dieter K. Fütterer

1.1 Introduction

The oceans of the world represent a natural depository for the dissolved and particulate products of continental weathering. After its input, the dissolved material consolidates by means of biological and geochemical processes and is deposited on the ocean floor along with the particulate matter from weathered rock. The ocean floor deposits therefore embody the history of the continents, the oceans and their pertaining water masses. They therefore provide the key for understanding Earth's history, especially valuable for the reconstruction of past environmental conditions of continents and oceans. In particular, the qualitative and quantitative composition of the sedimentary components reflect the conditions of their own formation. This situation may be more or less clear depending on preservation of primary sediment composition, but the processes of early diagenesis do alter the original sediment composition, and hence they alter or even wipe out the primary environmental signal. Hence, only an entire understanding of nature and sequence of processes in the course of sediment formation and its diagenetic alteration will enable us to infer the initial environmental signal from the altered composition of the sediments.

Looking at the sea-floor sediments from a geochemical point of view, the function of particles, or rather the sediment body as a whole, i.e. the solid phase, can be quite differently conceived and will vary with the perspective of the investigator. The "classical" approach – simply applying studies conducted on the continents to the oceans – usually commences with a geological-sedimentological investigation, whereafter the mineral composition is recorded in detail. Both methods lead to a more or less overall geochemical description of the entire system.

Another, more modern approach conceives the ocean sediments as part of a global system in which the sediments themselves represent a variable component between original rock source and deposition. In such a rather process-related and globalized concept of the ocean as a system, sediments attain special importance. First, they constitute the environment, a solid framework for the geochemical reactions during early diagenesis that occur in the pore space between the particles in the water-sediment boundary layer. Next to the aqueous phase, however, they are simultaneously starting material and reaction product, and procure, together with the porous interspaces, a more or less passive environment in which reactions take place during sediment formation.

1.2 Sources and Components of Marine Sediments

Ocean sediments are heterogeneous with regard to their composition and also display a considerable degree of geographical variation. Due to the origin and formation of the components various sediment types can be distinguished: Lithogenous sediments which are transported and dispersed into the ocean as detrital particles, either as terrigenous particles – which is most frequently the case – or as volcanogenic particles having only local importance; biogenous sediments which are directly produced by organisms or are formed by accumulation of skeletal fragments; hydrogenous or authigenic sediments which precipitate directly out of solution as new formations, or are formed de novo when the particles come into contact with the solution; finally, cosmogenic sediments which are only of secondary importance and will therefore not be considered in the following.

1

1.2.1   Lithogenous Sediments

The main sources of lithogenous sediments are ultimately continental rocks which have been broken up, crushed and dissolved by means of physical and chemical weathering, exposure to frost and heat, the effects of water and ice, and biological activity. The nature of the parent rock and the prevalent climatic conditions determine the intensity at which weathering takes place. Information about these processes can be stored within the remnant particulate weathered material, the terrigenous detritus, which is transported by various routes to the oceans, such as rivers, glaciers and icebergs, or wind. Volcanic activity also contributes to lithogenous sediment formation, however, to a lesser extent; volcanism is especially effective on the active boundaries of the lithospheric plates, the mid-ocean spreading ridges and the subduction zones.

The major proportion of weathered material is transported from the continents into the oceans by rivers as dissolved or suspension load, i.e. in the form of solid particulate material. Depending on the intensity of turbulent flow suspension load generally consists of particles smaller than 30 microns, finer grained than coarse silt. As the mineral composition depends on the type of parent rock and the weathering conditions of the catchment area, it will accordingly vary with each river system under study. Furthermore, the mineral composition is strongly determined by the grain-size distribution of the suspension load. This can be seen, for example, very clearly in the suspension load transported by the Amazon River (Fig. 1.1) which silt fraction (> 4 - 63 µm) predominantly consists of quartz and feldspars, whereas mica, kaolinite, and smectite predominate in the clay fraction.

It is not easy to quantify the amount of suspension load and traction load annually discharged by rivers into the oceans on a worldwide scale. In a conservative estimative approach which included 20 of the probably largest rivers, Milliman and Meade (1983) extrapolated this amount to comprise approximately $13 \cdot 10^9$ tons. Recent estimations (Milliman and Syvitski 1992) which included smaller rivers flowing directly into the ocean hold that an annual discharge of approximately $20 \cdot 10^9$ tons might even exist.

Under the certainly not very realistic assumption of an even distribution over a surface area of $362 \cdot 10^6$ km² which covers the global ocean floor, this amount is equivalent to an accumulation rate of 55.2 tons km^{-2} y^{-1}, or the deposition of an approximately 35 mm-thick sediment layer every 1000 years.

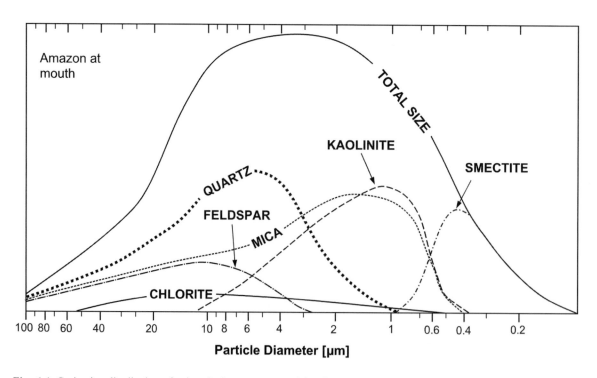

Fig. 1.1 Grain-size distribution of mineral phases transported by the Amazon River (after Gibbs 1977).

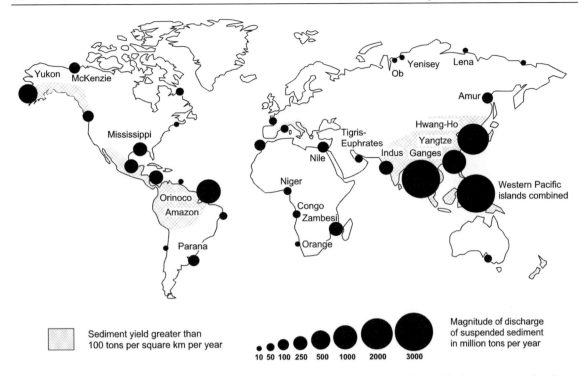

Fig. 1.2 Magnitude of annual particulate sediment discharge of the world's major rivers. The huge amount of sediment discharge in southeast Asia and the western Pacific islands is due to high relief, catchment, precipitation and human activity (Hillier 1995).

Most of the sediment transported to the coastline by the rivers today is deposited on protected coastal zones, in large estuaries, and on the shelves; only a rather small proportion of the sediment is transported beyond the shelf edge and reaches the bottom of the deep sea. The geographical distribution of the particulate discharge varies greatly worldwide, depending on the geographical distribution of the respective rivers, amount and concentration of the suspended material. According to Milliman and Syvitski (1992), the amount of suspension load is essentially a function of the surface area and the relief of the catchment region, and only secondarily does it depend on the climate and the water mass of the rivers. Apart from these influences, others like human activity, climate, and geological conditions are the essential factors for river systems in southeast Asia.

The southeast Asian rivers of China, Bangla Desh, India, and Pakistan that drain the high mountain region of the Himalayan, and the rivers of the western Pacific islands (Fig. 1.2), transport just about one half of the global suspension load discharged to the ocean annually. This must naturally also exert an effect on the sedimentation

rates in the adjoining oceanic region of the Indo-Pacific.

Sediment transport by icebergs which calve from glaciers and inland ice into the ocean at polar and subpolar latitudes is an important process for the discharge and dispersal of weathered coarse grained terrigenous material over vast distances. Due to the prevailing frost weathering in nival climate regions, the sedimentary material which is entrained by and transported by the ice is hardly altered chemically. Owing to the passive transport via glaciers the particles are hardly rounded and hardly sorted in fractions, instead, they comprise the whole spectrum of possible grain sizes, from meter thick boulders down to the clay-size fraction.

As they drift with the oceanic currents, melting icebergs are able to disperse weathered terrigenous material over the oceans. In the southern hemisphere, icebergs drift from Antarctica north to 40°S. In the Arctic, the iceberg-mediated transport is limited to the Atlantic Ocean; here, icebergs drift southwards to 45°N, which is about the latitude of Newfoundland. Coarse components released in the process of disintegration and melting leave behind "ice rafted detritus" (IRD), or "drop stones", which represent characteristic signals in the sediments and are of extremely high importance in paleoclimate reconstructions.

In certain regions, the transport and the distribution carried out by sea ice are important processes. This is especially true for the Arctic Ocean where specific processes in the shallow coastal areas of the Eurasian shelf induce the ice, in the course of its formation, to incorporate sediment material from the ocean floor and the water column. The Transpolar Drift distributes the sediment material across the Arctic Ocean all the way to the North Atlantic. Glacio-marine sedimentation covers one-fifth of present day's ocean floor (Lisitzin 1996).

Terrigenous material can be carried from the continents to the oceans in the form of mineral dust over great distances measuring hundreds to up to thousands of kilometers. This is accomplished by eolian transport. Wind, in contrast to ice and water, only carries particles of finer grain size, such as the silt and clay fraction. A grain size of approximately 80 μm is assumed to mark the highest degree of coarseness transportable by wind. Along wind trajectories, coarser grains such as fine sand and particles which as to their sizes are characteristic of continental loess soil (20-50 μm) usually fall out in the coastal areas, whereas the finer grains come to settle much farther away. The relevant sources for *eolian dust transport* are the semi-arid and arid regions, like the Sahel zone and the Sahara desert, the Central Asian deserts and the Chinese loess regions (Pye 1987). According to recent estimations (Prospero 1996), a total rate of approximately $1-2 \cdot 10^9$ tons y^{-1} dust is introduced into the atmosphere, of which about $0.91 \cdot 10^9$ tons y^{-1} is deposited into the oceans. This amount is, relative to the entire terrigenous amount of weathered material, not very significant; yet, it contributes considerably to sediment formation because the eolian transport of dust concentrates on few specific regions (Fig. 1.3). Dust from the Sahara contributes to sedimentation on the Antilles island Barbados at a rate of 0.6 mm y^{-1}, confirming that it is not at all justified to consider its contribution in building up deep-sea sediments in the tropical and sub-tropical zones of the North Atlantic as negligible. Similar conditions are to be found in the north-western Pacific and the Indian Ocean where great amounts of dust are introduced into the ocean from the Central Asian deserts and the Arabic desert.

According to rough estimates made by Lisitzin (1996), about 84 % of terrigenous sediment input into the ocean is effected by fluvial transport,

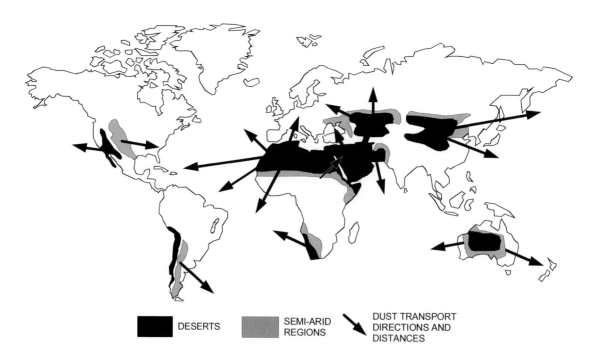

DESERTS   SEMI-ARID REGIONS   DUST TRANSPORT DIRECTIONS AND DISTANCES

Fig. 1.3 The world's major desert areas and semi-arid regions and potential long-distance eolian dust trajectories and oceanic depocenters (Hillier 1995).

somewhat more than 7 % by eolian transport, and less than 7 % is due to the activity of icebergs.

A distinctly less, but still not insignificant proportion of lithogenous sediment is formed by volcanic activity which is quite often coupled with processes of active subduction at the continental plate boundaries. A large proportion of pyroclastic fragments becomes wind-dispersed over large areas, whereafter they are usually retraced in oceanic sediments as finely distributed volcanic glass. Yet, the formation of single, distinct, cm-thick tephra layers might also occur in the deep sea where they represent genuine isochronous markers which can be used for correlation purposes and the time calibration of stratigraphic units. Layers of ash deposits in the eastern Mediterranean are prominent examples indicative of the eruption of the volcanic island Ischia in prehistoric times of more than 25,000 years ago, and Santorin about 3500 years ago.

Locally, there may be a frequent occurrence of tephra layers and significant concentrations of finely dispersed volcaniclastic material in deep-sea sediments especially in the proximal zones of volcanic activity, like in marginal zones of the modern Pacific Ocean. In a recent evaluation of tephra input into the Pacific Ocean sediments based on DSDP and ODP data Straub and Schmincke (1998) estimate that the minimum proportion of volcanic tephra corresponds to 23 vol.% of the existing Pacific oceanic sediments.

Lithogenous detrital components of marine sediments, despite all regional variability, include only few basic minerals (Table 1.1). With the exception of quartz, complete weathering, particularly the chemical weathering of metamorphic and igneous rock, leads to the formation of clay minerals. Consequently, this group represents, apart from the remaining quartz, the most important mineral constituent in sediments; clay minerals make up nearly 50 % of the entire terrigenous sediment. To a lesser degree, terrigenous detritus contains unweathered minerals, like feldspars. Furthermore, there are mica, non-biogenous calcite, dolomite in low quantities, as well as accessory heavy minerals, for instance, amphibole, pyroxene, apatite, disthene, garnet, rutile, anatase, zirconium, tourmaline, but they altogether seldom comprise more than 1 % of the sediment. Basically, each mineral found in continental rock - apart from their usually extreme low concentrations - may also be found in the oceanic sediments. The percentage in which the various minerals are present in a sediment markedly depends on the grain-size distribution.

The clay minerals are of special importance inasmuch as they not only constitute the largest proportion of fine-grained and non-biogenous sediment, but they also have the special geochemical property of absorbing and easily giving off ions, a property which affords more detailed observation. Clay minerals result foremost from the weathering of primary, rock forming aluminous silicates, like feldspar, hornblende and pyroxene, or even volcanic glass. Kaolinite, chlorite, illite, and smectite which represent the four most important groups of clay minerals are formed partly under very different conditions of weathering. Consequently, the analysis of their qualitative and quantitative distribution will enable us to draw essential conclusions on origin and transport, weathering and hydrolysis, and therefore on climate conditions of the rock's source region (Biscay 1965; Chamley 1989). The extremely fine-granular structure of clay minerals, which is likely to produce an active surface of 30 m^2g^{-1} sediment, as well as their ability to absorb ions internally within the crystal structure, or bind them superficially by means of reversible adsorption, as well as their capacity to temporarily bind larger amounts of water, all these properties are fundamental for us to consider clay sediments as a very active and effectively working "geochemical factory".

Clay minerals constitute a large part of the family of phyllosilicates. Their crystal structure is characterized by alternation of flat, parallel sheets, or layers of extreme thinness. For this reason clay minerals are called layer silicates. Two basic types of layers, or sheets make up any given clay mineral. One type of layer consists of tetrahedral sheets in which one silicon atom is surrounded by four oxygen atoms in tetrahedral configuration. The second type of layer is composed of octahedron sheets in which aluminum or magnesium is surrounded by hydroxyl groups and oxygen in a 6-fold coordinated arrangement (Fig. 1.4). Depending on the clay mineral under study, there is still enough space for other cations possessing a larger ionic radius, like potassium, sodium, calcium, or iron to fit in the gaps between the octahedrons and tetrahedrons. Some clay minerals - the so-called expanding or swelling clays - have a special property which allows them to incorporate hydrated cations into their structure. This process is reversible; the water changes the

Table 1.1 Mineralogy and relative importance of main lithogeneous sediment components.

	relative importance	idealized composition
Quarz	+++	SiO_2
Calcite	+	$CaCO_3$
Dolomite	+	$(Ca,Mg)CO_3$
Feldspars		
Plagioclase	++	$(Na,Ca)[Al(Si,Al)Si_2O_8]$
Orthoclase	++	$K[AlSi_3O_8]$
Muscovite	++	$KAl_2[(AlSi_3)O_{10}](OH)_2$
Clay minerals		
Kaolinite	+++	$Al_2Si_2O_5(OH)_4$
Mica Group		
e.g. Illite	+++	$K_{0.8-0.9}(Al,Fe,Mg)_2(Si,Al)_4O_{10}(OH)_2$
Chlorite Groupe		
e.g. Chlorite s.s.	+++	$(Mg_{3-y}Al_1Fe_y)Mg_3(Si_{4-x}Al)O_{10}(OH)_8$
Smektite Groupe		
e.g. Montmorillonite	+++	$Na_{0.33}(Al_{1.67}Mg_{0.33})Si_4O_{10}(OH)_2 \cdot nH_2O$
Heavy minerals, e.g.		
Amphiboles		
e.g. Hornblende	+	$Ca_2(Mg,Fe)_4Al[Si_7,Al_{22}](OH)_2$
Pyroxene		
e.g. Augite	+	$(Ca,Na)(Mg,Fe,Al)[(Si,Al)_2O_6]$
Magnetite	-	Fe_3O_4
Ilmenite	-	$FeTiO_3$
Rutile	-	TiO_2
Zircon	-	ZrO_2
Tourmaline	-	$(Na,Ca)(Mg,Fe,Al,Li)_3Al_6(BO_3)_3Si_6O_{18}(OH)_4$
Garnet		
e.g. Grossular	-	$Ca_3Al_2(SiO_4)_3$

volume of the clay particles significantly as it goes into or out of the clay structure. All in all, hydration can vary the volume of a clay particle by 95 %.

Kaolinite is the most important clay mineral of the two-layer group, also referred to as the 7-Ångstrom clay minerals (Fig. 1.4), which consist of interlinked tetrahedron-octahedron units. Illite and the smectites which have the capacity to bind water by swelling belong to the group of three-layered minerals, also referred to as 10-Ångstrom clay minerals. They are made of a combination of two tetrahedrally and one octahedrally coordi-nated sheets. Four-layer clay minerals, also known as 14-Ångstrom clay minerals, arise whenever a further autonomous octahedral layer emerges between the three-layered assemblies. This group comprises the chlorites and an array of various composites.

Apart from these types of clay minerals, there is a relatively large number of clay minerals that possess a mixed-layered structure made up of a composite of different basic structures. The result is a sheet by sheet chemical mixture on the scale of the crystallite. The most frequent mixed-layer structure consists of a substitution of illite and smectite layers.

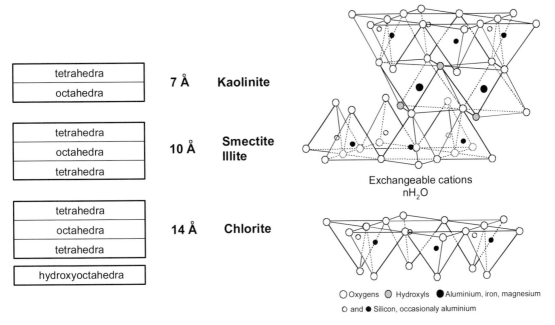

Fig. 1.4 Schematic diagram of clay mineral types: Left: According to the combination of tetrahedral- and octahedral-coordinated sheets; Right: Diagrammatic sketch of the structure of smectite (after Hillier 1995 and Grim 1968).

Kaolinite is a regularly structured di-octahedral two-layer mineral and preferentially develops under warm and humid conditions, by chemical weathering of feldspars in tropical soil. Good drainage is essential to assure the removal of cations released during hydrolysis. Abundance and distribution of kaolinite reflects soil-forming processes in the area of its origin which is optimal in lateritic weathering in the tropics. Owing to the fact that the occurrence of this mineral in ocean sediments is distinctly latitude-dependent, it is often referred to as the "mineral of low latitudes".

Illite is a three-layered mineral of the mica group and not really a specified mineral, instead the term illite refers to a group of mica-like minerals in the clay fraction; as such, it belongs to the most frequently encountered type of clay minerals. Illites are formed as detrital clay minerals by fragmentation in physical weathering. Chemical weathering (soil formation) in which potassium is released from muscovite also leads to the formation of illites. For this reason, illite is often referred to as incomplete mica or hydromica. The distribution of illites clearly reflects its terrestrial and detrital origin which is also corroborated by K/Ar-age determinations made on illites obtained from recent sediments. There are as yet no indications as to *in-situ* formations of illites in marine environments.

Similar to the illites are the cation-rich, expandable, three-layered minerals of the smectite group. The smectites, a product of weathering and pedogenic formation in temperate and sub-arid zones, hold an intermediate position with regard to their global distribution. However, smectites are often considered as indicative of volcanic environments, in fact smectite formation due to low-temperature chemical alteration of volcanic rocks is even a quite typical finding. Similarly, finely dispersed particles of volcanic glass may transform into smectite after a sufficiently long exposure to seawater. Smectites may also arise from muscovite after the release of potassium and its substitution by other cations. There is consequently no distinct pattern of smectite distribution discernible in the oceans.

The generally higher occurrence of smectite concentrations in the southern hemisphere can be explained with the relatively higher input of volcanic detritus. This is especially the case in the Southern Pacific.

Chlorite predominately displays a trioctahedral structure and is composed of a series of three layers resembling mica with an interlayered sheet of brucite (hydroxide interlayer). Chlorite is mainly released from altered magmatic rocks and from metamorphic rocks of the green schist facies as a result of physical weathering. It therefore characteristically depends on the type of parent rock. On account of its iron content, chlorite is prone to chemical weathering. Chlorite distinctly displays a distribution pattern of latitudinal zonation and due to its abundance in polar regions it is considered as the "mineral of high latitudes" (Griffin et al. 1968). The grain size of chlorite minerals is – similar to illite – not limited to the clay fraction (< 4 µm), but in addition encompasses the entire silt fraction (4 - 63 µm) as well.

1.2.2   Biogenous Sediments

Biogenous sediments generally refer to bioclastic sediments, hence sediments which are built of remnants and fragments of shells and tests produced by organisms – calcareous, siliceous or phosphatic particles. In a broader sense, biogenous sediments comprise all solid material formed in the biosphere, i.e. all the hard parts inclusive of the organic substance, the caustobioliths. The organic substance will be treated more comprehensively in Chapter 4, therefore they will not be discussed here.

The amount of carbonate which is deposited in the oceans today is almost exclusively of biogenous origin. The long controversy whether chemical precipitation of lime occurs directly in the shallow waters of the tropical seas, such as the banks of the Bahamas and in the Persian Gulf (Fig. 1.5), during the formation of calcareous ooids and oozes of acicular aragonite, has been settled in preference of the concept of biomineralization (Fabricius 1977). The tiny aragonite

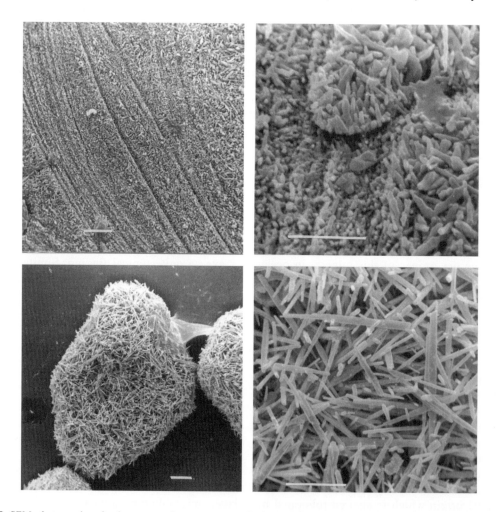

Fig. 1.5 SEM photographs of calcareous sediments composed of aragonite needles which are probably of biogenic origin. *Upper left:* slightly etched section of a calcareous ooid from the Bahamas showing subconcentric laminae of primary ooid coatings; *upper right*: close-up showing ooid laminae formed by small acicular aragonite needles. *Lower left:* silt-sized particle of aragonite mud from the Persian Gulf; *lower right:* close-up showing details of acicular aragonite needles measuring up to 10 µm in length, scalebar 5 µm.

needles (few micrometers) within the more or less concentric layers of calcareous ooids have been considered for a long time as primary precipitates from seawater. However, further investigations including the distribution of stable isotopes distinctly evidenced their biological origin as products of calcification by unicellular algae. Yet, one part of the acicular aragonite ooze might still originate from the mechanical disruption of shells and skeletal elements.

Although marine plants and animals are numerous and diverse, only relatively few groups produce hard parts capable of contributing to the formation of sediments, and only very few groups occur in an abundance relevant for sediment formation (Table 1.2). Relevant for sediment formation are only carbonate minerals in the form of aragonite, Mg-calcite and calcite, as well as biogenic

opal in the form of amorphous $SiO_2 \cdot nH_2O$. The sulfates of strontium and barium as well as various compounds of iron, manganese, and aluminum are of secondary importance, yet they are of geochemical interest, e.g. as tracers for the reconstruction of past environmental conditions. For example, the phosphatic particles formed by various organisms, such as teeth, bone, and shells of crustaceans, are major components of phosphorite rocks which permit us to draw conclusions about nutrient cycles in the ocean.

Large amounts of carbonate sediment accumulate on the relatively small surface of the shallow shelf seas – as compared to entire oceans surface – of the tropical and subtropical warm water regions, primarily by few lime-secreting benthic macrofossil groups. Scleritic corals, living in symbiosis with algae, and encrusting red algae consti-

Table 1.2 Major groups of marine organisms contributing to biogenic sediment formation and mineralogy of skeletal hard parts. Foraminifera and diatoms (underlined) are important groups of both plankton and benthos. x = common, (x) = rare (mainly after Flügel 1978, 2004 and Milliman 1974).

	Aragonite	Aragonite + Calcite	Mg-Calcite	Calcite	Calcite + Mg-Calcite	Opal	divers
Plankton							
Pteropods	x						
Radiolarians						x	celestite
Foraminifera	(x)		x	x	x		
Coccolithophores				x			
Dinoflagellates			x				organic
Silicoflagellates						x	
Diatoms						x	
Benthos							
Chlorophyta	x						
Rhodophyta	x		x		(x)		
Phaeophyta	x						
Sponges	x		x			x	celestite
Scleratinian corals	x						
Octocorals	x		x				
Bryozoens	x	x					
Brachiopods				x			phosphate
Gastropods	x	x					
Pelecypods	x	x		x			
Decapods			x				phosphate
Ostracods	(x)		x				
Barnacles	(x)		x				
Annelid worms	x	x	x			(x)	phosphate
Echinoderms			x				phosphate
Ascidians	x						

tute the major proportion of the massive structures of coral reefs. Together with calcifying green algae, foraminifera, and mollusks, these organisms participate in a highly productive ecosystem. Here, coarse-grained calcareous sands and gravel are essentially composed of various bioclasts attributable to the reef structure, of lime-secreting algae, mollusks, echinoderms and large foraminifera.

Fine-grained calcareous mud is produced by green algae and benthic foraminifera as well as by the mechanical abrasion of shells of the macrobenthos. Considerable amounts of sediment are formed by bioerosion, through the action of boring, grazing and browsing, and predating organisms. Not all the details have been elucidated as to which measure the chemical and biological decomposition of the organic matter in biogenic hard materials might lead to the formation of primary skeletal chrystalites on the micrometer scale, and consequently contribute to the fine-grained calcareous mud formation.

It is obvious that the various calcareous-shelled groups, especially of those organisms who secrete aragonite and Mg-calcite, contribute significantly to the sediment formation in the shallow seas, whereas greater deposits of biogenic opal are rather absent in the shallow shelf seas. The isostatically over-deepened shelf region of Antarctica, where locally a significant accumulation of siliceous sponge oozes occurs, however, makes a remarkable exception. The relatively low opal concentration in recent shelf deposits does not result from an eventual dilution with terrigenous material. The reason is rather that recent tropical shallow waters have low silicate concentrations, from which it follows that diatoms and sponges are only capable of forming slightly silicified skeletons that quickly remineralize in markedly silicate-deficient waters.

With an increasing distance from the coastal areas, out toward the open ocean, the relevance of planktonic shells and tests in the formation of sediments increases as well (Fig. 1.6). Planktonic lime-secreting algae and silica-secreting algae, coccolithophorids and diatoms that dwell as primary producers in the photic zone which thickness measures approximately 100 m, as well as the calcareous foraminifers and siliceous radiolarians, and silicoflagellates (Table 1.2), are the producers of the by far most widespread and essential deep-sea sediments: the calcareous and siliceous biogenic oozes. Apart from the groups mentioned,

planktonic mollusks, the aragonite-shelled pteropods, and some calcareous cysts forming dinoflagellates, also contribute to a considerable degree to sediment formation.

1.2.3   Hydrogenous Sediments

Hydrogenous sediments may be widely distributed, but as to their recent quantity they are relatively insignificant. They will be briefly mentioned in this context merely for reasons of being complete. According to Elderfield (1976), hydrogenous sediments can be subdivided into "precipitates", primary inorganic components which have precipitated directly from seawater, like sodium chloride, and "halmyrolysates", secondary components which are the reaction products of sediment particles with seawater, formed subsequent to *in-situ* weathering, but prior to diagenesis. Of these, manganese nodules give an example. In the scope of this book, these components are not conceived as being part of the "primary" solid phase sediment, but as "secondary" authigenic formations which only emerge in the course of diagenesis, as for instance some clay minerals like glauconite, zeolite, hydroxides of iron and manganese etc. In the subsequent Chapters 11 and 13 some aspects of these new formations will be more thoroughly discussed.

The distinction between detrital and newly formed, authigenic clay minerals is basically difficult to make on account of the small grain size and their amalgamation with quite similar detrital material. Yet it has been ascertained that the by far largest clay mineral proportion – probably more than 90 % – located in recent to subrecent sediments is of detrital origin (Chamley 1989; Hillier 1995). There are essentially three ways for smectites to be formed, which demand specific conditions as they are confined to local areas. Alterations produced in volcanic material is one way, especially by means of hydration of basaltic and volcanic glasses. This process is referred to as *palagonitization*. The probably best studied smectite formation consists in the vents of hydrothermal solutions and their admixture with seawater at the mid-oceanic mountain ranges. Should authigenic clay minerals form merely in recent surface sediments in very small amounts, their frequency during diagenesis (burial diagenesis), will demonstrate a distinct elevation. However, this aspect will not be considered any further beyond this point.

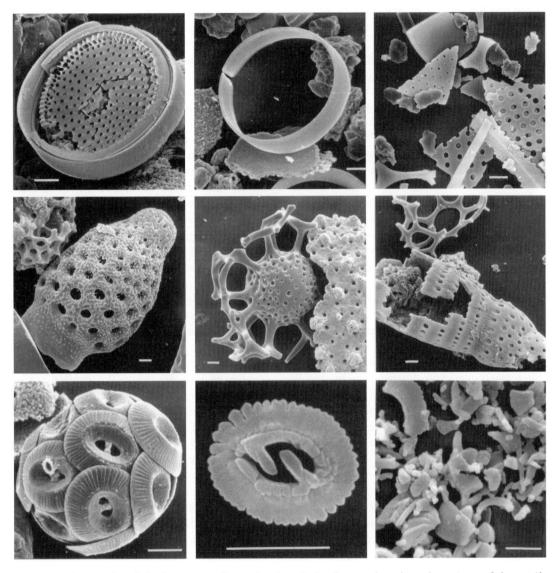

Fig. 1.6 SEM photographs of the important sediment-forming planktonic organisms in various stages of decay. *Above*: Siliceous centered diatoms. *Middle*: Siliceous radiolarian. *Below*: Calcareous coccolithophorid of single placoliths = coccoliths and lutitic abrasion of coccoliths. Identical scale of 5 μm.

1.3   Classification of Marine Sediments

As yet, there is no general classification scheme applicable to marine sediments that combines all the essential characteristics pertaining to a sediment. A large number of different schemes has been proposed in the literature which focus either on origin, grain-size distribution, chemical and mineralogical features of the sediment components, or the facial development of the sediments – all depending on the specific problem under study. The advantages and disadvantages of the various schemes will not be of any concern here. In the following, an attempt will be made to give a comprehensive concise summary, based on the combination of the various concepts.

Murray and Renard (1891) early introduced a basically simple concept which in its essentials differentiated according to the area of sediment deposition as well as to sediment sources. It distinguished (i) "shallow-water deposits from low-water mark to 100 fathoms", (ii) "terrigenous deposits in deep and shallow water close to land", i.e. the combined terrigenous deposits from the deep sea and the shelf seas, and (iii) "pelagic deposits in deep water removed from land". This scheme is very much appropriate to provide the basic framework for a simple classification scheme on the basis of terrigenous sediments, inclusive of the "shallow-water deposits" in the sense of Murray and Renard (1891), and the deep-sea sediments. The latter usually are subdivided into hemipelagic sediments and pelagic sediments.

1.3.1   Terrigenous Sediments

Terrigenous sediments, i.e. clastics consisting of material eroded from the land surface, are not only understood as nearshore shallow-water deposits on the shelf seas, but also comprise the deltaic foreset beds of continental margins, slump deposits at continental slopes produced by gravity transport, and the terrigenous-detrital shelf sediments redistributed into the deep sea by the activity of debris flows and turbidity currents.

The sand, silt, and clay containing shelf sediments primarily consist of terrigenous siliciclastic components transported downstream by rivers; they also contain various amounts of autochthonous biogenic shell material. Depending on the availability of terrigenous discharge, biogenic carbonate sedimentation might predominate in broad shelf regions. A great variety of grain sizes is typical for the sediments on the continental shelf, with very coarse sand or gravel accumulating in high-energy environments and very fine-grained material accumulating in low-energy environments. Coarse material in the terrigenous sediments of the deep sea is restricted to debris flow deposits on and near the continental margins and the proximal depocenters of episodic turbidites.

The development and the hydrodynamic history of terrigenous sediments is described in its essentials by the grain-size distribution and the derived sediment characteristics. Therefore, a classification on the basis of textural features appears suitable to describe the terrigenous sediments. The subdivision of sedimentary particles according the Udden-Wentworth scale encompasses four major categories: gravel (> 2 mm), sand (2 - 0.0625 mm), silt (0.0625 - 0.0039 mm) and clay (< 0.0039 mm), each further divided into a number of subcategories (Table 1.3). Plotting the percentages of these grain sizes in a ternary diagram results in a basically quite simple and clear classification of the terrigenous sediments (Fig. 1.7). Further subdivisions and classifications

within the various fields of the ternary diagram are made quite differently and manifold, so that a commonly accepted standard nomenclature has not yet been established. The reason for this lies, to some extent, in the fact that variable amounts of biogenic components may also be present in the sediment, next to the prevalent terrigenous, siliclastic components. Accordingly, sediment classifications are likely to vary and certainly will become confusing as well; this is especially true of mixed sediments. However, a certain degree of standardization has developed owing to the frequent and identical usage of terms descriptive of sediment cores, as employed in the international Ocean Drilling Program (ODP) (Mazullo and Graham 1987).

Although very imprecise, the collective term "mud" is often used in literature to describe the texture of fine-grained, mainly non-biogenic sediments which essentially consist of a mixture of silt and clay. The reason for this is that the differentiation of the grain-size fractions, silt and clay, is not easy to manage and is also very time-consuming with regard to the applied methods. However, it is of high importance for the genetic interpretation of sediments to acquire this information. For example, any sediment mainly consisting of silt can be distinguished, such as a distal turbidite or contourite, from a hemipelagic sediment mainly consisting of clay. The classification of clastic sediments as proposed by Folk (1980) works with this distinction, providing a more precise definition of *mud* as a term (Fig. 1.8).

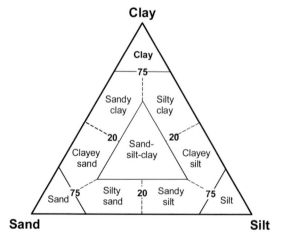

Fig. 1.7 Ternary diagram of sand-silt-clay grain-size distribution showing principal names for siliciclastic, terrigeneous sediments (from Shepard 1954).

mm	Udden-Wenthworth phi values (φ)	Terminology
1024	-10	Boulder
512	-9	
256	-8	Cobbles
128	-7	
64	-6	
32	-5	Pebble
16	-4	
8	-3	
4	-2	
3,36	-1,75	Granule
2,83	-1,5	
2,38	-1,25	
2,00	-1,0	
1,68	-0,75	Very coarse sand
1,41	-0,5	
1,19	-0,25	
1,00	0,0	
0,84	0,25	Coarse sand
0,71	0,50	
0,59	0,75	
0,50	1,00	
0,42	1,25	Medium sand
0,35	1,50	
0,3	1,75	
0,25	2,00	
0,21	2,25	Fine sand
0,177	2,50	
0,149	2,75	
0,125	3,00	
0,105	3,25	Very fine sand
0,088	3,50	
0,074	3,75	
0,0625	4,00	
0,053	4,25	Coarse silt
0,044	4,50	
0,037	4,75	
0,031	5,00	
	5,25	Medium silt
	5,50	
	5,75	
0,0156	6,00	
	6,25	Fine silt
	6,50	
	6,75	
0,0078	7,00	
	7,25	Very fine silt
	7,50	
	7,75	
0,0039	8,00	
0,00200	9,0	**Clay**
0,00098	10,0	
0,00049	11,0	

Table 1.3 Grain-size scales and textural classification following the Udden-Wentworth US Standard. The phi values (φ) according to Krumbein (1934) and Krumbein and Graybill (1965); $\phi = -\log_2 d/d_0$ where d_0 is the standard grain diameter (i.e. 1 mm).

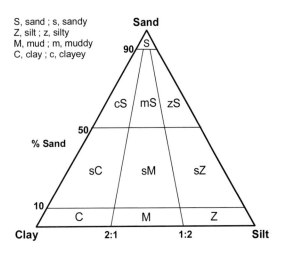

Fig. 1.8 Textural classification of clastic sediments (modified after Folk 1980).

It needs to be stated in this context, that, in order to compare the quantitative reports of published grain sizes, the international literature does not draw the line between silt and clay at 0.0039 mm – as the U.S standard and the Udden-Wentworth scale does (Table 1.3), or the French AFNOR-norm, but sets the limit at 0.002 mm, a value very often found in German literature and complying with the DIN-standards. To add to further diversity, modern publications from Russia mark the silt-clay transition at a grain size of 0.01 mm (e.g. Lisitzin 1996).

1.3.2   Deep-sea Sediments

The sediments in the deep sea consist of only few basic types which in their manifold combinations are suited for the description of a varied facial pattern (Table 1.4). The characteristic pelagic deep-sea sediment far from coastal areas is deep-sea red clay, an extremely fine-grained (median < 1 μm) red-brown clay sediment which covers the oceanic deep-sea basins below the Calcite Compensation Depth (CCD). More than 90 % is composed of clay minerals, other hydrogenous minerals, like zeolite, iron-manganese precipitates and volcanic debris. Such sediment composition demonstrates an authigenic origin. The small percentage of lithogenic minerals, such as quartz, feldspar and heavy minerals, confirms the existence of terrigenous components which in part should have originated from eolian transport processes. The biogenic oozes represent the most frequent type of deep-sea sediments; they mainly consist of shells and skeletal material from planktonic organisms living in the ocean where they drizzle from higher photic zones down to the ocean floor, like continuous rainfall, once they have died (Fig. 1.6). The fragments of the calcareous-shelled pteropods, foraminifera, and coccolithophorids constitute the calcareous oozes (pteropod ooze, foraminiferal ooze, or nannofossil ooze), whereas the siliceous radiolarians, silicoflagellates, and diatoms constitute the siliceous oozes (radiolarian ooze or diatomaceous ooze).

The hemipelagic sediments are basically made of the same components as the deep-sea red clay and the biogenic oozes, clay minerals and biogenic particles respectively, but they also contain an additional and sometimes dominating amount of terrigenous material, such as quartz, feldspars, detrital clay minerals, and some reworked biogenous components from the shelves (Table 1.4, Fig. 1.9).

Most deep-sea sediments can be described according to their composition or origin as a three-component system consisting of

(i)   biogenic carbonate,

(ii)   biogenic opal, and

(iii) non-biogenic mineral constituents.

The latter group comprises the components of the deep-sea red clay and the terrigenous siliciclastics. On the basis of experiences made in the Deep Sea Drilling Project, Dean et al. (1985) have

Table 1.4 Classification of deep-sea sediments according to Berger (1974).

I. (Eu-)pelagic deposits (oozes and clays)
    < 25 % of fraction > 5µm is of terrigenic, volcanogenic, and/or neritic origin
    Median grain size < 5µm (except in authigenic minerals and pelagic organisms)
    A. Pelagic clays. $CaCO_3$ and siliceous fossils < 30 %
        1. $CaCO_3$ 1 - 10 %. (Slightly) calcareous clay
        2. $CaCO_3$ 10 - 30 %. Very calcareous (or marl) clay
        3. Siliceous fossils 1 - 10 %. (Slightly) siliceous clay
        4. Siliceous fossils 10 - 30 %. Very siliceous clay
    B. Oozes. $CaCO_3$ or siliceous fossils > 30 %
        1. $CaCO_3$ > 30 %. < $^2/_3$ $CaCO_3$: marl ooze. > $^2/_3$ $CaCO_3$: chalk ooze
        2. $CaCO_3$ < 30 %. > 30 % siliceous fossils: diatom or radiolarian ooze

II. Hemipelagic deposits (muds)
    > 25 % of fraction > 5 µm is of terrigenic, volcanogenic, and/or neritic origin
    Median grain size >5 µm (except in authigenic minerals and pelagic organisms)
    A. Calcareous muds. $CaCO_3$ > 30 %
        1. < $^2/_3$ $CaCO_3$: marl mud. > $^2/_3$ $CaCO_3$: chalk mud
        2. Skeletal $CaCO_3$ > 30 %: foram ~, nanno ~, coquina ~
    B. Terrigenous muds, $CaCO_3$ < 30 %. Quarz, feldspar, mica dominant
        Prefixes: quartzose, arkosic, micaceous
    C. Volcanogenic muds. $CaCO_3$ < 30 %. Ash, palagonite, etc., dominant

III. Special pelagic and/or hemipelagic deposits
        1. Carbonate-sapropelite cycles (Cretaceous).
        2. Black (carbonaceous) clay and mud: sapropelites (e.g., Black Sea)
        3. Silicified claystones and mudstones: chert (pre-Neogene)
        4. Limestone (pre-Neogene)

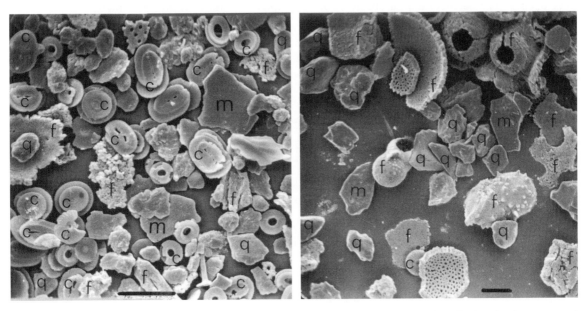

Fig. 1.9 Hemipelagic sediment from Sierra Leone Rise, tropical North Atlantic. *Left*: Fine silt-size fraction composed of coccoliths (c), foraminiferal fragments (f) and of detrital quartz (q) and mica (m). *Right*: Coarse silt-sized fraction predominately composed of foraminiferal fragments (f), some detrital quartz (q) and mica (m), scalebar 10 μm.

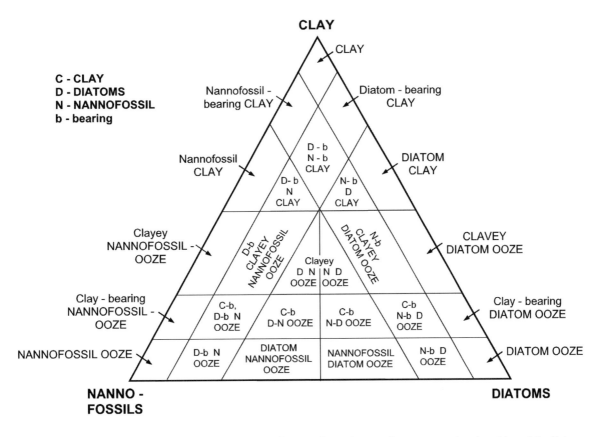

Fig. 1.10 Classification of deep-sea sediments according to the main constituents, e.g. clay (non-biogenic), diatoms (siliceous biogenic), and nannofossils (calcareous biogenic); (modified from Dean et al. 1985).

developed a very detailed and purely descriptive classification scheme (Fig. 1.10):

- The most frequently occurring component with a percentage higher than 50 % determines the designation of the sediment. Non-biogenic materials are specified accordingly on the basis of the grain-size fractions: sand, silt or clay. Biogenic material is referred to as "ooze" and preceded by the most abundant biogenic component: nannofossil ooze, foraminifera ooze, diatom ooze, and radiolarian ooze respectively.

- Each component measuring between 25-50 % is characterized by the following attributes: sandy, silty, clayey, or nannofossil, foraminiferal, diatomaceous, or radiolarian.

- Components with percentages between 10-25 % are referred to by adding the suffix "-bearing", as in "clay-bearing", "diatom-bearing".

- Components with percentages below 10 % are not expressed at all, but may be included by addition of the suffix "rich", as in "C_{org}-rich".

The thus established, four-divided nomenclature of deep-sea sediments with threshold limits of 10 %, 25 % and 50 % easily permits a quite detailed categorization of the sediment which is adaptable to generally rare, but locally frequent occurrences of components, e.g. zeolite, eventually important for a more complete description.

1.4   Global Patterns of Sediment Distribution

The overall distribution pattern of sediment types in the world's oceans depends on few elementary factors. The most important factor is the relative amount with which one particle species contributes to sediment formation. Particle preservation and eventual dilution with other sediment components will modify the basic pattern. The formation and dispersal of terrigenous constituents derived from weathering processes on the continents, as well as autochthonous oceanic-biogenic constituents, both strongly depend on the prevalent climate conditions, so that, in the oceans, a latitude-dependent and climate-related global pattern of sediment distribution will be the ultimate result.

1.4.1   Distribution Patterns of Shelf Sediments

The particulate terrigenous weathering products mainly transported from the continents by rivers are not homogeneously distributed over the ocean floor, but concentrate preferentially along the continental margins, captured either on the shelf or the continental slope (Fig. 1.11). Massive sediment layers are built where continental inputs are particularly high, and preferentially during glacial periods when sea levels were low.

Approximately 70 % of the continental shelf surface is covered with relict sediment, i.e. sediment deposited during the last glacial period under conditions different from today's, especially at times when the sea level was comparatively low (Emery 1968). It has to be assumed that there is a kind of textural equilibrium between these relict sediments and recent conditions. The fine-grained constituents of shelf sediments were eluted during the rise of the sea level in the Holocene and thereafter deposited, over the edge of the shelf onto the upper part of the continental slope, so that extended modern shelf surface areas became covered with sandy relict sediment (Milliman et al. 1972; Milliman and Summerhays 1975).

According to Emery (1968), the sediment distribution on recent shelves displays a plain and distinctly zonal pattern (Fig. 1.12):

Biogenic sediments with coarse-grained calcareous sediments predominate at lower latitudes,

Detrital sediments with riverine terrigenous siliciclastic material at moderate latitudes, and

Glacial sediments of terrigenous origin transported by ice are limited to high latitudes.

In detail, this well pronounced pattern may become strongly modified by local superimpositions. Coarse-grained biogenic carbonate sediments will be found at moderate and high latitudes as well, at places where the riverine terrigenous inputs are very low (Nelson 1988).

Today, as a result of the post-glacial high sea levels, most river-transported fine-grained material is deposited in the estuaries and on the flat inner shelves in the immediate proximity of river mouths. Only a small proportion is transported over the edge of the shelf onto the continental slope. These processes account for the development of mud belts on the shelf, of which 5 types

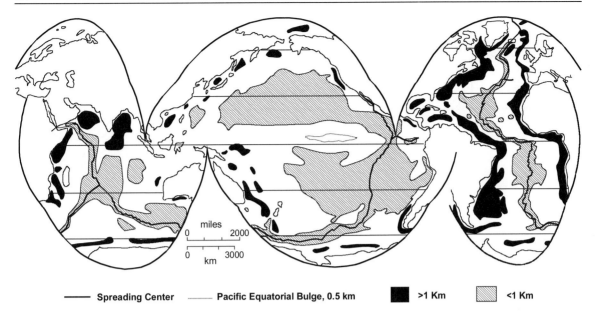

| ——— Spreading Center | ·········· Pacific Equatorial Bulge, 0.5 km | ■ >1 Km | ▨ <1 Km |

Fig. 1.11 Sediment thickness to acoustic basement in the world ocean (from Berger 1974).

of shelf mud accumulation can be distinguished (McCave 1972, 1985): "*muddy coasts, nearshore, mid- and outer-shelf mud belts and mud blankets*' (Fig. 1.13). Mud belts depend on the amount of discharged mud load, the prevalent tidal and/or current system, or the distribution of the sus-

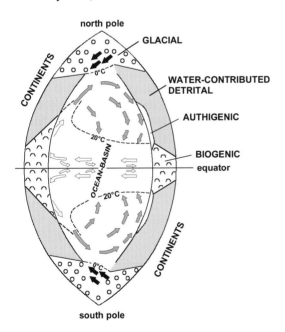

Fig. 1.12 Latitudinal distribution of sedimentary facies of the shallow marine environment of continental shelves in an idealized ocean. *Bold arrows* = cold water; *light arrows* = warm water; *grey arrows* = upwelling water (modified from Reineck and Singh 1973).

pended sediment. Muddy coasts will preferentially form near river mouths, whereas mid-shelf mud belts are characteristic of regions where wave and tidal activity are relatively lower than on the inner or outer shelf.

Especially in delta regions where the supply rates of terrigenous material are high – as in the tropics – even the entire shelf might become covered with a consistent blanket of mud, although the shelf represents a region of high energy conversion.

1.4.2 Distribution Patterns of Deep-sea Sediments

The two most essential boundary conditions in pelagic sedimentation are the nutrient content in the surface water which controls biogenic productivity and by this biogenic particle production, and the position of the calcite compensation depth (CCD) controlling the preservation of carbonate. The CCD, below which no calcite is found, describes a level at which the dissolution of biogenic carbonate is compensated for by its supply rate. The depth of the CCD is generally somewhere between 4 and 5 km below the surface, however, it varies rather strongly within the three great oceans due to differences in the water mass and the rates of carbonate production.

Calcareous ooze and pelagic clays are the predominant deep-sea sediments in offshore regions

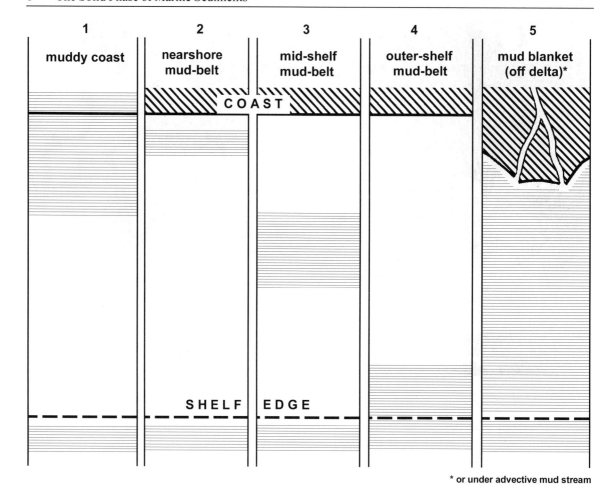

Fig. 1.13 Schematic representation of modern mud accumulation on continental shelves (modified from McCave (1972)).

(Table 1.5). The distribution of these sediments in the three great oceans shows a considerable degree of variation (Fig. 1.14). The distribution patterns strongly depend on the water depth, i.e. the position of the CCD. Calcareous ooze, primarily consisting of foraminiferal oozes and nannoplankton oozes, covers vast stretches of the sea-floor at water depths less than 3-4 km and roughly retraces the contours of the mid-oceanic ridges as well as other plateaus and islands, whereas pelagic clay covers the vast deep-sea plains in the form of deep-sea red clay. This particular pattern is especially obvious in the Atlantic Ocean.

Table 1.5 Relative areas of world oceans covered with pelagic sediments; area of deep-sea floor = $268.1 \cdot 10^6$ km^2 (from Berger (1976)).

Sediments (%)	Atlantic	Pacific	Indian	World
Calcareous ooze	65,1	36,2	54,3	47,1
Pteropod ooze	2,4	0,1	---	0,6
Diatom ooze	6,7	10,1	19,9	11,6
Radiolarian ooze	---	4,6	0,5	2,6
Red clays	25,8	49,1	25,3	38,1
Relative size of ocean (%)	23,0	53,4	23,6	100,0

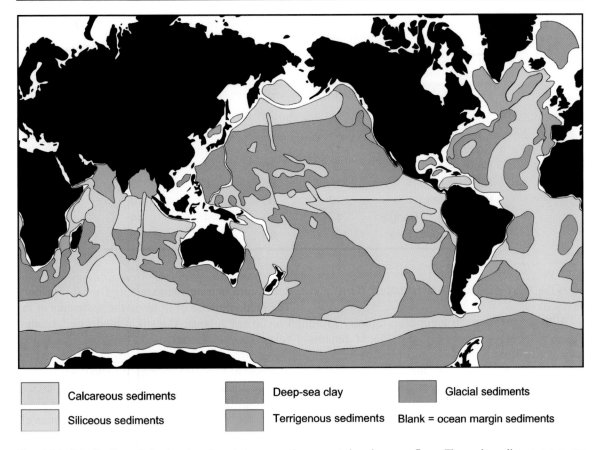

Calcareous sediments   Deep-sea clay   Glacial sediments

Siliceous sediments   Terrigenous sediments   Blank = ocean margin sediments

Fig, 1.14 Distribution of dominant sediment types on the present-day deep-sea floor. The main sediment types are deep-sea clay and calcareous oozes which patterns are predominately depth-controlled. (from Davies and Gorsline (1976)).

Siliceous oozes, mostly consisting of diatom ooze, form a conspicuous ring around Antarctica which clearly marks the zone of the Polar Front in the Antarctic Circumpolar Current. A broad band of radiolarian ooze covers the Pacific Ocean below the equatorial upwelling zone, whilst diatom ooze covers the oceanic margins of the Northern Pacific. Terrigenous sediments, especially in the form of mass flow deposits and turbidites cover vast stretches of the near-continental zones in the North Atlantic, the northeastern Pacific, and the broad deep-sea fans off the big river mouths in the northern Indian Ocean. Glacio-marine sediments are restricted to the continental margins of Antarctica and to the high latitudes of the North Atlantic.

1.4.3   Distribution Patterns of Clay Minerals

Apart from the pelagic clays of the abyssal plains and the terrigenous sediments deposited along the continental margins, primarily or exclusively consisting of clay minerals, various amounts of lithogenous clay minerals can be found in all types of ocean sediments (Table 1.6). The relative proportion of the various clay minerals in the sediments is a function of their original source, their mode of transport into the area of deposition – either by eolian or volcanic transport, or by means of water and ice – and finally the route of transportation (Petschick et al. 1996). In a global survey, it is easy to identify the particular interactions which climate, weathering on the continents, wind patterns, riverine transport, and oceanic currents have with regard to the relative distribution of the relevant groups of clay minerals (kaolinite, illite, smectite, and chlorite).

The distribution of kaolinite in marine sediments (Fig. 1.15) depends on the intensity of chemical weathering at the site of the rock's origin and the essential patterns of eolian and fluvial transport. Due to its concentration at equatorial and tropical latitudes, kaolinite is usually referred

19

Table 1.6 Average relative concentrations [wt.%] of the principal clay mineral groups in the < 2 □m carbonate-free fraction in sediments from the major ocean basins (data from Windom 1976).

	Kaolinite	Illite	Smectite	Chlorite
North Atlantic	20	56	16	10
Gulf of Mexico	12	25	45	18
Caribbian Sea	24	36	27	11
South Atlanic	17	47	26	11
North Pacific	8	40	35	18
South Pacific	8	26	53	13
Indian Ocean	16	30	47	10
Bay of Bengal	12	29	45	14
Arabian Sea	9	46	28	18

to as the "clay mineral of low latitudes" (Griffin et al. 1968).

Illite is the most frequent clay mineral to be found in ocean sediments (Fig. 1.16). It demonstrates a distinctly higher concentration in sediments at mid-latitudes of the northern oceans which are surrounded by great land masses. This follows particularly from its terrigenous origin and becomes evident when the Northern Pacific is compared with the Southern Pacific. The illite concentration impressively reflects the percentage and distribution of particles which were introduced into marine sediments by fluvial transport. The predominance of illites in the sediments of the Pacific and Atlantic Oceans at moderate latitudes, below the trajectories of the jet-stream, indicates the great importance of the wind system in the transport of fine-dispersed particulate matter.

The distribution pattern of smectite differs greatly in the three oceans (Fig. 1.17), and along with some other factors may be explained as an effect induced by dilution. Smectite is generally considered as an indicator of a "volcanic regime" (Griffin et al. 1968). Thus, high smectite concentra-

KAOLINITE

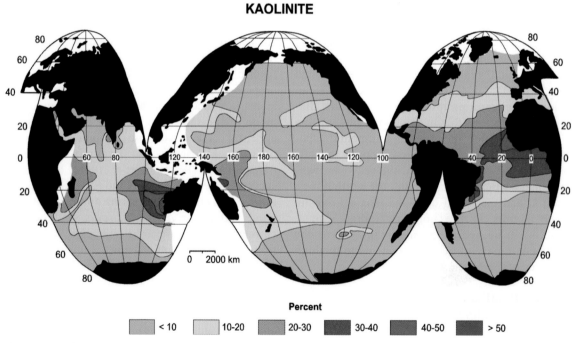

Fig. 1.15 Relative distribution of kaolinite in the world ocean, concentration in the carbonate-free < 2 μm size fraction (from Windom 1976).

ILLITE

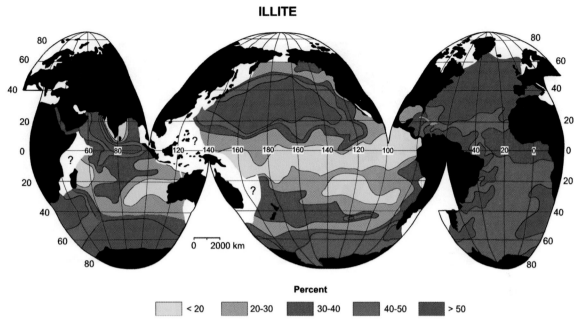

Fig. 1.16 Relative distribution of illite in the world ocean, concentration in the carbonate-free < 2 μm size fraction (from Windom 1976).

tions are usually observed in sediments of the Southern Pacific, in regions of high volcanic activity, where the sedimentation rates are very low due to great distances from the shoreline, and where the dilution with other clay minerals is low as well. The low smectite concentration in the North Atlantic results from terrigenous detritus inputs which are rich in illites and chlorites.

The distribution of chlorite in deep-sea sediments (Fig. 1.18) is essentially inversely related to

SMECTITE

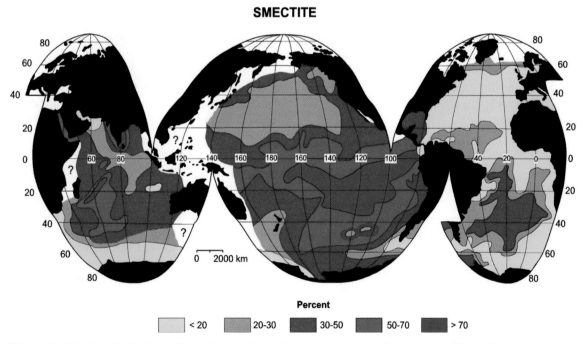

Fig. 1.17 Relative distribution of smectite in the world ocean, concentration in the carbonate-free < 2 μm size fraction (from Windom 1976).

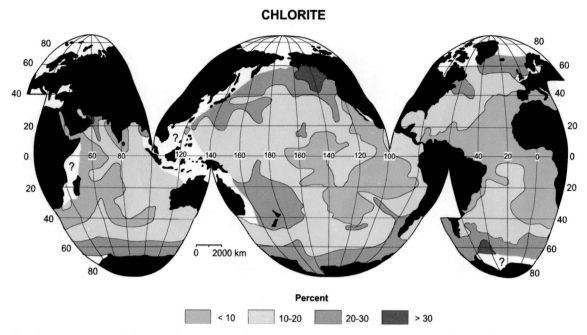

Fig. 1.18 Relative distribution of chlorite in the world ocean, concentration in the carbonate-free < 2 μm size fraction (from Windom 1976).

the pattern of kaolinite. Although chlorite is distributed homogeneously over the oceans, its highest concentration is measured in polar regions and therefore is referred to as the "high latitude mineral" (Griffin et al. 1968).

1.4.4   Sedimentation Rates

As can be seen in Table 1.8, the sedimentation rates of typical types of deep-sea sediments show a strong geographical variability which is based on the regionally unsteady import of terrigenous material and a highly variable biogenic productivity in the ocean.

Table 1.7 Sedimentation rates of red clay in various deep-sea basins of the world ocean (data from various sources, e.g. Berger 1974, Gross 1987).

| | Rate (mm ky⁻¹) | |
	Mean	Range
North Atlantic	1,8	0.5-6.2
South Atlantic	1,9	0.2-7.5
North Pacific	1,5	0.4-6.0
South Pacific	0,45	0.3-0.6

Basically, it can be stated that the sedimentation rate decreases with increasing distance from a sediment source, may this either be a continent or an area of high biogenic productivity. The highest rates of terrigenous mud formation are recorded on the shelf off river mouth's and on the continental slope, where sedimentation rates can amount up to several meters per one thousand years. Distinctly lower values are observed at detritus-starved continental margins, for example of Antarctica. The lowest sedimentation rates ever recorded lie between 1 and 3 mm ky⁻¹. and are connected to deep-sea red clay in the offshore deep-sea basins (Table 1.7), especially in the central Pacific Ocean.

Calcareous biogenic oozes demonstrate intermediate rates which frequently lie between 10 and 40 mm ky⁻¹. Their distribution pattern depends on the biogenic production and on the water depth, or the depth of the CCD. As yet, rate values between 2 and 10 mm ky⁻¹. were considered as normal for the sedimentation of siliceous oozes. Recent investigations in the region of the Antarctic Circumpolar Current have revealed that a very high biogenic production, in connection with lateral advection and "sediment focusing", can even give rise to sedimentation rates of more than 750 mm per thousand years (Fig. 1.19).

Table 1.8 Typical sedimentation rates of recent and subrecent marine sediments (data from various sources, e.g. Berger 1974, Gross 1987).

Facies	Area	Average Sedimentation Rate (mm ky^{-1})
Terrigeneous mud	California Borderland	50 - 2,000
	Ceara Abyssal Plain	200
	Antarctic Continental Margin	30-65
Calcareous ooze	North Atlantic (40-50 °N)	35-60
	North Atlantic (5-20 °N)	40-14
	Equatorial Atlantic	20-40
	Caribbean	~28
	Equatorial Pacific	5-18
	Eastern Equatorial Pacific	~30
	East Pacific Rise (0-20 °N)	20-40
	East Pacific Rise (~30 °N)	3-10
	East Pacific Rise (40-50 °N)	10-60
Siliceous ooze	Equatorial Pacific	2-5
	Antarctic, Indian Sector	2-10
	Antarctic, Atlantic Sector	25-750
Red clay	Northern North Pacific (muddy)	10-15
	Central North Pacific	1-2
	Tropical North Pacific	0-1
	South Pacific	0.3-0.6
	Antarctic, Atlantic Sector	1-2

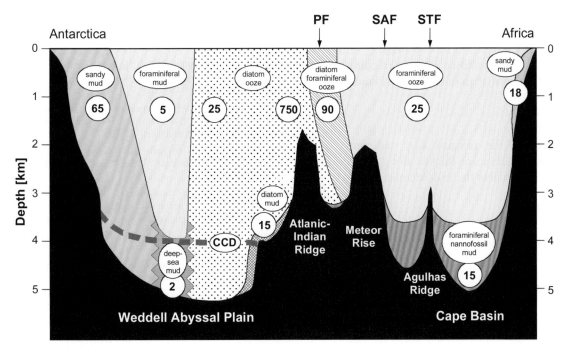

Fig. 1.19 Distribution of major sediment facies across the frontal system of the Antarctic Circumpolar Current (ACC) between Africa and Antarctica. Numbers are typical sedimentation rates in mm ky^{-1}. PF = Polar Front, SAF = Sub Antarctic Front, STF = Subtropical Front.

Velocity of sedimentation or deposition of material through time, i.e. sedimentation rate in this chapter is given solely as millimeter per 1000 years (mm/1000 y or mm ky^{-1}). This describes the vertical thickness of sediment deposited in a certain period of time. It assumes more or less linear and constant processes – not regarding any inconsistency hiatuses – between two datum levels and is, therefore, often referred to as linear sedimentation rate. This term has to be distinguished from mass accumulation rate which refers to a sediment mass deposited in a certain period of time within a defined area, usually given as grams per centimeter square per 1000 years (g cm^{-2} ky^{-1}). For further reading see Bruns and Hass (1999).

1.5   Problems

Problem 1

Is there any correlation between the distribution of calcareous and siliceous sediments, terrigenous sediments and deep-sea clay?

Problem 2

Explain the distribution pattern of clay minerals kaolinite, illite, chlorite and smectite in the world ocean. Why should be illite more common in the North Atlantic than in the South Atlantic?

Problem 3

What are the reasons why sediments of the Atlantic Ocean are thicker along the continental margin than near the mid-Atlantic ridge?

Problem 4

What causes the variability of sedimentation rates?

Problem 5

Summarize and explain the main aspects of deep sea sediment classification.

References

Berger, W.H., 1974. Deep-sea sedimentation. In: Burk, C.A. and Drake, C.L. (eds) The geology of continental margins. Springer Verlag, Berlin, Heidelberg, New York, pp 213-241.

Berger, W.H., 1976. Biogenic deep sea sediments: Production, preservation and interpretation. In: Riley, J.P. and Chester, R. (eds) Chemical Oceanography, Academic, 5. Press, London, NY, San Francisco, pp 266-388.

Berger, W.H. and Herguera, J.C., 1991. Reading the sedimentary record of the ocean's productivity. Pimary productivity and biogeochemical cycles in the sea. In: Falkowski, P.G. and Woodhead, E.D. (eds) Plenum Press, New York, London, pp 455-486.

Biscay, P.E., 1965. Mineraloy and sedimentation of recent deep-sea clay in the Atlantic Ocean and adjacent seas and oceans. Geol. Soc. Am. Bull., 76: 803-832.

Bruns, P. and Hass, H.C., 1999. On the determination of sediment accumulation rates. GeoResearch Forum Vol. 5, Trans Tech Publication, Zürich, 244 pp.

Chamley, H., 1989. Clay sedimentology. Springer Verlag, Berlin, Heidelberg, New York, 623 pp.

Davies, T.A. and Gorsline, D.S., 1976. Oceanic sediments and sedimentary processes. In: Riley, J.P. and Chester, R. (eds) Chemical oceanography, 5. Academic Press, London, New York, San Francisco, pp 1-80.

Dean, W.E., Leinen, M. and Stow, D.A.V., 1985. Classification of deep-sea fine-grained sediments. Journal of Sediment Petrology, 55: 250-256.

Elderfield, H., 1976. Hydrogeneous material in marine sediments; excluding manganese nodules. In: Riley, J.P. and Chester, R. (eds) Chemical Oceanography, 5. Academic Press, London, New York, San Francisco, pp 137-215.

Emery, K.O., 1968. Relict sediments on continental shelves of the world. Bull. Am. Assoc. Petrol. Geologists, 52: 445-464.

Fabricius, F., 1977. Origin of marine ooids and grapestones. Contributions to sedimentology, 7, Schweizerbart, Stuttgart, 113 pp.

Flügel, E., 1978. Mikrofazielle Untersuchungsmethoden von Kalken. Springer Verlag, Berlin, Heidelberg, New York, 454 pp.

Flügel, E., 2004. Microfacies of carbonate rocks: analysis, interpretation and application. Springer Verlag, Berlin, Heidelberg, New York, 976 pp.

Folk, R.L., 1980. Petrology of sedimentary rocks. Hemphill. Publ. Co., Austin, 182 pp.

Garrels, R.M. and Mackenzie, F.T., 1971. Evolution of sedimentary rocks. Norton & Co, New York, 397 pp.

Gibbs, R.J., 1977. Transport phase of transition metals in the Amazon and Yukon rivers. Geological Society of America Bulletin, 88: 829-843.

Gingele, F., 1992. Zur klimaabhängigen Bildung biogener und terrigener Sedimente und ihre Veränderungen durch die Frühdiagenese im zentralen und östlichen Südatlantic. Berichte, Fachbereich Geowissenschaften, Universität Bremen, 85, 202 pp.

Goldberg, E.D. and Griffin, J.J., 1970. The sediments of the northern Indian Ocean. Deep-Sea Research, 17: 513-537.

Griffin, J.J., Windom, H. and Goldberg, E.D., 1968. The distribution of clay minerals in the World Ocean. Deep-Sea Research, 15: 433-459.

Grim, R.E., 1968. Clay Mineralogy, McGraw-Hill Publ., NY, 596 pp.

Gross, M.G., 1987. Oceanography: A view of the Earth. Prentice Hall, Englewood Cliffs, NJ, 406 pp.

Hillier, S., 1995. Erosion, sedimentation and sedimentary origin of clays. In: Velde, B. (ed) Origin and mineralogy of clays. Springer Verlag, Berlin, Heidelberg, New York, pp 162-219.

Krumbein, W.C., 1934. Size frequency distribution of sediments. Journal of Sediment Petrology, 4: 65-77.

Krumbein, W.C. and Graybill, F.A., 1965. An introduction to statistical models in Geology. McGraw-Hill, New York, 328 pp.

Lisitzin, A.P., 1996. Oceanic sedimentation: Lithology and Geochemistry. Amer. Geophys. Union, Washington, D. C., 400 pp.

Mazullo, J. and Graham, A.G., 1987. Handbook for shipboard sedimentologists. Ocean Drilling Program, Technical Note, 8, 67 pp.

McCave, I.N., 1972. Transport and escape of fine-grained sediment from shelf areas. In: Swift, D.J.P., Duane, D.B. and Pilkey, O.H. (eds), Shelf sediment transport, process and pattern. Dowden, Hutchison & Ross, Stroudsburg, pp 225-248.

Milliman, J.D., Pilkey, O.H. and Ross, D.A., 1972. Sediments of the continental margins off the eastern United States. Geological Society of America Bulletin, 83: 1315-1334.

Milliman, J.D., 1974. Marine carbonates. Springer Verlag, Berlin, Heidelberg, New York, 375 pp.

Milliman, J.D. and Summerhays, C.P., 1975. Upper continental margin sedimentation of Brasil. Contribution to Sedimentology, 4, Schweizerbart, Stuttgart, 175 pp.

Milliman, J.D. and Meade, R.H., 1983. World-wide delivery of river sediment to the ocean. The Journal of Geology, 91: 1-21.

Milliman, J.D. and Syvitski, J.P.M., 1992. Geomorphic/tectonic control of sediment discharge to the ocean: The importance of small mountainous rivers. The Journal of Geology, 100: 525-544.

Murray, J. and Renard, A.F., 1891. Deep sea deposits - Report on deep sea deposits based on specimens collected during the voyage of H.M.S. Challenger in the years 1873-1876. „Challenger" Reports, Eyre & Spottiswood, London; J. Menzies & Co, Edinburgh; Hodges, Figgis & Co, Dublin, 525 pp.

Nelson, C.S., 1988. Non-tropical shelf carbonates - Modern and Ancient. Sedimentary Geology, 60, 1-367 pp.

Petschick, R., Kuhn, G. and Gingele, F., 1996. Clay mineral distribution in surface sediments of the South Atlantic: sources, transport, and relation to oceanography. Marine Geology, 130: 203-229.

Prospero, J.M., 1996. The atmospheric transport of particles in the ocean. In: Ittekkot, V., Schäfer, P.,

Honjo, S. and Depetris, P.J. (eds), Particle Flux in the Ocean. Wiley & Sons, Chichester, New York, Brisbane, Toronto, Singapore, pp 19-52.

Pye, K., 1987. Aeolian dust and dust deposits. Academic Press, London, New York, Toronto, 334 pp.

Reineck, H.E. and Singh, I.B., 1973. Depositional sedimentary environments. Springer Verlag, Berlin, Heidelberg, New York, 439 pp.

Seibold, E. and Berger, W.H., 1993. The sea floor - An introduction to marine geology. Springer Verlag, Berlin, Heidelberg, New York, 356 pp.

Shepard, F., 1954. Nomenclature based on sand-silt-clay ratios. Journal of Sediment Petrology, 24: 151-158.

Straub, S.M. and Schmincke, H.U., 1998. Evaluating the tephra input into Pacific Ocean sediments: distribution in space and time. Geologische Rundschau, 87: 461-476.

Velde, B., 1995. Origin and mineralogy of clays. Springer Verlag, Berlin, Heidelberg, New York, 334 pp.

Windom, H.L., 1976. Lithogeneous material in marine sediments. In: Riley, J.P. and Chester, R. (eds), Chemical Oceanography. Academic Press, London, New York, pp 103-135.

2 Physical Properties of Marine Sediments

Monika Breitzke

Generally, physical properties of marine sediments are good indicators for the composition, microstructure and environmental conditions during and after the depositional process. Their study is of high interdisciplinary interest and follows various geoscientific objectives.

Physical properties provide a *lithological and geotechnical description* of the sediment. Questions concerning the composition of a depositional regime, slope stability or nature of seismic reflectors are of particular interest within this context. Parameters like P- and S-wave velocity and attenuation, elastic moduli, wet bulk density and porosity contribute to their solution.

As most physical properties can be measured very quickly in contrast to parameters like carbonate content, grain size distribution or oxygen isotope ratio, they often serve as *proxy parameters* for geological or paleoceanographical processes and changes in the environmental conditions. Basis for this approach are correlations between different parameters which allow to derive regression equations of regional validity. Highly resolved physical properties are used as input parameters for such equations to interpolate coarsely sampled geological, geochemical or sedimentological core logs. Examples are the P-wave velocity, attenuation and wet bulk density as indices for the sand content or mean grain size and the magnetic susceptibility as an indicator for the carbonate content or amount of terrigenous components. Additionally, highly resolved physical property core logs are often used for time series analysis to derive quaternary *stratigraphies* by orbital tuning.

In marine geochemistry, physical parameters which quantify the amount and distribution of pore space or indicate alterations due to diagenesis are important. Porosities are necessary to calculate flow rates, permeabilities describe how easy a fluid flows through a porous sediment and increased magnetic susceptibilities can be interpreted by an increased iron content or, more generally, by a high amount of magnetic particles within a certain volume.

This high interdisciplinary value of physical property measurements results from the different effects of changed environmental conditions on sediment structure and composition. Apart from anthropogeneous influences variations in the environmental conditions result from climatic changes and tectonic events which affect the ocean circulation and related production and deposition of biogenic and terrigenous components. Enhanced or reduced current intensities during glacial or interglacial stages cause winnowing, erosion and redeposition of fine-grained particles and directly modify the sediment composition. These effects are subject of sedimentological and paleoceanographical studies. Internal processes like early diagenesis, newly built authigenic sediments or any other alterations of the solid and fluid constituents of marine sediments are of geochemical interest. Gravitational mass transports at continental slopes or turbidites in large submarine fan systems deposit as coarse-grained chaotic or graded beddings. They can be identified in high-resolution physical property logs and are often directly correlated with sea level changes, results valuable for sedimentological, seismic and sequence stratigraphic investigations. Generally, any changes in the sediment structure which modify the elastic properties of the sediment influence the reflection characteristics of the subsurface and can be imaged by remote sensing seismic or echosounder surveys.

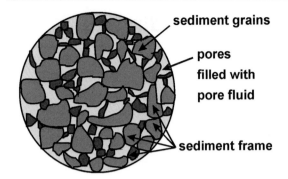

Fig. 2.1 Components of marine sediments. The single particles are the sediment grains. The voids between these particles – the pores – are filled with pore fluid, usually sea water. Welded particles and sediment grains in close contact build the sediment frame.

2.1   Introduction

Physical properties of marine sediments depend on the properties and arrangement of the solid and fluid constituents. To fully understand the image of geological and paleoceanographical processes in physical property core logs it is helpful to consider the single components in detail (Fig. 2.1).

By most general definition a sediment is a collection of particles - the sediment grains - which are loosely deposited on the sea floor and closely packed and consolidated under increasing lithostatic pressure. The voids between the sediment grains - the pores - form the pore space. In water-saturated sediments it is filled with pore water. Grains in close contact and welded particles build the sediment frame. Shape, arrangement, grain size distribution and packing of particles determine the elasticity of the frame and the relative amount of pore space.

Measurements of physical properties usually encompass the whole, undisturbed sediment. Two types of parameters can be distinguished: (1) bulk parameters and (2) acoustic and elastic parameters. Bulk parameters only depend on the relative amount of solid and fluid components within a defined sample volume. They can be approximated by a simple volume-oriented model (Fig. 2.2a). Examples are the wet bulk density and porosity. In contrast, acoustic and elastic parameters depend on the relative amount of solid and fluid components and on the sediment frame including arrangement, shape and grain size distribution of the solid particles. Viscoelastic wave propagation models simulate these complicated structures, take the elasticity of the frame into account and consider interactions between solid and fluid constituents. (Fig. 2.2b). Examples are the velocity and attenuation of P- and S-waves. Closely related parameters which mainly depend on the distribution and capillarity of the pore space are the permeability and electrical resistivity.

Various methods exist to measure the different physical properties of marine sediments. Some parameters can be measured directly, others

Sediment Models

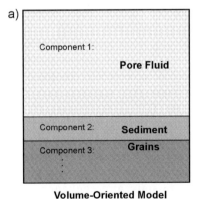

Volume-Oriented Model
for Bulk Parameters

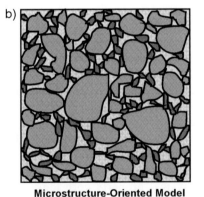

Microstructure-Oriented Model
for Acoustic, Elastic and Related
Parameters

Fig. 2.2 Two types of sediment models. (a) The layered, volume-oriented model for bulk parameters only depends on the relative amount of solid and fluid components. (b) The microstructure-oriented model for acoustic and elastic parameters takes the complicated shape and geometry of the particle and pore size distribution into account and considers interactions between the solid and fluid constituents during wave propagation.

indirectly by determining one or several related parameters and computing the desired property by empirical or model-based equations. For consolidated sedimentary, igneous and metamorphic rocks Schön (1996) described a large variety of such methods for all common physical properties. Some of these methods can also be applied to unconsolidated, water-saturated sediments, others have to be modified or are completely inappropriate. Principally, if empirical relations or sediment models are used, the implicit assumptions have to be checked carefully. An example is Archie's law (Archie 1942) which combines the electrical resistivities of the pore fluid and saturated sediment and with its porosity. An exponent and multiplier in this equation depend on the sediment type and composition and change from fine- to coarse-grained and from terrigenous to biogenic sediments. Another example is Wood's equation (Wood 1946) which relates P-wave velocities to porosities. This model approximates the sediment by a dilute suspension and neglects the sediment frame, any interactions between particles or particles and pore fluid and assumes a 'zero' frequency for acoustic measurements. Such assumptions are only valid for a very limited set of high porosity sediments so that for any comparisons these limitations should be kept in mind.

Traditional measuring techniques use small chunk samples taken from the split core. These techniques are rather time-consuming and could only be applied at coarse sampling intervals. The necessity to measure high-resolution physical property logs rapidly on a milli- to centimeter scale for a core-to-core or core-to-seismic data correlation, and the opportunity to use high-resolution physical property logs for stratigraphic purposes (e.g. orbital tuning) forced the development of non-destructive, automated logging systems. They record one or several physical properties almost continuously at arbitrary small increments under laboratory conditions. The most common tools are the <u>m</u>ulti <u>s</u>ensor <u>t</u>rack (MST) system for P-wave velocity, wet bulk density and magnetic susceptibility core logging onboard of the *Ocean Drilling Program* research vessel JOIDES Resolution (e.g. Shipboard Scientific Party 1995), and the commercially available <u>m</u>ulti <u>s</u>ensor <u>c</u>ore <u>l</u>ogger (MSCL) of GEOTEK™ (Schultheiss and McPhail 1989; Weaver and Schultheiss 1990; Gunn and Best 1998). Additionally, other core logging tools have simultaneously been developed for special research

interests which for instance record electrical resistivities (Bergmann 1996) or full waveform transmission seismograms on sediment cores (Breitzke and Spieß 1993). They are particularly discussed in this paper.

While studies on sediment cores only provide information on the local core position, lateral variations in physical properties can be imaged by remote sensing methods like high-resolution seismic or sediment echosounder profiling. They facilitate core-to-core correlations over large distances and allow to evaluate physical property logs within the local sedimentation environment.

In what follows the theoretical background of the most common physical properties and their measuring tools are described. Examples for the wet bulk density and porosity can be found in Section 2.2. For the acoustic and elastic parameters first the main aspects of Biot-Stoll's viscoelastic model which computes P- and S-wave velocities and attenuations for given sediment parameters (Biot 1956a, b, Stoll 1974, 1977, 1989) are summarized. Subsequently, analysis methods are described to derive these parameters from transmission seismograms recorded on sediment cores, to compute additional properties like elastic moduli and to derive the permeability as a related parameter by an inversion scheme (Sect. 2.4).

Examples from terrigenous and biogenic sedimentation provinces are presented (1) to illustrate the large variability of physical properties in different sediment types and (2) to establish a sediment classification which is only based on physical properties, in contrast to geological sediment classifications which mainly uses parameters like grain size distribution or mineralogical composition (Sect. 2.5).

Finally, some examples from high-resolution narrow-beam echosounder recordings present remote sensing images of terrigenous and biogenic sedimentation environments (Sect. 2.6).

2.2   Porosity and Wet Bulk Density

Porosity and wet bulk density are typical bulk parameters which are directly associated with the relative amount of solid and fluid components in marine sediments. After definition of both parameters this section first describes their traditional analysis method and then focuses on recently developed techniques which determine porosities and wet bulk densities by gamma ray attenuation and electrical resistivity measurements.

The porosity (ϕ) characterizes the relative amount of pore space within a sample volume. It is defined by the ratio

$$\phi = \frac{volume\ of\ pore\ space}{total\ sample\ volume} = \frac{V_f}{V} \quad (2.1)$$

Equation 2.1 describes the fractional porosity which ranges from 0 in case of none pore volume to 1 in case of a water sample. Multiplication with 100 gives the porosity in percent. Depending on the sediment type porosity occurs as *inter-* and *intraporosity*. Interporosity specifies the pore space between the sediment grains and is typical for terrigenous sediments. Intraporosity includes the voids within hollow sediment particles like foraminifera in calcareous ooze. In such sediments both inter- and intraporosity contribute to the total porosity.

The wet bulk density (ρ) is defined by the mass (m) of a water-saturated sample per sample volume (V)

$$\rho = \frac{mass\ of\ wet\ sample}{total\ sample\ volume} = \frac{m}{V} \quad (2.2)$$

Porosity and wet bulk density are closely related, and often porosity values are derived from wet bulk density measurements and vice versa. Basic assumption for this approach is a two-component model for the sediment with uniform grain and pore fluid densities (ρ_g) and (ρ_f). The wet bulk density can then be calculated using the porosity as a weighing factor

$$\rho = \phi \cdot \rho_f + (1 - \phi) \cdot \rho_g \quad (2.3)$$

If two or several mineral components with significantly different grain densities contribute to the sediment frame their densities are averaged in (ρ_g).

2.2.1   Analysis by Weight and Volume

The traditional way to determine porosity and wet bulk density is based on weight and volume measurements of small sediment samples. Usually they are taken from the centre of a split core by a syringe which has the end cut off and a definite volume of e.g. 10 ml. While weighing can be done very accurately in shore-based laboratories mea-

surements onboard of research vessels require special balance systems which compensate the shipboard motions (Childress and Mickel 1980). Volumes are measured precisely by Helium gas pycnometers. They mainly consist of a sample cell and a reference cell and employ the ideal gas law to determine the sample volume. In detail, the sediment sample is placed in the sample cell, and both cells are filled with Helium gas. After a valve connecting both cells are closed, the sample cell is pressurized to (P_1). When the valve is opened the pressure drops to (P_2) due to the increased cell volume. The sample volume (V) is calculated from the pressure ratio (P_1/P_2) and the volumes of the sample and reference cell, (V_s) and (V_{ref}) (Blum 1997).

$$V = V_s + \frac{V_{ref}}{1 - P_1/P_2} \quad (2.4)$$

While weights are measured on wet and dry samples having used an oven or freeze drying, volumes are preferentially determined on dry samples. A correction for the mass and volume of the salt precipitated from the pore water during drying must additionally be applied (Hamilton 1971; Gealy 1971) so that the total sample volume (V) consists of

$$V = V_{dry} - V_{salt} + V_f \quad (2.5)$$

The volumes (V_f) and (V_{salt}) of the pore space and salt result from the masses of the wet and dry sample, (m) and (m_{dry}), from the densities of the pore fluid and salt ($\rho_f = 1.024$ g cm^{-3} and $\rho_{salt} = 2.1$ g cm^{-3}) and from the fractional salinity (s)

$$V_f = \frac{m_f}{\rho_f} = \frac{m - m_{dry}}{(1-s) \cdot \rho_f} \quad (2.6)$$

with : $\quad m_f = \dfrac{m - m_{dry}}{1 - s}$

$$V_{salt} = \frac{m_{salt}}{\rho_{salt}} = \frac{m_f - (m - m_{dry})}{\rho_{salt}} \quad (2.7)$$

with : $\quad m_{salt} = m_f - (m - m_{dry})$

Together with the mass (m) of the wet sample equations 2.5 to 2.7 allow to compute the wet bulk density according to equation 2.2.

Wet bulk density computations according to equation 2.3 require the knowledge of grain densities (ρ_g). They can also be determined from weight and volume measurements on wet and dry samples (Blum 1997)

$$\rho_g = \frac{m_g}{V_g} = \frac{m_{dry} - m_{salt}}{V_{dry} - V_{salt}} \qquad (2.8)$$

Porosities are finally computed from the volumes of the pore space and sample as defined by the equations above.

2.2.2 Gamma Ray Attenuation

The attenuation of gamma rays passing radially through a sediment core is a widely used effect to analyze wet bulk densities and porosities by a non-destructive technique. Often ^{137}Cs is used as source which emits gamma rays of 662 keV energy. They are mainly attenuated by Compton scattering (Ellis 1987). The intensity I of the attenuated gamma ray beam depends on the source intensity (I_0), the wet bulk density (ρ) of the sediment, the ray path length (d) and the specific Compton mass attenuation coefficient (μ)

$$I = I_0 \cdot e^{-\mu \rho d} \qquad (2.9)$$

To determine the source intensity in practice and to correct for the attenuation in the liner walls first no core and then an empty core liner are placed between the gamma ray source and detector to measure the intensities (I_{air}) and (I_{liner}). The difference ($I_{air} - I_{liner}$) replaces the source intensity (I_0), and the ray path length (d) is substituted by the measured outer core diameter ($d_{outside}$) minus the double liner wall thickness ($2d_{liner}$). With these corrections the wet bulk density (ρ) can be computed according to

$$\rho = -\frac{1}{\mu \cdot (d_{outside} - 2d_{liner})} \cdot \ln\left(\frac{I}{I_{air} - I_{liner}}\right) \qquad (2.10)$$

Porosities are derived by rearranging equation 2.3 and assuming a grain density (ρ_g).

The specific Compton mass attenuation coefficient (μ) is a material constant. It depends on the energy of the gamma rays and on the ratio (Z/A) of the number of electrons (Z) to the atomic mass (A) of the material (Ellis 1987). For most sediment and rock forming minerals this ratio is about 0.5, and for a ^{137}Cs source the corresponding mass attenuation coefficient (μ_g) for sediment grains is 0.0774 cm^2 g^{-1}. However, for the hydro-

gen atom (Z/A) is close to 1.0 leading to a significantly different mass attenuation coefficient (μ_f) in sea water of 0.0850 cm^2 g^{-1} (Gerland and Villinger 1995). So, in water-saturated sediments the effective mass attenuation coefficient (μ) results from the sum of the mass weighted coefficients of the solid and fluid constituents (Bodwadkar and Reis 1994)

$$\mu = \phi \cdot \frac{\rho_f}{\rho} \cdot \mu_f + (1 - \phi) \cdot \frac{\rho_g}{\rho} \cdot \mu_g \qquad (2.11)$$

The wet bulk density (ρ) in the denominator is defined by equation 2.3. Unfortunately, equations 2.3 and 2.11 depend on the porosity (ϕ), a parameter which should actually be determined by gamma ray attenuation. Gerland (1993) used an average 'processing porosity' of 50% for terrigenous, 70% for biogenic and 60% for cores of mixed material to estimate the effective mass attenuation coefficient for a known grain density. Whitmarsh (1971) suggested an iterative scheme which improves the mass attenuation coefficient. It starts with an estimated 'processing porosity' and mass attenuation coefficient (eq. 2.11) to calculate the wet bulk density from the measured gamma ray intensity (eq. 2.10). Subsequently, an improved porosity (eq. 2.3) and mass attenuation coefficient (eq. 2.11) can be calculated for a given grain density. Using these optimized values in a second iteration wet bulk density, porosity and mass attenuation coefficient are re-evaluated. Gerland (1993) and Weber et al. (1997) showed that after few iterations (< 5) the values of two successive steps differ by less than 0.1‰ even if a 'processing porosity' of 0 or 100% and a mass attenuation coefficient of 0.0774 cm^2 g^{-1} or 0.0850 cm^2 g^{-1} for a purely solid or fluid 'sediment' are used as starting values.

Gamma ray attenuation is usually measured by automated logging systems, e.g. onboard of the *Ocean Drilling Program* research vessel JOIDES Resolution (Boyce 1973, 1976) by the multisensor core logger of GEOTEK™ (Schultheiss and McPhail 1989; Weaver and Schultheiss 1990; Gunn and Best 1998) or by specially developed systems (Gerland 1993; Bodwadkar and Reis 1994; Gerland and Villinger 1995). The emission of gamma rays is a random process which is quantified in counts per second, and which are converted to wet bulk densities by appropriate calibration curves (Weber et al. 1997). To get a representative value for the gamma ray attenuation gamma counts must be integrated over a sufficiently long time interval. How different integration times and measuring increments influence the quality, resolution and reproducibility of gamma ray logs illustrates Figure 2.3. A 1 m long section of gravity core PS2557-1 from the South African continental margin was repeatedly measured with increments from 1 to 0.2 cm and integration times from 10 to 120 s. Generally, the dominant features, which can be related to changes in the lithology, are reproduced in all core logs. The prominent peaks are more pronounced and have higher amplitudes if the integration time increases (from A to D). Additionally, longer integration times reduce the scatter (from A to B). Fine-scale lithological variations are best resolved by the shortest measuring increment (D).

A comparison of wet bulk densities derived from gamma ray attenuation with those measured on discrete samples is shown in Figure 2.4a for two gravity cores from the Arctic (PS1725-2) and Antarctic Ocean (PS1821-6). Wet bulk densities,

porosities and grain densities of the discrete samples were analyzed by their weight and volume. These grain densities and a constant 'processing porosity' of 50% were used to evaluate the gamma counts. Displayed versus each other both data sets essentially differ by less than ±5% (dashed lines). The lower density range (1.20 - 1.65 g cm^{-3}) is mainly covered by the data of the diatomaceous and terrigenous sediments of core PS1821-6 while the higher densities (1.65 - 2.10 g cm^{-3}) are characteristic for the terrigenous core PS1725-2. According to Gerland and Villinger (1995) the scatter in the correlation of both data sets probably results from (1) a slight shift in the depth scales, (2) the fact that both measurements do not consider identical samples volumes, and (3) artefacts like drainage of sandy layers due to core handling, transportation and storage.

A detailed comparison of both data sets is shown in Figure 2.4b for two segments of core PS1725-2. Wet bulk densities measured on discrete samples agree very well with the density log derived from gamma ray attenuation. Additionally,

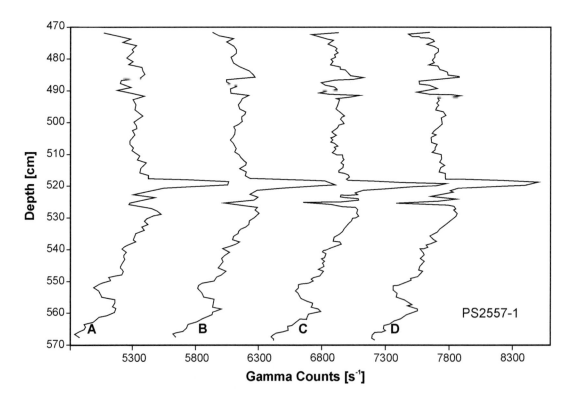

Fig. 2.3 Gamma counts for a 1 m long core section of gravity core PS2557-1 used to illustrate the influence of various integration times and measuring increments. Curve A is recorded with an increment of 1 cm and an integration time of 10 s, curve B with an increment of 1 cm and an integration time of 20 s, curve C with an increment of 0.5 cm and an integration time of 60 s and curve D with an increment of 0.2 cm and an integration time of 120 s. To facilitate the comparison each curve is set off by 800 counts. Modified after Weber et al. (1997).

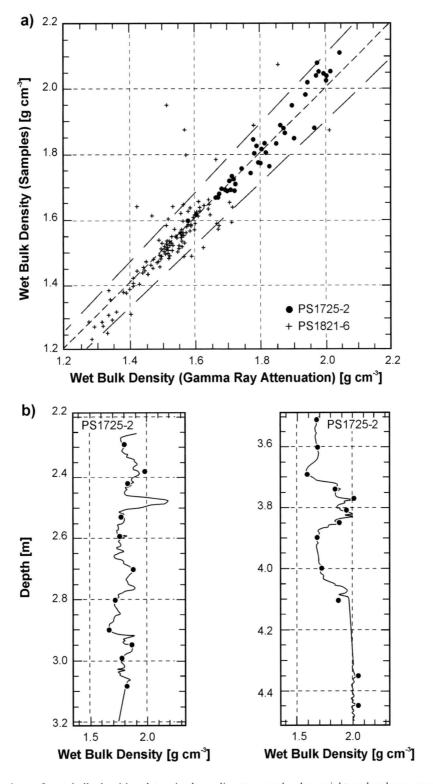

Fig. 2.4 Comparison of wet bulk densities determined on discrete samples by weight and volume measurements and calculated from gamma ray attenuation. (a) Cross plot of wet bulk densities of gravity cores PS1821-6 from the Antarctic and PS1725-2 from the Arctic Ocean. The dashed lines indicate a difference of ±5% between both data sets. (b) Wet bulk density logs derived from gamma ray attenuation for two 1 m long core sections of gravity core PS1725-2. Superimposed are density values measured on discrete samples. Modified after Gerland and Villinger (1995).

the striking advantage of the almost continuous wet bulk density log clearly becomes obvious. Measured with an increment of 5 mm a lot of fine-scale variations are defined, a resolution which could never be reached by the time-consuming analysis of discrete samples (Gerland 1993; Gerland and Villinger 1995).

The precision of wet bulk densities can be slightly improved, if the iterative scheme for the mass attenuation coefficient is applied (Fig. 2.5a). For core PS1725-2 wet bulk densities computed with a constant mass attenuation coefficient are compared with those derived from the iterative procedure. Below 1.9 g cm^{-3} the iteration produces slightly smaller densities than are determined with a constant mass attenuation coefficient and are

thus below the dotted 1:1 line. Above 1.9 g cm^{-3} densities based on the iterative procedure are slightly higher.

If both data sets are plotted versus the wet bulk densities of the discrete samples the optimization essentially becomes obvious for high densities (>2.0 g cm^{-3}, Fig. 2.5b). After iteration densities are slightly closer to the dotted 1:1 line.

While this improvement is usually small and here only on the order of 1.3% ($\approx$0.02 g cm^{-3}), differences between assumed and true grain density affect the iteration more distinctly (Fig. 2.5c). As an example wet bulk densities of core PS1725-2 were calculated with constant grain densities of 2.65, 2.75 and 2.10 g cm^{-3}, values which are typical for calcareous, terrigenous and diatomaceous

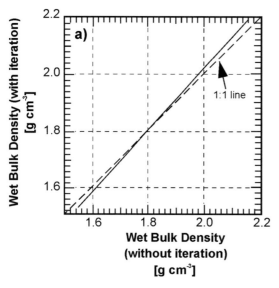

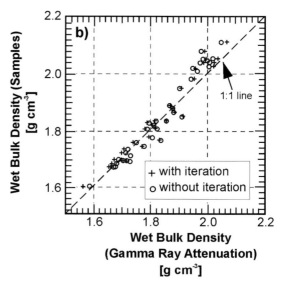

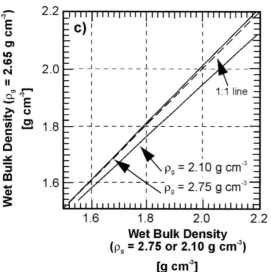

Fig. 2.5 Influence of an iterative mass attenuation coefficient determination on the precision of wet bulk densities. The gamma ray attenuation log of gravity core PS1725-2 was used as test data set. (a) Wet bulk densities calculated with a constant mass attenuation coefficient ('processing porosity' =50%) are displayed versus the data resulting from the iteration. A pore fluid density of 1.024 g cm^{-3} and a constant grain density of 2.7 g cm^{-3} were used, and the iteration was stopped if densities of two successive steps differed by less than 0.1‰ (b) Cross plot of wet bulk densities measured on discrete samples versus wet bulk densities calculated from gamma ray attenuation with a constant mass attenuation coefficient (O) and with the iterative scheme (+). (c) Influence of grain density on iteration. Three grain densities of 2.65, 2.75 and 2.1 g cm^{-3} were used to calculate wet bulk densities. Modified after Gerland (1993).

sediments. Displayed as cross-plot, calculations with grain densities of 2.65 and 2.75 g cm^{-3} almost coincide and follow the dotted 1:1 line. However, computations with a grain density of 2.10 g cm^{-3} differ by 1.3 - 3.6% (0.02 - 0.08 g cm^{-3}) from those with a grain density of 2.65 g cm^{-3}. This result is particularly important for cores composed of terrigenous and biogenic material of significantly different grain densities like the terrigenous and diatomaceous components in core PS1821-6. Here, an iteration with a constant grain density of 2.65 g cm^{-3} would give erroneous wet bulk densities in the diatomaceous parts. In such cores a depthdependent grain density profile should be applied in combination with the iterative procedure, or otherwise wet bulk densities would better be calculated with a constant mass attenuation coefficient. However, if the grain density is almost constant downcore the iterative scheme could be applied without any problems.

2.2.3 Electrical Resistivity (Galvanic Method)

The electrical resistivity of water-saturated sediments depends on the resistivity of its solid and fluid constituents. However, as the sediment grains are poor conductors an electrical current mainly propagates in the pore fluid. The dominant transport mechanism is an electrolytic conduction by ions and molecules with an excess or deficiency of electrons. Hence, current propagation in water-saturated sediments actually transports material through the pore space, so that the resistivity depends on both the conductivity of the pore water and the microstructure of the sediment. The conductivity of pore water varies with its salinity, and mobility and concentration of dissolved ions and molecules. The microstructure of the sediment is controlled by the amount and distribution of pore space and its capillarity and tortuosity. Thus, the electrical resistivity cannot be considered as a bulk parameter which strictly only depends on the relative amount of solid and fluid components, but as shown below, it can be used to derive porosity and wet bulk density as bulk parameters after calibration to a 'typical' sediment composition of a local sedimentation environment.

Several models were developed to describe current flow in rocks and water-saturated sediments theoretically. They encompass simple plane layered models (Waxman and Smits 1968) as well

as complex approximations of pore space by self similar models (Sen et al. 1981), effective medium theories (Sheng 1991) and fractal geometries (Ruffet et al. 1991). In practice these models are of minor importance because often only few of the required model parameters are known. Here, an empirical equation is preferred which relates the resistivity (R_s) of the wet sediment to its fractional porosity (ϕ) (Archie 1942)

$$F = \frac{R_s}{R_f} = a \cdot \phi^{-m} \qquad (2.12)$$

The ratio of the resistivity (R_s) in sediment to the resistivity (R_f) in pore water defines the formation (resistivity) factor (F). (a) and (m) are constants which characterize the sediment composition. As Archie (1942) assumed that (m) indicates the consolidation of the sediment it is also called cementation exponent (cf. Sect. 3.2.2). Several authors derived different values for (a) and (m). For an overview please refer to Schön (1996). In marine sediments often Boyce's (1968) values (a = 1.3, m = 1.45), determined by studies on diatomaceous, silty to sandy arctic sediments, are applied. Nevertheless, these values can only be rough estimates. For absolutely correct porosities both constants must be calibrated by an additional porosity measurement, either on discrete samples or by gamma ray attenuation. Such calibrations are strictly only valid for that specific data set but, with little loss of accuracy, can be transferred to regional environments with similar sediment compositions. Wet bulk densities can then be calculated using equation 2.3 and assuming a grain density (cf. also section 3.2.2).

Electrical resistivities can be measured on split cores by a half-automated logging system (Bergmann 1996). It measures the resistivity (R_s) and temperature (T) by a small probe which is manually inserted into the upper few millimeters of the sediment. The resistivity (R_f) of the interstitial pore water is simultaneously calculated from a calibration curve which defines the temperature-conductivity relation of standard sea water (35‰ salinity) by a fourth power law (Siedler and Peters 1986).

The accuracy and resolution that can be achieved compared to measurements on discrete samples were studied on the terrigenous square barrel kastenlot core PS2178-5 from the Arctic Ocean. If both data sets are displayed as cross plots porosities mainly range within the dashed 10% error lines, while densities mainly differ by

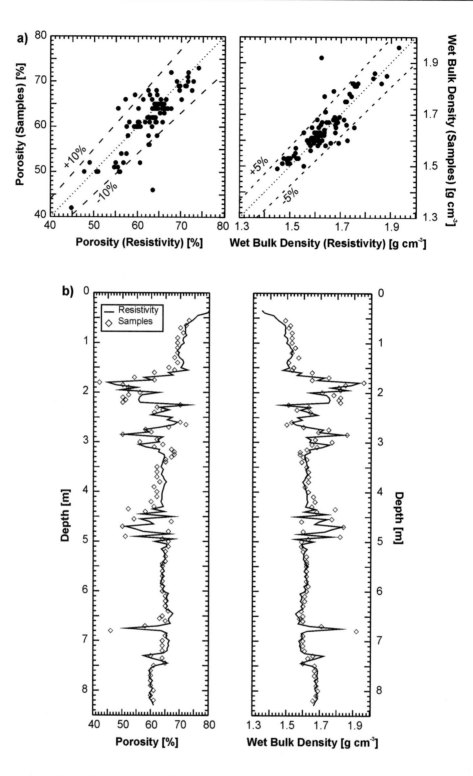

Fig. 2.6 Comparison of porosities and wet bulk densities measured on discrete samples and by electrical resistivities. Boyce's (1968) values for the coefficients (a) and (m) and pore fluid and grain densities of 1.024 g cm^{-3} and 2.67 g cm^{-3} were used to convert formation factors into porosities and wet bulk densities. Wet and dry weights and volumes were analyzed on discrete samples. (a) Cross plots of both data sets for square barrel kastenlot core PS2178-5. The dashed lines indicate an error of 10% for the porosity and 5% for the density data. (b) Porosity and wet bulk density logs of core PS2178-5 derived from resistivity measurements. Superimposed are porosity and density values measured on discrete samples. Data from Bergmann (1996).

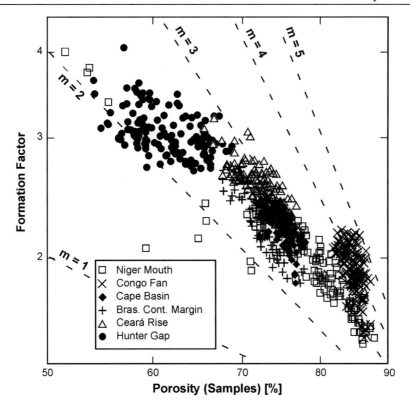

Fig. 2.7 Formation factor versus porosity for six gravity cores retrieved from different sedimentation provinces in the South Atlantic. Porosities were determined on discrete samples by wet and dry weights and volumes, formation factors by resistivity measurements. The dashed lines indicate Archie's law for a = 1 and cementation exponents (m) between 1 and 5. For a description of the sedimentation provinces, core numbers, coring locations, sediment compositions, water depths and constants (a) and (m) derived from linear least square fits please refer to Table 2.1. Unpublished data from M. Richter, University Bremen, Germany.

less than 5% (Fig. 2.6a). The core logs illustrate that the largest differences occur in the laminated sandy layers, particularly between 1.5 and 3.5 m and between 4.3 and 5.1 m depth (Fig. 2.6b). A drainage of the sandy layers, slight differences in the depth scale and different volumes considered by both methods probably cause this scatter. Nevertheless, in general both data sets agree very well having used Boyce's (1968) values for (a) and (m) and a typical terrigenous grain density of 2.67 g/cm³ for porosity and wet bulk density computations. This is valid because the sedimentation environment of core PS2178-5 from the Arctic Ocean is rather similar to the Bering Sea studied by Boyce (1968).

The influence of the sediment composition on the formation factor-porosity relation illustrates Figure 2.7 for six provinces in the South Atlantic. For each core porosities and formation factors were evaluated at the same core depths by wet and dry weights and volumes of discrete samples

and by electrical resistivities. Displayed as a cross plot on a log-log scale the data sets of each core show different trends and thus different cementation exponents. Jackson et al. (1978) tried to relate such variations in the cementation exponent to the sphericity of sediment particles. Based on studies on artificial samples they found higher cementation exponents if particles become less spherical. For natural sediments incorporating a large variety of terrigenous and biogenic particle sizes and shapes, it is difficult to verify similar relations. It is only obvious that coarse-grained calcareous foraminiferal oozes from the Hunter Gap show the lowest and fine-grained, diatom-bearing hemipelagic mud from the Congo Fan upwelling the highest cementation exponent. Simultaneously, the constant (a) decreases with increasing cementation exponent (Tab. 2.1). Possibly, in natural sediments the amount and distribution of pore space are more important than the particle shape. Generally, cementation exponent (m), constant (a) and formation factor (F) together characterize a specific environment. Table 2.1 summarizes the values of (a) and (m) derived from a linear least square fit to the log-log display of the data sets and shortly describes the sediment compositions.

With these values for (a) and (m) calibrated porosity logs are calculated which agree well with porosities determined on discrete samples (Fig. 2.8, black curves). The error in porosity that may result from using the 'standard' Boyce's (1968) values for (a) and (m) instead of those derived from the calibration appears in two different ways (gray curves). (1) The amplitude of the downcore porosity variations might be too large, as is illustrated by core GeoB2110-4 from the Brazilian Continental Margin. (2) The log might be shifted to higher or lower porosities, as is shown by core GeoB1517-1 from the Ceará Rise. Only if linear regression results in values for (a) and (m) close to Boyce's (1968) coefficients the error in the porosity log is negligible (core GeoB1701-4 from the Niger Mouth).

In order to compute absolutely correct wet bulk densities from calibrated porosity logs a grain density must be assumed. For most terrigenous and calcareous sediment cores this parameter is not very critical as it often only changes by few percent downcore (e.g. 2.6 - 2.7 g cm^{-3}). However, in cores from the Antarctic Polar Frontal Zone where an interlayering of diatomaceous and calcareous oozes indicates the advance and retreat of the oceanic front during glacial and interglacial stages grain densities may vary between about 2.0 and 2.8 g cm^{-3}. Here, depth-dependent values must either be known or modeled in order to get correct wet bulk density variations from resistivity measurements. An example for this approach are resistivity measurements on ODP core 690C from the Maud Rise (Fig. 2.9). While the carbonate log (b) clearly indicates calcareous layers with high and diatomaceous layers with zero $CaCO_3$ percentages (O'Connell 1990), the resistivity-based porosity log (a) only scarcely reflects these lithological changes. The reason is that calcareous and diatomaceous oozes are characterized by high inter- and intraporosities incorporated in and between hollow foraminifera and diatom shells. In contrast, the wet bulk density log measured onboard of JOIDES Resolution by gamma ray attenuation ((c), gray curve) reveals pronounced variations. They obviously correlate with the $CaCO_3$-content and can thus only be attributed to downcore changes in the grain density. So, a grain density model (d) was developed. It averages the densities of carbonate (ρ_{carb} = 2.8 g cm^{-3}) and biogenic opal (ρ_{opal} = 2.0 g cm^{-3}) (Barker, Kennnett et al. 1990) according to the fractional $CaCO_3$-content (C), $\rho_{model} = C \times \rho_{carb} + (1 - C) \times \rho_{opal}$. Based on this model wet bulk densities ((c), black curve) were derived from

Table 2.1 Geographical coordinates, water depth, core length, region and composition of the sediment cores considered in Figure 2.7. The cementation exponent (m) and the constant (a) are derived from the slope and intercept of a linear least square fit to the log-log display of formation factors versus porosities.

Core	Coordinates	Water Depth	Core Length	Region	Sediment Composition	Factor (a)	Cementation Exponent (m)
GeoB 1306-2	35°12.4'S 26°45.8'W	4058 m	6.97 m	Hunter Gap	foram.-nannofossil ooze, sandy	1,6	1,3
GeoB 1517-1	04°44.2'N 43°02.8'W	4001 m	6.89 m	Ceará Rise	nannofossil-foram. ooze	1,3	2,1
GeoB 1701-4	01°57.0'N 03°33.1'E	4162 m	7.92 m	Niger Mouth	clayey mud, foram. bearing	1,4	1,4
GeoB 1724-2	29°58.2'S 08°02.3'E	5084 m	7.45 m	Cape Basin	red clay	1,0	2,8
GeoB 2110-4	28°38.8'S 45°31.1'W	3011 m	8.41 m	Braz. Cont. Margin	pelagic clay. Foram. bearing	0,8	3,3
GeoB 2302-2	05°06.4'S 10°05.5'E	1830 m	14.18 m	Congo Fan	hemipelagic mud, diatom bearing, H$_2$S	0,8	5,3

the porosity log which agree well with the gamma ray attenuation densities ((c), gray curve) and show less scatter.

Though resistivities are only measured half-automatically including a manual insertion and removal of the probe, increments of 1 - 2 cm can be realized within an acceptable time so that more fine-scale structures can be resolved than by analysis of discrete samples. However, the real advantage compared to an automated gamma ray attenuation logging is that the resistivity measurement system can easily be transported, e.g.

onboard of research vessels or to core repositories, while the transport of radioactive sources requires special safety precautions.

2.2.4 Electrical Resistivity (Inductive Method)

A second, non-destructive technique to determine porosities by resistivity measurements uses a coil as sensor. A current flowing through the coil induces an electric field in the unsplit sediment core while it is automatically transported through

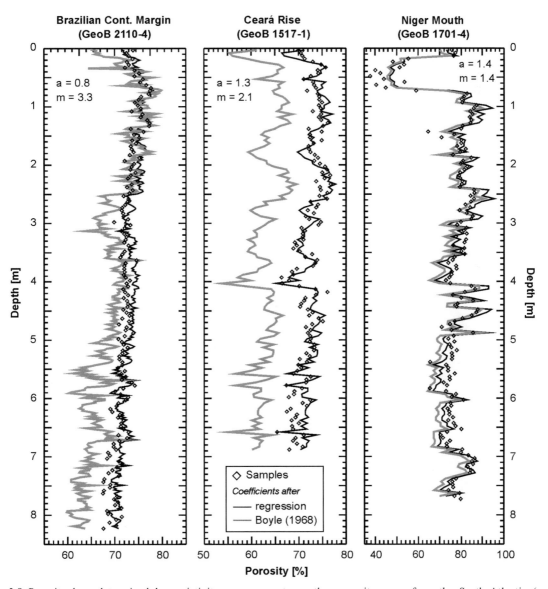

Fig. 2.8 Porosity logs determined by resistivity measurements on three gravity cores from the South Atlantic (see also Fig. 2.7 and Table 2.1). Gray curve: Boyce's (1968) values were used for the constants (a) and (m). Black curve: (a) and (m) were derived from the slope and intercept of a linear least square fit. These values are given at the top of each log. Superimposed are porosities determined on discrete samples by weight and volume measurements (unpublished data from P. Müller, University Bremen, Germany).

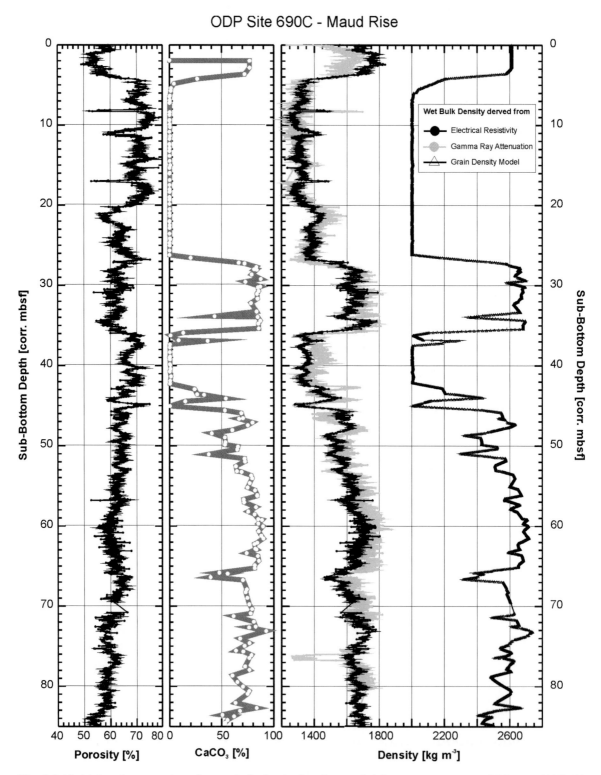

Fig. 2.9 Model based computation of a wet bulk density log from resistivity measurements on *ODP* core 690C. (a) Porosity log derived from formation factors having used Boyce's (1968) values for (a) and (m) in Archie's law. (b) Carbonate content (O'Conell 1990). (c) Wet bulk density log analyzed from gamma ray attenuation measurements onboard of JOIDES Resolution (gray curve). Superimposed is the wet bulk density log computed from electrical resistivity measurements on archive halves of the core (black curve) having used the grain density model shown in (d). Unpublished data from B. Laser and V. Spieß, University Bremen, Germany.

the centre of the coil (Gerland et al. 1993). This induced electric field contains information on the magnetic and electric properties of the sediment.

Generally, the coil characteristic is defined by the quality value(Q)

$$Q = \omega \cdot \frac{L(\omega)}{R(\omega)} \qquad (2.13)$$

(L(ω)) is the inductance, (R(ω)) the resistance and (ω) the (angular) frequency of the alternating current flowing through the coil (Chelkowski 1980). The inductance (L(ω)) depends on the number of windings, the length and diameter of the coil and the magnetic permeability of the coil material. The resistance (R(ω)) is a superposition of the resistance of the coil material and losses of the electric field induced in the core. It increases with decreasing resistivity in the sediment. Whether the inductance or the resistance is of major importance depends on the frequency of the current flowing through the coil. Changes in the inductance (L(ω)) can mainly be measured if currents of some kilohertz frequency or less are used. They simultaneously indicate variations in the magnetic susceptibility while the resistance (R(ω)) is insensitive to changes in the resistivity of the sediment. In contrast, operating with currents of several megahertz allows to measure the resistivity of the sediment by changes of the coil resistance (R(ω)) while variations in the

magnetic susceptibility do not affect the inductance (L(ω)). The examples presented here were measured with a commercial system (Scintrex CTU 2) which produces an output voltage that is proportional to the quality value (Q) of the coil at a frequency of 2.5 MHz, and after calibration is inversely proportional to the resistivity of the sediment.

The induced electric field is not confined to the coil position but extends over some sediment volume. Hence, measurements of the resistance (R(ω)) integrate over the resistivity distribution on both sides of the coil and provide a smoothed, low-pass filtered resistivity record. The amount of sediment volume affected by the induction process increases with larger coil diameters. The shape of the smoothing function can be measured from the impulse response of a thin metal plate glued in an empty plastic core liner. For a coil of about 14 cm diameter this gaussian-shaped function has a half-width of 4 cm (Fig. 2.10), so that the effect of an infinitely small resistivity anomaly is smeared over a depth range of 10 - 15 cm. This smoothing effect is equivalent to convolution of a source wavelet with a reflectivity function in seismic applications and can accordingly be removed by deconvolution algorithms. However, only few applications from longcore paleomagnetic studies are known up to now (Constable and Parker 1991; Weeks et al. 1993).

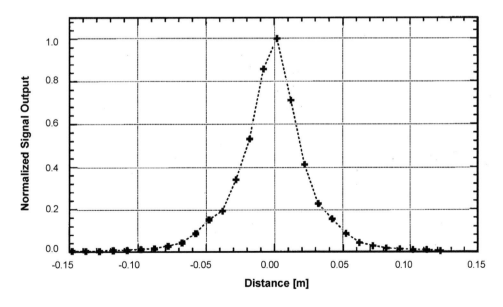

Fig. 2.10 Impulse response function of a thin metal plate measured by the inductive method with a coil of about 14 cm diameter. Modified after Gerland et al. (1993).

The low-pass filtering effect particularly becomes obvious if resistivity logs measured by the galvanic and inductive method are compared. Figure 2.11 displays such an example for a terrigenous core from the Weddell Sea (PS1635-1) and a biogenic foraminiferal and diatomaceous core from the Maud Rise (PS1836-3) in the Antarctic Ocean. Resistivities differ by maximum 15% (Fig. 2.11a), a rather high value which mainly results from core PS1635-1. Here, the downcore logs illustrate that the inductive methods produces lower resistivities than the galvanic method (Fig. 2.11b). In detail, the galvanic resistivity log reveals a lot of pronounced, fine-scale variations which cannot be resolved by induction measurements but are smeared along the core depth. For the biogenic core PS1836-3 this smoothing is not so important because lithology changes more gradually.

2.3   Permeability

Permeability describes how easy a fluid flows through a porous medium. Physically it is defined by Darcy's law

$$q = \frac{\kappa}{\eta} \cdot \frac{\partial p}{\partial x}$$ (2.14)

which relates the flow rate (q) to the permeability (κ) of the pore space, the viscosity (η) of the pore fluid and the pressure gradient ($\partial p / \partial x$) causing the fluid flow. Simultaneously, permeability depends on the porosity (ϕ) and grain size distribution of the sediment, approximated by the mean grain size (d_m). Assuming that fluid flow can be simulated by an idealized flow through a bunch of capillaries with uniform radius ($d_m/2$) (Hagen Poiseuille's flow) permeabilities can for instance be estimated from Kozeny-Carman's equation (Carman 1956; Schopper 1982)

$$\kappa = \frac{d_m^2}{36k} \cdot \frac{\phi^3}{(1-\phi)^2}$$ (2.15)

This relation is approximately valid for unconsolidated sediments of 30 - 80% porosity (Carman 1956). It is used for both geotechnical applications to estimate permeabilities of soil (Lambe and Whitman 1969) and seismic modeling of wave propagation in water-saturated sediments (Biot 1956a, b; Hovem and Ingram 1979; Hovem 1980;

Ogushwitz 1985). (κ) is a constant which depends on pore shape and tortuosity. In case of parallel, cylindrical capillaries it is about 2, for spherical sediment particles about 5, and in case of high porosities ≥ 10 (Carman 1956).

However, this is only one approach to estimate permeabilities from porosities and mean grain sizes. Other empirical relations exist, particularly for regions with hydrocarbon exploration (e.g. Gulf of Mexico, Bryant et al. 1975) or fluid venting (e.g. Middle Valley, Fisher et al. 1994) which compute depth-dependent permeabilities from porosity logs or take the grain size distribution and clay content into account.

Direct measurements of permeabilities in unconsolidated marine sediments are difficult, and only few examples are published. They confine to measurements on discrete samples with a specially developed tool (Lovell 1985), to indirect estimations by resistivity measurements (Lovell 1985), and to consolidation tests on ODP cores using a modified medical tool (Olsen et al. 1985). These measurements are necessary to correct for the elastic rebound (MacKillop et al. 1995) and to determine intrinsic permeabilities at the end of each consolidation step (Fisher et al. 1994). In Section 2.4.2 a numerical modeling and inversion scheme is described which estimates permeabilities from P-wave attenuation and dispersion curves (c.f. also section 3.6).

2.4   Acoustic and Elastic Properties

Acoustic and elastic properties are directly concerned with seismic wave propagation in marine sediments. They encompass P- and S-wave velocity and attenuation and elastic moduli of the sediment frame and wet sediment. The most important parameter which controls size and resolution of sedimentary structures by seismic studies is the frequency content of the source signal. If the dominant frequency and bandwidth are high, fine-scale structures associated with pore space and grain size distribution affect the elastic wave propagation. This is subject of ultrasonic transmission measurements on sediment cores (Sects. 2.4 and 2.5). At lower frequencies larger scale features like interfaces with different physical properties above and below and bedforms like mud waves, erosion zones and channel levee systems are the dominant structures imaged

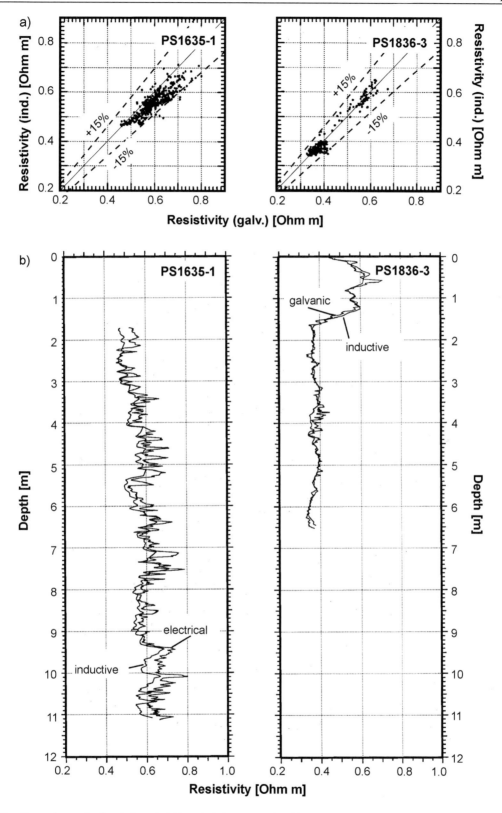

Fig. 2.11 Comparison of electrical resistivities (galv.) measured with the small hand-held probe and determined by the inductive method (ind.) for the gravity cores PS1635-1 and PS1836-3. (a) Cross plots of both data sets. The dashed lines indicate a difference of 15%. (b) Downcore resistivity logs determined by both methods. Modified after Gerland et al. (1993).

by sediment echosounder and multi-channel seismic surveys (Sect. 2.6).

In this section first Biot's viscoelastic model is summarized which simulates high- and low-frequency wave propagation in water-saturated sediments by computing phase velocity and attenuation curves. Subsequently, analysis techniques are introduced which derive P-wave velocities and attenuation coefficients from ultrasonic signals transmitted radially across sediment cores. Additional physical properties like S-wave velocity, elastic moduli and permeability are estimated by an inversion scheme.

2.4.1   Biot-Stoll Model

To describe wave propagation in marine sediments mathematically, various simple to complex models have been developed which approximate the sediment by a dilute suspension (Wood 1946) or an elastic, water-saturated frame (Gassmann 1951; Biot 1956a, b). The most common model which considers the microstructure of the sediment and simulates frequency-dependent wave propagation is based on Biot's theory (Biot 1956a, b). It includes Wood's suspension and Gassmann's elastic frame model as low-frequency approximations and combines acoustic and elastic parameters - P- and S-wave velocity and attenuation and elastic moduli - with physical and sedimentological parameters like mean grain size, porosity, density and permeability.

Based on Biot's fundamental work Stoll (e.g. 1974, 1977, 1989) reformulated the mathematical background of this theory with a simplified uniform nomenclature. Here, only the main physical principles and equations are summarized. For a detailed description please refer to one of Stoll's publications or Biot's original papers.

The theory starts with a description of the microstructure by 11 parameters. The sediment grains are characterized by their grain density (ρ_g) and bulk modulus (K_g), the pore fluid by its density (ρ_f), bulk modulus (K_f) and viscosity (η). The porosity (ϕ) quantifies the amount of pore space. Its shape and distribution are specified by the permeability (κ), a pore size parameter $a=d_m/3 \cdot \phi/(1-\phi)$, d_m = mean grain size (Hovem and Ingram 1979; Courtney and Mayer 1993), and structure factor $a'=1-r_0(1-\phi^{-1})$ ($0 \leq r_0 \leq 1$) indicating a tortuosity of the pore space (Berryman 1980). The elasticity of the sediment frame is considered by its bulk and shear modulus (K_m and μ_m).

An elastic wave propagating in water-saturated sediments causes different displacements of the pore fluid and sediment frame due to their different elastic properties. As a result (global) fluid motion relative to the frame occurs and can approximately be described as Poiseuille's flow. The flow rate follows Darcy's law and depends on the permeability and viscosity of the pore fluid. Viscous losses due to an interstitial pore water flow are the dominant damping mechanism. Intergranular friction or local fluid flow can additionally be included but are of minor importance in the frequency range considered here.

Based on generalized Hooke's law and Newton's 2. Axiom two equations of motions are necessary to quantify the different displacements of the sediment frame and pore fluid. For P-waves they are (Stoll 1989)

$$\nabla^2 (H \cdot e - C \cdot \zeta) = \frac{\partial^2}{\partial t^2} (\rho \cdot e - \rho_f \cdot \zeta)$$

(2.16a)

$$\nabla^2 (C \cdot e - M \cdot \zeta) = \frac{\partial^2}{\partial t^2} (\rho_f \cdot e - m \cdot \zeta) - \frac{\eta}{\kappa} \cdot \frac{\partial \zeta}{\partial t}$$

(2.16b)

Similar equations for S-waves are given by Stoll (1989). Equation 2.16a describes the motion of the sediment frame and Equation 2.16b the motion of the pore fluid relative to the frame.

$$e = div(\vec{u})$$ (2.16c)

and

$$\zeta = \phi \cdot div(\vec{u} - \vec{U})$$ (2.16d)

are the dilatations of the frame and between pore fluid and frame ($\vec{u}$ = displacement of the frame, $\vec{U}$ = displacement of the pore fluid). The term ($\eta/\kappa \cdot \partial \zeta/\partial t$) specifies the viscous losses due to global pore fluid flow, and the ratio (η/κ) the viscous flow resistance.

The coefficients (H), (C), and (M) define the elastic properties of the water-saturated model. They are associated with the bulk and shear moduli of the sediment grains, pore fluid and sediment frame (K_g), (K_f), (K_m), (μ_m) and with the porosity (ϕ) by

$$H = (K_g - K_m)^2 / (D - K_m) + K_m + {}^4/_3 \cdot \mu_m \quad (2.17a)$$

$$C = (K_g \cdot (K_g - K_m)) / (D - K_m) \quad (2.17b)$$

$$M = K_g^2 / (D - K_m) \qquad (2.17c)$$

$$D = K_g \cdot (1 + \phi \cdot (K_g/K_f - 1)) \qquad (2.17d)$$

The apparent mass factor $m = a' \cdot \rho_f / \phi$ ($a' \geq 1$) in Equation 2.16b considers that not all of the pore fluid moves along the maximum pressure gradient in case of tortuous, curvilinear capillaries. As a result the pore fluid seems to be more dense, with higher inertia. (a') is called structure factor and is equal to 1 in case of uniform parallel capillaries.

In the low frequency limit $H - 4/3\mu_m$ (Eq. 2.17a) represents the bulk modulus computed by Gassmann (1951) for a 'closed system' with no pore fluid flow. If the shear modulus (μ_m) of the frame is additionally zero, the sediment is approximated by a dilute suspension and Equation 2.17a reduces to the reciprocal bulk modulus of Wood's equation for 'zero' acoustic frequency, ($K^{-1} = \phi/K_f + (1-\phi)/K_g$; Wood 1946).

Additionally, Biot (1956a, b) introduced a complex correction function (F) which accounts for a frequency-dependent viscous flow resistance (η/κ). In fact, while the assumption of an ideal Poiseuille flow is valid for lower frequencies, deviations of this law occur at higher frequencies. For short wavelengths the influence of pore fluid viscosity confines to a thin skin depth close to the sediment frame, so that the pore fluid seems to be less viscous. To take these effects into account the complex function (F) modifies the viscous flow resistance (η/κ) as a function of pore size, pore fluid density, viscosity and frequency. A complete definition of (F) can be found in Stoll (1989).

The equations of motions 2.16 are solved by a plane wave approach which leads to a 2 x 2 determinant for P-waves

$$\det\begin{pmatrix} H \cdot k^2 - \rho\omega^2 & \rho_f\omega^2 - C \cdot k^2 \\ C \cdot k^2 - \rho_f\omega^2 & m\omega^2 - M \cdot k^2 - i\omega F\eta/\kappa \end{pmatrix} = 0$$

$$(2.18)$$

and a similar determinant for S-waves (Stoll 1989). The variable $k(\omega) = k_r(\omega) + ik_i(\omega)$ is the complex wavenumber. Computations of the complex zeroes of the determinant result in the phase velocity $c(\omega) = \omega/k_r(\omega)$ and attenuation coefficient $\alpha(\omega) = k_i(\omega)$ as real and imaginary parts. Generally, the determinant for P-waves has two and that for S-

waves one zero representing two P- and one S-wave propagating in the porous medium. The first P-wave (P-wave of first kind) and the S-wave are well known from conventional seismic wave propagation in homogeneous, isotropic media. The second P-wave (P-wave of second kind) is similar to a diffusion wave which is exponentially attenuated and can only be detected by specially arranged experiments (Plona 1980).

An example of such frequency-dependent phase velocity and attenuation curves presents Figure 2.12 for P- and S-waves together with the slope (power (n)) of the attenuation curves ($\alpha = k \cdot f^n$). Three sets of physical properties representing typical sand, silt and clay (Table 2.2) were used as model parameters. The attenuation coefficients show a significant change in their frequency dependence. They follow an $\alpha \sim f^2$ power law for low and an $\alpha \sim \sqrt{f}$ law for high frequencies and indicate a continuously decreasing power (n) (from 2 to 0.5) near a characteristic frequency $f_c = (\eta\phi)/(2\pi\kappa\rho_f)$. This characteristic frequency depends on the microstructure (porosity (ϕ), permeability (κ)) and pore fluid (viscosity (η), density (ρ_f)) of the sediment and is shifted to higher frequencies if porosity increases and permeability decreases. Below the characteristic frequency the sediment frame and pore fluid moves in phase and coupling between solid and fluid components is at maximum. This behaviour is typical for clayey sediments in which low permeability and viscous friction prevent any relative movement between pore fluid and frame up to several MHz, in spite of their high porosity. With increasing permeability and decreasing porosity the characteristic frequency diminishes so that in sandy sediments movements in phase only occur up to about 1 kHz. Above the characteristic frequency wavelengths are short enough to cause relative motions between pore fluid and frame.

Phase velocities are characterized by a low- and high-frequency plateau with constant values and a continuous velocity increase near the characteristic frequency. This dispersion is difficult to detect because it is confined to a small frequency band. Here, dispersion could only be detected from 1 - 10 kHz in sand, from 50 - 500 kHz in silt and above 100 MHz in clay. Generally, velocities in coarse-grained sands are higher than in fine-grained clays.

S-waves principally exhibit the same attenuation and velocity characteristics as P-waves. However, at the same frequency attenuation is

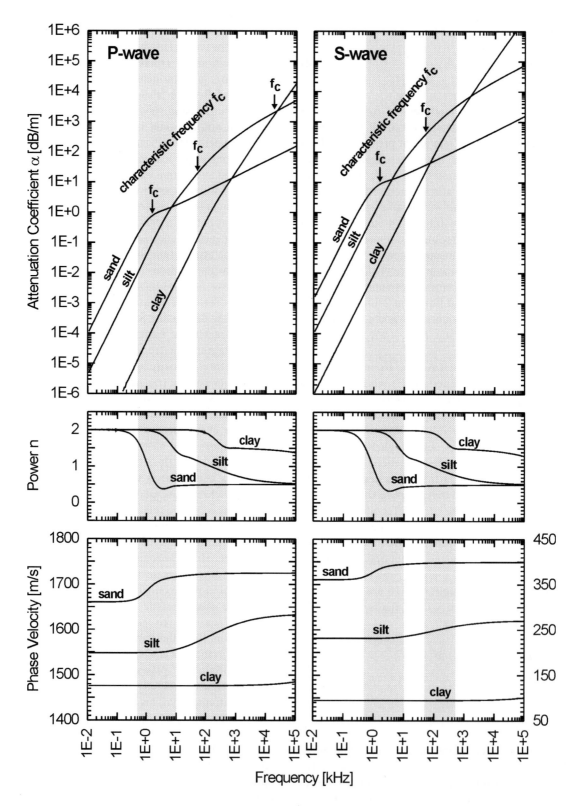

Fig. 2.12 Frequency-dependent attenuation and phase velocity curves and power (n) of the attenuation law $\alpha = k \cdot f^n$ computed for P-and S-waves in typical sand, silt and clay. The gray-shaded areas indicate frequency bands typical for ultrasonic measurements on sediment cores (50 - 500 kHz) and sediment echosounder surveys (0.5 - 10 kHz). Modified after Breitzke (1997).

about one magnitude higher, and velocities are significantly lower than for P-waves. The consequence is that S-waves are very difficult to record due to their high attenuation, though they are of great value for identifying fine-scale variations in the elasticity and microstructure of marine sediments. This is even valid if the low S-wave velocities are taken into account and lower frequencies are used for S-wave measurements than for P-wave recordings.

The two gray-shaded areas in Figure 2.12 mark two frequency bands typical for ultrasonic studies on sediment cores (50 - 500 kHz) and sediment echosounder surveys (0.5 - 10 kHz). They are displayed in order to point to one characteristic of acoustic measurements. Attenuation coefficients analyzed from ultrasonic measurements on sediment cores cannot directly be transferred to

sediment echosounder or seismic surveys. Primarily they only reflect the microstructure of the sediment. Rough estimates of the attenuation in seismic recordings from ultrasonic core measurements can be derived if ultrasonic attenuation is modeled, and attenuation coefficients are extrapolated to lower frequencies by such model curves.

2.4.2 Full Waveform Ultrasonic Core Logging

To measure the P-wave velocity and attenuation illustrated by Biot-Stoll's model an automated, PC-controlled logging system was developed which records and stores digital ultrasonic P-waveforms transmitted radially across marine sediment cores (Breitzke and Spieß 1993). These transmission measurements can be done at arbitrary small depth

Table 2.2 Physical properties of sediment grains, pore fluid and sediment frame used for the computation of attenuation and phase velocity curves according to Biot-Stoll's sediment model (Fig. 2.12).

Parameter	Sand	Silt	Clay
Sediment Grains			
Bulk Modulus K_g [10^9 Pa]	38	38	38
Density ρ_g [g cm^{-3}]	2.67	2.67	2.67
Pore Fluid			
Bulk Modulus K_f [10^9 Pa]	2.37	2.37	2.37
Density ρ_f [g cm^{-3}]	1.024	1.024	1.024
Viscosity η [10^{-3} Pa·s]	1.07	1.07	1.07
Sediment Frame			
Bulk Modulus K_m [10^6 Pa]	400	150	20
Shear Modulus μ_m [10^6 Pa]	240	90	12
Poisson Ratio σ_m	0.25	0.25	0.25
Pore Space			
Porosity ϕ [%]	50	60	80
Mean Grain Size d_m [10^{-6} m]	70	30	2
Permeability κ [m^2]	$5.4 \cdot 10^{-11}$	$2.3 \cdot 10^{-12}$	$7.1 \cdot 10^{-15}$
Pore Size Parameter $a = d_m / 3 \phi /(1-\phi)$ [10^{-6} m]	23	15	2.7
Ratio κ / a^2	0.1	0.01	0.001
Structure Factor $a' = 1 - r_0 (1-\phi^{-1})$	1.5	1.3	1.1
Constant r_0	0.5	0.5	0.5

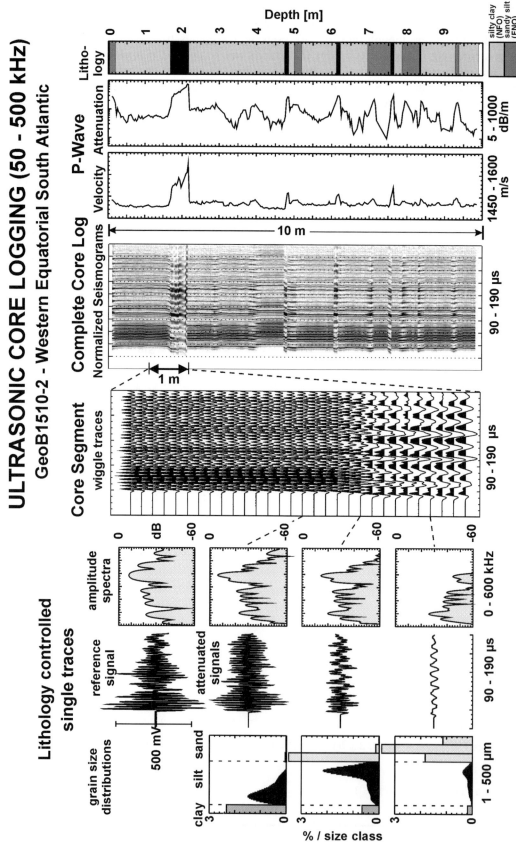

Fig. 2.13 Full waveform ultrasonic core logging, from lithology controlled single traces to the gray shaded pixel graphic of transmission seismograms, and P-wave velocity and attenuation logs derived from the transmission data. The single traces on the left-hand side reflect true amplitudes while the wiggle traces of the core segment and the pixel graphic are normalized to maximum values. Data from Breitzke et al. (1996).

increments so that the resulting seismogram sections can be combined to an ultrasonic image of the core.

Figure 2.13 displays the most prominent effects involved in full waveform ultrasonic core logging. Gravity core GeoB1510-2 from the western equatorial South Atlantic serves as an example. The lithology controlled single traces and amplitude spectra demonstrate the influence of increasing grain sizes on attenuation and frequency content of transmission seismograms. Compared to a reference signal in distilled water, the signal shape remains almost unchanged in case of wave propagation in fine-grained clayey sediments (1st attenuated trace). With an increasing amount of silty and sandy particles the signal amplitudes are reduced due to an enhanced attenuation of high-frequency components (2nd and 3rd attenuated trace). This attenuation is accompanied by a change in signal shape, an effect which is particularly obvious in the normalized wiggle trace display of the 1 m long core segment. While the upper part of this segment is composed of fine-grained nannofossil ooze, a calcareous foraminiferal turbidite occurs in the lower part. The downward coarsening of the graded bedding causes successively lower-frequency signals which can easily be distinguished from the high-frequency transmission seismograms in the upper fine-grained part. Additionally, first arrival times are lower in the coarse-grained turbidite than in fine-grained nannofossil oozes indicating higher velocities in silty and sandy sediments than in the clayey part. A conversion of the normalized wiggle traces to a gray-shaded pixel graphic allows us to present the full transmission seismogram information on a handy scale. In this ultrasonic image of the sediment core lithological changes appear as smooth or sharp phase discontinuities resulting from the low-frequency waveforms in silty and sandy layers. Some of these layers indicate a graded bedding by downward prograding phases (1.60 - 2.10 m, 4.70 - 4.80 m, 7.50 - 7.60 m). The P-wave velocity and attenuation log analyzed from the transmission seismograms support the interpretation. Coarse-grained sandy layers are characterized by high P-wave velocities and attenuation coefficients while fine-grained parts reveal low values in both parameters. Especially, attenuation coefficients reflect lithological changes much more sensitively than P-wave velocities.

While P-wave velocities are determined online during core logging using a cross-correlation technique for the first arrival detection (Breitzke and Spieß 1993)

$$v_P = \frac{d_{outside} - 2d_{liner}}{t - 2t_{liner}} \qquad (2.19)$$

($d_{outside}$ = outer core diameter, $2\ d_{liner}$ = double liner wall thickness, t = detected first arrival, $2t_{liner}$ = travel time across both liner walls), attenuation coefficients are analyzed by a post-processing routine. Several notches in the amplitude spectra of the transmission seismograms caused by the resonance characteristics of the ultrasonic transducers required a modification of standard attenuation analysis techniques (e.g. Jannsen et al. 1985; Tonn 1989, 1991). Here, a modification of the spectral ratio method is applied (Breitzke et al. 1996). It defines a window of bandwidth ($b_i = f_{ui} - f_{li}$) in which the spectral amplitudes are summed (Fig. 2.14a).

$$\overline{A}(f_{mi}, x) = \sum_{f_i = f_{li}}^{f_{ui}} A(f_i, x) \qquad (2.20)$$

The resulting value ($\overline{A}(f_{mi}, x)$) is related to that part of the frequency band which predominantly contributes to the spectral sum, i.e. to the arithmetic mean frequency (f_{mi}) of the spectral amplitude distribution within the ith band. Subsequently, for a continuously moving window a series of attenuation coefficients ($\alpha(f_{mi})$) is computed from the natural logarithm of the spectral ratio of the attenuated and reference signal

$$\alpha(f_{mi}) = \ln\left[\frac{\overline{A}(f_{mi}, x)}{\overline{A}_{ref}(f_{mi}, x)}\right] \Bigg/ x = k \cdot f^n \qquad (2.21)$$

Plotted in a log α - log f diagram the power (n) and logarithmic attenuation factor (log k) can be determined from the slope and intercept of a linear least square fit to the series of ($f_{mi}, \alpha(f_{mi})$) pairs (Fig. 2.14b). Finally, a smoothed attenuation coefficient $\alpha(f) = k \cdot f^n$ is calculated for the frequency (f) using these values for (k) and (n).

Figure 2.14b shows several attenuation curves ($\alpha(f_{mi})$) analyzed along the turbidite layer of core GeoB1510-2. With downward-coarsening grain sizes attenuation coefficients increase. Each linear regression to one of these curves provide a power (n) and attenuation factor (k), and thus one value $\alpha = k \cdot f^n$ on the attenuation log of the complete core (Fig. 2.14c).

ATTENUATION ANALYSIS

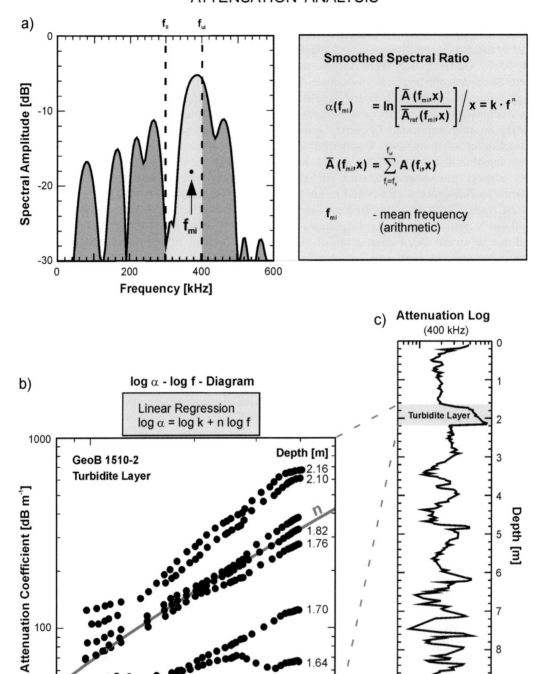

Fig. 2.14 Attenuation analysis by the smoothed spectral ratio method. (a) Definition of a moving window of bandwidth $b_i = f_{ui} - f_{li}$, in which the spectral amplitudes are summed. (b) Seven attenuation curves analyzed from the turbidite layer of gravity core GeoB1510-2 and linear regression to the attenuation curve in 1.82 m depth. (c) Attenuation log of gravity core GeoB1510-2 for 400 kHz frequency. Data from Breitzke et al. (1996).

As the P-wave attenuation coefficient obviously depends on the grain size distribution of the sediment it can be used as a proxy parameter for the mean grain size, i.e. for a sedimentological parameter which is usually only measured at coarse increments due to the time-consuming grain size analysis methods. For instance, this can be of major importance in current controlled sedimentation environments where high-resolution grain size logs might indicate reduced or enhanced current intensities.

If the attenuation coefficients and mean grain sizes analyzed on discrete samples of core GeoB1510-2 are displayed as a cross plot a second order polynomial can be derived from a least square fit (Fig. 2.15a). This regression curve then allows to predict mean grain sizes using the attenuation coefficient as proxy parameter. The accuracy of the predicted mean grain sizes illus-trates the core log in Figure 2.15b. The predicted gray shaded log agrees well with the superim-posed dots of the measured data.

Similarly, P-wave velocities can also be used as proxy parameters for a mean grain size prediction. However, as they cover only a small range (1450 - 1650 m/s) compared to attenuation coefficients (20 - 800 dB/m) they reflect grain size variations less sensitively.

Generally, it should be kept in mind, that the regression curve in Figure 2.15 is only an example. Its applicability is restricted to that range of attenuation coefficients for which the regression curve was determined and to similar sedimentation environments (calcareous foraminiferal and nannofossil ooze). For other sediment compo-sitions new regression curves must be determined, which are again only valid for that specific sedimentological setting.

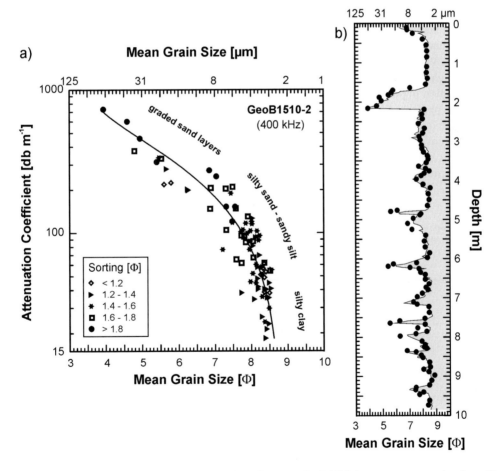

Fig. 2.15 (a) Attenuation coefficients (at 400 kHz) of gravity core GeoB1510-2 versus mean grain sizes. The solid line indicates a second degree polynomial used to predict mean grain sizes from attenuation coefficients. (b) Comparison of the predicted mean grain size log (gray shaded) with the data measured on discrete samples (solid dots). Mean grain sizes are given in $\Phi = -log_2\ d$, d = grain diameter in mm. Modified after Breitzke et al. (1996).

Biot-Stoll's theory allows us to model P-wave velocities and attenuation coefficients analyzed from transmission seismograms. As an example Figure 2.16 displays six data sets for the turbidite layer of core GeoB1510-2. While attenuation coefficients were analyzed as described above frequency dependent P-wave velocities were determined from successive bandpass filtered transmission seismograms (Courtney and Mayer 1993; Breitzke 1997). Porosities and mean grain sizes enter the modeling computations via the pore size parameter (a) and structure factor (a'). Physical properties of the pore fluid and sediment grains are the same as given in Table 2.2. A bulk and shear modulus of 10 and 6 MPa account for the elasticity of the frame. As the permeability is the parameter which is usually unknown but has the strongest influence on attenuation and velocity dispersion, model curves were computed for three constant ratios κ/a^2 = 0.030, 0.010 and 0.003 of permeability (κ) and pore size parameter (a). The resulting permeabilities are given in each diagram. These theoretical curves show that the attenuation and velocity data between 170 and 182 cm depth can consistently be modeled by an appropriate set of input parameters. Viscous losses due to a global pore water flow through the sediment are sufficient to explain the attenuation in these sediments. Only if the turbidite base is approached (188 - 210 cm depth) the attenuation and velocity dispersion data successively deviate from the model curves probably due to an increasing amount of coarse-grained foraminifera. An additional damping mechanism which

might either be scattering or resonance within the hollow foraminifera must be considered.

Based on this modeling computations an inversion scheme was developed which automatically iterates the permeability and minimizes the difference between measured and modeled attenuation and velocity data in a least square sense (Courtney and Mayer 1993; Breitzke 1997). As a result S-wave velocities and attenuation coefficients, permeabilities and elastic moduli of water-saturated sediments can be estimated. They are strictly only valid if attenuation and velocity dispersion can be explained by viscous losses. In coarse-grained parts deviations must be taken into account for the estimated parameters, too. Applied to the data of core GeoB1510-2 in 170, 176 and 182 cm depth S-wave velocities of 67, 68 and 74 m/s and permeabilities of $5 \cdot 10^{-13}$, $1 \cdot 10^{-12}$ and $3 \cdot 10^{-12}$ m^2 result from this inversion scheme.

2.5   Sediment Classification

Full waveform ultrasonic core logging was applied to terrigenous and biogenic sediment cores to analyze P-wave velocities and attenuation coefficients typical for the different settings. Together with the bulk parameters and the physical properties estimated by the inversion scheme they form the data base for a sediment classification which identifies different sediment types from their

Table 2.3 Geographical coordinates, water depth, core length, region and composition of the sediment cores considered for the sediment classification in Section 2.5.

Core	Coordinates	Water Depth	Core Length	Region	Sediment Composition
40KL	07°33.1'N 85°29.7'E	3814 m	8.46 m	Bengal Fan	terrigenous clay, silt, sand
47KL	11°10.9'N 88°24.9'E	3293 m	10.00 m	Bengal Fan	terrigenous clay, silt, sand; foram. and nannofossil ooze
GeoB 2821-1	30°27.1'S 38°48.9'W	3941 m	8.19 m	Rio Grande Rise	foram. and nannofossil ooze
PS 2567-2	46°56.1'S 06°15.4'E	4102 m	17.65 m	Meteor Rise	diatomaceous mud/ooze; few foram. and nannofossil layers

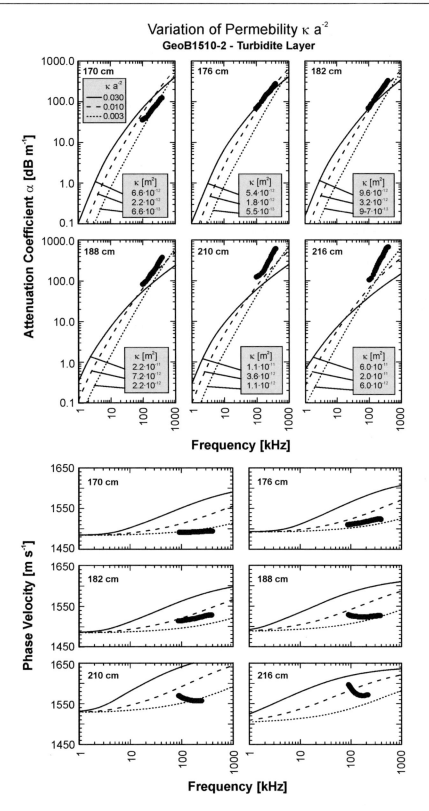

Fig. 2.16 Comparison of P-wave attenuation and velocity dispersion data derived from ultrasonic transmission seismograms with theoretical curves based on Biot-Stoll's model for six traces of the turbidite layer of gravity core GeoB1510-2. Permeabilities vary in the model curves according to constant ratios κ/a^2 = 0.030, 0.010, 0.003 (κ = permeability, a = pore size parameter). The resulting permeabilities are given in each diagram. Modified after Breitzke et al. (1996).

acoustic and elastic properties. Table 2.3 summarizes the cores used for this sediment classification.

2.5.1   Full Waveform Core Logs as Acoustic Images

That terrigenous, calcareous and biogenic siliceous sediments differ distinctly in their acoustic properties is shown by four transmission seismogram sections in Figure 2.17. Terrigenous sediments from the Bengal Fan (40KL, 47KL) are composed of upward-fining sequences of turbidites characterized by upward decreasing attenuations and P-wave velocities. Coarse-grained basal sandy layers can easily be located by low-frequency waveforms and high P-wave velocities.

Calcareous sediments from the Rio Grande Rise (GeoB2821-1) in the western South Atlantic also exhibit high- and low-frequency signals which scarcely differ in their P-wave velocities. In these sediments high-frequency signals indicate fine-grained nannofossil ooze while low-frequency signals image coarse-grained foraminiferal ooze.

The sediment core from the Meteor Rise (PS2567-2) in the Antarctic Ocean is composed of diatomaceous and foraminiferal-nannofossil ooze deposited during an advance and retreat of the Polar Frontal Zone in glacial and interglacial stages. Opal-rich, diatomaceous ooze can be identified from high-frequency signals while foraminiferal-nannofossil ooze causes higher attenuation and low-frequency signals. P-wave velocities again only show smooth variations.

Acoustic images of the complete core lithologies present the colour-encoded graphics of the transmission seismograms, in comparison to the lithology derived from visual core inspection (Fig. 2.18). Instead of normalized transmission seismograms instantaneous frequencies are displayed here (Taner et al. 1979). They reflect the dominant frequency of each transmission seismogram as time-dependent amplitude, and thus directly indicate the attenuation. Highly attenuated low-frequency seismograms appear as warm red to white colours while parts with low attenuation and high-frequency seismograms are represented by cool green to black colours.

In these attenuation images the sandy turbidite bases in the terrigenous cores from the Bengal Fan (40KL, 47KL) can easily be distinguished. Graded beddings can also be identified from slightly prograding phases and continuously

decreasing travel times. In contrast, the transition to calcareous, pelagic sediments in the upper part (> 5.6 m) of core 47KL is rather difficult to detect. Only above 3.2 m depth a slightly increased attenuation can be observed by slightly warmer colours at higher transmission times (> 140 µs). In this part of the core (> 3.2 m) sediments are mainly composed of coarse-grained foraminiferal ooze, while farther downcore (3.2 - 5.6 m) fine-grained nannofossils prevail in the pelagic sediments.

The acoustic image of the calcareous core from the Rio Grande Rise (GeoB2821-1) shows much more lithological changes than the visual core description. Cool colours between 1.5 - 2.5 m depth indicate unusually fine grain sizes (Breitzke 1997). Alternately yellow/red and blue/black colours in the lower part of the core (> 6.0 m) reflect an interlayering of fine-grained nannofossil and coarse-grained foraminiferal ooze. Dating by orbital tuning shows that this interlayering coincides with the 41 ky cycle of obliquity (von Dobeneck and Schmieder 1999) so that fine-grained oozes dominate during glacial and coarse-grained oozes during interglacial stages.

The opal-rich diatomaceous sediments in core PS2567-2 from the Meteor Rise are characterized by a very low attenuation. Only 2 - 3 calcareous layers with significantly higher attenuation (yellow and red colours) can be identified as prominent lithological changes.

2.5.2   P- and S-Wave Velocity, Attenuation, Elastic Moduli and Permeability

As the acoustic properties of water-saturated sediments are strongly controlled by the amount and distribution of pore space, cross plots of P-wave velocity and attenuation coefficient versus porosity clearly indicate the different bulk and elastic properties of terrigenous and biogenic sediments and can thus be used for an acoustic classification of the lithology. Additional S-wave velocities (and attenuation coefficients) and elastic moduli estimated by least-square inversion specify the amount of bulk and shear moduli which contribute to the P-wave velocity (Breitzke 2000).

The cross plots of the P-wave parameters of the four cores considered above illustrate that terrigenous, calcareous and diatomaceous sediments can uniquely be identified from their position in both diagrams (Fig. 2.19). In terri-

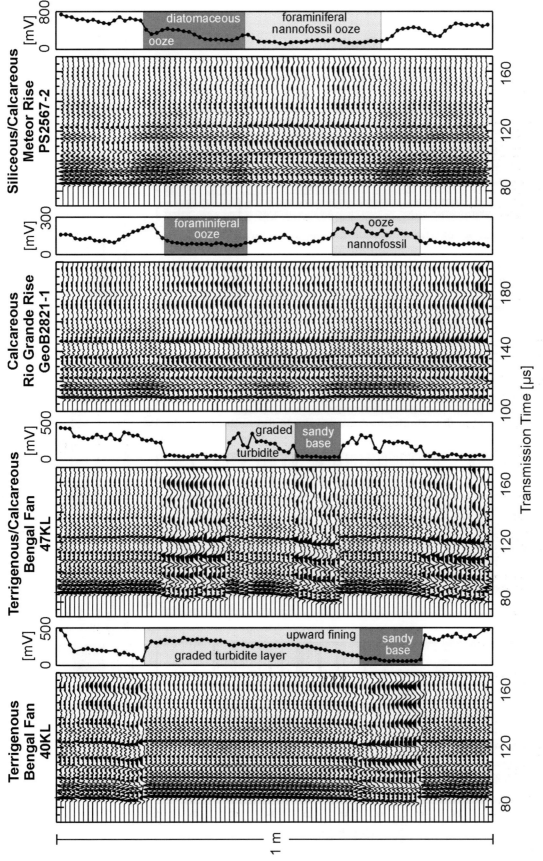

Fig. 2.17 Normalized transmission seismogram sections of four 1 m long core segments from different terrigenous and biogenic sedimentation environments. Seismograms were recorded with 1 cm spacing. Core depths are 5.49-6.43 m (40KL), 6.03-6.97 m (47KL), 5.26-6.14 m (GeoB2821-1), 12.77-13.60 m (PS2567-2). Maximum amplitudes of the transmission seismograms are plotted next to each section from 0 to ... mV, as given at the top.

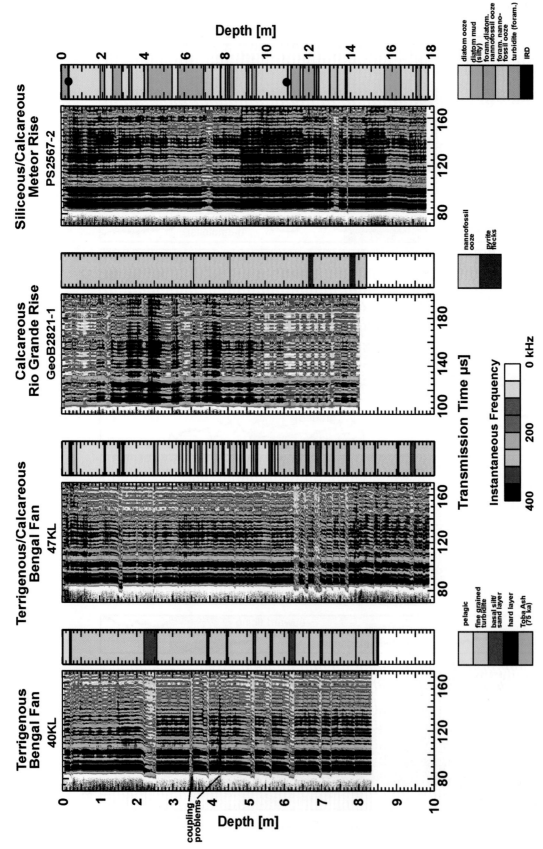

Fig. 2.18 Ultrasonic images of the transmission measurements on cores 40KL, 47KL, GeoB2821-1 and PS2567-2 retrieved from different terrigenous and biogenic sedimentation environments. Displayed are the colour-encoded instantaneous frequencies of the transmission seismograms and the lithology derived from visual core inspections.

56

genous sediments (40KL, 47KL) P-wave velocities and attenuation coefficients increase with decreasing porosities. Computed S-wave velocities are very low ($\approx$ 60 - 65 m s^{-1}) and almost independent of porosity (Fig. 2.20a) whereas computed S-wave attenuation coefficients (at 400 kHz) vary strongly from $4 \cdot 10^3$ dB m^{-1} in fine-grained sediments to $1.5 \cdot 10^4$ dB m^{-1} in coarse-grained sediments (Fig. 2.20b). Accordingly, shear moduli are also low and do not vary very much (Fig. 2.21b) so that higher P-wave velocities in terrigenous sediments mainly result from higher bulk moduli (Fig. 2.21a).

If calcareous, particularly foraminiferal components (FNO) are added to terrigenous sediments porosities become higher ($\approx$ 70 - 80%, 47KL). P-wave velocities slightly increase from their minimum of 1475 m s^{-1} at 70% porosity to 1490 m s^{-1} at 80% porosity (Fig. 2.19a) mainly due to an increase in the shear moduli from about 6.5 to 8.5 MPa whereas bulk moduli remain almost constant at about 3000 MPa (Fig. 2.21b). However, a much more pronounced increase can be observed in the P-wave attenuation coefficients (Fig. 2.19b). For porosities of about 80% they reach the same values (about 200 dB m^{-1}) as terrigenous sediments of about 55% porosity, but can easily be distinguished because of higher P-wave velocities in terrigenous sediments. S-wave attenuation coefficients increase as well in these hemipelagic sediments from about $1 \cdot 10^5$ dB m^{-1} at 65% porosity to about $2.5 \cdot 10^5$ dB m^{-1} at 80% porosity (Fig. 2.20b), and are thus even higher than in terrigenous sediments.

Calcareous foraminiferal and nannofossil oozes (NFO) show similar trends in both P-wave and S-wave parameters as terrigenous sediments, but are shifted to higher porosities due to their additional intraporosities (GeoB2821-1; Figs. 2.19, 2.20).

In diatomaceous oozes P- and S-wave wave velocities increase again though porosities are very high (>80%, PS2567-2; Figs. 2.19, 2.20). Here, the diatom shells build a very stiff frame which causes high shear moduli and S-wave velocities (Fig. 2.21). It is this increase in the shear moduli which only accounts for the higher P-wave velocities while the bulk moduli remain almost constant and are close to the bulk modulus of sea water (Fig. 2.21). P-wave attenuation coefficients are very low in these high-porosity sediments (Fig. 2.19b) whereas S-wave attenuation coefficients are highest (Fig. 2.20b).

Permeabilities estimated from the least square inversion mainly reflect the attenuation charac-

teristics of the different sediment types (Fig. 2.22). They reach lowest values of about $5 \cdot 10^{-14}$ m^2 in fine-grained clayey mud and nannofossil ooze. Highest values of about $5 \cdot 10^{-11}$ m^2 occur in diatomaceous ooze due to their high porosities. Nevertheless, it should be kept in mind that these permeabilities are only estimates based on the input parameters and assumptions incorporated in Biot-Stoll's model. For instance one of these assumptions is that only mean grain sizes are used, but the influence of grain size distributions is neglected. Additionally, the total porosity is usually used as input parameter for the inversion scheme without differentiation between inter- and intraporosities. Comparisons of these estimated permeabilities with direct measurements unfortunately do not exist up to know.

2.6   Sediment Echosounding

While ultrasonic measurements are used to study the structure and composition of sediment cores, sediment echosounders are hull-mounted acoustic systems which image the upper 10-200 m of sediment coverage by remote sensing surveys. They operate with frequencies around 3.5-4.0 kHz. The examples presented here were digitally recorded with the narrow-beam Parasound echosounder and ParaDIGMA recording system (Spieß 1993).

2.6.1   Synthetic Seismograms

Figure 2.23 displays a Parasound seismogram section recorded across an inactive channel of the Bengal Fan. The sediments of the terrace were sampled by a 10 m long piston core (47KL). Its acoustic and bulk properties can either be directly compared to the echosounder recordings or by computations of synthetic seismograms. Such modeling requires P-wave velocity and wet bulk density logs as input parameters. From the product of both parameters acoustic impedances $I = v_P \cdot \rho$ are calculated. Changes in the acoustic impedance cause reflections of the normally incident acoustic waves. The amplitude of such reflections is determined by the normal incidence reflection coefficient $R = (I_2 - I_1) / (I_2 + I_1)$, with (I_1) and (I_2) being the impedances above and below the interface. From the series of reflection coefficients the reflectivity can be computed as impulse response function, including all internal

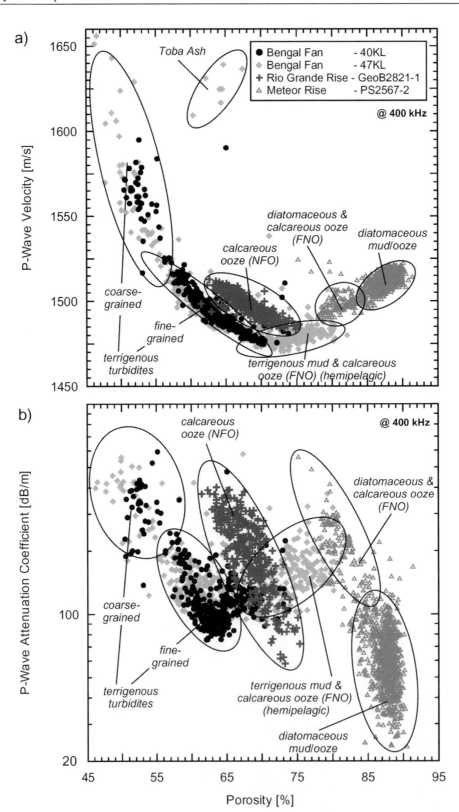

Fig. 2.19 (a) P-wave velocities and (b) attenuation coefficients (at 400 kHz) versus porosities for the four sediment cores 40KL, 47KL, GeoB2821-1, and PS2567-2. NFO is nannofossil foraminiferal ooze, FNO is foraminiferal nannofossil ooze (nomenclature after Mazullo et al. (1988)). Modified after Breitzke (2000).

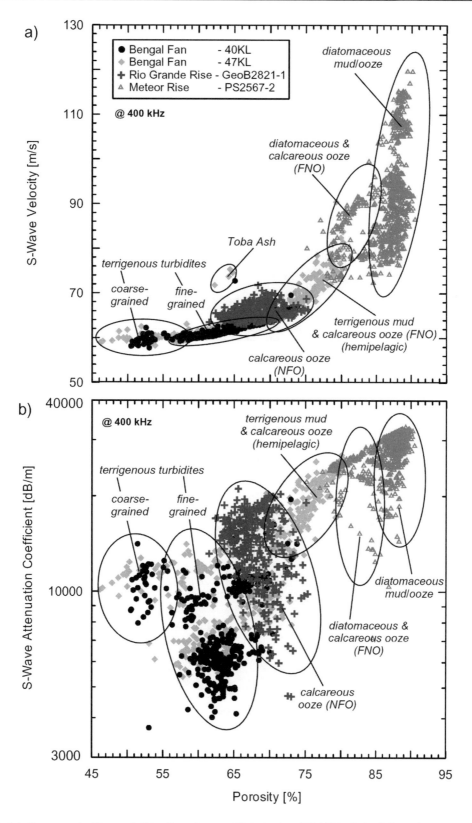

Fig. 2.20 (a) S-wave velocities and (b) attenuation coefficients (at 400 kHz) derived from least square inversion based on Biot-Stoll's theory versus porosities for the four sediment cores 40KL, 47KL, GeoB2821-1, and PS2567-2. NFO and FNO as in Figure 2.19. Modified after Breitzke (2000).

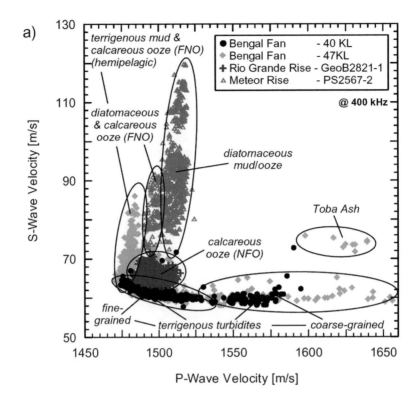

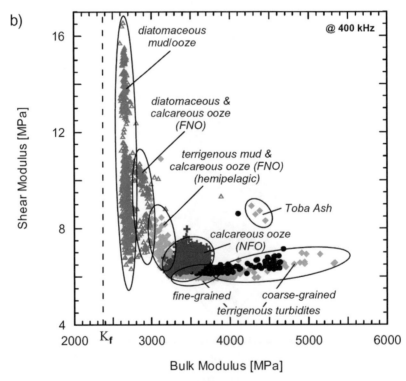

Fig. 2.21 (a) S-wave velocity versus P-wave velocity and (b) shear modulus versus bulk modulus derived from least square inversion based on Biot-Stoll's theory for the four sediment cores 40KL, 47KL, GeoB2821-1, and PS2567-2. The dotted line indicates an asymptote parallel to the shear modulus axis which crosses the value for the bulk modulus of seawater K_f = 2370 MPa on the bulk modulus axis. NFO and FNO as in Figure 2.19. Modified after Breitzke (2000).

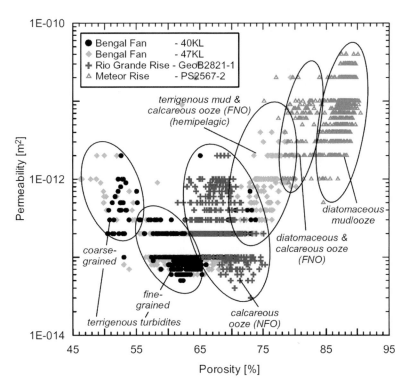

Fig. 2.22 Permeabilities derived from least square inversion based on Biot-Stoll's theory versus porosities for the four sediment cores 40KL, 47KL, GeoB2821-1, and PS2567-2. The regular spacing of the permeability values is due to the increment (10 permeability values per decade) used for the optimization process in the inversion scheme. NFO and FNO as in Figure 2.19. Modified after Breitzke (2000).

multiples. Convolution with a source wavelet finally provides the synthetic seismogram. In practice, different time- and frequency domain methods exist for synthetic seismogram computations. For an overview, please refer to books an theoretical seismology (e.g. Aki and Richards, 2002). Here, we used a time domain method called 'state space approach' (Mendel et al. 1979).

Figure 2.24 compares the synthetic seismogram computed for core 47KL with the Parasound seismograms recorded at the coring site. The P-wave velocity and wet bulk density logs used as input parameters are displayed on the right-hand side, together with the attenuation coefficient log as grain size indicator, the carbonate content and an oxygen isotope ($\delta^{18}O$-) stratigraphy. An enlarged part of the gray shaded Parasound seismogram section is shown on the left-hand side. The comparison of synthetic and Parasound data indicates some core deformations. If the synthetic seismograms are leveled to the prominent reflection caused by the Toba Ash in 1.6 m depth about 95 cm sediment are missing in the overlying younger part of the core. Deeper reflec-

tions, particularly caused by the series of turbidites below 6 m depth, can easily be correlated between synthetic and Parasound seismograms, though single turbidite layers cannot be resolved due to their short spacing. Slight core stretching or shortening are obvious in this lower part of the core, too.

From the comparison of the gray shaded Parasound seismogram section and the wiggle traces on the left-hand side with the synthetic seismogram, core logs and stratigraphy on the right-hand side the following interpretation can be derived. The first prominent reflection below sea floor is caused by the Toba Ash layer deposited after the explosion of volcano Toba (Sumatra) 75,000 years ago. The underlying series of reflection horizons (about 4 m thickness) result from an interlayering of very fine-grained thin terrigenous turbidites and pelagic sediments. They are younger than about 240,000 years (oxygen isotope stage 7). At that time the channel was already inactive. The following transparent zone between 4.5 - 6.0 m depth indicates the transition to the time when the channel was active, more than

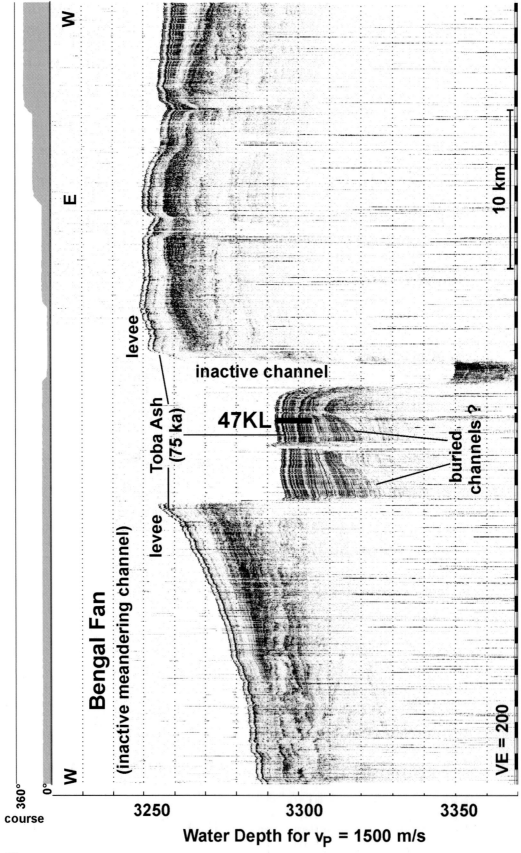

Fig. 2.23 Parasound seismogram section recorded across an inactive meandering channel in the Bengal Fan. The sediments of the terrace were sampled by piston core 47KL marked by the black bar. Vertical exaggeration (VE) of sedimentary structures is 200. The ship's course is displayed above the seismogram section. The terrace exhibits features which might be interpreted as old buried channels. Modified after Breitzke (1997).

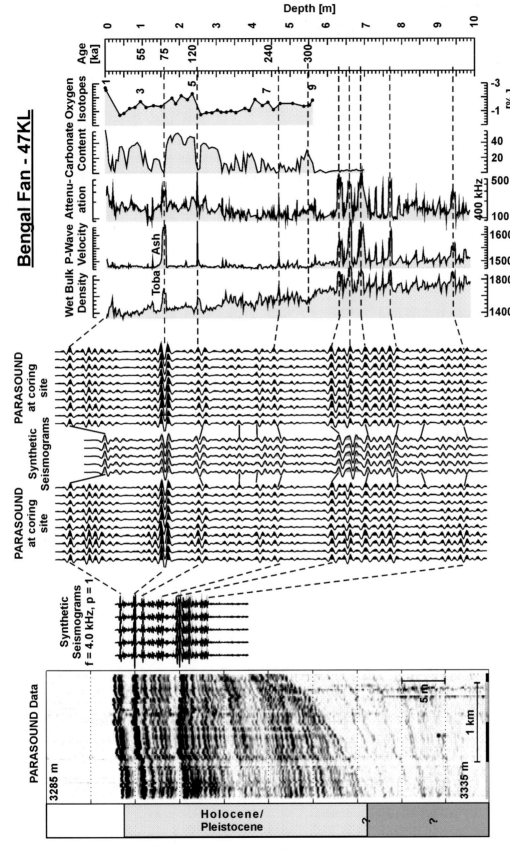

Fig. 2.24 Coring site of 47KL. *From left to right:* Enlarged part of the Parasound seismogram section recorded on approaching site 47KL. Comparison of synthetic seismograms computed on basis of P-wave velocity and wet bulk density measurements on 47KL with Parasound seismograms recorded at the coring site. Physical properties and age model of piston core 47KL. Wet bulk densities, carbonate content and oxygen isotope stratigraphy (*G. ruber pink*) are taken from (Kudrass 1994, 1996). Modified after Breitzke (1997).

300,000 years ago. Only terrigenous turbidites causing strong reflections below 6 m depth were deposited.

Transferred to the Parasound seismogram section in Figure 2.23 this interpretation means that the first prominent reflection horizon in the levees can also be attributed to the Toba Ash layer (75 ka). The following transparent zone indicates completely pelagic sediments deposited on the levees during the inactive time of the channel (< 240 ka). The suspension cloud was probably trapped between the channel walls so that only very thin, fine-grained terrigenous turbidites were deposited on the terrace but did not reach the levee crests. The high reflectivity part below the transparent zone in the levees is probably caused by terrigenous turbidites deposited here more than 300,000 years ago during the active time of the channel.

2.6.2 Narrow-Beam Parasound Echosounder Recordings

Finally, some examples of Parasound echosounder recordings from different terrigenous and biogenic provinces are presented to illustrate that characteristic sediment compositions can already be recognized from such remote sensing surveys prior to the sediment core retrieval.

The first example was recorded on the Rio Grande Rise in the western South Atlantic during RV Meteor cruise M29/2 (Bleil et al. 1994) and is typical for a calcareous environment (Fig. 2.25a). It is characterized by a strong sea bottom reflector caused by coarse-grained foraminiferal sands. Signal penetration is low and reaches only about 20 m.

In contrast, the second example from the Conrad Rise in the Antarctic Ocean, recorded during RV Polarstern cruise ANT XI/4 (Kuhn, pers. communication), displays a typical biogenic siliceous environment (Fig. 2.25b). The diatomaceous sediment coverage seems to be very transparent, and the Parasound signal penetrates to about 160 m depth. Reflection horizons are caused by calcareous foraminiferal and nanno-fossil layers deposited during a retreat of the Polar Frontal Zone during interglacial stages.

The third example recorded in the Weddell Sea/ Antarctic Ocean during RV Polarstern cruise ANT XI/4 as well (Kuhn, pers. communication) indicates a deep sea environment with clay sediments (Fig. 2.25c). Signal penetration again is high (about 140 m), and reflection horizons are very

sharp and distinct. Zones with upward curved reflection horizons might possibly be indicators for pore fluid migrations.

The fourth example from the distal Bengal Fan (Hübscher et al. 1997) was recorded during RV Sonne cruise SO125 and displays terrigenous features and sediments (Fig. 2.25d). Active and older abandoned channel levee systems are characterized by a diffuse reflection pattern which indicates sediments of coarse-grained turbidites. Signal penetration in such environments is rather low and reaches about 30 - 40 m.

Acknowledgment

We thank the captains, crews and scientists onboard of RV Meteor, RV Polarstern and RV Sonne for their efficient cooperation and help during the cruises in the South Atlantic, Antarctic and Indian Ocean. Additionally, special thanks are to M. Richter who provided a lot of unpublished electrical resistivity data and to F. Pototzki, B. Pioch and C. Hilgenfeld who gave much technical assistance and support during the development of the full waveform logging system. V. Spieß developed the ParaDIGMA data acquisition system for digital recording of the Parasound echosounder data and the program system for their processing and display. His help is greatly appreciated, too. H. Villinger critically read an early draft and improved the manuscript by many helpful discussions. The research was funded by the Deutsche Forschungsgemeinschaft (Special Research Project SFB 261 at Bremen University, contribution no. 251 and project no. Br 1476/2-1+2) and by the Federal Minister of Education, Science, Research and Technology (BMBF), grant no. 03G0093C.

Appendix

A: Physical Properties of Sediment Grains and Sea Water

The density (ρ_g) and bulk modulus (K_g) of the sediment grains are the most important physical properties which characterize the sediment type - terrigenous, calcareous and siliceous - and composition. Additionally, they are required as input parameters for wave propagation modeling,

a)

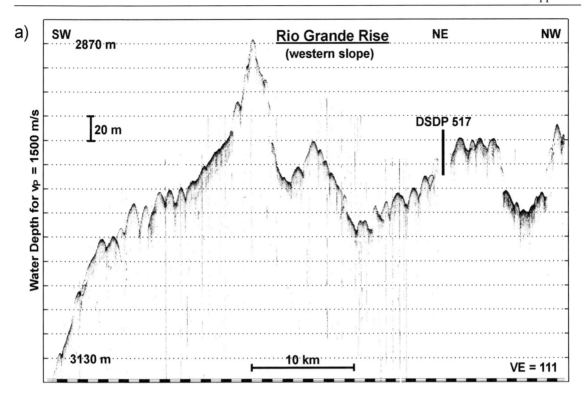

b)

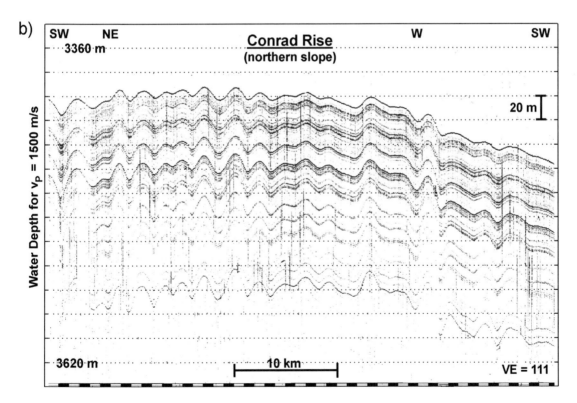

Fig. 2.25 a, b Parasound seismogram sections representing typical calcareous (Rio Grande Rise) and opal-rich diatomaceous (Conrad Rise) environments. Lengths and heights of both profiles amount to 50 km and 300 m, respectively. Vertical exaggeration (VE) is 111. The Parasound seismogram section from the Conrad Rise was kindly provided by G. Kuhn, Alfred-Wegener-Institute for Polar- and Marine Research, Bremerhaven, Germany and is an unpublished data set recorded during RV Polarstern cruise ANT IX/4. Modified after Breitzke (1997).

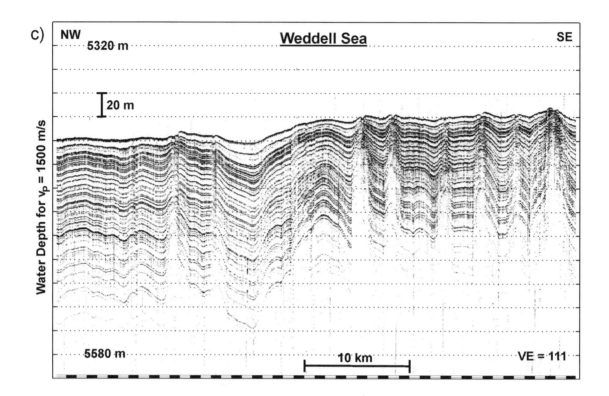

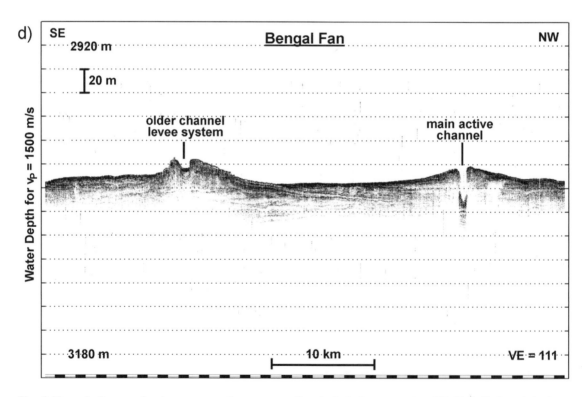

Fig. 2.25 c, d Parasound seismogram sections representing typical deep sea clay (Weddell Sea) and terrigenous sediments (Bengal Fan). Lengths and heights of both profiles amount to 50 km and 300 m, respectively. Vertical exaggeration (VE) is 111. The Parasound seismogram section from the Weddell Sea was kindly provided by G. Kuhn, Alfred-Wegener-Institute for Polar- and Marine Research, Bremerhaven, Germany and is an unpublished data set recorded during RV Polarstern cruise ANT IX/4. Modified after Breitzke (1997).

e.g. with Biot-Stoll's sediment model (Sect. 2.4.1). Table 2A-1 provides an overview on densities and bulk moduli of some typical sediment forming minerals.

Density (ρ_f), bulk modulus (K_f), viscosity (η), sound velocity (v_p), sound attenuation coefficient (α) and electrical resistivity (R_f) of sea water depend on salinity, temperature and pressure. Table 2A-2 summarizes their values at laboratory conditions, i.e. 20°C temperature, 1 at pressure and 35‰ salinity.

B: Corrections to Laboratory and In Situ Conditions

Measurements of physical properties are usually carried out under laboratory conditions, i.e. room temperature (20°C) and atmospheric pressure (1 at).

Table 2A-1 Physical properties of sediment forming minerals, at laboratory conditions (20°C temperature, 1 at pressure). After [1]Wohlenberg (1982) and [2]Gebrande (1982).

Mineral	Density[1] ρ_g [g cm^{-3}]	Bulk Modulus[2] K_g [10^9 Pa]
terrigenous sediments		
quartz	2.649 - 2.697	37.6 - 38.1 (trigonal)
biotite	2.692 - 3.160	42.0 - 60.0
muscovite	2.770 - 2.880	43.0 - 62.0
hornblende	3.000 - 3.500	84.0 - 90.0
calcareous sediments		
calcite	2.699 - 2.882	70.0 - 76.0
siliceous sediments		
opal	2.060 - 2.300	no data

Table 2A-2 Physical properties of sea water, at laboratory conditions (20°C temperature, 1 at pressure, 35 ‰ salinity). After [1]Wille (1986) and [2]Siedler and Peters (1986).

Parameter	Value
sound velocity[1] v_P [m s^{-1}]	1521 m s^{-1}
sound attenuation coefficient[1] α [10^{-3} dB m^{-1}]	
4 kHz	~0.25
10 kHz	~0.80
100 kHz	~ 40
400 kHz	~ 120
1000 kHz	~ 300
density[2] ρ_f [g/cm^{-3}]	1.024
bulk modulus[2] K_f [10^9 Pa]	2.37
viscosity[2] η [10^{-3} Pa·s]	1.07
electrical resistivity[2] R_f [Ω·m]	0.21

In order to correct for slight temperature variations in the laboratory and to transfer laboratory measurements to in situ conditions, usually temperature and in situ corrections are applied. Temperature variations mainly affect the pore fluid. Corrections to in situ conditions should consider both the influence of reduced temperature and increased hydrostatic pressure at the sea floor.

Porosity and Wet Bulk Density

In standard sea water of 35‰ salinity density increases by maximum $0.3 \cdot 10^{-3}$ g cm^{-3} per °C (Siedler and Peters 1986), i.e. by less than 0.1% per °C. Hence, temperature corrections can usually be neglected.

Differences between laboratory and in-situ porosities are less than 0.001% (Hamilton 1971) and can thus be disregarded, too. Sea floor wet bulk densities are slightly higher than the corresponding laboratory values due to the hydrostatic pressure and the resulting higher density of the pore water. For Central Pacific sediments of 75 - 85% porosity Hamilton (1971) estimated a density increase of maximum 0.01 g cm^{-3} for water depths between about 500 - 3000 m and of maximum 0.02 g cm^{-3} for water depths between about 3000 - 6000 m. Thus, corrections to sea floor conditions are usually of minor importance, too. However, for cores of several hundred meter length the effect of an increasing lithostatic pressure has to be taken into account.

Electrical Resistivity

If porosities and wet bulk densities are determined by galvanic resistivity measurements (Sect. 2.2.3) varying sediment temperatures are considered by computation of the formation factor (F) (see Eq. 2.12). While the resistivity (R_s) of the sediment is determined by the small hand-held probe (cf. Sect. 2.2.3) the resistivity (R_f) of the pore fluid is derived from a calibration curve which describes the temperature (T) - conductivity (c) relation by a fourth power law (Siedler and Peters 1986)

$$R_f^{-1} = c_o + c_1 T + c_2 T^2 + c_3 T^3 + c_4 T^4 \quad (2.22)$$

The coefficients (c_0) to (c_4) depend on the geometry of the probe and are determined by a least square fit to the calibration measurements in standard sea water.

P-wave velocity and attenuation

Bell and Shirley (1980) demonstrated that the P-wave velocity of marine sediments increases almost linearly by about 3 m s^{-1} per °C while the attenuation is independent of sediment temperature, similar to the temperature dependence of sound velocity and attenuation in sea water. Hence, to correct laboratory P-wave velocity measurements to a reference temperature of 20°C Schultheiss and McPhail's (1989) equation

$$v_{20} = v_T + 3 \cdot (20 - T) \quad (2.23)$$

can be applied, with (v_{20}) = P-wave velocity at 20°C (in m s^{-1}), (T) = sediment temperature (in °C) and (v_T) = P-wave velocity measured at temperature (T) (in m s^{-1}).

To correct laboratory P-wave velocity measurements to in situ conditions a modified time-average equation (Wyllie et al. 1956) can be used (Shipboard Scientific Party 1995)

$$\frac{1}{v_{insitu}} = \frac{1}{v_{lab}} + \phi \cdot \left(\frac{1}{c_{insitu}} - \frac{1}{c_{lab}} \right) \quad (2.24)$$

(v_{lab}) and ($v_{in\text{-}situ}$) are the measured laboratory (20°C, 1 at) and the corrected in situ P-wave velocities, (c_{lab}) and ($c_{in\text{-}situ}$) the sound velocity in sea water at laboratory and in situ conditions and (ϕ) the porosity. If laboratory and sea floor pressure, temperature and salinity are known from tables or CTD measurements (c_{lab}) and ($c_{in\text{-}situ}$) can be computed according to Wilson's (1960) equation

$$c = 1449.14 + c_T + c_P + c_s + c_{STP} \quad (2.25)$$

(c_T), (c_P), (c_S) are higher order polynomials which describe the influence of temperature (T), pressure (P) and salinity (S) on sound velocity. (c_{STP}) depends on all three parameters. Complete expressions for (c_T), (c_P), (c_S), (c_{STP}) can be found in Wilson (1960). For T = 20°C, P = 1 at and S = 0.035 Wilson's equation results in a sound velocity of 1521 m s^{-1}.

2.7 Problems

Problem 1

What is the difference between bulk parameters and acoustic/elastic parameters concerning the distribution of pore fluid and sediment grains ?

Problem 2

Which direct and which indirect methods do you know to determine the porosity and wet bulk density of marine sediment, which of both parameters is primarily determined by each method, and how can porosity values be converted to density values and vice versa ?

Problem 3

Assume that you have recovered a sediment core in the Antarctic Ocean close to the Polar Frontal Zone. The core is composed of an interlayering of calcareous and siliceous sediments. You would like to determine the downcore porosity and wet bulk density logs by electrical resitivity measurements. What do you have to consider during conversion of porosities to wet bulk densities ?

Problem 4

Which sediment property mainly influences the ultrasonic acoustic wave propagation (50 - 500 kHz), and how affects it the P-wave velocity and attenuation in terrigenous and calcareous sediments ?

Problem 5

Which parameter determines the amplitude or strength of reflection horizons during sediment echosounding, and which are the two physical properties that contribute to it ?

References

Aki K., Richards P.G., 2002. Quantitative Seismology. University Science Books, 700 pp

Archie G.E., 1942. The electrical resistivity log as an aid in determining some reservoir characteristics. Transactions of the American Institute of Mineralogical, Metalurgical and Petrological Engineering 146: 54-62

Barker P.F., Kennett J.P. et al., 1990. Proceedings of the Ocean Drilling Program, Scientific Results 113, College Station, TX (Ocean Drilling Program), 1033 pp

Bell D.W., Shirley D.J., 1980. Temperature variation of the acoustical properties of laboratory sediments. Journal of the Acoustical Society of America 68: 227-231

Bergmann U., 1996. Interpretation of digital Parasound echosounder records of the eastern Arctic Ocean on the basis of sediment physical properties. Rep. on Polar Research, Alfred-Wegener Institute for Polar and Marine Research, Bremerhaven 183: 164 pp

Berryman J.G., 1980. Confirmation of Biot's theory. Applied Physics Letters 37: 382-384

Biot M.A., 1956a. Theory of propagation of elastic waves in a fluid-saturated porous solid. II. Higher frequency range. Journal of the Acoustical Society of America 28: 179-191

Biot M.A., 1956b. Theory of wave propagation of elastic waves in a fluid-saturated porous solid. I. Low-frequency range. Journal of the Acoustical Society of America 28: 168-178

Bleil U. et al., 1994. Report and preliminary results of Meteor cruise M29/2 Montevideo - Rio de Janeiro, 15.07. - 08.08.1994. Ber. FB Geow. Univ. Bremen 59: 153 pp

Blum P., 1997. Physical properties handbook: a guide to the shipboard measurement of physical properties of deep-sea cores. Technical Note 26, Ocean Drilling Program, College Station, Texas

Bodwadkar S.V., Reis J.C., 1994. Porosity measurements of core samples using gamma-ray attenuation. Nuclear Geophysics 8: 61-78

Boyce R.E., 1968. Electrical resistivity of modern marine sediments from the Bering Sea. Journal of Geophysical Research 73: 4759-4766

Boyce R.E., 1973. Appendix I. Physical properties - Methods. In: Edgar NT, Sanders JB et al (eds) Initial Reports of the Deep Sea Drilling Project 15, Washington, US Government Printing Office, pp 1115-1127

Boyce R.E., 1976. Definitions and laboratory techniques of compressional sound velocity parameters and wet-water content, wet-bulk density, and porosity parameters by gravity and gamma ray attenuation techniques. In: Schlanger S.O., Jackson E.D. et al (eds) Initial Reports of the Deep Sea Drilling Project 33, Washington, US Government Printing Office, pp 931-958

Breitzke M., Spieß V., 1993. An automated full waveform logging system for high-resolution P-wave profiles in marine sediments. Marine Geophysical Researches 15: 297-321

Breitzke M., Grobe H., Kuhn G., Müller P., 1996. Full waveform ultrasonic transmission seismograms - a fast new method for the determination of physical and sedimentological parameters in marine sediment cores. Journal of Geophysical Research 101: 22123-22141

Breitzke M., 1997. Elastische Wellenausbreitung in marinen Sedimenten - Neue Entwicklungen der Ultraschall Sedimentphysik und Sedimentechographie. Ber. FB Geo Univ. Bremen 104: 298 pp

Breitzke M., 2000. Acoustic and elastic characterization of marine sediments by analysis, modeling, and inversion of ultrasonic P wave transmission seismograms, Journal of Geophysical Research 105: 21411-21430

Bryant W.R., Hottman W., Trabant P., 1975. Permeability of unconsolidated and consolidated marine sediments, Gulf of Mexico. Marine Geotechnology 1: 1-14

Carman P.C., 1956. Flow of gases through porous media. Butterworths Scientific Publications, London, 182 pp

Chelkowski A., 1980. Dielectric Physics. Elsevier, Amsterdam, 396 pp

Childress J.J., Mickel T.J., 1980. A motion compensated shipboard precision balance system. Deep Sea Research 27A: 965-970

Constable C., Parker R., 1991. Deconvolution of log-core paleomagnetic measurements - spline therapy for the linear problem. Geophysical Journal International 104: 453-468

Courtney R.C., Mayer L.A., 1993. Acoustic properties of fine-grained sediments from Emerald Basin: toward an inversion for physical properties using the Biot-Stoll model. Journal of the Acoustical Society of America 93: 3193-3200

Dobeneck v.T., Schmieder F., 1999. Using rock magnetic proxy records for orbital tuning and extended time series analyses into the super- and sub-Milankovitch bands. In: Fischer G and Wefer G (eds) Use of proxies in paleoceanography: examples from the South Atlantic. Springer Verlag Berlin, pp 601 - 633.

Ellis D.V., 1987. Well logging for earth scientists. Elsevier, Amsterdam, 532 pp

Fisher A.T., Fischer K., Lavoie D., Langseth M., Xu J., 1994. Geotechnical and hydrogeological properties of sediments from Middle Valley, northern Juan de Fuca Ridge. In: Mottle M.J., Davis E., Fisher A.T., Slack J.F. (eds) Proceedings of the Ocean Drilling Program, Scientific Results 139, College Station, TX (Ocean Drilling Program), pp 627-647

Gassmann F., 1951. Über die Elastizität poröser Medien. Vierteljahresschrift der Naturforschenden Gesellschaft in Zürich 96: 1-23

Gealy E.L., 1971. Saturated bulk density, grain density and porosity of sediment cores from western equatorial Pacific: Leg 7, Glomar Challenger. In: Winterer E.L. et al. (eds) Initial Reports of the Deep Sea Drilling Project 7, Washington, pp 1081-1104

Gebrande H., 1982. Elastic wave velocities and constants of elasticity at normal conditions. In: Hellwege K.H. (ed) Landolt-Börnstein. Numerical Data and Functional Relationships in Science and Technology. Group V: Geophysics and Space Research 1, Physical Properties of Rocks, Subvolume b, Springer Verlag, Berlin, pp 8-35

Gerland S., 1993. Non-destructive high resolution density measurements on marine sediments. Rep. on Polar Research, Alfred-Wegener Institute for Polar and Marine Research, Bremerhaven 123: 130 pp

Gerland S., Richter M., Villinger H., Kuhn G., 1993. Non-destructive porosity determination of Antarctic marine sediments derived from resistivity measurements with the inductive method. Marine Geophy-

sical Researches 15: 201-218

Gerland S., Villinger H., 1995. Nondestructive density determination on marine sediment cores from gamma-ray attenuation measurements. Geo-Marine Letters 15: 111-118

Gunn D.E., Best A.I., 1998. A new automated nondestructive system for high resolution multi-sensor logging of open sediment cores. Geo-Marine Letters 18: 70-77

Hamilton E.L., 1971. Prediction of in situ acoustic and elastic properties of marine sediments. Geophysics 36: 266-284

Hovem J.M., Ingram G.D., 1979. Viscous attenuation of sound in saturated sand. Journal of the Acoustical Society of America 66: 1807-1812

Hovem J.M., 1980. Viscous attenuation of sound in suspensions and high-porosity marine sediments. Journal of the Acoustical Society of America 67: 1559-1563

Hübscher C., Spieß V., Breitzke M., Weber M.E., 1997. The youngest channel-levee system of the Bengal Fan: results from digital echosounder data. Marine Geology 141: 125 - 145

Jackson P.D., Taylor-Smith D., Stanford P.N., 1978. Resistivity-porosity-particle shape relationships for marine sands. Geophysics 43: 1250-1268

Jannsen D., Voss J., Theilen F., 1985. Comparison of methods to determine Q in shallow marine sediments from vertical seismograms. Geophysical Prospecting 33: 479-497

Lambe T.W., Whitman R.V., 1969. Soil mechanics. John Wiley and Sons Inc, New York, 553 pp

Lovell M.A., 1985. Thermal conductivity and permeability assessment by electrical resistivity measurements in marine sediments. Marine Geotechnology 6: 205-240

MacKillop A.K., Moran K., Jarret K., Farrell J., Murray D., 1995. Consolidation properties of equatorial Pacific Ocean sediments and their relationship to stress history and offsets in the Leg 138 composite depth sections. In: Pisias N.G., Mayer L.A., Janecek T.R., Palmer-Julson A., van Andel T.H. (eds) Proceedings of the Ocean Drilling Program, Scientific Results 138, College Station, TX (Ocean Drilling Program), pp 357-369

Mazullo J.M., Meyer A., Kidd R.B., 1988. New sediment classification scheme for the Ocean Drilling Program. In: Mazullo JM, Graham AG (eds) Handbook for Shipboard Sedimentologists, Technical Note 8, Ocean Drilling Program, College Station, Texas, pp 45 - 67

Mendel J.M., Nahi N.E., Chan M., 1979. Synthetic seismograms using the state space approach. Geophysics 44: 880-895

O'Connell S.B., 1990. Variations in upper cretaceous and Cenozoic calcium carbonate percentages, Maud Rise, Weddell Sea, Antarctica. In: Barker P.F., Kennett J.P. et al. (eds) Proceedings of the Ocean Drilling Program, Scientific Results 113, College Station, TX (Ocean Program), pp 971-984

Ogushwitz P.R., 1985. Applicability of the Biot theory. II. Suspensions. Journal of the Acoustical Society of America 77: 441-452

Olsen H.W., Nichols R.W., Rice T.C., 1985. Low gradient permeability measurements in a triaxial system. Geotechnique 35: 145-157

Plona T.J., 1980. Oberservation of a second bulk compressional wave in a porous medium at ultrasonic frequencies. Applied Physics Letters 36: 159-261

Ruffet C., Gueguen Y., Darot M., 1991. Complex conductivity measurements and fractal nature of porosity. Geophysics 56: 758-768

Schopper J.R., 1982. Permeability of rocks. In: Hellwege KH (ed) Landolt-Börnstein. Numerical Data and Functional Relationships in Science and Technology, Group V: Geophysics and Space Research 1, Physical Properties of Rocks, Subvolume a, Berlin, Springer Verlag, Berlin, pp 278-303

Schön J.H., 1996. Physical Properties of Rocks - Fundamentals and Principles of Petrophysics. Handbook of Geophysical Exploration 18, Section I, Seismic Exploration. Pergamon Press, Oxford, 583 pp

Schultheiss P.J., McPhail S.D., 1989. An automated P-wave logger for recording fine-scale compressional wave velocity structures in sediments. In: Ruddiman W, Sarnthein M et al (eds), Proceedings of the Ocean Drilling Program, Scientific Results 108, College Station TX (Ocean Drilling Program), pp 407-413

Sen P.N., Scala C., Cohen M.H., 1981. A self-similar model from sedimentary rocks with application to dielectric constant of fused glass beads. Geophysics 46: 781-795

Sheng P., 1991. Consistent modeling of electrical and elastic properties of sedimentary rocks. Geophysics 56, 1236-1243

Shipboard Scientific Party, 1995. Explanatory Notes. In: Curry W.B., Shackleton N.J., Richter C. et al. (eds), Proceedings of the Ocean Drilling Program, Initial Reports 154, College Station TX (Ocean Drilling Program), pp 11-38

Siedler G., Peters H., 1986. Physical properties (general) of sea water. In: Hellwege K.H., Madelung O. (eds) Landolt-Börnstein. Numerical Data and Functional Relationships in Science and Technology. Group V: Geophysics and Space Research 3, Oceanography, Subvolume a, Springer Verlag, Berlin, pp 233-264

Spieß V., 1993. Digitale Sedimentechographie - Neue Wege zu einer hochauflösenden Akustostratigraphie. Ber. FB Geo Univ. Bremen 35: 199 pp

Stoll R.D., 1974. Acoustic waves in saturated sediments. In: Hampton L. (ed) Physics of sound in marine sediments. Plenum Press, New York, pp 19-39

Stoll R.D., 1977. Acoustic waves in ocean sediments. Geophysics 42: 715-725

Stoll R.D., 1989. Sediment Acoustics. Springer Verlag, Berlin, 149 pp

Taner M.T., Koehler F., Sheriff R.E., 1979. Complex seismic trace analysis. Geophysics 44: 1041 - 1063

Tonn R., 1989. Comparison of seven methods for the computation of Q. Physics of the Earth and Planetary Interiors 55: 259-268

Tonn R., 1991. The determination of the seismic quality factor Q from VSP data: a comparison of different computational methods. Geophysical Prospecting 39: 1-27

Waxman M.H., Smits L.J.M., 1968. Electrical conductivities in oil bearing shaly sandstones. Society of Petroleum Engineering 8: 107-122

Weaver, P.P.E., Schultheiss P.J., 1990. Current methods for obtaining, logging and splitting marine sediment cores. Marine Geophysical Researches 12: 85-100

Weber M.E., Niessen F., Kuhn G., Wiedicke M., 1997. Calibration and application of marine sedimentary physical properties using a multi-sensor core logger. Marine Geology 136: 151-172

Weeks R., Laj C., Endignoux L., Fuller M., Roberts A., Manganne R., Blanchard E., Goree W., 1993. Improvements in long-core measurement techniques: applications in paleomagnetism and paleoceanography. Geophysical Journal International 114: 651-662

Whitmarsh R.B., 1971. Precise sediment density determination by gamma-ray attenuation alone, Journal of Sedimentary Petrology 41: 882-883

Wille P., 1986. Acoustical properties of the ocean. In: Hellwege KH, Madelung O (eds) Landolt-Börnstein. Numerical Data and Functional Relationships in Science and Technology. Group V: Geophysics and Space Research 3, Oceanography, Subvolume a, Springer Verlag, Berlin, pp 265-382

Wilson W.D., 1960. Speed of sound in sea water as a function of temperature, pressure and salinity. Journal of the Acoustical Society of America 32: 641-644

Wohlenberg J., 1982. Density of minerals. In: Hellwege K.H. (ed) Landolt-Börnstein. Numerical Data and Functional Relationships in Science and Technology. Group V: Geophysics and Space Research 1, Physical Properties of Rocks, Subvolume a, Springer Verlag, Berlin, pp 66-113

Wood A.B., 1946. A textbook of sound. G Bell and Sons Ltd, London, 578 pp

Wyllie M.R., Gregory A.R., Gardner L.W., 1956. Elastic wave velocities in heterogeneous and porous media. Geophysics 21: 41-70

3 Quantification of Early Diagenesis: Dissolved Constituents in Pore Water and Signals in the Solid Phase

Horst D. Schulz

A chapter on the pore water of sediments and the processes of early diagenesis, reflected by the concentration profiles therein, can certainly be structured in different ways. One possibility is, for example, to start with the sample-taking strategies, followed by the analytical treatment of the samples, then a model presentation to substantiate the processes, and finally a quantitative evaluation of the measured profiles with regard to material fluxes and reaction rates. The reader who would prefer this sequence is recommended to begin with the Sections 3.3, 3.4, 3.5 and to consult the Sections 3.1 and 3.2 later. However, in our opinion, marine geochemistry departed from its initial development stage of sample collection and anlysis some years ago. This book is therefore written with a different structure from earlier texts. The well understood theoretical knowledge on concentration profiles has to be introduced from the beginning. Only then do problems arise which necessitate the investigation of particular substances in pore water and the application of specific methods of sampling and analysis.

Particularly, two fundamental approaches are conceivable for investigating diagenesis processes, both having their advantages and disadvantages. The classical way leads to a sedimentological, mineralogical, and in the broadest sense geochemical examination of the solid phase. The advantage of this approach is that sample withdrawal and analytical procedures are generally easier to carry out and produce more reliable results. On the other hand, in most cases this approach will not provide us with any information on the timing at which specific diagenetical alterations in the sediment took place. Nothing will be learnt about reaction rates, reaction kinetics, and mostly we do not even know whether the processes, interpreted on account of certain measurements and/or observations, occur today or whether they were terminated some time before. Furthermore, the distinction between the primary signals from the water column and the results of diagenetic reactions, often remains unclear.

A section dealing with the concentrations of elements in the solid phase of the sediment should not be missing despite the mentioned disadvantages and problems that come along with a geochemical description of a sediment/pore water system. Section 3.7 therefore contains a brief presentation of the various sample-processing methods and analyses, next to some examples of measured element profiles. One of the greatest problems encountered in interpreting element concentrations in the sediment's solid phase will always be one's ability to distinguish between a primary signal from the water column, i.e. from the introduction into the sediment, and the diagenetic alteration of the material.

If the processes of diagenesis and their time-dependency are to be recorded directly, then a second approach appears to be more reasonable; one that includes the measurement of concentration profiles in pore water. Figure 3.1 shows various concentration profiles of substances dissolved in the pore water of sediments taken at a depth of approximately 4000 m, off the estuary of the River Congo. The differently shaped curves shown here directly reflect the reactions that are happening today. The investigation of pore water thus constitutes the only way to determine the reaction rates and the fluxes of material which they produce. Yet, this procedure is rather laborious and unfortunately contains numerous opportunities for errors. Most errors are caused as soon as the sediment sample is taken, others are caused upon preparing the pore water from the sediment, or in the course of

GeoB 1401

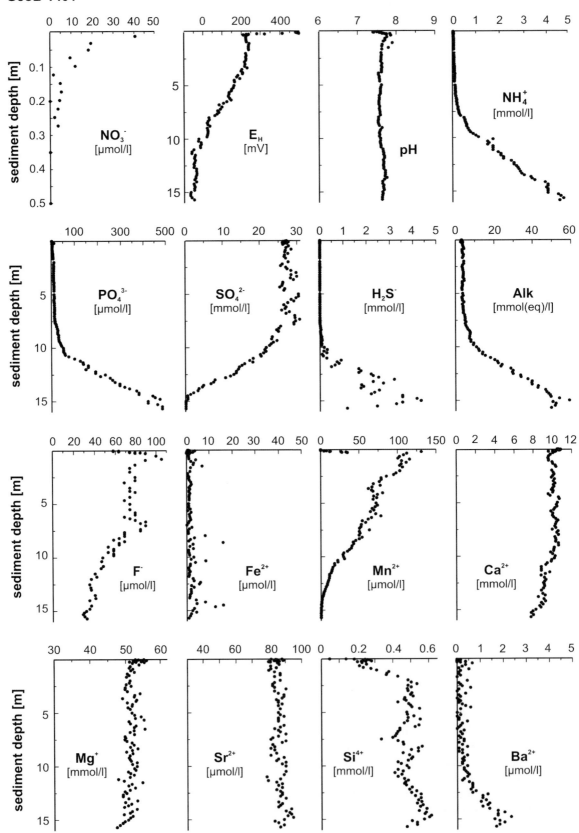

Fig. 3.1 Concentration profiles in the pore water fractions of sediments obtained off the estuary of the River Congo, at a depth of approximately 4000 m. The sediments contain a relatively high amount of TOC. Values ranging from 1 to 3.5 wt. % indicate that this sediment is characterized by the high reaction rates of various early diagenesis processes. These processes are reflected by diffusion fluxes over gradients and by reaction rates determined by gradient changes (after Schulz et al. 1994).

analyzing low concentrations obtained from small sample amounts. Actually, one would prefer to gain all results by *in situ* measurements at the bottom of the ocean, which is feasible, however, only in very few exceptional cases.

3.1 Introduction: How to Read Pore Water Concentration Profiles

The processes of early diagenesis - irrespective of whether we are dealing with microbiological redox-reactions or predominantly abiotic reactions of dissolution and precipitation - are invariably reflected in the pore water of sediments. The aquatic phase of the sediments is the site where the reactions occur, and where they become visible as time-dependent or spatial distributions of concentrations. Thus, if the early diagenesis processes are to be examined and quantified in the young marine sediment, the foremost step will always consist of measuring the concentration profiles in the pore water fraction.

For reasons of simplification and enabling an easy understanding of the matter, only three fundamental processes will be considered in this section, along with their manifestation in the pore water concentration profiles:

- Consumption of a reactant in pore water (e.g. consumption of oxygen in the course of oxidizing organic material)
- Release of a substance from the solid phase into the pore water fraction (e.g. release of silica due to the dissolution of opal).
- Diffusion transport of dissolved substances in pore water and across the sediment/ bottom water boundary.

Figure 3.2 shows schematic diagrams of some possible concentration profiles occurring in ma-

rine pore water. Here, the profile shown in Figure 3.2a is quite easy to understand. In the event of a substance that is not subject to any early diagenetic process, the pore water as the formation water contains exactly the same concentration as the supernatant ocean water. If the concentration in the ocean floor water changes, this change will

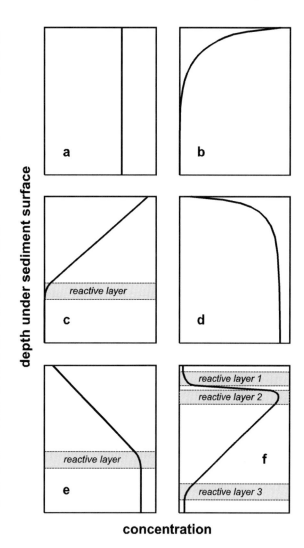

concentration

Fig. 3.2 Principal concentration profile shapes as in measured pore water:
a: Profile of a non-reacting substance (e.g. chloride); b: profile of a substance that is depleted in the upper layers of the sediment (e.g. oxygen); c: profile of a substance that is consumed in a particular reactive layer; d: profile of a substance that is released into pore water in the upper layers of the sediment (e.g. silica); e: profile of a substance that is released into the pore water in specific reactive layers; f: profile of a substance that is released into pore water in one discrete depth (reactive layer 2), and, in two other depths (reactive layers 1 and 3) is removed from the pore water by consumption (e.g. manganese).

be reproduced in the pore water fraction on account of diffusion processes. After a definite time of non-stationary conditions, which depends on the depth of the profile under study and the required accuracy of the measurement, the concentration of the ocean floor water prevails again, stationary and constant within the entire profile. The details and the problems concerning stationary and non-stationary conditions in pore water will be thoroughly discussed in Section 3.2.

Figure 3.2b shows the concentration profile of a substance which is consumed by early diagenetic reactions in the upper layers of the sediment. For example, the profiles of dissolved oxygen often look like this (cf. Chap. 6). Assuming steady-state conditions (cf. Sect. 3.2), the concentration gradient near the sediment surface will display its highest increases, because whatever is consumed in the deeper layers will arrive there by means of diffusion. With increasing depths the concentration gradient and the diffusive material transport will decline, until, at a particular depth, concentration, gradient, and diffusion simultaneously approach zero. This sediment depth, referred to as the penetration depth of a substance, often represents a steady-state condition composed of the diffusive reinforcement from above and the depletion of the substance in the sediment by the reactions of early diagenesis.

In principle, the situation depicted in Figure 3.2c is quite the same. Here, the process is just limited to one reactive layer. With an example of oxygen as the dissolved substance, this could be a layer containing easily degradable organic matter, or the sedimentary depth in which oxygen encounters and reacts with a reductive solute species coming from below (e.g. Mn^{2+} or Fe^{2+}). At any rate, the space above the reactive layer is characterized by a constant gradient in the concentration profile and thus by a diffusive transport which is everywhere the same. Within the reactive layer the dissolved substance is brought to zero concentration by depletion.

The reverse case, which is in principle quite similar, is demonstrated in Figure 3.2d and e. Here, a substance is released anywhere into the layers near the sediment surface (Fig. 3.2d), whereas it might be released into the pore water only in one particular layer (Fig. 3.2e). The solute could be, for example, silica that often displays

such concentration profiles on account of the dissolution of sedimentary opal.

The concentration profile shown in Figure 3.2f is much more complex than the others in Figure 3.2 and only comprehensible if the interactions of several processes taking place at various depths below the sediment surface are regarded. Such a profile is characteristic, for example, of the concentrations of divalent manganese in marine pore water (see also Figure 3.1, Figure 3.8, and Chapter 11). Manganese oxide functions as an electron acceptor and reacts with organic matter in reactive layer 2, whereupon chemically reduced divalent manganese dissolves. A small percentage diffuses downwards (flat gradient) and precipitates as manganese carbonate or manganese sulfide in reactive layer 3. The major proportion diffuses upwards (much steeper gradient), encounters dissolved oxygen from the sediment surface once again in reactive layer 1, is reoxidized and precipitates back to manganese oxide.

The described relations can be summarized by stating a few rules for reading and understanding of pore water concentration profiles:

- Diffusive material fluxes always occur in the form of concentration gradients; concentration gradients always represent diffusive material fluxes.
- Reactions occurring in pore water in most cases constitute changes in the concentration gradient; changes in the concentration gradient always represent reactions occurring in pore water.
- A concave-shaped alteration in the concentration gradient profile (cf. Fig. 3.2b, c) signifies the depletion of a substance from pore water; conversely, a convex-shaped concentration gradient profile (cf. Fig. 3.2d, e) always depicts the release of a substance into the pore water.

However, the following also needs to be observed:

- If a substance is involved in two reactions taking place at the same depth, and is released into the pore water fraction by one reaction and then withdrawn again by the other, neither of the participating reactions will become evident in the pore-water concentration profile.

In the following Section 3.2, the laws of diffusion and the particularities of their application to sediment pore waters will be treated in more detail. In this context, the problem of steady state and non-steady state situations will have to be covered, since the simple examples described above have anticipated steady state situations. Moreover, as a further simplification, the examples of this section have by necessity neglected advection, bioirrigation, and bioturbation. These processes will be discussed in detail in Section 3.6. Section 3.7 will then cover the investigations of the sediment's solid phase and thereby disclose the result of one or the other process of early diagenesis.

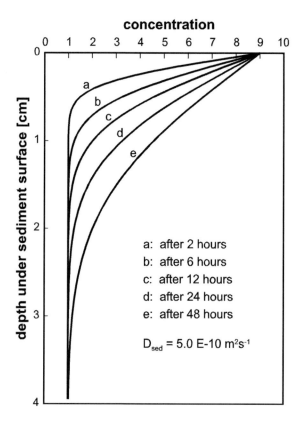

Fig. 3.3 Calculated non-steady state concentration profiles in pore water of a young sediment. It was assumed that the concentration of '1' has been previously constant in the pore water of the sediment as well as in the supernatant bottom water for a long period of time, so that a steady state situation was prevalent. Then the concentration of bottom water changed shortly to '9'. The concentration profiles a to e are non-steady states after 2 and up to 48 hours. The calculation of such non-steady state concentration profiles can be performed , for example, with the aid of the model program CoTAM (Hamer and Sieger 1994) or CoTReM (cf. Chap. 15).

3.2 Calculation of Diffusive Fluxes and Diagenetic Reaction Rates

3.2.1 Steady State and Non-Steady State Situations

In the preceding section and especially in all the following chapters, one pair of terms will assume extraordinary importance for describing and understanding biogeochemical processes: the steady state and non-steady state situation.

Let us first consider the steady state situation as its description is more straight forward in a model concept. In Figure 3.2c a concentration profile is shown in which a substance is continually consumed at a specific rate of reaction and within a reactive layer. At the same time, a constant concentration is prevalent in the bottom water above the sediment surface, an infinite reservoir as compared to the consumption in the sediment. It follows therefore that a constant concentration gradient exists between the sediment surface and the reactive layer, and thus everywhere the same diffusive flux. Such a concentration gradient is referred to as being in steady state. It remains in this condition as long as its determining factors - turn-over rates in the reactive layer, concentration at the sea-floor, dimension and properties of the space between the reactive layer and sediment surface - are not changed.

Any change of the conditions that is liable to exert any kind of influence on the concentration profile, terminates the steady state situation. A non-steady state emerges, which is a time-dependent situation occurring in the pore water. If the system remains unperturbed in the changed situation for a sufficient length of time, a new steady state situation can become established, different from the first and reflecting the novel configuration of conditions.

Strictly speaking, there are no real steady-state situations in nature. Even the sun had begun to shine at a certain time, the earth and, upon her, the oceans have come into existence at a certain time, and all things must pass sooner or later. Hence, the term referring to the steady-state condition also depends on the particular stretch of time which is under study, as well as on the dimension of the system, and, not least, on the accuracy of the measurements with which we examine the parameters that describe the system.

A given concentration in the pore water of a pelagic sediment, several meters below the sediment surface, can be measured today, next month, next year, and after 10 years. Within the margins of reasonable analytical precision, we will always measure just about the same value and rightfully declare the situation to be steady-state (That there are also exceptions to the rule may be concluded from the Examples 2 and 3 described in Section 15.3.2). At the same time, pore water concentrations in sedimentary surface areas near the same pelagic sediment could be subject to considerable seasonal variation (such as residual deposits of algal bloom periods) and thus be classified as being in a non-steady state.

A calculated example for pore water concentrations in a non-steady state condition is shown in Figure 3.3. Details concerning the calculation procedure will not be discussed here. The conceptual model employed will be described in Section 3.2.4, a suitable computer model is described in Chapter 15. A typical diffusion coefficient characteristic of young marine sediments and a characteristic porosity coefficient were used in the calculation.

In the calculated example shown in Figure 3.3 the assumption was made that a constant concentration of '1' has been prevalent for a long time in the pore water of the young sediment and in the bottom water above it. Then, the bottom water underwent a momentary change to yield a concentration of '9'. By means of diffusion, the new concentration gradually spreads into the pore water. After 2 hours it reaches a depth of less than 1 cm, after 48 hours, respectively, a depth of about 3 to 4 cm below the sediment surface. If the concentration of '9' remains constant long enough in the bottom water above the sediment, this concentration will theoretically penetrate into an infinitely great depth. To what extent, and into what depths of pore water, these concentrations are to be assigned to steady state or non-steady states can only be determined in each particular case, having its own concentration in the bottom water over a given period of time. At any rate, the allocation of one or the other state can only be done separately for each system, each parameter, and each time interval. Calculations as shown in Figure 3.3 are likely to produce valuable preliminary concepts, for evaluating real measured pore water profiles.

In the next example, the influence of seasonal variation in the bottom water lying above the

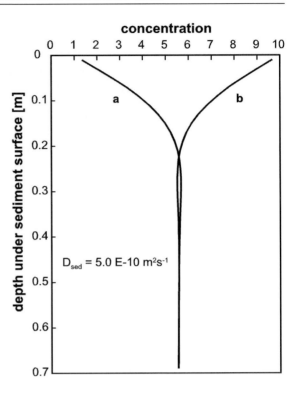

Fig. 3.4 Calculated non-steady state concentration profiles in the pore water of a young sediment. The calculation assumes that concentrations of '1' and '10' were prevailing alternately in the bottom water over the sediment, each over a period of half a year. After several years, the curve a. reflects the situation of concentration '1' at the end of one half year, whereas curve b. reflects the situation of concentration '10' at the end of one half year. The calculation of such non-steady state concentration profiles can be performed, for example, with the aid of the model programs EXPLICIT or CoTReM (cf. Chapter 15).

sediment will be examined, as well as the resulting non-steady states in the pore water fraction. To this end, the following boundary conditions are selected: A substance concentration of '1' is supposed to be prevalent in the bottom water over one half year, afterwards the concentration changes to '10' for one half year, then it changes back to a value of '1', and so on, continually changing. Figure 3.4 shows the result of such an oscillatory situation after several years. The curve denoted 'a' demonstrates the situation in which the bottom water concentration was '1' after half a year; the curve denoted 'b' reflects the situation in which the bottom water concentration was '10' after half a year. It is evident that essential effects of such changes can only be observed down to a depth of less than 0.2 m below the sediment surface. In this model calculation, the effects of the preceding half year still remain visible in a depth between about 0.22 m and 0.35 m below the sediment surface. Below a depth of 0.4 m no further non-steady states can be seen, a constant and steady-state mean concentration of '5.5' prevails.

The conditions proposed in Figure 3.4 describe a very extreme situation. On the one hand, the half-yearly alternating concentrations differ by a whole order of magnitude, on the other hand, no transitory periods with intermediate concentrations were anticipated. Seasonal variations in nature usually display smaller concentration differences and are also likely to possess transitory intermediate concentrations. Both diminish the effects of non-steady state formation in sedimentary depths as were shown in the example of Figure 3.4.

Figure 3.4 also demonstrates clearly the effects of the analytical precision on differentiating steady state and non-steady states in pore water. In the theoretical calculation shown in Figure 3.4, the difference between both curves is still visible in form of a non-steady state at a depth between 0.22 and 0.4 m. However, if the possible errors are taken into consideration that occur during sampling and in the analytic treatment of pore water samples (Sects. 3.3 - 3.5), then there is likely to be no current parameter with which these differences could be discovered practically. Below a depth of 0.2 m, one would always measure the same concentration and consequently judge the situation as stationary, or as a steady state.

3.2.2 The Steady State Situation and Fick's First Law of Diffusion

According to Fick's first law of diffusion, the diffusive flux (J) is directly proportional to concentration gradient ($\partial C/\partial x$) under steady state conditions. The factor of proportionality is the temperature-dependent and substance-dependent diffusion coefficient (D^0):

$$J = -D^0 \cdot \frac{\partial C}{\partial x} \qquad (3.1)$$

The negative sign indicates that the diffusive flux runs in opposition to the gradient's direction from high concentrations to lower concentrations. In these terms, increasing concentrations in greater depths will yield negative gradients in the sediment along with an upwards directed positive flux, and vice-versa.

The relations shown in Equation 3.1 are only valid for free solutions, thus without the 'disturbing' sedimentary solid phase. The diffusion coefficient D^0 is only valid for infinitely dilute solutions. Boudreau (1997) and Iversen and Jørgensen (1993) have summarized the current state of knowledge concerning the application of the diffusion laws to the pore water volume of sediments, accomplished on the basis of an extensive literature survey, and have thoroughly discussed the general problem. As nothing further needs to be added at this point, the calculation of the values shown in the following Tables 3.1 and 3.2 were performed by using the relations published by Boudreau (1997). Table 3.1 contains the diffusion coefficients for a number of ions, gases, and uncharged complex compounds dissolved in seawater at various temperatures.

Looking at diffusion processes in the pore water volume of sediments, it must be taken into consideration that diffusion can only take place within the pore water volume (porosity ϕ), hence a diffusive flux can only be proportionally effective with regard to this spatial compartment. Beyond this limitation, the diffusion coefficient is distinctly lower in the pore water volume of a sediment (D_{sed}) than in free solution. The diffusive flux in the sediment (J_{sed}) is calculated as:

$$J_{sed} = -\phi \cdot D_{sed} \cdot \frac{\partial C}{\partial x} \qquad (3.2)$$

The diffusion coefficient in the pore water volume of sediments differs from the diffusion coefficient of free solutions in such a manner that diffusion in the pore water volume cannot follow a straight course, but must take 'deviations' around each single grain. The degree of deviation around particles is called tortuosity (θ). It describes the mean ratio between the real length of the pathway and the straight-line distance. Tortuosity can be quantified directly by measuring the electrical resistivity (R) (McDuff and Ellis 1979) and employing a related 'formation factor' (F).

$$F = R_s / R_f \qquad (3.3)$$

In this equation, R_s is the specific electrical resistivity for the whole system composed of sediment and pore water, and R_f denotes the electrical resistivity for pore water only. Since the electric cur-

Table. 3.1 Diffusion coefficient in free solution for sea-water at various temperatures. The calculations were performed on the basis of relations derived from a comprehensive literature study carried out by Boudreau (1997).

Ion	0 °C	5 °C	10 °C	15 °C	20 °C	25 °C
			D^{sw} $[m^2 s^{-1}]$			
H^+	5.17E-09	5.89E-09	6.60E-09	7.30E-09	8.01E-09	8.69E-09
D^+	3.03E-09	3.68E-09	4.33E-09	4.97E-09	5.61E-09	6.24E-09
Li^+	4.21E-10	5.34E-10	6.45E-10	7.56E-10	8.66E-10	9.74E-10
Na^+	5.76E-10	7.15E-10	8.52E-10	9.88E-10	1.12E-09	1.26E-09
K^+	9.08E-10	1.10E-09	1.29E-09	1.47E-09	1.66E-09	1.84E-09
Rb^+	9.70E-10	1.17E-09	1.36E-09	1.56E-09	1.75E-09	1.94E-09
Cs^+	9.79E-10	1.17E-09	1.36E-09	1.55E-09	1.74E-09	1.93E-09
Ag^+	7.43E-10	9.11E-10	1.08E-09	1.24E-09	1.40E-09	1.57E-09
NH_4^+	9.03E-10	1.10E-09	1.29E-09	1.47E-09	1.66E-09	1.85E-09
Ba^{2+}	3.86E-10	4.68E-10	5.49E-10	6.30E-10	7.10E-10	7.88E-10
Be^{2+}	2.44E-10	3.10E-10	3.75E-10	4.39E-10	5.03E-10	5.66E-10
Ca^{2+}	3.42E-10	4.26E-10	5.09E-10	5.91E-10	6.72E-10	7.53E-10
Cd^{2+}	3.15E-10	3.85E-10	4.56E-10	5.25E-10	5.95E-10	6.63E-10
Co^{2+}	3.15E-10	3.85E-10	4.56E-10	5.25E-10	5.95E-10	6.63E-10
Cu^{2+}	3.22E-10	3.96E-10	4.69E-10	5.41E-10	6.13E-10	6.84E-10
Fe^{2+}	3.15E-10	3.84E-10	4.54E-10	5.22E-10	5.91E-10	6.58E-10
Hg^{2+}	3.19E-10	4.17E-10	5.13E-10	6.09E-10	7.04E-10	7.98E-10
Mg^{2+}	3.26E-10	3.93E-10	4.60E-10	5.25E-10	5.91E-10	6.55E-10
Mn^{2+}	3.02E-10	3.75E-10	4.46E-10	5.17E-10	5.88E-10	6.57E-10
Ni^{2+}	3.19E-10	3.80E-10	4.40E-10	4.99E-10	5.58E-10	6.16E-10
Sr^{2+}	3.51E-10	4.29E-10	5.08E-10	5.85E-10	6.62E-10	7.38E-10
Pb^{2+}	4.24E-10	5.16E-10	6.08E-10	6.98E-10	7.88E-10	8.77E-10
Ra^{2+}	3.72E-10	4.65E-10	5.57E-10	6.48E-10	7.39E-10	8.28E-10
Zn^{2+}	3.15E-10	3.85E-10	4.55E-10	5.24E-10	5.93E-10	6.60E-10
Al^{3+}	2.65E-10	3.46E-10	4.26E-10	5.05E-10	5.83E-10	6.61E-10
Ce^{3+}	2.80E-10	3.41E-10	4.02E-10	4.62E-10	5.22E-10	5.80E-10
La^{3+}	2.64E-10	3.28E-10	3.91E-10	4.53E-10	5.15E-10	5.76E-10
Pu^{3+}	2.58E-10	3.13E-10	3.69E-10	4.24E-10	4.79E-10	5.32E-10
OH^-	2.46E-09	2.97E-09	3.48E-09	3.98E-09	4.47E-09	4.96E-09
OD^-	1.45E-09	1.76E-09	2.07E-09	2.38E-09	2.68E-09	2.98E-09
$Al(OH)_4^-$	4.24E-10	5.37E-10	6.50E-10	7.62E-10	8.73E-10	9.82E-10
Br^-	9.51E-10	1.16E-09	1.36E-09	1.56E-09	1.76E-09	1.96E-09
Cl^-	9.13E-10	1.12E-09	1.32E-09	1.52E-09	1.72E-09	1.92E-09
F^-	5.98E-10	7.58E-10	9.17E-10	1.07E-09	1.23E-09	1.39E-09
HCO_3^-	4.81E-10	6.09E-10	7.37E-10	8.63E-10	9.89E-10	1.11E-09
$H_2PO_4^-$	3.82E-10	4.86E-10	5.90E-10	6.92E-10	7.94E-10	8.94E-10
HS^-	9.89E-10	1.11E-09	1.24E-09	1.36E-09	1.49E-09	1.61E-09
HSO_3^-	6.04E-10	7.34E-10	8.63E-10	9.91E-10	1.12E-09	1.24E-09
HSO_4^-	5.69E-10	7.13E-10	8.55E-10	9.96E-10	1.14E-09	1.27E-09
I^-	9.33E-10	1.13E-09	1.33E-09	1.53E-09	1.73E-09	1.92E-09

Ion	0 °C	5 °C	10 °C	15 °C	20 °C	25 °C
			D^{sw} $[m^2s^{-1}]$			
IO_3^-	4.43E-10	5.61E-10	6.78E-10	7.93E-10	9.08E-10	1.02E-09
NO_2^-	9.79E-10	1.13E-09	1.28E-09	1.43E-09	1.58E-09	1.73E-09
NO_3^-	9.03E-10	1.08E-09	1.26E-09	1.44E-09	1.62E-09	1.79E-09
Acetate$^-$	4.56E-10	5.71E-10	6.84E-10	7.96E-10	9.08E-10	1.02E-09
Lactate$^-$	4.19E-10	5.32E-10	6.44E-10	7.54E-10	8.64E-10	9.72E-10
CO_3^{2-}	4.12E-10	5.04E-10	5.96E-10	6.87E-10	7.78E-10	8.67E-10
HPO_4^{2-}	3.10E-10	3.93E-10	4.75E-10	5.56E-10	6.37E-10	7.16E-10
SO_3^{2-}	4.31E-10	5.47E-10	6.62E-10	7.77E-10	8.91E-10	1.00E-09
SO_4^{2-}	4.64E-10	5.72E-10	6.79E-10	7.86E-10	8.91E-10	9.95E-10
$S_2O_3^{2-}$	4.58E-10	5.82E-10	7.06E-10	8.28E-10	9.50E-10	1.07E-09
$S_2O_4^{2-}$	3.75E-10	4.63E-10	5.49E-10	6.35E-10	7.20E-10	8.04E-10
$S_2O_6^{2-}$	5.04E-10	6.40E-10	7.75E-10	9.08E-10	1.04E-09	1.17E-09
$S_2O_8^{2-}$	4.64E-10	5.89E-10	7.12E-10	8.35E-10	9.57E-10	1.08E-09
Malate^{2-}	3.19E-10	4.05E-10	4.90E-10	5.74E-10	6.57E-10	7.39E-10
PO_4^{3-}	2.49E-10	3.16E-10	3.82E-10	4.48E-10	5.13E-10	5.77E-10
Citrate^{3-}	2.54E-10	3.22E-10	3.90E-10	4.57E-10	5.23E-10	5.89E-10
H_2	1.92E-09	2.36E-09	2.81E-09	3.24E-09	3.67E-09	4.10E-09
He	2.74E-09	3.37E-09	4.00E-09	4.63E-09	5.24E-09	5.85E-09
NO	1.02E-09	1.26E-09	1.49E-09	1.72E-09	1.95E-09	2.18E-09
N_2O	9.17E-10	1.13E-09	1.34E-09	1.55E-09	1.75E-09	1.96E-09
N_2	8.69E-10	1.07E-09	1.27E-09	1.47E-09	1.66E-09	1.85E-09
NH_3	9.96E-10	1.23E-09	1.45E-09	1.68E-09	1.90E-09	2.12E-09
O_2	1.00E-09	1.23E-09	1.46E-09	1.69E-09	1.91E-09	2.13E-09
CO	1.00E-09	1.24E-09	1.47E-09	1.69E-09	1.92E-09	2.14E-09
CO_2	8.38E-10	1.03E-09	1.22E-09	1.41E-09	1.60E-09	1.79E-09
SO_2	6.94E-10	8.54E-10	1.01E-09	1.17E-09	1.33E-09	1.48E-09
H_2S	9.17E-10	1.13E-09	1.34E-09	1.55E-09	1.75E-09	1.96E-09
Ar	8.65E-10	1.06E-09	1.26E-09	1.46E-09	1.65E-09	1.85E-09
Kr	8.38E-10	1.03E-09	1.22E-09	1.41E-09	1.60E-09	1.79E-09
Ne	1.31E-09	1.62E-09	1.92E-09	2.22E-09	2.51E-09	2.81E-09
CH_4	7.29E-10	8.97E-10	1.06E-09	1.23E-09	1.39E-09	1.56E-09
CH_3Cl	6.51E-10	8.01E-10	9.50E-10	1.10E-09	1.24E-09	1.39E-09
C_2H_6	6.03E-10	7.42E-10	8.80E-10	1.02E-09	1.15E-09	1.29E-09
C_2H_4	6.77E-10	8.33E-10	9.88E-10	1.14E-09	1.29E-09	1.44E-09
C_3H_8	5.07E-10	6.23E-10	7.40E-10	8.54E-10	9.69E-10	1.08E-09
C_3H_6	6.29E-10	7.74E-10	9.18E-10	1.06E-09	1.20E-09	1.34E-09
H_4SiO_4	4.59E-10	5.67E-10	6.76E-10	7.84E-10	8.92E-10	1.00E-09
$B(OH)_3$	5.14E-10	6.36E-10	7.57E-10	8.78E-10	9.99E-10	1.12E-09

rent in the pore water volume of sediments is bound to the presence of charged ions in solution, the same deviations are valid (cf. Sect. 2.1.2). The tortuosity (θ) is then calculated by applying the porosity (ϕ) to the equation according to (Berner 1980):

$$\theta^2 = \phi \cdot F \qquad (3.4)$$

The diffusion coefficient in sediments (D_{sed}) can be calculated on the basis of a dimensionless tortuosity and the diffusion coefficient in free solutions of sea-water (D^{sw} in Table 3.1):

$$D_{sed} = D^{sw} / \theta^2 \qquad (3.5)$$

If the tortuosity is not known on account of electric conductivity measurements and the 'formation factor' (F), its value may be estimated, a bit less accurately, by its empirical relationship to the porosity. About a dozen different empirical equations are known from literature. The most frequently used is Archie's Law (Archie 1942):

$$\theta^2 = \phi^{(1-m)} \qquad (3.6)$$

The adaptation to specific data is done by means of (m) as long as parallel values are available for sediments obtained from direct measurements of their electrical conductivity. Boudreau (1997) shows, however, that this relation does not have any advantage as compared to:

$$\theta^2 = 1 - \ln(\phi^2) \qquad (3.7)$$

Table 3.2 Tortuosity expressed in terms of a sediment's porosity. The calculation was performed by using the Equation 3.7 as published by Boudreau (1997).

ϕ	θ^2	ϕ	θ^2	ϕ	θ^2
0,20	4,22	0,44	2,64	0,68	1,77
0,22	4,03	0,46	2,55	0,70	1,71
0,24	3,85	0,48	2,47	0,72	1,66
0,26	3,69	0,50	2,39	0,74	1,60
0,28	3,55	0,52	2,31	0,76	1,55
0,30	3,41	0,54	2,23	0,78	1,50
0,32	3,28	0,56	2,16	0,80	1,45
0,34	3,16	0,58	2,09	0,82	1,40
0,36	3,04	0,60	2,02	0,84	1,35
0,38	2,94	0,62	1,96	0,86	1,30
0,40	2,83	0,64	1,89	0,88	1,26
0,42	2,74	0,66	1,83	0,90	1,21

This relation (Boudreau's law) has been used for the calculation of various porosity values prevalent in marine sediments as listed in Table 3.2.

By applying the contents of the Tables 3.1 and 3.2 as well as the relation expressed in Equation 3.5, the various examples of the following section are quantifiable. Notwithstanding, it should be emphasized that tortuosity values obtained by electrical conductivity measurements should always be, if available at all, favored to estimated values deduced on account of an empirical relation to porosity.

3.2.3 Quantitative Evaluation of Steady State Concentration Profiles

This section intends to demonstrate the application of Fick's first law of diffusion to some selected examples of concentration profiles which were derived from marine pore water samples. It needs to be stressed that all these calculations require that steady-state conditions are present. The calculation of non-steady state conditions will only be dealt with later in Section 3.2.4.

Figure 3.5 shows an oxygen profile measured *in-situ*. The part of concentration gradient exhibiting the highest inclination is clearly located directly below the sediment surface. This gradient of 22.1 mol $m^{-3}m^{-1}$ is identified in Figure 3.5 (cf. Sect. 12.2). A relatively high degree of porosity will have to be assumed for this sediment near the sediment surface. A porosity of $\phi = 0.80$ yields a tortuosity value (θ^2) of 1.45, according to Table 3.2. The diffusion coefficient for oxygen dissolved in free seawater at 5 °C is shown in Table 3.1 to amount to $D^{sw} = 1.23 \cdot 10^{-9}$ m^2s^{-1}. Applying Equation 3.5, it follows that the diffusion coefficient for oxygen in sediments is $D_{sed} = 8.5 \cdot 10^{-10}$ m^2s^{-1}. This yields the diffusive oxygen flux from the bottom water into the sediment $J_{sed,\,oxygen}$ as:

$$\begin{aligned} J_{sed,oxygen} &= -0.80 \cdot 8.5 \cdot 10^{-10} \cdot 22.1 \\ &= -1.5 \cdot 10^{-8}\ [\text{mol m}^{-2}\text{s}^{-1}] \end{aligned} \qquad (3.8)$$

To arrive at less complicated and more imaginable figures, and in order to compare this value with, for instance, sedimentological data, we multiply with the number of seconds in a year ($365 \cdot 24 \cdot 60 \cdot 60 = 31,536,000$) and then we obtain:

$$\begin{aligned} J_{sed,oxygen} &= -1.5 \cdot 10^{-8} \cdot 31,536,000 \\ &= -0.47\ [\text{mol m}^{-2}\text{a}^{-1}] \end{aligned} \qquad (3.9)$$

If we now assume that oxygen reacts in the sediment exclusively with C_{org} in a ratio of 106:138 as Froelich et al. (1979) have indicated (cf. Section 3.2.5), then we obtain the amount of C_{org} which is annually oxidized per m^2:

$$R_{ox,Corg} = 0.47 \cdot (106/138) \cdot 12 = 4.4 \; [gC \, m^{-2} a^{-1}] \quad (3.10)$$

It must be pointed out that we have to differentiate very distinctly between two very different statements. On the one hand, the profile inescapably proves that 0.47 mol m^{-2}a^{-1} oxygen are consumed in the sediment. On the other hand, the calculation that 4.4 g m^{-2}a^{-1} C_{org} is equivalent to this amount requires that all oxygen is, in fact, used in the oxidation of organic matter. However, it is imaginable that at least a fraction of oxygen is consumed by the oxidation of other reduced inorganic solute species (e.g. Fe^{2+}, Mn^{2+}, or NH_4^+; cf. also with example shown in Fig. 3.8). There is

indeed evidence that, depending on the specific conditions of the various marine environments, one or the other reaction contributes more or less to the consumption of oxygen. At any rate, this needs to be verified by other measurements, for instance, by recording the concentration profiles of the reducing solute species.

Figure 3.6 shows the concentration profile of dissolved sulfate obtained from the pore water of sediments sampled from the Amazon deep sea fan. The pore water was extracted by compression of sediment sampled with the gravity corer, and was immediately afterwards analyzed by ion-chromatography (compare Sects. 3.3 and 3.4). As compared to the previous example, a depth range comprising more than two orders of magnitude is dealt with here. The paths for diffusion are hence considerably longer in this example. Yet, the sulfate concentration in sea-water is also two orders of magnitude higher than the concentration of oxygen so that, in total, a similar gradient is formed nevertheless.

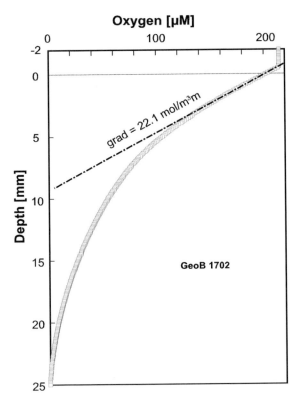

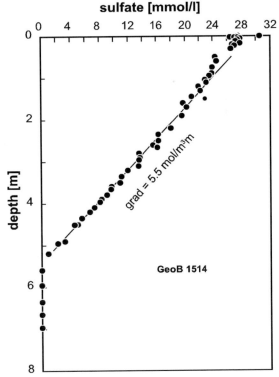

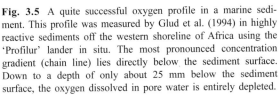

Fig. 3.5 A quite successful oxygen profile in a marine sediment. This profile was measured by Glud et al. (1994) in highly reactive sediments off the western shoreline of Africa using the 'Profilur' lander in situ. The most pronounced concentration gradient (chain line) lies directly below the sediment surface. Down to a depth of only about 25 mm below the sediment surface, the oxygen dissolved in pore water is entirely depleted.

Fig. 3.6 Sulfate profile in pore water from sediments of the Amazon deep sea fan at a water depth of about 3500 m. A linear concentration gradient can be distinctly derived from the sediment surface down to a depth of about 5.4 m. The gradient change, and thus a change in the diffusive flux, is strongly limited to a depth interval of at the most 10 to 20 cm (after Schulz et al. 1994).

In this case a concentration gradient of 5.5 mol m^{-3} m^{-1} was derived as a mean value for the depths ranging from 0 to 5.4 m. Upon examining the curve in more detail, it is obvious that the gradient is less pronounced in the upper 2 meters which is probably explained by a somewhat higher porosity and a concurrently unchanged diffusive flux. As the discovered gradient within the sulfate profile is located distinctly deeper under the sediment surface than in the previous example, it is reasonable to assume a lower degree of porosity. A porosity of $\phi = 0.60$ yields, according to Table 3.2, a tortuosity value (θ^2) of 2.02. On consulting Table 3.1 we find that the diffusion coefficient for sulfate in sea-water at 5 °C is $D^{sw} = 5.72 \cdot 10^{-10}$ m^{2}s^{-1}. Using Equation 3.5 it follows that the diffusion coefficient in the sediment amounts to $D_{sed} = 2.8 \cdot 10^{-10}$ m^{2}s^{-1}. Thus, the diffusive sulfate flux from the bottom water into the sediment corresponds to $J_{sed, sulfate}$:

$$J_{sed,sulfate} = -0.60 \cdot 2.8 \cdot 10^{-10} \cdot 5.5$$
$$= -9.2 \cdot 10^{-10} \; [\text{mol m}^{-2}\text{s}^{-1}] \qquad (3.11)$$

To arrive at more relevant figures and for reasons of comparison with other sedimentologic data, we multiply this value with the number of seconds in one year (31,536,000) and obtain:

$$J_{sed,sulfate} = -9.2 \cdot 10^{-10} \cdot 31,536,000$$
$$= -0.029 \; [\text{mol m}^{-2}\text{a}^{-1}] \qquad (3.12)$$

If we assume that sulfate reacts in the sediment exclusively with C_{org}, in a ratio of 106:53 as Froelich et al. (1979) have reported, then for the amount of C_{org} that is annually oxidized per square meter amounts to:

$$R_{ox,Corg} = 0.029 \cdot (106/53) \cdot 12$$
$$= 0.70 \; [\text{gC m}^{-2}\text{a}^{-1}] \qquad (3.13)$$

It should be indicated at this point as well that the calculated diffusive sulfate flux from the bottom water into the sediment, and from there into a depth of about 5.4 m, is the unequivocal consequence of the profile shown in Figure 3.6. It also follows that this sulfate is degraded in the depth of 5.4 m within a depth interval of at the most 10 to 20 cm thickness. The calculated C_{org} amount that undergoes conversion again depends on the assumption made by Froelich et al. (1979) that indeed the whole of sulfate reacts with organic carbon. Several studies demonstrated that this must

not be generally the case. For sediments obtained from the Skagerak, Iversen and Jørgensen (1985) showed that an essential proportion of sulfate is consumed in the oxidation of methane. At different locations of the upwelling area off the shores of Namibia and Angola, Niewöhner et al. (1998) could even prove that the entire amount of sulfate is consumed due to the oxidation of methane which diffuses upwards in an according gradient.

Figure 3.7 shows a nitrate profile obtained from sediments of the upwelling area off Namibia which is rather characteristic of marine pore water. The processes behind such nitrate profiles are now well understood. The details of these reactions are described in the chapters 5 and 6; here, they will be discussed only briefly as much is necessary for the comprehension of calculated substance fluxes.

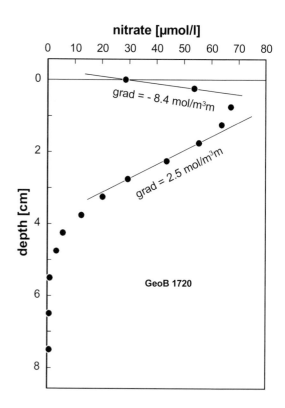

Fig. 3.7 Nitrate profile of pore water obtained from sediments of the upwelling area off the coast of Namibia. The profile displays the shape which is typical of nitrate profiles, with a maximum at a depth which is determined by the decomposition of organic material and the oxidized nitrogen released from it after having reacted with the dissolved oxygen. The gradients indicated document a flux upward into the bottom water and a flux downward into zones where nitrate functions as an electron acceptor in the oxidation of other substances.

The maximum at a specific depth below sea-level is the result of oxidation of organic material by the oxygen that diffuses into the sediment zone from above. Here, the nitrogen of the organic material is converted into nitrate. Mostly, a more pronounced gradient transports the major proportion of nitrate upwards into the bottom water. The smaller proportion travels downwards along a shallow gradient where it is finally consumed as an electron acceptor in the oxidation of other substances.

All these processes can be derived directly and quantitatively from Figure 3.7. Since they are bound to a sedimentary zone which lies very close to the surface, a high degree of porosity can be assumed. According to Table 3.2, the porosity $\phi = 0.80$ corresponds to a tortuosity (θ^2) of 1.45. Table 3.1 shows that the diffusion coefficient for nitrate in free sea-water at 5 °C amounts to $D^{sw} = 1.08 \cdot 10^{-9}$ m^2s^{-1}. Applying Equation 3.5 yields a sedimentary diffusion coefficient of $D_{sed} = 7.4 \cdot 10^{-10}$ m^2s^{-1}. Hence, the diffusive nitrate flux from the sediment to the bottom water compartment is calculated as:

$$\begin{aligned} J_{nitrate,up} &= -0.80 \cdot 7.4 \cdot 10^{-10} \cdot (-8.4) \\ &= 5.0 \cdot 10^{-9} \, [\text{mol m}^{-2}\text{s}^{-1}] \end{aligned} \qquad (3.14)$$

or as

$$\begin{aligned} J_{nitrate,up} &= 5.0 \cdot 10^{-9} \cdot 31{,}536{,}000 \\ &= 0.16 \, [\text{mol m}^{-2}\text{a}^{-1}] \end{aligned} \qquad (3.15)$$

The nitrate flux in a downward direction is calculated accordingly:

$$\begin{aligned} J_{nitrate,down} &= -0.80 \cdot 7.4 \cdot 10^{-10} \cdot 2.5 \\ &= -1.5 \cdot 10^{-9} \, [\text{mol m}^{-2}\text{s}^{-1}] \end{aligned} \qquad (3.16)$$

or as

$$\begin{aligned} J_{nitrate,down} &= -1.5 \cdot 10^{-9} \cdot 31{,}536{,}000 \\ &= -0.047 \, [\text{mol m}^{-2}\text{a}^{-1}] \end{aligned} \qquad (3.17)$$

The sum of both fluxes yields a minimal estimate value of the total nitrate concentration released from organic matter due to its reaction with oxygen. However, the real value for the total amount of released nitrate must be higher than the sum of both calculated fluxes. The gradient of the downward directed flux may be quite reliably calculated from numerous points, however, the more pronounced gradient of the flux leading upward into the bottom water consists only of two points, one of which merely represents the concentration in the bottom water whereas the other represents the pore water of the uppermost 0.5 cm of sediment. It is probably correct to assume that the gradient is more pronounced in closer proximity to the sediment surface, and hence the flux should also prove to be more enhanced. A more accurate statement would only be possible under *in-situ* conditions with a depth resolution similar to the oxygen profile shown in Figure 3.5. As for measurements performed under *ex-situ* conditions, a better depth resolution than shown in Figure 3.7 is very hard to obtain.

If we assume that the released nitrate exclusively originates from the oxidation of organic matter, and that the organic matter is oxidized in a manner in which the C:N ratio corresponds to the Redfield-ratio of 106:16 (cf. Sect. 3.2.5), then we can also calculate the conversion rate of organic matter on the basis of the nitrate profile:

$$\begin{aligned} R_{ox,Corg} = [\text{abs}(J_{nitrate,up}) \\ + \text{abs}(J_{nitrate,down})] \cdot (106/16) \end{aligned} \qquad (3.18)$$

or by employing the values of the above example:

$$\begin{aligned} R_{ox,Corg} &= (0.16 + 0.047) \cdot (106/16) \\ &= 1.37 \, [\text{mol m}^{-2}\text{a}^{-1}] \end{aligned} \qquad (3.19)$$

or

$$R_{ox,Corg} = 16.5 \, [\text{gC m}^{-2}\text{a}^{-1}] \qquad (3.20)$$

At any rate, such a value might serve only as a rough estimation, since apart from the aforementioned error, the calculation procedure implicitly contains some specific assumptions. It has been already mentioned that the C:N ratio of the oxidized material is supposed to be (106/16 = 6.625). Publications of Hensen et al. (1997), however, indicate that especially in sediments with a rich abundance of organic matter a distinctly lower ratio of almost 3 is imaginable (cf. Chap. 6). It must also be considered that the measured profiles are not just influenced by diffusion in the surface zones, but also by the processes of bioturbation and bioirrigation (cf. Sect. 3.6.2).

In principle, the shape of the manganese profile shown in Figure 3.8 is not dissimilar to the nitrate profile of Figure 3.7. Here, we again identify the zones of maximum concentrations – in this particular case about 0.5 m below the sea-floor level – as the site of dissolved manganese release into the pore water. Again, a pronounced negative gradient transports the dissolved manganese upwards away from the zone of its release. This time, however, the substance flux does not reach up to the bottom water above the sediment, instead, very low manganese concentrations are measured at a depth which is only few centimeters below the sediment surface. A flat positive gradient leads a smaller fraction of the released manganese to greater depth where it is withdrawn from the pore water at about 14 to 15 m below the sediment surface.

Here as well, similar to the previously discussed example, both fluxes can be calculated. The upwardly directed flux from the manganese release zone ($J_{manganese, up}$) is obtained from the gradient (-0.5 mol m^{-3}m^{-1}), where in the upper zone the sediment possesses an assumed porosity of 0.80. Considering the values in Table 3.1 and 3.2, as well as the Equation 3.5, a sedimentary diffusion coefficient of $D_{sed} = 2.6 \cdot 10^{-10}$ m^{2}s^{-1} yields the fol-

lowing manganese flux:

$$
\begin{aligned}
J_{manganese, up} &= -0.80 \cdot 2.6 \cdot 10^{-10} \cdot (-0.5) \\
&= 1.0 \cdot 10^{-10} \ [\text{mol m}^{-2}\text{s}^{-1}]
\end{aligned}
\tag{3.21}
$$

or

$$
\begin{aligned}
J_{manganese, up} &= 1.0 \cdot 10^{-10} \cdot 31,536,000 \\
&= 3.2 \cdot 10^{-3} \ [\text{mol m}^{-2}\text{a}^{-1}]
\end{aligned}
\tag{3.22}
$$

Likewise, the downward-directed manganese flux is calculated, however, taking a porosity degree of $\phi = 0.60$ into account and an accordingly calculated $D_{sed} = 1.9$ E-10 m^{2}s^{-1}:

$$
\begin{aligned}
J_{manganese, down} &= -0.60 \cdot 1.9 \cdot 10^{-10} \cdot 0.0084 \\
&= -9.6 \cdot 10^{-13} \ [\text{mol}^1\text{m}^{-2}\text{s}^{-1}]
\end{aligned}
\tag{3.23}
$$

or

$$
\begin{aligned}
J_{manganese, down} &= -9.6 \cdot 10^{-13} \cdot 31,536,000 \\
&= -3.0 \cdot 10^{-5} \ [\text{mol m}^{-2}\text{a}^{-1}]
\end{aligned}
\tag{3.24}
$$

Again, both fluxes added together constitute the total release of dissolved manganese from the sediment into the pore water. Since the downward-directed flux is in this case almost two orders of magnitude lower than the upward-directed flux, it may be neglected considering the possible errors occurring in the determination of the upward stream. The release of manganese is equivalent to the conversion of oxidized tetravalent manganese into the soluble divalent manganese. Which substance is the electron donor in this reaction cannot be concluded from the manganese profile. It could be organic matter as proposed by Froelich et al. (1979). In this case we should not overlook the fact that the converted substance amounts are one order of magnitude lower than they are, for instance, in sulfate fluxes (Fig. 3.6), and more than two orders of magnitude lower than they are in the flux of oxygen. (Fig. 3.5). It should also be noted that upon reducing one mole of Mn(IV) to Mn(II) only two moles of electrons are exchanged, whereas it amounts to 4 moles of electrons per mole oxygen, and even 8 moles of electrons per mole sulfate. Even the 'impressive' gradient of the manganese profile shown in Figure 3.8 does not represent an essential fraction of the overall diagenetic processes in the sediment, involved in the oxidation of organic matter.

The upward-directed manganese flux does not reach into the bottom water, instead, the Mn^{2+} is re-oxidized in a depth of only few centimeters below the sediment surface. Generally, one would ex-

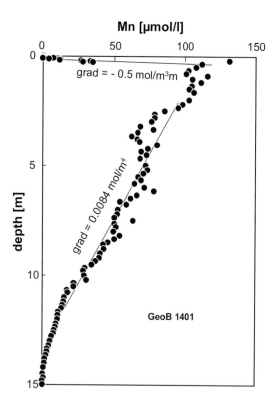

Fig. 3.8 A manganese profile in pore water of sediments off the Congo River estuary, in a water depth of approx. 4000 m. The profile is, in principle, quite similar to the profile of nitrate previously shown in Figure 3.7. Here, manganese is released into the pore water at a specific depth below the sediment surface. A gradient with a high negative slope leads most of the Mn^{2+} upwards; another gradient, positive and more level, conveys manganese into a precipitation zone, which was just included in the lowest core meter (between 14 and 15 m).

pect this re-oxidation to occur by the action of dissolved oxygen (cf. Chap. 11).

It is not within the scope of the present discussion to decide whether the variations in the concentration profile that were averaged upon calculating the gradient of $0.0084 \ mol \ m^{-3}m^{-1}$ shown in Figure 3.8, merely reflect the inaccuracy pertinent to the sampling technique and/or the analytical procedure, or whether they represent discrete processes of their own. Since they do not vary independently, and since several points appear to constitute smaller minima and maxima, it may be suggested that these measurements do not simply represent analytical variations, but reasonable and true values. At any rate, the whole curve leads in its course to very low values that are reached at a depth of 14 m along with a distinct change of the gradient's slope. In many cases, a precipitation of Mn(II)-carbonate is to be expected. The identification of such diagenetic phases anticipated in geochemical modeling is discussed in more detail in Section 15.1.

3.2.4   The Non-Steady State Situation and Fick's Second Law of Diffusion

All examples of the preceding Sections 3.2.2 and 3.2.3 are strictly only valid under steady-state conditions when concentrations, and hence the gradients and diffusive fluxes, are constant over time. Fick's second law of diffusion is applicable in non-steady state situations:

$$\frac{\partial C}{\partial t} = D_{sed} \cdot \frac{\partial^2 C}{\partial x^2} \qquad (3.25)$$

The diffusion coefficient in free solution (D) or the diffusion coefficient in sedimentary pore water (D_{sed}) are used due to the conditions of the system. In contrast to Fick's first law of diffusion, the time co-ordinate (t) appears here next to the local co-ordinate (x). The concentrations and thus the gradients and fluxes are variable for these co-ordinates. Such a partial differential equation cannot be solved without determining a specific configuration of boundary conditions. On the other hand, most of the known solutions are without great practical value for the geochemist due to their very specifically chosen sets of boundary conditions that are seldom related to real situations. In the following only one solution will therefore be presented in detail. The following boundary conditions are to be considered as valid: In the whole sediment profile below the sediment surface the same diffusion coefficient (D_{sed}) is assumed to

prevail, the same concentration (C_0) prevails in the sediment's pore volume. After a certain point in time (t=0) the bottom water attains another concentration (C_{bw}) at the sediment surface. The concentrations ($C_{x,t}$) in the depth profile (co-ordinate x = depth below the sediment surface) at a specific time-point (t) are for these conditions described as:

$$C_{x,t} = C_0 + (C_{bw} - C_0) \cdot erfc\left\{ x \Big/ \left(2 \cdot \sqrt{D_{sed} \cdot t}\right) \right\}$$

$$(3.26)$$

The error function (erfc(a)), related to the Gauss-function, can be approximated according to Kinzelbach (1986) with the relation:

$$erfc(a) = exp(-a^2) \cdot (b_1 \cdot c + b_2 \cdot c^2 + b_3 \cdot c^3 \\ + b_4 \cdot c^4 + b_5 \cdot c^5) \qquad (3.27)$$

with:

$$\begin{aligned} b_1 &= 0.254829592 \\ b_2 &= -0.284496736 \\ b_3 &= 1.421413741 \\ b_4 &= -1.453152027 \\ b_5 &= 1.061405429 \end{aligned}$$

and:

$$c = 1 \, / \, [1 + 0.327591117 \cdot abs(a)]$$

for negative values for (a) it follows:

$$erfc(a) = 2 - erfc(a)$$

Figure 3.9 shows the graphical representation of the error function according to the approximation published by Kinzelbach (1986). This is the function complementary to the error function of Boudreau (1997):

$$erfc(a) = 1 - erfc(a) \qquad (3.28)$$

With this analytical solution of Fick's Second Law of Diffusion, the various curves in Figure 3.3 can now be calculated. On doing this, one will find that the outcome is exactly the same as in the corresponding calculations with the different numerical solutions presented in Chapter 15. Yet, the components of Figure 3.4, with the multiple change of the concentration in bottom water and the 'memory' of which is preserved over several cycles in the pore water fraction, is not accessible with this rather simple analytical solution.

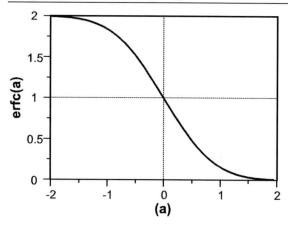

(a)

Fig. 3.9 Graphical representation of the error function in the approximated form after Kinzelbach (1986). This is the function complementary to the error function of Boudreau (1997). Additionally, in the form published by Boudreau (1997), the range of negative values for (a) is omitted, as this has little relevance for sediments.

3.2.5 The Primary Redox-Reactions: Degradation of Organic Matter

Nearly all biogeochemical processes in young marine sediments during early diagenesis are directly or indirectly connected with the degradation of organic matter. This organic matter is produced by algae in the euphotic zone of the water column by photosynthesis. Usually, only a small part of the primary production reaches the sediment surface and of which only a small part is incorporated into the sediment where it becomes the driving force for most of the primary diagenetic redox-reactions (cf. Fig. 12.1).

The conceptual model for the degradation of organic matter in marine sediments was first proposed by Froelich et al. (1979). Although many more details, variations and specific pathways of these redox-reactions have become known in the meantime, this 'Froelich-model' of the primary redox-reactions in marine sediments is still valid

Another example that can be assessed with this analytical solution results from the following considerations: At the beginning of the Holocene, about 10,000 years ago, the sea level rose more than 100 m as a result of thawing ice, which is equivalent to 3 % of the entire water column. This means that seawater had been previously about 3 % higher in concentration. If we assume a chloride concentration of 20,000 mg/l in the seawater today, and thus a mean concentration of 20,600 mg/l in seawater of the ice age, then we are able to calculate the non-steady state chloride profile in pore water with the application of the analytical solution of Equation 3.26.

From the result of this calculation (shown in Fig. 3.10) it follows that we will find just about one half of the ice age seawater concentration (20,300 mg/l) at a depth of 12 m below the sediment surface. If we consider that the reliability of our analytical methods lies at best somewhere around 1.5 %, the exemplary calculation reveals that the effect in pore water is almost at the limit of detection.

Other applications of such analytical solutions hardly make any sense, since, with the exception of chloride, practically all other parameters of pore water are strongly influenced by complex biogeochemical processes. In order to retrace these processes appropriately, analytical solutions for non-steady states in pore water are usually not sufficiently flexible. Hence, numeric solutions are mostly employed. These will be discussed later in Chapter 15 with regard to connection to biogeochemical reactions.

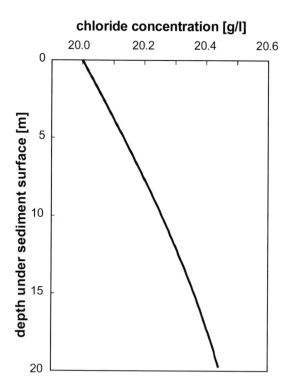

chloride concentration [g/l]

Fig. 3.10 Calculated concentration profile in the pore water of a marine sediment according to an analytical solution of Fick's Second Law of Diffusion. For reasons of simplification it was assumed that the seawater contained 3 % less chloride concentration since the beginning of the Holocene as a result of thawing ice. This lower concentration (20,000 mg/l) had enough time over 10,000 years to replace the higher concentration (20,600 mg/l) from the sediment.

(Fig. 3.11). Usually, these reactions are based on a very simplified organic matter with a C:N:P-ratio of 106:16:1 ('Redfield-ratio') as described by Redfield (1958) (cf. Sects. 5.4.4 and 6.2). The standard free energies for these model reactions are also listed in Figure 3.11 (cf. Sect. 5.4.1). The reactions in this figure are listed in order of declining energy yield from top to bottom.

Based on these reactions, a succession of different redox zones is established:

- Close to the sediment surface, dissolved oxygen is usually transported from the bottom water into the sediment either by molecular diffusion or as a result of biological activity. In this upper zone (the oxic zone), dissolved oxygen is the electron acceptor for the degradation of organic matter. Products of this reaction are carbonate, nitrate and phosphate derived from the nitrogen and phosphorus in the organic matter. For details of these reactions, see Chapter 6.

- Below the oxic zone follows a zone where manganese(IV) oxides in the solid phase of the sediment serve as electron acceptors. Products of this reaction are usually carbonate, nitrogen, phosphate and dissolved Mn^{2+}-ions in the pore water. These dissolved Mn^{2+}-ions are usually transported either by diffusion or by bio-activities to the oxid zone where manganese is re-oxidized and precipitates again as manganese(IV) oxide. These reactions belong to a manganese cycle, by which oxygen is transported into deeper parts of the sediment. For details of these reactions, see Chapter 11.

- Below this zone, nitrate serves as an electron acceptor, which is a product of the redox-

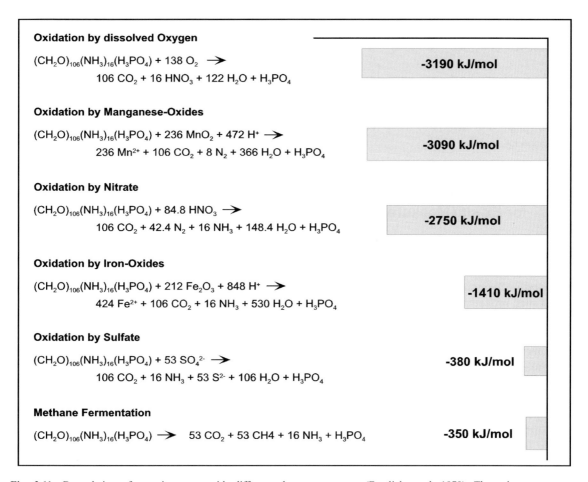

Fig. 3.11 Degradation of organic matter with different electron acceptors (Froelich et al. 1979). The columns represent the different amounts of resulting energy. The oxidation of organig matter by nitrate describes a reaction of the nitrate to elementary nitrogen and leaves the nitrogen of the organic matter as ammonium. Other reactions are possible (cf. section 6.3).

reactions in the oxic zone. Carbonate, phosphate and nitrogen, as well as ammonia are produced. The oxygen for the oxidation of organic matter in this zone is derived from the nitrate, produced in the oxic zone. In most cases, this process oxidizes at least one order of magnitude less organic matter than the reactions in the oxic zone. For details of these reactions, see Chapter 6.

- Below this zone, iron(III) oxides or iron(III) hydroxides in the solid phase of the sediments act as electron acceptors. For details of these reactions, see Chapter 7.

- Below this zone, dissolved sulfate serves as electron acceptor for the oxidation of organic matter, according to the 'Froelich-model'. Recent publications however showed that, in most cases, not organic matter is oxidized in this zone, but predominantly methane, which diffuses up from the deeper parts of the sediment (Niewöhner et al. 1998). For details of these reactions, see Chapter 8.

- The reaction with the lowest yield of standard free energy is methane fermentation with the products carbonate, methane, ammonia and phosphate. For details of these reactions, see Chapter 4.

3.3   Sampling of Pore Water for *Ex situ* Measurements

Ideally, one would prefer to analyze pore water exclusively under *in situ* conditions, as will be explained in Section 3.5, for the parameters that permit such procedure. The pressure change, and frequently the change of temperature as well, are usually coupled to *ex situ* measurement and exert a number of influences of varying potential. However, the *ex situ* measurement will certainly remain a necessity for quite a long time, with regard to most of the substances dissolved in pore water and especially for great depths below the sediment surface. This book is not the place to give a general review on sediment sampling techniques. Rather, the more common procedures for sediment sampling will be introduced with an emphasis put on pore water analysis. Then, the particularities, the possible errors as well as problems arising in

the application of these sampling techniques, will be discussed.

The following section is concerned with the separation of the aquatic pore water phase from the solid sediment phase. As with all particularly problematic and error-inducing procedures, these various techniques have, depending on the case at hand, their specific advantages and disadvantages.

As a matter of course, one would want to analyze the obtained pore water as soon as it has been separated from the sediment to quantify the dissolved substances therein. In daily routine proceedings, compromises must be made since not all analyses can be carried out simultaneously, and since each and every analytical instrument is not present on board a ship. Thus, it will be necessary to describe the state of knowledge concerning pore water storage, transport and preservation.

3.3.1   Obtaining Samples of Sediment for the Analysis of Pore Water

A number of different techniques are available to withdraw samples from the marine sediment. Depending on the scientific question under study, a tool may be chosen that is either capable of taking the sample without harming the sediment surface or disturbing the supernatant bottom water (e.g. multicorer), or that punches out a large as possible sample from the sediment surface area (e.g. box corer), or one that yields cores from the upper less solid meters of the sediment which are as long as possible and most undisturbed (gravity corer, piston corer, box corer).

The Box Corer

The generic term 'box corer' denotes a number of tools of different size and design used in sampling marine sediments, mostly lowered from ships by means of steel wire rope to the bottom. All have a square or rectangular metal box in common which is pressed into the sediment by their own weight, or perhaps by additionally mounted weights. Upon lifting the tool from the sea-floor the first pull on the steel ropes is used to close the box by means of a shovel while it is still situated in the sediment. At the same time, an opening at the top is shut, more or less tightly, so that the bottom water immediately above the sediment is entrapped. The lateral dimensions of small metal box corers are about 10 cm and they reach to that extent into the sediment. Larger box corers ('giant box corer') possess lateral dimensions and penetration depths of 50 cm, respectively.

Apart from the large sample volume that is collected, the giant box corer has the advantage that the *in situ* temperature is kept stable at least in the central zones of the sample, even if the sample is raised at the equator, from a depth of several thousand meters where the sediment is about 2° C cold. This is about the only advantage the box corer has with regard to the geochemical pore water analysis, whereas various disadvantages are to be considered. The shutter on top of the box corer is often not very tightly sealed, and thus the entrapped bottom water might become uncontrollably adulterated upon being raised upwards through the water column. On hoisting the loaded, heavy box corer out of the water, and during its later transportation, onto the deck of a ship, the sediment surface is mostly destroyed to an extent that at least the upper 1-2 cm become worthless for the subsequent pore water analysis.

The Multicorer

The multicorer is also employed from the ship using steel wire ropes and can also be used for all depths under water. On applying this tool, up to 12 plastic tubes (mostly acrylic polymers), each measuring a length of about 60 cm and about 5-10 cm in diameter, are simultaneously inserted approximately 30 cm into the sediment. As with the box corer, the first pull of the steel rope on lifting the appliance is used to seal the plastic tubes on both ends. These shutters are usually tight enough to ensure that the entrapped water will later represent the genuine bottom water.

Variations to the *in situ* conditions are caused in greater depths (about 1000 m and more) by the expansion of the pore water, which happens relative to the sediment, when the pressure diminishes. The uppermost millimeters of the profile are then distorted. Moreover, the temperature rise gives cause for disturbances upon raising the samples upwards out of great depths, in the course of which microbial activity is activated within the sediment sample, which is distinctly higher than under *in situ* conditions.

On the other hand, the multicorer provides, at present, the best solution for *ex situ* sampling of sediments from the sediment/water interface. The nitrate profile shown in Figure 3.7 was measured in pore water extracted from a sediment sample which had been obtained by using the multicorer. It shows clearly that the concentration profile can be measured in pore water with an almost undisturbed depth resolution of 0.5 cm per each sample. The reliable sampling technique using the multicorer also becomes evident upon comparing the *in situ* measured oxygen profiles (Holby and Riess 1996) with the *ex situ* measured oxygen profiles of a multicorer sample (Enneking et al. 1996). Both measurements (Figure 3.12) were conducted at the same location, at the same time, and lead to the same oxygen penetration depth and nearly identical oxygen concentration profiles. In both cases, the oxygen was measured with micro-electrodes.

Light-weight, High-momentum Gravity Corer

A light-weight, high-momentum gravity corer was first described by Meischner and Ruhmohr (1974) (cf. Figure 3.13). Somewhat modified variants

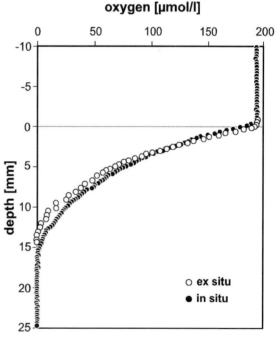

Fig. 3.12 *In situ* measured oxygen profile (dots, Holby and Riess 1996). In addition, an oxygen profile which was also measured ex situ with a micro-electrode is shown (circles, Enneking et al. 1996). The sample was obtained with a multicorer tube. Both profiles were measured in the course of the Meteor M 34/2 expedition at the same time and from the same location in an upwelling area off Namibia, at a depth of approx. 1,300 m below sea level.

made by various manufacturers are in use, whereby 'Rumohr corer' is a model frequently used. In most cases, a plastic tube (mostly an acrylic polymerization product) with a diameter of about 6 cm and a variable weight, is pressed 0.5 m to 1 m deep into the sediment. Upon pulling the tube out, the top aperture is shut whereby the bottom water that just covers the sediment becomes entrapped. The tube is not sealed at the bottom because (mostly) the closure at the top suffices to prevent the sediment from falling out. But, as soon as the high-momentum gravity corer is held out of the water, the bottom aperture of the tube must be sealed with an adequate rubber stopper.

The high-momentum gravity corer is most frequently used in shallow waters, on the shelf and from the decks of smaller ships, since it does happen from time to time that the bulk sample slides out of the bottom prior to its closure with the rubber stopper, and owing to the fact that the corer has only one tube that is filled with sample of sediment. In greater depths, and whenever a larger ship is available, the multicorer will be used in preference. Otherwise, the quality of the samples is entirely comparable. Furthermore, the high-momentum gravity corer is capable of extracting sample cores of up to 1 m in length.

The Gravity Corer and the Piston Corer

Upon using the gravity corer, a steel tube of about 13 cm in diameter is pressed between 6 m and 23 m deep into the sediment with the aid of a lead weight weighing about 3 - 4 t. Inside the steel tube, a plastic liner made of HD-PVC is situated that encompasses the core. Upon extraction from the sediment and raising the tube upwards through the water column, the tube's top aperture is kept shut by a valve. Below, another valve-like shutter - the core catcher - prevents the core material from falling out. Generally, cores of about 10 - 15 m in length are obtained in pelagic sediments, although at times they reach up to 20 m.

In a piston corer the same steel tubes and plastic liners are employed, but with a lighter lead weight. However, this corer is additionally equipped with a shear-action mechanism that moves a piston upwards through the tube when the tool is immersed into the sediment. The piston stroke produces a vacuum that facilitates the penetration of core into the tube. By using the piston corer, core lengths can be achieved which are frequently 20 - 30 m in length and thus several meters longer than the ones obtained with the gravity corer. Yet, it is observed that the cores obtained with the

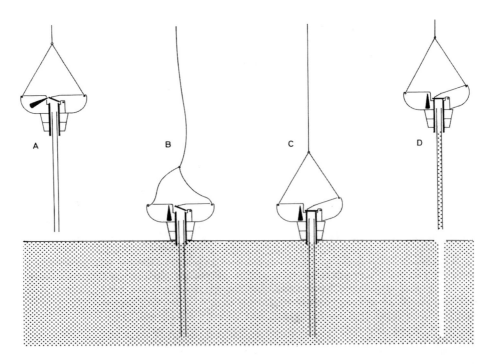

Fig. 3.13 Application of a high-momentum gravity corer (Meischner and Rumohr 1974) to obtain samples from marine sediments. The device can also be stationed on smaller vessels and is suited to extract almost unperturbed cores measuring 1 m in length to be applied in pore water analysis.

gravity corer and with the piston corer display the same stratigraphical depth zones of the sediment. It is therefore assumed that the cores sampled with the gravity corer are 'somewhat compressed' especially in the lower parts, whereas the piston corer produces cores that are 'somewhat extended' in length. How this comes into effect and what conclusions are to be drawn with regard to the pore water samples is yet completely uncertain. In the examined pore water profiles of sediments extracted with the gravity corer, there were no indications as to any noticeable compression to date.

Both sampling tools can only be operated from the decks of larger ships that are equipped with steel ropes and winches designed for managing weights of 10 - 15 tons. The author would personally always prefer using the gravity corer because its handling is easier, safer and faster; and because, especially in low latitudes, every minute counts in which the core is unnecessarily exposed to high temperatures on board the deck of the ship.

At least the upper 10 to 30 cm of the core length obtained with either tool is usually adulterated in that it is not appropriate for pore water analysis. The multicorer, high-momentum gravity corer, or at least the box corer should be employed in a parallel procedure to ensure that this layer will also be included as part of the sample. It should not be overlooked that, especially in the deep sea, sampling with two different tools 'at the same site' might imply a distance of several 100 m on the ocean floor. From this deviation considerable differences in pore water composition, and in some of the biogeochemical reactions close to sediment surface, are likely to result. Hence the specification as to 'same site' must be acknowledged with caution.

After the usage of either tool - the gravity corer and the piston corer - the sediment core obtained is immediately dissected into pieces of 1 m in length within the tube. Usually, the one meter long tubular pieces, tightly sealed with caps, are stored at *in situ* temperature prior to the subsequent further processing which is to be carried out as quickly as possible.

The Box-shaped Gravity Corer (Kastenlot)

In principle, the box-shaped gravity corer is not dissimilar to the above mentioned gravity corer. Here, a metal box with a core length of about 10 m and a lateral dimension of about 0.1 m up to 0.3 m is used instead of a steel tube with a plastic liner.

The core box is shut by valves upon being pulled out of the sediment and is then hoisted upwards through the water column. The advantage of using the box-shaped gravity corer consists in providing a large core diameter, less perturbations due to the metal box's thinner walls, and by removing one side of the box it allows the sediment stratigraphy to be examined. The essential disadvantage in examining pore water samples consists in the fact that the core is distinctly less accessible, and that processing of the core under an inert atmosphere (glove box) is hardly feasible.

The Harpoon sampler

Sayles et al. (1976) have described a device that allows collection of pore water samples under *in situ* conditions. To achieve this, a tube is pressed about 2 meters deep into the sediment, like a harpoon. Then the pore water is withdrawn from various sections of the tube, through opened valves, and concomitantly passed through filters. The crucial step of separating sediment and pore water thus happens under *in situ* conditions. The obtained water samples are then ready to be analyzed *ex situ*. The concentrations which were measured by Sayles et al. (1976), however, do not differ greatly from those which were obtained 'from the same location' by expressing pore water under *ex situ* conditions.

Jahnke et al. (1982) used the same (or, at least a quite similar) tool in 4450 m deep waters of the equatorial Pacific, at approximately $0°$ to $10°$ N, and approximately $140°$ W (MANOP-site). Here, the distances between the locations in which the harpoon sampler was used and in which, for reasons of comparison, pore water was obtained by compression after the sampling was carried out with the box corer, amounted to 300 m to up to 3000 m. In measurements, which were corroborated several times, Jahnke et al. (1982) found similar concentrations of nitrate, nitrite, silica, pH, and manganese. Yet, the concentrations of ΣCO_2, alkalinity and phosphate were distinctly higher and displayed statistical significance.

3.3.2 Pore Water Extraction from the Sediment

In the following section, the separation of pore water from the sediment will be described. The necessary procedures will be described in the sequence in which they are employed. This is to say, that we will begin with the filled tubes of the multicorer, the high-momentum gravity corer, or the meter-sized pieces of the gravity corer or piston corer. The entire processing steps described in the following for the core material should be performed at temperatures which should be kept as close as possible to the temperatures prevailing under *in situ* conditions.

In this context, the problem needs to be dealt with as to how long a tightly sealed core, which is exposed to the *in situ* temperature all the while, can be left to itself before it is processed without the occurrence of any essential perturbation in the pore water fraction. In order to extract pore water from a sediment core, such as is shown in Figure 3.1, and subsequently analyze and preserve it, even a practiced team would require several days. In most cases, a compromise needs to be found therefore, between the highest number of samples and most rapid processing.

A special situation prevailed in the case of the core shown in Figure 3.1, as the ship was cruising for a relatively long time after the core had been taken. Then the experiment was conducted that led to the results shown in Figure 3.1. The core was analyzed with an almost unusual high sampling density. As these procedures afforded plenty of time, and since it was unsure whether the pore water of the sediment core, which was stored at *in situ* temperature in the meantime, would change within the prospective processing time of ten days, the single pieces measuring one meter were intentionally *not* analyzed in a depth sequence, but in a *randomized* sequence. Thereby, the variations in the processing time had to become reflected as discontinuities in the corresponding data at the end of each meter interval. This had not been the case at either interval, from which it follows that a processing time of 10 days was obviously quite innocuous to the quality of the samples.

Analysis of Dissolved Gases

For some substances dissolved in pore water everything will be too late for a reliable analysis as soon as the sediment core lies freshly, but in a decompressed state, on the deck of the ship. This holds true predominantly for the dissolved gases.

In this respect, dissolved oxygen is relatively easy to manage, a circumstance which becomes evident upon comparing the *in situ* oxygen profile with the *ex situ* profile, both measured at the same sites (shown in Fig. 3.12). The reason for this similarity is that the relatively low concentrations, that are often below the saturation level, do not significantly assume a condition of oversaturation even after decompression. Notwithstanding, the results shown in Figure 3.12 reflect the situation too favourably, because differences of up to a factor of 2 were also observed between *in situ* and *ex situ* conditions with regard to the penetration depth of oxygen, and thus to the corresponding reaction rates as well. As to what extent these differences between *in situ* and *ex situ* conditions really exist, or whether such differences result from measurements carried out at not exactly identical sites, has not yet been sufficiently investigated.

The case is obviously similar for dissolved carbon dioxide whose concentration is essentially determined by the equilibrium of the aquatic carbonate phases. As for the alkalinity, no remarkable variations were found in measured values, even at high concentrations, when the measurements were performed successively on adjacent parts of the same sediment core (cf. alkalinity profile shown in Fig. 3.1).

The measurement of sulfide is much more troublesome, especially at high concentrations (It should be noted that H_2S is strongly toxic and that one can become quickly accustomed to its smell after prolonged presence in the laboratory. When working with sulfide-containing core material the laboratory should be ventilated thoroughly at all times!). As a general rule, everything already perceived by its smell is already lost to analysis. Since sulfide is readily analyzed by various methods in aqueous solution, most errors arise from decompression of the core and the subsequent separation of pore water from the sediment. As soon as decompression begins, a great quantity starts to degas, initially forming a finely distributed effervescence. In this condition, measurements can mostly still be carried out, however with less satisfactory results, provided that a specific volume of water-containing degassed sediment is punched out with a syringe and immediately brought into an alkaline environment (SAOB = <u>S</u>ulfur <u>A</u>nti-<u>O</u>xidizing <u>B</u>uffer with pH > 13, after Cornwell and Morse 1987).

Even more difficult is the sampling and the analysis of sediments with a marked content of methane gas. In this particular case, a considerable amount of degassing of the sample occurs immediately upon decompression, thus not

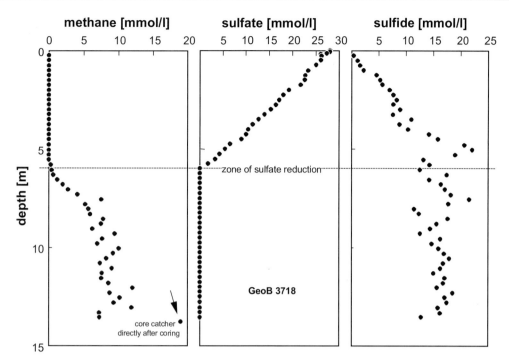

Fig. 3.14 Concentration profiles of pore water from anoxic sediments obtained from an upwelling area off Namibia at a water depth of approximately 1300 m. The analysis of sulfide and methane was carried out in samples that were punched out with syringes from small and quickly sawed-out 'windows' in the fresh sediment core. As for sulfide, these syringe-drawn samples were brought into an alkaline environment, whilst for methane analysis the samples were stored in head space vials for subsequent gas-chromatography analysis. The arrow points to a methane sample that originated from a sealed sediment core obtained by using a sample from the 'core catcher' (after Niewöhner et al. 1998).

permitting the measurement of the concentrations afterwards which are oversaturated under conditions of normal atmospheric pressure.

The results of measurements carried out on a core with high sulfide and methane concentrations are shown in Figure 3.14. The material was obtained in 1300 m deep water, in a high productivity zone of upwelling off the coast of Namibia. It can clearly be seen that in the depth zone of sulfate reduction, methane moving upwards meets with the downwards-diffusing sulfate, both substances displaying almost identical flux rates. These processes are dealt with more thoroughly in Chapter 8; at this point, just the sampling at the site of the core and the analytical sample treatment will be discussed.

As for both substances, sulfide and methane, concentrations found in the core are similar to that of sulfate. For the purpose of sampling, small 'windows' (2 x 3 cm) were cut into the plastic liners with a saw, immediately after the meter-long segments from the gravity corer were available. In each of these windows, 2-3 ml samples of fresh sediment were punched out with a syringe. For the

determination of sulfide some of these sufficiently undisturbed partial samples were placed into a prepared alkaline milieu (see above, SAOB); for the determination of methane others were transferred directly to head space vials. The head space vials comprising a volume of 50 ml contained 20 ml of a previously prepared solution of 1.2 M NaCl + 0.3 M HgCl$_2$. The analysis of sulfide was performed by measuring the sample with an ion-selective electrode, whereas methane analysis was done in the gaseous volume of the headspace vials by gas-chromatography.

It is noticeable that the values in the profiles hardly scatter at the somewhat lower concentrations (about 4 mmol/l for methane and 7 mmol/l for sulfide, respectively). The values of the higher concentrations obtained in greater depths scatter more strongly and are altogether far too low. This became evident since the highest methane concentration is to be found in a sample that was not obtained from an extra sawed-out 'window', but from one that was previously taken directly from the core catcher at the lower open end of the core. As to the question concerning the potential electron donor for sulfate reduction it was, however, of particular importance that methane could be reliably determined up to a concentration of 2-3 mmol/l, and that the upwards directed gradient reached into the zone of the reaction.

Opening and Sampling a Core under
Inert Gas inside the 'Glove Box'

In order to extract the pore water from the sediment, the next step consists in opening, or rather cutting the meter-long segments previously excised from the sediment in a lengthwise fashion. In the case of multicorer or high-momentum gravity corer, the cores are pushed upwards by using an appropriate piston. Caution should be taken that anoxic sediments (easy to identify: if uncertain, everything that is not distinctly and purely brown in color) must not, by any means, come into contact with an oxygen-containing atmosphere in the course of sampling. Therefore the opening of the core segments must happen in an inert atmosphere, or, as the case may be, the core must be pushed from below into the 'glove box' where

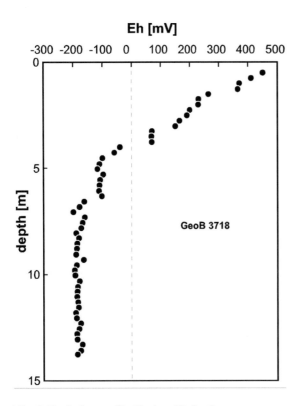

Fig. 3.15 Redox profile (E_H in mV) for the same core as shown in Figure 3.14. Values varying approximately +/- 30 mV must be allowed, whereas an otherwise plausible profile is measured in coherence to the prevailing redox processes.

such an atmosphere is maintained. A 'glove box' that suits the purpose and that can also be easily self-made and adapted to the requirements of the prevailing conditions is described by De Lange (1988).

The inert gases of higher molecular weight, as well as nitrogen, are suited for maintaining the desired inert gas atmosphere. As for nitrogen it must be considered that the least expensive version of the gas does not contain sufficiently low amounts of oxygen. Argon is often more useful, as ordinary argon that is used for welding usually has sufficiently low oxygen. A slight excess pressure in the 'glove box' maintains that, in the event of small leakage, the dissipation of gas might increase, yet the invasion of oxygen is securely prevented.

Measurement of E_H and pH
with Punch-in Electrodes

In the glove box, the redox potential (E_H-value) and the pH value should be measured by applying punch-in electrodes directly to the freshly cut core surface. Suitable electrodes are available today from most suppliers because they are, for example, quite often used in the examination of cheese. However, there is a risk that the electrodes become damaged when larger particles or solidified kernels are present in the sediment sample.

The measurement of the pH-value is relatively easy to do since the function of the electrode can be checked repeatedly with the aid of appropriate calibration solutions. However it must merely be taken into consideration that the comparison with *in situ* measurements close to the sediment surface (cf. Sect. 3.5) may not necessarily lead to corroborated values, because the decompression within the core material has an immediate effect on the pH-value as well.

The redox potential (E_H-value in mV) is considered as a quite troublesome and controversial parameter, with regard to the actual measurement and to its consequences as well. The electrodes available for this purpose cannot be checked with calibration solutions. The occasionally mentioned calibration solutions that function on the basis of divalent and trivalent Fe-ions are not satisfactory, since they actually just confirm the electrode's electric functioning and, at the same time, 'shock' the electrode to such a degree that it may remember the value of the calibration solution for several days to come. The method of choice can

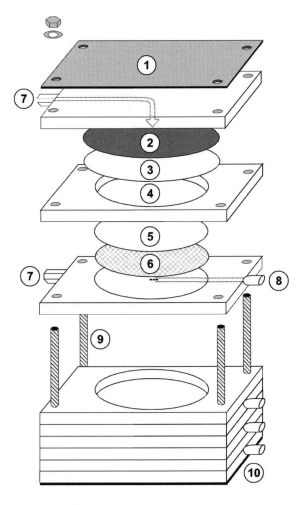

(1) Top sheet (V2A)

(2) Rubberdisk

(3) Parafilm

(4) Chamber for the sample

(5) Filter membrane(0.2 µm)

(6) Tissue

(7) Pressure supply

(8) Leak

(9) Bottom sheet (V2A)

(10) Thread (V2A)

Fig. 3.16 Extractor used for the preparation of pore water from low density sediments (modified after Schlüter 1990).

only consist of using well polished Pt-electrodes and of being continually aware of the plausibility and variation of the measured values. Whenever necessary, the electrode must be substituted for another well-polished Pt-electrode. Useful recommendations as to the redox potential and its measurement are described by Kölling (1986, 2000) and by Seeburger and Käss (1989). A compilation of fundamentals, processes and applications of redox measurements in natural aquatic systems was published by Schüring et al. (2000). A measured E_H-profile is shown in Figure 3.15 for the same core as was shown in the previous Figure 3.14. It can be seen that the values scatter by approximately +/- 30 mV, but that otherwise a reasonable profile has been measured that is likely to contribute to an understanding of the redox processes occurring in the sediment.

Extraction of Pore water

For extracting pore water from the sediment, various authors have developed appliances that all have a common design (e.g. Reeburgh 1967; De Lange 1988; Schlüter 1990). In each case, a sediment sample volume of 100 - 200 ml is transferred into a container made of PE or PTFE which has on the bottom side a piece of round filter foil (pore size: 0.1 - 0.2 µm) measuring 5-10 cm in diameter. On top, the sample is sealed with a rubber cover onto which argon or nitrogen is applied at a pressure of 5-15 atm. The sediment sample is thus compacted and a part of the pore water passes downwards through the filter layer. By this method, a pore water sample comprising 30 - 50 ml can easily be obtained from the water-rich and less solid layers lying close to the sediment surface (Figure 3.16).

It might make sense, in case of anoxic sediments, to load the extractor in an inert atmosphere, or even better, to carry out the whole procedure in the glove box under these conditions. This becomes inevitable if, for instance, divalent iron is to be analyzed. The question of how fast this pore water sample needs to be analyzed for single constituents, will be discussed further in Section 3.4.

Certain losses of CO_2 from the sample into the inert atmosphere of the glove box, and due to that, an increase of pH and a precipitation of calcite seem to be inevitable. Only the combined *in situ* measurement of pH, CO_2 and Ca^{2+} will lead to a reliable measurement of the calcite-carbonate-system.

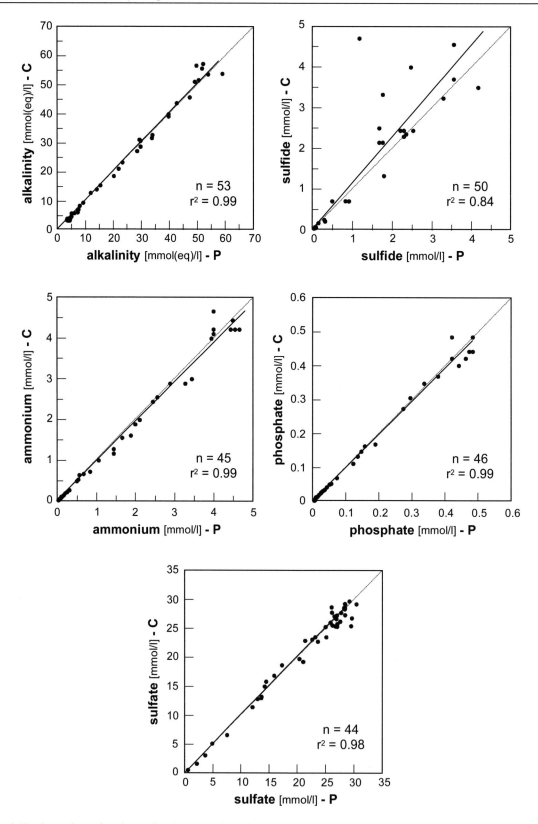

Fig. 3.17 Comparison of analyses after the preparation of pore water by compression and centrifugal extraction. In principle, there are no observable differences. The larger variation of the sulfide values might be attributed to the decompression of the core and/or might have occurred upon mounting the sediment into the press or into the centrifuge (after Schulz et al. 1994).

It should be noted that the choice of the filter used in the extraction procedure will determine what will be evaluated as dissolved constituent and what will be evaluated as particle or solid sedimentary phase. The entire field of colloids which had not yet been investigated thoroughly for marine sediments will therefore be either evaluated as part of the solid phase, or as dissolved constituents, depending on the analytical method employed, for instance, whenever this fraction is eliminated by the addition of hydrochloric acid. With respect to the available and customary methods, a lot of work still remains to be done on this subject.

Centrifugal Extraction of Pore water

Centrifugal extraction of pore water from sediments is a procedure not very frequently performed although it appears rather easy at first examination. Anoxic sediments must be loaded into tightly sealed centrifugation tubes, to be filled and capped under the inert atmosphere conditions of a glove box. But as these plastic tubes are often not sufficiently air-tight, the invasion of ambient air might affect the sample. Moreover, a cooling centrifuge must be regularly employed, since heating the sediment above the *in situ* temperature is by no means desirable. According to the opinion of most workers, the operation of a centrifuge on board a ship is rather troublesome and, most frequently, leads in various ways to the reception of high repair bills.

In the course of centrifugation, the sediment will be compacted in such a manner that a supernatant of pore water can be decanted. Generally, less pore water is gained as compared to pressure exertion. At any rate, the water must be passed through a filter possessing a pore-size of 0.1 to 0.2 μm. As for anoxic sediments, all these steps must be carried out inside the glove box and under an inert atmosphere. The measurements comparing centrifugal extraction and pressing techniques, which have been done so far, indicate that both methods yield quite identical values (see Fig. 3.17).

Whole Core Squeezing Method

A simple method of pore water preparation from a sediment core that is still kept in a plastic liner has been described by Jahnke (1988). In this method, which is also suitable for employing on anoxic sediments, holes are drilled into the tubes used in the multicorer, high-momentum gravity corer, or for a partial sample derived from the box corer. These holes are sealed prior to core sampling with plastic bolts and O-rings. As soon as the tube is filled with the sediment core, the device is mounted solidly permitting a piston from above and below to hold the core into position. Then, the prepared drainage outlets are opened and sample ports mounted to them, through which the pore water is drained off and passed through a filter. If the core is compressed either by the mechanical force of the pistons, or by infusion of gas, the filtered pore water runs off through the prefixed apertures and can be collected without coming into contact with the atmosphere, provided that such precautions were taken. Jahnke (1988) compares concentration profile examples of pore water obtained by the application of this method with pore water extracted centrifugally. Within the margins of attainable precision, these profiles must be conceived as truly 'identical'.

Pore-Water Extraction with Rhizons

Rhizons were originally developed to gain pore water samples which were then used to analyze the moisture of water in unsaturated soils (Meijboom and van Nordwijk 1992). A rhizon is supposed to function like an artificial root of a plant, hence its name is derived from 'rhiza', the Greek word for root. The device consists of thin tubes of various lengths (several cm) measuring approx. 2.5 mm in diameter. They are closed on one end and possess a connector on the other, over which a negative pressure is built up inside the tube. The tube material consists of a hydrophilic porous polymer, with a pore diameter of about 0.1 μm. The actually quite instable tube is reinforced by an inserted piece of wire made of inert metal or durable plastic, allowing it to easily penetrate the soil or a sediment from the outside. In case of hard soils or sediments, a hole must first be punctured with a similarly shaped piece of metal or durable plastic.

The special property of the hydrophilic porous polymer is that, after short soaking in water, it becomes permeable to water - but not to air - once a vacuum is applied inside. Wherever the rhizon and the pore water come into contact, the negative pressure will draw the water into the interior, without allowing air to enter or destroy the vacuum. The water thus drawn inside will be simul-

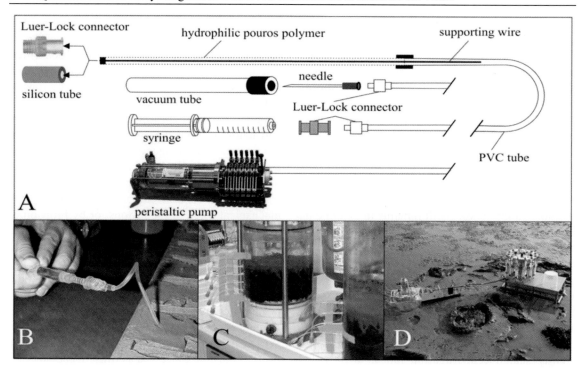

Fig. 3.18 Extraction of pore water from marine sediments with rhizons. (A) Basic construction of a rhizon including three different techniques to apply a vacuum. (B) Insertion into an opened core section. (C) Insertion into a multicorer core. (D) Application in the intertidal flat region, from the sediment surface (from: Seeberg-Elverfeldt et al. 2005).

taneously filtered through the 0.1μm wide pores, permitting its analysis without further treatment.

Seeberg-Elverfeldt et al. (2005) applied rhizons for the first time in marine environments and obtained very good results. Figure 3.18 shows two different application modes. The upper part of Figure (A) outlines three different possibilities how the vacuum may be generated. First, the outside connector of the rhizon can be equipped with a hypodermic needle which is punctured through a rubber seal into a previously evacuated glass vessel. This glass vessel will later contain the pore water being drawn into it. The vacuum can also be generated by means of a disposable syringe or a peristaltic pump. The lower part of Figure 3.18 demonstrates various modes of application, (B) on an open core section, (C) laterally inserted into a multicorer tube through small apertures, and (D) directly inserted into a naked sandflat sediment at low tide.

Previous applications of rhizons produced pore-water analyses which were very similar to those of compressed pore-water samples, except for the phosphate concentrations which were significantly higher when rhizons were used. This agrees with the observations made by Jahnke et al. (1982) upon drawing *in situ* samples with a harpoon sampler (cf. Section 3.3.1) and indicates that using rhizons to extract pore water obviously comes closer to the conditions *in situ* than by pressing or centrifugation. The method of extracting pore water with rhizons not only seems to be easier, but better as well.

Rhizons are manufactured and distributed by Rhizosphere Research Products (Dolderstraat 62, NL-6706 JG Wageningen, The Netherlands) or Eijkelkamp (P.O. Box 4, NL-6987 ZG Giesbeek, The Netherlands). Their price is markedly higher as compared to filters used for pressing pore water out of a sediment sample, however, they can be used more than just once. Special sizes and connectors will be supplied, if a larger number of pieces is ordered.

Sediments as yet Impossible to Sample

Whatever has been said so far as to the separation of pore water from sediments holds true only for fine-grained material that is, in most cases, still quite rich in water. Fortunately, this is what most marine sediments and almost all deep-sea sediments are. However, sediments consisting of pure sand might give rise to problems (e.g. sands from the shelf or purely foraminiferal sands on the mid-ocean ridge). The reason for this is, first of all, that these sediments cannot be compacted by pressure exertion or centrifugation. Consequently, practically no water can be obtained from them at all. It is of even more importance that these sediments possess a high degree of permeability so that as soon as the corer is raised out of the water, the pore water (and sometimes the core material as well) is spilled out from the bottom. A simple method for the extraction of pore water from coarse, sandy material by centrifugation is described by Saager et al. (1990). However, this technique does not solve the problem of how to bring a specimen of sand and its pore water into a centrifuge unperturbed. Only the water which is present at the sampling site at a certain point in time can be collected, even if the abovementioned rhizons are applied. However, it must not be the same water which had been *in situ* at this location.

In situ measurements of pH-values and oxygen with the aid of landers are certainly still possible (cf. Sect. 3.5). But since the thin electrodes are likely to break off upon examining such material, these sediments are not very much favoured. Fortunately, this particular sediment type often contains only a small amount of decomposable organic matter, so that the biogeochemical processes of diagenesis do not display reaction rates as high as other sediment types. Upon sparing these sediments, the pore water geochemist usually suffers no great loss.

3.3.3   Storage, Transport and Preservation of Pore Water

In this section, the vessels in which the pressed or centrifuged pore water is kept until the time for analysis arrives will be briefly discussed. In Section 3.4 it will also be considered, in connection with the analytical methods, how much time one may allow to pass between pore water preparation and analysis. Hence the decision depends on whether immediate analysis is necessary, or whether the analyses can be made at a later time, provided that an appropriate preservation of the material has been accomplished.

Vessels, Bottles

Bottles made of polyethylene have proven to be quite useful for the storage of filtered pore water samples. In most cases the disposable 20 ml PE-bottles which are used in scintillation measure-ments are very much suited to the purpose. These bottles need not be cleaned prior to use, but are just briefly rinsed with the first milliliters that run out of the press or the rhizon. A second use does not really make sense considering the low price of these bottles and the obligatory amount of work effort and detergent to be invested in their cleaning. If several bottles are to be used for the storage of one and same sample, for instance when the sample volume is rather large due to water-rich sediments, precautions should be taken to be sure that the entire sample amount is homogenized prior to the analytical measurements, since differences in contrations between the first and the last drop of pore water are possible.

Preservation of Pore water Samples

Only the alkali-metals, sulfate, and the halogenides do not require any preservation of the samples prior to their analytical measurement at a later date. Otherwise, pore water samples that are not to be analyzed immediately can be preserved with regard to some dissolved substances, allowing a delayed analysis.

The concentrations of most cations - as long as different valences of iron, for instance, are not to be determined - are measurable after a longer period of time when the samples' pH is lowered by the addition of acid (mostly HNO_3) to a value of at least pH 2 or pH 3. Care should be taken that a sufficient amount of 'suprapur' quality acid is added. The required amounts can vary considerably depending on the alkalinity of the sample. Usually, concentrated acid will be used in order to prevent unnecessary dilution of the sample. It is also important that the sample has been previously passed through a filter of 0.1 - 0.2 μm poresize, since the acid will dissolve all particles (and certainly the colloids that pass the filter as well).

As for the substances that are altered in their concentration by microbial activity, the toxification of sample is recommended. Mercury (II) salts are not very much favored since the handling of these extremely toxic substances - which are toxic also to man - requires special precaution measures. Chloroform is banned from most laboratories because it is considered as particularly carcinogenic. Thus, only the solvent TTE (trifluorotrichloro-ethane) remains which is frequently used in water analyses. The addition of one droplet usually suffices since only a small but sufficient proportion dissolves whereas the rest remains at the bottom phase of the bottle. Even though the compound is not as toxic in humans as those previously mentioned, no one would think of pipetting the sample thus (or otherwise) treated by mouth. It is anticipated that these toxified samples have a shelf-life of about 4 weeks.

Previously, water samples were often frozen. This procedure is not at all suited for marine pore water because mineral precipitations might occur. These precipitations do not become soluble upon thawing the samples, hence the pore water samples have become irreversibly damaged.

3.4   Analyzing Constituents in Pore Water, Typical Profiles

This section does not claim to re-write the analytical treatment of sea water anew, nor is it intended to present this aspect as a mere recapitulation of facts which will be far from being complete. This task would by far exceed the possibilities and, above all, the scope of this book. However, analytical methods are subject to continual improvement and relatively fast changes so that many of the methods to be introduced here would soon become outdated. Therefore, only a short account of the current methods will be given, especially of those that demand comment on the specificity to marine pore water.

The analytical details of methods that are currently used in the examination of marine pore water can be looked up in the internet under the following address:

http://www.geochemie.uni-bremen.de

We will take it upon ourselves to continually update the internet page, to evaluate the suggestions we receive for improvement, and eventually make this information available to other researchers.

If we take for granted that, in each single case, the determination of oxygen dissolved in marine pore water, as well as the measurement of the pH-value, E_H-value, and certainly the measurement of the temperature as well, has already been performed, the following sequence of single analytical steps is applied to the pressed or centrifuged pore water samples:

- The titration of alkalinity must be carried out as quickly as possible, i.e., within minutes or maximally within one hour, as a replacement to the direct determination of carbonate. Especially when high alkalinity values are present (maximal values up to 100 mmol(eq) l^{-1} are known to occur in pore water, whereas sea water contains only about 2.5 mmol(eq) l^{-1}), the adjustment to atmospheric pCO_2 might induce too low alkalinity values due to precipitation of calcite within the sample.

- If sampling for methane and sulfide has been performed as described above (see Sect. 3.3.2 'Analyses of Dissolved Gases'), the analytical measurement of the already preserved samples does not have to take place immediately within the first few minutes. Yet, the analysis should not be deferred for more than a couple of hours.

- Analyzing compounds that are sensitive to microbial action (phosphate, ammonium, nitrate, nitrite, silica) should always be completed approximately within one day, if the samples were not preserved by the addition of TTE (trifluorotrichloroethane). Even in its preserved state, it is not recommended to store the sample for longer than four weeks. Storage is best in the dark at about 4°C; freezing the samples is not recommended.

- The determination of sulfate and halogenides is not coupled to any particular time. It must be ascertained that no evaporation of the sample is likely to occur. Other instances of sample perturbation are not known. Storage is best at about 4°C; freezing the samples is not recommended.

Photometrical Determinations, Auto-Analyzer

Most procedures for photometric analysis of sea water samples were described by Grasshoff et al. (1999). Owing to the quicker processing speed, and especially the need to apply lower amounts of sample solution, one would generally employ the auto-analyzer method to this end. The method is also described in the aforementioned textbook. The auto-analyzer can be either self-built or may be purchased from diverse commercial suppliers as a ready-to-use appliance. The analysis of marine pore water is generally characterized by small sample volumes and, as for some parameters, concentrations that are distinctly different from sea water. Allowing for these limitations, the recipes summarized under the above internet-address have proven quite useful for the determination of phosphate, ammonium, nitrate, nitrite, and iron(II) in pore water.

Figure 3.1 shows the characteristic profiles of ammonium and phosphate concentrations that were measured by photometrical analysis of pore water obtained from reactive sediments possessing a high amount of organic matter. Both parameters demonstrate quite similarly shaped curves compared to alkalinity. Here, the ratio of the concentrations derived from both profiles lies close to the C:N:P Redfield-ratio of 106:16:1 and clearly documents their release into the pore water due to the decomposition of organic matter. A typical photometric nitrate profile in reactive sediment zones near the sediment surface is shown in the quantitative evaluation of fluxes and reaction rates presented in Figure 3.7.

Alkalinity

Usually, alkalinity is actually the ultimate parameter that is subject to a procedure of 'genuinely chemical' titration, in this regard as a proxy parameter for carbonate. A new spectrophotometric method for the determination of alkalinity was proposed by Sarazin et al. (1999). Mostly, alkalinity will be, for reasons of simplification, set equal to the total carbonate concentration, although a number of other substances in pore water will contribute to the titration of alkalinity as well. Most geochemical model programs (cf. Chap. 15) foresee the input of titrated alkalinity as an alternative to the input of carbonate. The model program will than calculate the proportion allocatable to the different carbonate species.

Another problem arising from titration of alkalinity in a pore water sample usually consists in the fact that samples with very small volumes cannot easily be used. Most ocean chemists will be accustomed to the titration of a volume of 100 ml, or at least 10 ml. Under certain circumstances, a pore water sample obtained from a definite depth might contain in total not more than 10 ml, hence, at best, merely 1 ml needs to be sacrificed for the titration of alkalinity. This requires that markedly pointed and thin (thus easily breakable) pH-electrodes are used. These are immersed, together with an electronically controlled micropipette, into a small vial so that a tiny magnetic stirrer bead still has enough room to fit inside as well.

Mostly, an adequate amount of 0.001 M hydrochloric acid will be added to a previously pipetted volume of 1 ml, thus the pH will be brought to a value of about 3.5. Since the dilution of the pre-pipetted volume must be considered, the alkalinity (Alk) is calculated according to the following equation:

$$Alk = [(V_{HCl} \cdot C_{HCl}) - 10^{-pH} \cdot (V_0 + V_{HCl}) \cdot f_{H^+}^{-1}] \cdot V_0^{-1}$$

(3.29)

in which V_{HCl} describes the volume of added hydrochloric acid, C_{HCl} represents the molality of the added hydrochloric acid, pH denotes the pH value the solution attains after the addition of hydrochloric acid, V_0 is the pre-pipetted sample volume, and f_{H^+} the activity coefficient for H^+-Ions in solution. The activity coefficient f_{H^+} can be determined according to the method described by Grasshoff et al. (1999), or calculated by employing a geochemical model (e.g. PHREEQC, cf. Chap. 15). The Equation 3.29 was also formulated in the first edition of the textbook published by Grasshoff et al. (1983) [equation 8-25 on page 108]. It must be pointed out that a very misleading error has unfortunately found its way into the equation in this reference. This error is hardly noticeable when low concentrations are prevalent in ocean water, or when the sample volumes are comparably large. In the case of high concentrations in combination with small sample volumes, however, alkalinity values will be calculated that are prone to an error of more than a factor of 2. For this reason, the equation is presented here in its correct form.

Flow Injection Analysis

A very interesting method to analyze total carbon dioxide and ammonium in pore water was introduced by Hall and Aller (1992). In this method, a sample carrier stream and a gas receiver stream flow past one another, separated only by a gas-permeable PTFE (Teflon®) membrane. For determining the total CO_2, the sample carrier stream consists of 10-30 mM HCl. To this stream, the sample in a volume of about 20 µl is added via a HPLC injection valve. The carbon dioxide traverses the PFTE membrane and enters the gas receiver stream which, in this case, consists of 10 mM NaOH. The CO_2 taken up by the gas receiver stream causes an electrical conductivity change that can be determined exactly in a micro-sized continous flow cell.

For the analysis of ammonium, the sample carrier stream consists of 10 mM NaOH + 0.2 M Na-citrate. This converts ammonium into NH_3-gas which penetrates the PTFE membrane and travels into the gas receiver stream, in this case consisting of 50 μm HCl. Here as well, an electric conductivity cell is used for the measurements. To provide a steady and impulse-free flow of the solutions, a multichannel peristaltic pump is used. The flow rates in both cases were set to approx. 1.4 ml min[-1].

These very reliable and easy to establish analytical procedures for the determination of these two important parameters of marine pore water are of special interest, because they require only small sample amounts and because a reliable analysis is obtained over a broad range of concentrations.

Ion-selective Electrodes

Ion-selective electrodes were accepted in water analysis only with great reluctance, despite of the advantages that had been expected on their introduction about 20 years ago. The reasons for this were manifold. The analytical procedure is inexpensive only at first sight. As has been confirmed for some electrodes (especially when its handling is not always appropriate), aging sets in soon after initial use which becomes noticeable with decreased sensitivity, low stability of measured values, and prolonged adjustment times.

With regard to marine pore water, ion-selective electrodes were successfully applied in the determination of fluoride (standard addition method) and in the analysis of sulfide within a mixture of sediment and pore water processed into SAOB-buffer (cf. Sect. 3.3.2). Figure 3.1 shows a typical profile of fluoride measured with an ion-selective electrode in pore water. The total sulfide profile shown in Figure 3.14 was measured in the way described above, by using an ion-selective electrode. The measured values occasionally display considerable variations, probably on account of the sediment's decompression and thereafter they might be produced upon withdrawing the sample from the core. They do not necessarily come about during the course of the analytical measurement.

Ion-Chromatography

The analysis of chloride and sulfate in marine pore water samples after an approximately 20-fold dilution is a standard method for normal ion-chroma-tography (HPLC), so that no further discussion is needed here. Considering the very dilute sample solution and the low sample amount required in ion-chromatography, the applied quantity of pressed or centrifuged pore water is negligibly small. However, the high background of chloride and sulfate which is due to the salt content in sea water prevents, with or without dilution, the determination of all other anion species by ion-chromatography. The sulfate profiles measured with this method are shown in the Figures 3.1, 3.6, and 3.14. Since the chloride profiles are mostly not very interesting, because they reflect practically no early diagenesis reactions at all, they can be consulted for control and eventually for correction of the sulfate profiles, whenever analytical errors have emerged in the course of their concomitant determination, e.g. due to faulty dilutions, or mishaps occurring in the injection valve of the machinery.

ICP-AES, AAS, ICP-MS

Generally the alkali metals (Li, Na, K, Rb, Cs), the alkali earth metals (Mg, Ca, Sr, Ba) and other metals (e.g. Fe, Mn, Si, Al, Cu, Zn, Cr, Co, Ni) are determined in the acid-preserved pore water samples. Analytical details depend on the special conditions provided by the applied analytical instruments ICP-AES (Inductively Coupled Plasma Atomic Emission Spectrometer), ICP-MS (Inductively Coupled Plasma Mass Spectrometer), or AAS (Atomic Absorption Spectrometer) of the respective laboratory. A discussion of details is not appropriate in this chapter. In most cases dilutions of 1:10 or 1:100 will be measured depending on the salt content in marine environments, so that the required sample amounts are rather low.

Gas-Chromatography

The quantification of methane with the aid of gas chromatography (FID-detection) is an excellent standard method, which therefore does not need any further discussion. The main importance is the immediate withdrawal of a sediment/pore water sample as already mentioned in Section 3.3.2. This sample is placed into a 50 ml-headspace vial with a syringe in which the sample is combined with 20 ml of a prepared solution of 1.2 M NaCl + 0.3 M $HgCl_2$. After equilibrium is reached in the closed bottle between the methane concentration in the gaseous phase and its concentration present in aqueous solution (at least one hour), the gas phase is ready to be sampled by puncturing the rubber cap that seals the vial with a syringe.

techniques applied to capture the biogeochemical processes in the water/sediment transition zone has been provided by Viollier et al. (2003).

3.5 *In-situ* Measurements

The desire to measure the properties of a natural system, or profiles of properties, *in situ*, appears to be obvious whenever the system becomes subject to a change as a result of the sample collection procedure. In principle, this is applicable to many aquatic geosystems. Especially where solid and aquatic phases come together in a limited space, this boundary constitutes the site of essential reactions. Such a site is the surface boundary between bottom water and sediment. An essential biogeochemical process of early diagenesis is exemplified by the permeation of oxygen into the young sediment by diffusion, but also by the activity of organisms. The oxygen in the sediment reacts in many ways (cf. Chaps. 5 and 6). In part, the oxygen is depleted in the course of oxidizing organic matter, in part, upon re-oxidizing various reduced inorganic species (e.g. Fe^{2+}, Mn^{2+}).

In organic-rich and highly reactive sediments, the depletion of oxygen previously imported by means of diffusion already occurs in the uppermost millimeters of the sediment (cf. Sect. 3.5). However, these organic-rich and young sediments are mostly extremely rich in water (porosity values up to 0.9) and very soft, so that every sample removal will most likely induce a perturbation of the boundary zone which is, in a crucial way, maintained by diffusion.

Moreover, the actively mediated import of oxygen into the young sediment, which runs parallel to diffusion, and which is transported by the organisms living therein, will function 'naturally' only as long as these organism are not 'unnaturally' treated. Any interference caused by sampling is certain to be regularly coupled to a change of temperature, pressure and the quality of the bottom water, and thus means a change in the 'normal' living conditions. Here as well, it is of special interest to measure the active import of oxygen into the sediment governed by biological activity *in situ*. (cf. Sect. 3.6). An overview of the literature references and the state of knowledge related to the *in situ* measurement methods and

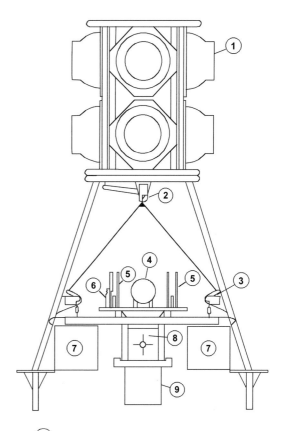

(1)	Glass floats (1 of 8)
(2)	Main ballast release
(3)	Secondary ballast release (1 of 3)
(4)	Electronic pressure case
(5)	Sampling racks
(6)	Syringes
(7)	Expendable ballast
(8)	Chamber lid and stirring mechanism
(9)	Chamber & chamber scoop

Fig. 3.19 Schematic representation of a Lander System which is also used in the deep sea. The machinery sinks freely, without any attachment, to the ocean floor where it carries out measurements and returns back to the surface after the release of ballast. The depicted version stands about 2-3 m in height. It is designed to conduct incubation experiments at the ocean floor, yet similar landers are used to record/ monitor in situ microprofiles of the oxygen concentration through the sediment/bottom water boundary layer. The perspective shows only two of the three feet upon which the lander stands (after Jahnke and Christiansen 1989).

Lander Systems

In some very shallow waters the *in situ* measurement at the sediment/bottom water boundary might be achieved from a ship, or by using light diving equipment. However, so-called 'Lander systems' have been developed for applications in greater depths and, above all, in the deep sea. Lander systems are deployed from the ship and sink freely, without being attached to the ship, down to the ocean floor where they softly 'land'. Standing on the bottom of the ocean they conduct measurements (profiles with microelectrodes), experiments (*in situ* incubations), and/or remove samples from the sediment or the bottom water. When the designated tasks are completed, ballast is released and the whole system emerges back to the surface to be taken on board.

The lander technique described above is, especially in the deep sea, more easily outlined in brief theory than to put into operation, as the involved workers need to invest a considerable amount of construction effort and operational experience in the lander systems so far used.

The problems which cannot be discussed here in detail consist of, for example, deploying the device into the water without imparting any damage to it, conducting the subsequent performance of sinking and landing at the proper speed, keeping the lander in a proper position, recording and storing the measured values, the timing of ballast release and thus applying the correct measure of buoyancy all the way up to the ocean's surface, and finally, the lander's retrieval and ultimate recovery on board of the ship. The correct performance of the *in situ* measurements is not even mentioned in this enumeration. In Figure 3.19 such an appliance is shown according to Jahnke and Christiansen (1989) which is very similar to the lander used by Gundersen and Jørgensen (1990) as well as Glud et al. (1994). A survey and comparative evaluation of 28 different lander systems is given by Tengberg et al. (1995) who also delivers an historical account on the development of landers. One of the first publications

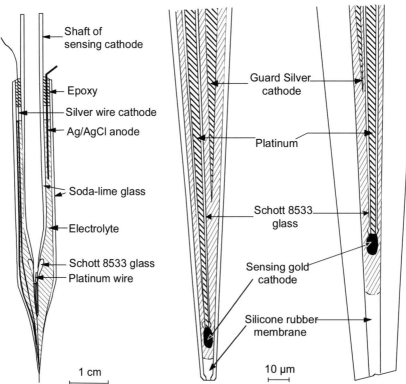

Fig. 3.20   Construction of an oxygen electrode equipped with a Guard-cathode (after Revsbech, 1989).

on the measurements of oxygen microprofiles in sediment/bottom water boundary layers of the deep sea is the report by Reimers (1978). Jørgensen and Revsbech (1985), subsequently, Archer et al. (1989) and Gundersen and Jørgensen (1990) were the first to measure a diffusive boundary layer using landers. The first *in situ* incubations using a lander were carried out by Smith Jr. and Teal (1973).

Profiles with Microelectrodes

A profile measured *in situ* with an oxygen microelectrode has already been presented and quantitatively evaluated in Figure 3.5. Such measurements became feasible only after oxygen electrodes had been developed that could measure oxygen concentrations at the sediment/bottom water boundary layers with a depth resolution of about 20 to 50 μm. It is also important to note that the measurement is carried out without being influenced by the oxygen depletion of the electrode. The development of the electrodes is closely linked to the name of N.P. Revsbech (Aarhus, Denmark) and is described in the publications of Revsbech et al. (1980), Jørgensen and Revsbech (1985), Revsbech and Jørgensen (1986), Revsbech (1989) and Kühl and Revsbech (2001).

The basic construction principle underlying the function of an oxygen microelectrode is shown in Figure 3.20, after a publication by Revsbech (1989). It is evident that such an electrode cannot be built without considerable practiced skill and experience - and they have to be either self-made or purchased at a rather high price. It is furthermore evident that such microelectrodes are extremely sensitive and break easily as, when they either hit a rather solid zone in the sediment, contain slight imperfections from manufacturing, or are handled with the slightest degree of ineptitude.

A construction unit is shown in Figure 3.21 that permits the integration of microelectrodes in a lander along with the necessary mechanical system, the electronic control and data monitoring equipment. It is of utmost importance that this unit is built into the lander in such a manner that the measurements will begin above the sediment surface, then insert the electrodes deep enough into the sediment, allowing later derivations of the essential processes from the concentration profile. This requires, among other pre-requisites, the correct estimation as to how

deep the feet of the lander will sink into the sediment. If the sediment is firmer than expected, only the bottom water might be measured; whereas, if the sediment is softer than expected, the profile might begin within the sediment. Thus,

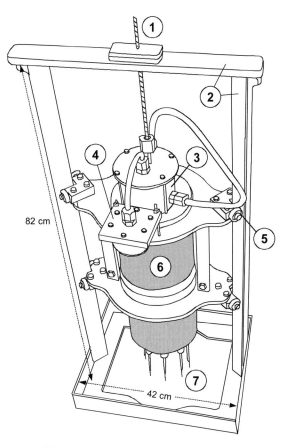

1. Thredded rod
2. Supporting Al - frame and tracks
3. DC - motor hausing
4. Oil - filled bladder
5. Radial ball bearings
6. Pressure cylinder
7. Microelectrodes

Fig. 3.21 Schematic representation of the mechanic and electronic unit applied to microelectrodes in a lander system. The round pressure cylinder contains the electronic control equipment and the data monitoring device, to the bottom of which various vertically positioned microelectrodes are mounted. The entire cylinder is lowered with the aid of a stepper motor and a corresponding mechanical system in pre-adjustable intervals so that the electrodes begin measuring in the bottom water zone and then become immersed into the sediment (after Reimers 1978).

much experience and many aborted profiles stand behind those profiles which are now shown in Figures 3.5 and 3.22.

By using such electrodes it is now possible to describe a diffusive boundary layer at the transition zone between the sediment and the bottom water with a sufficient degree of depth resolution (Fig. 3.22). This layer characterizes a zone which lies directly over the sediment and is not influenced by any current. Instead, a diffusion controlled transport of material characterizes the exchange between bottom water and sediment. This layer which possesses a thickness of 0.2 - 1 mm still belongs to the bottom water with regard to its porosity (100 % water), but due to its diffusion controlled material transport, it belongs instead to the sediment. In both oxygen profiles shown on the left in Figure 3.22, the location of the diffusive boundary layer is indicated. Above this layer, the bottom water is agitated, resulting in an almost constant oxygen concentration. A constant gradient is found within this layer because this is where only diffusive transport takes place, but no depletion of oxygen occurs. It may be of interest in this regard that the better known diffusion coefficient for water (D^{sw} in Equation 3.5 and in Table 3.1) must be applied for this diffusive transport, and not the diffusion coefficient for the

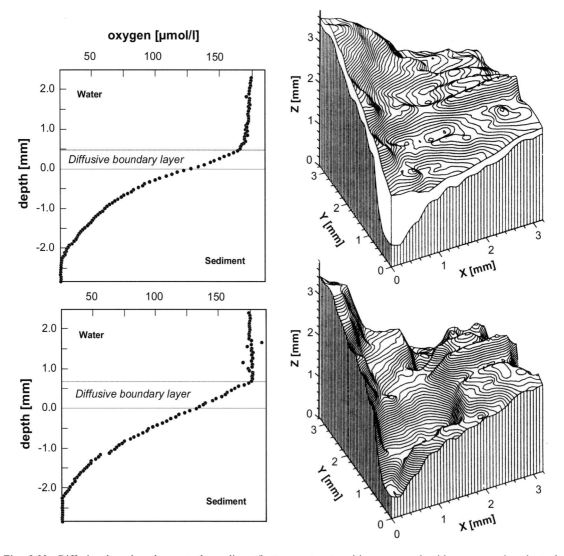

Fig. 3.22 Diffusive boundary layer at the sediment/bottom water transition measured with oxygen microelectrodes. Profiles are shown in both representations on the left. The two representations on the right show the respective spatial patterns. In the above right corner, the surface is shown, whereas the subsurface of the diffusive boundary layer is shown below (after Gundersen and Jørgensen 1990).

sediment (D_{sed}). Since the entire diffusive transport from the bottom water to the sediment crosses this transitory layer, measurements within the layer allow for especially reliable estimations of the oxygen consumption in the sediment.

To date, extremely good results have been obtained with oxygen microelectrodes in sediments with high reaction rates, because the essential processes can be analyzed only here, with profiles most often reaching a few centimeters into the sediment. The lower ends of these profiles are frequently marked by breakage of the electrodes. In sediments displaying less vigorous turnovers, the measurement mostly includes just one single oxygen concentration gradient, not reaching into the penetration depth for oxygen.

Less frequently, pH-profiles are published in literature when compared to the oxygen profiles. The details as to their shapes and their underlying processes are not yet sufficiently understood. Additionally, a reliable calibration under the given pressure conditions has not been accomplished for *in situ* pH-measurements. Consequently, only relative values are measured within one profile. Electrodes measuring H_2S can also be built as microelectrodes. However, they are only suited for *in situ* measurements when H_2S reaches close enough to the sediment surface.

Optodes

Apart from the great advantages of electrodes - they have practically opened a domain of the aquatic environment which was previously not accessible to measurement - the disadvantages should be mentioned as well. On the one hand, only the three aforementioned types are applicable (O_2, pH, H_2S), on the other hand, these electrodes are commercially available only at a rather high price or must be self-made. The manufacturing of one electrode demands - if one has the necessary experience – half a day up to one full day of labor. Additionally, the electrodes tend to break easily. Thus, in the experimental work that led to the publication by Glud et al. (1994) about 50 to 60 (!) microelectrodes were 'used up'.

It therefore appears to be reasonable to look for other methods and other measurement principles. Such an alternative principle consists of the construction of the optode (by analogy: electrodes as functioning electrically, optodes functioning optically). The basic principle (Fig. 3.23) consists of the conductance of light possessing a specific wavelength (450 nm) via glass fibers to the site of measurement. The tip of the glass fiber is surrounded by a thin layer of epoxy resin that contains a dye (ruthenium(II)-tris-4,7-diphenyl-1,1-phenanthroline).

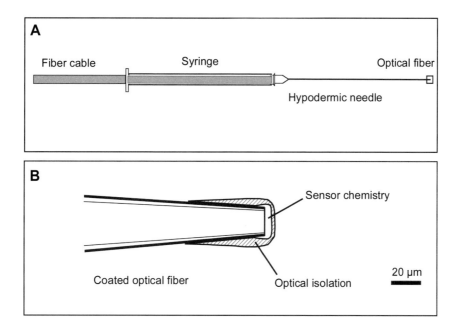

Fig. 3.23 Schematic representation of an oxygen microoptode. One end of the glass fiber can be mounted into a steel cannula for protection and stabilization (A). At the site of measurement, the tip of the glass fiber is surrounded by a thin layer of epoxy resin (B) which contains a fluorescent dye. The fluorescence properties of the dye depend on the concentration of the compound to be measured (after Klimant et al. 1995)

The dye emits a fluorescent light at a different wavelength (610 nm). The intensity of this fluorescence depends on the concentration of the substance to be measured that penetrates into the resin by diffusion, directly at the site of measurement. The construction of such microoptodes, and the first results of measuring oxygen at the sediment/bottom water boundary were described by Klimant et al. (1995, 1997), Glud et al. (1999) and Kühl and Revsbech (2001). According to Klimant et al. (1995), optodes can be developed for a number of substances (e.g. pH, CO_2, NH_3, O_2). As yet, this new technology has only been applied to oxygen measurements, i.e. in the form of the microoptode. Presumably, the construction details and the alchemy of the fluorescent dyes will still be the object of current development.

So-called planar optodes were developed in recent years, which, for example, allow us to draw two-dimensional maps showing the oxygen distribution in a sediment. The fluorescence induced by oxygen can be measured and made visible on a particularly processed stretch of foil (e.g. Glud et al. 2001).

in-situ Incubations

The diffusive input of oxygen into the sediment contributes to only one part of the whole oxygen budget. An input of oxygen into the sediment conducted by macroscopic organisms must be considered to occur especially in the sediments of shallow waters, and when the proportion of organic matter in the sediment is high. These proces-

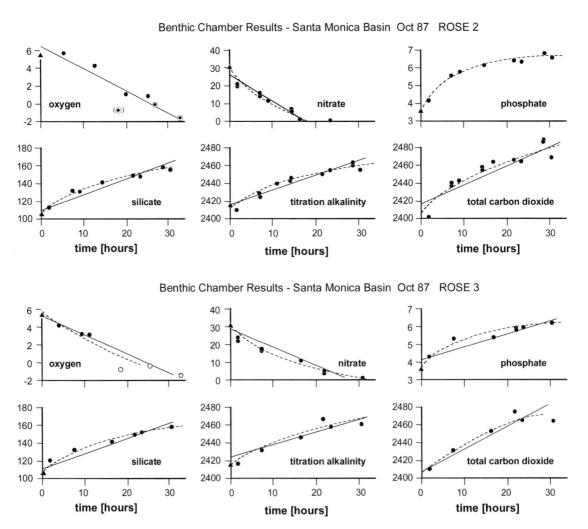

Fig. 3.24 Results of in situ incubation measurements. In both examples, a depletion of oxygen and nitrate and a concomitant release of phosphate, silica, and carbonate was observed in incubation time lasting 30 hours. The concentration unit is μmol/kg, alkalinity is expressed in μmol(eq)/kg. (after Jahnke and Christiansen 1989).

ses will be discussed more thoroughly in the following Section 3.6 concerned with bioirrigation.

The total oxygen consumption at the transition
between sediment and bottom water is measured
in so-called incubation experiments which are best
to be carried out under *in situ* conditions. To this
end, a distinct area of the sediment surface area
(usually several dm²) is covered with a closed box
in such a manner that a part of the bottom water is
entrapped. Inside the closed box, the concentrations of oxygen and/or nutrients are repeatedly
measured over a sufficient length of time. The
total exchange of substances is deduced from the
measured concentration changes in relation to the
volume of entrapped bottom water and the area
under study.

Figure 3.24 shows the result of such an *in situ*
experiment taken from a study conducted by
Jahnke and Christiansen (1989). The sediment's
uptake of the electron acceptors oxygen and
nitrate is easy to quantify, as well as the concomitant release of phosphate, silica, and carbonate
into the bottom water.

Glud et al. (1994) also carried out *in situ* incubation experiments in the upwelling area off the
coast of Namibia and Angola, in a parallel procedure to their recording diffusion controlled
oxygen profiles by means of microelectrodes.
Thus, they were able to obtain values for the
sediment's total oxygen uptake and, at about the

same site, simultaneously measured the exclusively diffusion-controlled oxygen uptake. The ratio
of total uptake and diffusive uptake yielded
values between 1.1 and 4. The higher ratios were
coupled to the altogether higher reaction rates
observable in organic-rich sediments at shallower
depths. Similarly a good correlation between the
difference of total oxygen uptake minus diffusive
oxygen uptake on the one hand, and the abundance of macroscopic fauna on the other hand
was observed.

Peepers, Dialysis, and Thin Film Techniques

At first glance, dialysis would appear to be a very
useful technique for obtaining pore water samples
under *in situ* conditions which can be analyzed
for nearly all constituents. Different types of
'peepers' are presently used which can be pressed
into the sediment from the sediment surface. Such
a peeper usually consists of a PE-column 0.5 - 2 m
long with 'peepholes' on its outside. Cells filled
with de-ionized water inside the peeper are
allowed to equilibrate via dialysis through a filter
membrane with the pore water outside the peeper.
However, an equilibration time of many days or
even a few weeks may be necessary if a volume of
fluid greater than 5 - 10 cm³ is necessary for
analysis and if a small peephole is required in
order to achieve a certain depth resolution. This

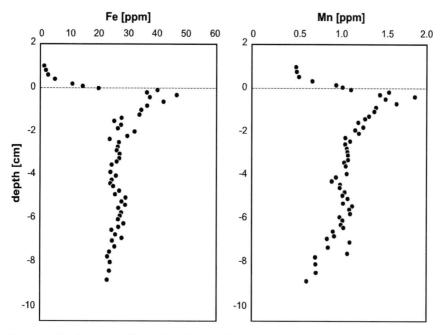

Fig. 3.25 Concentration-depth profiles for Mn and Fe through the sediment/water interface. For these measurements, a DET gel was deployed for 24 hours in situ. (after Davison et al. 1994)

restricts the use of dialysis-peepers to shallow water environments where diving is possible, and thus makes it especially useful in tidal areas. It should be pointed out that dialysis does not simply result in the sampling of pore water under *in situ* conditions. Only those aquatic species that are small enough to pass through the membrane can find their way through the peephole into the cell. The species distribution is therefore dependent on the nature of the membrane used. A review of different techniques is given by Davison et al. (2000). More relevant to the study of deep sea sediments are *in situ* techniques using the diffusive equilibration in thin films (DET) and diffusive gradients in thin films (DGT) (Davison et al. 1991; Davison and Zhang 1994; Davison et al. 1997; Davison et al. 2000).

Figure 3.25 shows concentration-depth profiles for Mn and Fe through the sediment/water interface with a depth resolution which would be impossible to obtain using any other technique. Measurements such as these have resulted in meaningful measurements of reaction rates within the uppermost layers of young sediments and of diffusive fluxes of various metals through the sediment/water interface for the first time. However, the preparation of the thin films, and their handling under field conditions is still quite troublesome and much work needs to be undertaken to further develop these techniques. Nonetheless, this seems to be a most promising approach for obtaining reliable *in situ* determinations of elements in sediment pore waters.

3.6   Influence of Bioturbation, Bioirrigation, and Advection

For the pure geochemist it would be most convenient if this chapter would close at this point. The concomitance of diffusive transport and reactions determinable on the basis of gradient alterations - although they are rather numerous - results in a sufficiently sophisticated system of pore water and sediment, quite well understood and calculable. For deeper zones below the sediment surface, beyond approximately 0.5 to 1 m, a description of the system in terms of diffusion and reactions probably seems to be quite acceptable, with a certain degree of accuracy obtained. The closer the zones under study are located to the sediment surface, the more inhomogeneities and

non-steady state conditions become of quantitative importance. These are mainly determined by the activity of organisms living in the sediment.

Depending on whether the effect on the sediment or the pore water is considered, this phenomenon is referred to as bioturbation or bioirrigation, respectively. Advection (sometimes also called convection) is the term used for the motion of water coupled to a pressure gradient which partly overlaps with bioirrigation, when permeabilities are influenced by the habitats of the organisms. At the end of this section, model concepts will be outlined that allow for a quantitative description of this heterogeneous, yet very reactive domain of the sediment.

Bioturbation

Bioturbation refers to the spatial rearrangement of the sediment's solid phase by diverse organisms, at least temporarily living in the sediment. This process implies that all sedimentary particles in the upper layers, which are inhabited by macroorganisms, are subject to a continual

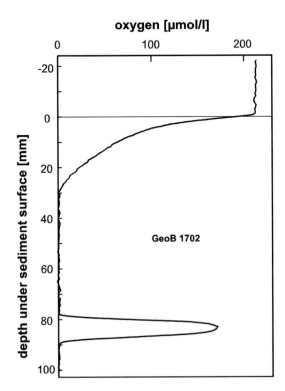

Fig. 3.26 In situ measurement of an oxygen profile in a water depth of 3100 m, off the mouth of the Congo River. The microelectrode detected an open cavity flooded with oxygen-rich bottom water in a depth between 80 mm and 90 mm below the sediment surface (after Glud et al. 1994).

transfer, i.e. seen in the long run, are integrated into a permanent cycle. From the publications of many authors (e.g. Aller 1988, 1990, 1994; Dicke 1986) it can be concluded that this layer, essentially influenced by bioturbation, generally reaches 5 to 10 cm, rarely even up to 15 cm, below the sediment surface. Only when a sediment particle has reached a greater depth in the course of continued sedimentation, will the 'recycling' process become a rare event until it ultimately comes to a final standstill.

This continued recycling of the sediment's solid phase is of great importance for a number of geobiochemical processes. By the action of this cycle, the electron donor (organic matter) and the electron acceptors (e.g. iron and manganese oxides) of the biogeochemical redox processes in the sediment are continually procured from the sediment surface. The results obtained by Fossing and Jørgensen (1990) can only be understood when considerably more sulfate is chemically reduced in the upper layers of the sediment than is accounted for by the sedimentation rate of directly imported organic matter. The re-oxidation of once released sulfide back to sulfate, which is also of importance in this regard, can only proceed when the oxides of iron and manganese, the electron acceptors, are continually supplied from the surface in an oxidized state, and return there in a chemically reduced form. Such a cycle has been described for iron and manganese, for instance, by Aller (1990); Van Cappellen and Wang (1996) as well as Haese (1997).

Bioirrigation

The process in which living organisms in the sediment actively transport bottom water through their habitats is known as bioirrigation. In this process, oxygen-rich water is usually pumped into the sediment, and water with less oxygen is pumped out. Figure 3.26 which is derived from the publication by Glud et al. (1994) demonstrates an oxygen profile measured in situ with the aid of a microelectrode. The electrode recorded a normal profile reaching about 30 mm below the sediment surface. Beyond, from about 30 mm to 80 mm below the sediment surface, the oxygen concentration was practically zero, whereas an open cavity was detected between 80 and 90 mm flooded with oxygen-rich bottom water.

In the publications submitted by Archer and Devol (1992), a comparative study on the purely diffusive oxygen flux (based on measurements with microelectrodes) and the total oxygen flux (based on incubations) was conducted for the shelf and the continental slope off the State of Washington. In a similar study, Glud et al. (1994) investigated the continental slope off Angola and Namibia. It was shown that the flux induced by bioirrigation was several times larger than the diffusive flux. However, such high fluxes only occur in densely populated sediments, mostly on the shelf. For deep sea conditions the total oxygen uptake was only slightly higher than the diffusive oxygen uptake of the sediment.

Advection of Pore Water

Except for bioirrigation, advection within the pore water fraction may only result from pressure gradients. Advection is sometimes referred to as convection, yet the terms are used rather synonymously. There are, in principle, three different potential causes leading to such pressure gradients, and thus to an advective flux:

- Sediment compaction and a resulting flow of water towards the sediment surface.

- Seafloor areas with warmer currents and a resulting upward-directed water flow, that corresponds with water flows directed downwards at other locations.

- Currents in bottom water that induce pressure differences at luv and lee sides of uneven patches on the sedimentary surface.

The first case can be quite easily assessed quantitatively on the basis of porosity measurements: one might imagine a fresh and water-rich deposited sediment column and observe the compaction as a further descent of the solid phase relative to the water that remains at the same place. Hence, the upwards directed advective flux results as the movement of water relative to the further sinking sediment. If the sediment exhibits a water content of approximately 0.9 at its surface, and even a value of 0.5 in several meters depth, and provided that the boundary between sediment and bottom water moves upwards due to the accumulation of new sediment, then it will inevitably follow that the compaction induces an advective flux which will not exceed values similar to the rate of sedimentation. As for deep sea sediments, this process can produce values in the range of a few centimeters per millennium.

Some examples for the second case were provided by Schultheiss and McPhail (1986). By using an analytical instrument that freely sinks to the ocean floor, they were able to measure pressure differences between pore water, located 4 m below the sediment surface, and the bottom water directly. At some locations, deep sea sediments of the Madeira Abyssal Plain displayed pressure differences of about 0 Pa, at other locations, however, the pressure in pore water 4 m below the sediment surface was significantly lower (120 or 450 Pa) than in bottom water. With regard to the porosity (ϕ) and the permeability coefficient of the sediment (k), the advective flux is calculated according to the following equation:

$$v_a = \left(k \cdot \frac{\Delta p}{\Delta x} \right) \Big/ \phi \qquad (3.30)$$

In this equation, known as the *Darcy Equation*, and which is applied in hydrogeology for calculating advective fluxes in groundwater, v_a [m s^{-1}] denotes the velocity with which a particle/solute crosses a definite distance in aqueous sediments. Δp refers to the pressure altitude measured in meters of water column (10^5 Pa = 1 bar $\approx$ 750 mm Hg $\approx$ 10.2 m water column); and Δx [m] is the distance across which the pressure difference is measured. In the example shown above (ϕ = 0.77, k = 7·10^{-9} m s^{-1}), this distance amounts to 4 m. Insertion into Equation 3.30 yields:

$$v_a = \left(7 \cdot 10^{-9} \cdot \frac{0.012}{4} \right) \Big/ 0.77 = 2.7 \cdot 10^{-11} \text{ [m s}^{-1}]$$

$$(3.31)$$

or:

$$v_a = 0.86 \text{ [mm a}^{-1}]$$

Applying the other value specified by Schultheiss and McPhail (1986) as 450 Pa, a velocity of v_a = 3.2 [mm a^{-1}] ensues. (These authors end up with the same values, yet they do so by using a different, unnecessarily complicated procedure for calculation). Such values would thus be estimated as one, or rather two orders of magnitude greater than those resulting from compaction. Actually, such values should distinctly be reflected by the concentration profiles in pore water. However, despite of the

huge number of published profiles, this consequence has not yet been demonstrated.

Upon applying the Darcy Equation 3.30 which is designed for permeable, sandy groundwaters, we must take into consideration that this equation is only applicable to purely laminar fluxes. Furthermore, it does not account for any forces effective near the grain's surface. A permeability value of k = 7·10^{-9} m s^{-1} implies that grain size of the sediment particles is almost identical to clay. Within such material, the forces working on the grain surfaces impose a limitation on the advective flux, so that the calculation of Equation 3.31, based on the pressure gradient, will surely lead to a considerable overestimation of the real advective flux.

The third case, a pressure gradient induced by the bottom current and a supposed advective movement of the pore water, is of importance mainly on the shelf where shallow waters, fast currents, an uneven underground, and high permeability values in coarse sediments are encountered. Ziebis et al. (1996) as well as Forster et al. (1996) were able to demonstrate by *in situ* measurements and in flume experiments that the influence of bottom currents may be indeed crucial for the superficial pore water of coarse sand sediments near to the coast.

At flow rates of about 10 cm s^{-1} over an uneven sediment surface (mounds up to 1 cm high), the oxygen measured by means of microelectrodes had penetrated to a maximum depth of 40 mm, whereas a penetration depth of only 4 mm was measured under comparable conditions when the sediment surface was even (Fig 3.27). Huettel et al. (1996) were able to show in similar flume experiments that not only solutes, but, in the uppermost centimeters, even fine particulate matter was likewise transported into the pore water of coarsely grained sediments. Similar processes with marked advective fluxes are, however, not to be expected in the finely grained sediments predominant in the deep sea.

Advection of Sediment

An advection of the sediment's solid phase does not, at first sight, seem to make any sense, because it implies - in contrast to bioturbation - a movement of sediment particles which favor a specific direction. There is no known process that describes an advection of a solid phase, instead, the definition of such a process actually only results from the definition of the position of the system of coordinates, and thus already indicates the model boundary conditions which will be later dealt with in more detail. If the zero-point of the coordinate-system is set by definition to the boundary between sediment and bottom water, it follows that the coordinate-system will travel upwards with the velocity of the sedimentation rate. Concomitantly, this also implies that sediment particles move downwards relative to the coordinate-system with the velocity of the sedimentation rate. This definition may appear somewhat formalistic, at first, but it becomes more real and relevant from a quantitative point of view when diagenetic reactions are studied. Then most turnovers are limited by the solid phase components that are introduced into the system by this form of 'advection'.

Conceptual Models

Conceptual models and their applications will be discussed in more detail in Chapter 15 where computer models will be presented and examples taken from various fields of application will be calculated. At this point, conceptual models will just be mentioned in brief summary and will contain the most essential statements of this chapter.

- The description of the diffusive transport by Fick's first and second law of diffusion includes the transport of soluble substances in pore water, which is not mediated by macroorganisms. In this regard, diffusive fluxes are always produced by gradients; and fluxes are always reflected by gradients. Upon assessment of concentration profiles with respect to material fluxes, it needs to be considered whether the depth zones of the investigated profile are sufficiently (quasi) stationary with regard to the studied parameter. Non-steady state conditions are appropriately described only by Fick's second law of diffusion.

- If no biogeochemical processes can be found in sediments at all, the pore water should display constant concentrations from the top down in a stationary way. Any reactions taking place in the pore water fraction, or between pore water and solid phase, will become visible as a change of the involved concentration gradients extending across the depth zone; any changes of the concentration gradients document the processes in which pore water has been involved.

- In principle, bioturbation as a macrobiological process should not be described in terms of easily manageable model concepts as can be done for molecular diffusion. Only if it is assured that the expansion of a given volume under study is large enough, and/or provided that the time-span necessary to make the

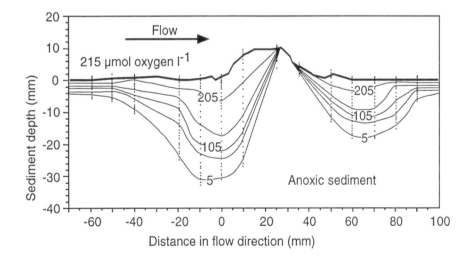

Fig. 3.27 Advective flow of pore water induced by bottom water flow and documented by O_2 penetration into the sediment around a small sediment mound (after Ziebis et al. 1996).

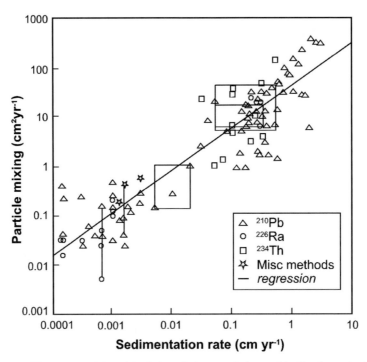

Fig. 3.28 Biodiffusion coefficients (particle mixing) in relation to their empirically determined dependence on the rate of sedimentation (after Tromp et al. 1995).

observation is sufficiently long, can we conceive of the bioturbation processes, on average, as identically effective at all places. In such cases, a model concept of 'biodiffusion' is frequently applied in which Fick's laws of diffusion are applied upon introducing an analogous 'biodiffusion coefficient'. Aller and De Master (1984) reported on having determined ^{234}Th and ^{210}Pb and drew conclusions using a model concept of bioturbation. Figure 3.28 shows an empirically determined biodiffusion constant according to Tromp et al. (1995) in its dependence on the sedimentation rate. Deep sea sediments with sedimentation rates in the range of 0.001 and 0.01 cm a^{-1} yield biodiffusion coefficients in the range of 0.1 and 1 cm^{2}a^{-1}, or, upon applying the dimensions contained in Table 3.1, values that lie between $3 \cdot 10^{-13}$ and $3 \cdot 10^{-12}$ m^{2}s^{-1}. Such biodiffusion coefficients are thus more than one order of magnitude smaller then the coefficients of molecular diffusion in sediment pore waters. The bioturbation process (= biodiffusion) is therefore without immediate relevance for pore water because, if we judge the matter realistically, its effects

are by far less than the accuracy of the molecular diffusion coefficient. Bioturbation (= biodiffusion) is, however, very important for the dynamics and the new organization of the sediment's solid phase. By this mechanism, and adjunct biogeochemical reactions, bioturbation consequently also exerts an indirect but nevertheless important influence on the distribution of concentrations in pore water.

- Upon introducing a related coefficient, Fick's laws of diffusion have also been applied to bioirrigation. However, this surely reflects a rather poor model concept. Figure 3.26 already showed this in two ways: firstly, this process is indeed of relevance and applicable to pore water. Secondly, both processes, molecular diffusion and biorrigation, take place simultaneously and are effective in parallel action at the same location. A workable model concept to deal with bioirrigation is therefore still missing.

- The advection of pore water has the same order of magnitude as the sedimentation rate and is, therefore, compared to molecular diffusion, not of great importance. However, advection as a model concept in itself may be quite easily included into computer models by the employment of the Darcy Equation 3.30. The advection of the sediment's solid phase results from the formal logic that the zero-point of the system of coordinates is always defined at the sediment surface, and that the sediment particles stream downward relative to the coordination system. This process is of great importance for balancing the diagenetical effects in the sediment.

3.7   Signals in the Sediment Solid Phase

As already described in the introductory section to this Chapter, the 'classic geological' approach to the processes of diagenesis was distinguished by the examination of the sediment's solid phase. And whenever a reconstruction of diagenetic processes in previous sediments - which nowadays consist of rock and have lost their original pore water content a long time ago -, is attempted, one will have to resort to an investigation of the solid phase. Analyzing the solid phase element profiles is much more reliable than is the case with pore water concentration profiles. However, the interpretation of such element profiles proves to be much more difficult, since the signals of the entire consecutive diagenetic processes interfere with those of sedimentation. Some of the element profiles measured in sediments can only be interpreted on account of the new understanding of diagenetic processes which was gained, in particular, after many pore water analyses had been carried out during the last two decades. The ratios of the sedimentary contents primarily originating from sedimentation compared to those originating from diagenetic processes might differ greatly from each other, depending on the element, the sedimentation area, and the type of diagenesis. In such case, it is always advisable to initially perform an estimative, rough calculation of the contents measured in the sediment in relation to the contents potentially altered in the sediment by the processes of diagenesis.

3.7.1   Analysis of the Sediment's Solid Phase

A variety of different analytical approaches lead to the profiles of sedimentary element contents. The two essential and most commonly used methods are nowadays the following:

- Laborious, yet very reliable, is the complete extraction of the dried, mortared and homogenized sediment sample using appropriate acids. Usually a microwave pressure extraction in a closed system composed of PTFE or a similarly inert and resistant material is applied for this purpose nowadays. With this method, all the sedimentary constituents are dissolved without exception by applying various mixtures of HF, HCl and HNO_3 ($HCLO_4$ can be avoided in most cases) at temperatures between 250 and 260 °C and a pressure ranging from 30 to 50 bars. In the ultimately received, slight nitric solution - hydrofluoric acid is left to evaporate in the course of the procedure - practically all elements (including the rare earth elements) can then be analyzed markedly above their limit of detection by atom absorption spectrometry (AAS), optical plasma emission spectrometry (ICP-OES) and/or plasma mass spectrometry (ICP-MS).

- X-ray fluorescence spectroscopy (XRF) is much easier in its performance but by no means as reliable particularly in case of low element contents. Here, the dried, mortared and homogenized samples can be used directly as powder, or better in this shape of pressed tablets, or even better still, as molten tablets. With this method, good detection limits are obtained for most of the quantitatively relevant elements, the access to trace metals, however, is not feasible to the same extent as in complete extraction and subsequent analysis of the extract solution.

The following comparison illustrates the greatly differing time and work effort associated with these two methods:

In order to analyze the quantitative element profiles, consisting of approximately 250 samples derived from one single core which was extracted with the gravity corer, several weeks are required for drying, mortaring, homogenization, weighing,

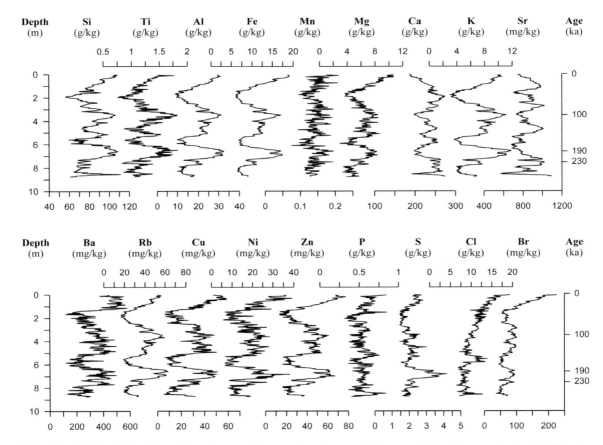

Fig. 3.29 Profile of the elemental contents within a sediment core which was obtained from a depth of 1952 m in the Cape Basin. Analysis was performed with a transportable X-ray fluorescnce anayzer (XRF) on board of a research vessel. Results were obtained within 24 hours after extracting the core (Wien et al. 2005a).

chemical extraction and analysis - provided that all steps of extraction and analysis are automated.

The element profiles of a sediment core from the Cape Basin (Wien et al. 2005a), which is shown in Figure 3.29, were obtained with a trans-portable XRF on board of a research vessel, the results were obtained within a period of 24 hours after core extraction with the gravity corer.

The energy-dispersive analysis was conducted with polarized X-rays. Powdered samples were measured because of the short processing time desired in this case, although lower detection limits and measured values displaying less scattering would certainly have been obtained by applying pressed or even molten tablets. An essential further improvement of the quality of measurements would have been accomplished if the more laborious wavelength-dispersive X-ray fluorescence (WDX), currently not available as transportable device, had been applied instead of the energy-dispersive X-ray fluorescence (EDX).

3.7.2   Interpretation of Element Profiles

The profiles of elemental contents shown in Figure 3.29 may serve as an example to briefly discuss the informational content, i.e. the various signals in these profiles. Core GeoB 8301 shown here originated from the Cape Basin at 34°46.0′S and 17°41.5′E and a water depth of 1952 m. The following signals can be distinctly recognized in the element profiles:

- The profiles of Ti, Al, Fe, K, Rb and also, to a limited extent, Mg are obviously determined by an input of terrestrial substance which is consequential to erosion. The curves which are very similar to each other represent the intensity and the type of erosion on the continent, formation and degradation of soil, as controlled by climate changes.
- The elements Ba, Cu, Ni und Zn produced similar curves, somewhat overlain by distur-bing measurement variances. However, espe-

cially in the case of barium at a depth of somewhat more than 1 m below the sediment surface, a much steeper rise of the measured values is recognizable than is characteristic of the other elements also controlled by terrestric processes. A similar steep rise at this depth is otherwise only encountered by strontium which is predominately controlled by marine processes. Various authors conceive barium as a proxy parameter of paleo-productivity in the surface water of oceans (e.g. Kasten et al. 2001). Also the elements copper, nickel and zinc which produced curves similar to barium reach the sediment when they are bound to organic matter, and thus create similar imprints in the sediment.

- The earth alkali metals calcium and strontium are almost exclusively determined by marine routes of entry. They predominately reach the sediment in the shape of calcareous shells originating from perished marine organisms. In this case, various organisms contain slightly different amounts of strontium, while others incorporate strontium to a variable degree, a process which depends on the surrounding temperature. For this core, and other sediment cores obtained from the Cape Basin, it could be demonstrated that a profile of the Sr/Ca ratio matched perfectly with the corresponding δ^{18}O-curve (SPECMAP curve, Imbrie et al. 1984), permitting an immediate deduction of an age model (cf. Wien et al. 2005b).

- The two elements in this core, calcium and strontium, were not even essentially influenced by diagenesis. Wherever the sediment's pCO_2 and concentration of dissolved carbonate changes in course of the degradation of organic matter, the respective dissolution/precipitation reactions of carbonate minerals can also be concomitantly determined (Pfeifer et al. 2002, Wenzhöfer et al. 2001).

- Large parts of the silicon curves show a resemblance to the elements which are introduced from land. This element obviously also reaches the marine sediment on similar routes of entry. However, individual peaks display a similarity to the elements which are controlled by marine conditions and thus provide evidence in favor of biogenic opal as

the route of entry, which is also quantitatively relevant.

- Sulfur reaches the sediment essentially only by means of diagenetic processes. Here, the fixation of sulfate-bound sulfur in the form of barite in the sulfate-methane transition (SMT) zone produces an interesting signal, but is quantitatively irrelevant on account of the low barium concentrations in the pore water. Sulfur precipitates in much greater amounts in shape of sulfide (Mackinawite FeS, pyrite FeS_2) wherever sulfide resulting from the reduction of sulfate comes into contact with divalent iron by means of diffusion. In the core shown in Figure 3.29 this obviously proceeded at a depth of about 7 m below the sediment surface over a relatively long period of time.

- The two halogens Cl and Br also provide some interesting pieces of information. Since a high concentration of chlorine prevails in ocean water, but is practically not involved in any diagenetic processes, chlorine can make its way into the XRF sample only by means of pore water extraction. The curve of chlorine shown in Figure 3.29 represents the porosity of the sediment material so precisely that its concentration could be reconstructed. This influence can also be seen in the case of bromine which is, however, overlain by a similar signal of nearly the same intensity as those of the elements Ba, Cu, Ni, Zn which are controlled by organic matter.

3.7.3 Correlation of Sediment Cores by the Contents of Elements

A stratigraphic correlation of different sediment cores only makes sense if the element applied in a correlation of a given core is practically not influenced by any diagenetic process at all, and when its curve is practically only influenced by its input into the sediment.

Diagenetic influences would not be coupled to the stratigraphy, but would reach various depths at various locations instead, totally independent of the diagenetic processes. Especially suited for correlations are elements that do not display too low concentrations and whose input into the sediment is essentially determined by one single route of entry. This can be, for example, either one of elements aluminum or iron for the terrestrial input, or calcium or strontium for the marine

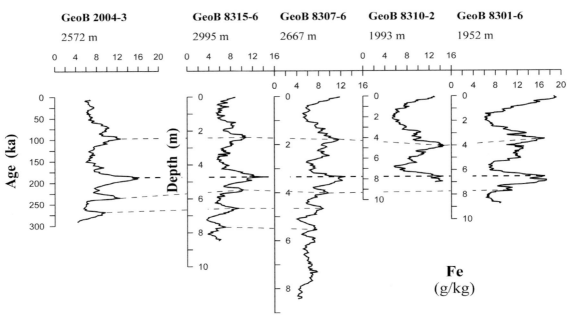

Fig. 3.30 Corelation of various sediment cores from the Cape Basin by means of sedimentary iron content profiles. This method permitted a safe projection of an age model available for one core to several other cores. (Wien et al. 2005a).

input. In Figure 3.30, core GeoB 8301 from the Cape Basin, shown in Figure 3.29, is correlated with several other cores derived from the Cape Basin by means of their iron profiles. Since a reliable age model was available for one of these cores (GeoB 2004), it could therefore also be applied to the other cores.

Graphical representations of element profiles frequently also show the ratios of elements. Above, the Sr/Ca ratio was mentioned as an example, along with its correlation with the SPECMAP curve in a certain sea area (Wien et al. 2005b). Especially popular is the representation of elemental contents reported as element/aluminum ratios. In this case, it is assumed that aluminum depends exclusively on a terrestrial route of entry. The element/aluminum ratio reveals the deviation of a given element from the behavior displayed by aluminum. An element/titanium ratio is used for the same purpose, however, it is not applied as often on account of the fact that titanium is not as reliable as aluminum and more difficult to analyze.

However, such element/aluminum or element/titanium ratios should never be applied in general and only with caution, as such procedure in fact systematically removes the terrestrial input signal from the data and, with it, all information which is attached to this course and running parallel to it.

A good approach is given if only the marine signal is to be examined, and somebody has started it at some time. Some workers may be encountered occasionally who, as a matter of principle, looking at their data study nothing but the element/aluminum ratios anymore. They will never learn anything about the terrestrial routes of input, because they simply will never see them. Studying the element/titanium ratios is particularly nonsensical in this regard. The ratio of one element to another should only be used, if it is exactly known why this ratio is studied, which part of the statement is underscored, and which one will be suppressed.

Acknowledgements

This is contribution No 0330 of the Research Center Ocean Margins (RCOM) which is financed by the Deutsche Forschungsgemeinschaft (DFG) at Bremen University, Germany.

3.8 Problems

Problem 1

Which of the following parameters have to be measured in the fresh sediment, which as soon as possible in squeezed or centrifuged pore water, which after preservation within some weeks, and which without any preservation even after long time: alkalinity, ammonia, chloride, manganese, nitrate, oxygen, pH-value, potassium, sodium, sulfate, sulfide, total iron?

Problem 2

In Figure 3.14 a sulfate reduction zone is documented by measured profiles of sulfate and methane in pore water. Estimate the diffusive methane flux $[mol\ m^{-2}a^{-1}]$ and use for the calculation the more reliable profile of sulfate concentrations. Use a temperature of 5°C and a sediment porosity of $\phi = 0.7$.

Problem 3

Figure 3.12 shows two very similar profiles of oxygen concentrations in pore water. Estimate the total oxygen consumption of the sediment and the rate of DIC production $[mol\ m^{-2}a^{-1}]$. Assume a C:N:P ratio of 106:16:1 for the organic matter. Use a temperature of 5°C and a sediment porosity of $\phi = 0.7$.

Problem 4

For which type of sediments, for which parameters, and for which steps of sampling, pore water extraction, and analyses a glove box with an inert gas atmosphere is necessary? Which inert gas would you prefer? Explain your decisions.

Problem 5

The contents of different elements in the sediment solid phase are determined by their different pathways into the sediment and by diagenetic processes. Which elements mainly document a marine input, which elements document a terrestrial input, which elements are mainly influenced by diagenetic processes?

References

Aller, R.C. and DeMaster, D.J., 1984. Estimates of particle flux and reworking at the deep-sea floor using 234Th/238U disequilibrium. Earth and Planetary Science Letters, 67: 308-318.

Aller, R.C., 1988. Benthic fauna and biogeochemical processes in marine sediments: The role of burrow structures. In: Blackburn, T.H. and Sørensen, J. (eds), Nitrogen cycling coastal marine environments. SCOPE. Wiley & Sons Ltd., pp. 301-338.

Aller, R.C., 1990. Bioturbation and manganese cycling in hemipelagic sediments. Phil. Trans. R. Soc. Lond., 331: 51-68.

Aller, R.C., 1994. The sedimentary Mn cycle in Long Island Sound: Its role as intermediate oxidant and the influence of bioturbation, O2, and Corg flux on diagenetic reaction balances. Journal of Marine Research, 52: 259-293.

Archer, D., Emerson, S. and Smith, C.R., 1989. Direct measurements of the diffusive sublayer at the deep sea floor using oxygen microelectrodes. Nature, 340: 623-626.

Archer, D. and Devol, A., 1992. Benthic oxygen fluxes on the Washington shelf and slope: A comparison of in situ microelectrode and chamber flux measurements. Limnology and Oceanography, 37: 614 - 629.

Archie, G.E., 1942. The electrical resistivity log as an aid in determining some reservoir characteristics. Trans. Am. Inst. Min. Metall., 146: 54 - 62.

Berner, R.A., 1980. Early diagenesis: A theoretical approach. Princton Univ. Press, Princton, NY, 241 pp.

Boudreau, B.P., 1997. Diagenetic models and their impletation: modelling transport and reactions in aquatic sediments. Springer, Berlin, Heidelberg, NY, 414 pp.

Cornwell, J.C. and Morse, J.W., 1987. The characterization of iron sulfide minerals in anoxic marine sediments. Marine Chemistry, 22: 193-206.

Davison, W., Grime, G.W., Morgan, J.A.W. and Clarke, K., 1991. Distribution of dissolved iron in sediment pore waters at submillimetre resolution. Nature, 352: 323-324.

Davison, W. and Zhang, H., 1994. In situ speciation measurements of trace components in natural waters using thin-film gels. Nature, 367: 546-548.

Davison, W., Zhang, H. and Grime, G.W., 1994. Performance characteristics of gel probes used for measuring the chemistry of pore waters. Environmental Science & Technology, 28: 1623-1632.

Davison, W., Fones, G.R. and Grime, G.W., 1997. Dissolved metals in surface sediment and a microbial mat at 100 μm resolution. Nature, 387: 885-888.

Davison, W., Fones, G., Harper, M., Teasdale, P. and Zhang, H., 2000. Dialysis, DET and DGT: in Situ Diffusional Techniques for Studying Water, Sediments and Soils, In: In Situ Chemical Analysis in Aquatic Systems, J. Buffle and G. Horvai, Eds.: Wiley, pp 495-569.

De Lange, G.J., 1988. Geochemical and early diagenetic aspects of interbedded pelagic/turbiditic sediments in two North Atlantic abyssal plains. Geologica Ultraiectina, Mededelingen van het Instituut vor Aardwetenschappen der Rijksuniversiteit te Utrecht, 57, 190 pp.

De Lange, G.J., Cranston, R.E., Hydes, D.H. and Boust, D., 1992. Extraction of pore water from marine sediments: A review of possible artifacts with pertinent examples from the North Atlantic. Marine Geology, 109: 53-76.

De Lange, G.J., 1992a. Shipboard routine and pressure-filtration system for pore-water extraction from suboxic sediments. Marine Geology, 109: 77-81.

De Lange, G.J., 1992b. Distribution of exchangeable, fixed, organic and total nitrogen in interbedded turbiditic/pelagic sediments of the Madeira Abyssal Plain, eastern North Atlantic. Marine Geology, 109: 95-114.

De Lange, G.J., 1992c. Distribution of various extracted phosphorus compounds in the interbedded turbiditic/pelagic sediments of the Madeira Abyssal Plain, eastern North Atlantic. Marine Geology, 109: 115-139.

Dicke, M., 1986. Vertikale Austauschkoeffizienten und Porenwasserfluß an der Sediment/Wasser Grenzfläche. Berichte aus dem Institut für Meereskunde an der Univ. Kiel, 155, 165 pp.

Enneking, C., Hensen, C., Hinrichs, S., Niewöhner, C., Siemer, S. and Steinmetz, E., 1996. Poor water chemistry. In: Schulz, H.D. and cruise participants (eds), Report and preliminary results of Meteor cruise M34/2 Walvis Bay-Walvis Bay, 29.01.1996 - 18.02.1996. Berichte, Fachbereich Geowissenschaften, Univ. Bremen, 78: 87-102.

Forster, S., Huettel, M. and Ziebis, W., 1996. Impact of boundary layer flow velocity on oxygen utilisation in coastal sediments. Mar. Ecol. Prog. Ser., 143: 173-185.

Fossing, H. and Jørgensen, B.B., 1990. Oxidation and reduction of radiolabeled inorganic sulfur compounds in an estuarine sediment, Kysing Fjord, Denmark. Geochimica et Cosmochimica Acta, 54: 2731-2742.

Froelich, P.N., Klinkhammer, G.P., Bender, M.L., Luedtke, N.A., Heath, G.R., Cullen, D., Dauphin, P., Hammond, D. and Hartman, B., 1979. Early oxidation of organic matter in pelagic sediments of the eastern equatorial Atlantic: suboxic diagenesis. Geochimica et Cosmochimica Acta, 43: 1075-1090.

Glud, R.N., Gundersen, J.K., Jørgensen, B.B., Revsbech, N.P. and Schulz, H.D., 1994. Diffusive and total oxygen uptake of deep-sea sediments in the eastern South Atlantic Ocean: in situ and laboratory measurements. Deep-Sea Research, 41: 1767-1788.

Glud, R.N., Klimant, I., Holst, G., Kohls, O., Meyer, V., Kühl, M. and Gundersen, J.K., 1999. Adaptation, Test and in situ Measurements with O_2 Micropt(r)odes on Benthic Landers. Deep Sea Research, 46: 171-183.

Glud, R.N., Tengberg, A., Kühl, M., Hall, P.O.J. and Klimant, I., 2001. An in situ instrument for planar O_2 optode measurements at benthic interfaces. Limnol. Oceanogr., 46: 2073-2080.

Grasshoff, K., Kremling, K. and Ehrhardt, M., 1999 (1st edition 1983). Methods of Seawater Analysis. 2nd edition, Wiley-VCH, Weinheim, NY, 600 pp.

Gundersen, J.K. and Jørgensen, B.B., 1990. Microstructure of diffusive boundary layer and the oxygen uptake of the sea floor. Nature, 345: 604-607.

Haese, R.R., 1997. Beschreibung und Quantifizierung frühdiagenetischer Reaktionen des Eisens in Sedimenten des Südatlantiks. Berichte, Fachbereich Geowissenschaften, Univ. Bremen, 99, 118 pp.

Hall, P.O.J. and Aller, R.C., 1992. Rapid, small-volume, flow injection analysis for CO_2 and NH_4^+ in marine and freshwaters. Limnology and Oceanography, 35: 1113-1119.

Hensen C., Landenberger H., Zabel M., Gundersen J.K., Glud R.N. and Schulz H.D. (1997) Simulation of early diagenetic processes in continental slope sediments off Southwest Africa: The copmputer model CoTAM tested. Marine Geology, 144: 191-210

Holby, O. and Riess, W., 1996. In Situ Oxygen Dynamics and pH-Profiles. In: Schulz, H.D. and cruise participants (eds), Report and preliminary results of Meteor cruise M34/2 Walvis Bay-Walvis Bay, 29.01.1996 - 18.02.1996. Berichte, Fachbereich Geowissenschaften, Univ. Bremen, 78: 85-87.

Huettel, M., Ziebis, W. and Forster, S., 1996. Flow-induced uptake of particulate matter in permeable sediments. Limnology and Oceanography, 41: 309-322.

Imbrie, J., Hays, J.D., Martinson, D.G., McIntyre, A., Mix, A.C., Morley, J.J., Pisias, N.G., Prell, W.L. and Shackleton, N.J., 1984. The orbital theory of Pleistocene climate: Support from a revised chronology of the marine $\delta^{18}O$ record. In: Berger, A.L., Imbrie, J., Hays, J., Kukla, G. Saltzman, B., Reidel, D. (Eds.) Milankovitch and Climate. Dordrecht. 1: 269-305

Iversen, N. and Jørgensen, B.B., 1985. Anaerobic methane oxidation rates at the sulfate-methane transition in marine sediments from Kattegat and Skagerrak (Denmark). Limnology and Oceanography, 30: 944-955.

Iversen, N. and Jørgensen, B.B., 1993. Diffusion coefficients of sulfate and methane in marine sediments: Influence of porosity. Geochimica et Cosmochimica Acta, 57: 571-578.

Jahnke, R.A., Heggie, D., Emerson, S. and Grundmanis, V., 1982. Pore waters of the central Pacific Ocean:

nutrient results. Earth and Planetary Science Letters, 61: 233-256.

Jahnke, R.A., 1988. A simple, reliable, and inexpensive pore-water sampler. Limnology and Oceanography, 33: 483-487.

Jahnke, R.A. and Christiansen, M.B., 1989. A free-vehicle benthic chamber instrument for sea floor studies. Deep-Sea Research, 36: 625-637.

Jørgensen, B.B. and Revsbech, N.P., 1985. Diffusive boundary layers and the oxygen uptake of sediments and detritus. Limnology and Oceanography, 30: 111-122.

Jørgensen, B.B., Bang, M. and Blackburn, T.H., 1990. Anaerobic mineralization in marine sediments from the Baltic Sea-North Sea transition. Marine Ecology Progress Series, 59: 39-54.

Kasten, S., Haese, R.R., Zabel, M., Rühlemann, C. and Schulz, H.D., 2001. Barium peaks at glacial terminations in sediments of the equatorial Atlantic Ocean – relict of deglacial productivity pulses?. Chemical Geology, 175: 635-651.

Kinzelbach, W., 1986. Goundwater Modeling - An Introducion with Sample Programs in BASIC. Elsevier, Amsterdam, Oxford, NY, Tokyo, 333 pp.

Klimant, I., Meyer, V. and Kühl, M., 1995. Fiber-oxic oxygen microsensors, a new tool in aquatic biology. Limnology and Oceanography, 40: 1159-1165.

Klimant, I., Kühl, M., Glud, R.N. and Holst, G., 1997. Optical Measurement of Oxygen and Temperature in Microscale: Strategies and Biological Applications. Sensors and Actuators B 38: 29-37.

Kölling, M., 1986. Vergleich verschiedener Methoden zur Bestimmung des Redoxpotentials natürlicher Gewässer. Meyniana, 38: 1-19.

Kölling, M., 2000. Comparison of Different Methods for Redox Potential Determination in Natural Waters. In: Schüring et al. (Eds.): Redox – Fundamentals, Processes and Applications. Springer, Berlin, Heidelberg etc., pp. 42-54.

Kühl, M. and Revsbech, N.P., 2001. Biogeochemical Microsensors for Boundary Layer Studies, In Boudreau, B.P. and Jorgensen, B.B. (Eds.) The Benthic Boundary Layer - Transport Processes and Biogeochemistry. Oxford University Press, pp. 180-210.

McDuff, R.E. and Ellis, R.A., 1979. Determining diffusion coefficients in marine sediments: A laboratory study of the validity of resistivity technique. American Journal of Science, 279: 66-675.

Meijboom, F.W. and van Nordwijk, M., 1992. Rhizon Soil Solution Samplers as artificial roots. In: Kutschera, L., Hübl, E., Lichtenegger, E., Persson, H. and Sobotnik, M., (Eds.): Root Ecology and its practical Application.- Proc 3rd ISSR Symp., Verein für Wurzelforschung, Klagenfurt, Austria; pp. 793-795.

Meischner, D. and Rumohr, J., 1974. A Light-weight, High-momentum Gravity Corer for Subaqueous Sediments. Senckenbergiana marit., 6: 105-117.

Niewöhner, C., Hensen, C., Kasten, S., Zabel, M. and Schulz, H.D., 1998. Deep sulfate reduction completely mediated by anaerobic methane oxi-dation in sediments of the upwelling area off Namibia. Geochimica et Cosmochimica Acta, 62: 455-464.

Pfeifer K., Hensen C., Adler M., Wenzhöfer F., Weber B. and Schulz H.D., 2002. Modeling of subsurface calcite dissolution, including the respiration and reoxidation processes of marine sediments in the region of equatorial upwelling off Gabon. Geochimica and Cosmochimica Acta, 66: 4247-4259

Redfield, A.C., 1958. The biological control of chemical factors in the environment. Am. Sci., 46: 206-226.

Reeburgh, W.S., 1967. An improved interstitial water sampler. Limnology and Oceanography, 12: 163-165.

Reimers, C.E., 1987. An in situ microprofiling instrument for measuring interfacial pore water gradients: methods and oxygen profiles from the North Pacific Ocean. Deep-Sea Research, 34: 2019-2035.

Revsbech, N.P., Jørgensen, B.B. and Blackburn, T.H., 1980. Oxygen in the sea bottom measured with a microelektrode. Science, 207: 1355-1356.

Revsbech, N.P. and Jørgensen, B.B., 1986. Microelectrodes: Their use in microbial ecology. Advances in Microbial Ecology, 9: 293-352.

Revsbech, N.P., 1989. An oxygen microsensor with a guard cathode. Limnology and Oceanography, 34: 474-478.

Saager, P.M., Sweerts, J.P. and Ellermeijer, H.J., 1990. A simple pore-water sampler for coarse, sandy sediments of low porosity. Limnology and Oceanography., 35: 747-751.

Sarazin, G., Michard, G. and Prevot, F., 1999. A rapid and accurate spectroscopic for alkalinity measurements in sea water samples. Wat. Res., 33: 290-294.

Sayles, F.L., Mangelsdorf, P.C., Wilson, T.R.S. and Hume, D.N., 1976. A sampler for the in situ collection of marine sedimentary pore waters. Deep-Sea Research, 23: 259-264.

Schüring, J., Schulz, H.D., Fischer, W.R., Böttcher, J. and Duijnisveld, W.H.M. (Eds.), 2000. Redox – Fundamentals, Processes and Applications. Springer, Berlin, Heidelberg etc., 251 pp.

Schlüter, M., 1990. Zur Frühdiagenese von organischem Kohlenstoff und Opal in Sedimenten des südlichen und östlichen Weddelmeeres. Berichte zur Polarforschung, Bremerhaven, 73, 156 pp.

Schultheiss, P.J. and McPhail, S.D., 1986. Direct indication of pore-water advection from pore pressure measurements in Madeira Abyssal Plain sediments. Nature, 320: 348-350.

Schulz, H.D., Dahmke, A., Schinzel, U., Wallmann, K. and Zabel, M., 1994. Early diagenetic processes, fluxes and reaction rates in sediments of the South Atlantic. Geochimica et Cosmochimica Acta, 58: 2041-2060.

Seeberg-Elverfeldt, Schlüter, M. and Kölling, M. 2005. Rhizon sampling of porewaters near the sediment-water interface of aquatic systems. Limnol. Oceanogr. Methods 3: 361-371.

Seeburger, I. and Käss, W., 1989. Grundwasser - Redoxpontentialmessung und Probennahmegeräte. DVWK-Schriften, Bonn, 84, 182 pp.

Smith, K.L.J. and Teal, J.M., 1973. Deep-sea benthic community respiration: An in-situ study at 1850 meters. Science, 179: 282-283.

Tengberg, A., De Bovee, F., Hall, P., Berelson, W., Chadwick, D., Ciceri, G., Crassous, P., Devol, A., Emerson, S., Gage, J., Glud, R., Graziottini, F., Gundersen, J., Hammond, D., Helder, W., Hinga, K., Holby, O., Jahnke, R., Khripounoff, A., Lieberman, S., Nuppenau, V., Pfannkuche, O., Reimers, C., Rowe, G., Sahami, A., Sayles, F., Schurter, M., Smallman, D., Wehrli, B. and De Wilde, P., 1995. Benthic chamber and profiling landers in oceanography - A review of design, technical solutions and function. Progress in Oceanography, 35: 253-292.

Tromp, T.K., van Cappellen, P. and Key, R.M., 1995. A global model for the early diagenetisis of organic carbon and organic phosphorus in marine sediments. Geochimica et Cosmochimica Acta, 59: 1259-1284.

Van Cappellen, P. and Wang, Y., 1996. Cycling of iron and manganese in surface sediments: a general theory for the coupled transport and reaction of carbon, oxygen, nitrogen, sulfur, iron, and manganese. American Journal of Science, 296: 197-243.

Viollier, E., Rabouille, C., Apitz, S.E., Breuer, E., Chaillou, G., Dedieu, K., Furukawa, Y, Grenz, C., Hall, P., Janssen, F., Morford, J.L., Poggiale, J.-C., Roberts, S., Shimmield, T., Taillefert, M., Tengberg, A., Wenzhöfer, F. and Witte, U., 2003. Benthic biogeochemistry: state of the art technologies and guidelines for the future of in situ survey. Journal of Experimental Marine Biology and Ecology, 285/286: 5-31.

Wenzhöfer F., Adler M., Kohls O., Hensen C., Strotmann B., Boehme S. and Schulz H.D., 2001. Calcite dissolution driven by benthic mineralization in the deep-sea: In situ measurements of Ca^{2+}, pH, pCO_2 and O_2. Geochimica et Cosmochimica Acta, 65: 2677-2690

Wien, K., Wissmann, D., Kölling, M. and Schulz, H.D. 2005a. Fast application of X-ray fluorescence spectrometry aboard ship: how good is the new portable Spectro Xepos analyser. Geomarine Letters, in press.

Wien, K., Kölling, M., and Schulz, H.D. 2005b. Close correlation between Sr/Ca ratio and SPECMAP record in bulk sediments from the Southern Cape Basin. Geo-Marine Letters, 25: 265-271.

Ziebis, W. and Forster, S., 1996. Impact of biogenic sediment topography on oxygen fluxes in permeable seabeds. Mar. Ecol. Prog. Ser., 140: 227-237.

4 Organic Matter: The Driving Force for Early Diagenesis

Jürgen Rullkötter

4.1 The Organic Carbon Cycle

The organic carbon cycle on Earth is divided into two parts (Fig. 4.1; Tissot and Welte 1984). The biological cycle starts with photosynthesis of organic matter from atmospheric carbon dioxide or carbon dioxide/ bicarbonate in the surface waters of oceans or lakes. It continues through the different trophic levels of the biosphere and ends with the metabolic or chemical oxidation of decayed biomass to carbon dioxide. The half-life is usually days to tens of years depending on the age of the organisms. The geological organic carbon cycle has a carbon reservoir several orders of magnitude larger than that of the biological organic carbon cycle ($6.4 \cdot 10^{15}$ t C compared with $3 \cdot 10^{12}$ t C in the biological cycle) and a turn-over time of millions of years. It begins with the incorporation of biogenic organic matter into sediments or soils. It then leads to the formation of natural gas, petroleum and coal or metamorphic forms of carbon (e.g. graphite), which may

be reoxidized to carbon dioxide after erosion of sedimentary rocks or by combustion of fossil fuels.

The tiny leak from the biological to the geological organic carbon cycle, particularly if seen from the point of view of a petroleum geochemist in the context of the formation of petroleum source rocks (see Littke et al. 1997a for an overview), is represented by the deposition and burial of organic matter into sediments. If looked at in detail, the transition from the biosphere to the geosphere is less well defined. The transformation of biogenic organic matter to fossil material starts immediately after the decay of living organisms. It may involve processes during transport, e.g. sinking through a water column, and alteration at the sediment surface or in the upper sediment layers where epi- and endobenthic organisms thrive. Furthermore, it may extend deeply into the sedimentary column where bacteria during the last decade or so were found to be still active at several hundreds of meters depth in layers deposited millions of years ago (Parkes et al. 1994; Wellsbury et al. 2002).

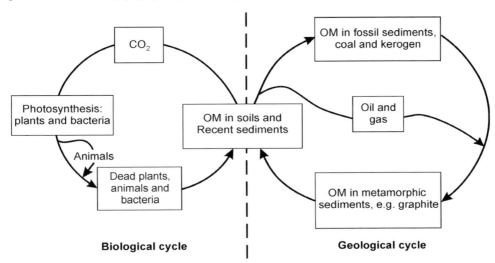

Fig. 4.1 The two major parts of the organic carbon cycle on Earth (after Tissot and Welte 1984). OM = Organic matter.

4.2   Organic Matter Accumulation in Sediments

In the fossil record, dark-colored organic-matter-rich layers (black shales, sapropels, petroleum source rocks in general) witness periods of time when conditions for organic matter accumulation in sediments apparently were particularly favorable. As the other extreme, massive sequences of white- or red-colored sedimentary rocks are devoid of organic matter. Although these rocks may contain abundant calcareous or siliceous plankton fossils, the organic matter of the organisms apparently was destroyed before it could be buried in the sediments.

Biogenic organic matter is considered labile (or metastable) under most sedimentary conditions due to its sensitivity to oxidative degradation, either chemically or biologically mediated. This is particularly true in well-oxygenated oceanic waters as they presently occur in the oceans almost worldwide. Thus, abundant accumulation of organic matter today is the exception rather than the rule. It is mainly restricted to the upwelling areas on the western continental margins and a few rather small oceanic deeps with anoxic bottom waters (such as the Cariaco Trench off Venezuela). In the geological past, more sluggish circulation in the deep ocean or in shallow epicontinental seas, accompanied by water column stratification, was probably one of the main causes leading to the deposition of massive organic-matter-

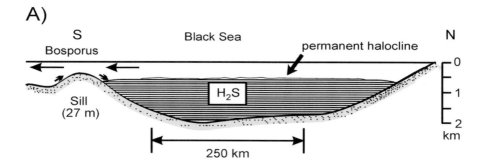

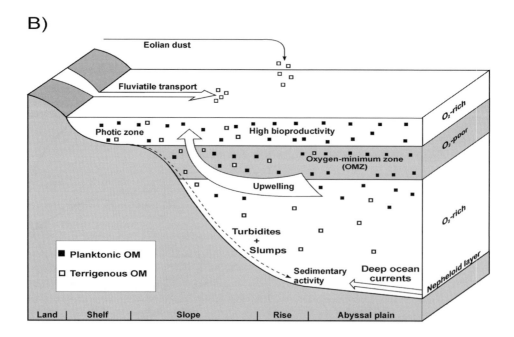

Fig. 4.2 Schematic models for organic matter accumulation in sediments. A) Stagnant basin or Black Sea model (after Demaison and Moore 1980 and Stein 1991); B) Productivity model (after Rullkötter et al. 1983). OM = Organic matter.

Table 4.1   Terminology for regimes of low oxygen concentrations and the resulting biofacies according to Tyson and Pearson (1991)

Oxygen (ml/l)	Environments	Biofacies	Physiological regime
8.0-2.0	Oxic	Aerobic	Normoxic
2.0-0.2	Dysoxic	Dysaerobic	Hypoxic
2.0-1.0	Moderate		
1.0-0.5	Severe		
0.5-0.2	Extreme		
0.2-0.0	Suboxic	Quasi-anaerobic	
0.0 (H$_2$S)	Anoxic	Anaerobic	Anoxic

rich rocks. A few examples are the Jurassic Posidonia Shales or Kimmeridge Clays in northwestern Europe, the Cretaceous black shales of the Atlantic Ocean and other oceanic areas of the world, and the Pliocene to Holocene sapropels of the Mediterranean Sea.

Stagnant oceanic bottom waters with a low concentration or absence of oxygen (anoxia) for a long time were considered the main prerequisite for the accumulation of high amounts of organic matter in sediments (Demaison and Moore 1980). More recently, a controversy developed about two contrasting models to explain the deposition of organic-matter-rich sediments in the marine realm, either (1) by preservation under anoxic conditions in a static situation or (2) by high primary productivity in a dynamic system (Fig. 4.2; Pedersen and Calvert 1990, 1991; Demaison 1991). The relative importance of these two dominant controlling parameters is still being heavily debated, although Stein (1986a) already conceived that either one of these parameters could play a decisive role in different oceanographic situations. Another parameter brought into discussion more recently is the protective role of organic matter adsorption on mineral surfaces and its influence on organic matter accumulation in marine sediments (Keil et al. 1994a, b; Mayer 1994, 1999; 2005; Ransom et al. 1998).

4.2.1   Productivity *Versus* Preservation

Recognition of the sensitivity of organic matter toward oxidative destruction led to the idea that the concentration of free oxygen in the water column and particularly at the sediment/water interface is the most important factor determining the amount of organic matter that is incorporated into sediments (e.g. Demaison and Moore 1980). The stagnant basin or Black Sea model (Fig. 4.2A), developed from this idea, is based on the observation that lack of replenishment of oxygen by restricted circulation in the bottom part of larger

water bodies can lead to longer-term oxygen-depleted (anoxic, suboxic; see Table 4.1 for definition) conditions in the water column and at the sediment/water interface. In the Black Sea (exceeding 2000 m water depth in the center), this is caused by the development of a very stable halocline (preventing vertical mixing) at about 100 m to 150 m water depth. The surface layer is fed by relatively light riverine freshwater from the continent, and denser saline deep water is flowing in at a low rate from the Mediterranean Sea over the shallow Bosporus sill. Over time, oxidation of sinking remnants of decayed organisms consumed all of the free oxygen in the deeper water, which was not effectively replenished by Mediterranean water. Instead, the deep water in the Black Sea (like in Lake Tanganyika, an analogous contempora-neous example of a large stratified lake; Huc 1988) contains hydrogen sulfide restricting life to anaerobic microorganisms that are commonly thought to degrade organic matter less rapidly than aerobic bacteria, although there are also opposing views (see, e.g., discussion by Kristensen et al. 1995; Hulthe et al. 1998). Lack of intense organic matter degradation under anoxic conditions would then not necessarily require high surface water bioproductivity for high organic carbon concentrations to occur in the sediment.

The proponents of primary productivity as the decisive factor controlling organic matter accumulation (e.g. Calvert 1987; Pedersen and Calvert 1990; Bailey 1991) suggested that changes in primary productivity with time in different areas of the world, induced by climatic and related oceanographic changes, explain the distribution of Cretaceous black shales and more recent (Quaternary) organic-matter-rich sediments better than the occurrence of anoxic conditions in oceanic bottom waters. Reduced oxygen concentrations in the water column, according to these authors, are a consequence of large amounts of decaying biomass settling toward the ocean bottom and consuming the dissolved oxygen.

The productivity model (Fig. 4.2B) is based on high primary bioproductivity in the photic zone of the ocean as it presently occurs in areas of coastal upwelling primarily on the western continental margins, along the equator and as a monsoon-driven phenomenon in the Arabian Sea and along the Oman and Somali coasts. Upwelling brings high amounts of nutrients to the surface, which stimulate phytoplanktonic growth (e.g. Suess and Thiede 1983; Thiede and Suess 1983; Summerhayes et al. 1992). On continental margins, the formation of oxygen-depleted water masses (oxygen minimum zones) usually implies that they impinge on the ocean bottom where they create depositional conditions similar to those in a stagnant basin. To which extent the reduction in oxygen concentration at the sediment/water interface enhances, or is required for, the preservation of organic matter in the sediments, is the main subject of debate between the proponents of the productivity and the preservation models (e.g. Pedersen and Calvert 1990, 1991; Demaison 1991). For the geological past, e.g. the Cretaceous, it is also conceivable that the equable climate on our planet led to a more sluggish circulation of ocean water worldwide. The lack of turnover in the water column may have caused the development of anoxic bottom water conditions in some parts of the ocean that may explain the formation of Cretaceous black shales almost synchronously in different areas (Sinninghe Damsté and Köster 1998 and references therein). Also, transgression as a consequence of eustatic sealevel rise may have enhanced accumulation of organic matter in shelf areas (Wenger and Baker 1986) whereas times of regression may have promoted organic matter accumulation in prograding delta fans in deeper water. A detailed discussion of organic matter accumulation in different oceanic settings, including deep marine silled basins, progradational submarine fans, upwelling areas, anoxic continental shelves and fluviodeltaic systems was provided by Littke et al. (1997a).

4.2.2   Primary Production of Organic Matter and Export to the Ocean Bottom

The total annual primary production by photosynthetic planktonic organisms in the modern world oceans has been estimated to be in the order of $30\text{-}50 \cdot 10^9$ tons of carbon (Berger et al. 1989; Hedges and Keil 1995). Oceanic carbon fixation is not evenly distributed, but displays zones of higher activity on continental margins

(several hundred g $C_{org} m^{-2} yr^{-1}$), whereas the central ocean gyres are mostly characterized by low primary production (about 25 g $C_{org} m^{-2} yr^{-1}$; e.g. Romankevitch 1984; Berger 1989). Due to plate tectonics and the variable distribution of land masses together with climatic developments, the global amount and distribution of organic matter production in the ocean is likely to have undergone significant changes with geological time.

Of the total biomass newly formed in the photic zone of the ocean, only a very small portion reaches the underlying seafloor and is ultimately buried in the sediment (for reviews of water column processes see Emerson and Hedges 1988; Wakeham and Lee 1989). Most of the organic matter enters the biological food web in the surface waters and is respired or used for new heterotrophic biomass production. Because of this intense recycling, it is difficult to determine the organic matter flux at different levels of the photic zone. Oceanographers and biogeochemists usually consider only the flux of organic matter through the lower boundary of the photic zone and term it 'new' production (Fig. 4.3). It equals 100 % of export production (see below) and is not to be confused with net photosynthesis which is gross photosynthesis minus algal respiration. Below this boundary, the content of organic matter in the water column decreases due to consumption in the food web and to microbiological and chemical degradation as observed from the analysis of material in sediment traps deployed at different water depths.

The water depth-dependent flux is termed export production (Fig. 4.3). Export production decreases rapidly just below the photic zone. Then, there is mostly a quasi-linear, slower decline at greater water depths until the organic matter reaches the benthic boundary (nepheloid) layer close to the sediment/water interface where the activity of epibenthic organisms enhances organic matter consumption again. This enhanced consumption continues in the upper sediment layer where burrowing organisms depend on the supply from the water column. Organic matter degradation eventually extends deeply into the sediment pile as became evident from the detection of a so-called deep biosphere at several hundred meters below the seafloor (e.g. Parkes et al. 1994). However, the rate of organic matter degradation apparently decreases significantly with increasing depth of burial. Overall, it is estimated that only 1 to 0.01 % of the primary production is buried deeply in marine sediments (cf. Fig. 12.1). The fraction strongly depends on a number of parameters including level of primary productivity, water depth, (probably) oxygen content in the water column and surface sediments (the latter affects benthic activity), particle size, adsorption to mineral surfaces and sediment porosity.

Although the fraction of organic matter burried in sediments is small relative to the amount produced by photosynthesis in oceanic surface waters, empirical relationships were derived from the analysis of sedimentary organic matter to estimate oceanic paleoproductivity (PaP [g C m^{-2}yr^{-1}]) in the geological past:

$$PaP = C \cdot \rho (100 - \phi/100)/0.003 \cdot S^{0.3} \qquad (4.1)$$

In this equation of Müller and Suess (1979), the Pleistocene paleoproductivity of the (oxic) ocean is related to the organic carbon content of a sediment in percent dry weight (C), the density of the dry sediment (ρ; g cm^{-3}), its porosity (ϕ; %) and the linear bulk

sedimentation rate (S; cm of total sediment per 1000 yr). The exponential factor was obtained from calibration with data from the present ocean. Because it was shown later that the organic carbon accumulation rate is a function of the carbon flux near the seafloor, which is related to both productivity and water depth, Stein (1986a, 1991) derived a more complex empirical relationship using flux data of Betzer et al. (1984):

$$PaP = 5.31[C(\rho_{WB} - 1.026\phi/100)]^{0.71} \cdot S^{0.07} \cdot D^{0.45}$$
$$(4.2)$$

with ρ_{WB} being the wet bulk density of the sediment and D the water depth. Values from this equation are proportional, although numerically not identical, to the results obtained from a third empirical relationship developed by Sarnthein et al. (1987, 1988):

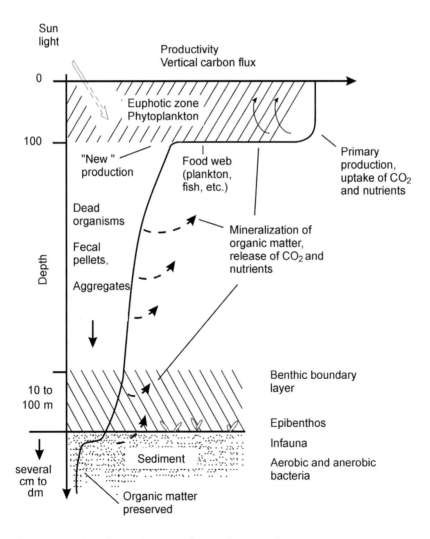

Fig.4.3 Schematic representation of organic matter flux to the ocean bottom.

$$PaP = [15.9C \cdot S \cdot \rho(1-\phi)]^{0.66} \cdot S_{B-C}^{-0.71} \cdot D^{0.32} \quad (4.3)^1$$

where S_{B-C} is the organic-carbon-free linear bulk sedimentation rate. Given the complexity of the sedimentation and burial processes of organic matter, the wide range of chemical and physical properties of organic matter from different organisms and the effects of organic matter alteration during diagenesis after incorporation in the sediment, the equations can only be considered rough estimates. It can be expected that the investigation of more oceanic sediment profiles in the future will result in further modification of the equations (cf. also Sarnthein et al. 1992; Stein and Macdonald 2005 and references therein). In particular, the extent of mixing of autochthonous marine organic matter with allochthonous organic matter from the continents (see 4.2.5) is difficult to estimate. Stein (1986a) used data from Rock-Eval pyrolysis, calibrated by organic petrographic data from microscopic analysis, to derive the marine organic matter fraction of a sediment from bulk pyrolysis measurements, but the correlation displays substantial scatter and can only be regarded a crude approximation.

Müller and Suess (1979) applied their paleoproductivity relationship (Eq. 4.1) to sediments from the deep ocean off Northwest Africa. They found that the Pleistocene interglacial periods had about the same productivity as that measured in the present-day ocean. It was three times higher during glacial periods, probably due to a higher nutrient supply by more intense mixing of water masses or stronger coastal upwelling. The more sophisticated Equation 4.3 of Sarnthein et al. (1987, 1988) yielded essentially the same results for the last 500,000 years as those from Equation 4.1. Typical Pleistocene productivities in the upwelling area off Northwest Africa ranged between 150 and 300 g $C_{org}m^{-2}yr^{-1}$, whereas the values were 20 to 50 g $C_{org}m^{-2}yr^{-1}$ in the central Atlantic Ocean (Stein et al., 1989).

The paleoproductivity equations above describe the relationship between surface-water productivity and organic carbon accumulation, specifically under conditions of an oxic water column. A different relationship for anoxic depositional settings, suggested by Bralower and Thierstein (1987), implies that at least 2 % of the organic carbon in the gross photosynthetic production is preserved in the sediments:

$$PaP = 5C \cdot S(\rho_{WB} - 1.026\phi/100) \quad (4.4)$$

Stein (1986a) used this equation to calculate paleoproductivity in the Mesozoic Atlantic Ocean. Interpretation was considered preliminary due to the difficulty of obtaining precise age information, and thus sedimentation rate data, for the older and more compacted Mesozoic sediments lean in microfossils. The estimated productivity appeared to have been low off Northwest Africa in the Jurassic, to have increased during the Early Cretaceous and to have reached maximum values similar to those today during Aptian-Albian times (about 110 million years ago). Interestingly, low paleoproductivity was calculated for the time of deposition of black shales at the Cenomanian-Turonian boundary (90 million years ago) indicating that preservation may have played a more important role for organic matter accumulation than productivity.

The empirical relationships for paleoproductivity assessment illustrate how organic matter accumulation is related to primary productivity through factors such as organic carbon flux through the water column and bulk sedimentation rate. In addition, there is evidence that reduced oxygen concentrations in the water column enhance organic matter preservation. Thus, organic-carbon-rich sediments and sedimentary rocks are likely to be formed by the mutually enhancing effects of oxygen depletion (static or dynamic; anoxia), and productivity. In view of this, it appears to be too restrictive to assign a single controlling factor (Pederson and Calvert 1990). For example, an anoxic water column in the Holocene Black Sea is in itself apparently not sufficient to lead to black shale formation, whereas the enhanced primary productivity in equatorial upwelling areas is not reflected in a high organic carbon content of the underlying sediments due to oxidation of the sinking organic matter in the deep oxic waters below the oxygen-minimum zone.

4.2.3 Transport of Organic Matter Through the Water Column

The extent of degradation of particulate organic matter as it sinks through the water column is influenced by the residence time of organic matter particles in the water column. A measure of vertical transport is the sinking velocity (vs; m s^{-1}), which for a spherical particle follows Stokes' law:

$$vs = [(\rho_2 - \rho_1) \cdot g \cdot D^2]/18\eta \quad (4.5)$$

ρ_2 and ρ_1 are the densities (g cm^{-3}) of the particle and the water, respectively, g is the acceleration due to

---

[1]   Numerical values in this equation were rounded because the author of this chapter believes that the number of decimals in the original publication suggests more accuracy than is both justified by, and required for, this empirical estimative approach.

gravity (m s^{-2}), D is the particle diameter (cm) and η is the dynamic viscosity (g cm^{-1}s^{-1}). Representative travel times of different idealized organic particles through a water column of 1000 m cover a wide range (Table 4.2). They reflect the effects of different densities and diameters. Smaller, less dense particles clearly settle very slowly.

The vastly different travel times of the particle types range from hours to years. They would be only slightly higher in saline ocean water. The density values in the table imply some association of the organic matter with mineral matter, either biogenic calcareous or siliceous (plankton) frustules or detrital mineral matter like clay. Pure organic matter would have a lower density than water and not sink to the ocean bottom at all. Densities, e.g., of coal macerals usually vary between 1.1 and 1.7 g cm^{-3} (van Krevelen 1961), and zooplankton fecal pellets often contain more mineral than organic matter. Degens and Ittekott (1987) strongly favored the transfer of organic matter by fecal pellets "which are jetted to the seafloor at velocities of about 500 m day^{-1}." Mineral-poor algal particles may have a very long residence time in the water column and a high chance of being metabolized before reaching the ocean floor. In deep oxic ocean water, the "fecal pellet express" may be an important mechanism of transporting marine organic matter to the seafloor. Microscopic analysis often revealed that all of the labile marine organic matter in such sediments occurred as 'amorphous' degraded material in rounded bodies which were ascribed to fecal pellets (e.g. Rullkötter et al. 1987). On the other hand, Plough et al. (1997) measured rapid rates of mineralization of fecal pellets (relative to their sinking time toward the seafloor). This

is consistent, however, with the observation that only intact fecal pellets occur in deep-sea sediments (PK Mukhopadhyay, personal communication 1987). Obviously, lysis of fecal pellet walls leads to rapid mineralization of the entire organic content, and only those fecal pellets reach the seafloor and are embedded in the sediment which escape this degradative process in the water column.

Other than noted in Table 4.2, real sinking velocities strongly depend on particle shape. For example, von Engelhardt (1973) showed that the sinking velocity of quartz grains of 10-100 μm diameter is greater by a factor of about one hundred than that of muscovite plates of equal diameter. Most organic matter particles, apart from fecal pellets, are not spherical or well rounded and thus have a lower sinking velocity than indicated in Table 4.2. The typical shape of terrigenous organic particles in young open-marine sediments is irregularly cylindrical with the longest axis being about twice the length of the shortest axis (Littke et al. 1991a).

4.2.4 The Influence of Sedimentation Rate on Organic Matter Burial

Müller and Suess (1979) demonstrated the influence of sedimentation rate on organic carbon accumulation under oxic open-ocean conditions. They found that the organic carbon content of marine sediments increases by a factor of about two for every tenfold increase in sedimentation rate. The underlying mechanism was believed to be the more rapid removal and protection of organic matter from oxic respiration and benthic digestion at the sediment/water interface by increasingly rapid burial (cf. Sect. 12.3.3). Also

Table 4.2   Sinking velocity and travel time for spherical particles in nonturbulent freshwater (after von Engelhardt 1973, Littke et al. 1997a)

Particle type	Sinking velocity (m·s^{-1})	Travel time (1000 m water)
Large terrigenous organic matter particle ∅ = 100 μm, ρ = 1.3 g·cm^{-3}	1.6·10^{-3}	7.1 days
Small terrigenous organic matter particle ∅ = 10 μm, ρ = 1.3 g·cm^{-3}	1.6·10^{-5}	1.9 years
Fecal pellet ∅ = 1 mm, ρ = 1.3 g·cm^{-3}	1.6·10^{-1}	2 hours[a]
Quartz grain ∅ = 10 μm, ρ = 2.6 g·cm^{-3}	8.7·10^{-5}	133 days

[a]Measured travel times for real fecal pellets: 4-20 days/1000 m (JK Volkman, pers. com. 1998)

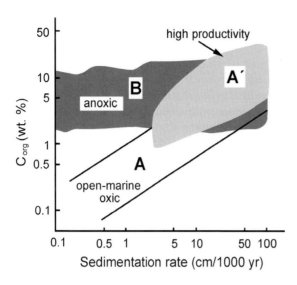

Fig. 4.4 Correlation between marine organic carbon content and sedimentation rate (after Stein 1986b, 1990). The distinction between fields A, A' and B is based on data derived from Recent to Miocene sediments deposited in normal open-ocean environments (field A), upwelling high-productivity areas (field A') and anoxic environments (field B).

According to Stein (1986b, 1990), the effects of oxic and anoxic conditions on marine organic matter preservation in oceanic sediments can be illustrated by a simple diagram of sedimentation rate versus organic carbon content (Fig. 4.4). Field A inside the diagonal lines represents the sedimentation rate-controlled accumulation of organic matter under open-marine oxic conditions. The hatched area B indicates anoxic or strongly oxygen-depleted conditions over a wide range of sedimentation rates with low rates being typical for stagnant basins like the Black Sea. The shaded area A', where areas A and B overlap at high sedimentation rates and high organic carbon contents, is typical of upwelling areas with high primary productivity on continental margins where the oxygen-minimum zone impinges on the shelf and upper slope. Strong dilution with mineral matter would place a sediment to the right of area A. Interestingly, the highly organic-matter-rich Atlantic Ocean black shales from a so-called 'world-wide anoxic event' at the Cenomanian-Turonian boundary (about 90 million years ago; see Herbin et al. 1986) all fall in the left part of area B, i.e. they appear to have been deposited at low sedimentation rates under anoxic conditions (Stein 1986b).

4.2.5   Allochthonous Organic Matter in Marine Sediments

As schematically indicated in Fig. 4.2B, marine sediments do not only accumulate organic matter from the (mainly planktonic) productivity in the overlying water column (autochthonous organic matter). Allochthonous organic matter originates from two other sources. One of them involves redeposition of marine sediments after erosion, often from a nearby location. Typical examples are contour currents along continental margins or downslope transport events on (steep) continental margins. In these cases, sediment initially deposited at shallow(er) water depth is eroded by currents, mechanical instability (oversteepening), earthquakes or other tectonic movements. The eroded sediment is transported down the continental slope and redeposited at a deeper location. This may occur as a turbidity current by which sediment is suspended in the near-bottom water column and then settles again. This process often involves particle size fractionation, i.e. at the new site the larger particles are deposited first and become overlain by a sequence of progressively finer particles (Bouma series). Alternatively, a massive package of sediment material (slump) of variable size, from very small to cubic kilometers, may be redeposited as a whole, usually in a deep slope or continental rise setting. The effect on the organic

conceivable, however, is that the relationship between sedimentation rate and organic carbon content is based on the protective effect of organic matter adsorption on mineral (particularly clay) surfaces, so that organic matter preservation increases with the increase of mineral surface available for adsorption (Keil et al. 1994a, b; Mayer 1994,1999, 2005; Collins et al. 1995; Ransom et al. 1998).

There is general agreement on the positive relationship between sedimentation rate and organic carbon content (e.g. Heath et al. 1977; Ibach 1982; Stein 1986a, b; Bralower and Thierstein 1987; Littke et al. 1991b). However, in cases where biostratigraphy provides accurate time control, it has been noted by Tyson (1987) that deposition of marine sediments with very high organic matter contents (petroleum source rocks) often appears to be associated with low rather than high sedimentation rates. Very high sedimentation rates at some point may lead to low organic matter concentrations in sediments due to dilution even if much of the sinking organic matter is preserved (Note difference between linear sedimentation rate [cm kyr^{-1}] and sediment accumulation rate [g cm^{-2}yr^{-1}], i.e. in a highly diluted sediment with a moderate to low organic carbon content deposited at a high linear bulk sedimentation rate, organic matter accumulation (or preservation) with time may still be high).

matter is different in these two cases. Turbidity currents may expose the (fossil) organic matter to an oxygen-rich water mass causing further degradation during resettling. Compact slump masses may transport significant amounts of labile organic matter from an initial depositional setting, favorable for organic matter preservation (e.g. in an oxygen-minimum zone), to an oxygen-rich deep-water environment. There is no enhanced mineralization in this case due to the undisturbed embedding of organic matter in the sediment matrix (cf., e.g., Cornford et al. 1979; Rullkötter et al. 1982).

Redeposition may occur almost synsedimentarily, i.e. the redeposited allochthonous sediment will differ only very little in age from the underlying autoch-thonous sediment. The organic matter content of both may reflect more or less contemporaneous primary productivity with the only exception that the remains of shallow(er)-water species are relocated to a deep-water environment. This may be more evident in the mineral fossils than in the organic matter assemblage, however. Alternatively, redeposition may occur a long time after initial sedimentation took place. Deep-sea drilling on the Northwest African continental margin, for example, has recovered extended Miocene sediment series which evidently contained slumps a few centimeters to several meters thick and of various age, from Tertiary to Middle Cretaceous (von Rad et al. 1979; Hinz et al. 1984). Investigation of the organic matter on a molecular level clearly demonstrated the corres-pondence between slump clasts embedded in the Lower Miocene sequence and underlying autoch-thonous Cenomanian series, whereas paleontological analysis showed a difference between outer shelf/upper slope species in the slumps and pelagic species in the autochthonous Cenomanian sediment (Rullkötter et al. 1984).

The second main source of allochthonous organic matter in marine sediments are the continents. Wind, rivers and glaciers transport large amounts of land-derived organic matter into the ocean (e.g. Romankevitch 1984; Hedges and Keil 1995; Hedges et al. 1997). Again, two principal types of organic matter have to be distinguished: (1) fresh or (in geological terms) recently biosynthesized land plant material and (2) organic matter contained in older sediments that were weathered and eroded on the continent in areas ranging from mountains to coastal swamps. Organic matter in older sedimentary rocks that are being eroded may have had an extended history of geothermal heating at great burial depth into the stages of oil or bituminous coal formation and beyond. This organic matter carries a distinct signal of geochemical matu-

ration that can easily be detected by bulk (e.g. atomic composition, pyrolysis yields, vitrinite reflectance; see Tissot and Welte 1984) and molecular geochemical parameters (e.g. compound ratios of specific geoche-mical fossils or biomarkers; see Peters et al. 2005). Due to its advanced level of diagenetic alteration, even peat can be distinguished from fresh organic matter in marine sediments. Only when sediments have been buried to a depth corresponding to the temperature which the eroded organic matter earlier experienced do both fresh and prematured organic matter continue diagenesis or maturation at the same rate. Geochemical reactions are virtually stopped (i.e. reaction rates become very low) as soon as sediments (and their organic matter contents) in the course of tectonic uplift are cooled to a temperature of about 15° C lower than their previous maximum temperature. During transport to the ocean, oxidation of organic matter eroded on land – and of terrestrial organic matter in general – has an effect similar to maturation. The product is a highly refractory, inert material , which in organic petrography is termed inertinite and which is easily recognizable under the microscope by its high reflectance (see Taylor et al. 1998 for details). Nevertheless, a substantial fraction of the terrigenous organic matter is reactive and metabolizable in the ocean (Hedges et al. 1997).

Wind-driven dust and aerosols carry terrigenous organic matter over long distances into the oceans and are estimated to contribute a total annual amount of $3.2 \cdot 10^8$ t carbon each year (Romankevitch 1984). Entire organoclasts, like pollen and spores, are blown offshore as are lipids from the waxy coatings of plant cuticles adsorbed to mineral grains (e.g. Rommerskirchen et al. 2003 and references therein). This wind-blown terrestrial material may comprise the bulk of the organic matter in open-ocean sediments where very little of the primary marine organic matter reaches the deep ocean floor. The higher resistance of terrestrial organic matter toward oxidation has been invoked to explain this selective enrichment. Summerhayes (1981) estimated that most organic-matter-rich sediments in the Atlantic Ocean, including the Cretaceous and Jurassic black shales, contain a 'background level' of 1 % terrestrial organic matter in total dry sediment.

Not all of this terrigenous material is brought into the ocean by winds. The most important contributors are the rivers draining into the ocean. Each year rivers transport approximately $0.4 \cdot 10^9$ t of dissolved and particulate organic carbon from continents to oceans (Schlesinger and Melack 1981). About 60 % of the river run-off derives from forested catchments, and the ratio of the contribution of tropical to temperate and boreal forests is about two to one (Schlesinger and Melack 1981). Much of the organic matter discharged by rivers appears to be soil-derived (Meybeck 1982; Hedges et al. 1986a, b) and highly degraded (Ittekott 1988; Hedges et al. 1994).

4.3   Early Diagenesis

4.3.1   The Organic Carbon Content of Marine Sediments

The content of organic carbon in marine surface sediments from different environments of the present-day oceans varies over several orders of magnitude depending on the extent of supply of organic matter, preservation conditions and dilution by mineral matter. The results of organic carbon measurements are usually expressed as TOC (total organic carbon) or C_{org} values in percent of dry sediment. Romankevich (1984) compiled data from a great number of analyses of ocean bottom sediments by various authors and from all over the world. The TOC values range from 0.01 % to more than 10 % C_{org} in a few cases. A statistical evaluation of data from the early phase of deep-sea drilling (Deep Sea Drilling Project Legs 1-31) showed

deep-sea sediments to have a mean organic carbon content of 0.3 %, with a median value of 0.1 % (McIver 1975). However, the range of samples was certainly not representative, and the organic carbon contents are probably biased toward higher values. For example, in the vast abyssal plains and other deep-water regions far away from the continents, an organic carbon content as high as 0.05 % is the exception rather than the rule.

In contrast to this, nearshore sediments on continental shelves and slopes usually have considerably higher organic carbon contents. Typical hemipelagic sediments on outer shelves and continental slopes range between 0.3 and 1 % C_{org}. Sediments within the oxygen-minimum zone of upwelling areas contain several percent of organic carbon with exceptional values exceeding 10 % C_{org} where upwelling is very intense, e.g. off Peru and southwest Africa, and where the oxygen-minimum zone extends into the shallow waters of the shelf. Still, generalizations are difficult to make because sedimentation conditions are highly variable in space.

Generalizations are also difficult to make with respect to variation of organic carbon content with time. For long periods of the geological record the present-day conditions of organic carbon burial can be projected to the past. There were times, however, mostly relatively

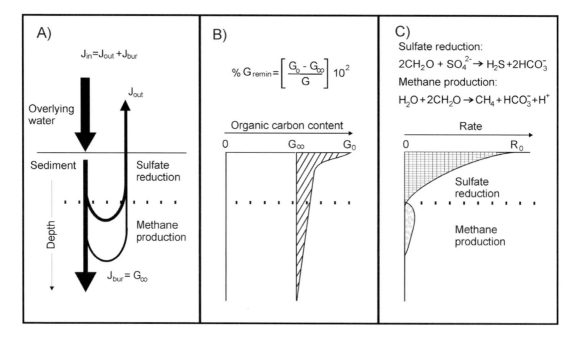

Fig. 4.5 Three independent approaches used to quantify organic matter degradation in anoxic coastal sediments: (A) carbon budget based on measurement of recycled (J_{out}; CO_2 and methane) and calculation of buried (J_{bur}) carbon fluxes; (B) kinetic modeling of the concentration/depth distribution of organic carbon (G; G_0 and G_∞ are organic carbon contents at the top and the bottom of the studied depth interval, respectively); (C) calculated organic carbon remineralization based on modeled or measured rates of sulfate reduction and methanogenesis (CH_2O stands for organic matter) (after Martens and Klump 1984).

short and often termed an event, when high amounts of organic matter were preserved, not only in shallow epicontinental seas, but also in deep-sea sediments. Examples are the Jurassic and particularly Cretaceous black shales of the Atlantic and Pacific Oceans with extreme organic carbon contents of 20-30 % and more (e.g. Herbin et al. 1986; Dumitrescu and Brassell 2005). Specific oceanographic conditions prevailed during the younger geological past in semi-enclosed basins like the Mediterranean Sea. Plio-Pleistocene sapropels in the eastern Mediterranean Sea were deposited at regular time intervals due to climatic changes induced by orbital forces, in this case the 23,000 year cycle of precession of the Earth's axis. Some of the Mediter-ranean sapropels, recovered during Leg 160 of the international Ocean Drilling Program, yielded more than 30 % C_{org} (Emeis et al. 1996).

The range of organic carbon contents in sediments and the associated variation in conditions for organic matter preservation imply that the amount of biogenic information incorporated in sediments as organic matter may vary drastically. In the same way, the extent to which the preserved organic matter is representative of the ecosystem in the water column above, may be vastly different. It is not surprising then that organic geochemists have preferentially investigated sediments with high organic carbon contents particularly when emphasis was on the formation of fossils fuels (petroleum or natural gas) or on molecular organic geochemical analysis which – at least in its early days – required relatively large amounts of material. It has to be kept in mind that this bias has certainly also influenced the choice of examples given throughout this chapter, although attempts are made to contrast case studies representing different environmental conditions in the oceanic realm.

Within a sediment, the organic carbon content decreases with increasing depth due to mostly microbiological remineralization, but possibly also due to abiological oxidation, during early (and later) diagenesis. The entire process takes place in a complex redox system where organic matter is the electron donor and a variety of substrates are electron acceptors. In other words, whenever organic matter is destroyed or altered by oxidation, a reaction partner has to be reduced. In an extended investigation of the biogeo-chemical cycling in an organic-matter-rich coastal marine basin, Martens and Klump (1984) schematically illustrated three independent approaches to quantify organic matter degradation in sediments with anoxic surface layers (Fig. 4.5). These involve (a) a mass balance of incoming, recycled and buried carbon fluxes, (b) kinetic modeling of the concentration/depth

distribution of organic carbon and (c) measurement of degradation rates in the sediment column. The redox zones in the example given in Fig. 4.5 are restricted to a depositional environment with anoxic surface sediment and comprise only sulfate reduction and methano-genesis. In the case of oxic conditions in the upper sediment layer, there would be additional oxygen, nitrate, Mn(IV) and Fe(III) reduction zones (Froelich et al. 1979; cf. also Fig. 4.6, where these zones are indicated, and Chap. 5).

Martens et al. (1992) applied the approaches in Fig. 4.5 to study the composition and fate of organic matter in coastal sediments of Cape Lookout Bight[2]. In separate studies it had previously been established that organic matter was mostly supplied from back-barrier island lagoons and marshes landward of the bight at a steady rate. Furthermore, the organic matter was extensively physically and biologically recycled in the lagoon before it ultimately accumulated in the sediments. Thus, systematic downcore decreases in amount of labile organic matter had to result from early diagenesis rather than variations of supply. The authors tried to answer the question of what fraction of the incoming particulate organic carbon (POC) is remine-ralized during early diagenesis under the conditions described by solving the simple mass balance equa-tion.

$$POC\ input = POC\ remineralized + POC\ buried$$

$$(4.6)$$

In their experience it has proven easiest to measure fluxes resulting from POC remineralization and burial and then to calculate POC input by adding these fluxes together. Numerical values of the fluxes are given in Figure 4.6. In this model, the incoming POC is either remineralized to CO_2, CH_4 and DOC (dissolved organic carbon) or buried. The CO_2, CH_4 and DOC produced during remineralization are either lost to the water column via sediment-water chemical exchange or buried as carbonate and dissolved components of sediment pore waters. Using ^{210}Pb-based sedimentation rates, the POC burial rate was found to be 117±19 mol C $m^{-2}yr^{-1}$. Sediment-water chemical exchange accounts for losses of 40.6±6.6 mol C $m^{-2}yr^{-1}$ as CO_2, CH_4 and DOC, whereas 7.0±1.1 mol C $m^{-2}yr^{-1}$ of these species, including

---

[2] Cape Lookout Bight (North Carolina, U.S.A.) is an "end member" environment with respect to sedimenta-tion rate (10 cm per year!), organic matter composi-tion, domination of anoxic degradation processes and direct ebullition of methane gas, i.e. not typical for present-day open-ocean marine sediments.

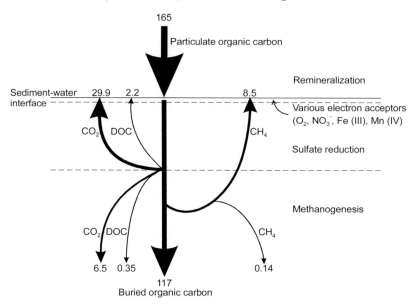

Fig. 4.6 Fluxes of carbon associated with organic matter supply, degradation and burial in Cape Lookout Bight sediments. The unit of all numerical flux values is moles C m^{-2}yr^{-1} (after Martens et al. 1992).

carbonate formed from CO_2, are buried (Fig. 4.6). The dissolved sediment-water exchange and burial fluxes sum to a total POC remineralization rate of 47.6±5.7 mol C m^{-2}yr^{-1}. When this value is added to the POC burial rate, a total POC input of 165±20 mol C m^{-2}yr^{-1} can be calculated from equation 4.6. From this result it follows that 29±5 % of the incoming POC is remineralized as an average over the first about ten years after sedimentation in Cape Lookout Bight.

Using a similar approach, Alperin et al. (1992) determined the POC remineralization rate for sediments of Skan Bay, Alaska, a pristine embayment with oxygen-depleted bottom water (<0.4 ml O_2/l water) and sulfidic surface sediments and with a shallow sill limiting advection of oxygen-rich water from the Bering Sea. Total sediment remineralization rate was calculated by three independent approaches: (1) the difference between POC deposition and preservation; (2) the quantity of carbon recycled to the water column and buried at depth; (3) depth-integrated rates of bacterial metabolism. The budget indicates that 84±3 % of the organic carbon deposited is remineralized in the upper meter of the sediment column representing approximately 100 years. A steady state is nearly reached, however, at a depth of about 70 cm, i.e. remineralization is already very slow after approximately 70 years of burial in Skan Bay. The initial content of more than 9 % organic carbon at the sediment surface dropped to less

than 2 % of dry sediment at 0.7 to 1 m depth and most of that would survive deeper burial.

A third case study of organic carbon recycling and preservation in coastal environments, including a comprehensive budget of inorganic reactants and products, is from the Aarhus Bay (Denmark), a shallow embayment in the Kattegat that connects the North Sea with the Baltic Sea (Fig. 4.7; Jørgensen 1996). The bulk annual sedimentation rate in Aarhus Bay is about 2 mm yr^{-1}. Photosynthesis is in the upper mesotrophic range and annually produces organic matter corresponding to 21.8 mol C m^{-2}yr^{-1}. Planktonic oxygen respiration corresponds to mineralization of 68 % of the primary productivity and 32% sedimentation, whereas direct sediment trap measurements accounted for 45% deposition. Of these 9.9 mol C m^{-2}yr^{-1}, about 2.2 mol C m^{-2}yr^{-1} are buried below the bioturbated zone. Metabolization in the sediment mainly occurs by oxygen and sulfate as electron acceptors, whereas nitrate, Mn(IV) and Fe(III) play a subordinate role. Methanogenesis was not included in the study of the carbon budget because only the water column and the shallow surface sediment were studied.

The three case studies show that organic matter preservation and, thus, organic carbon contents strongly depend on the specific local environmental conditions. The extent of remineralization in these three cases ranges from about 30 to 85 % in the upper sediment layers comprising, however, different time ranges. They correspond to the range quoted for continental shelf and estuarine sediments (20 to 90 %) by Henrichs and Reeburgh (1987). Apparently, organic carbon flux, bulk sedimentation rate, water depth, oxygen concentration in the bottom water and related extent of bioturbation of surface sediments all have an influence on the intensity of organic matter remineralization during early diagenesis and on how much organic matter is buried to a sediment depth where further remineralization only proceeds very slowly. Only that organic matter fraction can be considered to become fossilized in a strict sense and to enter the geological organic carbon cycle.

Methanogenesis

As indicated in Figure 4.6, the last step in the sedimentary metabolic pathway is methanogenesis. The formation of methane at a depth level, where all sulfate has been consumed, involves a group of strictly anaerobic archaea, collectively called methanogens. They use a small number of different low-molecular-weight substances for the biosynthesis of methane which, together with elemental hydrogen, in turn are formed

by bacterial fermentation from more complex organic substances during early diagenesis. The terminal electron acceptor in methanogenesis is carbon. The most prominent pathways are the reduction of carbon dioxide with molecular hydrogen and the transformation of acetic acid into methane and carbon dioxide (Eq. 4.7), although formic acid or methanol may be used as substrates as well.

$$CO_2 + 4\,H_2 \rightarrow CH_4 + 2\,H_2O$$
$$CH_3COOH \rightarrow CH_4 + CO_2 \qquad (4.7)$$

Methanogenesis is widespread in the marine environment, particularly on continental margins or in stagnant basins, where sufficient organic matter is deposited so that anoxic conditions occur at shallow depth below the seafloor (e.g. D'Hondt et al. 2002). The amounts of methane formed can be enormous in certain areas. Under suitable conditions of low temperature and high pressure this biogenic methane and pore water may form a solid, ice-like substance called methane clathrate or, more generally, gas hydrate (see Chap. 14). Another important process which is related to methanogenesis and of which many microbiological and biogeochemical details have only been revealed in recent years and still are being investigated,

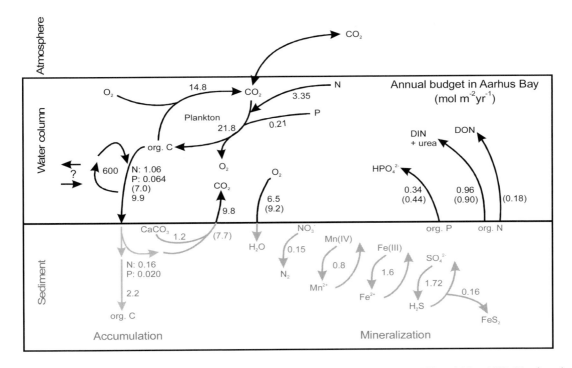

Fig. 4.7 Summary of fluxes and process rates measured in Aarhus Bay between May 1990 and May 1991. Numbers in parentheses were derived by difference while the others are based on independent rate measurements and calculations. Unit are given in mol m⁻²yr⁻¹ for each component. DIN = dissolved inorganic nitrogen; DON = dissolved organic nitrogen (after Jørgensen 1996).

is the anaerobic methane oxidation by consortia of archaea and sulfate-reducing bacteria, formally the reverse of methanogenesis (AOM; see Chap. 8).

4.3.2   Chemical Composition of Biomass

Apart from considering the fate of bulk organic matter (or organic carbon) during diagenesis, organic geo-chemistry has developed a more sophisticated under-standing of diagenetic organic matter transformation down to the molecular level. Fundamental to this understanding is a comparison of the organic consti-tuents of geological samples with the inventory of extant organisms. This was, and still partly is, hampered by the limited knowledge of the natural product chemistry particularly of unicellular marine algae, protozoans and bacteria.

The simplest way of describing the chemical nature of biomass is by its elemental composition. For marine phytoplankton as primary producers a relationship was found to the nutrients available in seawater which led to the definition of the Redfield ratio as $C:N:P = 106:16:1$ (Redfield et al. 1963). Derived from this is an average molecular formula of phytoplankton organic matter related to the general process of phytosynthesis (of which the reverse signifies remineralization):

$$106\ CO_2 + 106\ H_2O + 16\ NH_3 + H_3PO_4 \rightarrow$$
$$(CH_2O)_{106}(NH_3)_{16}H_3PO_4 + 106\ O_2$$

$$(4.8)$$

The formula of the organic matter product is often reduced to the summary version of $C_{106}H_{263}N_{16}O_{110}P$. It has no real chemical meaning in terms of molecular structure because it contains more hydrogen than the bonds of all the other atoms can account for. The reason is that the generalized average formula (i.e. the product in Eq. 4.8) is just the sum of separate neutral molecules which are involved in biosynthesis of organic matter. The formation of a molecular structure requires the formal loss of a number of molecules of water for condensation. Whereas the formula does not represent the correct elemental organic matter compo-sition of marine phytoplankton, at least not for hydrogen and oxygen, it has to be kept in mind that it is a crude generalisation in itself. It would certainly vary with nutrient conditions and planktonic species as has been observed, e.g., by Takahasi et al. (1985) in a study of plankton biomass from the Atlantic and Indian Oceans which resulted in a modified Redfield ratio of $122(\pm18):16:1$. There are quite a number of more recent studies that confirm this kind of deviation

from the Redfield ratio or extend the ratio by inclusion of trace metals (e.g. Leonardos and Geider 2004; Ho et al. 2003; cf. also Chap. 6).

Food chain and early diagenetic processes change the initial elemental composition drastically. Organic matter in sediments relative to the primary producers is enriched particularly in carbon and hydrogen, whereas it is depleted in oxygen (but the degree depends on the extent of oxidation of sedimentary organic matter), nitrogen and phosphorus. Depletion in phosphorus is due to the facile hydrolytic cleavage of bound phosphate groups. Loss of nitrogen occurs by prefe-rential degradation of organic nitrogen compounds as discussed later (see Sect. 4.4 for a discussion of C/N ratios). Sulfur, not originally included in the general formula, would be equal to or less in content than phosphorus. The enrichment of sulfur in fossil organic matter is, however, not due to a relative enrichment in the course of preferential loss of other elements (as is the case for C and H). Sulfur enrichment rather is a consequence of diagenetic incorporation of reduced inorganic sulfur species (like HS^- or corresponding polysulfides) which are formed from seawater sulfate by sulfate-reducing microorganisms in shallow sediments under anoxic conditions (see Chap. 8).

On the next higher level, the chemical composition of living organisms in the biosphere, despite their diversity, can be confined to a limited number of principal compound classes. Their proportions vary in the different groups of organisms as is evident from the estimates of Romankevitch (1984) for a few types of marine organisms (Table 4.3). Also, within the groups the compound class composition is highly variable (Table 4.4). It may even depend on the growth stage for a single species. The compound classes in turn comprise a very large number of single compounds with different individual chemical structures, although enzymatic systems limit the potential chemical diversity (that is why there are chemical biomarkers of taxonomic significance). Many of the compound classes are also represented in fossil organic matter, although not in the same proportions as they occur in the biosphere because of their different stabilities toward degradation and modification of original structures during sedimentation and diagenesis.

Nucleic Acids and Proteins

Nucleic acids, as ribonucleic acids (RNA) or desoxy-ribonucleic acids (DNA), are biological macromolecules carrying genetic information. They consist of a regular sequence of phosphate, sugar (pentose) and a small variety of base units, i.e. nitrogen-bearing heterocyclic compounds of the purine or pyrimidine type. During biosynthesis, the genetic information is transcribed into sequences of amino acids, which occur as peptides, proteins or enzymes in the living cell. These macromolecules vary widely in the number of amino acids and thus in molecular weight. They account for most of the nitrogen-bearing compounds in the cell and serve in such different functions as catalysis of biochemical reactions and formation of skeletal structures (e.g. shells, fibers, muscles).

During sedimentation of decayed organisms, nucleic acids and proteins are readily hydrolyzed chemically or enzymatically into smaller, water-soluble units. Refined analytical techniques, however, allow traces of DNA in sedimentary sequences, in combination with lipids, to be used to trace changes in phytoplanktonic populations with geological time (e.g. Coolen et al. 2004). Amino acids occur in rapidly decreasing concentrations in recent and subrecent sediments, but may also survive in small concentrations in older sediments, particularly if they are protected, e.g., by the calcareous frustules or shells of marine organisms. Nitrogen-bearing aromatic organic compounds in sediments and crude oils may relate to the purine and pyrimidine bases in nucleic acids, but this awaits unequivocal confirmation. A certain fraction of the nucleic acids and proteins reaching the sediment surface may be bound into the macromolecular organic matter network (humic substances, kerogen) of the sediments and there become protected against further rapid hydrolysis. Experiments in the laboratory have shown that kerogen-like material (melanoidins) can be obtained by heating amino acids with sugar.

Saccharides, Lignin, Cutin, Suberin

Sugars are polyhydroxylated hydrocarbons that together with their polymeric forms (oligosaccharides, polysaccharides) constitute an abundant proportion of the biological material, particularly in the plant kingdom. Polysaccharides occur as supporting units in skeletal tissues (cellulose, pectin, chitin) or serve as an energy depot, for example, in seeds (starch). Although polysaccharides are largely insoluble in water, they are easily converted to soluble C_5 (pentoses) and C_6 sugars (hexoses) by hydrolysis and, thus, in the sedimentary environment will have a short-term fate similar to that of the proteins.

Lignin is a structural component of plant tissues where it occurs as a three-dimensional network together with cellulose. Lignin is a macromolecular condensation product of three different propenyl (C_3-substituted) phenols (one type of few biogenic aromatic compounds). It is preserved, even during transport from land to ocean and during sedimentation to the seafloor where it occurs predominantly in humic organic matter of deltaic environments.

Cutin and suberin are lipid biopolymers of variable composition which are part of the protective outer coatings of all higher plants. Chemically, cutin and suberin are closely related polyesters composed of long-chain fatty and hydroxy fatty acid monomers. Both types of biopolymers represent labile, easily metabolizable terrigenous organic matter because they are sensitive to hydrolysis. After sedimentation, they have only a moderate preservation potential.

Table 4.3 Biochemical composition of marine organisms (after Romankevitch 1984).

Organism	Proteins (%)	Carbohydrates (%)	Lipids (%)	Ash (%)
Phytoplankton	30	20	5	45
Phytobenthos	15	60	0.5	25
Zooplankton	60	15	15	10
Zoobenthos	27	8	3	62

Table 4.4 The main chemical constituents of marine plankton in percent of dry weight (after Krey 1970).

Organism	Proteins (%)	Carbohydrates (%)	Lipids (%)	Ash (%)
Diatoms	24-48	0-31	2-10	30-59
Dinoflagellates	41-48	6-36	2-6	12-77
Copepods	71-77	0-4	5-19	4-6

Insoluble, Nonhydrolyzable Highly Aliphatic Biopolymers

Insoluble, nonhydrolyzable aliphatic biopolymers were discovered in algae and higher plant cell walls as well as in their fossil remnants in sediments (see de Leeuw and Largeau 1993; van Bergen et al. 2004 for overviews). These substances are called algaenan, cutan or suberan according to their origin or co-occurrence with cutin and suberin in extant organisms. They consist of aliphatic polyester chains cross-linked with ether bridges (Blokker et al. 1998, 2000) which render them very stable toward degradation. Pyrolysis and other rigorous methods are needed to decompose these highly aliphatic biopolymers. This explains why they are preferentially preserved in sediments.

Monomeric lipids

Biologically produced compounds that are insoluble in water but soluble in organic solvents such as chloroform, ether or acetone are called lipids. In a wider sense, these also include membrane components and certain pigments. They are common in naturally occurring fats, waxes, resins and essential oils. The low water solubility of the lipids derives from their hydrocarbon-like structures which are responsible for their higher survival rates during sedimentation compared to other biogenic compound classes like amino acids or sugars.

Various saturated and unsaturated fatty acids are the lipid components bound to glycerol in the triglyceride esters of fats (see Fig. 4.8 for examples of chemical structures of lipid molecules). Cell membranes consist to a large extent of fatty acid diglycerides with the third hydroxyl group of glycerol bound to phosphate or another hydrophilic group. In waxes, fatty acids are esterified with long-chain alcohols instead of glycerol. Plant waxes contain unbranched, long-chain saturated hydrocarbons (*n*-alkanes) with a predominance of odd carbon numbers (e.g. C_{27}, C_{29}, C_{31}) in contrast to the acids and alcohols which show an even-carbon-number predominance.

Isoprene (2-methylbuta-1,3-diene), a branched diunsaturated C_5 hydrocarbon, is the building block of a large family of open-chain and cyclic isoprenoids and terpenoids (Fig. 4.8). Essential oils of higher plants are enriched in monoterpenes (C_{10}) with two isoprene units. Farnesol, an unsaturated C_{15} alcohol, is an example of a sesquiterpene with three isoprene units. The acyclic diterpene phytol is probably the most abundant isoprenoid on Earth. It occurs esterified to chlorophyll *a* and some bacteriochlorophylls and is,

thus, widely distributed in the green pigments of aquatic and subaerial plants. Sesterterpenes (C_{25}) are of relatively minor importance except in some methanogenic bacteria (cf. Volkman and Maxwell 1986).

Cyclization of squalene (or its epoxide) is the biochemical pathway to the formation of a variety of pentacyclic triterpenes (C_{30}) consisting of six isoprene units. Triterpenoids of the oleanane, ursane, lupane and other less common types are restricted to higher plants, and in exceptional cases may dominate the extractable organic constituents of deep-sea sediments like in Baffin Bay (ten Haven et al. 1992). The geochemically most important and widespread triterpenes are from the hopane series, like diploptene which occurs in ferns, cyanobacteria and other eubacteria. The predominant source of hopanoids are bacterial cell membranes, however, which contain bacteriohopanetetrol (and closely related molecular species) as rigidifiers. This C_{35} compound has a sugar moiety attached to the triterpane skeleton via a carbon-carbon bond (Fig. 4.8). The widespread distribution of bacteria on Earth through time makes the hopanoids ubiquitous constituents of all organic-matter assemblages (Rohmer et al. 1992).

Steroids are tetracyclic compounds that are also biochemically derived from squalene epoxide cyclization, but have lost, in most cases, up to three methyl groups. Cholesterol (C_{27}) is the most important sterol of animals and occurs in some plants as well. Higher plants frequently contain C_{29} sterols (e.g. sitosterol) as the most abundant compound of this group. Steroids together with terpenoids are typical examples of biological markers (chemical fossils) because they contain a high degree of structural information that is retained in the carbon skeleton after sedimentation (e.g. Poynter and Eglinton 1991; Peters et al. 2005) and often provides a chemotaxonomic link between the sedimentary organic matter and the precursor organisms in the biosphere.

Carotenoids, red and yellow pigments of algae and land plants, are the most important representatives of the tetraterpenes (C_{40}). Due to their extended chain of conjugated double bonds (e.g. β-carotene; Fig. 4.8) they are labile in most depositional environments and are found widespread but in low concentrations in marine surface sediments. Aromatization probably is one of the dominating diagenetic pathways in the alteration of the original structure of carotenoids in the sediment. Diagenetic intermolecular cross-linking by sulfur bridges may preserve the carotenoid carbon skeletons to a certain extent.

A second pigment type of geochemical significance are the chlorophylls and their derivatives that during

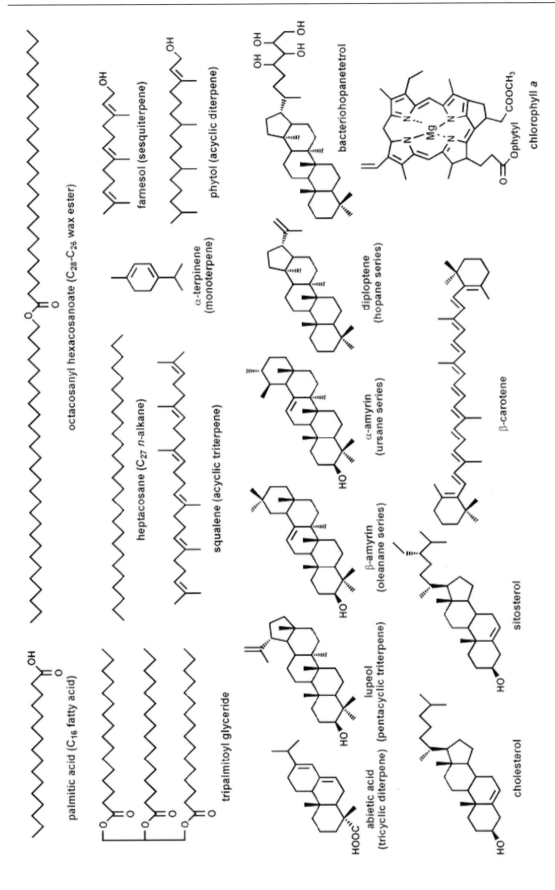

Fig. 4.8 Structural formulae of representative low-molecular-weight lipids in living organisms and surface sediments (after Rullkötter 2001).

diagenesis are converted into the fully aromatized porphyrins. Most porphyrins in sediments and crude oils are derived from the green plant pigment chlorophyll *a* and from bacteriochlorophylls.

4.3.3   The Principle of Selective Preservation

Organic compounds and compound classes differ in their potential to be preserved in sediments and to survive early diagenesis. As a general rule, water-soluble organic compounds, or organic macromolecules, which are easily hydrolyzed to water-soluble monomers, have a low preservation potential. In contrast to this, compounds with a low solubility in water such as lipids and hydrolysis-resistant macromolecules are selectively enriched in the sedimentary organic matter. Table 4.5 is a compilation of the source and preservation potential of some common organic compound types. It is based on anticipated chemical stabilities related to

structures, reported biodegradability and reported presence in the geosphere, but not on mechanisms of preservation such as mechanical protection or bacteriostatic activities of certain chemicals in the (paleo)environment (de Leeuw and Largeau 1993).

The near-surface sediment layers represent the transition zone where biological organic matter is transformed into fossil organic matter. There are two slightly differing views about the nature of this process. The classical view (Fig. 4.9; Tissot and Welte 1984) implies that biopolymers are (mainly) enzymatically degraded into the corresponding biomonomers. The monomers then are either used by sediment bacteria and archaea to synthesize their own biomass or as a source of energy. Alternatively, they may randomly recombine by condensation or polymerization to geomacromolecules (see Sect. 4.3.4). The discovery of nonhydrolyzable, highly aliphatic biopolymers in extant organisms and geological samples has led to a reappraisal of the processes involved in the formation

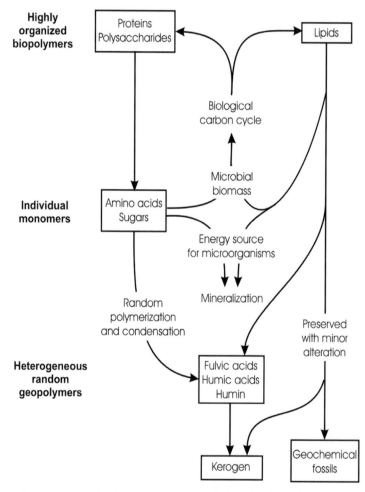

Fig. 4.9 From biomass to geomacromolecules - a summary of the classical view of processes involved in the transformation of biogenic organic matter into kerogen and geochemical fossils (after Tissot and Welte 1984).

of geomacromolecules (Tegelaar et al. 1989; de Leeuw and Largeau 1993). In a scheme modified from that of Tissot and Welte (1984), more emphasis is placed on the selective preservation of biopolymers (Fig. 4.10). This means that the role of consecutive and random polymerisation and polycondensation reactions of biomonomers formed by hydrolysis or other degradative pathways may be less important than previously thought. Support to this view is given by the detection of the close morphological relationship between some fossil 'ultralaminae' and the thin resistant outer cell walls of green microalgae (Largeau et al. 1990).

To complete the modified view of geomacromolecule formation, the process of 'natural vulcanisation' (Fig. 4.10) has been proposed to play a major role under suitable conditions (e.g. Sinninghe Damsté et al. 1989a, 1990; de Leeuw and Sinninghe Damsté 1990). Many marine sediments contain high-molecular-weight organic sulfur substances that are thought to be derived from intermolecular incorporation of inorganic sulfur species (HS⁻, polysulfides) into functionalized lipids during early diagenesis. This requires the reduction of seawater sulfate by sulfate-reducing microorganisms under anoxic conditions (see Chap. 8). Sulfur incorporation into organic matter is further enhanced in depositional systems that are iron-limited, i.e. organosulfur compounds are particularly abundant in areas that receive little continental detritus with clays enriched in iron and where instead biogenic carbonate or opal is the dominant mineral component of the sediment.

As a consequence of the discussion of organomineral interaction for the preservation of organic matter in sediments (see Sect. 4.2), Collins et al. (1995) raised the question if sorption of organic matter on mineral surfaces did not lead to a rebirth of the classical

Table 4.5 Inventory of selected biomacromolecules, their occurrence in extant organisms, and their potential for survival during sedimentation and diagenesis (after Tegelaar et al. 1989 and de Leeuw and Largeau 1993; see there for chemical structures). The 'preservation potential' ranges from - (extensive degradation under all depositional conditions) to ++++ (little degradation under any depositional conditions).

Biomacromolecules	Occurrence	'Preservation potential'
Starch	Vascular plants; some algae; bacteria	-
Glycogen	Animals	-
Poly-β-hydroxyalkanoates	Eubacteria	-
Cellulose	Vascular plants; some fungi	-/+
Xylans	Vascular plants; some algae	-/+
Galactans	Vascular plants; algae	-/+
Gums	Vascular plants	+
Alginic acids	Brown algae	-/+
Dextrans	Eubacteria; fungi	+
Xanthans	Eubacteria	+
Chitin	Arthropods; copepods; crustacea; fungi; algae	+
Proteins	All organisms	-/+
Mureins	Eubacteria	+
Teichoic acids	Eubacteria	+
Bacterial lipopolysaccharides	Gram-positive eubacteria	++
DNA, RNA	All organisms	-
Glycolipids	Plants; algae; eubacteria	+/++
Polyisoprenoids (rubber, gutta)	Vascular plants	+
Polyprenols and dolichols	Vascular plants; bacteria; animals	+
Resinous polyterpenoids	Vascular plants	+/++
Cutins, suberins	Vascular plants	+/++
Lignins	Vascular plants	++++
Sporopollenins	Vascular plants	+++
Algaenans	Algae	++++
Cutans	Vascular plants	++++
Suberans	Vascular plants	++++

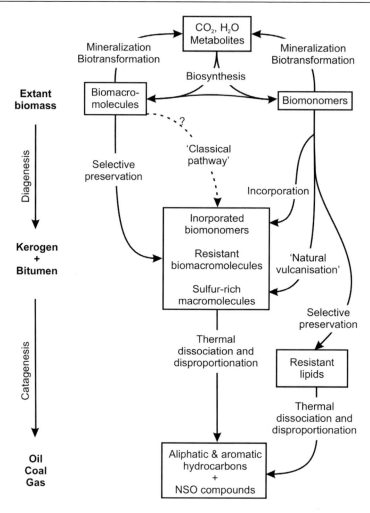

Fig. 4.10   The selective preservation pathway model of kerogen formation (after de Leeuw and Largeau 1993 and Tegelaar et al. 1989).

condensation pathway for geomacromolecule formation. Neither adsorption nor condensation alone may be a satisfactory process for preservation of labile organic substances. Adsorption of monomers can merely retard their biodegradation, and condensation is not favored in (pore water) solution. However, if the processes operate in concert – adsorption promoting condensation and condensation enhancing adsorption of further reactants – a plausible mechanism for the preservation of organic matter arises. Condensation reactions between adsorbed compounds would lead to the formation of very strongly bound macromolecules resulting in a marked divergence in the diagenetic history of the adsorbed monolayer and nonmineral-bound organic matter. More direct evidence is, however, still required to establish the quantitative importance of this process relative to other processes, such as selective preservation.

4.3.4   The Formation of Fossil Organic Matter and its Bulk Composition

The geomacromolecular organic matter surviving microbial degradation in the early phase of diagenesis consists of three fractions, termed fulvic acids, humic acids and humin. They are ill-defined in their chemical structures, but are operationally distinguished by their solubilities in bases (humic and fulvic acids) and acids (fulvic acids only). All three are considered to be potential precursors of kerogen, which designates the type of geomacromolecules formed at a later stage of diagenesis. Kerogen is also only operationally defined as being insoluble in non-oxidizing acids, bases and organic solvents (Durand 1980; Tissot and Welte 1984). Besides this high-molecular-weight organic material, sediments contain low-molecular-weight organic substances that are collectively called bitumen and are extractable with organic solvents. Bitumen consists of nonpolar hydrocarbons and a variety of polar lipids with a great number of different functional groups, such as ketones, ethers, esters, alcohols, fatty acids and corresponding sulfur-bearing compounds.

Fulvic acids and subsequently humic acids decrease in their concentrations over time as a result of progressive combination reactions with increasing diagenesis. This process of kerogen formation concurrently involves the elimination of small molecules like water, carbon dioxide, ammonia or hydrogen sulfide (Huc and Durand 1977). As a consequence, the degree of condensation of the macromolecular kerogen increases. In terms of elemental composition it becomes enriched in carbon and hydrogen and depleted in oxygen, nitrogen and sulfur. This carbon- and hydrogen-rich kerogen ultimately is the source material for the formation of petroleum and natural gas which starts when burial is deep and temperatures are so high that the carbon-carbon bonds in the kerogen are thermally cracked. The main phase of oil formation typically occurs between 90 °C and 120 °C, but the range largely depends on the heating rate (slow or rapid burial) and on the chemical structure of the kerogen. For example, kerogens rich in sulfur start oil formation earlier because carbon-sulfur bonds are broken more easily than carbon-carbon bonds. The product, however, is a heavy oil (high density, high viscosity) rich in sulfur

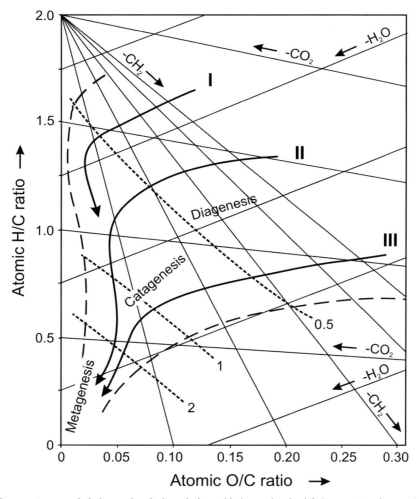

Fig. 4.11   Kerogen types and their geochemical evolution with increasing burial (temperature increase) in a van Krevelen diagram of atomic H/C versus O/C ratios from elemental analysis (after Tissot and Welte 1984). Roman numbers indicate kerogen types, bold trend lines are idealized average values from a large number of data points. In organic geochemistry, diagenesis is the early (low-temperature) range of this evolution, catagenesis signifies the phase of petroleum and wet gas formation by thermal cracking, and metagenesis is the high-temperature range where still some dry gas (methane, ethane) is formed. Numbers associated with broken lines in the diagram (0.5, 1, 2) indicate approximate vitrinite reflectance (R_o) values commonly measured to indicate thermal maturity. Vitrinites are fossil woody organic particles of characteristic shape which increase their reflectivity as a function of geothermal history; the amount of reflected light can be quantitatively measured with a light microscope under defined conditions (see Sect. 4.5.3 and Taylor et al. 1998 for more details).

and economically less valuable than light and sulfur-lean crude oil. The difference between an oil shale and a rock actively generating petroleum is only the thermal history. An oil shale simply has not completed the phase of diagenetic release of small molecules described above. For more details on petroleum formation consult the textbooks of Tissot and Welte (1984), Hunt (1996) and Welte et al. (1997).

Molecular structural information about kerogen can be inferred from elemental analysis, spectroscopic methods and the results of pyrolysis and selected chemical degradation experiments (see Rullkötter and Michaelis 1990 for an overview). Yet, with the understanding that kerogen is a complex heterogeneous macromolecular substance with contributions from a variety of organisms and a wide range of chemical alterations that occurred during diagenesis, it becomes clear that there will be no single molecular structure of kerogen, and only certain characteristic units can be described at the molecular level.

Kerogens have been classified into types derived from H/C and O/C atomic ratios in a van Krevelen diagram (Fig. 4.11). The types indicated are related to the hydrogen and oxygen richness, relative to carbon, of the biogenic precursor material. Roughly, kerogen Type I is related to hydrogen-rich organic matter as occurring, e.g., in waxes and algal mats, kerogen Type II represents typical oceanic plankton material, and kerogen Type III is typical of land-derived organic matter which has been transformed into lignite or coal. A kerogen type IV not indicated in the diagram has occasionally been defined to have very low H/C ratios and to represent highly oxidized, largely inert organic matter. The bold solid trend lines indicated in Figure 4.11 then represent the changes in elemental composition initially occuring during diagenesis (evolution grossly parallel to the x-axis due to loss of oxygen functionalities) and later during oil and gas formation (catagenesis; evolution grossly parallel to y-axis due to loss of hydrogen-rich petroleum hydrocarbons), until a carbon-rich residue is the ultimate product (near origin in xy diagram). For more details see Tissot and Welte (1984).

4.3.5   Early Diagenesis at the Molecular Level

A small portion of sedimentary organic matter is soluble in organic solvents and contains lipid compounds that are either directly inherited from the biological precursor organisms or cleaved by hydrolysis from larger cellular units like cell walls or membranes (cf. Figs. 4.9 and 4.10). The compounds include individual

substances as well as homologous series of structurally related compounds. Most of them are functionalized polar lipids that undergo decarboxylation (organic acids) and dehydration reactions (alcohols) during early diagenesis to produce saturated and olefinic hydrocarbons, of which the latter are progressively hydrogenated into their saturated analogs during later diagenesis. Alternatively, aromatic hydrocarbons are formed by the loss of hydrogen. If these hydrocarbons essentially have the same carbon skeletons and steric configurations as their functionalized biogenic precursors, they are called biological markers or molecular fossils (see Sections 4.3.2 and 4.3.5). Parallel to retention of the biogenic carbon skeleton, structural rearrangements, catalyzed by clay minerals, partial cleavage of substituents or ring opening may occur as side reactions during diagenetic transformation of biogenic lipids. During the earlier phases of diagenesis, including processes occurring in the water column, such alterations appear to be mediated by microbial activity. With increasing burial they are more and more driven by thermodynamic constraints as temperature increases.

The following discussion of biological marker reactions of course only applies to that fraction of lipid compounds that have escaped the highly efficient degradation in the uppermost sediment layer. It has been established through quantitative assessment of transformation reactions that degradation in this zone may occur over timescales of days and that reaction rates have often been underestimated by an order of magnitude (Canuel and Martens 1996). It was furthermore demonstrated in this study that the degradation processes can be highly selective and depend on the origin of the compounds (marine, bacterial or terrestrial). Fatty acids apparently are particularly sensitive to degradation whereas sterols and hydrocarbons have a higher chance of entering the deeper sediment. As a consequence, quantitative assessment of the source and fate of organic matter based on biological markers will be strongly limited as long as diagenetic effects cannot be separated from variations in organic matter supply.

4.3.6   Biological Markers (Molecular Fossils)

Molecules with a high degree of structural complexity are particularly informative and thus suitable for studying geochemical reactions because they provide the possibility of relating a certain product to a specific precursor. For example, specific biomarkers have been assigned to some common groups of microalgae. These compounds include long-chain (C_{37}-C_{39}) n-alkenones, highly-branched isoprenoid alkenes, long-chain *n*-alkandiols, 24-methylenecholesterol and dinosterol. They have been found to be unique for, or obviously be preferentially biosynthesized by, haptophytes, diatoms, eustigmatophytes, diatoms and dinoflagellates, respectively (see Volkman et al. 1998; Volkman 2005 for comprehensive overviews). Other long-chain *n*-alkyl lipids (e.g. Eglinton and Hamilton 1967), diterpenoids and 3-oxygenated triterpenoids (e.g. Simoneit 1986) are considered useful tracers for organic matter from vascular land plants.

Although certain biological markers may be chemotaxonomically very specific, care has to be taken when using relative biomarker concentrations in geological samples to derive quantitative figures of the biological species that have contributed to the total organic matter. First of all, different types of biological markers may have different reactivities and, thus, may be selectively preserved during diagenesis (Hedges and Prahl 1993). In this respect, sequestering of reactive

biomarkers by the formation of high-molecular-weight organic sulfur compounds may play an important role (e.g. Sinninghe Damsté et al 1989b; see Fig. 4.10). Furthermore, there may be a fractionation between high- and low-molecular-weight compounds. An extreme example is the (lacustrine) Messel oil shale. In its organic matter fraction, the residues of dinoflagellates are represented by abundant 4-methyl steroids in the bitumen whereas the labile cell walls were not preserved. On the other hand, certain green algae are clearly identifiable under the electron microscope due to the highly aliphatic biopolymers in their cell walls, but no biomarkers specific for green algae were found in the extractable organic matter (Goth et al. 1988).

The scheme in Figure 4.12 is an example of extensive and variable biomarker reactions after sedimentation. It illustrates the fate of sterols particularly during diagenesis. Although the scheme looks complex, it shows only a few selected structures out of more than 300 biogenic steroids and geochemical conversion

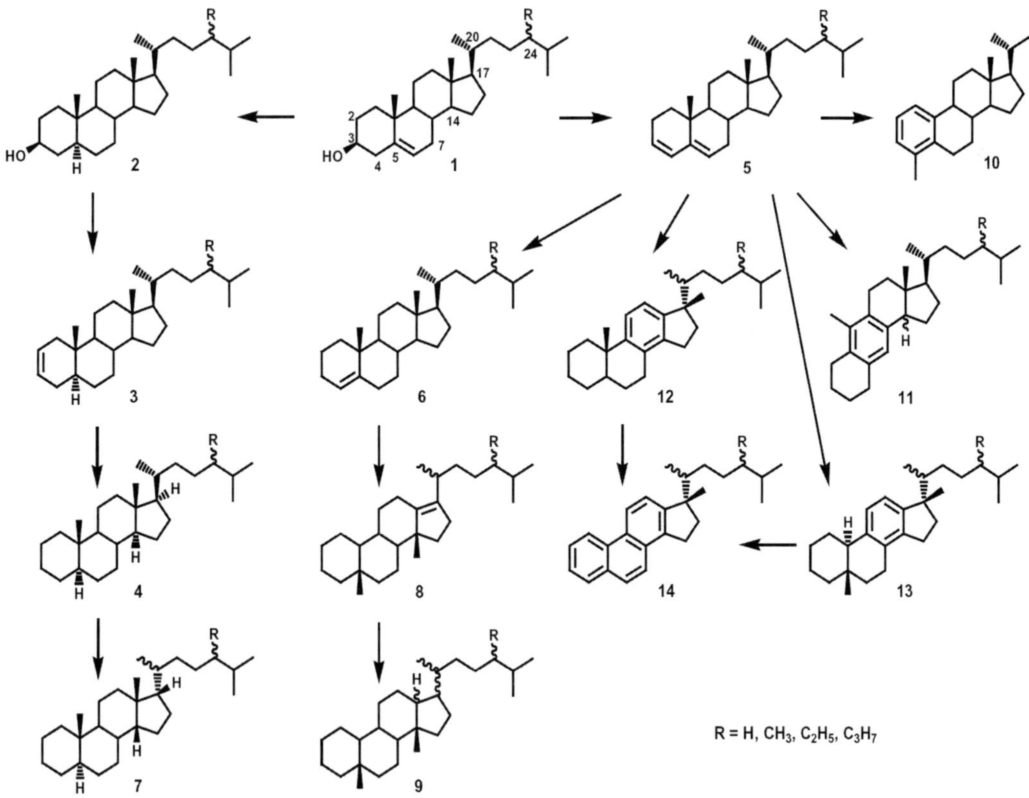

Fig. 4.12 Diagenetic and catagenetic transformation of steroids. The precursor sterols are gradually transformed during diagenesis into saturated hydrocarbons by dehydration (elimination of water) and hydrogenation of the double bonds. At higher temperatures, during catagenesis, the thermodynamically most stable stereoisomers are formed. Alternatively, dehydration leads to aromatic steroid hydrocarbons which are stable enough to occur in crude oils (after Rullkötter 2001). See text for detailed description of the reaction sequences.

products presently known to occur in sediments. The biogenic precursor chosen as an example in Figure 4.12 is cholesterol (structure 1, R=H), a widely distributed steroid in a variety of plants, but more typical of animals (e.g. zooplankton). Hydrogenation of the double bond leads to the formation of the saturated cholestanol (2). This reaction occurs in the uppermost sediment layers soon after deposition and is believed to involve microbial activity. Elimination of water gives rise to the unsaturated hydrocarbon 3. At the end of the diagenetic stage, the former unsaturated steroid alcohol 1 will have been transformed to the saturated steroid hydrocarbon 4 after a further hydrogenation step. An alternative route to the saturated sterane 4 is via dehydration of cholesterol (1, R=H), which yields the diunsaturated

compound 5. Hydrogenation leads to a mixture of two isomeric sterenes (6; isomer with double bond in position 5 like in the starting material (1) not shown in Fig. 4.12). This compound cannot be formed from 3 as suggested for a long time, because such a double bond migration would require more energy than is available under the diagenetic conditions in sediments (de Leeuw et al. 1989). Further hydrogenation of 6 affords the saturated hydrocarbon 4. A change in steric configuration of this molecule, e.g. to form 7, occurs only during the catagenesis stage at elevated temperatures. A side reaction from sterene 6 is a skeletal rearrangement leading to diasterene 8 where the double bond has moved to the five-membered ring and two methyl groups (represented by the bold bonds) are now bound

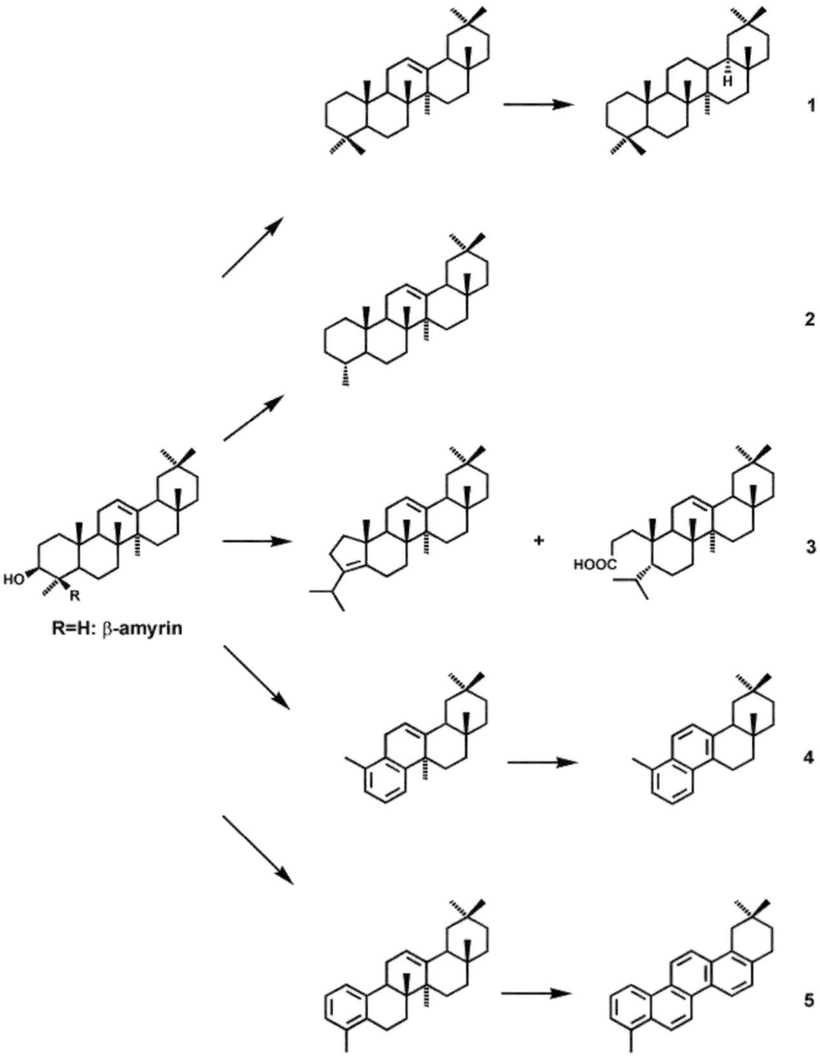

Fig. 4.13 Schematic representation of five different (mostly oxidative) diagenetic reactions of triterpenoids from higher plants. R in the starting material should be an oxygen function, at least in the second pathway (after Rullkötter et al. 1994).

to the bottom part of the ring system. This reaction has been shown in the laboratory to be catalyzed by acidic clays. Thus, diasterenes (8) and the corresponding diasteranes (9), formed from diasterenes by hydrogenation during late diagenesis, are found in shales but not in those carbonates that lack acidic clays.

An additional alternative diagenetic transformation pathway of steroids leads to aromatic instead of saturated hydrocarbons. The diolefin 5 is a likely intermediate on the way to the aromatic steroid hydrocarbons 10-14. Compounds 10 and 11 are those detected first in the shallow sediment layers. They obviously are labile and do not survive diagenesis. During late diagenesis, the aromatic steroid hydrocarbon 12 with the aromatic ring next to the five-membered ring cooccurs with compounds 10 and 11 in the sediments, but is also stable enough to survive elevated temperatures and thus to be found in crude oils. There is also a corresponding rearranged monoaromatic steroid hydrocarbon (13). During catagenesis, monoaromatic steroid hydrocarbons are progressively transformed into triaromatic hydrocarbons (14) before the steroid record is completely lost by total destruction of the carbon skeleton at higher temperatures.

As a second example, Figure 4.13 shows five different diagenetic reaction pathways for pentacyclic triterpenoids of terrestrial origin that were found to be abundant, e.g., in Tertiary deep-sea sediments of Baffin Bay (ten Haven et al. 1992). Diagenetic alteration with full retention of the carbon skeleton (e.g. in the case of β-amyrin; R=H) leads to an olefinic hydrocarbon after elimination of the oxygen functionality in the A-ring and later to the fully saturated hydrocarbon. If the substituent group R is a hydroxyl or carboxylic acid group, oxidation would yield an unstable ketocarboxylic acid, which instantaneously eliminates CO_2 leading to a carbon skeleton with one carbon atom less than the starting molecule (second pathway; see Rullkötter et al. 1994). Direct chemical elimination of the hydroxyl group in ring A causes ring contraction, and eventually the ring is opened by oxidative cleavage of the double bond (third pathway). If the carbon atoms of the A-ring are completely lost during degradation, then subsequent aromatization may lead into the fourth pathway. Alternatively, aromatization may start with the intact carbon skeleton giving rise to a series of partly or fully aromatized pentacyclic hydrocarbons (fifth pathway). All these alterations are typical for terrigenous triterpenoids. They probably start soon after the decay of the organisms (or parts thereof, e.g. leaves) and continue during transport into the ocean. The compounds described and several others have been found in numerous marine sediments (see Corbet et al. 1980 and Rullkötter et al. 1994 for overviews).

4.4 Organic Geochemical Proxies

4.4.1 Total Organic Carbon and Sulfur

Organic carbon profiles in a sedimentary sequence, particularly if they are obtained with high stratigraphic resolution (e.g. Stein and Rack 1995), provide direct evidence for changes in depositional patterns. An in-depth interpretation, however, usually requires additional information on the quality of the organic matter, i.e. on its origin (marine versus terrigenous) and/or its degree of oxidation during deposition. The relationship between organic carbon and sedimentation rate may help to distinguish different depositional environments or to determine paleoproductivity as already discussed in Section 4.2.

Furthermore, the relationship between organic carbon and sulfur is also characteristic of the paleoenvironment. Leventhal (1983) and Berner and Raiswell (1983) observed an increase in pyrite sulfur content in marine sediments with increasing amount of total organic carbon (Fig. 4.14). The rationale behind this is that the amount of metabolizable organic matter available to support sulfate-reducing bacteria increases

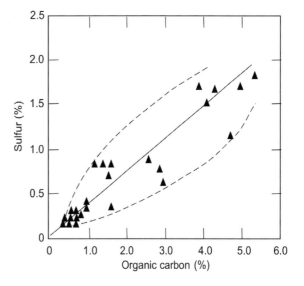

Fig. 4.14 Plot of weight percent organic carbon vs. weight percent pyrite sulfur for normal-marine modern sediments. Each plotted point represents the average value of samples in a given core, taken at a sediment depth where contents of organic carbon and pyrite have attained quasi-steady-state values, i.e. where early diagenesis of carbon and sulfur is (essentially) complete. The dashed lines enclose data from a variety of other studies (after Berner and Raiswell 1983). Sediments deposited under anoxic (euxinic) conditions would plot above the trend line, freshwater sediments significantly below.

with the total amount of organic matter arriving at the sediment-water interface. As a consequence, the sedimentary pyrite sulfide content is positively correlated with the non-metabolized (resistant or unused) organic matter content (TOC). The trendline in Figure 4.14 is considered representative of normal marine (oxic) environments. Data from the Black Sea plot above the trendline (higher S/C ratios) because consumption of organic matter by sulfate-reducing bacteria leads to excess hydrogen sulfide, available for pyrite formation, in the water column. In contrast to this, freshwater sediments have very low S/C (or high C/S) ratios because of the low sulfate concentrations in most freshwater bodies. Although the trendline is based on pyrite sulfur, it is not important whether the reduced sulfur is present as metal sulfide (mostly pyrite) or bound to the organic matter. This is a question of iron limitation rather than sulfate reduction.

As outlined in Section 4.3.1 (and discussed more extensively in Chap. 8) there is a close connection between organic matter remineralization during early diagenesis and microbial sulfate reduction. If all of the sulfate reduction products were precipitated as pyrite or bound into (immobile) organic matter, measuring the amount of sulfur in these species would provide an easy way for determining the amount of organic matter that has been remineralized and was not preserved in the sediment. However, the main product of sulfate reduction, hydrogen sulfide, is volatile and can escape from the sediment, particularly when bioturbation of the surface sediment supports this transport.

Release of hydrogen sulfide from the sediment plays a less important role under strictly anoxic conditions where fine lamination indicates that the environment is hostile to burrowing organisms and bioturbation does not occur. It has been shown that in these cases the initial amount of organic matter deposited can be estimated by measuring concentrations of reduced sulfur in such sediments. Considering the amount of organic matter consumed by sulfate reduction, Lallier-Vergès (1993) defined a sulfate reduction index (SRI) as

$$\text{SRI} = \text{\% initial organic carbon} \, / \, \text{\% preserved organic carbon} \qquad (4.9)$$

The amount of initial organic carbon then is the sum of the preserved organic carbon (measured as TOC) and the metabolized organic carbon (determined from the sulfur content with stoichiometric correction for the sulfate reduction process). Furthermore, the diffusive loss has to be taken into account. With a

correction factor of 0.75 and a term $1/(1-qH_2S)$ for the diffusive loss, Vetö et al. (1994) calculated the initial (or original) organic carbon content of a sediment before sulfate reduction as

$$\text{TOC}_{orig} = \text{TOC} + 0.75S \cdot 1/(1-qH_2S) \qquad (4.10)$$

where TOC and S are the measured values of total organic carbon and total sulfur. The authors estimate that the diffusive H_2S loss in non-bioturbated sediments usually is less than 45 % and that this value can only be reached in cases of very high organic matter supply, high reactivity of this organic matter and iron limitation. Littke et al. (1991b) and Lückge et al. (1996) calculated that sulfate reduction consumed between 20 and 50 % of the initially sedimented organic matter (or 1-3 % of primary productivity) both from ancient rocks (Posidonia Shale) and recent sediments (Oman Margin and Peru upwelling systems). In a study of the Pakistan continental margin in the northern Arabian Sea (Littke et al. 1997b), they clearly demonstrated that the carbon-sulfur relationship only holds in the laminated sections of the sediment profile whereas it fails (strongly underestimates sulfate reduction) in the intercalated homogeneous, i.e. bioturbated, sediments for the reason explained before.

4.4.2   Marine *Versus* Terrigenous Organic Matter

As pointed out in Section 4.2.5, even deep-sea sediments deposited in areas remote from continents may contain a mixture of marine and terrigenous organic matter. For any investigation of autochthonous marine organic matter preservation or marine paleoproductivity, these two sources of organic matter have to be distinguished. Furthermore, global or regional climate fluctuations have changed the pattern of continental run-off and ocean currents in the geological past (see Sect. 4.4.3). Being able to recognize variations in marine and terrigenous organic matter proportions may, thus, be of great significance in paleoclimatic and paleoceanographic studies.

A variety of parameters are used to assess organic matter sources. Bulk parameters have the advantage that they are representative of total organic matter, whereas molecular parameters address only part of the extractable organic matter, which in turn is only a small portion of total organic matter. Some successful applications of molecular parameters show that the small bitumen fraction may be representative of the total, but there are many other examples where this is not the case. On the other hand, oxidation of marine organic matter has the same effect on some bulk parameters as an admixture of terrigenous organic matter, because the latter is commonly enriched in oxygen through biosynthesis. It is, therefore, advisable to rely on more than one parameter, and to obtain complementary information.

C/N Ratio

Carbon/nitrogen (C/N) ratios of phytoplankton and zooplankton are around 6, freshly deposited marine organic matter ranges around 10, whereas terrigenous organic matter has C/N ratios of 20 and above (e.g. Meyers 1994, 1997 and references therein). This difference can be ascribed to the absence of cellulose in algae and its abundance in vascular plants and to the fact that algae are instead rich in proteins. Both weight and atomic ratios are used by various authors, but due to the small difference in atomic mass of carbon and nitrogen, absolute numbers of ratios do not deviate greatly.

Selective degradation of organic matter components during early diagenesis has the tendency to modify (usually increase) C/N values already in the water column. Still, these ratios are sometimes sufficiently well preserved in shallow-marine sediments to allow a rough assessment of terrigenous organic matter contribution (e.g. Jasper and Gagosian 1990; Prahl et al. 1994). A different trend exists in deep oceanic sediments with low organic carbon contents. Inorganic nitrogen (ammonia) is released during organic matter decomposition and adsorbed to the mineral matrix (particularly clays) where it adds significantly to the total nitrogen. The C/N ratio is then changed to values below those of normal marine/terrigenous organic matter proportions (Müller 1977; Meyers 1994). This effect should be small in sediments containing more than 0.3 % organic carbon. On the other hand, many sapropels from the eastern Mediterranean Sea and organic-matter-rich sediments underlying upwelling areas have conspicuously high C/N ratios (>15), i.e. well in the range of land plants despite a dominance of marine organic matter, for reasons yet to be determined (see Bouloubassi et al. 1999 for an overview). Because such "atypical" C/N ratios were determined in quite a number of sediments more recently it is advised not to place too much emphasis on the significance of this bulk parameter.

Hydrogen and Oxygen Indices

Hydrogen Index (HI) values from Rock-Eval pyrolysis (see Sect. 4.5.2) below about 150 mg HC/g TOC are typical of terrigenous organic matter, whereas HI values of 300 to 800 mg HC/g TOC are typical of marine organic matter. Deep-sea sediments rich in organic matter usually show values of only 200-400 mg HC/g TOC, even if marine organic matter strongly dominates. Oxidation has lowered the hydrogen content of the organic matter in this case. It should also be mentioned that Rock-Eval pyrolysis was developed as a screening method for rapidly determining the hydrocarbon generation potential of petroleum source rocks (Espitalié et al. 1985) and that a range of complications may occur with sediments buried only to shallow depth. For example, unstable carbonates may decompose below the shut-off temperature of 390 °C (cf. Sect. 4.5.2) which increases the Oxygen Index and falsely indicates a high oxygen content of the organic matter. Furthermore, Rock-Eval pyrolysis cannot be used for sediments with TOC < 0.3 % because of the so-called mineral matrix effect. If sediments with low organic carbon contents are pyrolyzed, a significant amount of the products may be adsorbed to the sediment minerals and are not recorded by the flame ionization detector, thus lowering the Hydrogen Index (Espitalié et al. 1977).

Maceral Composition

If the morphological structure of organic matter is well preserved in sediments, organic petrographic investigation under the microscope is probably the most informative method to distinguish marine and terrestrial organic matter contributions to marine sediments by the relative amounts of macerals (organoclasts) derived from marine biomass and land plants (see Sect. 4.5.3). Many marine sediments, however, contain an abundance of non-structured organic matter (e.g. in upwelling areas; Lückge et al. 1996) which cannot easily be assigned to one source or the other. Furthermore, comprehensive microscopic studies are time-consuming. In his paleoproductivity assessments (see Sect. 4.2), Stein (1986a) calibrated Hydrogen Indices from Rock-Eval pyrolysis of marine sediments with microscopic data and suggested to use the more readily available pyrolysis data as a proxy for marine/terrigenous organic matter proportions.

Stable Carbon and Hydrogen Isotope Ratios

Carbon isotope ratios are principally useful to distinguish between marine and terrestrial organic matter sources in sediments and to identify organic matter from different types of land plants. The stable carbon isotopic composition of organic matter reflects the isotopic composition of the carbon source as well as the discrimination (fractionation) between ^{12}C and

[13]C during photosynthesis (e.g. Hayes 1993). Most plants, including phytoplankton and almost all trees and most shrubs, incorporate carbon into their biomass using the Calvin (C_3) pathway which discriminates against ^{13}C to produce a shift in $\delta^{13}C$ values of about -20 ‰ from the isotope ratio of the inorganic carbon source. Some plants use the Hatch-Slack (C_4) pathway (many subtropical savannah grasses and sedges) which leads to an isotope shift of about -7 ‰. Other plants, mostly succulents, utilize the CAM (crassulacean acid metabolism) pathway, which more or less switches between the C_3 and C_4 pathways and causes the $\delta^{13}C$ values to depend on the growth dynamics.

Organic matter produced from atmospheric carbon dioxide ($\delta^{13}C \approx$ -7 ‰) by land plants using the C_3 pathway has an average $\delta^{13}C$ value of approximately -27 ‰ and by those using the C_4 pathway approximately -14 ‰. Marine algae use dissolved bicarbonate, which has a $\delta^{13}C$ value of approximately 0 ‰. As a consequence, marine organic matter typically has $\delta^{13}C$ values varying between -20 ‰ and -22 ‰. Isotopic fractionation, among others, is also temperature dependent which, e.g., in cold polar waters may lead to carbon isotope values for marine organic matter of -26‰ or lower (e.g. Rau et al. 1991). The fact that a number of environmental and biological variables

influence the stable carbon isotope composition of plants may cause variations from the generalized numbers given above (see Killops and Killops 2005 for an overview). The 'typical' difference of about 7 ‰ between organic matter of marine primary producers and land plants has nevertheless been successfully used to trace the sources and distributions of organic matter in coastal oceanic sediments (e.g. Westerhausen et al. 1993; Prahl et al. 1994; Rommerskirchen et al. 2003 and references therein). Figure 4.15 shows data for a continental margin setting with the Congo Fan as an example. The carbon isotope ratios are measured values compiled from various sources and show the pathway from the different biological systems to the mixed data in the sediments in a transect from the shallow continental margin to the deep ocean.

The availability of dissolved CO_2 in ocean water has an influence on the carbon isotopic composition of algal organic matter because isotopic discrimination toward ^{12}C increases when the partial pressure of carbon dioxide (pCO_2) is high and decreases when it is low (see Fogel and Cifuentes 1993 for an overview). Organic-matter $\delta^{13}C$ values, therefore, become indicators not only of origins of organic matter but also of changing paleoenvironmental conditions on both short- and long-term scales. For example, the $\delta^{13}C$ values

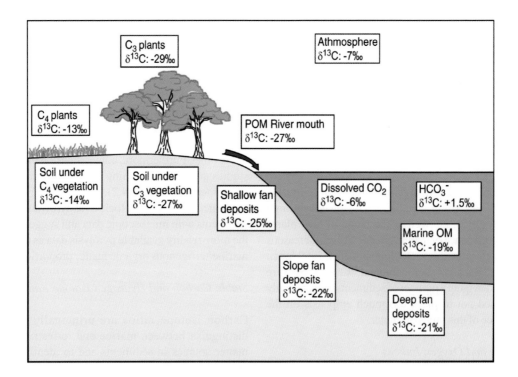

Fig. 4.15 Carbon stable isotope data in different compartments of a continental margin settings with the Congo Fan as an example. Data for $\delta^{13}C$ values of biota and sediments are from measurements in the Congo Fan and the Congo River catchment area (Mariotti et al. 1991, Müller et al. 1994, Muzuka 1999, Schwartz et al. 1986, Westerhausen et al. 1993).

of dissolved inorganic carbon (DIC; CO_2, bicarbonate, carbonate) available for photosynthesis in the cells varies over the year with the balance between photosynthetic uptake and respiratory production. During summer and spring, when rates of photosynthesis are high, the isotope ratio of remaining DIC is enriched in ^{13}C. In fall, when respiration is the dominant process, the $\delta^{13}C$ of DIC becomes more negative because organic matter is remineralized.

Fluctuations that have been measured in the $\delta^{13}C$ values of sedimentary organic matter over the Earth's history (e.g. Schidlowski 1988) can thus be interpreted in terms of the productivity in the water column and the availability of DIC in a particular geological time period. In a study of sediments from the central equatorial Pacific Ocean spanning the last 255,000 years it has been demonstrated that the carbon isotopic composition of fossil organic matter depends on the exchange between atmospheric and oceanic CO_2. Changes with time can then be used to estimate past atmospheric carbon dioxide concentrations (Jasper et al. 1994).

Other than stable carbon isotopes the stable hydrogen isotopes do not reflect the fixation processes in photosynthesis, but rather the hydrological cycle under which the respective plants have grown (Sessions et al. 1999; Sauer et al. 2001). This means that lake biota will differ in hydrogen isotope composition from marine organisms, and land plants thriving under humid or arid conditions will show a large spread in hydrogen isotope ratios. This is schematically illustrated in the $\delta^{13}C$ vs. δ^2H diagram in Figure 4.16 where C_4 and CAM plants are clearly separated from each other, which is often not the case when only carbon isotope ratios are determined. Due to the large atomic weight difference between hydrogen and deuterium (^2H), δ^2H values extend over a much larger range than $\delta^{13}C$ values, and the former may require specific mathematical algorithms to calculate hydrogen isotopic fractionations in biogeochemical systems (Sessions and Hayes 2005).

Carbon fixation during photosynthesis is not the only process causing isotope fractionation. Further fractionation occurs during the subsequent biosynthetic reactions leading to a deviation of the carbon isotopic signature of important groups of natural products from the bulk isotopic value of the organism. Lipids have the lightest $\delta^{13}C$ values, whereas proteins and carbohydrates are significantly heavier (more enriched in ^{13}C). Even within the lipids, e.g., compound groups may differ in their carbon isotope ratios as a consequence of their individual biosynthetic pathways. Still, the general difference in carbon isotope signatures

between plants of different origins is largely maintained. In a diagram like that in Figure 4.16, $\delta^{13}C$ values of individual lipids would just be collectively shifted to lighter values on the x-axis compared to the total carbon values of their source organisms.

Compound-specific isotope analysis allows stable carbon (and hydrogen) isotope analysis of individual biomarkers for the assessment of organic matter sources and paleoenvironmental reconstruction (e.g. Hayes et al. 1987, 1990). Since the implementation of the new analytical technique in the last 15 years, numerous applications have been published. Examples are the analysis of plant wax lipids in Atlantic Ocean deep sea sediments to reconstruct organic matter flux from the African continent and the effect of climatic change on the vegetation in Africa (e.g., Huang et al. 2000; Rommerskirchen et al. 2003; Schefuß et al. 2004). Molecular carbon isotope analysis has recently become particularly important in the investigation of the methane cycle. Biogenic methane produced by archaea is isotopically particularly light ($\delta^{13}C = -60$ to -100 ‰). Methanotrophic organisms using this biogenic methane as their carbon source biosynthesize, among others, membrane lipids which may isotopically be as light as -120 ‰. The combination of molecular and isotope analysis has, thus, helped a great deal in unraveling the complex processes particularly involved in anaero-

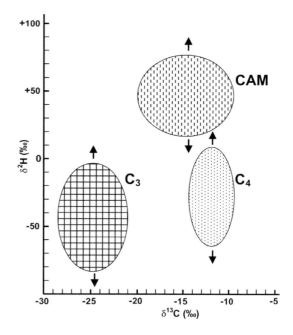

Fig. 4.16 Schematic representation of carbon and hydrogen stable isotope ratios of plants with different biosynthetic pathways. See text for more details. Arrows indicate that hydrogen isotope ratios may cover an even larger range than indicated by the encircled areas.

bic methane oxidation (e.g. Orphan et al. 2001, 2002; Elvert et al. 2003; Wakeham et al. 2003).

Marine sediments may not only be a mixture from autochthonous and allochthonous sources, the different biogenic components may also be of different age. Whereas mixing by redeposition of continental slope sediments by slumping or turbidite flow are well known and often recognizable on a bulk level, there may also be more intimate mixing of components of diverse age when, e.g., terrestrial organic matter is transported to the ocean through rivers taking a significant amount of time. Radiocarbon analysis of individual biomarkers has revealed that their ^{14}C ages can differ significantly within a given sediment sample, but some of these differences are not easy to explain (e.g. Eglinton et al. 1997; Pearson et al. 2001). Mollenhauer et al. (2003) compared the ^{14}C ages of carbonate frustrules of planktonic foraminifera with those of long-chain alkenones (cf. Sect. 4.4.3), both widely used for reconstructing the chronostratigraphy of near-surface marine sediments, in hemipelagic muds from the Namibia continental margin in the South Atlantic Ocean. They found the alkenones to be systematically and up to 2000 years older than the foraminifera. Among several possible alternative explanations for this discrepancy they favored long-range transport of the alkenones, possibly from the Argentine basin across the Atlantic Ocean, before they were ultimately deposited together with the autochthonous, younger foraminifera.

4.4.3 Molecular Paleo-Seawater Temperature and Climate Indicators

Past Sea-Surface Temperatures (SST) Based on Long-Chain Alkenones

Paleoceanographic studies have taken advantage of the fact that biosynthesis of a major family of organic compounds by certain microalgae depends on the water temperature during growth. The microalgae belong to the class of *Haptophyceae* (often also named *Prymnesiophyceae*) and notably comprise the marine coccolithophorids *Emiliania huxleyi* and *Gephyrocapsa oceanica*. The whole family of compounds, which are found in marine sediments of Recent to Lower Cretaceous age throughout the world ocean, is a complex assemblage of aliphatic straight-chain ketones and esters with 37 to 41 carbon atoms and two to four double bonds (see Brassell 1993 and Brassell et al. 2004 for more details). Principally only the C_{37} methylketones with 2 and 3 double bonds are used for past sea-surface temperature assessment (Fig. 4.17), although the relationship of the tetra-unsaturated C_{37} alkenone, which is commonly more abundant in lacustrine systems, to salinity and temperature was recently given special attention (Sikes and Sicre 2002).

It was found from the analysis of laboratory cultures and field samples that the extent of unsaturation (number of double bonds) in these long-chain ketones varies linearly with growth temperature of the algae over a wide temperature range (Brassell et al. 1986; Prahl and Wakeham 1987). To describe this, an unsaturation index was suggested, which in its simplified form is defined by the concentration ratio of the two C_{37} ketones:

$$U_{37}^{K'} = [C_{37:2}]/[C_{37:2} + C_{37:3}] \qquad (4.11)$$

Calibration was then made with the growth temperatures of laboratory cultures of different haptophyte species and with ocean water temperatures at which plankton samples had been collected. From these data sets, a number of different calibration curves evolved for different species and different parts of the world ocean so that some doubts arose as to the universal applicability of the unsaturation index. In a major analytical effort, Müller et al. (1998) resolved the complications and arrived at a uniform calibration for the global ocean from 60°N to 60°S. The resulting relationship,

$C_{37:2}$ heptatriaconta-15E,22E-dien-2-one

$C_{37:3}$ heptatriaconta-8E,15E,22E-trien-2-one

Fig. 4.17 Structural formulae of long-chain alkenones used for paleo-sea surface temperature assessment.

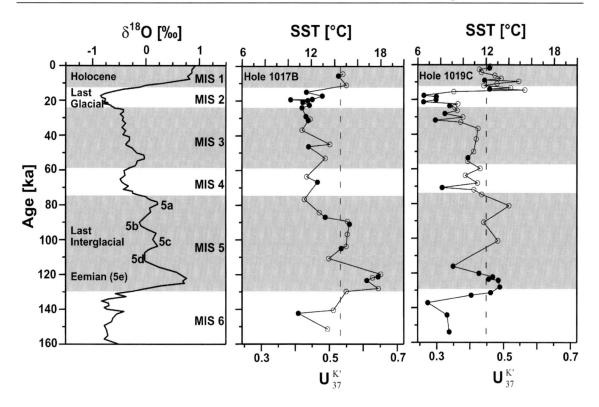

Fig. 4.18 Alkenone-based sea-surface temperature (SST) reconstruction of two sediment sections from Ocean Drilling Program (ODP) Holes 1017B (Southern California, 50 km west of Point Arguello) and 1019C (Northern California, 60 km west of Crescent City in the Eel River Basin) on the California continental margin (after Mangelsdorf et al. 2000) compared to the global $\delta^{18}O$ chronostratigraphy (SPECMAP; after Martinson et al. 1987). MIS = Marine isotope stage.

$$U^{K'}_{37} = 0.033T + 0.044 \qquad (4.12)$$

is identical within error limits with the widely used calibrations of Prahl and Wakeham (1987) and Prahl et al. (1988) based on *Emiliania huxleyi* cultures ($U^{K'}_{37} = 0.033T + 0.043$). Müller et al. (1998) also found that the best correlations were obtained using ocean water temperatures from 0 to 10 m water depth, suggesting that the sedimentary $U^{K'}_{37}$ ratio reflects mixed-layer temperatures and that the production of alkenones within or below the thermocline was not high enough to significantly bias the mixed-layer temperature signal. Regional variations in the seasonality of primary production also have only a negligible effect on the U signal in the sediments. Furthermore, the strong linear relationships obtained for the South Atlantic Ocean and the global ocean indicate that U values of the sediments are neither affected to a measurable degree by changing species compositions nor by growth rate of algae and nutrient availability, other than expected from culture experiments. Significant effort was undertaken to intercalibrate the analysis of the alkenone parameter world-wide (Rosell-

Melé et al. 2001). Continuing research on the alkenone parameter, however, still reveals additional calibrations for extreme environments (e.g. Sicre et al. 2002) or the temperature range over which the alkenone index can be applied (Pelejero and Calvo 2003).

The alkenone index presently is one of the most frequently used molecular organic geochemical parameters in the geosciences. A simple example of an application is shown in Figure 4.18. The global oxygen isotope stratigraphy (SPECMAP; after Martinson et al. 1987) is compared with the sea-surface temperature curve reconstructed from the sedimentary alkenone ratio ($U^{K'}_{37}$) for two holes drilled by the Ocean Drilling Program into deep-sea sediments off the coast of California (Mangelsdorf et al. 2000). The temperature follows the variations of the $\delta^{18}O$ values reasonably well. They both illustrate the repetitive change from cold to warm climatic stages and *vice versa*. It is also noteworthy that the absolute temperatures, as expected from the oceanographic setting, are higher at the location of Hole 1017B (50 km west of Point Arguello, just north of the Santa Barbara Channel) than at the Northern Californian location 60 km west of Crescent City in the Eel River basin (Hole 1019C).

155

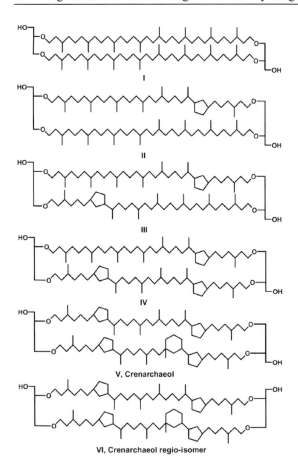

Fig. 4.19 Crenarchaeotal membrane lipid moieties used for TEX_{86} paleothermometry (after Schouten et al. 2002).

TEX$_{86}$ Paleothermometry

Recently, a new geochemical temperature proxy, the TEX_{86} index, was introduced by Schouten et al. (2002). It is based on the number of cyclopentane moieties in the glycerol dialkyl tetraethers (GDGTs; Fig. 4.19) of the membrane lipids of a group of marine archaea called Crenarchaeota, which changes as a response to temperature in the growth medium as demonstrated both by core top samples of marine sediments (Schouten et al. 2002) and in culture experiments (Wuchter et al. 2004). It is defined as follows:

$$TEX_{86} = (III + IV + VI)/(II + III + IV + VI) \quad (4.13)$$

where the Roman numbers represent the relative contents of four different membrane lipids (Fig. 4.19). With increasing temperature, the number of cyclopentane rings in the Crenarchaeotal membrane lipids increases. The TEX_{86} index showed a significant linear

correlation with growth temperature in the incubation experiments (Eq. 4.14) and with the annual mean sea-surface temperature in the surface sediments (Eq. 4.15).

$$TEX_{86} = 0.015 \cdot T + 0.10 \quad (r^2 = 0.79) \quad (4.14)$$
$$TEX_{86} = 0.015 \cdot T + 0.29 \quad (r^2 = 0.92). \quad (4.15)$$

Both equations exhibit identical slopes but differ slightly in the intercept with the y-axis. Two additional aspects make this new proxy parameter particularly promising. The temperature range (0-35 °C) for the TEX_{86} appears to be wider than for the alkenone index (5-28 °C), and the Crenarchaeota are likely to significantly predate the first occurrence of the alkenone-producing haptophytes in Earth history.

ACL Index Based on Land Plant Wax Alkanes

In marine sediments, higher-plant organic matter can be an indicator of climate variations both by the total amount indicating enhanced continental run-off during times of low sea level or of humid climate on the continent and by specific marker compounds indicating a change in terrestrial vegetation as a consequence of regional or global climatic variations. Long-chain n-alkanes are commonly used as the most stable and significant biological markers of terrigenous organic matter supply (e.g. Eglinton and Hamilton 1967). The odd carbon-numbered C_{27}, C_{29}, C_{31} and C_{33} n-alkanes are major components of the epicuticular waxes of higher plants. These terrestrial biomarkers are often preferentially enriched in the marine environment, particularly under oligotrophic surface waters, because the compounds are protected by the resistant character of the plant particles and in part by the highly water-insoluble nature of the waxes themselves (Kolattukudy 1976).

The carbon number distribution patterns of n-alkanes in leaf waxes of higher land plants depend on the climate under which they grow. The distributions show a trend of increasing chain length nearer to the equator, i.e. at lower latitude (Gagosian et al. 1987), but they are also influenced by humidity (Hinrichs et al. 1998). In addition, waxes of tree leaves have molecular distributions different from those of grasses with either the C_{27} or the C_{29} n-alkane having the highest relative concentration in trees and the C_{31} n-alkane in grasses (Cranwell 1973). Poynter (1989) defined an Average Chain Length (ACL) index to describe the chain length variations of n-alkanes,

$$ACL_{27-33} = (27[C_{27}] + 29[C_{29}] + 31[C_{31}] + 33[C_{33}])/$$
$$([C_{27}] + [C_{29}] + [C_{31}] + [C_{33}]) \quad (4.16)$$

in which $[C_x]$ signifies the content of the *n*-alkane with x carbon atoms. Poynter (1989) demonstrated the sensitivity of sedimentary *n*-alkane ACL values to past climatic changes. In Santa Barbara basin (offshore California) sediments from the last 160,000 years, Hinrichs et al. (1998) found the highest ACL values in the Eemian climate optimum (about 125,000 yr B.P.). Over the entire sediment section, the ACL values were higher in homogeneous sediment layers deposited in periods of more humid climate than in laminated sediments deposited under a semi-arid continental climate like that of today (Fig. 4.20). The sedimentary ACL variations most probably recorded the climatic changes on the continent for the following two reasons:

- Vegetation patterns on the continent responded rapidly to climatic oscillations, which were often characterized by drastic changes of temperature and precipitation. During relatively warm and dry periods when mainly laminated sediments were accumulated in the Santa Barbara basin, smaller proportions of grass-derived biomass may have contributed to the sedimentary organic matter.

- Changes in continental precipitation significantly affected the degree of erosion and the transport of terrigenous detritus to the ocean, enhancing the proportion of biomass from other source areas (probably over longer distances from higher altitudes). This explains best the almost parallel and abrupt changes of oceanic conditions (e.g. bottom-water oxygen concentrations affecting sediment texture) and terrestrial signals recorded in the sediments (e.g. ACL indices).

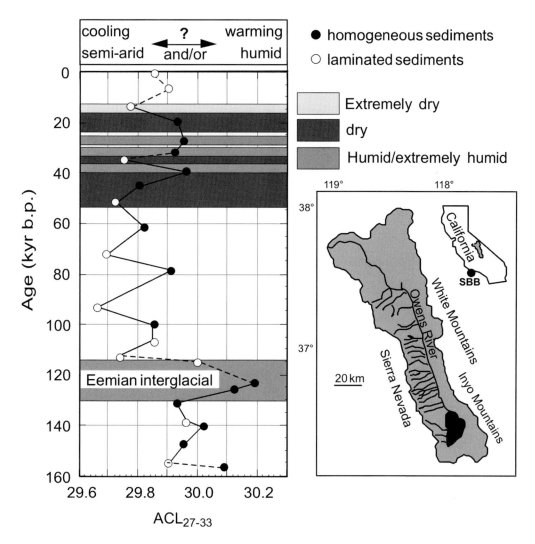

Fig. 4.20   Average chain length (ACL) time-series in relation to sediment texture (homogeneous versus laminated) for a core in the central Santa Barbara basin offshore California (after Hinrichs et al. 1998). Information on continental humidity from a sediment core in the Owens Lake basin was taken from Benson et al. (1996).

4.5   Analytical Techniques

Organic geochemical analyses of marine sediments may range from the rapid determination of a few bulk parameters to a high level of sophistication if trace organic constituents are to be investigated at the molecular level. The analytical scheme in Figure 4.21 is one of many that are currently being used in different laboratories. It is certainly not complete nor is it fully applied to each sample, particularly not when large series are studied. For example, the analysis of amino acids (e.g. Mitterer 1993), sugars (Cowie and Hedges 1984; Moers et al. 1993), lignin (Goñi and Hedges 1992), or humic substances (Brüchert 1998; Senesi and Miano 1994; Rashid 1985) requires a modification of the scheme in Figure 4.21. Other than in inorganic geochemical analysis, where modern instrumentation allows the simultaneous determination of a wide range of element concentrations,

the millions of possible organic compounds require an *a priori* selectivity, but even the analysis of selected lipids can be quite time-consuming. In addition, the amount of sample material required for molecular organic geochemical studies is higher than for many other types of analyses, and this limits stratigraphic (time) resolution.

4.5.1   Sample Requirements

Most organic geochemical methods require a well homogenized, pulverized sample aliquot. An exception is reflected-light microscopy in organic petrography where the association of organic matter with the sediment matrix can be quite informative and is, therefore, preserved. Before homogenization, the sample needs drying either at ambient temperature or by freeze-drying. Higher temperatures are to be avoided due to the thermal lability of the organic matter. For the same reason, sediments – particularly those of young

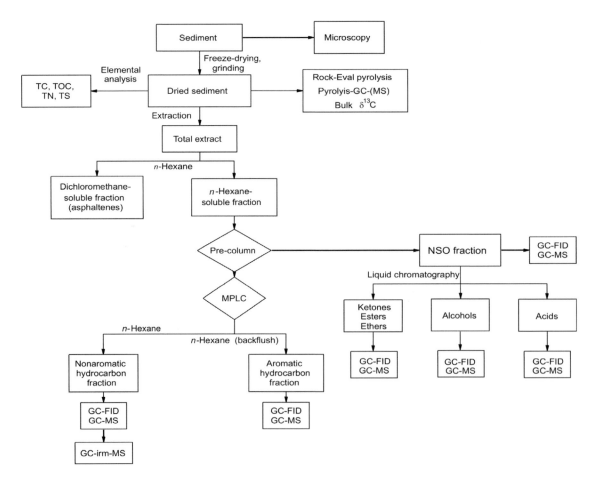

Fig. 4.21   Example of an analytical scheme of organic geochemical analyses of marine sediments comprising bulk and molecular parameters.

age in which bacteria may still be active – should be stored deep-frozen (-18°C or lower) between sampling and analysis. Grinding can be done by mortar and pestle or in an electrical ball or disc mill, but excessive grinding should be avoided due to the associated rise in temperature in the sample.

4.5.2 Elemental and Bulk Isotope Analysis

The basic parameter determined in most organic geochemical studies is the total organic carbon (TOC, C_{org}) content. Most marine sediments and sedimentary rocks contain carbon both as carbonates (C_{inorg}, C_{carb}, C_{min}) and as organic matter. There are numerous methods for quantifying carbon, most of them are based on heating solid samples in an oxygen atmosphere and detection of the evolving CO_2 by coulometric or spectrometric techniques or by a thermal conductivity detector. Commonly used instruments are elemental analyzers, which determine carbon, nitrogen, hydrogen (only applicable to pure organic matter), and sulfur (CHN, CNS, CS analyzers). Organic carbon is either determined directly, after destruction of carbonate with mineral acids before combustion in the elemental analyzer, or as the difference between total carbon (combustion) and mineral carbon (measurement of CO_2 released upon acid treatment).

For the determination of bulk stable carbon isotope ratios ($^{13}C/^{12}C$) the organic matter is converted to carbon dioxide by oxidation following digestion of the sediment with mineral acid to remove carbonates. Traditionally, oxidation of organic matter was performed off-line, CO_2 was separated from other gaseous oxidation products, and the purified gas introduced into an isotope ratio mass spectrometer. Modern instruments provide on-line combustion isotope-ratio measurement facilities. In this case, an elemental analyzer is connected to an isotope ratio mass spectrometer *via* a special interface that allows removal of gases other than CO_2. This configuration can also be used to separate sulfur and nitrogen oxides which, after on-line conversion to a suitable single species (SO_2 and N_2, respectively), can be used to determine stable sulfur ($^{34}S/^{32}S$) and nitrogen ($^{15}N/^{14}N$) isotope ratios. A special technical configuration of the mass spectrometer is required for hydrogen isotope ($^2H/^1H$) ratio measurement. Isotope ratios are not determined directly, but relative to an internationally accepted standard. The results are reported in the delta notation ($\delta^{13}C$, $\delta^{34}S$, $\delta^{15}N$, δ^2H [or commonly δD for deuterium]) relative to this standard. For details see, e.g., Fogel and Cifuentes (1993) and references therein.

4.5.3 Rock-Eval Pyrolysis and Pyrolysis Gas Chromatography

Rock-Eval pyrolysis (Espitalié et al. 1985) is conducted using bulk sediment samples to determine, (1) the amount of hydrocarbon-type compounds already present in the sample (S1 peak [mg hydrocarbons per g sediment]; compounds released at low temperature and roughly equivalent to the amount of organic matter extractable with organic solvents), (2) the amount of hydrocarbon-type compounds generated by pyrolytic degradation of the macromolecular organic matter during heating up to 550°C (S2 peak [mg hydrocarbons per g sediment]), (3) the amount of carbon dioxide released from the organic matter up to 390°C, i.e. before carbonates decompose (S3 peak [mg CO_2 per gram sediment]), and (4) the temperature of maximum pyrolysis yield (Tmax [°C]). The Hydrogen Index (HI) and Oxygen Index (OI) derived from the S2 and S3 values correspond to the pyrolysis yield normalized to the content of organic carbon (mg hydrocarbons and mg CO_2 per g TOC, respectively). The results of Rock-Eval pyrolysis are usually displayed in a van-Krevelen-type diagram of HI *versus* OI values which roughly corresponds to an H/C versus O/C atomic ratio van Krevelen diagram (see Fig. 4.11; Tissot and Welte 1984).

One of the methods of studying the composition of macromolecular sedimentary organic matter in more detail is the molecular analysis of pyrolysis products. For this purpose, the pyrolysis products are transferred to a gas chromatographic column and analyzed as described for extractable organic matter in Sect. 4.5.5, with or without the combination with a mass spectrometer. Both flash pyrolysis (Curie-point pyrolysis; samples are heated on a magnetic wire by electrical induction almost instantaneously, e.g., to 610°C) or off-line pyrolysis at various heating rates have been applied to geological samples (see Larter and Horsfield 1993 for an overview of various pyrolysis techniques).

4.5.4 Organic Petrography

Organic petrography is the study of the macroscopically and, more importantly, microscopically recognizable organic matter components initially of coal, but meanwhile more generally in sediments and sedimentary rocks. Organic-matter-rich rocks and coal are usually studied as polished blocks in reflected light under a petrographic microscope. When sediments are lean in organic matter, the organic particles have to be concentrated by dissolution of the mineral matrix in consecutive treatments with hydrochloric and hydrofluoric acid. The concentrates are then analyzed as smear slides in transmitted light or embedded in araldite® resin and subsequently studied as polished blocks similar to whole-rock samples.

Organic particles visible under the microscope (>1 μm) are called macerals. The most important groups in the order of increasing reflectance are liptinite, vitrinite and inertinite. Liptinites are lipid-rich parts of aquatic (e.g. alginite) or land plants (e.g. cutinite, suberinite, sporinite, resinite), the terms indicating the origin of these organoclasts. Many liptinites are probably related to nonhydrolyzable, highly aliphatic biopolymers found in algae (algaenan) and land plants (cutinan, suberan) (see Sect. 4.3.3). These biopolymers serve as cell wall components of the organisms and their stability allows the morphological shapes of plant material to be preserved after sedimentation and burial so that they can be identified under the microscope. Vitrinites derive from the woody parts of higher plants. Inertinites are highly reflecting particles of strongly oxidized or geothermally heated organic matter of various origin, most commonly from higher plants. Non-structured, often (incorrectly) called amorphous, organic matter is known as bituminite or sapropelinite. Lipid-rich organic matter, even if finely dispersed, can be recognized under the microscope after UV irradiation by its bright fluorescence.

In addition to maceral distribution, organic petrographers determine vitrinite reflectance as a measure of geothermal evolution of sedimentary organic matter. For more details of microscopic analysis see Taylor et al. (1998).

4.5.5   Bitumen Analysis

The larger part of sedimentary organic matter is insoluble in organic solvents. The proportion of the soluble fraction (bitumen) can be relatively high in surficial sediments, then decreases in amount with increasing depth of burial due to formation of humic substances and kerogen. It only increases again when temperatures become high enough for thermal kerogen cracking to generate petroleum (Tissot and Welte 1984).

The most common solvent used for extraction of bitumen from dried sediments is dichloromethane (CH_2Cl_2) with a small admixture (e.g. 1%) of methanol,

although more polar mixtures like chloroform/methanol or chloroform/toluene are also used. Occasionally, when very polar lipids from surficial sediments are to be extracted, wet sediment samples are preferred, and extraction starts with acetone or methanol or a mixture of these two because they mix with water. Extraction is then repeated with dichloromethane or a solvent of similar polarity. Extraction in a Soxhlet apparatus usually takes one to two days, whereas reasonably complete extracts can be obtained within minutes with the support of ultrasonication or blending. After filtration, the solvent is removed by rotary evaporation and the total extract yield determined gravimetrically, as is commonly done for any subfraction after further separation. Internal standards for quantitation of single compounds are added either before or after extraction, occasionally only after liquid chromatographic separation (see later).

The most polar and highest-molecular-weight extract components (called asphaltenes, a term derived from petroleum geochemistry, but often also applied to surficial sediments or even biological material) may interfere with many subsequent separations and analyses. For this reason, these components are frequently removed from the total extract by dissolving it in a small amount of, e.g., dichloromethane and adding a large excess of a nonpolar solvent like n-hexane (or n-pentane, n-heptane). The n-hexane-soluble extract fraction can be further separated into compound classes (or fractions of similar polarities) either by column liquid chromatography, medium-pressure liquid chromatography (MPLC: Radke et al. 1980; HMPLC: Willsch et al. 1997), high-performance liquid chromatography (HPLC), or thin-layer chromatography (TLC), depending on the sample quantity used and the sophistication of the separation required. In MPLC separation, as indicated in Figure 4.21, the polar lipids are withheld by a pre-column filled with deactivated silica gel, and only the hydrocarbons are separated on the main silica gel column into nonaromatic and aromatic hydrocarbon fractions. Because the main separation system is only operated with n-hexane, the difficult separation between the two hydrocarbon fractions is very reproducible, an important prerequisite for some applications, particularly in petroleum geochemistry. The polar lipids can be removed from the pre-column with a polar solvent. If emphasis is placed on the subfractionation of the polar heterocompounds, then HMPLC may be the method of choice. There are, however, many variations of this separation scheme. The nature of the geological samples and the scientific objectives will determine the type and extent of separation required.

Total extracts and/or liquid chromatographic subfractions are then analyzed by capillary column gas chromatography using a flame ionization detector. Except for hydrocarbon fractions, derivatization is commonly applied to render polar lipids more volatile in order to reduce gas chromatographic retention times and to improve peak shape at the detector. Carboxylic acids are usually transformed into their methyl esters, and hydroxyl or amine groups into their trimethylsilyl ether derivatives. Alternatively, both acid and hydroxyl groups can be silylated. Acetate formation is another common derivatization method. A variety of derivatization reagents are commercially available for this purpose.

Only a few major compound series can be recognized at the level of their molecular structures based on relative retention times and distribution patterns by gas chromatography alone. This applies to *n*-alkanes in the nonaromatic hydrocarbon fraction, *n*-fatty acids in the carboxylic acid fraction and in some cases *n*-alkanols in the neutral polar fraction. High abundance of a few single compounds (pristane, phytane, long-chain alkenones) sometimes also allow their direct identification from gas chromatograms.

The most powerful technique for assigning molecular structures to constituents of complex mixtures as they are found in the lipid extracts of geological samples is the combination of capillary column gas chromatography and mass spectrometry (GC-MS). Although the expression "identification" is frequently used, GC-MS alone is insufficient to fully characterize a new compound whose gas chromatographic and mass spectrometric behavior has not been described before. Normally, GC-MS analysis relies on a comparison with GC and MS data published in the (geochemical) literature or with data of standards, commercially available or synthesized in the laboratory, or on the interpretation of mass spectral fragmentation patterns following common empirical rules (e.g. McLafferty and Turecek 1993). Unfortunately, there is no comprehensive compilation of mass spectra of geochemically relevant organic compounds, although a significant number of spectra was recently published by de Leeuw (2004). Peters et al. (2005) have provided a detailed coverage of hydrocarbons and selected polar compounds of significance in petroleum geochemistry and other fields of organic geochemistry.

The youngest, revolutionary development in analytical organic geochemistry is the on-line coupling of a gas chromatograph to an isotope ratio mass spectrometer via a combustion interface (GC-irm-MS; Hayes et al. 1990; Freeman et al. 1994). This instrument allows the determination of stable carbon and hydrogen isotope ratios of single organic compounds in complex mixtures provided they are gas chromatographically reasonably well separated. Chemotaxonomic relations to specific precursor organisms are then possible if these are distinct from other organisms in their carbon isotope fractionation behavior during photosynthesis and biosynthesis. Sample preparation and analysis are similar to those for GC-MS analysis with the provision that the isotopic composition of derivatizing agents is accounted for in data interpretation.

4.6 The Future of Marine Geochemistry of Organic Matter

The evolution of organic geochemistry has always been closely connected to the developments in instrumental techniques for the analysis of organic compounds in geological samples. Now that the instrumentation has reached a high level of sophistication and particularly sensitivity, one of the future targets will certainly be higher stratigraphic (time) resolution which is particularly important for climate research. Advancement in the fundamental understanding particularly of the early part of the geological organic carbon cycle will depend on the cooperation between organic geochemists and microbiologists. They will have to refine the knowledge of the biological effects on the early diagenesis of organic matter arriving at the sediment-water interface and becoming buried in the uppermost sediment. It will be necessary to broaden the natural product inventory of microorganisms, both of sedimentary bacteria and archaea and of unicellular algae, protozoans and other organisms at the lower end of the food chain in order to arrive at solid chemotaxonomic relationships between source organisms and molecular fossils.

Carefully designed laboratory simulation experiments together with high-resolution field studies will refine the mass balance approaches of organic matter exchange between sediment and water column related to early diagenetic processes. Finally, mathematical modeling of transport and reaction processes will become an increasingly important tool in marine geochemistry of organic matter.

Acknowledgment

Dr. Rüdiger Stein (Alfred Wegener Institute of Polar and Marine Research, Bremerhaven, Germany) and Dr. Ute Güntner (ICBM, University of Oldenburg) kindly provided material for preparation of figures. Dörte Gramberg, Sabine Henne, Kai Mangelsdorf, Joachim Rinna, Elke Spannhacke, Adelheid Wegner-Demmer and Holger Winkler (ICBM, University of Oldenburg, Germany) were helpful in drafting the figures.

4.7   Problems

Problem 1

What are the factors influencing organic matter preservation in marine sediments?

Problem 2

What are the main electron acceptors in the metabolization of organic matter during early diagenesis?

Problem 3

What is meant with selective preservation of organic matter?

Problem 4

What is the biological marker concept?

Problem 5

Can you name a few applications of (molecular) organic geochemistry?

References

Alperin, M.J., Reeburgh, W.S. and Devol, A.H., 1992. Organic carbon remineralization and preservation in sediments of Scan Bay, Alaska. In: Whelan, J.K. and Farrington, J.W. (eds), Organic matter: Productivity, accumulation and preservation in recent and ancient sediments. Columbia University Press, NY, pp. 99-122.

Bailey, G.W., 1991. Organic carbon flux and development of oxygen deficiency on the northern Benguela continental shelf south of 22°S: Spatial and temporal variability. In: Tyson, R.V. and Pearson, T.H. (eds), Modern and ancient continental shelf anoxia. Geol. Soc. Spec. Publ., 58, Blackwell, Oxford, pp. 171-183.

Benson, L.V., Burdett, J.W., Kashgarian, M., Lund, S.P., Phillips, F.M. and Rye, R.O., 1996. Climatic and hydrologic oscillations in the Owens Lake basin and adjacent Sierra Nevada, California. Science, 274: 746-749.

Berger, W.H., 1989. Global maps of ocean productivity. In: Berger, W.H., Smetacek, V.S. and Wefer, G. (eds), Productivity of the ocean: Present and past. Dahlem Workshop Rep., Life Sci. Res. Rep., 44, Wiley, Chichester, pp. 429-456.

Berger, W.H., Smetacek, V.S. and Wefer, G. (eds) 1989. Productivity of the ocean: Present and past. Dahlem Workshop Rep., Life Sci .Res. Rep., 44. Wiley, Chichester, 471 pp.

Berner, R. and Raiswell, R., 1983. Burial of organic carbon and pyrite sulphur in sediments over Phanaerozoic time: A new theory. Geochim. Cosmochim. Acta, 47: 855-862.

Betzer, P.R., Showers, W.J., Laws, E.A., Winn, C.D., Ditullo, G.R. and Kroopnick, P.M., 1984. Primary productivity and particle fluxes on a transect of the equator at 153°W in the Pacific Ocean. Deep-Sea Res., 31: 1-11.

Blokker, P., Schouten, S., van den Ende, H., de Leeuw, J.W., Hatcher, P.G. and Sinninghe Damsté, J.S., 1998. Chemical structure of algaenans from the fresh water algae *Tetraedron minimum, Scenedesmus communis and Pediastrum boryanum*. Org. Geochem., 29: 1453-1468.

Blokker, P., Schouten, S., de Leeuw, J.W., Sinninghe Damsté, J.S. and van den Ende, H., 2000. A comparative study of fossil and extant algaenans using ruthenium tetroxide degradation. Geochim. Cosmochim. Acta, 64: 2055-2065.

Bouloubassi, I., Rullkötter, J. and Meyers, P.A., 1999. Origin and transformation of organic matter in Pliocene-Pleistocene Mediterranean sapropels: Organic geochemical evidence reviewed. Mar. Geol. 153: 177-199.

Bralower, T.J. and Thierstein, H.R., 1987. Organic carbon and metal accumulation in Holocene and mid-Cretaceous marine sediments: Paleoceanographic significance. In: Brooks, J. and Fleet, A.J. (eds), Marine petroleum source rocks. Geol. Soc. Spec. Publ., 26, Blackwell, Oxford, pp. 345-369.

Brassell, S.C., Eglinton, G., Marlowe, I.T., Pflaumann, U. and Sarnthein, M., 1986. Molecular stratigraphy: A new tool for climatic assessment. Nature, 320: 129-133.

Brassell, S.C., 1993. Applications of biomarkers for delineating marine paleoclimatic fluctuations during the Pleistocene. In: Engel, M.H. and Macko, S.A. (eds), Organic geochemistry. Principles and applications. Plenum Press, NY, pp. 699-738.

Brassell, S.C., Dumitrescu, M. and the ODP Leg 198 Shipboard Scientific Party, 2004. Recognition of alkenones in a lower

Aptian porcellanite from the west-central Pacific. Org. Geochem., 35: 181-188.

Brüchert, V., 1998. Early diagenesis of sulfur in estuarine sediments: The role of sedimentary humic and fulvic acids. Geochim. Cosmochim. Acta, 62: 1567-1586.

Calvert, S.E., 1987. Oceanic controls on the accumulation of organic matter in marine sediments. In: Brooks, J. and Fleet, A.J. (eds), Marine petroleum source rocks. Geol. Soc. Spec. Publ., 26, Blackwell, Oxford, pp. 137-151.

Canuel, E.A. and Martens, C.S., 1996. Reactivity of recently deposited organic matter: Degradation of lipid compounds near the sediment-water interface. Geochim. Cosmochim. Acta, 60: 1793-1806.

Collins, M.J., Bishop, A.N. and Farrimond, P., 1995. Sorption by mineral surfaces: Rebirth of the classical condensation pathway for kerogen formation? Geochim. Cosmochim. Acta, 59: 2387-2391.

Coolen, M.J.L., Muyzer, G., Rijpstra, W.I.C., Schouten, S., Volkman, J.K. and Sinninghe Damsté, J.S., 2004. Combined DNA and lipid analyses of sediments reveal changes in Holocene haptophyte and diatom populations in an Antarctic lake. Earth Planet. Sci. Lett., 223: 225-239.

Corbet, B., Albrecht, A. and Ourisson, G., 1980. Photochemical or photomimetic fossil triterpenoids in sediments and petroleum. J. Am. Chem. Soc., 78: 183-188.

Cornford, C., Rullkötter, J. and Welte, D., 1979. Organic geochemistry of DSDP Leg 47a, Site 397, eastern North Atlantic: Organic petrography and extractable hydrocarbons. In: von Rad, U., Ryan, W.B.F. et al. (eds), Initial Reports DSDP, 47, US Government Printing Office, Washington, DC, pp. 511-522.

Cowie, G.L. and Hedges, J.I., 1984. Determination of neutral sugars in plankton, sediments, and wood by capillary gas chromatography of equilibrated isomeric mixtures. Anal. Chem., 56: 497-504.

Cranwell, P.A., 1973. Chain-length distribution of n-alkanes from lake sediments in relation to postglacial environments. Freshw. Biol., 3: 259-265.

de Leeuw, J.W., Cox, H.C., van Graas, G., van de Meer, F.W., Peakman, T.M., Baas, J.M.A. and van de Graaf, B., 1989. Limited double bond isomerisation and selective hydrogenation of sterenes during early diagenesis. Geochim. Cosmochim. Acta, 53: 903-909.

de Leeuw, J.W. and Sinninghe Damsté, J.S., 1990. Organic sulphur compounds and other biomarkers as indicators of palaeosalinity: A critical evaluation. In: Orr, W.L. and White, C.M. (eds), Geochemistry of sulphur in fossil fuels, ACS symposium series, 429, American Chemical Society, Washington, DC, pp. 417-443.

de Leeuw, J.W. and Largeau, C., 1993. A review of macromolecular organic compounds that comprise living organisms and their role in kerogen, coal and petroleum formation. In: Engel, M.H. and Macko, S.A. (eds), Organic geochemistry. Principles and applications. Plenum Press, NY, pp 23-72.

de Leeuw, J.W., 2004. Mass Spectra of Geochemicals, Petrochemicals and Biomarkers. Wiley, CD-ROM.

Degens, E.T. and Ittekot, V., 1987. The carbon cycle - tracking the path of organic particles from sea to sediment. , In: Brooks, J. and Fleet, A. (eds), Marine petroleum source rocks. Geol. Soc. Spec. Publ. No. 26, Blackwell, Oxford, pp. 121-135.

Demaison, G.J., 1991. Anoxia vs productivity: What controls the formation of organic-carbon-rich sediments and sedimentary rocks? Discussion. Bull. Am. Assoc. Petrol. Geol., 75: 499.

Demaison, G.J. and Moore, G.T., 1980. Anoxic environments and oil source bed genesis. Bull. Am. Assoc. Petrol. Geol., 64: 1179-1209.

D'Hondt., S., Rutherford., S. and Spivack, A.J., 2002. Metabolic activity of subsurface life in deep-sea sediments. Science, 295: 2067-2070.

Dumitrescu, M. and Brassell, S.C., 2005. Biogeochemical assessment of sources of organic matter and paleoproductivity during the Early Aptian Oceanic Anoxic Event at Shatsky Rise, ODP Leg 198. Org. Geochem., 36: 1002-1022.

Durand, B., 1980. Kerogen. Insoluble organic matter from sedimentary rocks. Editions Technip, Paris, 519 pp.

Eglinton, G. and Hamilton, R.J., 1967. Leaf epicuticular waxes. Science, 156: 1322-1335.

Eglinton, T.I., Benitez-Nelson, B.C., Pearson, A., McNichol, A.P., Bauer, J.E. and Druffel, E.R.M., 1997. Variability in radiocarbon ages of individual organic compounds from marine sediments. Science, 277: 796-799.

Elvert, M., Boetius, A., Knittel, K. and Jørgensen, B.B., 2003. Characterization of specific membrane fatty acids as chemotaxonomic markers for sulfate-reducing bacteria involved in anaerobic oxidation of methane. Geomicrobiol. J., 20: 403-419.

Emeis, K.-C., Robertson, A.H.F., Richter, C. et al. (eds) 1996. Proceedings of the Ocean Drilling Program. Initial Reports, 160, ODP, College Station (TX), 972 pp.

Emerson, S. and Hedges, J.I., 1988. Processes controlling the organic carbon content of open ocean sediments. Paleoceanography, 3: 621-634.

Espitalié, J., LaPorte, J.L., Madec, M., Marquis, F., Leplat, P., Paulet, J. and Boutefeu, A., 1977. Méthode rapide de caractérisation des roches mères, de leur potentiel pétrolier et de leur degré d'évolution. Rev. Inst. Fr. Pét., 32: 23-42.

Espitalié, J., Deroo, G. and Marquis, F., 1985. La pyrolyse Rock-Eval et ses applications. Rev. Inst. Fr. Pét., 40: 755-784.

Fogel, M.L. and Cifuentes, L.A., 1993. Isotope fractionation during primary production. In: Engel, M.H. and Macko, S.A. (eds), Organic geochemistry. Principles and applications. Plenum Press, NY, pp 73-98.

Freeman, K.H., Boreham, C.J., Summons, R.E. and Hayes, J.M., 1994. The effect of aromatization on the isotopic compositions of hydrocarbons during early diagenesis. Org. Geochem., 21: 1037-1049.

Froelich, P.N., Klinkhammer, G.P., Bender, M.L., Luedtke, N.A., Heath, G.R., Cullen, D., Dauphin, P., Hammond, D., Hartmann, B. and Maynard, V., 1979. Early oxidation of organic matter in pelagic sediments of eastern equatorial Atlantic: Suboxic diagenesis. Geochim. Cosmochim. Acta, 43: 1075-1090.

Gagosian, R.B., Peltzer, E.T. and Merrill, J.T., 1987. Long-range transport of terrestrially derived lipids in aerosols from the South Pacific. Nature, 325: 800-803.

Goñi, M.A. and Hedges, J.I., 1992. Lignin dimers: Structures, distribution and potential geochemical applications. Geochim. Cosmochim. Acta, 56: 4025-4043.

Goth, K., de Leeuw, J.W., Püttmann, W. and Tegelaar, E.W., 1988. The origin of Messel shale kerogen. Nature, 336: 759-761.

Hayes. J.M., Takigiku, R., Ocampo, R., Callot H.J. and Albrecht, P., 1987. Isotopic compositions and probable origins of organic molecules in the Eocene Messel shale. Nature, 329: 48-51.

Hayes, J.M., Freeman, K.H., Popp, B.N. and Hoham, C.H., 1990. Compound-specific isotope analyses, a novel tool for reconstruction of ancient biogeochemical processes. Org. Geochem., 16: 1115-1128.

Hayes, J.M., 1993. Factors controlling ^{13}C contents of sedimentary organic compounds: Principles and evidence. Mar. Geol., 113: 111-125.

Heath, G.R., Moore, T.C. and Dauphin, J.P., 1977. Organic carbon in deep-sea sediments. In: Anderson, N.R. and Malahoff, A. (eds), The fate of fossil fuel CO_2 in the oceans. Plenum Press, NY, pp. 605-625.

Hedges, J.I., Clark, W.A., Quay, P.D., Richey, J.E., Devol, A.H. and Santos, U.M., 1986a. Composition and fluxes of particulate organic material in the Amazon River. Limnol. Oceanogr., 31: 717-738.

Hedges, J.I., Cowie, G.L., Quay, P.D., Grootes, P.M., Richey, J.E., Devol, A.H., Farwell, G.W., Schmidt, F.W. and Salati, E., 1986b. Organic carbon-14 in the Amazon river system. Nature, 321: 1129-1131.

Hedges, J.I. and Prahl, F.G., 1993. Early diagenesis: Consequences for applications of molecular biomarkers. In: Engel, M.H. and Macko, S.A. (eds), Organic geochemistry. Principles and applications. Plenum Press, NY, pp. 237-253.

Hedges, J.I., Ertel, J.R., Richey, J.E., Quay, P.D., Benner, R., Strom, M. and Forsberg, B., 1994. Origin and processing of organic matter in the Amazon River as indicated by carbohydrates and amino acids. Limnol. Oceanogr., 39: 743-761.

Hedges, J.I. and Keil, R.G., 1995. Sedimentary organic matter preservation: An assessment and speculative synthesis. Mar. Chem., 49: 81-115.

Hedges, J.I., Keil, R.G. and Benner, R., 1997. What happens to terrestrial organic matter in the ocean? Org. Geochem., 27: 195-212.

Henrichs, S.M. and Reeburgh, W.S., 1987. Anaerobic mineralization of marine sediment organic matter: Rates and the role of anaerobic processes in the oceanic carbon economy. Geomicrobiol. J., 5: 191-237.

Herbin, J.P., Montadert, L.O., Müller, C., Gomez, R., Thurow, J. and Wiedmann, J., 1986. Organic-rich sedimentation at the Cenomanian/Turonian boundary in oceanic and coastal basins in the North Atlantic and Tethys. In: Summerhayes, C.P. and Shackleton, N.J. (eds), North Atlantic paleoceanography, Geol. Soc. Spec. Publ., 21, Blackwell, Oxford, 389-422.

Hinrichs, K.-U., Rinna, J. and Rullkötter, J., 1998. Late Quaternary paleoenvironmental conditions indicated by marine and terrestrial molecular biomarkers in sediments from the Santa Barbara basin. In: Wilson, R.C. and Tharp, V.L. (eds), Proceedings of the fourteenth annual Pacific climate (PACLIM) conference, April 6-9, 1997. Interagency Ecological Program, Technical Report, 57, California Department of Water Resources, Marysville (CA), pp. 125-133.

Hinz, K., Winterer. E.L. et al. (eds) 1984. Initial Reports of the Deep Sea Drilling Project, 79, US Government Printing Office, Washington, DC, 934 pp.

Ho, T.Y., Quigg, A., Zinkel, Z.V., Milligan, A.J., Wyman, K., Falkowski, P.G. and Morel, F.M.M., 2003. The elemental composition of some marine phytoplankton. J. Phycol., 39: 1145-1159.

Huang, Y., Dupont, L., Sarnthein, M., Hayes, J.M. and Eglinton, G., 2000. Mapping of C_4 plant input from North West Africa into North East Atlantic sediments. Geochim. Cosmochim. Acta, 64: 3505-3513.

Huc, A.Y. and Durand, B., 1977. Occurrence and significance of humic acids in ancient sediments. Fuel, 56: 73-80.

Huc, A.Y., 1988. Aspects of depositional processes of organic matter in sedimentary basins. Org. Geochem., 13: 263-272.

Hulthe, G., Hulth, S. and Hall, P.O.J., 1998. Effect of oxygen on degradation rate of refractory and labile organic matter in continental margin sediments. Geochim. Cosmochim. Acta, 62: 1319-1328.

Hunt, J.M., 1996. Petroleum geochemistry and geology. Freeman, NY, 743 pp.

Ibach, L.E., 1982. Relationships between sedimentation rate and total organic carbon content in ancient marine sediments. Bull. Am. Assoc. Petrol. Geol., 66: 170-188.

Ittekott, V., 1988. Global trends in the nature of organic matter in river suspensions. Nature, 332: 436-438.

Jasper, J.P. and Gagosian, R.B., 1989. Alkenone molecular stratigraphy in an oceanic environment affected by glacial fresh-water events. Paleoceanography, 4: 603-614.

Jasper, J.P. and Gagosian, R.B., 1990. The sources and deposition of organic matter in the Late Quaternary Pigmy Basin, Gulf of California. Geochim. Cosmochim. Acta, 54: 117-132.

Jasper, J.P., Hayes, J.M., Mix, A.C. and Prahl, F.G., 1994. Photosynthetic fractionation of ^{13}C and concentrations of dissolved CO_2 in the central equatorial Pacific during the last 255,000 years. Paleoceanography, 9: 781-798.

Jørgensen, B.B., 1996. Case study - Aarhus Bay. In: Jørgensen, B.B. and Richardson, K. (eds), Eutrophication in coastal marine ecosystems. Coastal and estuarine studies, 52, American Geophysical Union, Washington, DC, pp. 137-154.

Keil, R.G., Tsamakis, E., Fuh, C.B., Giddings, J.C. and Hedges, J.I., 1994a. Mineralogical and textural controls on the organic matter composition of coastal marine sediments: hydrodynamic separation using SPLITT-fractionation. Geochim. Cosmochim. Acta, 58: 879-893.

Keil, R.G., Montluçon, D.B., Prahl, F.G. and Hedges, J.I., 1994b. Sorptive preservation of labile organic matter in marine sediments. Nature, 370: 549-552.

Killops, S. and Killops, V., 2005. Introduction to organic geochemistry. Blackwell, Oxford, 393 pp.

Kolattukudy, P.E., 1976. Chemistry and biochemistry of natural waxes. Elsevier, Amsterdam, 459 pp.

Krey, J., 1970. Die Urproduktion des Meeres (in German). In: Dietrich, G. (ed), Erforschung des Meeres. Umschau, Frankfurt, pp. 183-195.

Kristensen, E., Ahmed, S.A. and Devol, A.H., 1995. Aerobic and anaerobic decomposition of organic matter in marine sediments. Which is faster? Limnol. Oceanogr., 40: 1430-1437.

Lallier-Vergès, E., Bertrand, P. and Desprairies, P., 1993. Organic matter composition and sulfate reduction intensity in Oman margin sediments. Mar. Geol., 112: 57-69.

Largeau. C., Derenne, S., Casadevall, E., Berkaloff, C., Corolleur, M., Lugardon, B., Raynaud, J.F. and Connan, J., 1990. Occurrence and origin of "ultralaminar" structures in "amorphous" kerogens of various source rocks and oil shales. Org. Geochem., 16: 889-895.

Larter, S.R. and Horsfield, B., 1993. Determination of structural components of kerogens by the use of analytical pyrolysis methods. In: Engel, M.H. and Macko, S.A. (eds), Organic geochemistry. Principles and applications. Plenum Press, NY, pp. 271-287.

Leonardos, N. and Geider, R.J., 2004. Limnol. Oceanogr., 49: 2105-2114.

Leventhal, J.S., 1983. An interpretation of carbon and sulphur relationships in Black Sea sediments as indicators of environments of deposition. Geochim. Cosmochim. Acta, 47: 133-138.

Littke, R., Rullkötter, J. and Schaefer, R.G., 1991a. Organic and carbonate carbon accumulation on Broken Ridge and Ninetyeast Ridge, central Indian Ocean. In: Weissel, J., Pierce, J. et al. (eds), Proceedings of the Ocean Drilling Program, Sci. Res., 121, ODP, College Station (TX), pp. 467-487.

Littke, R., Baker, D.R., Leythaeuser, D. and Rullkötter, J., 1991b. Keys to the depositional history of the Posidonia Shale (Toarcian) in the Hils syncline, northern Germany. In: Tyson, R.V. and Pearson, T.H. (eds), Modern and ancient continental shelf anoxia. Geol. Soc. Spec. Publ., 58, The Geological Society, London, pp. 311-334.

Littke, R., Baker, D.R. and Rullkötter, J., 1997a. Deposition of petroleum source rocks. In: Welte, D.H., Horsfield, B. and Baker, D.R. (eds), Petroleum and basin evolution. Springer-Verlag, Heidelberg, pp. 271-333.

Littke, R., Lückge, A. and Welte, D.H., 1997b. Quantification of organic matter degradation by microbial sulphate reduction for Quaternary sediments from the northern Arabian Sea. Naturwissenschaften, 84: 312-315.

Lückge, A., Boussafir, M., Lallier-Vergès, E. and Littke, R., 1996. Comparative study of organic matter preservation in immature sediments along the continental margins of Peru and Oman. Part I: Results of petrographical and bulk geochemical data. Org. Geochem., 24: 437-451.

Mangelsdorf, K., Güntner, U. and Rullkötter, J., 2000. Climatic and oceanographic variations on the California continental margin during the last 160 kyr. Org. Geochem., 31: 829-846.

Mariotti, A., Gadel, F., Giresse, P. and Kinga-Mouzeo, 1991. Carbon isotope composition and geochemistry of particulate organic matter in the Congo River (Central Africa): Application to the study of Quaternary sediments off the mouth of the river. Chem. Geol., 86: 345-357.

Martens, C.S. and Klump, J.V., 1984. Biogeochemical cycling in an organic-rich coastal marine basin - 4. An organic carbon budget for sediments dominated by sulfate reduction and methanogenesis. Geochim. Cosmochim. Acta, 48: 1987-2004.

Martens, C.S., Haddad, R.I. and Chanton, J.P., 1992. Organic matter accumulation, remineralization, and burial in an anoxic coastal sediment. In: Whelan, J.K. and Farrington, J.W. (eds), Organic matter: Productivity, accumulation and preservation in recent and ancient sediments, Columbia University Press, NY, pp 82-98.

Martinson, D.G., Pisias, N.G., Hays, J.D., Imbrie, J., Moore, T.C.J. and Shackleton, N.J., 1987. Age dating and the orbital theory of the ice ages: Development of a high-resolution 0 to 300,000-year chronostratigraphy. Quat. Res., 27: 1-29.

Mayer, L.M., 1994. Surface area control of organic carbon accumulation in continental shelf sediments. Geochim. Cosmochim. Acta, 58: 1271-1284.

Mayer, L.M., 1999. Extent of coverage of mineral surfaces by organic matter in marine sediments. Geochim. Cosmochim. Acta, 63: 207-215.

Mayer, L.M., 2005. Erratum to L. M. Mayer (1999) "Extent of coverage of mineral surfaces by organic matter in marine sediments" *Geochimica et Cosmochimica Acta* 63, 207-215. Geochim. Cosmochim. Acta, 69: 1375.

McIver, R., 1975. Hydrocarbon occurrences from JOIDES Deep Sea Drilling Project. Proceedings of the 9th World Petroleum Congress (Tokyo), 2, Applied Science Publishers, Barking, pp. 269-280.

McLafferty, F.W. and Turecek, F., 1993. Interpretation of mass spectra. University Science Books, Mill Valley (CA), 371 pp.

Meybeck, M., 1982. Carbon, nitrogen, and phosphorus transport by world rivers. Am. J. Sci., 282: 401-450.

Meyers, P.A., 1994. Preservation of elemental and isotopic source identification of sedimentary organic matter. Chem. Geol., 144: 289-302.

Meyers, P.A., 1997. Organic geochemical proxies of paleoceanographic, paleolimnologic, and paleoclimatic processes. Org. Geochem., 27: 213-250.

Mitterer, R.M., 1993. The diagenesis of proteins and amino acids in fossil shells. In: Engel, M.H. and Macko, S.A. (eds), Organic geochemistry. Principles and applications. Plenum Press, NY, pp 739-753.

Moers, M.E.C., Jones, D.M., Eakin, P.A., Fallick, A.E., Griffiths, H. and Larter, S.R., 1993. Carbohydrate diagenesis in hypersaline environments: Application of GC-IRMS to the stable isotope analysis of derivatives of saccharides from surficial and buried sediments. Org. Geochem., 20: 927-933.

Mollenhauer, G., Eglinton, T.I., Ohkouchi, N., Schneider, R.R., Müller, P.J., Grootes, P.M. and Rullkötter, J., 2003. Asynchronous alkenone and foraminifera records from the Benguela Upwelling System. Geochim. Cosmochim. Acta, 67: 2157-2171.

Müller, P.J., 1977. C/N ratios in Pacific deep-sea sediments: Effect of inorganic ammonium and organic nitrogen compounds sorbed by clays. Geochim. Cosmochim. Acta, 41: 765-776.

Müller, P.J. and Suess, E., 1979. Productivity, sedimentation rate and sedimentary organic matter in the oceans - organic carbon preservation. Deep-Sea Res., 27A: 1347-1362.

Müller, P.J., Schneider, R. and Ruhland, G., 1994. Late Quarternary pCO₂ variations in the Angola Current: Evidence from organic carbon δ^{13}C and alkenone temperatures. In: Zahn, R. et al. (eds.), Carbon cycling in the glacial ocean: Constraints on the ocean's role in global change, Nato ASI Series I, Vol. 17. Springer, Heidelberg, pp. 343-366.

Müller, P.J., Kirst, G., Ruhland, G., von Storch, I. and Rosell-Melé, A. 1998. Calibration of the alkenone paleo-temperature index $U^{K'}_{37}$ based on core-tops from the eastern South Atlantic and the global ocean (60°N-60°S). Geochim. Cosmochim. Acta, 62: 1757-1772.

Muzuka, A.N.N., 1999. Isotopic compositions of tropical East African flora and their potential as source indicators of organic matter in coastal marine sediments. J. Afr. Earth Sci., 28: 757-766.

Orphan, V.J., House C.H., Hinrichs, K.-U., McKeegan, K.D. and DeLong, E.F., 2001. Methane-consuming archaea revealed by direct coupled isotopic and phylogenetic analysis. Science, 293: 484-487.

Orphan, V.J., House, C.H., Hinrichs, K.-U., McKeegan, K.D. and DeLong, E.F., 2002. Multiple archaeal groups mediate methane oxidation in anoxic cold seep sediments. PNAS, 99: 7663-7668.

Parkes, R.J., Cragg, B.A., Bale, S.J., Getliff, J.M., Goodman, K., Rochelle, P.C., Fry, J.C., Weightman, A.J. and Harvey, S.M., 1994. Deep bacterial biosphere in Pacific ocean sediments. Nature, 371: 410-413.

Pearson, A., McNichol, A.P., Benitez-Nelson, B.C., Hayes, J.M. and Eglinton, T.I., 2001. Origins of lipid biomarkers in Santa Monica Basin surface sediments: A case study using compound-specific δ^{14}C analysis. Geochim. Cosmochim. Acta, 65: 3123-3137.

Pedersen, T.F. and Calvert, S.E., 1990. Anoxia versus productivity: What controls the formation of organic-carbon-rich sediments and sedimentary rocks? Bull. Am. Assoc. Petrol. Geol., 74: 454-466.

Pedersen, T.F. and Calvert, S.E., 1991. Anoxia vs productivity: What controls the formation of organic-carbon-rich sediments and sedimentary rocks? Reply. Bull. Am. Assoc. Petrol. Geol., 75: 500-501.

Pelejero, C. and Calvo, E., 2003. The upper end of the $U^{K'}_{37}$ temperature calibration revisited. Geochem. Geophys. Geosyst., 4: 1014, doi:10.1029/2002GC000431.

Peters, K.E., Walters, C.C. and Moldowan, J.M., 2005. The biomarker guide (2 volumes). Cambridge University Press, Cambridge, 1190 pp.

Plough, H., Kühl, M., Buchholz-Cleven, B. and Jørgensen, B.B., 1997. Anoxic aggregates - an ephemeral phenomenon in the pelagic environment? Aqu. Microb. Ecol., 13: 285-294.

Poynter, J., 1989. Molecular stratigraphy: The recognition of palaeoclimate signals in organic geochemical data. PhD Thesis, University of Bristol, 324 pp.

Poynter, J. and Eglinton, G., 1991. The biomarker concept - strengths and weaknesses. Fresenius' J. Anal. Chem., 339: 725-731.

Prahl, F.G. and Wakeham, S.G., 1987. Calibration of unsaturation patterns in long-chain ketone compositions for paleotemperature assessment. Nature, 330: 367-369.

Prahl, F.G., Muehlhausen, L.A. and Zahnle, D.L., 1988. Further evaluation of long-chain alkenones as indicators of paleoceanographic conditions. Geochim. Cosmochim. Acta, 52: 2303-2310.

Prahl, F.G., Ertel, J.R., Goñi, M.A., Sparrow, M.A. and Eversmeyer, B., 1994. Terrestrial organic carbon contributions to sediments on the Washington margin. Geochim. Cosmochim. Acta, 58: 3035-3048.

Radke, M., Willsch, H. und Welte, D.H., 1980. Preparative hydrocarbon group type determination by automated medium pressure liquid chromatography. Anal. Chem., 52: 406-411.

Ransom, B., Kim, D., Kastner, M. and Wainwright, S., 1998. Organic matter preservation on continental slopes: Importance of mineralogy and surface area. Geochim. Cosmochim. Acta, 62: 1329-1345.

Rashid, M.A., 1985. Geochemistry of marine humic compounds. Springer-Verlag, Berlin, 300 pp.

Rau, G.H., Takahashi, T., Des Marais, D.J. and Sullivan, C.W., 1991. Particulate organic matter δ^{13}C variations across the Drake Passage. J. Geophys. Res., 96: 15131-15135.

Redfield, A.C., Ketchum, B.H. and Richards, F.W., 1963. The influence of organisms on the composition of sea water. In: Hill, M.N. (ed), The sea, 2, Wiley Interscience, NY, pp. 26-77.

Rohmer, M., Bisseret, P. and Neunlist, S., 1992. The hopanoids, prokaryotic triterpenoids and precursors of ubiquitous molecular fossils. In: Moldowan, J.M., Albrecht, P. and Philp, R.P. (eds), Biological markers in sediments and petroleum. Prentice Hall, Englewood Cliffs (NJ), pp. 1-17.

Romankevitch, E.A., 1984. Geochemistry of organic matter in the ocean. Springer-Verlag, Heidelberg, 334 pp.

Rommerskirchen, F., Eglinton, G., Dupont, L., Güntner, U., Wenzel, C. and Rullkötter, J., 2003. A north to south transect of Holocene Southeast Atlantic continental margin sediments: Relationship between aerosol transport and compound-specific δ^{13}C land plant biomarker and pollen records. Geochem. Geosyst., 4 (12), 1101, DOI 10.1029/2003GC000541.

Rosell-Melé, A., Bard, E., Emeis, K.-C., Grimalt, J.O., Müller, P., Schneider, R., Bouloubassi, I., Epstein, B., Fahl, K., Fluegge, A., Freeman, K., Goñi, M., Güntner, U., Hartz, D., Hellebust, S., Herbert, T., Ikehara, M., Ishiwatari, R., Kawamura, K., Kenig, F., de Leeuw, J.W., Lehman, S., Mejanell, L., Ohkouchi, N., Pancost, R.D., Pelejero, C., Prahl, F., Quinn, J., Rontani, J.-F., Rostek, F., Rullkötter, J., Sachs, J., Blanz, T., Sawada, K., Schulz-Bull, D., Sikes, E., Sonzogni, C., Ternois, Y., Versteegh, G., Volkman, J.K. and Wakeham, S., 2001. Precision of the current methods to measure the alkenone proxy $U^{K'}_{37}$ and absolute alkenone abundance in sediments: Results of an interlaboratory comparison study. Geochem. Geophys. Geosyst., 2: Paper number 2000GC000141.

Rullkötter, J., Cornford, C. and Welte, D.H., 1982. Geochemistry and petrography of organic matter in Northwest African continental margin sediments: quantity, provenance, depositional environment and temperature history. In: von Rad, U., Hinz, K., Sarnthein, M. and Seibold, E. (eds), Geology of the Northwest African continental margin, Springer-Verlag, Heidelberg, pp. 686-703.

Rullkötter, J., Vuchev, V., Hinz, K., Winterer, E.L., Baumgartner, P.O., Bradshaw, M.L., Channel, J.E.T., Jaffrezo, M., Jansa,

L.F., Leckie, R.M., Moore, J.M., Schaftenaar, C., Steiger, T.H. and Wiegand, G.E., 1983. Potential deep sea petroleum source beds related to coastal upwelling. In: Thiede, J. and Suess, E. (eds), Coastal upwelling: Its sediment record, Part B: Sedimentary records of ancient coastal upwelling. Plenum Press, NY, pp. 467-483.

Rullkötter, J., Mukhopadhyay, P.K., Schaefer, R.G. and Welte, D.H., 1984. Geochemistry and petrography of organic matter in sediments from Deep Sea Drilling Project Sites 545 and 547, Mazagan Escarpment. In: Hinz, K., Winterer, E.L. et al. (eds), Initial Reports DSDP, 79, US Government Printing Office, Washington, DC, pp. 775-806.

Rullkötter, J., Mukhopadhyay, P.K. and Welte, D.H., 1987. Geochemistry and petrography of organic matter from Deep Sea Drilling Project Site 603, lower continental rise off Cape Hatteras. In: van Hinte, J.E., Wise, S.E. Jr et al. (eds), Initial Reports DSDP, 92, US Government Printing Office, Washington DC, pp. 1163-1176.

Rullkötter, J. and Michaelis, W., 1990. The structure of kerogen and related materials. A review of recent progress and future trends. Org. Geochem., 16: 829-852.

Rullkötter, J., Peakman, T.M. and ten Haven, H.L., 1994. Early diagenesis of terrigenous triterpenoids and its implications for petroleum geochemistry. Org. Geochem., 21: 215-233.

Rullkötter, J., 2001. Geochemistry, organic. In: Meyers, R.A. (ed), The encyclopedia of physical science and technology, 6, Academic Press, San Diego, pp. 547-572.

Sarnthein, M., Winn, K. and Zahn, R., 1987. Paleoproductivity of oceanic upwelling and the effect of atmospheric CO_2 and climatic change during deglaciation times. In: Berger, W.H. and Labeyrie, L.D. (eds), Abrupt climatic change. Reidel, Dordrecht, pp 311-337.

Sarnthein, M., Winn, K., Duplessy, J.C. and Fontugne, M.R., 1988. Global variations of surface water productivity in low- and mid-latitudes: Influence of CO_2 reservoirs of the deep ocean and atmosphere during the last 21,000 years. Paleoceanography, 3: 361-399.

Sarnthein, M., Pflaumann, U., Ross, R., Tiedemann, R. and Winn, K., 1992. Transfer functions to reconstruct ocean paleoproductivity: A comparison. In: Summerhayes, C.P., Prell, W.L. and Emeis, K.C. (eds), Upwelling systems. Evolution since the early Miocene. Geol. Soc. Spec. Publ., 64, Blackwell, Oxford, pp 411-427.

Sauer, P.E., Eglinton, T.I., Hayes, J.M., Schimmelmann, A. and Sessions, A.L., 2001. Compound-specific D/H ratios of lipid biomarkers from sediments as a proxy for environmental and climatic conditions. Geochim. Cosmochim. Acta, 65: 213-222.

Schefuß, E., Versteegh, G.J.M., Jansen, J.H.F. and Sinninghe Damsté, J.S., 2004. Lipid biomarkers as major source and preservation indicators in SE Atlantic surface sediments. Deep-Sea Res. I, 51: 1199-1228.

Schidlowski, M. 1988. A 3,800-million-year isotopic record of life from carbon in sedimentary rocks. Nature, 333: 313-318.

Schlesinger, W.H. and Melack, J.M., 1981. Transport of organic carbon in the world's rivers. Tellus, 33: 172-187.

Schouten, S., Hopmans, E.C., Schefuß, E. and Sinninghe Damsté, J.S., 2002. Distributional variations in marine crenarchaeotal lipids: A new tool for reconstructing ancient sea water temperatures? Earth Planet. Sci. Lett., 204: 265-274.

Schwartz, D., Mariotti, A., Lanfranchi, R. and Guillet, B., 1986. $^{13}C/^{12}C$ ratios of soil organic matter as indicators of vegetation changes in the Congo. Geoderma, 39: 97-103.

Senesi, N. and Miano, T.M. (eds) 1994. Humic substances in the global environment and implications on human health. Elsevier, Amsterdam, 1390 pp.

Sessions, A.L., Burgoyne, T.W., Schimmelmann, A. and Hayes, J.M., 1999. Fractionation of hydrogen isotopes in lipid biosynthesis. Org. Geochem., 30: 1193-1200.

Sessions, A.L. and Hayes, J.M., 2005. Calculation of hydrogen isotopic fractionations in biogeochemical systems. Geochim. Cosmochim. Acta, 69: 593-597.

Sicre, M.-A., Bard, E., Ezat, U. and Rostek, F., 2002. Alkenone distributions in the North Atlantic and Nordic sea surface waters. Geochem. Geophys. Geosyst., 3: doi:10.1029/2001GC000159.

Sikes, E.L. and Sicre M.-A., 2002. Relationship of the tetra-unsaturated C_{37} alkenone to salinity and temperature: Implications for paleoproxy applications. Geochem. Geophys. Geosyst., 3: 1063, doi:10.1029/2002GC000345.

Simoneit, B.R.T., 1986. Cyclic terpenoids of the geosphere. In: Johns, R.B. (ed), Biological markers in the sedimentary record. Elsevier, Amsterdam, pp. 43-99.

Sinninghe Damsté, J.S., Eglinton, T.I., de Leeuw, J.W. and Schenck, P.A., 1989a. Organic sulphur in macromolecular sedimentary organic matter I. Structure and origin of sulphur-containing moieties in kerogen, asphaltenes and coal as revealed by flash pyrolysis. Geochim. Cosmochim. Acta, 53: 873-899.

Sinninghe Damsté, J.S., Rijpstra, W.I.C., Kock-van Dalen, A.C., de Leeuw, J.W. and Schenck, P.A., 1989b. Quenching of labile functionalized lipids by inorganic sulphur species: Evidence for the formation of sedimentary organic sulphur compounds at the early stages of diagenesis. Geochim. Cosmochim. Acta, 53: 1343-1355.

Sinninghe Damsté, J.S., Eglinton, T.I., Rijpstra, W.I.C. and de Leeuw, J.W., 1990. Characterization of organically bound sulphur in high-molecular-weight sedimentary organic matter using flash pyrolysis and Raney Ni desulphurisation. In: Orr, W.L. and White, C.M. (eds), Geochemistry of sulphur in fossil fuels, ACS symposium series, 429, American Chemical Society, Washington, DC, pp. 486-528.

Sinninghe Damsté, J.S. and Köster, J., 1998. A euxinic North Atlantic ocean during the Cenomanian/Turonian oceanic anoxic event. Earth Planet. Sci. Lett., 158: 165-173.

Stein, R., 1986a. Surface-water paleo-productivity as inferred from sediment deposited in oxic and anoxic deep-water environments of the Mesozoic Atlantic Ocean. Mitt. Geol.-Paläont. Inst. Univ. Hamburg, 60: 55-70.

Stein, R., 1986b. Organic carbon and sedimentation rate - further evidence for anoxic deep-water conditions in the Cenomanian/Turonian Atlantic Ocean. Mar. Geol., 72: 199-209.

Stein, R., 1990. Organic carbon content/sedimentation rate relationship and its paleoenvironmental significance for marine sediments. Geo-Mar. Lett., 10: 37-40.

Stein, R., 1991. Accumulation of organic carbon in marine sediments. Lect. Notes Earth Science, 34: 1-217.

Stein, R. and Rack, F., 1995. A 160,000-year high-resolution record of quantity and composition of organic carbon in the Santa Barbara basin (Site 893). In: Kennett, J.P.,

Baldauf, J. and Lyle, M. (eds), Proceedings of the Ocean Drilling Program, Sci. Res., 146, ODP, College Station (TX), pp. 125-138.

Stein, R. and Macdonald, R.W. (eds) 2005. The organic carbon cycle in the Arctic Ocean. Springer-Verlag, Berlin, 363 pp.

Suess, E. and Thiede, J. (eds) 1983. Coastal upwelling: Its sediment record. Part A: Responses of the sedimentary regime to present coastal upwelling. Plenum Press, NY, 604 pp.

Summerhayes, C.P., 1981. Organic facies of middle Cretaceous black shales in deep North Atlantic. Bull. Am. Assoc. Petrol. Geol., 65: 2364-2380.

Summerhayes, C.P., Prell, P.I. and Emeis, K.C. (eds) 1992. Upwelling systems: Evolution since the early Miocene. Geol. Soc. Spec. Publ., 64, Blackwell, Oxford, 519 pp.

Takahasi, T., Broecker, W.S. and Langer, S., 1985. Redfield ratio based on chemical data from isopycnal surfaces. J. Geophys. Res., 90: 6907-6924.

Taylor, G.H., Teichmüller, M., Davis, A., Diessel, C.F.K., Littke, R. and Robert, P. (eds) 1998. Organic petrology: A new handbook incorporating some revised parts of Stach's textbook of coal petrology. Gebrüder Borntraeger, Berlin, 704 pp.

Tegelaar, E.W., Derenne, S., Largeau, C. and de Leeuw, J.W., 1989. A reappraisal of kerogen formation. Geochim. Cosmochim. Acta, 53: 3103-3107.

ten Haven, H.L., Peakman, T.M. and Rullkötter, J., 1992. Early diagenetic transformation of higher plant triterpenoids in deep sea sediments from Baffin Bay. Geochim. Cosmochim. Acta, 56: 2001-2024.

Thiede, J. and Suess, E. (eds) 1983. Coastal upwelling: Its sediment record. Part B: Sedimentary records of ancient coastal upwelling. Plenum Press, NY, 610 pp.

Tissot, B.P. and Welte, D.H., 1984. Petroleum formation and occurrence. Springer-Verlag, Heidelberg, 699 pp.

Tyson, R.V., 1987. The genesis and palynofacies characteristics of marine petroleum source rocks. In: Brooks, J. and Fleet, A.J. (eds), Marine petroleum source rocks. Geol. Soc. Spec. Publ., 26, Blackwell, Oxford, pp. 47-67.

Tyson, R.V. and Pearson, T.H. (eds) 1991. Modern and ancient continental shelf anoxia. Geol. Soc. Spec. Publ., 58, Blackwell, Oxford, 470 pp.

van Bergen, P.F., Blokker, P., Collinson, M.E., Sinninghe Damsté, J.S. and de Leeuw, J.W., 2004. Structural biomacromolecules in plants: What can be learnt from the fossil record. In: Hemsley, A.R. and Poole, I. (eds), Evolution of plant physiology. Elsevier, Amsterdam, pp. 133-154.

van Krevelen, D.W., 1961. Coal typology – chemistry – physics – constitution. Elsevier, Amsterdam, 513 pp.

Vetö, I., Hetényi, M., Demény, A. and Hertelendi, E., 1994. Hydrogen index as reflecting intensity of sulphide diagenesis in non-bioturbated, shaly sediments. Org. Geochem., 22: 299-310.

Volkman, J.K. and Maxwell, J.R., 1986. Acyclic isoprenoids as biological markers. In: Johns, R.B. (ed), Biological markers in the sedimentary record. Elsevier, Amsterdam, pp. 1-42.

Volkman, J.K., Barrett, S.M., Blackburn, S.I., Mansour, M.P., Sikes, E. and Gelin, F., 1998. Microalgal biomarkers: A review of recent research developments. Org. Geochem., 29: 1163-1179.

Volkman, J.K., 2005. Sterols and other triterpenoids: Source specificity and evolution of biosynthetic pathways. Org. Geochem. 36: 139-159.

von Engelhardt, W., 1973. Sedimentpetrologie, Teil III: Die Bildung von Sedimenten und Sedimentgesteinen (in German). Schweizerbarth, Stuttgart, 378 pp.

von Rad, U., Ryan, W.B.F. et al. (eds) 1979. Initial Reports DSDP 47, US Government Printing Office, Washington, D.C., 835 pp.

Wakeham, S.G. and Lee, C., 1989. Organic geochemistry of particulate matter in the ocean: The role of particles in oceanic sedimentary cycles. Org Geochem., 14: 83-96.

Wakeham, S.G., Lewis, C.M., Hopmans, E.C., Schouten, S. and Sinninghe Damsté, J.S., 2003. Archaea mediate anaerobic oxidation of methane in deep euxinic waters of the Black Sea. Geochim. Cosmochim. Acta, 67: 1359-1374.

Wellsbury, P., Mather, I.D. and Parkes, R.J., 2002. Geomicrobiology of deep, low organic carbon sediments in the Woodlark Basin, Pacific Ocean. FEMS Microbiol. Ecol., 42: 59-70.

Welte, D.H., Horsfield, B. and Baker, D.R. (eds) 1997. Petroleum and basin evolution. Springer-Verlag, Heidelberg, 535 pp.

Wenger, L.M. and Baker, D.R., 1986. Variations in organic geochemistry of anoxic-oxic black shale-carbonate sequences in the Pennsylvanian of the Midcontinent, U.S.A. Org. Geochem., 10: 85-92.

Westerhausen, L., Poynter, J., Eglinton, G., Erlenkeuser, H. and Sarnthein, M., 1993. Marine and terrigenous origin of organic matter in modern sediments of the equatorial East Atlantic: The $\delta^{13}C$ and molecular record. Deep-Sea Res. I, 40: 1087-1121.

Willsch, H., Clegg, H., Horsfield, B., Radke, M. and Wilkes, H., 1997. Liquid chromatographic separation of sediment, rock, and coal extracts and crude oils into compound classes. Anal. Chem., 69: 4203-4209.

Wuchter, C., Schouten, S., Coolen, M.J.L. and Sinninghe Damsté, J.S., 2004. Temperature-dependent variation in the distribution of tetraether membrane lipids of marine Crenarchaeota: Implications for TEX_{86} paleothermometry. Paleoceanography, 19: PA4028, doi:10.1029/2004PA001041.

5 Bacteria and Marine Biogeochemistry

Bo Barker Jørgensen

Geochemical cycles on Earth follow the basic laws of thermodynamics and proceed towards a state of maximal entropy and the most stable mineral phases. Redox reactions between oxidants such as atmospheric oxygen or manganese oxide and reductants such as ammonium or sulfide may proceed by chemical reaction, but they are most often accelerated by many orders of magnitude through enzymatic catalysis in living organisms. Throughout Earth's history, prokaryotic physiology has evolved towards a versatile use of chemical energy available from this multitude of potential reactions. Biology, thereby, to a large extent regulates the rate at which the elements are cycled in the environment and affects where and in which chemical form the elements accumulate. By coupling very specifically certain reactions through their energy metabolism, the organisms also direct the pathways of transformation and the ways in which the element cycles are coupled. Microorganisms possess an enormous diversity of catalytic capabilities which is still only incompletely explored and which appears to continuously expand as new organisms are discovered. A basic understanding of the microbial world and of bacterial energy metabolism is therefore a prerequisite for a proper interpretation of marine geochemistry - a motivation for this chapter on biogeochemistry.

The role of microorganisms in modern biogeochemical cycles is the result of a long evolutionary history. The earliest fossil evidence of prokaryotic organisms dates back three and a half billion years (Schopf and Klein 1992; Brasier et al. 2002). Only some 1.5 billion years later did the evolution of oxygenic photosynthesis lead to a build-up of oxygen on the surface of our planet and it may have taken another 1.5 billion years before the oxygen level in the atmosphere and ocean rose to the present-day level, thus triggering the rapid evolutionary radiation of metazoans at the end of the Proterozoic era. Through the two billion years that Earth was inhabited exclusively by microorganisms, the main element cycles and biogeochemical processes known today evolved. The microscopic prokaryotes developed the complex enzymatic machinery required for these processes and are even today much more versatile with respect to basic types of metabolism than plants and animals which developed over the last 600 million years. In spite of their uniformly small size and mostly inconspicuous morphology, the prokaryotes are thus physiologically much more diverse than the metazoans. In the great phylogenetic tree of all living organisms, humans are more closely related to slime molds than the sulfate reducing bacteria are to the methanogenic archaea. The latter two belong to separate domains of prokaryotic organisms, the Bacteria and the Archaea, respectively (the term 'prokaryote' is used rather than 'bacteria' when also the archaea are included). Animals and plants, including all the eukaryotic microorganisms, belong to the third domain, Eukarya.

5.1 Role of Microorganisms

5.1.1 From Geochemistry to Microbiology – and back

Due to the close coupling between geochemistry and microbiology, progress in one of the fields has often led to progress in the other. Thus, analyses of chemical gradients in the pore water of marine sediments indicate where certain chemical species are formed and where they react with each other. The question is then, is the reaction biologically catalyzed and which microorganisms may be involved?

An example is the distribution of dissolved ferrous iron and nitrate, which in deep sea sediments often show a diffusional interface between the two species (Fig. 5.1A). Based on such gradients, Froelich et al. (1979), Klinkhammer (1980) and others suggested that Fe^{2+} may be readily oxidized by nitrate, presumably catalyzed by bacteria. Marine microbiologists, thus, had the background information to start searching for nitrate reducing bacteria which use ferrous iron as a reductant and energy source. It took, however, nearly two decades before such bacteria were isolated for the first time and could be studied in pure culture (Straub et al. 1996; Benz et al. 1998; Fig. 5.1B). The bacteria appear to occur widespread in aquatic sediments but their quantitative importance is still not clear (Straub and Buchholz-Cleven 1998). The bacteria oxidize ferrous iron (here ferrous carbonate) according to the following stoichiometry:

$$10FeCO_3 + 2NO_3^- + 24H_2O \rightarrow$$
$$10Fe(OH)_3 + N_2 + 10HCO_3^- + 8H^+ \qquad (5.1)$$

The observation of a deep diffusional interface between sulfate and methane in marine sediments also led to a long-lasting search by microbiologists for methane oxidizing sulfate reducers. The geochemical data and experiments demonstrated clearly that methane is oxidized to CO_2 below the sediment surface, at a depth where no potential oxidant other than sulfate seems to remain (Reeburgh 1969; Iversen and Jørgensen 1985; Alperin and Reeburgh 1985; Chaps. 3, 8 and 14). Only recently were the microorganisms discovered that can carry out the complete oxidation of methane with sulfate. These consist of unique syntrophic aggregates of archaea and sulfate reducing bacteria (Boetius et al. 2000; Orphan et al. 2001). The archaea apparently oxidize the methane through a partial reversal of the metabolic pathway of methane formation. Thermodynamic calculations show that the reversed direction of methane formation from H_2 and CO_2 (Eq. 5.2) is exergonic if the H_2 partial pressure is kept extremely low (Hoehler et al. 1994, 1998). The sulfate reducing

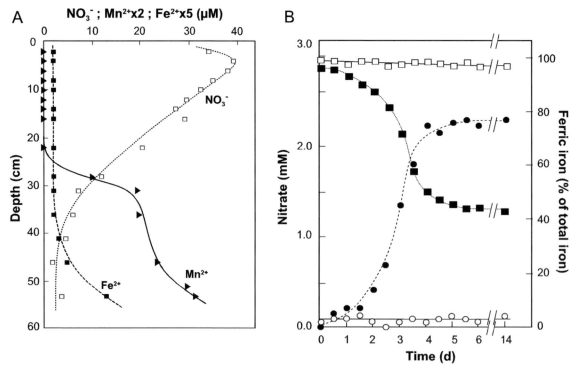

Fig. 5.1 A) Pore water gradients of nitrate, dissolved manganese, and iron in sediments from the eastern equatorial Atlantic at 5000 m depth. The gradients indicate that Fe^{2+} is oxidized by NO_3^-, whereas Mn^{2+} may be oxidized by O_2 (no data). (Data from Froelich et al. 1979; Station 10GC1). B) Anaerobic bacterial oxidation of ferrous to ferric iron with nitrate in an enrichment culture. Filled symbols show results from a growing culture, open symbols shows a control experiment with killed cells (no concentration changes). Bacteria are clearly needed for the fast iron oxidation with nitrate. Symbols show ferric iron ($\bullet + \circ$) and nitrate ($\blacksquare + \square$). (Data from Straub et al. 1996).

bacteria are known to be highly efficient H_2 scavengers (Eq. 5.3):

$$CH_4 + 2H_2O \rightarrow CO_2 + 4H_2 \qquad (5.2)$$

$$4H_2 + SO_4^{2-} + 2H^+ \rightarrow H_2S + 4H_2O \qquad (5.3)$$

There are also many examples of the reverse inspiration, that progress in microbiology has led to a new understanding of geochemistry. One such example was the discovery of a widespread ability among laboratory cultures of sulfate reducing and other anaerobic bacteria to disproportionate inorganic sulfur compounds of intermediate oxidation state (Bak and Cypionka 1987). By such a disproportionation, which can be considered an inorganic fermentation, elemental sulfur (S^0) or thiosulfate ($S_2O_3^{2-}$) may be simultaneously reduced to sulfide and oxidized to sulfate:

$$4S^0 + 4H_2O \rightarrow 3H_2S + SO_4^{2-} + 2H^+ \qquad (5.4)$$

$$S_2O_3^{2-} + H_2O \rightarrow H_2S + SO_4^{2-} \qquad (5.5)$$

A search for the activity of such bacteria, by the use of radiotracers and sediment incubation experiments, revealed their widespread ocurrence in the seabed and their great significance for the marine sulfur cycle (Jørgensen 1990; Jørgensen and Bak 1991; Thamdrup et al. 1993). Disproportionation reactions also cause a strong fractionation of sulfur isotopes, which has recently led to a novel interpretation of stable sulfur isotope signals in modern sediments and sedimentary rocks with interesting implications for the evolution of oxygen in the global atmosphere and ocean (Canfield and Teske 1996). The working hypothesis is that the large isotopic fractionations of sulfur, which in the geological record started some 600-800 million years ago, are the result of disproportionation reactions. From modern sediments we know that these conditions require two things: bioturbation and an efficient oxidative sulfur cycle. Both point towards a coupled evolution of metazoans and a rise in the global oxygen level towards the end of the Proterozoic and start of the Cambrian.

5.1.2 Approaches in Marine Biogeochemistry

The approaches applied in marine biogeochemistry are diverse, as indicated in Figure 5.2,

and range from pure geochemistry to experimental ecology, microbiology and molecular biology. Mineral phases and soluble constituents are analyzed and the data used for modeling of the diagenetic reactions and mass balances (Berner 1980; Boudreau 1997; Chap. 15). Dynamic processes are studied in retrieved sediment cores, which are used to analyze solute fluxes across the sediment-water interface or to measure process rates by experimental approaches using, e.g. radiotracers, stable isotopes or inhibitors. Studies are also carried out directly on the sea floor using advanced instrumentation such as autonomous benthic landers, remotely operated vehicles (ROV's), or manned submersibles. Benthic landers have been constructed which can be deployed on the open ocean from a ship, sink freely to the deep sea floor, carry out pre-programmed measurements while storing data or samples, release ballast and ascend again to the sea surface to be finally retrieved by the ship (Tengberg et al. 1995). Such in situ instruments may be equipped with A) samplers and analytical instruments for studying the benthic boundary layer (Thomsen et al. 1996), B) microsensors for high-resolution measurements of chemical gradients in the sediment (Reimers 1987; Gundersen and Jørgensen 1990), C) flux chambers or eddy correlation instrumentation for measurements of the exchange of dissolved species across the sediment-water interface (Smith et al. 1976; Berelson et al. 1987; Berg et al. 2004), D) coring devices for tracer injection and measurements of processes down to 0.5 meter sediment depth (Greeff et al. 1998).

Progress in microsensor technology has also stimulated research on the interaction between processes, bacteria and environment at a high spatial resolution (cf. Sect. 3.4). Microsensors currently used in marine research can analyze O_2, CO_2, pH, NO_3^-, Ca^{2+}, S^{2-}, H_2S, CH_4 and N_2O as well as physical parameters such as temperature, light, flow or position of the solid-water interface (Kühl and Revsbech 2001). Sensors have mostly been based on electrochemical principles. However, microsensors based on optical fibers (optodes) or sensors based on enzymes or bacteria (biosensors) are gaining importance. Two-dimensional optical sensors, planar optodes, for oxygen, pH or CO_2 can be used to map the dynamic distribution of chemical species at the sediment-water interface (Glud et al. 2001). Examples of microsensor data from marine sediments are given in Chapter 3.

Among the important contributions of micro-biologists to biogeochemistry is to quantify the populations of bacteria, isolate new types of microorganisms from the marine environment, study their physiology and biochemistry in labo-ratory cultures, and thereby describe the microbial diversity and metabolic potential of natural microbial communities. There are currently about a thousand species of prokaryotes described from the sea and the number is steadily increasing. The rather time-consuming task of isolating and describing new bacterial species, however, sets a limit to the rate of progress. The classical micro-biological approaches also have shortcomings, among others that only the viable or culturable bacteria are recognized, and that the counting procedures for these viable bacteria, furthermore, underestimate the true population size by several orders of magnitude.

In recent years, the rapid developments in molecular biology have opened new possibilities for the study of bacterial populations and their metabolic activity without cultivation, even at the level of individual cells. A new classification of

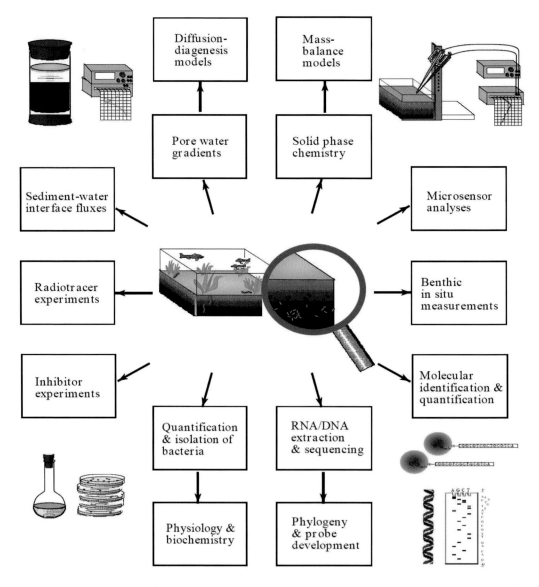

Fig. 5.2 Approaches of marine biogeochemistry. Top: Geochemical methods based on solute and solid phase analyses and modelling. Left: Experimental methods for the analyses of process rates. Bottom: Identification, quantification and characterization of the microbial populations. Right: High resolution and in situ methods for the analyses of microbial populations and their microenvironment. Graphics by Niels Birger Ramsing.

the prokaryotic organisms has developed at a genetic level, based on sequence analyses of small-subunit ribosomal RNA (16S rRNA), which now allows a phylogenetic identification of microorganisms, even of those that have not yet been isolated and studied in laboratory cultures. This has revealed much greater species diversity than had been anticipated just a decade ago and the number of "species" according to this genetic diversity has reached ca 10.000, or ten-fold more than the isolated species, and it is rapidly growing. The RNA and DNA based methods have for the first time enabled a true quantification of defined bacterial groups in nature. By the use of molecular probes that bind specifically to ribosomal RNA of selected target organisms both cultured and uncultured bacteria can now be identified phylogenetically. When such probes are fluorescently labelled, a sediment sample may be stained by the probes and individual cells of the target bacteria can then be observed and counted directly under a fluorescense microscope. This technique is called Fluorescence in situ Hybridization (FISH) and has become an indispensable tool in biogeochemistry.

As an example of the quantification of bacteria, Llobet-Brossa et al. (1998) found that the total cell density of microorganisms in sediment from the German Wadden Sea was up to $4 \cdot 10^9$ cells cm^{-3}. Such a population density of several billion microorganisms in a teaspoon of mud is typical of coastal marine sediments. Of all these cells, up to 73% could be identified by FISH as eubacteria and up to 45% were categorized to known groups of eubacteria. The sulfate reducing bacteria comprised some 10-15% of the identified microorganisms, while other members of the group proteobacteria comprised 25-30%. Surprisingly, members of the *Cytophaga-Flavobacterium* group were the most abundant in all sediment layers. These bacteria are specialized in the degradation of complex macromolecules and their presence in high numbers indicates their role in the initial hydrolytic degradation of organic matter.

Such studies of the spatial distribution of bacteria provide information on where they may be actively transforming organic or inorganic compounds, and thereby indicate which processes are likely to take place. This is particularly important when chemical species are rapidly recycled so that the dynamics of their production or consumption is not easily revealed by geochemical analyses. The introduction of high

resolution tools such as microsensors and molecular probes has helped to overcome one of the classical problems in biogeochemistry, namely to identify the relationship between the processes that biogeochemists analyze, and the microorganisms who carry them out. The magnitude of this problem may be appreciated when comparing the scale of bacteria with that of humans. The bacteria are generally 1-2 µm large, while we are 1-2 m. As careful biogeochemists we may use a sediment sample of only 1 cm^3 to study the metabolism of, e.g. methane producing bacteria. For the study of the metabolism of humans, this sample size would by isometric analogy correspond to a soil volume of 1000 km^3. Thus, it is not surprising that very sensitive methods, such as the use of radiotracers, are often necessary when bacterial processes are to be demonstrated over short periods of time (hours-days). It is also obvious, that a 1 cm^3 sample will include a great diversity of prokaryotic organisms and metabolic reactions and that a much higher resolution is required to sort out the activities of individual cells or clusters of organisms.

5.2 Life and Environments at Small Scale

The size spectrum of living organisms and of their environments is so vast that it is difficult to comprehend (Fig. 5.3). The smallest marine bacteria with a size of <0.4 µm are at the limit of resolution of the light microscope, whereas the largest whales may grow to 30 m in length, eight orders of magnitude larger. The span in biomass is nearly the third power of this difference, less than $(10^8)^3$ or about 10^{22} (since whales are not spherical), which is comparable to the mass ratio between humans and the entire Earth. It is therefore not surprising that the world as it appears in the microscale of bacteria is also vastly different from the world we humans can perceive and from which we have learned to appreciate the physical laws of nature. These are the classical laws of Newton, relating mass, force and time with mass movement and flow and with properties such as acceleration, inertia and gravitation. As we go down in scale and into the microenvironment of marine bacteria, these properties lose their significance. Instead, viscosity becomes the strongest force and molecular diffusion the fastest transport.

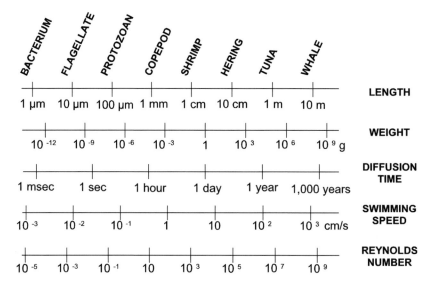

Fig. 5.3 Relationships between size, diffusion time and Reynolds number. Representative organisms of the different scales and their length and biomass are indicated. The diffusion times were calculated for small solutes with a molecular diffusion coefficient of 10^{-5} cm^2 s^{-1}. The Reynolds numbers were calculated for organisms swimming at a speed typical for their size or for water parcels of similar size and moving at similar speed.

5.2.1   Hydrodynamics of Low Reynolds Numbers

Water movement on a large scale is characterized by inertial flows and turbulence. These are the dominant mechanisms of transport and mixing, yet they are inefficient in bringing substrates to the microorganisms living in the water. This is because the fundamental property required for turbulence, namely the inertial forces associated with mass, plays no role for very small masses, or at very small scales, relative to the viscosity or internal friction of the fluid. In the microscale, time is insignificant for the movement of water, particles or organisms and only instantaneous forces are important. Furthermore, fluid flow is rather simple and predictable, the water sticking to any solid surface and adjacent water volumes slipping past it in a smooth pattern of laminar flow. The transition from laminar to turbulent flow depends on the scale and the flow velocity and is described by the dimensionless Reynolds number, Re, which expresses the ratio between inertial and viscous forces affecting the fluid or the particle considered:

$$Re = uL/v \qquad (5.6)$$

where u is the velocity, L is the characteristic dimension of the water parcel or particle, and v is

the kinematic viscosity of the seawater (ca. 0.01 cm^2 s^{-1} at 20°C). The transition from low (<1) to high Reynolds numbers for swimming organisms, for sinking particles, or for hydrodynamics in general lies in the size range of 0.1-1 mm.

5.2.2   Diffusion at Small Scale

Diffusion is a random movement of molecules due to collision with water molecules which leads to a net displacement over time, a 'random walk'. When a large number of molecules is considered, the mean deviation, L (more precisely: the root mean square of deviations), from the starting position is described by a simple but very important equation, which holds the secret of diffusion:

$$L = \sqrt{2Dt} \qquad (5.7)$$

where D is the diffusion coefficient and t is the time. This equation says that the distance molecules are likely to travel by diffusion increases only with the square root of time, not with time itself as in the locomotion of objects and fluids which we generally know from our macroworld. Expressed in a different way, the time needed for diffusion increases with the square of the distance:

$$t = \frac{L^2}{2D} \qquad (5.8)$$

These are very counter-intuitive relations which have some surprising consequences. From Eq. 5.7 one can calculate the 'diffusion velocity', which is distance divided by time:

$$\text{'Diffusion velocity'} = \frac{L}{t} = \sqrt{2D/t} \qquad (5.9)$$

This leads to the curious conclusion that the shorter the period over which we measure diffusion, the larger is its velocity and vice versa. This is critical to keep in mind when working with diagenetic models, because the different chemical species have time and length scales of diffusion and reaction which vary over many orders of magnitude and which are therefore correspondingly difficult to compare.

Calculations from Eq. 5.8 of the diffusion times of oxygen molecules at 10°C show that it takes one hour for a mean diffusion distance of 3-4 mm, whereas it takes a day to diffuse 2 cm and 1000 years for 10 meters (Table 5.1). For a small organic molecule such as glucose, these diffusion times are about three times longer. Over the scale of a bacterium, however, diffusion takes only 1/1000 second. Thus, for bacteria of 1 μm size, one could hardly envision a transport mechanism which would outrun diffusion within a millisecond. The transition between predominantly diffusive to predominantly advective or turbulent transport of solutes lies in the range of 0.1 mm for actively swimming organisms and somewhat higher for passively sinking marine aggregates (Fig. 5.3).

5.2.3 Diffusive Boundary Layers

The transition from a turbulent flow regime with advective and eddy transport to a small scale dominated by viscosity and diffusional transport is apparent when an impermeable solid-water interface such as the sediment surface is approached (Fig. 5.4). According to the classical eddy diffusion theory, the vertical component of the eddy diffusivity, E, decreases as a solid interface is approached according to: $E = A \nu Z^{3-4}$, where A is a constant, ν is the kinematic viscosity, and Z is the height above the bottom. An exponent of 3-4

Table 5.1 Mean diffusion times for O_2 and glucose over distances ranging from 1 μm to 10 m.

Diffusion distance	Time (10°C)	
	Oxygen	Glucose
1 μm	0.34 ms	1.1 ms
3 μm	3.1 ms	10 ms
10 μm	34 ms	110 ms
30 μm	0.31 s	1.0 s
100 μm	3.4 s	10 s
300 μm	31 s	100 s
600 μm	2.1 min	6.9 min
1 mm	5.7 min	19 min
3 mm	0.8 h	2.8 h
1 cm	9.5 h	1.3 d
3 cm	3.6 d	12 d
10 cm	40 d	130 d
30 cm	1.0 yr	3.3 yr
1 m	10.8 yr	35 yr
3 m	98 yr	320 yr
10 m	1090 yr	3600 yr

shows that the eddy diffusivity drops very steeply as the sediment surface is approached. In the viscous sublayer, which is typically about 1 cm thick in the deep sea, the eddy diffusivity falls below the kinematic viscosity of ca 10^{-2} cm^2 s^{-1}. Even closer to the sediment surface, the vertical eddy diffusion coefficient for mass falls below the molecular diffusion coefficient, D, which is constant for a given solute and temperature, and which for small dissolved molecules is in the order of 10^{-5} cm^2 s^{-1}. The level where E becomes smaller than D defines the diffusive boundary layer, δ_e, which is typically about 0.5 mm thick. In this layer, molecular diffusion is the predominant transport mechanism, provided that the sediment is impermeable and stable.

The diffusive boundary layer plays an important role for the exchange of solutes across the sediment-water interface (Jørgensen 2001). For chemical species which have a very steep gradient in the diffusive boundary layer it may limit the flux and thereby the rate of chemical reaction. This may be the case for the precipitation of manganese on iron-manganese nodules (Boudreau 1988) or for the dissolution of carbonate shells and other minerals such as alabaster in the deep sea (Santschi et al. 1991). For chemical species with a

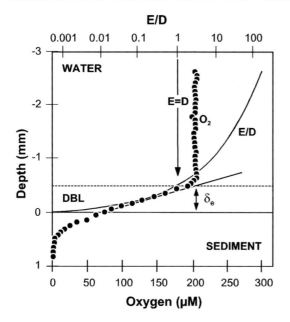

Fig. 5.4 Oxygen microgradient (data points) at the sediment-water interface compared to the ratio, E/D (logarithmic scale), between the vertical eddy diffusion coefficient, E, and the molecular diffusion coefficient, D. Oxygen concentration was constant in the overflowing seawater. It decreased linearly within the diffusive boundary layer (DBL), and penetrated only 0.7 mm into the sediment. The DBL had a thickness of 0.45 mm. Its effective thickness, δ_e is defined by the intersection between the linear DBL gradient and the constant bulk water concentration. The diffusive boundary layer occurs where E becomes smaller than D, i.e. where E/D = 1 (arrow). Data from Aarhus Bay, Denmark, at 15 m water depth during fall 1990 (Gundersen et al. 1995).

weak gradient, such as sulfate, the diffusive boundary layer plays no role since the uptake of sulfate is totally governed by diffusion-reaction within the sediment. The function of the diffusive boundary layer as a barrier for solute exchange is also reflected in the mean diffusion time of molecules through the layer, which is about 1 min over a 0.5 mm distance (Table 5.1).

The existence of a diffusive boundary layer is apparent from microsensor measurements of oxygen and other solutes at the sediment-water interface. In a Danish coastal sediment, the concentration of oxygen dropped steeply over the 0.5 mm thick boundary layer (Fig. 5.4) which consequently had a significant influence on the regulation of oxygen uptake in this sediment of high organic-matter turnover. Figure 5.5 shows, as another example, oxygen penetration down to 13 mm below the surface in fine-grained sediment. The profile was measured in situ in the seabed by a free-falling benthic lander operating with a

100 µm depth resolution, which was just sufficient to resolve the diffusive boundary layer. Based on the boundary layer gradient, the vertical diffusion flux of oxygen across the water-sediment interface can be calculated (cf. Chap. 3).

5.3   Regulation and Limits of Microbial Processes

Bacteria and other microorganisms are the great biological catalysts of element cycling at the sea floor. The degradation and remineralization of organic matter and many redox processes among inorganic species are dependent on bacterial catalysis, which may accelerate such processes up to 10^{20}-fold relative to the non-biological reaction rate. It is, however, important to keep in mind that this biological catalysis is based on living organisms, each of which has its special

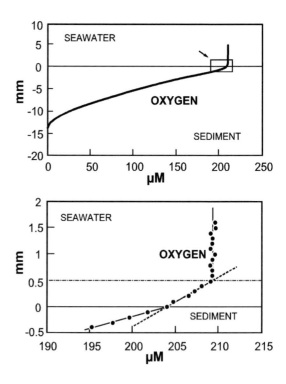

Fig. 5.5 Oxygen gradient measured in situ by a benthic lander in Skagerrak at the transition between the Baltic Sea and the North Sea at 700 m water depth. Due to the high depth resolution of the microelectrode measurements it was possible to analyze the O_2 microgradient within the 0.5 mm thick diffusive boundary layer. The framed part in the upper graph is blown up in the lower graph. (Data from Gundersen et al. 1995).

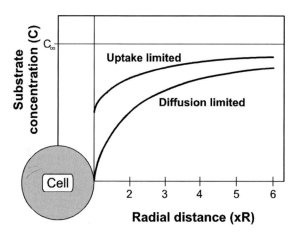

Fig. 5.6 Theoretical concentration gradient of substrate molecules around a spherical cell at different radial distances from its center (R is the radius of the cell). The concentration of substrate in the bulk water is C_∞ and concentration curves were calculated from Eq. 5.10 for a cell limited in its substrate uptake by external diffusion ('Diffusion limited') or by the uptake capacity across its own cell membrane ('Uptake limited'). Note how the substrate concentration only gradually approaches the bulk concentration with increasing distance from the cell.

requirements and limits for its physical and chemical environment, its supply of nutrients, growth rate and mortality, interaction with other organisms, etc..

5.3.1   Substrate Uptake by Microorganisms

Among the many reasons for the importance of prokaryotic organisms in biogeochemical cycling are:

A) Their metabolic versatility, especially the many types of anaerobic metabolism and the ability to degrade complex polymeric substances or to catalyze reactions among inorganic compounds.

B) Their small size which allows them to inhabit nearly any environment on the surface of the Earth and which strongly enhances the efficiency of their catalytic activity.

C) The wide range of environmental conditions under which they thrive, including temperature, salinity, pH, hydrostatic pressure, etc..

The metabolic versatility of prokaryotic organisms is discussed in Section 5.4. The small size of the

individual cells is related to their nutrition exclusively on solutes such as small organic molecules, inorganic ions or gases. The uptake of food by the prokaryotic cells thus takes place principally by molecular diffusion of small molecules to the cell surface and their transport through the cytoplasmic membrane into the cell. This constrains the relationships between cell size, metabolic rate, substrate concentration and molecular diffusion coefficients (D) (e.g. Koch 1990; 1996; Karp-Boss et al. 1996). The concentration gradient around the spherical cell is:

$$C_r = (R/r) \cdot (C_0 - C_\infty) + C_\infty \ , r > R \qquad (5.10)$$

where C_r is the concentration at the radial distance, r, C_0 is the substrate concentration at the cell surface, C_∞ is the ambient substrate concentration, and R is the radius of the cell (Fig. 5.6). The maximal substrate uptake rate of a cell is reached when the substrate concentration at the cell surface is zero. The total diffusion flux, J, to the cell is then:

$$J = 4 \pi D R \, C_\infty \qquad (5.11)$$

This flux provides the maximal substrate supply to the diffusion-limited cell, which has a volume of $4/3 \pi R^3$. The flux thus determines the maximal specific rate of bacterial metabolism of the substrate molecules, i.e. the metabolic rate per volume of biomass:

$$\text{Specific metabolic rate} = (4 \pi D R \, C_\infty)/(^4/_3 \pi R^3) = (3D/R^2)C_\infty \qquad (5.12)$$

Eq. 5.12 shows that the biomass-specific metabolic rate of the diffusion-limited cell varies inversely with the square of its size. This means that the cell could potentially increase its specific rate of metabolism 4-fold if the cell diameter were only half as large. The smaller the cell, the less likely it is that its substrate uptake will reach diffusion limitation. Thus, at the low substrate concentrations normally found in marine environments, microorganisms avoid substrate limitation by forming small cells of <1 μm size. Thereby, the bacteria become limited by their transport efficiency of molecules across the cell membrane rather than by diffusion from their surroundings (Fig. 5.6). In the nutrient-poor seawater, where substrates are available only in sub-micromolar and even nanomolar concentrations, free-living bacteria may have an impressingly high substrate uptake efficiency which corresponds to each second clearing the substrate from a seawater volume that is several hundred times their own volume. In sediments, on the other hand, bacteria often form microcolonies or their diffusion supply is impeded by sediment structures so that the microorganisms may at times be diffusion limited in their substrate uptake. It is not clear, however, how this affects the kinetics of substrate turnover in sediments.

The recent discovery of 'giant' bacteria of up to ¾ mm diameter in the seabed off Namibia seems contradictory to the constraints on bacterial size discussed above (Schulz et al. 2001; Chap. 8). These sulfide-oxidizing and nitrate-reducing bacteria of a new genus, *Thiomargarita namibiensis*, have a unique intracellular organelle, a liquid vacuole, which fills more than 90% of the cellular volume. The cytoplasm is thus only a thin peripheral layer a few μm thick that is not diffusion limited.

5.3.2   Substrate Limitation in the Deep Sub-surface

The theoretical limit for substrate availability required to sustain bacterial growth and survival is of great biogeochemical significance, e.g. in relation to the exceedingly slow degradation of organic material in million-year old sediments and oil reservoirs deep under the sea floor (Stetter et al. 1993; Parkes et al. 1994). Until a few decades ago, the microbial world was thought to be limited to the upper meters of the seabed, while the deeper sediments were thought to be sterile in spite of significant amounts of organic material buried at much greater depths. The exploration of the deep sub-seafloor biosphere started in the early 1980'ies when the first evidence of microbial activity was provided by studies of methane formation and sulfate reduction in cores obtained from the Deep Sea Drilling Program. Systematic counts of fluorescently stained cells in cores of the Ocean Drilling Program have since then led to a large data base on the population size of deep biosphere microorganisms (Parkes et al. 2000; D'Hondt et al. 2004). Fig. 5.7 shows in a double-logarithmic plot the depth distribution of these prokaryotic cells, from the top cm to the greatest sampling depth of 800 mbsf. The plot shows a large scatter in population density between different sites cored but also a systematic

decrease in cell numbers from $>10^9$ cells cm^{-3} at the surface to $<10^6$ cells cm^{-3} at depth. Whitman et al. (1998) made a global extrapolation based on the available data and came to the astonishing conclusion that the prokaryotes of sub-seafloor sediments constitute a "hidden majority" equivalent to 1/2 to 5/6 of Earth's prokaryotic biomass and 1/10 to 1/3 of Earth's total living biomass.

This vast prokaryotic population plays a critical role in global carbon cycling by controlling the amount of deposited organic material that becomes buried to great depth in the seabed and stored there for many millions of years. The gradual geothermal heating of sediments as they become buried to many hundred meters depth enhances the availability of even highly refractory

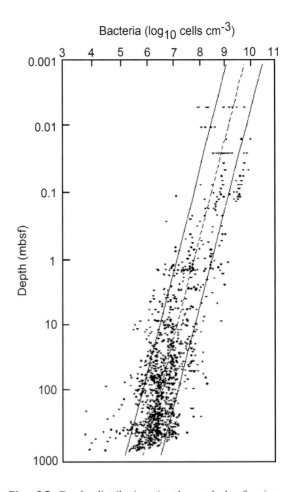

Fig. 5.7 Depth distribution in the seabed of micro-organisms enumerated by direct microscopic counts of cells stained by a fluorescent DNA stain. The graph shows the global data from ODP cores in a double-log plot, from the sediment surface to 800 m subsurface. Data from Parkes et al. (2000).

organic material to microbial attack and leads to further mineralization with the ultimate formation of methane (Wellsbury et al. 1997). The methane slowly diffuses upwards from the deep deposits and becomes oxidized to CO_2 as it reaches the bottom of the sulfate zone (Chap. 8), or it accumulates as gas hydrate within the upper sediment strata (Borowski et al. 1996). The gas hydrates constitute a reduced carbon reservoir that greatly exceeds the amount of carbon in all living organisms on Earth (Kvenvolden, 1993; Chap. 14).

It remains an open question how the microorganisms in subsurface sediments are able to survive on the extremely low energy and carbon flux available for each cell. By DNA/RNA based techniques it could be shown that a significant fraction of the bacteria are alive and active rather than dormant or even dead (Schippers et al. 2005). Yet, calculations from cell numbers and measured or modeled mineralization rates show that the organic carbon flux available in million year old sediments allows only generation times of years to thousands of years (Whitman et al. 1998; Schippers et al. 2005).

5.3.3 Temperature as a Regulating Factor

The sea floor is mostly a cold environment with 85% of the global ocean having temperatures below 5°C. At the other extreme, hydrothermal

vents along the mid-oceanic ridges have temperatures reaching above 350°C. In each temperature range from the freezing point to an upper limit of around 110°C there appear to be prokaryotic organisms that are well adapted and even thrive optimally at that temperature. Thus, extremely warm-adapted (hyperthermophilic) methane producing or elemental-sulfur reducing bacteria, which live at the boiling point of water, are unable to grow at temperatures below 60-70°C because it is too cold for them (Stetter 1996).

It is well known that chemical processes as well as bacterial metabolism are slowed down by low temperature. Yet, the biogeochemical recycling of deposited organic material in marine sediments does not appear to be less efficient or less complete in polar regions than in temperate or tropical environments. Sulfate reduction rates in marine sediments of below 0°C around Svalbard at 78° north in the Arctic Ocean are comparable to those at similar water depths along the European coast (Sagemann et al. 1998). Figure 5.8A shows the short-term temperature dependence of sulfate reduction rates in such Svalbard sediments. As is typical for microbial processes, the optimum temperature of 25-30°C is high above the in situ temperature, which was -1.7°C during summer. Above the optimum, the process rate dropped steeply, which is due to enzymatic denaturation and other physiological malfunctioning of the cells and which shows that this is a

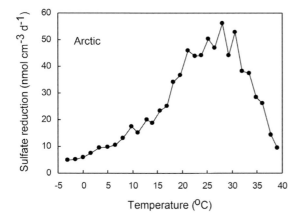
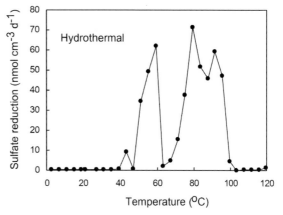

Fig. 5.8 Temperature regulation of bacterial sulfate reduction in different marine sediments. The data show the rates of sulfate reduction in a homogenized sample from the upper few tens of cm of sediment measured by short-term incubations in a temperature gradient block with radiolabelled sulfate as a tracer. A) Arctic sediment from 175 m depth in Storfjorden on Svalbard at 78°N where the in situ temperature was –1.7°C. (Data from Sagemann et al. 1998). B) Hydrothermal sediment from the Guaymas Basin, Gulf of California, at 2000 m depth. The hydrothermal fluid here seeps up through fine-grained, organic-matter rich sediment which was collected at 12-18 cm depth where the *in situ* temperature was 54-71°C. (Data from Weber and Jørgensen 2002).

biologically catalyzed process. Chemically catalyzed processes, in contrast, tend to have a continuous increase in rate with increasing temperature. Although the sulfate reduction was slow at <0°C this does not mean that the bacteria do not function well at low temperature. On the contrary, the Arctic sulfate reducers had their highest growth efficiency (i.e. highest biomass production per amount of substrate consumed) at around 0°C, in contrast to sulfate reducers from temperate environments which have their highest growth efficiency at the warm temperatures experienced during summer (Isaksen and Jørgensen 1996; Knoblauch and Jørgensen 1999).

The radiotracer method of measuring sulfate reduction in sediments (see Sect. 5.6) is a sensitive tool to demonstrate the temperature strains among the environmental microorganisms. In 54-71°C warm sediment from the hydrothermal sediments of the Guaymas Basin, Gulf of California, two main groups of sulfate reducers could be discriminated from their temperature optima: a) moderate thermophiles with an optimum at 60°C, b) extreme thermophiles with optimum at 80-90°C (Fig. 5.8B). Moderately thermophilic sulfate reducers are well known from pure cultures of *Desulfotomaculum*, whereas extreme thermophiles with optimum at 85°C are known among the genus, *Archaeoglobus* (Stetter 1988). Such prokaryotic organisms have been isolated from hydrothermal environments and have also been found in oil reservoirs 3000 m below the seabed where the temperature is up to 110°C and the hydrostatic pressure up to 420 bar.

Surprisingly, sulfate reducers adapted to the normal low temperatures of the main sea floor were also unknown until recently. Such cold-adapted (psychrophilic) bacteria are generally scarce among the culture collections in spite of their major biogeochemical significance. This is partly because of their slow growth which makes them difficult to isolate and cumbersome to study. A number of psychrophilic sulfate reducers were recently isolated which have temperature optima down to 7°C and are unable to grow above 10-15°C because it is too hot (Knoblauch et al. 1999).

These examples demonstrate a general problem in marine microbiology, namely that the large prokaryotic diversity comprises adaptations to very diverse environments and that only a very small fraction, maybe less than one percent, of the bacterial species in the ocean is known to science today. Many of the unknown microorganisms may be among the biogeochemically very important species. The estimate is based on recent methodological advances in molecular biology that have made it possible to analyze the diversity of natural prokaryotic populations, including the large majority that has still not been isolated or studied.

5.3.4 Other Regulating Factors

The major area of the sea floor lies in the deep sea below several thousand meters of water where an enormous hydrostatic pressure prevails. Since the pressure increases by ca 1 bar (1 atm) for every 10 meters, bacteria living at 5000 m depth must be able to withstand a pressure of 500 bar (50 MPa). Microorganisms isolated from sediments down to 3000-4000 m have been found to be preferentially barotolerant, i.e. they grow equally well at sea surface pressure as at their in situ pressure. At depths exceeding 4000 m the isolated bacteria become increasingly barophilic, i.e. they grow optimally at high pressures, and bacteria isolated from deep sea trenches at 10,000 m depth were found to grow optimally at 700-1000 bar (Yayanos 1986). Many barophilic bacteria are also psychrophilic in accordance with the low temperature of 1-4°C prevailing in the deep sea (DeLong et al. 1997). They appear to grow relatively slowly which may be an adaptation to low temperature and low nutrient availability rather than a direct effect of high pressure. The degradation of organic material appears to be just as efficient in the deep sea as in shallower water, since only a few percent of sedimenting detritus resist mineralization and are buried deep down into the sediment.

5.4 Energy Metabolism of Prokaryotes

Microorganisms can be considered 1 μm large bags of enzymes in which the important biogeochemical processes are catalyzed. This analogy, however, is too crude to understand how and why these microscopic organisms drive and regulate the major cycles of elements in the ocean. This requires a basic knowledge of their energy metabolism and physiology.

The cells use the chemical energy of organic or inorganic compounds for cell functions which

Table 5.2 Pathways of organic matter oxidation, hydrogen transformation and fermentation in the sea floor and their standard free energy yields, ΔG^0, per mol of organic carbon. [CH_2O] symbolizes organic matter of unspecified composition. ΔG^0 values at pH 7 according to Thauer et al. (1977), Conrad et al. (1986), and Fenchel et al. (1998). (cf. Fig. 3.11).

Pathway and stoichiometry of reaction	ΔG^0 (kJ mol^{-1})
Oxic respiration:	
[CH_2O] + $O_2 \rightarrow CO_2 + H_2O$	-479
Denitrification:	
$5[CH_2O] + 4NO_3^- \rightarrow 2N_2 + 4HCO_3^- + CO_2 + 3H_2O$	-453
Mn(IV) reduction:	
[CH_2O] + $3CO_2 + H_2O + 2MnO_2 \rightarrow 2Mn^{2+} + 4HCO_3^-$	-349
Fe(III) reduction:	
[CH_2O] + $7CO_2 + 4Fe(OH)_3 \rightarrow 4Fe^{2+} + 8HCO_3^- + 3H_2O$	-114
Sulfate reduction:	
$2[CH_2O] + SO_4^{2-} \rightarrow H_2S + 2HCO_3^-$	-77
$4H_2 + SO_4^{2-} + H^+ \rightarrow HS^- + 4H_2O$	-152
$CH_3COO^- + SO_4^{2-} + 2H^+ \rightarrow 2CO_2 + HS^- + 2H_2O$	-41
Methane production:	
$4H_2 + HCO_3^- + H^+ \rightarrow CH_4 + 3H_2O$	-136
$CH_3COO^- + H^+ \rightarrow CH_4 + CO_2$	-28
Acetogenesis:	
$4H_2 + 2CO_3^- + H^+ \rightarrow CH_3COO^- + 4H_2O$	-105
Fermentation:	
$CH_3CH_2OH + H_2O \rightarrow CH_3COO^- + 2H_2 + H^+$	10
$CH_3CH_2COO^- + 3H_2O \rightarrow CH_3COO^- + HCO_3^- + 3H_2 + H^+$	77

require work: growth and division, synthesis of macromolecules, transport of solutes across the cell membrane, secretion of exoenzymes or exopolymers, movement etc. The organisms catalyze reduction-oxidation (redox) processes from which they conserve a part of the energy and couple it to the formation of a proton gradient across the cytoplasmic membrane. The socalled *proton motive force* established by this gradient is comparable to the electron motive force of an ordinary battery. It is created by the difference in electrical charge and H^+ concentration between the inside (negative and low H^+, i.e. alkaline) and the outside (positive and high H^+) of the membrane. By reentry of protons into the cell through membrane-bound ATPase protein complexes, it drives the formation of high energy phosphate bonds in compounds such as *ATP* (adenosine triphosphate), which functions as a transient storage of the energy and is continuously recycled as the phosphate bond is cleaved in energy-requiring processes. ATP is utilized very widely in organisms as the fuel to drive energy requiring processes, and a fundamental question in all cellular processes is, how much ATP do they produce or consume?

Redox processes, whether biological or chemical, involve a transfer of one or more electrons between the chemical reactants. An example is the oxidation of ferrous iron to ferric iron by oxygen at low pH:

$$2Fe^{2+} + \tfrac{1}{2}O_2 + 2H^+ \rightarrow 2Fe^{3+} + H_2O \qquad (5.13)$$

By this reaction, a single electron is transferred from each Fe^{2+} ion to the O_2 molecule. The ferrous iron is thereby oxidized to ferric iron, whereas the oxygen is reduced to water. In the geochemical literature, O_2 in such a reaction is termed the *oxidant* and Fe^{2+} the *reductant*. In the biological literature, the terms *electron acceptor* for O_2 and *electron donor* for Fe^{2+} are used.

5.4.1   Free Energy

Chemical reactions catalyzed by microorganisms yield highly variable amounts of energy and some are directly energy consuming. The term *free energy*, G, of a reaction is used to express the energy released per mol of reactant, which is available to do useful work. The change in free energy is conventionally expressed as ΔG^0, where the symbol Δ should be read as 'change in' and the superscript ^{0} indicates the following standard conditions: pH 7, 25°C, a 1 M concentration of all reactants and products and a 1 bar partial pressure of gases. The value of ΔG^0 for a given reaction is expressed in units of kilojoule (kJ) per mol of reactant. If there is a net decrease in free energy (ΔG^0 is negative) then the process is *exergonic* and may proceed spontaneously or biologically catalyzed. If ΔG^0 is positive, the process is *endergonic* and energy from ATP or from an accompanying process is required to drive the reaction. The change in free energy of several metabolic processes in prokaryotic organisms is listed in Table 5.2.

A thorough discussion of the theory and calculation of ΔG^0 for a variety of anaerobic microbial processes is given by Thauer et al. (1977). As a general rule, processes for which the release of energy is very small, < ca 20 kJ mol^{-1}, are insufficient for the formation of ATP and are thus unable to serve the energy metabolism of microorganisms (Hoehler et al. 1998). It is important to note, however, that standard conditions are seldom met in the marine environment and that the actual conditions of pH, temperature and substrate/product concentrations must be known before the energetics of a certain reaction can be realistically calculated. Several processes, which under standard conditions would be endergonic, may be exergonic in the normal marine sediment.

An important example of this is the formation of H_2 in several bacterial fermentation processes which is exergonic only under low H_2 partial pressure. The hydrogen cycling in sediments is therefore dependent on the immediate consumption of H_2 by other organisms, such as the sulfate reducing bacteria, which keep the partial pressure of H_2 extremely low. The H_2-producing and the H_2-consuming bacteria thus tend to grow in close proximity to each other, thereby facilitating the diffusional transfer of H_2 at low concentration from one organism to the other, a so-called 'inter-species hydrogen transfer' (Conrad et al. 1986; Schink 1997).

5.4.2   Reduction-Oxidation Processes

The electron transfer in redox processes is often accompanied by a transfer of protons, H^+. The simplest example is the oxidation of H_2 with O_2 by the so-called 'Knallgas-bacteria', which occur widespread in aquatic sediments:

$$H_2 + \tfrac{1}{2}O_2 \rightarrow H_2O \qquad (5.14)$$

In the energy metabolism of cells, an intermediate carrier of the electrons (and protons) is commonly required. Such an *electron carrier* is, for instance, NAD^+ (nicotinamid adenin dinucleotide), which formally accepts two electrons and one proton and is thereby reduced to NADH. The NADH may give off the electrons again to specialized electron acceptors and the protons are released in the cell sap. Thereby, the NADH, which must be used repeatedly, is recycled.

A redox process such as Eq. 5.14 formally consists of two reversible half-reactions. The first is the oxidation of H_2 to release electrons and protons:

$$H_2 \Leftrightarrow 2e^- + 2H^+ \qquad (5.15)$$

The second is the reduction of oxygen by the transfer of electrons (and protons):

$$\tfrac{1}{2}O_2 + 2e^- + 2H^+ \Leftrightarrow H_2O \qquad (5.16)$$

Compounds such as H_2 and O_2 vary strongly in their tendency to either give off electrons and thereby become oxidized (H_2) or to accept electrons and thereby become reduced (O_2). This tendency is expressed as the redox potential, E_0', of the compounds. This potential is expressed in volts and is measured electrically in reference to a standard compound, namely H_2. By convention, redox potentials are expressed for half reactions

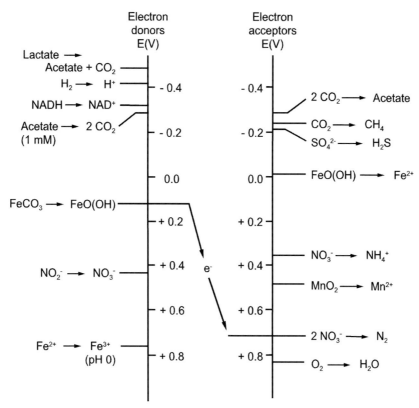

Fig. 5.9 The 'electron towers' of redox processes in biogeochemistry. By the half-reaction on the left side, electrons are released from an electron donor and are transferred to an electron acceptor in the half-reaction on the right side. The drop in redox potential between donor and acceptor is a measure of the chemical energy released by the process. The redox potentials are here calculated for standard conditions at pH 7 and 1 mM concentrations of substrates and products. As an example, the electron transfer (arrow between electron towers) is shown for the oxidation of ferrous carbonate with nitrate (see text).

which are reductions, i.e. 'oxidized form + e⁻ → reduced form'. When protons are involved in the reaction, the redox potential in the biological literature is expressed at pH 7, because the interior of cells is approximately neutral in pH. A reducing compound such as H_2, with a strong tendency to give off electrons (Eq. 5.15), has a strongly negative E_0' of -0.414 V. An oxidizing compound such as O_2, with a strong tendency to accept electrons (Eq. 5.16), has a strongly positive E_0' of +0.816 V. The free energy yield, ΔG^0, of a process is proportional to the difference in E_0' of its two half-reactions, $\Delta E_0'$, and the redox potentials of electron donor and electron acceptor in a microbial energy metabolism therefore provide important information on whether they can serve as useful substrates for energy metabolism:

$$\Delta G^0 = -n\, F\, \Delta E_0'\; [\text{kJ mol}^{-1}] \qquad (5.17)$$

where n is the number of electrons transferred by the reaction and F is Faraday's constant (96,485 Coulomb mol⁻¹).

The redox potentials of different half-reactions, that are of importance in biogeochemistry, are shown in Figure 5.9. The two 'electron towers' show the strongest reductants (most negative E_0') at the top and the strongest oxidants at the bottom. Reactions between electron donors of more negative E_0' with electron acceptors of more positive E_0' are exergonic ('downhill') and may provide the basis for biological energy metabolism. The larger the drop in E_0' between electron donor and acceptor, the more energy is released. As an example, the oxidation of ferrous iron in the form of $FeCO_3$ with NO_3^- is shown, a process that was discussed in Section 5.1.1 (Fig. 5.1, Straub et al. 1996). The oxidation of organic compounds such as lactate or acetate by O_2 releases maximum amounts of energy, and aerobic respiration is, accordingly, the basis for metazoan life and for many microorganisms.

183

5.4.3   Relations to Oxygen

Before discussing the biological reactions further, a few basic concepts should be clarified. One is the relation of living organisms to O_2. Those organisms, who live in the presence of oxygen, are termed *aerobic* (meaning 'living with oxygen'), whereas those who live in the absence of oxygen are *anaerobic*. This discrimination is both physiologically and biogeochemically important. In all aerobic organisms, toxic forms of oxygen, such as peroxide or superoxide, are formed as by-products of the aerobic metabolism. Aerobic organisms rapidly degrade these aggressive species with enzymes such as catalase, peroxidase or superoxide dismutase. Many organisms, which are obligately anaerobic, lack these enzymes and are not able to grow in the presence of oxygen for this and other reasons. Some organisms need oxygen for their respiration, yet they are killed by higher O_2 concentrations. They are *microaerophilic* and thrive at the lower boundary of the O_2 zone, between the *oxic* (containing oxygen) and the *anoxic* (without oxygen) environments. Other, non-obligately (facultatively) anaerobic bacteria may live both in *oxidized* and in *reduced* zones of the sediment. The oxidized sediment is characterized by redox potentials, as measured by a naked platinum electrode, of $E_H > 0\text{-}100$ mV and up to about +400 mV. Oxidized sediments are generally brown to olive because iron minerals are in oxidized forms. In most marine sediments the O_2 penetration is small relative to the depth of oxidized iron minerals. Most of the oxidized zone is therefore anoxic and has been termed *suboxic* (Froelich et al. 1979). The reducing sediment below the suboxic zone has E_H below 0-100 mV and down to -200 mV and may be black or gray from different forms of iron sulfide minerals.

5.4.4   Definitions of Energy Metabolism

We are now ready to explore the main types of energy metabolism. Microbiologists use a terminology for the different types based on three criteria: a) the energy source, b) the electron donor and c) the carbon source (Table 5.3). To the term for each of these criteria or their combination is added '-*troph*', meaning 'nutrition'. For each of the three criteria there are two alternatives, thus leading to $2^3 = 8$ possibilities. The combinations are, however, partly coupled and only five of them are of biogeochemical significance.

The energy source of the *phototrophic* organisms such as plants, algae and photosynthetic bacteria is light. For the *chemotrophic* organisms such as animals, fungi and many bacteria it is chemical energy, e.g. of glucose, methane or ammonium. Many phototrophic microorganisms have the capacity to switch between a phototrophic mode of life in the light and a heterotrophic mode, where they take up organic substrates.

The electron donor of green plants is an inorganic compound, H_2O, and the final electron acceptor is CO_2:

$$CO_2 + H_2O \rightarrow CH_2O + O_2 \tag{5.18}$$

where CH_2O symbolizes the organic matter in plant biomass. A more accurate stoichiometry, originally suggested by Redfield (1958) for marine phytoplankton assimilating also nitrate and phosphate as nutrients, is:

$$106CO_2 + 16NO_3^- + HPO_4^{2-} + 122H_2O + 18H^+ \rightarrow$$
$$C_{106}H_{263}O_{110}N_{16}P + 138O_2 \tag{5.19}$$

The chemical composition shows that biomass is more reduced than CH_2O, in particular due to the lipid fraction.

Table 5.3 Microbiological terminology for different types of energy metabolism.

Energy source		Electron donor		Carbon source	
Light:	*photo-*	Inorganic:	*litho-*	CO_2:	*auto-troph*
Chemical:	*chemo-*	Organic:	*organo-*	Organic C:	*hetero-troph*

The electron donor of many biogeochemically important, phototrophic or chemotrophic bacteria is also inorganic, such as H_2, Fe^{2+} or H_2S. All these organisms are called *lithotrophic* from 'lithos' meaning rock. In contrast, *organotrophic* organisms such as animals use an organic electron donor:

$$CH_2O + O_2 \rightarrow CO_2 + H_2O \qquad (5.20)$$

Finally, organisms such as the green plants, but also many lithotrophic bacteria, are *autotrophs*, i.e. they are able to build up their biomass from CO_2 and other inorganic nutrients. Animals and many bacteria living on organic substrates instead incorporate organic carbon into their biomass and are termed *heterotrophs*. Among the prokaryotes, there is not a strict discrimination between autotrophy and heterotrophy. Thus, heterotrophs generally incorporate 4-6% CO_2 (mostly to convert C_3 compounds to C_4 compounds in anaplerotic reactions, in order to compensate for C_4 compounds which were consumed for biosynthetic purposes in the cell). This heterotrophic CO_2 assimilation has in fact been used to estimate the total heterotrophic metabolism of bacterioplankton based on their dark $^{14}CO_2$ incorporation. As another example, sulfate reducing bacteria living on acetate are in principle heterotrophs but derive about 30% of their cell carbon from CO_2. Aerobic methane oxidizing bacteria, which strictly speaking are also living on an organic compound, incorporate 30-90% CO_2.

The three criteria in Table 5.3 can now be used in combination to specify the main types of energy metabolism of organisms. Green plants are photolithoautotrophs while animals are chemoorganoheterotrophs. This detail of taxonomy may seem an exaggeration for these organisms which are in daily terms called photoautotrophs and heterotrophs. However, the usefulness becomes apparent when we need to understand the function of, e.g. chemolithoautotrophs or chemolithoheterotrophs for the cycling of nitrogen, manganese, iron or sulfur in marine sediments (see Sect. 5.5).

5.4.5 Energy Metabolism of Microorganisms

The basic types of energy metabolism and representative organisms of each group are compiled in Table 5.4. Further information can be found in Fenchel et al. (1998), Ehrlich (1996), Madigan et al. (1997), Canfield et al. (2005) and other textbooks. Most photoautotrophic organisms use water to reduce CO_2 according to the highly endergonic reaction of Eq. 5.18. Since water has a high redox potential it requires more energy than is available in single photons of visible light to transfer electrons from water to an electron carrier, $NADP^+$, which can subsequently reduce CO_2 through the complex pathway of the Calvin-Benson cycle. Modern plants and algae, which use H_2O as an electron donor, consequently transfer the electrons in two steps through two photocenters. In photosystem II, electrons are transferred from H_2O which is oxidized to O_2. In photosystem I these electrons are transferred to a highly reducing primary acceptor and from there to $NADP^+$. More primitive phototrophic bacteria, which predominated on Earth before the evolution of oxygenic photosynthesis, have only one photocenter and are therefore dependent on an electron donor of a lower redox potential than water. In the purple and green sulfur bacteria or cyanobacteria, H_2S serves as such a low-E_H electron donor. The H_2S is oxidized to S^0 and mostly further to sulfate. Some purple bacteria are able to use Fe^{2+} as an electron donor to reduce CO_2. Their existence had been suggested many years ago, but they were discovered and isolated only recently (Widdel et al. 1993; Ehrenreich and Widdel 1994). This group is geologically interesting, since direct phototrophic Fe(II) oxidation with CO_2 opens the theoretical possibility for iron oxidation in the absence of O_2 some 2.0-2.5 billion years ago, at the time when the great deposits of banded iron formations were formed on Earth.

Some of the purple and green bacteria are able to grow photoheterotrophically, which may under some environmental conditions be advantageous as they cover their energy requirements from light but can assimilate organic substrates instead of spending most of the light energy on the assimilation of CO_2. The organisms may either take up or excrete CO_2 in order to balance the redox state of their substrate with that of their biomass.

5.4.6 Chemolithotrophs

The chemolithotrophs comprise a large and diverse group of exclusively prokaryotic organisms, which play important roles for mineral cycling in marine sediments (Table 5.4). They conserve energy from the oxidation of a range of

Table 5.4 Main types of energy metabolism with examples of representative organisms.

Metabolism		Energy source	Carbon source	Electron donor	Organisms
Photoautotroph			CO_2	H_2O	Green plants, algae, cyanobacteria
		Light		H_2S, S^0, Fe^{2+}	Purple and green sulfur bact. (*Chromatium*, *Chlorobium*), Cyanobacteria
Photoheterotroph			Org. C $\pm$ CO_2		Purple and green non-sulfur bact. (*Rhodospirillum*, *Chloroflexus*)
Chemo-lithotroph	Chemolitho-autotroph	Oxidation of inorganic compounds	CO_2	Aerobic: H_2S, S^0, $S_2O_3^{2-}$, FeS_2 NH_4^+, NO_2^- H_2 Fe^{2+}, Mn^{2+} Anaerobic: $H_2 + SO_4^{2-}$ $H_2S/S^0/S_2O_3^{2-} + NO_3^-$ $H_2 + CO_2 \rightarrow CH_4$ $H_2 + CO_2 \rightarrow$ acetate	Aerobic: Colorless sulfur bact. (*Thiobacillus*, *Beggiatoa*) Nitrifying bact. (*Thiobacillus*, *Nitrobacter*) Hydrogen bact. (*Hydrogenomonas*) Iron bact. (*Ferrobacillus*, *Shewanella*) Anaerobic: Some sulfate reducing bact. (*Desulfovibrio spp.*) Denitrifying sulfur bact. (*Thiobac. denitrificans*) Nitrate reducing sulfur bacteria (*Thiomargarita*) Methanogenic archaea Acetogenic bact.
	Mixotroph		CO_2 or Org. C	H_2S, S^0, $S_2O_3^{2-}$	Colorless sulfur bact. (some *Thiobacillus*)
	Chemolitho-heterotroph		Org. C	H_2S, S^0, $S_2O_3^{2-}$ H_2	Colorless sulfur bact. (some *Thiobacillus*, *Beggiatoa*) Some sulfate reducing bact.
Heterotroph (=chemoorganotroph)		Oxidation of organic compounds	Org. C (max. 30% CO_2)	Org. C	Aerobic: Animals, fungi, many bacteria Anaerobic: Denitrifying bacteria Mn- or Fe-reducing bacteria Sulfate reducing bacteria Fermenting bacteria
			Org. C_1 (30-90% CO2)	$CH_4 + O_2$ $CH_4 + SO_4^{2-}$	Methane oxidizing bacteria Methane oxidizing archaea

inorganic compounds. Many are autotrophic and assimilate CO_2 via the Calvin cycle similar to the plants. Some are heterotrophic and assimilate organic carbon, whereas some can alter between these carbon sources and are termed *mixotrophs*. Many respire with O_2, while others are anaerobic and use nitrate, sulfate, CO_2 or metal oxides as electron acceptors.

It is important to note that the autotrophic chemolithotrophs have a rather low growth yield (amount of biomass produced per mol of substrate consumed). For example, the sulfur bacteria use most of the electrons and energy from sulfide oxidation to generate ATP. They need this transient energy storage for CO_2 assimilation via the Calvin cycle (reductive pentose phosphate cycle) which energetically is a rather inefficient pathway in spite of its widespread ocurrence among autotrophic organisms. Only some 10-20% of the electrons derived from H_2S oxidation flow into CO_2 and are used for autotrophic growth (Kelly 1982). As a consequence, it has a limited significance for the overall organic carbon budget of marine sediments whether sulfide is oxidized autocatalytically without bacterial involvement or biologically with a resulting formation of new bacterial biomass (Jørgensen and Nelson 2004). In spite of this low growth yield, filamentous sulfur bacteria such as *Beggiatoa* or *Thioploca* may often form mats of large biomass at the sediment surface and thereby influence the whole community of benthic organisms as well as the chemistry of the sediment (Fossing et al. 1995).

Among the aerobic chemolithotrophs, colorless sulfur bacteria may oxidize H_2S, S^0, $S_2O_3^{2-}$ or FeS_2 to sulfate. Many H_2S oxidizers, such as *Beggiatoa*, accumulate S^0 in the cells and are conspicuous due to the highly light-refractive S^0-globules which give *Beggiatoa* mats a bright white appearence. Nitrifying bacteria consist of two groups, those which oxidize NH_4^+ to NO_2^- and those which oxidize NO_2^- to NO_3^-. The hydrogen oxidizing 'Knallgas-bacteria' mentioned before (Eq. 5.14) oxidize H_2 to water. They have a higher efficiency of ATP formation than other chemolithotrophs because their electron donor, H_2, is highly reducing which provides a large energy yield when feeding electrons into the respiratory chain (see below). A variety of iron and manganese oxidizing bacteria play an important role for metal cycling in the suboxic zone, but this group of organisms is still very incompletely known. Much research has been done on the important acidophilic forms, e.g. *Thiobacillus ferrooxidans*, associated with acid mine drainage, due to their significance and because it is possible to grow these in a chemically stable medium without excessive precipitation of iron oxides.

The anaerobic chemolithotrophs use alternative electron acceptors such as nitrate, sulfate or CO_2 for the oxidation of their electron donor. Thus, several sulfur bacteria can respire with nitrate (denitrifiers) and thereby oxidize reduced sulfur species such as H_2S, S^0 or $S_2O_3^{2-}$ to sulfate. Also the above mentioned iron oxidizing nitrate reducing bacteria belong in this group. Many sulfate reducing bacteria can use H_2 as electron donor and thus live as anaerobic chemolithotrophs. Two very important groups of organisms are the methanogens and the acetogens. The former, which are archaea and not bacteria, form methane from CO_2 and H_2:

$$CO_2 + 4H_2 \rightarrow CH_4 + H_2O \qquad (5.21)$$

while the latter form acetate:

$$2CO_2 + 4H_2 \rightarrow CH_3COO^- + H^+ \qquad (5.22)$$

In marine sediments below the sulfate zone, methanogenesis is the predominant terminal pathway of organic carbon degradation. Methane may also be formed from acetate or from organic C_1-compounds such as methanol or methylamines:

$$CH_3COO^- + H^+ \rightarrow CH_4 + CO_2 \qquad (5.23)$$

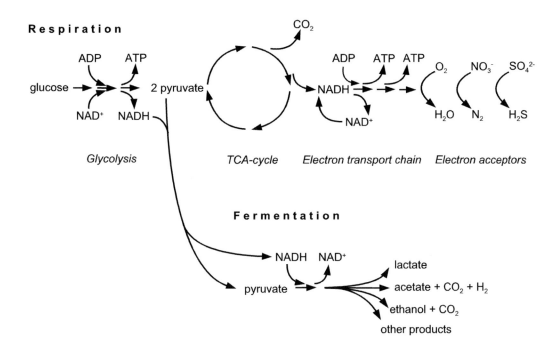

Fig. 5.10 Main pathways of catabolic metabolism in respiring and fermenting heterotrophic organisms (see text).

5.4.7   Respiration and Fermentation

The best known type of energy metabolism in the seabed is the aerobic respiration by heterotrophic organisms such as animals and many bacteria. Heterotrophic bacteria take up small organic molecules such as glucose, break them down into smaller units, and ultimately oxidize them to CO_2 with oxygen. The first pathway inside the cell, called *glycolysis*, converts glucose to pyruvate and conserves only a small amount of potential energy in ATP (Fig. 5.10). Only few electrons are transferred to NAD^+ to form reduced NADH. Through the complex cyclic pathway called the *tricarboxylic acid cycle (TCA cycle)*, the rest of the available electrons of the organic substrate is transferred, again mostly to form NADH, and CO_2 is released. The electron carrier, NADH, is recycled by transferring its electrons via a membrane-bound *electron transport chain* to the *terminal electron acceptor*, O_2, which is thereby reduced to H_2O. Through the electron transport chain, the proton-motive force across the cell membrane is maintained and most of the energy carrier, ATP, is generated. The energy yield of aerobic respiration is large, and up to 38 mol of ATP may be formed per mol of glucose oxidized. This corresponds to a 43% efficiency of energy utilization of the organic substrate, the rest being lost as entropy (heat). By the transfer of electrons from the electron carrier, NADH, to the terminal electron acceptor, O_2, the reduced form of the carrier, NAD^+, is regenerated and the electron flow can proceed in a cyclic manner.

Many heterotrophic prokaryotes are anaerobic and have the ability to use terminal electron acceptors other than O_2 for the oxidation of their organic substrates (Fig. 5.10). The denitrifying bacteria respire with NO_3^- which they reduce to N_2 and release into the large atmospheric nitrogen pool. The sulfate reducing bacteria can use a range of organic molecules as substrates and oxidize these with the concomitant reduction of SO_4^{2-} to H_2S. Some bacteria can use manganese(III, IV) or iron(III) oxides as electron acceptors in their heterotrophic metabolism and reduce these to Mn^{2+} or Fe^{2+}.

Many bacteria do not possess a complete TCA-cycle and electron transport chain and are thus not able to use an external electron acceptor as terminal oxidant. This prevents the formation of large amounts of ATP via the electron transport chain and thus leads to a low energy yield of the metabolism. The organisms may still form a small amount of ATP by so-called substrate level phosphorylation through the glycolysis, and this is sufficient to enable growth of these *fermentative* organisms. The cells must, however, still be able to recycle the reduced electron carrier, NADH, in order to continue the glycolytic pathway. In principle, they do this by transferring the electrons from NADH to an intermediate such as pyruvate (CH_3COCOO^-). An example of a fermentation reaction is the formation of lactate ($CH_3CHOHCOO^-$) by lactic acid bacteria (brackets around the pyruvate indicate that this is an intermediate and not an external substrate being assimilated and transformed):

$$[CH_3COCOO^-] + NADH + H^+ \rightarrow$$
$$CH_3CHOHCOO^- + NAD^+ \qquad (5.24)$$

It is seen that fermentation does not require an external electron acceptor and there is no net oxidation of the organic substrate, glucose. There is rather a reallocation of electrons and hydrogen atoms within the cleaved molecule, whereby a small amount of energy is released. Fermentations often involve a cleavage of the C_3 compound to a C_2 compound plus CO_2. The pyruvate is then coupled to coenzyme-A and cleaved in the form of acetyl-CoA and energy is subsequently conserved by the release of acetate (CH_3COO^-):

$$[CH_3COCOO^-] \rightarrow CH_3COO^- + CO_2 + H_2 \qquad (5.25)$$

The H_2 is formed by a transfer of electrons from pyruvate via the enzyme ferredoxin to NAD^+:

$$NADH + H^+ \rightarrow NAD^+ + H_2 \qquad (5.26)$$

The degradation of organic matter via fermentative pathways to small organic molecules such as lactate, butyrate, propionate, acetate, formate, H_2 and CO_2 is very important in marine sediments, since these compounds are the main substrates for sulfate reduction and partly for methane formation.

A form of inorganic fermentation of sulfur compounds was discovered in recent years in several sulfate reducers and other anaerobic bacteria (Bak and Cypionka 1987). These organisms may carry out a disproportionation of S^0, $S_2O_3^{2-}$ or SO_3^{2-} by which H_2S and SO_4^{2-} are formed simultaneously (Eqs. 5.4 and 5.5). Disproportionation reactions have turned out to

Table 5.5 Diversity of sulfur metabolism among the sulfate reducing bacteria. The changes in free energy, ΔG^0, have been calculated for standard conditions. The data show that the disproportionation of elemental sulfur is an endergonic process under standard conditions and therefore requires an efficient removal of the formed HS^- to pull the reaction and make it exergonic. After Cypionka (1994).

Pathway and stoichiometry	ΔG^0 (kJ mol^{-1})
Reduction of sulfur compounds:	
$SO_4^{2-} + 4H_2 + H^+ \rightarrow HS^- + 4H_2O$	-155
$SO_3^{2-} + 3H_2 + H^+ \rightarrow HS^- + 3H_2O$	-175
$S_2O_3^{2-} + 4H_2 \rightarrow 2HS^- + 3H_2O$	-179
$S^0 + H_2 \rightarrow HS^- + H^+$	-30
Incomplete sulfate reduction:	
$SO_4^{2-} + 2H_2 + H^+ \rightarrow S_2O_3^{2-} + 5H_2O$	-65
Disproportionation:	
$S_2O_3^{2-} + H_2O \rightarrow SO_4^{2-} + HS^- + H^+$	-25
$4SO_3^{2-} + H^+ \rightarrow 3SO_4^{2-} + HS^-$	-236
$4S^0 + 4H_2O \rightarrow SO_4^{2-} + 3HS^- + 5H^+$	+33*
Oxidation of sulfur compounds:	
$HS^- + 2O_2 \rightarrow SO_4^{2-} + H^+$	-794
$HS^- + NO_3^- + H^+ + H_2O \rightarrow SO_4^{2-} + NH_4^+$	-445
$S_2O_3^{2-} + 2O_2 + H_2O \rightarrow 2SO_4^{2-} + 2H^+$	-976
$SO_3^{2-} + {}^1/_2O_2 \rightarrow SO_4^{2-}$	-257

play an important role in the sulfur cycle of marine sediments and for the isotope geochemistry of sulfur (Jørgensen 1990; Thamdrup et al. 1993; Canfield and Teske 1996).

In comparison to all other heterotrophs, the microorganisms oxidizing methane and other C_1 compounds such as methanol, have a unique metabolic pathway which involves oxygenase enzymes and thus requires O_2. Only aerobic methane-oxidizing bacteria have been isolated and studied in laboratory culture, yet methane oxidation in marine sediments is known to take place mostly anaerobically at the transition to the sulfate zone. Microbial consortia that oxidize methane with sulfate have in particular been studied at methane seeps on the sea floor and the communities can now also be grown in the laboratory (Boetius et al. 2000; Orphan et al. 2001; Nauhaus et al. 2002) Anaerobic methane oxidation is catalyzed by archaea that use a key enzyme related to the coenzyme-M reductase of methanogens, to attack the methane molecule (Krüger et al. 2003; see Sect. 5.1). The best studied of these ANME (ANaerobic MEthane

oxidizers) archaea depend on a symbiosis with sulfate reducing bacteria for a complete methane oxidation to CO_2. It is an interesting question whether some of the ANME archaea can also reduce sulfate and thus carry out the entire process within a single cell.

Although the different groups of prokaryotes may be catagorized in the scheme of Table 5.4, it should be noted that they are often very diverse and flexible and may thus fit into different categories according to their immediate environmental conditions and mode of life. A good example are the sulfate reducing bacteria which are typically obligate anaerobes specialized on the oxidation of a limited range of small organic molecules with sulfate (Postgate 1984; Widdel 1988). This group has, however, a great diversity in the types of sulfur metabolism, as listed in Table 5.5 (Bak and Cypionka 1987; Dannenberg et al. 1992; Krekeler and Cypionka 1995). Different sulfate reducing bacteria may, alternatively to SO_4^{2-}, use SO_3^{2-}, $S_2O_3^{2-}$ or S^0 as electron acceptors or may disproportionate these in the absence of an appropriate electron donor such as H_2. They may even respire with oxygen or nitrate and may oxidize reduced sulfur compounds with oxygen. There is also evidence from marine sediments and pure cultures that sulfate reducing bacteria may reduce oxidized iron minerals (Coleman et al. 1993) and are even able to grow with ferric hydroxide as electron acceptor (Knoblauch et al. 1999; cf. Sect. 7.4.2.2):

$$8Fe(OH)_3 + CH_3COO^- \rightarrow$$
$$8Fe^{2+} + 2HCO_3^- + 5H_2O + 15OH^- \quad (5.27)$$

5.5   Pathways of Organic Matter Degradation

Organic material is deposited on the sea floor principally as aggregates which sink down through the water column as a continuous particle rain (Chap. 4). This particulate organic flux is related to the primary productivity of the overlying plankton community and to the water depth through which the detritus sinks while being gradually decomposed. As a mean value, some 25-50% of the primary productivity reaches the sea floor in coastal seas whereas the fraction is only about 1% in the deep sea (Suess 1980; cf. Fig. 12.1). Within the sediment, most organic material remains associated with the particles or is sorbed to inorganic sediment grains. Freshly arrived material often forms a thin detritus layer at the sediment-water interface, in particular after the sedimentation of a phytoplankton bloom. This detritus is a site of high microbial activity and rapid organic matter degradation, even in the deep sea where the bacterial metabolism is otherwise strongly limited by the low organic influx (Lochte and Turley 1988). Due to the feeding by benthic invertebrates and to the mixing of sediment by their burrowing activity (bioturbation) the deposited organic particles are gradually buried into the sediment and become an integral part of the sedimentary organic matter.

5.5.1   Depolymerization of Macromolecules

The organic detritus is a composite of macromolecular compounds such as structural carbohydrates, proteins, nucleic acids and lipid complexes. Prokaryotic organisms are unable to take up particles or even larger organic molecules and are restricted to a molecular size less than ca 600 daltons (Weiss et al. 1991). A simple organic molecule such as sucrose has a molecular size of 342 daltons. Thus, although some bacteria are able to degrade and metabolize fibers of cellulose, lignin, chitin or other structural polymers, they must first degrade the polymers before they can assimilate the monomeric products (Fig. 5.11). This depolymerization is caused by exoenzymes produced by the bacteria and either released freely into their environment or associated with the outer membrane or cell wall. The latter requires a direct contact between the bacterial cells and the particulate substrate and many sediment bacteria are indeed associated with solid surfaces in the sediment. The excretion of enzymes is a loss of carbon and nitrogen and thus an energy investment of the individual bacterial cells, but these have strategies of optimizing the return of monomeric products (Vetter et al. 1998) and may regulate the enzyme production according to the presence of the polymeric target compound in their environment ('substrate induction'). Thus, there is a positive correlation between the availability of specific polymeric substances in the sediment and the concentration of free enzymes which may degrade them (Boetius and Damm 1998; Boetius and Lochte 1996).

The depolymerization of sedimentary organic matter is generally the rate-limiting step in the

sequence of mineralization processes (Arnosti 2004). This may be concluded from the observation that the monomeric compounds and the products of further bacterial degradation do not accumulate in the sediment but are rapidly assimilated and metabolized. Thus, the concentrations of individual free sugars, amino acids or lipids in the pore water are generally low. Dissolved organic matter (DOM) is a product of partial degradation and may be released from the sediment to the overlying water. It mainly consists of complex dissolved polymers and oligomers rather than of monomers. Dissolved organic molecules may not only be taken up by bacteria but may also be removed from the pore water by adsorption onto sediment particles or by condensation reactions, such as monosaccharides and amino acids forming melanoidins (Sansone et al. 1987; Hedges 1978). Organic matter that is not mineralized may be adsorbed to mineral surfaces and transformed to 'geomacromolecules' which are highly resistant to enzymatic hydrolysis and bacterial degradation and which become more permanently buried with the sediment (Henrichs 1992; Keil et al. 1994). This buried fraction also contains some of the original organic matter which was deposited as refractory biomolecules such as lignins, tannins, algaenan, cutan and suberan (Tegelaar et al. 1989). The burial of organic matter deep in the sediment shows a positive correlation with the rate of deposition and constitutes in the order of 5-20% of the initially deposited organic matter in shelf sediments and 0.5-3% in deep-sea sediments (Henrichs and Reeburgh 1987)

5.5.2   Aerobic and Anaerobic Mineralization

The particulate detritus consumed by metazoans, and the small organic molecules taken up by microorganisms in the oxic zone, may be mineralized completely to CO_2 through aerobic respiration within the individual organisms. The aerobic food chain consists of organisms of very diverse feeding biology and size, but of a uniform type of energy metabolism, namely the aerobic respiration. The oxic zone is, however, generally only mm-to-cm thick in finegrained shelf sediments as shown by measurements with O_2 microsensors (Chap. 3). In slope and deep-sea sediments the oxic zone expands to reach depths of many cm or dm (Reimers 1987; Wenzhöfer and Glud 2002).

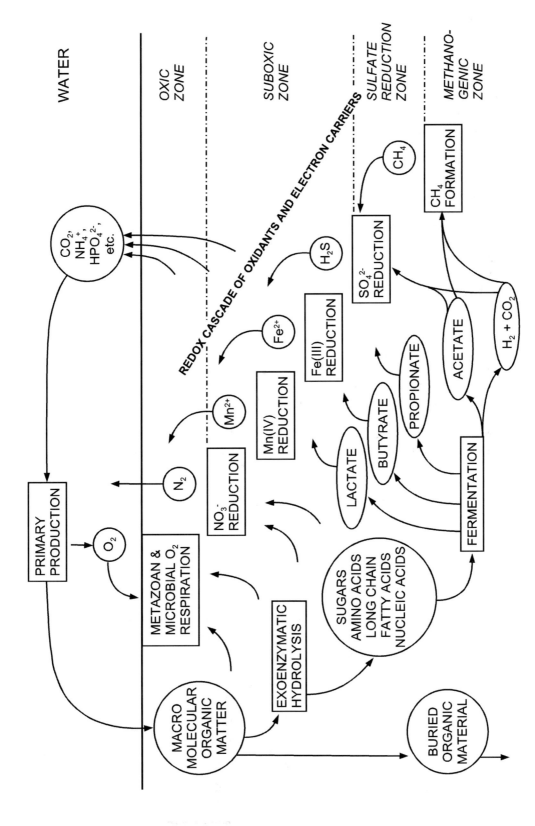

Fig. 5.11 Pathways of organic carbon degradation in marine sediments and their relation to the geochemical zonations and the consumption of oxidants. (After Fenchel and Jørgensen 1977).

Much of the organic mineralization thus takes place within the anoxic sediment. This anoxic world is inhabited primarily by prokaryotic organisms that have a high diversity of metabolic types (Fig. 5.11). There are denitrifiers and metal oxide reducers in the suboxic zone that can utilize a wide range of monomeric organic substances and can respire these to CO_2. With depth into the sediment, however, the energy yield of bacterial metabolism becomes gradually smaller and the organisms become narrower in the spectrum of substrates which they can use. While denitrifiers are still very versatile with respect to usable substrates, the sulfate reducers are mostly unable to respire, for example sugars or amino acids. Instead, these monomeric compounds are taken up by fermenting bacteria and converted into a narrower spectrum of fermentation products that include primarily volatile fatty acids such as formate, acetate, propionate and butyrate, as well as H_2, lactate, some alcohols and CO_2. Through a second fermentation step the products may be focused even further towards the key products: acetate, H_2 and CO_2. The sulfate reducers depend on these products of fermentation which they can respire to CO_2. Several well-known sulfate reducers such as *Desulfovibrio* spp. can only carry out an incomplete oxidation of substrates such as lactate, and they excrete acetate as a product. Other sulfate reducers have specialized on acetate and catalyze the complete oxidation to CO_2. The degradative capacity of anaerobic bacteria seems, however, to be broader than previously expected and new physiological types

continue to be discovered. As an example, sulfate reducing bacteria able to oxidize aromatic and aliphatic hydrocarbons have now been isolated, which shows that anaerobic prokaryotes are able to degrade important components of crude oil in the absence of oxygen (Rueter et al. 1994; Rabus et al. 1996).

The methane forming (methanogenic) archaea can use only a narrow spectrum of substrates, primarily H_2 plus CO_2 and acetate (Eqs. 5.21 and 5.23). Within the sulfate zone, which generally penetrates several meters down into the seabed, the sulfate reducing bacteria compete successfully with the methanogens for these few substrates and methanogenesis is, therefore, of little significance in the sulfate zone. Only a few 'non-competitive' substrates such as methylamines are used only by the methanogens and not by the sulfate reducers (Oremland and Polcin 1982). Below the sulfate zone, however, there are no available electron acceptors left other than CO_2, and methane accumulates here as the main terminal product of organic matter degradation.

5.5.3   Depth Zonation of Oxidants

The general depth sequence of oxidants used in the mineralization of organic matter is
$$O_2 \rightarrow NO_3^- \rightarrow Mn(IV) \rightarrow Fe(III) \rightarrow SO_4^{2-} \rightarrow CO_2.$$
This sequence corresponds to a gradual decrease in redox potential of the oxidant (Fig. 5.9) and thus to a decrease in the free energy available by respiration with the different electron acceptors (Table 5.2; cf. Sect. 3.2.5). The ΔG^0 of oxic

Table 5.6 Annual budget for the mineralization of organic carbon and the consumption of oxidants in a Danish coastal sediment, Aarhus Bay, at 15 m water depth. The basic reaction and the change in oxidation step are shown for the elements involved. The rates of processes were determined for one m^2 of sediment and were all recalculated to carbon equivalents. From data compiled in Jørgensen (1996).

Reaction	Δ Oxidation steps	Measured rate	Estimated carbon equivalents
		mol m^{-2} yr^{-1}	
$[CH_2O] \rightarrow CO_2$	C: $0 \rightarrow +4 = 4$	9,9	9,9
$O_2 \rightarrow H_2O$	O: $0 \rightarrow -2 = 2 \cdot 2$	9,2	9,2
$NO_3^- \rightarrow N_2$	N: $+5 \rightarrow 0 = 5$	0,15	0,19
$Mn(IV) \rightarrow Mn^{2+}$	Mn: $+4 \rightarrow +2 = 2$	0,8	0,4
$Fe(III) \rightarrow Fe^{2+}$	Fe: $+3 \rightarrow +2 = 1$	1,6	0,4
$SO_4^{2-} \rightarrow HS^-$	S: $+6 \rightarrow -2 = 8$	1,7	3,4

respiration is the highest, -479 kJ mol^{-1}, and that of denitrification is nearly as high. Table 5.2 shows that respiration of one mol of organic carbon with sulfate yields only a fraction of the energy of respiration with oxygen. The remaining potential chemical energy is not lost but is mostly bound in the product, H_2S. This energy may become available to other microorganisms, such as chemolithotrophic sulfur bacteria, when the sulfide is transported back up towards the sediment surface and comes into contact with potential oxidants.

The quantitative importance of the different oxidants for mineralization of organic carbon has been studied intensely over the last few decades, both by diagenetic modeling and by incubation experiments. It is generally found that oxygen and sulfate play the major role in shelf sediments, where 25-50% of the organic carbon may be mineralized anaerobically by sulfate reducing bacteria (Jørgensen 1982). With increasing water depth and decreasing organic influx down the continental slope and into the deep sea, the depth of oxygen penetration increases and sulfate reduction gradually looses significance. Nitrate seems to play a minor role as an oxidant of organic matter. Manganese oxides occur in shelf sediments mostly in lower concentrations than iron oxides and, expectedly, the Mn(IV) reduction rates should be lower than those of Fe(III) reduction. Manganese, however, is recycled nearer to the sediment surface and relatively faster than iron. These solid-phase oxidants are both dependent on bioturbation as the mechanism to bring them from the sediment surface down to their zone of reaction. The shorter this distance, the faster can the metal oxide be recycled.

Table 5.6 shows an example of process rates in a coastal marine sediment. In this comparison, it is important to keep in mind that the different oxidants are not equivalent in their oxidation capacity. When for example the iron in Fe(III) is reduced, the product is Fe^{2+} and the iron atoms have been reduced only one oxidation step from +3 to +2. Sulfur atoms in sulfate, in contrast, are reduced eight oxidation steps from +6 in SO_4^{2-} to $-$2 in H_2S. One mol of sulfate, therefore, has 8-fold higher oxidation capacity than one mol of iron oxide. In order to compare the different oxidants and their role for organic carbon oxidation, their reduction rates were recalculated to carbon equivalents on the basis of their change in oxidation step. It is then clear that oxygen and

sulfate were the predominant oxidants in this sediment, sulfate oxidizing about 44% of the organic carbon in the sediment. Of the H_2S formed, 15% was buried as pyrite, while the rest was reoxidized and could potentially consume one third of the total oxygen uptake. This redox balance is typical for coastal sediments where up to half of the oxygen taken up by the sediment is used for the direct or indirect reoxidation of sulfide and reduced manganese and iron (Jørgensen and Nelson 2004).

Manganese and iron behave differently with respect to their reactivity towards sulfide. Iron binds strongly to sulfide as FeS or FeS_2, whereas manganese does not. Furthermore, the reduced iron is more reactive than reduced manganese when it reaches the oxic zone, and Fe^{2+} generally does not diffuse out of the sediment, although it may escape by advective pore water flow (Huettel et al. 1998). The Mn^{2+}, in contrast, easily recycles via the overlying water column and can thereby be transported from the shelf out into deeper water. This mechanism leads to the accumulation of manganese oxides in some continental slope sediments (see Chap. 11). The important role of manganese and iron in slope sediments is evident from Figure 5.12. Manganese constituted about 5% of the total dry weight of this sediment. Below the manganese zone, iron oxides and then sulfate took over the main role as oxidants. In the upper few cm of the sediment, manganese oxide was the dominant oxidant below the O_2 zone (Canfield 1993).

The consecutive reduction of oxidants with depth in the marine sediment and the complex reoxidation of their products constitute the 'redox cascade' (Fig. 5.11). An important function of this sequence of reactions is the transport of electrons from organic carbon via inorganic species back to oxygen. The potential energy transferred from the organic carbon to the inorganic products is thereby released and may support a variety of lithotrophic microorganisms. These may make a living from the oxidation of sulfide with Fe(III), Mn(IV), NO_3^- or O_2. Others may be involved in the oxidation of reduced iron with Mn(IV), NO_3^- or O_2 etc. The organisms responsible for these reactions are only partly known and new types continue to be isolated.

The depth sequence of electron acceptors in marine sediments, from oxygen to sulfate, is accompanied by a decrease in the degradability of the organic material remaining at that depth. This

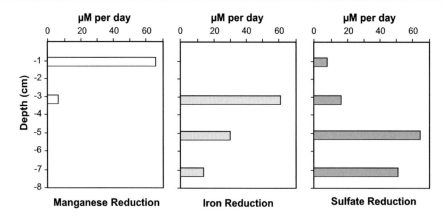

Fig. 5.12 Depth zonation of reduction rates for the oxidants, Mn(IV), Fe(III) and SO_4^{2-} in a marine sediment from Skagerrak (Denmark) at 700 m water depth. This sediment was particularly rich in manganese oxide. Data from Canfield (1993).

decrease can be formally considered to be the result of a number of different organic carbon pools, each with its own degradation characteristics and each being degraded exponentially with time and depth (Westrich and Berner 1984). The consecutive depletion of these pools leads to a steep decrease in organic carbon reactivity, which follows the sum of several exponential decays, and which can be demonstrated from the depth distribution of oxidant consumption rates. Thus, the O_2 consumption rate per volume in the oxic zone of a coastal sediment was found to vary between 3,000 and 30,000 nmol O_2 cm^{-3} d^{-1} as the oxygen penetration depth varied between 5 mm in winter and 1 mm in summer (Gundersen et al. 1995). Some five cm deeper into the sediment, where sulfate reduction predominated and reached maximal activity, the rates of carbon mineralization had decreased 100-fold to 25-150 nmol cm^{-3} d^{-1} between winter and summer (Thamdrup et al. 1994). A few meters deep into the sediment, where methanogenesis predominated below the sulfate zone, carbon mineralization rates were again about a 100-fold lower. Although the rates down there may seem insignificant, the slow organic matter decomposition to methane proceeds to great depth in the sediment and is therefore important on an areal basis. This methane diffuses back up towards the sulfate zone where it is oxidized to CO_2 at the expense of sulfate (Chap. 8). It is equivalent to 10% or more of the total sulfate reduction in coastal sediments, depending on the depth of sulfate penetration (Iversen and Jørgensen 1985; Niewöhner et al. 1998; Chap. 8).

Whereas Fig. 5.12 illustrates highly active processes in shelf sediment, deep sea sediments have orders of magnitude lower process rates and very different biogeochemistry. In large provinces of the ocean floor the organic flux is so low that sulfate does not become depleted throughout the entire sediment column and iron and manganese reduction may be the predominant anaerobic mineralization processes. Fig. 5.13 shows an example from the eastern tropical Pacific at 3,300 m water depth where Miocene to Holocene carbonate and ciliceous oozes were cored during the Ocean Drilling Program Leg 201 (D'Hondt, Jørgensen, Miller et al. 2003). The sediment deposit is here 420 m thick and the porewater profiles reach down to the basaltic ocean crust with an age of 16 million years. Prokaryotic cells were found throughout the sediment, from the surface to the bottom, with a large diversity of organisms, mostly of unknown phylogenetic types with no representatives among current laboratory cultures (Parkes et al. 2005). The distributions of dissolved inorganic carbon (DIC) and sulfate are nearly mirror images reflecting the oxidation of organic carbon to CO_2 by sulfate reduction, whereby sulfate is consumed and DIC is produced:

$$2[CH_2O] + SO_4^{2-} \rightarrow 2HCO_3^- + H_2S \qquad (5.28)$$

At the bottom of the sediment column, the DIC and sulfate concentrations surprisingly return to near seawater values. This is due to a very slow flow of sea water through the fractured crust that exchanges solutes with the overlying sediment by

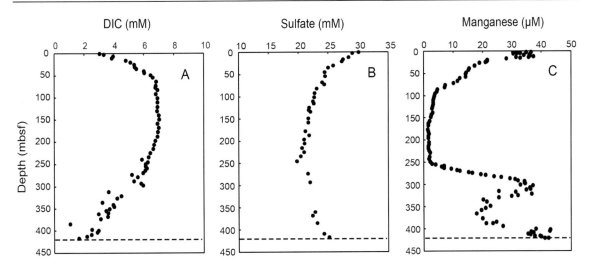

Fig. 5.13 Porewater profiles of dissolved inorganic carbon (DIC), sulfate and manganese (Mn^{2+}) in the pore water measured during ODP Leg 201 at Site 1226 in the eastern tropical Pacific Ocean. Data from D'Hondt, Jørgensen, Miller et al. (2003).

molecular diffusion. This seawater even transports nitrate and oxygen and enables the existence of aerobic microorganisms hundreds of meters beneath the ocean floor (D'Hondt et al. 2004). The distribution of Mn^{2+} shows that manganese reduction takes place in the upper 100 to 150 m of the sediment column and then again below 250 m. The lower zone of manganese reduction is due to an excess of manganese oxide buried during a low-productivity period in the Miocene before 7 Ma. This example shows that microbial processes taking place in the deep subsurface sediments are even today affected by the past oceanographic history millions of years ago.

5.6 Methods in Biogeochemistry

A diversity of approaches, of which only a few examples can be discussed here, is used in marine biogeochemistry (Fig. 5.2). Among the important goals is to quantify the rate at which biological and chemical processes take place in different depth zones of the sediment. These are often fast processes for which the reactants may have turnover times in the order of days or hours or even minutes. Data on dissolved species in the pore water and on solid-phase geochemistry can be used in diagenetic models to calculate such rates, as discussed in Chapter 3 (e.g. Schulz et al. 1994). The dynamic process is then derived from the chemical gradients in the pore water and from knowledge about the diffusion coefficient of the chemical species. Such diffusion-diagenesis models work best for either very steep gradients, e.g. of oxygen for which diffusion is rapid, or very deep gradients, e.g. of sulfate which penetrates deep below the zone affected by biological transport of pore water (bioirrigation) or of solid-phase sediment (bioturbation). For intermediate depth scales, e.g. in the suboxic zone, the burrowing fauna influences the transport processes so strongly that advection and bioirrigation tend to dominate over molecular diffusion. If the transport factor is enhanced to an unknown degree, a rate calculation based on molecular diffusion would be correspondingly wrong. In such cases, it may be more realistic to model the solid phase combined with estimates of the rate of burial and of mixing by the infauna (bioturbation). Burial and mixing rates are most often estimated from the vertical distribution and decay of natural radionuclides, such as ^{210}Pb, in the sediment.

Another problem is that many compounds are not only consumed but also recycled in a chemical zone. Thus, sulfate in the upper sediment layers is both consumed by sulfate reduction and produced by sulfide oxidation so that the net sulfate removal may be insignificant relative to the total reduction rate. The net removal of sulfate in the whole sediment column is determined by the amount of sulfide trapped in pyrite and is often only 10% of the gross sulfate reduction.

5.6.1   Incubation Experiments

In addition to modeling, process rates can be measured experimentally in sediment samples by following the concentration changes of the chemical species over time, either in the pore water or in the solid phase. By such incubation experiments it is critical that the physico-chemical conditions and the biology of the sediment remain close to the natural situation. Since gradual changes with time are difficult to avoid, the duration of the experiment should preferably be restricted to several hours up to a day. This is generally too short a time interval for significant changes in the populations of anaerobic bacteria. A careful way of containing the sediment samples is to distribute the sediment from different depths into gas-tight plastic bags under an atmosphere of nitrogen and keep these at the natural temperature. The sediment can then be mixed without dilution or contact with a gas phase, and sub-samples can be taken over time for the analyses of chemical concentrations. Such a technique has been used to measure the rates of organic matter mineralization from the evolution of CO_2 or NH_4^+ or to measure the reduction rates of manganese and iron oxides (Fig. 5.12, Canfield et al. 1993; Hansen et al. 2000).

5.6.2   Radioactive Tracers

Radioactive tracers are used to identify and measure microbial or chemical processes in marine sediments. Radiotracers are mostly applied when chemical analysis alone is too insensitive or if the pathway of processes is more complex or cyclic. Thus, if a process is very slow relative to the pool sizes of the reactants or products, a long-term experiment of many days or months would be required to detect a chemical change. Since this would lead to changing sediment conditions and thus to non-natural process rates, higher sensitivity is required. By the use of a radiotracer, hundred- or thousand-fold shorter experiment durations may often be achieved. An example is the measurement of sulfate reduction rates (see Sect. 5.6.3). Sulfate can be analyzed in pore water samples with a precision of about $\pm 2\%$, and a reduction of 5-10% or more is, therefore, required to determine its rate with a reasonable confidence. If radioactive, ^{35}S-labelled sulfate is used to trace the process, it is possible to detect the reduction of only 1/100,000 of the SO_4^{2-} by analyzing the radioactivity of the sulfide formed (Kallmeyer et al. 2004). The sensitivity by using radiotracer is thus improved >1000-fold over the chemical analysis.

The radioisotopes most often used in biogeochemistry are shown in Table 5.7, together with examples of their application. ^3H is a weak ß-emitter, whereas ^{32}P is a hard ß-emitter. ^{14}C and ^{35}S have similar intermediate energies. The ^{14}C is most widely used, either to study the synthesis of new organic biomass through photo- or chemosynthesis ($^{14}CO_2$ assimilation), or to study the transformations and mineralization of organic material such as plankton detritus or specific compounds such as glucose, lactate, acetate and other organic molecules. The ^3H-labelled substrates have in particular been used in connection with autoradiography, where the incorporation of label is quantified and mapped at high spatial resolution by radioactive exposure of a photographic film emulsion or a ß-imager. By this technique it may be possible to demonstrate which microorganisms are actively

Table 5.7   Radio-isotopes most often used as tracers in biogeochemistry. The measurement of radioactivity is based on the emission of β-radiation (high-energy electrons) or γ-radiation (electromagnetic) with the specified maximum energy.

Isotope	Emission mode, E_{max}	Half-life	Examples of application
^3H	β, 29 keV	12 years	Turnover of org. compounds, autoradiography
^{14}C	β, 156 keV	5730 years	Turnover of org. compounds, CO_2 assimilation
^{32}P	β, 1709 keV	14 days	Phosphate turnover & assimilation
^{35}S	β, 167 keV	87 days	Sulfate reduction
^{55}Fe	γ, 6 keV	2.7 years	Iron oxidation and reduction

assimilating the labeled compound. Since phosphorus does not undergo redox processes in sediments, ^{32}P (or ^{33}P) is mostly used to study the dynamics and uptake of phosphate by microorganisms. ^{35}S has been used widely to study processes of the sulfur cycle in sediments, in particular to measure sulfate reduction.

Although radiotracer techniques may offer many advantages, they also have inherent problems. For example, applications of ^{35}S to trace the pathways of H_2S oxidation have been flawed by isotope exchange reactions between the inorganic reduced sulfur species such as elemental sulfur, polysulfides and iron sulfide (Fossing et al. 1992; Fossing 1995). This isotope exchange means that the sulfur atoms swap places between two compounds without a concomitant net chemical reaction between them. If a ^{35}S atom in H_2S changes place with a ^{32}S atom in S^0, the resulting change in radioactivity will appear as if ^{35}S-labelled H_2S were oxidized to S^0, although there may be no change in the concentrations of H_2S or S^0. When the distribution of ^{35}S radiolabel is followed with time the results can hardly be distinguished from a true net process and may be incorrectly interpreted as such. There is no similar isotopic exchange with sulfate at normal environmental temperatures, which would otherwise prevent its use for the measurement of sulfate reduction rates.

The radioisotopes of iron or manganese, ^{55}Fe, ^{59}Fe and ^{54}Mn, have had only limited application as tracers in biogeochemical studies. Experiments with ^{55}Fe (or ^{59}Fe) as a tracer for Fe(III) reduction in marine sediments showed problems of unspecific binding and possibly isotope exchange (King 1983; Roden and Lovley 1993), which has discouraged other researchers from the use of this isotope. Similar problems complicate experiments with manganese as a radiotracer. The iron isotope, ^{57}Fe, has been used in a completely different manner for the analysis of iron speciation in marine sediments. The isotope, ^{57}Fe, is added to a sediment and is allowed to equilibrate with the iron species. It can then be used to analyze the oxidation state and the mineralogy of iron by Mössbauer spectroscopy and thus to study the oxidation and reduction of iron minerals.

It has been a serious draw-back in studies of nitrogen transformations that a useful radioisotope of nitrogen does not exist. The isotope, ^{13}N, is available only at accelerator facilities and has a half-life of 5 min, which strongly limits its applicability. Instead, the stable isotope, ^{15}N, has been used successfully as a tracer in studies of nitrogen transformations in the marine environment. For example, the use of $^{15}NO_3^-$ by the 'isotope pairing' technique has offered possibilities to study the process of denitrification of $^{15}NO_3^-$ to N_2 (Nielsen 1992). The $^{15}NO_3^-$ is added to the water phase over the sediment and is allowed to diffuse into the sediment and gradually equilibrate with $^{14}NO_3^-$ in the pore water. By analyzing the isotopic composition of the formed N_2, i.e. $^{14}N^{14}N$, $^{14}N^{15}N$

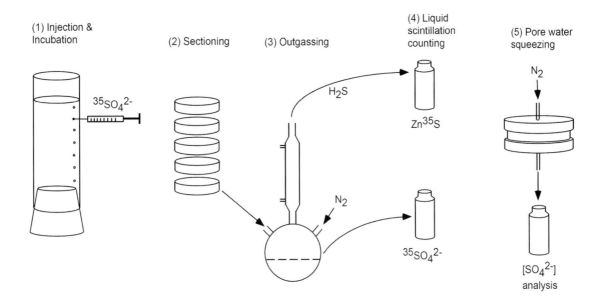

Fig. 5.14 Principle of sulfate reduction measurement in sediment using $^{35}SO_4^{2-}$ as a tracer (see text).

and $^{15}N^{15}N$, it has been possible to calculate not only the rate of denitrification, but even to discriminate between denitrification from nitrate diffusing down from the overlying seawater and denitrification from nitrate formed by nitrification within the sediment. The results have shown that at the normal, low concentrations of nitrate in seawater, <10-20 µM, the main source of nitrate for denitrification is the internally formed nitrate derived from nitrification.

5.6.3   Example: Sulfate Reduction

As an example of a radio-tracer method, the measurement of sulfate reduction rates using $^{35}SO_4^{2-}$ according to Jørgensen (1978); Fossing and Jørgensen (1989); Kallmeyer et al. (2004) is described in brief (Fig. 5.14). The $^{35}SO_4^{2-}$ is injected with a microsyringe (1) into whole, intact sediment cores in quantities of a few microliters which contain ca 100 kBq (1 becquerel = 1 radioactive disintegration per second; 37 kBq = 1 microcurie). After 4-8 hours the core is sectioned (2), and the sediment samples are fixed with zinc acetate. The zinc binds the sulfide and prevents its oxidation and it also kills the bacteria and prevents further sulfate reduction. The reduced ^{35}S is then separated from the sediment by acidification (3), and the released $H_2^{35}S$ is transferred in a stream of N_2 to a trap containing zinc acetate where $Zn^{35}S$ precipitates. The radioactivities of the $Zn^{35}S$ and of the remaining $^{35}SO_4^{2-}$ are then analyzed by liquid scintillation counting (4). In a sediment core the concentration gradient of sulfate in the pore water is analyzed (5), and the rates of sulfate reduction can then be calculated according to the equation:

Sulfate reduction rate =

$$\phi\left(SO_4^{2-}\right)\cdot\frac{H_2^{35}S}{^{35}SO_4^{2-}}\cdot\frac{24}{t}\cdot1.06 \; \text{nmol } SO_4^{2-}\text{cm}^{-3}\text{day}^{-1}$$

(5.29)

where ϕ is porosity, (SO_4^{2-}) is sulfate concentration in the pore water ($\phi(SO_4^{2-})$ is then sulfate concentration per volume sediment, $H_2^{35}S$ is radioactivity of total reduced sulfur), $^{35}SO_4^{2-}$ is radioactivity of added sulfate tracer, t is experiment time in hours, and 1.06 is a correction factor for the small dynamic isotope discrimination by the bacteria against the heavier isotope. One nmol (nanomol) is 10^{-9} mol. This formula is a good approximation as long as only a small fraction of the labeled sulfate is reduced during incubation, a condition normally fulfilled in marine sediments.

Sulfate reduction rates in marine shelf sediments commonly lie in the range of 1-100 nmol SO_4^{2-} cm^{-3} day^{-1} (Jørgensen 1982). Since the sulfate concentration in the pore water is around 28 mM or 20 µmol cm^{-3}, the turn-over time of the sulfate pool is in the order of 1-50 years. A purely chemical experiment would thus require a month to several years of incubation. This clearly illustrates why a radiotracer technique is required to measure the rate within several hours. In sediment cores from the open eastern equatorial Pacific obtained by the Ocean Drilling Program it has been possible to push the detectability of this radiotracer method to its physical limit by measuring sulfate reduction rates of <0.001 nmol cm^{-3} d^{-1} in 9 million year old sediments at 300 m below the seafloor (Parkes et al. 2005).

Similar principles as described here are used in measurements of, e.g. the oxidation of ^{14}C-labelled methane or the formation of methane from ^{14}C-labelled CO_2 or acetate. Often there are no gaseous substrates or products, however, and it is not possible to separate the radioisotopes as efficiently as by the measurement of sulfate reduction or methane formation and oxidation. This is the case for radiotracer studies of intermediates in the mineralization of organic matter in sediments, e.g. of ^{14}C-labelled sugars, amino acids, or volatile fatty acids, studies which have been important for the understanding of the pathways of organic degradation (Fig. 5.11). These organic compounds, however, occur at low concentrations and have a fast turn-over of minutes to hours, so the sensitivity is less important. More important is the fact that these compounds are at a steady-state between production and consumption, which means that their concentration may not change during incubation in spite of their fast turnover. A simple chemical experiment would thus not be able to detect their natural dynamics, but a radiotracer experiment may.

5.6.4   Specific Inhibitors

This limitation of chemical experiments is often overcome by the use of a specific inhibitor (Oremland and Capone 1988). The principle of an inhibitor technique may be, a) to block a sequence of processes at a certain step in order to observe the accumulation of an intermediate compound, b) to inhibit a certain

Table 5.8 Some inhibitors commonly used in biogeochemistry and microbial ecology.

Inhibitor	Process	Principle of function
BES	Methanogenesis	Blocks CH_4 formation (methyl-CoM reductase)
MoO_4^{2-}	Sulfate reduction	Blocks SO_4^{2-} reduction (depletes ATP pool)
Nitrapyrin	Nitrification	Blocks autotrophic NH_4^+ oxidation
Acetylene	Denitrification	Blocks $N_2O \rightarrow N_2$ (also blocks nitrification)
Acetylene	N_2-fixation	C_2H_2 is reduced to C_2H_4 instead of $N_2 \rightarrow NH_4^+$
DCMU	Photosynthesis	Blocks electron flow between Photosystem II $\rightarrow$ I
Cyanide	Respiration	Blocks respiratory enzymes
β-fluorolactate	Lactate metabol.	Blocks heterotrophic metabolism of lactate
Chloramphenicol	Growth	Blocks prokaryotic protein synthesis
Cycloheximide	Growth	Blocks eukaryotic protein synthesis

group of organisms in order to observe whether the process is taken over by another group or may proceed autocatalytically, c) to let the inhibitor substitute the substrate in the process and measure the transformation of the inhibitor with higher sensitivity. Ideally, the inhibitor should be specific for only the relevant target organisms or metabolic reaction. In reality, most inhibitors have side effects and it is necessary through appropriate control experiments to determine these and to find the minimum inhibitor dose required to obtain the required effect. Some examples are given in Table 5.8.

BES (2-bromoethanesulfonic acid) is a structural analogue of mercaptoethanesulfonic acid, also known as coenzyme-M in methanogenic bacteria, a coenzyme associated with the terminal methylation reactions in methanogenesis. BES inhibits this methylation and thus the formation of methane. It belongs to the near-ideal inhibitors because its effect is specific to the target group of organisms. BES has been used to determine the substrates for methane formation in aquatic sediments and to show that some substrates such as acetate and H_2 are shared in competition between the methanogens and the sulfate reducers, whereas others such as methylamines are 'non-competitive' substrates which are used by the methanogens alone (Oremland and Polcin 1982; Oremland et al. 1982).

Similarly, molybdate (MoO_4^{2-}) together with other group VI oxyanions are analogues of sulfate and inhibit sulfate reduction competitively.

Molybdate specifically interferes with the initial 'activation' of sulfate through reaction with ATP and tends to deplete the ATP pool, thus leading to cell death of sulfate reducing bacteria. Molybdate has also been important for clearing up the substrate interactions between methanogens and sulfate reducers in sediments. Molybdate has been used to demonstrate quantitatively which substrates play a role for sulfate reduction in marine sediments (Fig. 5.15). When molybdate is added to sediment at a concentration similar to that of seawater sulfate, 20 mM, sulfate reduction stops. The organic substrates, which were utilized by the sulfate reducers in the uninhibited sediment and which were kept at a minimum concentration as long as these bacteria were active, then suddenly start to accumulate because they are no longer consumed. Since the bacterial processes leading to the formation of these substrates are not inhibited, the substrates will accumulate at a similar rate at which they were used by the sulfate reducers before inhibition. Such experiments have demonstrated that acetate, propionate, butyrate, isobutyrate and H_2 are among the most important substrates for sulfate reducing bacteria in marine sediments (Sørensen et al. 1981; Christensen 1984).

Nitrapyrin (N-serve) was first introduced in agriculture as a means to inhibit the conversion of ammonium fertilizer to nitrate with subsequent wash-out of the nitrate. The nitrapyrin blocks the copper-containing cytochrome oxidase involved in the initial enzymatic oxidation of ammonium to

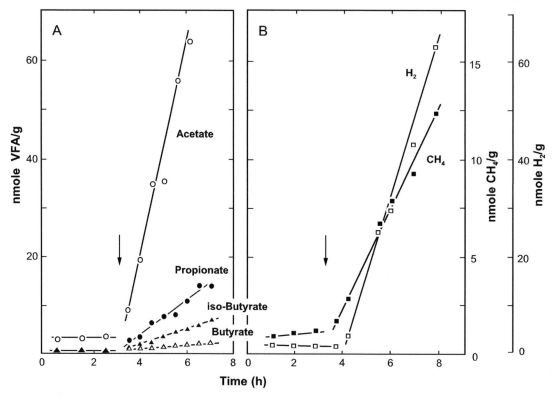

Fig. 5.15 Inhibitor experiment for the demonstration of substrates used by sulfate reducing bacteria in a coastal marine sediment. The concentrations of volatile fatty acids, hydrogen and methane are followed during a time-course experiment over 8 hours. At 3.5 hours (arrow) molybdate was added and the substrates accumulate at a rate corresponding to their rate of consumption before inhibition. The formation of methane shows the release of competition for the common substrates for methanogenesis and sulfate reduction (H_2 and acetate). Data from Sørensen et al. (1981).

hydroxylamine. The rate of ammonium accumulation after nitrapyrin inhibition is thus a measure of the nitrification rate before inhibition (Henriksen 1980).

Acetylene (C_2H_2) has been used to study the process of denitrification in which it blocks the last step from N_2O to N_2 (Sørensen 1978; Seitzinger et al. 1993). It causes an accumulation of N_2O which can be analyzed by a gas chromatograph equipped with an electron capture detector or by a N_2O microelectrode (Revsbech et al. 1988). The N_2O accumulation rate is equal to the denitrification rate before inhibition. Acetylene is also used for studies of N_2 fixation. Acetylene has a triple bond analogous to N_2 ($HC{\equiv}CH$ vs. $N{\equiv}N$) and can substitute N_2 competitively. Instead of reducing N_2 to NH_4^+, the key enzyme of nitrogen fixation, nitrogenase, reduces acetylene to ethylene ($H_2C{=}CH_2$). The formation rate of ethylene, which is easily analyzed in the headspace by a gas chromatograph with flame ionization detector, is thus a measure of the nitrogen fixation rate.

Other inhibitors are applied in studies of photosynthesis (DCMU) and respiration (cyanide) or are used to distinguish activities of prokaryotic (chloramphenicol) versus eukaryotic (cycloheximide) microorganisms (Table 5.8).

5.6.5   Other Methods

Information on the hydrolytic activity in marine sediments has been obtained from the use of model substrates labeled with fluorescent dyes such as methylumbelliferone (MUF) or fluorescein. These substrates may be small dimeric molecules, the hydrolytic cleavage of which releases the fluorescence signal, which is then indicative of the activity of specific enzymes such as glucosidase, chitobiase, lipase, aminopeptidase or esterase (Chrost 1991). Also large fluorescently labeled polymers such as the polysaccharides laminarin or pullulan have been used in experiments to demonstrate the mechanism and kinetics of bacterial degradation (Arnosti 1996).

Acknowledgements

This is contribution No 0331 of the Research Center Ocean Margins (RCOM) which is financed by the Deutsche Forschungsgemeinschaft (DFG) at Bremen University, Germany.

5.7 Problems

Problem 1

Following the last glaciation, the salinity of the oceans decreased due to melting of the polar ice caps. This decrease in chloride and sodium and other major ions of sea water has led to a lowered concentration in the pore water of the upper seabed. By using the equation for diffusion distance versus time (ca. 10,000 years), calculate how deep this concentration decrease has diffused into the seabed. Diffusion coefficients are given in Chapter 3. Make sure that similar units are being used.

Problem 2

Which of the following processes are energetically feasible (exergonic)?
  1) Fe(II) oxidation with SO_4^{2-}
  2) CH_4 oxidation with NO_3^-
  3) Fe(II) oxidation with MnO_2
  4) H_2S oxidation with SO_4^{2-}
  5) H_2 oxidation with CO_2

Problem 3

The filamentous sulfur bacteria, *Thioploca*, oxidize sulfide with nitrate in their energy metabolism. Alternatively, they can assimilate CO_2 or acetate as a carbon source. What kind(s) of "-troph" is *Thioploca*? (use Table 5.3 and 5.4).

Problem 4

Bacteria in the open ocean tend to be smaller than bacteria in nutrient-rich coastal ecosystems. What could be the reason?

Problem 5

A dead phytoplankton cell sinks to the sea floor. What must happen to it (burial, degradation) before its organic carbon can be converted to methane by autotrophic methanotrophs?

Problem 6

Methane is produced from acetate at extremely low rates in deep sub-seafloor sediments. The process can be measured experimentally using a radioactive isotope, ^{14}C-acetate. How can the experiment be designed to obtain the highest sensitivity and detect the low rate?
  1) increase/decrease in sample size?
  2) increase/decrease in amount of radioactivity?
  3) increase/decrease in incubation time?
  4) other?

References

Alperin, M.J. and Reeburgh, W.S., 1985. Inhibition Experiments on Anaerobic Methane Oxidation. Applied and Environmental Microbiology, 50: 940-945.

Arnosti, C., 1996. A new method for measuring polysaccharide hydrolysis rates in marine environments. Organic Geochemistry, 25: 105-115.

Arnosti, C., 2004. Speed bumps and barricades in the carbon cycle: Substrate structural effects on carbon cycling. Marine Chemistry, 92: 263-273.

Bak, F. and Cypionka, H., 1987. A novel type of energy metabolism involving fermentation of inorganic sulphur compounds. Nature, 326: 891-892.

Benz, M., Brune, A. and Schink, B., 1998. Anaerobic and aerobic oxidation of ferrous iron and neutral pH by chemoheterotrophic nitrate-reduction bacteria. Archives of Microbiology, 169: 159-165.

Berelson, W.M., Hammond, D.E., Smith, K.L. Jr; Jahnke, R.A., Devol, A.H., Hinge, K.R., Rowe, G.T. and Sayles, F. (eds), 1987. In situ benthic flux measurement devices: bottom lander technology. MTS Journal, 21: 26-32.

Berg, P., Røy, H., Janssen, F., Meyer, V., Jørgensen, B.B., Hüttel, M. and De Beer, D., 2003. Oxygen uptake by aquatic sediments measured with a novel non-invasive EDDY-correlation technique. Marine Ecology Progress Series, 261: 75-83.

Berner, R.A., 1980. Early diagenesis: A theoretical approach. Princton Univ. Press, Princton, NY, 241 pp.

Boetius, A. and Lochte, K., 1996. Effect of organic enrichments on hydrolytic potentials and growth of bacteria in deep-sea sediments. Marine Ecology Progress Series, 140: 239-250.

Boetius, A. and Damm, E., 1998. Benthic oxygen uptake, hydrolytic potentials and microbial biomass at the Arctic continental slope. Deep-Sea Research I, 45: 239-275.

Boetius, A., Ravenschlag, K., Schubert, C.J., Rickert, D., Widdel, F., Gieseke, A., Amann, R., Jørgensen, B.B., Witte, U. and Pfannkuche, O., 2000. A marine microbial consortium apparently mediating anaerobic oxidation of methane. Nature, 407: 623-626.

Borowski, W.S., Paull, C.K. and Ussler, W., 1996. Marine pore-water sulfate profiles indicate in situ methane flux from underlying gas hydrate. Geology, 24: 655-658.

Boudreau, B.P., 1988. Mass-transport constraints on the growth of discoidal ferromanganese nodules. American Journal of Science, 288: 777-797.

Boudreau, B.P., 1997. Diagenetic models and their impletation: modelling transport and reactions in aquatic sediments. Springer, Berlin, Heidelberg, NY, 414 pp.

Brasier, M.D., Green, O.R., Jephcoat, A.P., Kleppe, A.K., van Krankendonk, M.J., Lindsay, J.F., Steele, A. and Grassineau, N.V., 2002. Questioning the evidence for Earth's oldest fossils. Nature, 416: 76-81.

Canfield, D.E., 1993. Organic matter oxidation in marine sediments. In: Wollast, R., Mackenzie, F.T. and Chou, L. (eds), Interactions of C, N, P and S biogeochemical cycles. NATO ASI Series, 14. Springer, Berlin, Heidelberg, NY, pp. 333-363.

Canfield, D.E., Jørgensen; B.B., Fossing, H., Glud, R.N., Gundersen, J., Ramsing, N.B., Thamdrup, B., Hansen, J.W., Nielsen, L.P. and Hall, P.O.J., 1993b. Pathways of organic carbon oxidation in three continental margin sediments. Marine Geology, 113: 27-40.

Canfield, D.E. and Teske, A., 1996. Late Proterozoic rise in atmospheric oxygen concentration inferred from phylogenetic and sulphur-isotope studies. Nature, 382: 127-132.

Canfield, D.E., Kristensen, E. and Thamdrup, B., 2005. Aquatic Geomicrobiology. Elsevier, San Diego, 640 pp.

Christensen, D., 1984. Determination of substrates oxidized by sulfate reduction in intact cores of marine sediments. Limnology and Oceanography, 29: 189-192.

Chrost, R.J., 1991. Microbial enzymes in aquatic environments. Springer, Berlin, Heidelberg, NY, 317 pp.

Coleman, M.L., Hedrick, D.B., Lovley, D.R., White, D.C. and Pye, K., 1993. Reduction of Fe(III) in sediments by sulphate-reducing bacteria. Nature, 361: 436-438.

Conrad, R., Schink, B. and Phelps, T.J., 1986. Thermodynamics of H2-consuming and H2-producing metabolic reactions in diverse methanogenic environments under in situ conditions. FEMS Microbiology Ecology, 38: 353-360.

Cypionka, H., 1994. Novel metabolic capacities of sulfate-reducing bacteria, and their activities in microbial mats. In: Stal, L.J. and Caumette, P.(eds), Microbial mats, NATO ASI Series, 35, Springer, Berlin, Heidelberg, NY, pp. 367-376.

Dannenberg, S., Kroder, M., Dilling, W. and Cypionka, H., 1992. Oxidation of H2, organic compounds and inorganic sulfur compounds coupled to reduction of O2 or nitrate by sulfate-reducing bacteria. Archives of Microbiology, 158: 93-99.

DeLong, E.F., Franks, D.G., Yayanos, A.A., 1997. Evolutionary relationships of cultivated psychrophilic and barophilic deep-sea bacteria. Applied and Environmental Microbiology 63: 2105-2108.

D'Hondt, S.L., Jørgensen, B.B., Miller, D.J., et al., 2003. *Proceedings of the Ocean Drilling Program, Initial Reports*, 201 [CD-ROM]. Available from: Ocean Drilling Program, Texas A&M University, College Station TX 77845-9547, USA.

D'Hondt, S., Jørgensen, B.B., Miller, D.J., Batzke, A., Blake, R., Cragg, B.A., Cypionka, H., Dickens, G.R., Ferdelman, T., Hinrichs, K.-U., Holm, N.G., Mitterer, R., Spivack, A., Wang, G., Bekins, B., Engelen, B., Ford, K., Gettemy, G., Rutherford, S.D., Sass, H., Skilbeck, C.G., Aiello, I.W., Guèrin, G., House, C.H., Inagaki, F., Meister, P., Naehr, T., Niitsuma, S., Parkes, R.J., Schippers, A., Smith, D.C., Teske, A., Wiegel, J., Padilla, C.N. and Acosta, J.L.S., 2004. Distributions of microbial activities in deep subseafloor sediments. Science, 306: 2216-2221.

Ehrenreich, A. and Widdel, F., 1994. Anaerobic oxidation of ferrous iron by purple bacteria, a new type of phototrophic metabolism. Applied and Environmental Microbiology, 60: 4517-4526.

Ehrlich, H.L., 1996. Geomicrobiology. Marcel Dekker, NY, 719 pp.

Fenchel, T.M. and Jørgensen, B.B., 1977. Detritus food chains of aquatic ecosystems: The role of bacteria. In: Alexander, M. (ed), Advances in Microbial Ecology, 1, Plenum Press, NY, pp. 1-58.

Fenchel, T., King, G.M. and Blackburn, T.H., 1998. Bacterial biogeochemistry: The ecophysiology of mineral cycling. Academic Press, London, 307 pp.

Fossing, H. and Jørgensen, B.B., 1989. Measurement of bacterial sulfate reduction in sediments: evaluation of a single-step chromium reduction method. Biogeochemistry, 8: 205-222.

Fossing, H., Thode-Andersen, S. and Jørgensen, B.B., 1992. Sulfur isotope exchange between 35S-labeled inorganic sulfur compounds in anoxic marine sediments. Marine Chemistry, 38: 117-132.

Fossing, H., Gallardo, V.A., Jørgensen, B.B., Hüttel, M., Nielsen, L.P., Schulz, H., Canfield, D.E., Forster, S., Glud, R.N., Gundersen, J.K., Küfer, J., Ramsing, N.B., Teske, A., Thamdrup, B. and Ulloa, O., 1995. Concentration and transport of nitrate by the mat-forming sulphur bacterium Thioploca. Nature, 374: 713-715.

Fossing, H., 1995. 35S-radiolabeling to probe biogeochemical cycling of sulfur. In:. Vairavamurthy, M.A and. Schoonen, M.A.A (eds), Geochemical transformations of sedimentary sulfur. ACS Symposium Series, 612, American Chemical Society, Washington, DC, pp. 348-364.

Froelich, P.N., Klinkhammer, G.P., Bender, M.L., Luedtke, N.A., Heath, G.R., Cullen, D., Dauphin, P., Hammond,

D., Hartman, B. and Maynard, V., 1979. Early oxidation of organic matter in pelagic sediments of the eastern equatorial Atlantic: suboxic diagenesis. Geochimica et Cosmochimica Acta, 43: 1075-1088.

Glud, R.N., Gundersen, J.K., Jørgensen, B.B., Revsbech, N.P. and Schulz, H.D., 1994. Diffusive and total oxygen uptake of deep-sea sediments in the eastern South Atlantic Ocean: in situ and laboratory measurements. Deep-Sea Research, 41: 1767-1788.

Glud, R. N., Tengberg, A., Kühl, M., Hall, P.O.J. and Klimant, I., 2001. An in situ instrument for planar O_2 optode measurements at benthic interfaces. Limnology and Oceanography, 46: 2073-2080.

Greeff, O., Glud, R.N., Gundersen, J., Holby, O. and Jørgensen, B.B., 1998. A benthic lander for tracer studies in the sea bed: in situ measurements of sulfate reduction. Continental Shelf Research, 18: 1581-1594.

Gundersen, J.K. and Jørgensen, B.B., 1990. Microstructure of diffusive boundary layers and the oxygen uptake of the sea floor. Nature, 345: 604-607.

Gundersen, J.K., Glud, R.N. and Jørgensen, B.B., 1995. Oxygen turnover in the sea bed (in Danish), Vol. 57. Danish Ministry of Environment and Energy, Copenhagen, 155 pp.

Hansen, J.W., Thamdrup, B., and Jørgensen, B.B., 2000. Anoxic incubation of sediment in gas-tight plastic bags: A method for biogeochemical process studies. Marine Ecology Progress Series, 208: 273-282.

Hedges, J.I., 1978. The formation and clay mineral reactions of melanoidins. Geochimica et Cosmochimica Acta, 42: 69-76.

Henrichs, S.M. and Reeburgh, W.S., 1987. Anaerobic mineralization of marine sediment organic matter: rates and the role of anaerobic processes in the oceanic carbon economy. Geomicrobiological Journal, 5: 191-237.

Henrichs, S.M., 1992. Early diagenesis of organic matter in marine sediments: progress and perplexity. Marine Chemistry, 39: 119-149.

Henriksen, K., 1980. Measurement of in situ rates of nitrification in sediment. Microbial Ecology, 6: 329-337.

Hoehler, T.M., Alperin, M.J., Albert, D.B. and Martens, C.S., 1994. Field and laboratory studies of methane oxidation in an anoxic marine sediment: Evidence for a methanogen-sulfate reducer consortium. Global Biogeochemical Cycles, 8: 451-463.

Hoehler, T.M., Alperin, M.J., Albert, D.B. and Martens, C.S., 1998. Thermodynamic control on hydrogen concentrations in anoxic sediments. Geochimica et Cosmochimica Acta, 62: 1745-1756.

Huettel, M., Ziebis, W., Forster, S. and Luther, G.W., 1998. Advective transport affecting metal and nutrient distributions and interfacial fluxes in permeable sediments. Geochimica et Cosmochimica Acta, 62: 613-631.

Isaksen, M.F. and Jørgensen, B.B., 1996. Adaptation of psychrophilic and psychrotrophic sulfate-reducing bacteria to permanently cold marine environments. Applied and Environmental Microbiology, 62: 408-414.

Iversen, N. and Jørgensen, B.B., 1985. Anaerobic methane oxidation rates at the sulfate-methane transition in marine sediments from Kattegat and Skagerrak (Denmark). Limnology and Oceanography, 30: 944-955.

Jørgensen, B.B., 1978. A comparison of methods for the quantification of bacterial sulfate reduction in coastal marine sediments. I. Measurement with radiotracer techniques. Geomicrobiology Journal, 1: 11-27.

Jørgensen, B.B., 1982. Mineralization of organic matter in the sea bed - the role of sulphate reduction. Nature, 296: 643-645.

Jørgensen, B.B., 1990. A thiosulfate shunt in the sulfur cycle of marine sediments. Science, 249: 152-154.

Jørgensen, B.B. and Bak, F., 1991. Pathways and microbiology of thiosulfate transformation and sulfate reduction in a marine sediment (Kattegat, Denmark). Applied and Environmental Microbiology, 57: 847-856.

Jørgensen, B.B., 2001. Microbial life in the diffusive boundary layer. In: Boudreau, B.P. and Jørgensen, B.B. (eds), The benethic boundary layer: transport processes and biogeochemistry. Oxford University Press, Oxford, pp. 348-373.

Jørgensen, B. B. and Nelson, D.C, 2004. Sulfide oxidation in marine sediments: Geochemistry meets microbiology. In Amend, J.P., Edwards, K.J. and Lyons, T.W. (eds), Sulfur Biogeochemistry – Past and Present. Geological Society of America, pp. 63-81.

Kallmeyer, J., Ferdelman, T.G., Weber, A., Fossing, H. and Jørgensen, B.B., 2004. A cold chromium distillation procedure for radiolabeled sulfide applied to sulfate reduction measurements. Limnology and Oceanography: Methods, 2: 171-180.

Karp-Boss, L., Boss, E. and Jumars, P.A., 1996. Nutrient fluxes to planktonic osmotrophs in the presence of fluid motion. Oceanography and Marine Biology Annual Reviews, 34: 71-107.

Keil, R.G., Montlucon, Prahl, F.G. and Hedges, J.I., 1994b. Sorptive preparation of labile organic matter in marine sediments. Nature, 370: 549-552.

Kelly, D.P., 1982. Biochemistry of the chemolithotrophic oxidation of inorganic sulphur. Philosophic Transactions of the Royal Society of London, 298: 499-528.

King, G.M., 1983. Sulfate reduction in Georgia salt marsh soils: An evaluation of pyrite formation by use of 35S and 55Fe tracers. Limnology and Oceanography, 28: 987-995.

Klinkhammer, G.P., 1980. Early diagenesis in sediments from the eastern equatorial Pacific, II. Pore water metal results. Earth and Planetary Science Letters, 49: 81-101.

Knoblauch, C. and Jørgensen, B.B., 1999. Effect of temperature on sulfate reduction, growth rate, and growth yield in five psychrophilic sulfate-reducing bacteria from Arctic sediments. Environmental Microbiology, 1: 457-467.

Knoblauch, C., Sahm, K. and Jørgensen, B.B., 1999. Psychrophilic sulphate reducing bacteria isolated from permanently cold arctic marine sediments: description of *Desulfofrigus oceanense* gen. nov., sp. nov., *Desulfofrigus fragile* sp. nov., *Desulfofaba gelida* gen. nov., sp. nov., *Desulfotalea psychrophila* gen. nov., sp. nov., and *Desulfotalea arctica* sp. nov. Journal of Systematic Bacteriology, 49: 1631-1643.

Koch, A.L., 1990. Diffusion, the crucial process in many aspects of the biology of bacteria. In: Marshall, K.C. (ed), Advances in microbial ecology, 11, Plenum, NY, pp. 37-70.

Koch, A.L., 1996. What size should a bacterium be? A question of scale. Annual Reviews of Microbiology, 50: 317-348.

Krekeler, D. and Cypionka, H., 1995. The preferred electron acceptor of *Desulfovibrio desulfuricans* CSN. FEMS Microbiology Ecology, 17: 271-278.

Krüger, M., Meyerdierks, A., Glöckner, F.O., Amann, R., Widdel, F., Kube, M., Reinhardt, R., Kahnt, J., Böcher, R., Thauer, R.K. and Shima, S, 2003. A conspicuous nickel protein in microbial mats that oxidize methane anaerobically. Nature, 426: 878-881.

Kühl, M. and Revsbech, N.P., 2001. Biogeochemical microsensors for boundary layer studies. In: Boudreau, B.P. and Jørgensen, B.B. (eds.), The benthic boundary layer: transport processes and biogeochemistry. Oxford University Press, Oxford, pp. 180-210.

Kvenvolden, K.A., 1993. Gas hydrates - geological perspective and global change. Reviews in Geophysics, 31: 173-187.

Llobet-Brossa, E., Roselló-Mora, R. and Ammann, R., 1998. Microbial community composition of Wadden Sea sediments as revealed by fluorescence in situ hybridization. Applied and Environmental Microbiology, 64: 2691-2696.

Lochte, K. and Turley, C.M., 1988. Bacteria and cyanobacteria associated with phytodetritus in the deep sea. Nature, 333: 67-69.

Madigan, M.T., Martinko, J.M. and Parker, J., 1997. Biology of microorganisms. Prentice Hall, London, 986 pp.

Nauhaus, K., Boetius, A., Krüger, M. and Widdel, F., 2002. *In vitro* demonstration of anaerobic oxidation of methane coupled to sulphate reduction in sediment from a marine gas hydrate area. Environmental Microbiology 4: 296-305.

Nielsen, L.P., 1992. Denitrification in sediment determined from nitrogen isotope pairing. FEMS Microbiology Ecology, 86: 357-362.

Niewöhner, C., Hensen, C., Kasten, S., Zabel, M. and Schulz, H.D., 1998. Deep sulfate reduction completely mediated by anaerobic methane oxidation in sediments of the upwelling area off Namibia. Geochimica et Cosmochimica Acta, 62: 455-464.

Oremland, R.S. and Polcin, S., 1982. Methanogenesis and sulfate reduction: Competitive and noncompetetive substrates in estuarine sediments. Applied and Environmental Microbiology, 44: 1270-1276.

Oremland, R.S., Marsh, L.M. and Polcin, S., 1982. Methane production and simultaneous sulphate reduction in anoxic saltmarsh sediments. Nature, 296: 143-145.

Oremland, R.S. and Capone, D.G., 1988. Use of „specific" inhibitors in biogeochemistry and microbial ecology. In: Marshall, K.C. (ed), Advances in microbial ecology, 10, Plenum Press, NY, pp. 285-383.

Orphan, V.J., House, C.H., Hinrichs, K.-U., McKeegan, K.D. and DeLong, E.F., 2001. Methane-consuming archaea revealed by directly coupled isotopic and phylogenetic analysis. Science, 293: 484-487.

Parkes, R.J., Cragg, B.A. and Wellsbury, P., 2000. Recent studies on bacterial populations and processes in marine sediments: a review. Hydrogeology Journal, 8: 11-28.

Parkes, J.R., Cragg, B.A., Bale, S.J., Getliff, J.M., Goodman, K., Rochell, P.A., Fry, J.C., Weightman, A.J. and Harvey, S.M., 1994. Deep bacterial biosphere in Pacific Ocean sediments. Nature, 371: 410-413.

Parkes, R.J., Webster, G., Cragg, B.A., Weightman, A.J., Newberry, C.J., Ferdelman, T.G., Kallmeyer, J., Jørgensen, B.B., Aiello, I.W., and Fry, J.C., 2005. Deep sub-seafloor prokaryotes stimulated at interfaces over geological time. Nature, 436: 390-394.

Postgate, J.R., 1984. The sulfate-reducing bacteria. Cambridge University Press, London, 151 pp.

Rabus, F., Fukui, M., Wilkes, H. and Widdel, F., 1996. Degradative capacities and 16S rRNA-targeted whole-cell hybridization of sulfate-reducing bacteria in an anaerobic enrichment culture utilizing alkylbenzenes from crude oil. Applied and Environmental Microbiology, 62: 3605-3613.

Redfield, A.C., 1958. The biological control of chemical factors in the environment. American Scientist, 46: 206-222.

Reeburgh, W.S., 1969. Observations of gases in Chesapeake Bay sediments. Limnology and Oceanography, 14: 368-375.

Reimers, C.E., 1987. An in situ microprofiling instrument for measuring interfacial pore water gradients: methods and oxygen profiles from the North Pacific Ocean. Deep-Sea Research, 34: 2019-2035.

Revsbech, N.P., Nielsen, L.P., Christensen, P.B. and Sørensen, J., 1988. Combined oxygen and nitrous oxide microsensor for denitrification studies. Applied and Environmental Microbiology, 54: 2245-2249.

Roden, E.E. and Lovley, D.R., 1993. Evaluation of 55Fe as a tracer of Fe(III) reduction in aquatic sediments. Geomicrobiology Journal, 11: 49-56.

Rueter, P., Rabus, R., Wilkes, H., Aeckersberg, F., Rainey, F.A., Jannasch, H.W. and Widdel, F., 1994. Anaerobic oxidation of hydrocarbons in crude oil by new types of sulphate-reducing bacteria. Nature, 372: 455-458.

Sagemann, J., Jørgensen, B.B. and Greeff, O., 1998. Temperature dependence and rates of sulfate reduction in cold sediments of Svalbard, Arctic Ocean. Geomicrobiology Journal, 15: 83-98.

Sansone, F.J., Andrews, C.C. and Okamoto, M.Y., 1987. Adsorption of short-chain organic acids onto nearshore marine sediments. Geochimica et Cosmochimica Acta, 51: 1889-1896.

Santschi, P.H., Anderson R.F., Fleisher, M.Q. and Bowles, W., 1991. Measurements of diffusive sublayer thicknesses in the ocean by alabaster dissolution, and their implications for the measurements of benthic fluxes. Journal of Geophysical Research, 96: 10.641-10.657.

Schink, B., 1997. Energetics of syntrophic cooperation in methanogenic degradation. Microbiology and Molecular Biology Reviews, 61: 262-280.

Schippers, A., Neretin, L.N., Kallmeyer, J., Ferdelman, T.G., Cragg, B.A., Parkes, R.J. and Jørgensen, B.B., 2005. Prokaryotic cells of the deep sub-seafloor biosphere identified as living bacteria. Nature, 433: 861-864.

Schopf, J.W. and Klein, C. (eds), 1992. The proterozoic biosphere. Cambridge University Press, Cambridge, 1348 pp.

Schulz, H.D., Dahmke, A., Schinzel, U., Wallmann, K. and Zabel, M., 1994. Early diagenetic processes, fluxes and reaction rates in sediments of the South Atlantic. Geochimica et Cosmochimica Acta, 58: 2041-2060.

Schulz, H., Brinkhoff, T., Ferdelman, T.G., Hernandez Marine, M., Teske, A. and Jørgensen, B.B., 1999. Dense populations of a giant sulfur bacterium in Namibian shelf sediments. Science, 284: 493-495.

Seitzinger, S. P., Nielsen, L.P., Caffrey, J. and Christensen, P.B., 1993. Denitrification measurements in aquatic sediments: A comparison of three methods. Biogeochemistry, 23: 147-167.

Smith, K.L.Jr., Clifford, C.H. Eliason, A.h., Walden, B., Rowe, G.T. and Teal, J.M., 1976. A free vehicle for measuring benthic community metabolism. Limnology and Oceanography, 21: 164-170.

Sørensen, J., 1978. Denitrification rates in a marine sediment as measured by the acetylene inhibition technique. Applied and Environmental Microbiology, 35: 301-305.

Sørensen, J., Christensen, D. and Jørgensen, B.B., 1981. Volatile fatty acids and hydrogen as substrates for sulfate-reducing bacteria in anaerobic marine sediment. Applied and Environmental Microbiology, 42: 5-11.

Stetter, K.O., 1988. *Archaeoglobus fulgidus* gen. nov., sp. nov.: a new taxon of extremely thermophilic Archaebacteria. Systematic and Applied Microbiology, 10: 172-173.

Stetter, K.O., Huber, R., Blöchl, E., Knurr, M., Eden, R.D., Fielder, M., Cash, H. and Vance, I., 1993. Hyperthermophilic archaea are thriving in deep North Sea and Alaskan oil reservoirs. Nature, 365: 743-745.

Stetter, K.O., 1996. Hyperthermophilic procaryotes. FEMS Microbiology Revue, 18: 149-158.

Straub, K.L., Benz, M., Schink, B. and Widdel, F., 1996. Anaerobic, nitrate-dependent microbial oxidation of ferrous iron. Applied and Environmental Microbiology, 62: 1458-1460.

Straub, K.L. and Buchholz-Cleven, B.E.E., 1998. Enumeration and detection of anaerobic ferrous iron-oxidizing, nitrate-reducing bacteria from diverse European sediments. Applied and Environmental Microbiology, 64: 4846-4856.

Suess, E., 1980. Particulate organic carbon flux in the oceans-surface productivity and oxygen utilization. Nature, 288: 260-263.

Tegelaar, E.W., de Leeuw, J.W., Derenne, S. and Largeau, C., 1989. A reappraisal of kerogen formation. Geochimica et Cosmochimica Acta, 53: 3103-3106.

Tengberg, A., de Bovee, F., Hall, P, Berelson, W., Chadwick, D., Ciceri, G., Crassous, P., Devol, A., Emerson, s., Gage, J., Glud, R., Graziottin, F., Gundersen, J., Hammond, D., Helder, W., Hinga, K., Holby, O., Jahnke, R., Khripounoff, A., Lieberman, S., Nuppenau, V., Pfannkuche, O., Reimers, C., Rowe, G., Sahami, A., Sayles, F., Schurter, M., Smallman, D., Wehrli, B. and de Wilde, P., 1995. Benthic chamber and profiling landers in oceanography - A review of design, technical solutions and function. Progress in Oceanography, 35: 253-292.

Thamdrup, B., Finster, K., Hansen, J.W. and Bak, F., 1993. Bacterial disproportionation of elemental sulfur coupled to chemical reduction of iron or manganese. Applied and Environmental Microbiology, 59: 101-108.

Thamdrup, B., Fossing, H. and Jørgensen, B.B., 1994. Manganese, iron, and sulfur cycling in a coastal marine sediment, Aarhus Bay, Denmark. Geochimica et Cosmochimica Acta, 58: 5115-5129.

Thauer, R.K., Jungermann, K. and Decker, K., 1977. Energy conservation in chemotrophic anaerobic bacteria. Bacterial Reviews, 41: 100-180.

Thomsen, L., Jähmlich, S., Graf, G., Friedrichs, M., Wanner, S. and Springer, B., 1996. An instrument for aggregate studies in the benthic boundary layer. Marine Geology, 135: 153-157.

Vetter, Y.A., Deming, J.W., Jumars, P.A. and Kriegerbrockett, B.B., 1998. A predictive model of bacterial foraging by means of freely released extracellular enzymes. Microbiology Ecology, 36: 75-92.

Weber, A. and Jørgensen, B.B., 2002. Bacterial sulfate reduction in hydrothermal sediments of the Guaymas Basin, Gulf of California, Mexico. Deep-Sea Research, 49: 827-841.

Weiss, M.S., Abele, U., Weckesser, J., Welte, W. und Schulz, G.E., 1991. Molecular architecture and electrostatic properties of a bacterial porin. Science, 254: 1627-1630.

Wellsbury, P., Goodman, K., Barth, T., Cragg, B.A., Barnes, S.P. and Parkes R.J., 1997. Deep marine biosphere fuelled by increasing organic matter availability during burial and heating. Nature, 388: 573-576.

Wenzhöfer, F. and Glud, R.N., 2002. Benthic carbon mineralization in the Atlantic: a synthesis based on in situ data from the last decade. Deep-Sea Research I, 49: 1255-1279.

Westrich, J.T. and Berner, R.A., 1984. The role of sedimentary organic matter in bacterial sulfate reduction: The G model tested. Limnology and Oceanography, 29: 236-249.

Whitman, W.B., Coleman, D.C. and Wiebe, W.J., 1998. Prokaryotes: The unseen majority. Proceedings of the National. Academy of Sciences, USA, 95: 6578-6583.

Widdel, F., 1988. Microbiology and ecology of sulfate- and sulfur-reducing bacteria. In: Zehnder, A.J.B. (ed). Biology of anaerobic microorganisms. Wiley & Sons, NY, pp. 469-585.

Widdel, F., Schnell, S., Heising, S., Ehrenreich, A., Assmus, B. and Schink, B., 1993. Ferrous iron oxidation by anoxygenic phototrophic bacteria. Nature, 362: 834-836.

Yayanos, A.A., 1986. Evolutional and ecological implications of the properties of deep-sea barophilic bacteria. Proc. Natl. Acad. Sci., 83: 9542-9546.

6 Benthic Cycling of Oxygen, Nitrogen and Phosphorus

CHRISTIAN HENSEN, MATTHIAS ZABEL AND HEIDE N. SCHULZ

6.1 Introduction

All particles settling at the sea floor continuously undergo diagenetic alteration due to physical and chemical processes in the sediment (e.g. particle mixing, compaction, redox reactions). Marine sediments are the primary long-term repository of organic matter and the systematic analysis of the control mechanisms and processes modifying the original input signal is of key importance for the understanding and the reconstruction of biogeochemical cycles in the ocean. This chapter mainly focuses on processes occurring at the transition zone between sediments and bottom water where fresh, bio-available organic material is subject to intense bacterially mediated oxidation. Oxygen and nitrate are treated together because they are thermodynamically the most favorable electron acceptors in the diagenetic sequence of organic matter decomposition and their pathways are coupled through oxidation of reduced nitrogen species (nitrification) in oxic systems (cf. Section 3.2.3). Generally, their involvement in the biogeochemical cycles of the ocean is much more complex than only seen from this point of view and therefore, the combined examination of both parameters is for reasons of convenience and follows the general concept of this book. Both oxygen and nitrate pathways are very important and inherent for the understanding of the oceanic carbon cycle. Oxygen is introduced into surface waters by photosynthesis and, even more important, by exchange with the atmosphere. Conversely, it is consumed in the course of the degradation of organic matter. The latter occurs throughout the water column and in the sediments resulting in the release of carbon dioxide, nitrate,

and phosphate. Nitrate itself is then used as the "next" suitable electron acceptor for organic matter degradation in environments where oxygen availability is limited, such as oxygen minimum zones or below the oxygen penetration depth in the sediments. Nitrate and phosphate are the most important limiting nutrients for driving primary productivity in the surface water. There is, however, a still ongoing debate on whether one or the other is the limiting nutrient on different time and spatial scales. The arguments are often called the "biologist's view" (nitrogen limitation) and the "geochemist's view" (phosphorus regulation), because they implicate two different perspectives on looking at the oceanic biogeochemical cycles. The typical geochemist would argue that over long time scales nitrogen fixers (converting N_2 to organic nitrogen) may be able to balance the nitrate deficit by using the huge reservoir of dissolved N_2 and allowing certain levels of productivity even if nitrate becomes exhausted relative to phosphorus. The supply of phosphorus depends solely on the riverine input and the remineralisation in the water column and in the sediments. Once it is exhausted, there is no other means of replenishing the reservoir. The view of the biologist might be supported by the fact that in certain areas of the global ocean small residues of phosphorus exist while nitrate is undetectable and the inorganic dissolved N:P ratio of the global ocean is often below 16:1. A sort of compromising theory has been proposed by Tyrrell (1999), which differentiates between nitrogen as the "proximate limiting nutrient" (with regard to immediate growth rates) and phosphate as the "ultimate limiting nutrient" (with regard to the whole system productivity over long timescales). However, since this is not the final conclusion and a more

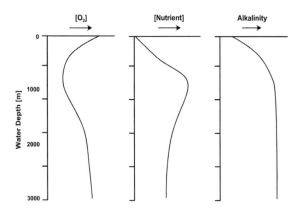

Fig. 6.1 Idealized vertical sections of dissolved oxygen, nutrients and alkalinity through the water column. The nutrients are completely consumed in surface waters. With increasing depth nutrients and CO_2 are produced while oxygen is consumed during organic matter decomposition.

overlaying water, adsorb to iron minerals (see Section 7.4.4.3), or precipitate as phosphate bearing minerals like apatite.

In the following sections, we will first give a short overview concerning the distribution and the geochemistry of oxygen nitrate in the modern oceans followed by a more detailed description of the relevant geochemical processes in marine sediments.

6.2   Distribution of Oxygen, Nitrate and Phosphate in Seawater

The distribution of dissolved oxygen in seawater results from the interaction of different factors. Those are (a) the input of oxygen across the atmosphere-ocean interface and the oxygen production by phytoplankton, (b) the microbially catalyzed degradation of organic matter and oxidation of other reduced substances, and (c) physical transport and mixing processes in the ocean. Theoretically, the oxygen concentration in seawater is limited by its solubility, but in fact the saturation concentration is only reached in surface waters. At some places, surface waters are even supersaturated with respect to oxygen,

detailed discussion is beyond the scope of this chapter we would like refer to general compilations of this topic (i.e. Falkowski 1997; Tyrrell 1999; Wallmann 2003; Mills et al. 2004).

Phosphate, in contrast to nitrate, does not change between different redox states. Phosphate is released into the pore water from degrading organic material and by polyphosphate accumulating bacteria, and may either diffuse into the

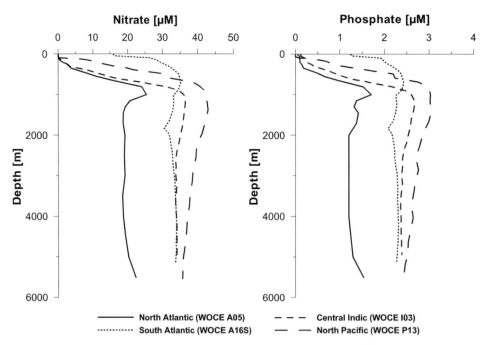

Fig. 6.2 Vertical sections of nitrate and phosphate in different ocean basins. Older water masses in the Pacific and Indian Ocean are generally more enriched in nitrate and phosphate (data from WOCE-sections A05, A16S, I03, and P13; WOCE 2002).

which is thought to be caused by entrapment of air bubbles (cf. Chester 1990). Below the productive mixed layer oxygen is depleted by bacterial respiration processes. Numerous studies show that this process is most intense within the upper 1,000 m of the water column, which marks approximately the lower end of the permanent thermocline. As a consequence significant oxygen minimum zones can be observed in areas with high surface water productivity (see Fig. 4.2). Such areas exist where upwelling water masses significantly enhance the supply of nutrients, mainly at the west coasts of the continents (trade wind belts of America and Africa), but also along the equatorial divergence zones and related to the Polar Fronts (e.g. Antoine et al. 1996; Behrenfeld and Falkowski 1997; see Fig. 12.5). Generally, the distribution of oxygen in the water column is dependent on how much the oxygen depletion exceeds the supply by vertical and lateral advection and diffusion at a certain depth. Since carbon oxidation is the main reason for oxygen depletion, a syngenetical increase of dissolved nutrients (nitrate and phosphate) and a decrease of particulate organic carbon with increasing water depth is the typical feature. Figure 6.1 shows idealized depth profiles of oxygen, a typical nutrient (like phosphate or nitrate), and alkalinity. Alkalinity, in this case, has to be understood as a sum parameter for dissolved carbon species which increase with depth due to the continued decay of organic material. These patterns, however, may deviate between the ocean basins depending on the oceanographic setting (currents, mixing of water masses), the particle-transport through the water column, and the composition of mineralized organic matter. Figure 6.2 shows some examples of the nitrate distribution in different ocean basins. The deep waters of the Pacific and the Indian Ocean are enriched in nitrate relative to the North Atlantic due to deep-water transport through the ocean basins (see below). As pointed out above, both nutrients are depleted in the surface water.

The investigation of the key processes has largely benefited from the invention of sediment traps measuring the particle flux through the water column over long periods of time. A number of researchers have attempted to quantify export fluxes of organic carbon from surface waters and their transition to the bottom by empirical formulations (cf. Section 4.2; e.g. Betzer et al. 1984; Berger et al. 1987; Martin et al. 1987). Generally, these are exponential and power equa-

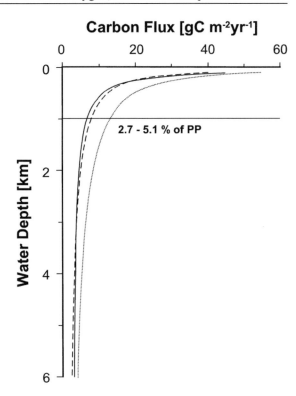

Fig. 6.3 Example calculations for the transit flux of organic carbon after empirical equations assuming a surface primary productivity of 250 gC m^{-2} yr^{-1}. More than 95% of the organic carbon is oxidized above the 1,000 m horizon.

Solid line: J_{Corg} [gC m^{-2} yr^{-1}] = 17 PP/z + PP/100 (Berger et al. 1987)
Broken line: J_{Corg} [gC m^{-2} yr^{-1}] = 9 PP/z + 0.7 PP/z$^{0.5}$ (Berger et al. 1987)
Dotted line: J_{Corg} [gC m^{-2} yr^{-1}] = 0.409 PP$^{1.41}$/z$^{0.628}$ (Betzer et al. 1984)

tions predicting about 90% of the remineralization within the upper hundreds of meters of the water column (cf. Fig 12.1). To illustrate this process Figure 6.3 shows the application of three different equations to predict the vertical (transit) carbon flux. In all cases a surface primary productivity of 250 gC m^{-2}yr^{-1} is assumed. At a depth horizon of 1,000 m only less than 5% of primarily produced organic carbon remain.

Comparative studies of the relation between primary productivity and benthic mineralization processes (Jahnke et al. 1990; Rowe et al. 1994; Hensen et al. 2000) show, however, that these empirical formulations are restricted to a limited, regional use (cf. Section 12.3). A further important factor in this regard is the reaction stoichiometry of organic matter degradation, since it determines the proportional release of CO_2, NO_3 and PO_4. Based on planktonic decomposition studies,

Redfield et al. (1963) suggested an overall C/N/P/ O_2 ratio of 106/16/1/138 (see Eq. 6.1, cf. Fig 3.11). However, although widely used, subsequent investigations have put this formulation into question. Deviating ratios were formulated as 140/ 16/1/172 (Takahashi et al. 1985) or 117/16/1/170 (Anderson and Sarmiento 1994) implying that there is still debate on the general validity of the use of one "Redfield ratio" for all ocean basins and all water depths. Instead, C/N/P ratios seem to be subject to regional variation.

Oxygen depth profiles show an opposite trend to alkalinity and nitrate resulting from low mineralization rates in the deep ocean waters and input of oxygen-rich water masses by advective transport. Figure 6.4 shows a meridional transect of oxygen concentration through the Atlantic Ocean compiled by Reid (1994). The most prominent pattern is the southward flow of oxygen-rich North Atlantic Deep Water raising the oxygen concentration along its flow path into the equatorial South Atlantic at water depths of 3,000 – 4,000 m. Less prominent, but still significant, is also the northward flow of oxygen-rich Antarctic Intermediate Water which is marked by elevated oxygen concentrations on a south-north path (50°S – 20°S) between 0 – 1,000 m water depth. The distribution of oxygen in ocean water is therefore strongly dependent on large-scale circulation patterns. The same is valid for the distribution of nutrients: Vertical concentration profiles are always a mixture of *in situ* decomposition and advective transport processes. The global deep-water circulation pattern shows a general flow path from the North Atlantic to the North Pacific and Indian Oceans. As a result, "older" deep waters in the Pacific and Indian Oceans are depleted of oxygen and enriched in nutrients (Fig. 6.2) and CO_2 (Broecker and Peng 1982; Kennett 1982; Chester 1990; see Chapter 9).

When studying the early diagenesis of deep-sea sediments, it is very important to consider all features of the oceanic environment and consequently of the composition of the overlying deep-ocean water, since it determines the availability of any solute, i.e. oxygen or nitrate, as possible oxidants for organic carbon mineralization. As a consequence of intense deep water mixing and limited supply of degradable material to the sediments within the big central gyres, oxidant limitation is the exceptional case, which has, however, not always been the case in earth history (cf. Section 4.1).

6.3   The Role of Oxygen, Nitrogen and Phosphorus in Marine Sediments

To estimate the role of oxygen and nitrate we have to describe the general processes occurring close to the sediment water interface, the methods how to measure concentrations, fluxes, and consumption rates, and how to relate them to organic matter degradation and other processes. Subsequently, we will show examples from case studies to characterize the magnitude of fluxes and different environments in deep-sea sediments. The early diagenetic processes at the sediment-water interface are of special interest in global biogeochemical cycles because it is decided at this separation line between ocean water and sediment if any substance is recycled or buried for a long period of time in a geological sense.

6.3.1   Respiration and Redox Processes

6.3.1.1   Nitrification and Denitrification

In principle, the sequence of oxidants is determined by the energy yield for the microorganisms. When oxygen and nitrate are depleted reduction of Mn and Fe (oxo)hydroxides and sulfate as well as methane fermentation follow in the sequence with decreasing yield of energy (Froelich et al. 1979; cf. Section 3.2.5 and subsequent chapters). This sequence is generally valid, even though numerous studies have identified an overlap of carbon oxidation pathways within the sediment resulting from competition between microbial populations (Canfield 1993) and the presence of microenvironments (e.g. Jørgensen 1977; cf. Chapters 7, 8, 12).

The general equations of coupled oxic respiration and nitrification (6.1) and denitrification (6.2) describing the "top" of the diagenetic sequence are given as:

Oxic respiration and nitrification
$$(CH_2O)_{106}(NH_3)_{16}(H_3PO_4) + 138\ O_2 \rightarrow$$
$$106\ CO_2 + 16\ HNO_3 + H_3PO_4 + 122\ H_2O$$
$$\Delta G^0 = -3190\ kJ\ mol^{-1} \tag{6.1}$$

Denitrification
$$(CH_2O)_{106}(NH_3)_{16}(H_3PO_4) + 94.4\ HNO_3 \rightarrow$$
$$106\ CO_2 + 55.2\ N_2 + H_3PO_4 + 177.2\ H_2O$$
$$\Delta G^0 = -3030\ kJ\ mol^{-1} \tag{6.2}$$

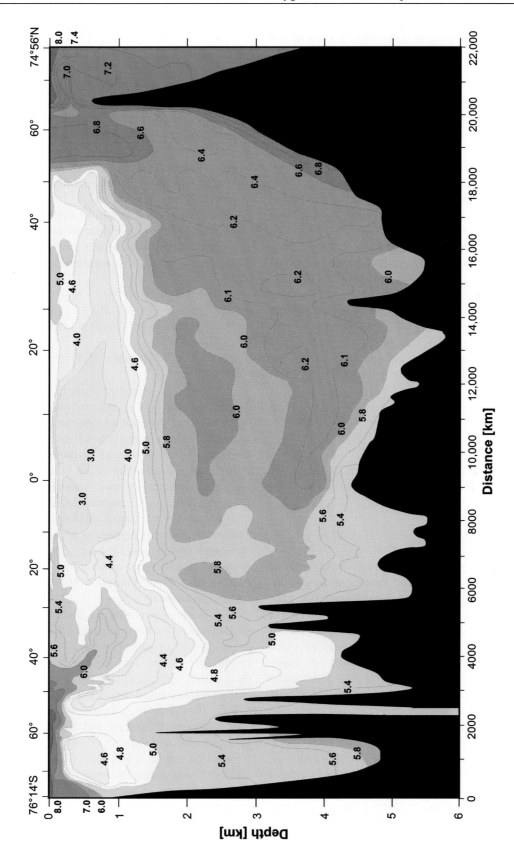

Fig. 6.4 North-South section of dissolved oxygen (ml/l) across the Atlantic Ocean (after Reid 1994). See text for explanation.

The ΔG^0-values indicate the higher energy yield for using oxygen rather than nitrate as the terminal electron acceptor. In these equations, it is assumed that organic matter with a C/N/P ratio of 106/16/1 is oxidized. Equation (6.1) requires that ammonia released during oxic respiration is quantitatively oxidized to nitrate. The process of nitrification is, however, more complex than described above and is known as a stepwise oxidation of nitrogen species by different microbes. Those can be grouped into ammonia oxidizers, generally with a genus name starting with the prefix Nitroso- and nitrite oxidizers, starting with the prefix Nitro-. Although ammonia and nitrite oxidizing bacteria are physiologically depending on each other and usually occur in close proximity, they are phylogenetically two distinct groups of bacteria, which are not closely related (Bock and Wagner 2001). The oxidation of ammonia to nitrite is also a two step process in which hydroxylamine is formed as an intermediate product (Eq. 6.3). The second step yields the biogeochemically useful energy (Eq. 6.4):

$$2\,NH_3 + O_2 \rightarrow 2\,NH_2OH \tag{6.3}$$

$$NH_2OH + O_2 \rightarrow NO_2^- + H_2O + H^+ \tag{6.4}$$

In the following step nitrite is oxidized by lithotrophs like *Nitrobacter* or *Nitrococcus* to nitrate (Eq. 6.5):

$$NO_2^- + \tfrac{1}{2}\,O_2 \rightarrow NO_3^- \tag{6.5}$$

In summary, nitrifying bacteria are considered to be strictly aerobic and therefore depend on adequate oxygen supply for their energy gain (Painter 1970). Experimental results of Henriksen et al. (1981) on control factors of nitrification rates in shallow water sediments from Denmark revealed that nitrification is strongly dependent on temperature, oxygen availability (oxygen penetration depth), ammonia supply and the number of nitrifying bacteria. These complex interactions are more thoroughly discussed in comprehensive reviews on nitrification in coastal marine environments by Kaplan (1983) and Henriksen and Kemp (1988).

Denitrification starts when oxygen is almost depleted (below the oxygen penetration depth) by inducing an enzymatic system of nitrate reductase and nitrite reductase by facultative aerobic bacteria, which can only use nitrogenous oxides if oxygen is - nearly - absent. Measurements carried out with a combined microsensor for O_2 and N_2O indicated that denitrification is restricted to a thin anoxic layer below the oxic zone (Christensen et al. 1989). Denitrification is the only biological process that produces free nitrogen. It removes fixed nitrogen compounds and, therefore, exerts a negative feedback on eutrophication making it a crucial process for the preservation of life on earth. For example, denitrification in rivers and coastal environments reduces the supply of fixed nitrogen from the continents by about 40%. (Seitzinger 1988).

The reduction of nitrate to dinitrogen occurs first as a reduction of nitrate to nitrite (Eq. 6.6) and then a stepwise reduction to nitrogen oxide, dinitrogen oxide (Eq. 6.7) and dinitrogen (Eq. 6.8):

$$NO_3^- + 2\,H^+ + 2\,e^- \rightarrow NO_2^- + H_2O \tag{6.6}$$

$$2\,NO_2^- + 6\,H^+ + 4\,e^- \rightarrow N_2O + 3\,H_2O \tag{6.7}$$

$$N_2O + 2\,H^+ + 2\,e^- \rightarrow N_2 + H_2O \tag{6.8}$$

The last reduction step from nitrous oxide to dinitrogen (Eq. 6.8) is not always completed so that the final product of denitrification is not necessarily dinitrogen. Nitrous oxide may therefore be produced or consumed during denitrification. A compilation of Seitzinger (1998) for coastal marine environments shows, however, that in most cases the net ratios between $N_2O:N_2$ production rates are usually very small (between 0.0002-0.06). The total amount of dinitrogen produced obviously depends on the partial pressure of oxygen (higher oxygen contents seem to be suitable for the production of N_2O; Jørgensen et al. 1984), the pH, and the presence of H_2S.

The major prerequisite for denitrification is the availability of nitrate (including nitrite). In marine sediments the dominant sources of nitrate are the production during nitrification and the supply from overlying bottom water by means of bioturbation, bioirrigation, and diffusion (see Section 6.3.2). Furthermore, denitrification is strongly dependent on temperature, but also on the oxygen concentration and the availability of organic matter are limiting for the process (Middelburg et al. 1996a). There is also evidence that denitrification may be reduced at high sulfate reduction rates, because low sulfide concentrations completely inhibit nitrification which in turn is necessary for denitrification (Seitzinger 1988). Generally, most suitable conditions for denitrification are obtained

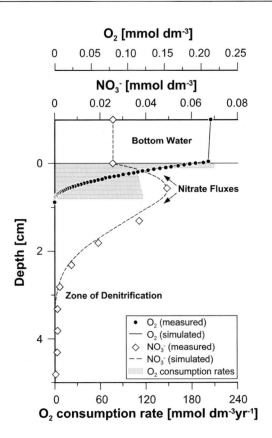

Fig. 6.5 Measured and simulated profiles of oxygen and nitrate of station GeoB 1711 from the continental slope off Namibia at a water depth of approximately 2000 m. Degradation of organic matter with a C/N ration of 3.7 was assumed for simulation. Bars indicate oxygen consumption rates required for model fit (after Hensen et al. 1997).

The ability of some bacteria to reduce nitrite further to ammonia (Jørgensen and Sørensen 1985; Sørensen 1987) can be expressed by the following equation:

$$NO_2^- + 8\,H^+ + 6\,e^- \rightarrow NH_4^+ + 2\,H_2O \qquad (6.9)$$

Figure 6.5 shows typical pore water profiles of oxygen and nitrate measured in organic rich sediments off Namibia summing up the net reactions described above. Due to nitrification, the highest nitrate concentrations are reached approximately at the oxygen penetration depth. At about 3 cm depth, nitrate is consumed in the process of denitrification. The nitrate profile indicates an upward flux into the bottom water and a downward flux to the zone of denitrification. Both profiles are verified by the application of Equations 6.1 and 6.2 within the computer model CoTAM/CoTReM (cf. Chapter 15) as indicated by the solid and dashed lines.

It could be shown that the processes described above are of key importance for the distribution of oxygen and nitrate in the sediment column and can directly be related to the degradation of organic matter. The reduction by any other reduced species (e.g. H_2S, Fe^{2+}, Mn^{2+}) can, however, also be an important pathway (see below and cf. Chapters 7, 8). Generally, most of these processes are microbially mediated (cf. Chapter 5; e.g. Chapelle 1993; Stumm and Morgan 1996). This also includes the (re-oxidation) of upward diffusing reduced nitrogen species (mainly ammonia) during oxic respiration. In the example of Figure 6.5 the C/N ratio of decomposed organic matter had to be decreased to a factor of 3.7 (instead of 6.625; cf. Eq. 6.1) to reproduce the measured nitrate concentration profile. This is a reasonable explanation because fresh organic matter (with a high fraction of N-rich amino acids) could be supplied in this specific marine environment (cf. Section 4.4), but the oxidation of upward diffusing reduced nitrogen species and/or artificially increased nitrate concentrations (see Section 6.3.2) have to be considered for interpretation.

The general reaction principle and the coupling of all processes described in this section is illustrated in Figure 6.6. All dissolved species created during nitrification and denitrification may either escape into the bottom water, mainly by diffusion, or are involved in subsequent redox processes. Ammonium is generally re-oxidized in

at intermediate carbon availability levels when carbon is not limiting for oxic respiration, but sulfate reduction is still low or absent. Studies of Jørgensen and Sørensen (1985) along a salinity gradient in a Danish estuary revealed evidence that denitrification and nitrate reduction (Eq. 6.9) became a more important pathway as sulfate became limited due to increased freshwater input close to the river outlet.

Many bacteria are able to reduce nitrate to ammonia in order to use nitrate as a nitrogen source for the build up of biomass (assimilation). Also the dissimilatoty nitrate reduction to ammonia (DNRA), where nitrate is respired to ammonia, is a widespread physiological ability among bacteria. Nevertheless, until recently it was thought to have small ecological significance, because the potential energy gain of nitrate reduction to nitrogen is higher (compare Fig. 5.9).

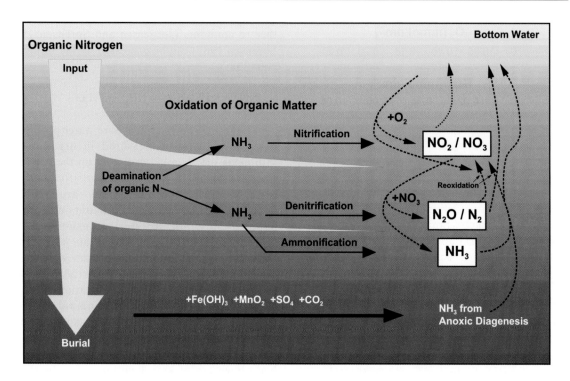

Fig. 6.6 Pathways of nitrogen in marine surface sediments. Arrows: black, organic matter degradation; gray, particulate organic nitrogen; dotted, diffusion of solutes.

oxygenated sediments (cf. Section 6.3.1.3), but it may also diffuse into the bottom water, if oxygen or other suitable oxidants are either limited or absent.

Nitrate as terminal electron acceptor is especially important for the oxidation of ferrous iron (Straub et al. 1996) or hydrogen sulfide. Particularly among the sulfide oxidizing bacteria a unique adaptation for the use of nitrate as electron acceptor has evolved: a central vacuole used for the storage of nitrate (Fossing et al. 1995). The vacuolated sulfur bacteria contain nitrate in concentrations ranging from several tens to hundreds of mmolar, which is up to five orders of magnitude higher than the ambient environmental concentrations (Fossing et al. 1995). High internal nitrate concentrations are found in three genera of sulfur bacteria (Fig. 6.7) called *Beggiatoa* (McHatton et al. 1996), *Thioploca* (Fossing et al. 1995) and *Thiomargarita* (Schulz et al. 1999). All of these bacteria are also storing the electron donor sulfide in the form of sulfur globules. The vacuolated sulfur bacteria can be unusually large (up to several 100 μm diameter), because the volume of the vacuole is not metabolically active. Their sulfur inclusions scatter the light giving these

bacteria a bright white appearance, which makes it possible to see them with the naked eye.

Beggiatoa and *Thioploca* are filamentous bacteria, meaning that the cells occur in a row and are connected with each other. The filaments are motile by gliding. *Thioploca* filaments are found as very dense mats in sediments of the upwelling areas off Chile and Peru. The filaments live as bundles in vertical sheaths reaching up to 20 cm into the sediment (Fig. 6.7). In the sheaths they glide between the surface of the sediment, where they take up nitrate, and deeper parts of the sediment, where sulfide produced by sulfate reducing bacteria is available (Fossing et al. 1995). The larger nitrate storing forms of *Beggiatoa* are frequently encountered in areas with locally enhanced sulfide flux, such as hydrothermal vents and seeps or methane hydrates. In contrast to *Thioploca* the filaments are not forming bundles. They are usually found as white or more seldom orange mats at the sediment surface (Fig. 6.7). In the Benguela upwelling region, a large area of the seafloor covered with loose diatome ooze, is populated by *Thiomargarita* cells (Schulz et al. 1999). In contrast to their close relatives *Beggiatoa* and *Thioploca* these sulfur bacteria are not motile and do not form true filaments. The individual, spherical cells are hold together in strings by a slime sheath (Fig. 6.7). They are the largest bacteria known so far (>700 μm diameter) and may survive up to several years in sulfidic sediments without getting into contact with nitrate or oxygen. Instead of actively gliding in between different sediment horizons they seem to rely on getting passively transported into nitrate containing water by sediment suspension.

Even though it is quite obvious that the nitrate storing sulfur bacteria use nitrate as electron acceptor for the oxidation of sulfide, it is still not completely clear, whether the end product of nitrate respiration is ammonia or nitrogen. As this group of bacteria cannot be grown in pure cultures yet, all conclusions on their metabolism are drawn from indirect evidences and remain to some degree speculative. Earlier reports on denitrification in *Beggiatoa* (Sweerts et al. 1990) have been doubted since Otte et al. (1999) could show ammonia production in cleaned *Thioploca* filaments. At first glance, it seems more likely to assume denitrification, as the energy yield per mole sulfide oxidized with nitrate is much higher if

nitrate is reduced to nitrogen (-714.7 kJ mol[-1]) and not to ammonia (-436.2 kJ mol[-1]) (Jørgensen and Nelson 2004). Nevertheless, per mole nitrate the energy yields of both processes are quite comparable with -446.7 kJ mol[-1] for denitrification and 436.2 kJ mol[-1] for dissimilatory reduction of nitrate to ammonia (DRNA). Furthermore, the last step of denitrification is easily blocked by sulfide, which suggests that DNRA might be a more efficient way of anaerobic sulfide oxidation than denitrification (Jørgensen and Nelson 2004).

6.3.1.2 Coupling of Oxygen and Nitrate to other Redox Pathways

Below the oxygenated zone anoxic diagenesis is stimulated, if biodegradable organic matter is still sufficiently available. Anoxic mineralization processes result in a release of reduced species, like ammonia (Fig. 6.6), into pore water which might be re-oxidized again as they diffuse back to the surface layers. Stimulation of anoxic diagenesis requires high input of organic matter to the sediment surface which is usually combined with high sediment advection restricted to coastal and high

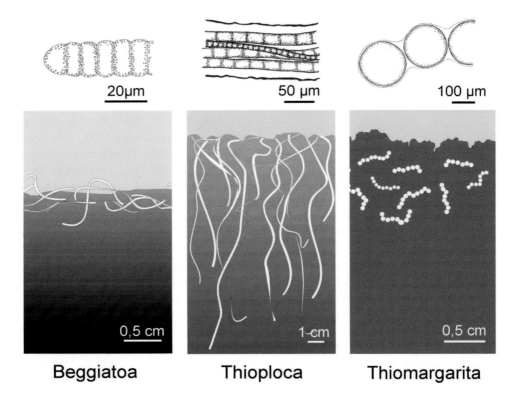

Fig. 6.7 Morphology of the three genera of nitrate storing sulfur bacteria. Above: appearance in the light microscope. Below: distribution in the sediment.

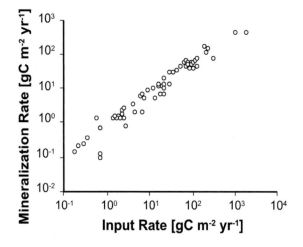

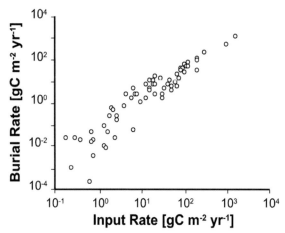

Fig. 6.8 Input rates of organic carbon plotted against carbon mineralization and carbon burial from different data sources (after Henrichs 1992).

production areas adjacent to continental margins. The general coupling between sediment advection, deposition (burial) and mineralization of organic matter is depicted in Figure 6.8 and has been documented in numerous studies (cf. Sections 4.2, 4.3; e.g. Henrichs and Reeburgh 1987; Henrichs 1992; van Cappellen et al. 1993; Tromp et al. 1995).

The amount of fresh organic matter arriving at the sediment surface also constitutes a control parameter for the population density of benthic macrofauna responsible for the biological mixing of the sediment. Biological mixing is known to be much more important for the transport of labile organic particles to deeper sediment layers than sedimentation (cf. Section 7.4.4). The strong correlation between sedimentation rate and particle mixing was recently compiled by Tromp et al. (1995) and is shown in Figure 3.28. Based on this

compilation Tromp et al. (1995) derived the following regression equation:

$$\log D_{bio} = 1.63 + 0.85 \log \varpi \qquad (6.10)$$

where D_{bio} is the bioturbation coefficient ($cm^2\ yr^{-1}$) and ϖ is the sedimentation rate ($cm\ yr^{-1}$). A closer look at the data shown in Figure 3.28 and Equation 6.10 makes clear that there is a difference between both parameters of up to three orders of magnitude. The correlation is, however, only applicable when other environmental factors, like bottom water anoxia, extreme sedimentation rates, or current action, are not effective. Bottom water deficiency of oxygen (below 20% sat.) has been shown to seriously decrease the bioturbation intensity (Rhoads and Morse 1971) and below an oxygen saturation of about 5% nearly no macrofauna will survive (Baden et al. 1990).

For some practical reasons, oxygen is often used as a measure for total respiration of a sediment, mainly in the marine environment (cf. Jahnke et al. 1996; Seiter et al. 2005; Section 12.5.2; Figs. 12.17 - 12.19). Because of all subsequent mineralization processes occurring below, this is of course not strictly correct. Rather a complete net-reoxidation of all reduced species produced during anoxic diagenesis is required – ultimately by means of oxygen (Pamatmat 1971).

Any fixation and burial of reduced species (e.g. the formation of sulfides or carbonates; pyrite, siderite,...) or the escape of reduced solutes across the sediment-water interface (e.g. CH_4, NH_4^+; N_2O, N_2, Mn^{2+}, Fe^{2+}; Bartlett et al. 1987; Seitzinger 1988; Devol 1991; Tebo et al. 1991; Johnson et al. 1992; Thamdrup and Canfield 1996; Wenzhöfer et al. 2002) results in an underestimation of the total respiration. The evaluation whether a reoxidation is complete is generally very difficult and is limited by the availability of measurements of all key species. The main questions are: (1) How big is the contribution of each pathway compared to the total mineralization? (2) To which amount and by which processes are these pathways interrelated? Since in most studies a lack of information on certain parameters remains, or different methods are applied to determine one pathway (e.g. differences resulting from *in situ* / *ex situ* determination of a species, or different methods to determine for example denitrification and sulfate reduction rates; see Section 6.4), the conclusion remains at least to some extent arbitrary. Reimers et al.

(1992) calculated for rapidly accumulating sediments on the continental margin off California that aerobic respiration is the major pathway of organic matter oxidation and more than 90% of the oxygen flux into the sediments is used for organic carbon oxidation. Since respiration by anoxic processes is estimated to exceed 30% of the total mineralization in the sediment this indicates an incomplete reoxidation cycle. Results of Canfield et al. (1993 a,b) and Wenzhöfer et al. (2002) from continental shelf sediments of the Baltic Sea indicate that oxygen consumption by reoxidation processes is quantitatively more important than aerobic respiration. In these environments sulfate reduction seems to be the dominant respiration process (Wenzhöfer et al. 2002; Thamdrup et al. 1994), although metal oxides can also play a significant role in particular cases. Canfield et al. (1993 a,b) have shown that the importance of metal oxides in the diagenetic sequence is strongly dependent on bioturbation activity. Mineralization rates simply derived from pore water gradients might underestimate the true rates by one order of magnitude (Haese 1997). Such complex interactions between different pathways of organic matter decomposition and redox reactions are restricted to coastal marine environments and highly accumulating upwelling regions. In oligotrophic regions of the deep sea, 100 to 1,000-times more organic carbon is oxidized by oxygen than by sulfate reduction and other pathways (Canfield 1989). For the major part of the world oceans the oxygen flux into the sediment provides a good approximation to the total rate of organic carbon oxidation.

6.3.1.3 Anaerobic Oxidation of Ammonium with Nitrate (Anammox)

The usual way of transferring biological bound nitrogen to dinitrogen is a series of microbial activities starting with the release of ammonia from degraded organic material (ammonification), the oxidation of ammonia with oxygen to nitrate (nitrification) and the reduction of nitrate to dinitrogen, when nitrate is used as electron acceptor for the oxidation of organic material (Fig. 6.9, cf. Section 6.3.1.1). The last step, denitrification, was until recently thought to be the most important microbial process releasing gaseous nitrogen and thereby counteracting eutrophication. Nevertheless, an anaerobic microbial process removing ammonia has been proposed early on, because in the absence of oxygen, ammonia is not accumulating in rates corresponding to the break down of organic material (Richards et al. 1965) and thermo-dynamical calculations suggest that it would be possible for bacteria to gain energy by the oxidation of ammonia with nitrate or nitrite (Broda 1977). The first direct evidence for the occurrence of anaerobic ammonia oxidation derived from wastewater bioreactors, where bacterial populations oxidizing ammonia with nitrite and producing dinitrogen could be grown in enrichment cultures (Mulder et al. 1995; van de Graaf et al. 1995).

Until now none of the bacteria carrying out the anammox reaction could be isolated into pure culture, but much information could be gained from enrichment cultures. All anammox bacteria

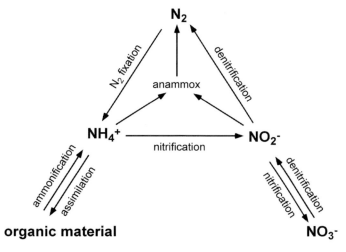

Fig. 6.9 Schematic nitrogen cycle (modified after Shapleigh 2000).

known so far seem to belong to the *Plancto-mycales* (Strous et al. 1999) a group of bacteria, which harbors relatively few isolated pure cultures, but seems to be quite abundant in marine environments. The process of anammox depends on nitrite, which occurs as an intermediate product during nitrification and denitrification (Fig. 6.9). Thus, anammox should not be seen as an alternative process to denitrification, but rather as a short cut in the nitrogen cycle, which still depends on the production of nitrite from denitrification. During the oxidation of ammonia with nitrite the highly toxic gas hydrazine (N_2H_4) is formed as an intermediate. To avoid a toxification of the cells with hydrazine the oxidation of ammonia to dinitrogen is restricted to a special cell compartment called anammoxosome. In the membrane of the anammoxosome a very unusual type of lipids called ladderanes is found that seem to be typical for anammox bacteria.

Rates of anaerobic ammonia oxidation in the environment can be determined by anaerobic incubation of samples with ^{15}N-labelled ammonia and recording the formation of labeled dinitrogen over time. For detection of anammox bacteria in the environment the ladderane lipids can be used as biomarkers for the detection of anammox bacteria. As all anammox bacteria known so far are phylogenetically closely related, it is also possible to search for anammox bacteria with molecular techniques such as fluorescent in situ hybridization. Although anammox bacteria were first described from wastewater reactors they were now also detected in various marine environments like the Skagerrak (Thamdrup and Dalsgaard 2002), the Black Sea (Kuypers et al. 2003), the Golfo Dulce (Dalsgaard et al. 2003) and the Benguela upwelling area (Kuypers et al. 2005). Thus, we have to assume that a considerable part of the total bacterial production of dinitrogen is carried out by anaerobic ammonia oxidation.

6.3.1.4 Nitrogen Isotopes in Marine Sediments

The stable isotope composition of sedimentary organic matter has widely been used to describe the state of the oceanic nitrogen cycle as it allows conclusions on nitrogen sources and transformation processes (i.e. assimilation or denitrification), which are subject to isotopic fractionation (i.e. Francois et al. 1992; Altabet and Francois 1994; Altabet et al. 1999; Thunell et al. 2004). We will briefly discuss this issue because early diagenesis

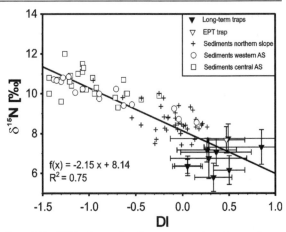

Fig. 6.10 $\delta^{15}N$-values of sediment trap and surface sediment samples (Arabian Sea) plotted vs. the degradation index (DI). The correlation indicates isotopic fractionation in the course of organic matter degradation (Gaye-Haake et al. 2005).

plays a certain, but still somewhat unconstrained role affecting the isotopic composition of organic nitrogen compounds. Stable nitrogen isotope ratios are usually expressed as

$$\delta^{15}N = \left[\left(\frac{^{15}N/^{14}N_{sample}}{^{15}N/^{14}N_{standard}} \right) - 1 \right] \cdot 1000 \qquad (6.11)$$

where the standard, atmospheric nitrogen, has a $\delta^{15}N$ of 0‰.

Deep water nitrate seems to have a fairly constant $\delta^{15}N$-level of about 5‰ (Sigman et al. 2000), which must be the result of almost balanced inputs and losses. The major input of nitrate is via nitrogen fixation – the ability of certain prokaryotes to transform dinitrogen to ammonia – which is then incorporated into new biomass. Although the controls are not completely understood, the process is of utmost significance in ocean biogeochemistry, because it may enable certain levels of primary productivity, even in nitrate-starved regions of the surface ocean. Since nitrogen fixation is considered to cause only little fractionation it produces fixed nitrogen with about the same $\delta^{15}N$ as atmospheric nitrogen (i.e. Codispoti et al. 2001). At present, we have to assume that the major losses of nitrate are caused by denitrification in the water column and the sediments. Due to the preferential use of the light isotope in the course of mineralization processes deep water nitrate becomes more enriched in ^{15}N. In suboxic zones of the water column the total nitrate pool is usually not depleted considerably by denitrification, which allows a significant fractionation and δ^{15}N-enrichment of the residual nitrate of up to 20‰ (Brandes et al. 1998). On annual time scales, the uptake of nitrate in the euphotic zone is generally believed to be complete (Thunell et al. 2004), so that upwelling of ^{15}N-enriched water masses causes the production of relatively ^{15}N-rich organic material. This process is believed to be particularly enhanced during interglacial rather than glacial times and gave rise to the use of δ^{15}N values of sedimentary material as a recorder of nutrient utilization and paleoproductivity (Altabet et al. 1995; Thunell and Kepple 2004). The applicability of this proxy has, however, been disputed, because of potential alterations of the primary signal during nitrification and denitrification in the sediments. Variations in the δ^{15}N-record are considered to be dependent on the preservation of the organic material, hence burial potential controlled by the sedimentation rate and the oxygenation of the water column (cf. Sections 6.3.1.2 and 6.5.2). Whereas in regions characterized by high particle fluxes and low bottom water oxygen levels δ^{15}N-values are indistinguishable from sediment trap material, poor preservation of organic matter may lead to increased ^{15}N/^{14}N-ratios of up to 5‰ (i.e. Altabet et al. 1994; 1999; Thunell et al. 2004). The shift to higher δ^{15}N-values is mainly attributed to the decomposition of amino acids, which are the main carriers of nitrogen in fresh marine detritus and more susceptible to degradation. Based on the observation that amino acids are rapidly consumed in surface sediments Dauwe et al. (1999) developed the so-called degradation index (DI), which has recently been used by Gaye-Haake et al. (2005) in order to demonstrate the dependence of the nitrogen isotope composition on the degradation of the sediments. Their data from the Arabian Sea clearly show the shift from lower (sediment traps) to higher δ^{15}N-values (sediments) with increasing degradation (Fig. 6.10).

6.3.2 Input and Redistribution of Phosphate in Marine Sediments

6.3.2.1 P-Species and Forms of Bonding

Particulate phosphorus reaches marine sediments in various portions of inorganic and organic fractions. In a recent study Faul et al. (2005) have investigated the phosphorus distribution in sinking oceanic particulate matter from a wide range of oceanic regimes. Correspondingly, the P flux to the sediment is typically dominated by reactive P components including organic P (~40%), authigenic P (~25%), and labile P and/or phases which are associated with iron oxyhydroxides (~21%). With only about 13%, the non-reactive detrital fraction seems to be less important. While the most important carrier of inorganic P are iron oxyhydroxides, which mostly

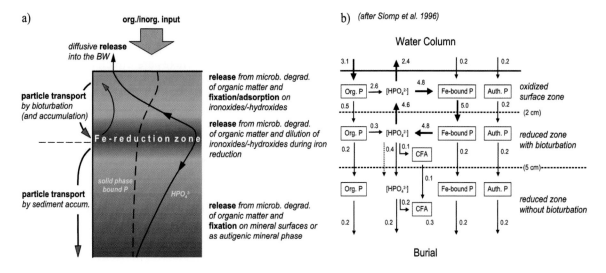

Fig. 6.11 The benthic phosphorus cycle in deep-sea sediments. a) generalized processes of particulate transport, release and fixation; b) example of fluxes of P (in 10^{-4} µmol cm^{-2}d^{-1}) between the pore water and the sediment P reservoirs as calculated with a model for a deep-water location at the western European continental platform Goban Spur (after Slomp et al. 1996).

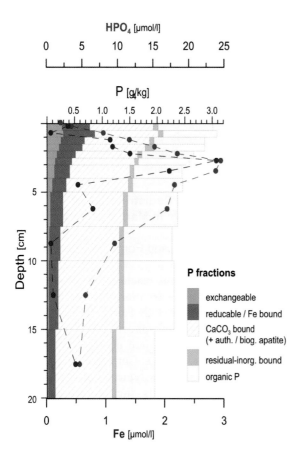

Fig. 6.12 Reservoirs of phosphorus in surface sediments and accompanying pore water profile of iron and phosphate. An example from the continental margin off Namibia (Zabel unpubl. data).

occur in form of aggregated colloidal suspensions or coating surfaces around suspended clay particles (e.g. Krom and Berner 1980; Froelich 1988), the organic fraction consists of organic matter and fish debris. After their accumulation on the sediment surface, the primary phases undergo an intensive redistribution due to early diagenetic modification, which is mainly affected by the close coupling of the P-cycling with the geochemistry of redox-sensitive iron phases (e.g. Froelich et al. 1982; Ruttenberg and Berner 1993; Slomp et al. 1996, 2002, 2004; Filippelli 1997, 2001; Schenau and de Lange 2001). A simplified illustration of most relevant processes in the generalised benthic P cycle is given in Figure 6.11.

The major species of dissolved phosphorus in the marine environment is HPO_4^{2-}. Phosphate is mainly released to the (pore) water either during microbial degradation of organic matter and/or

concomitant with the reduction of ferric iron. Since phosphate cannot be used as an electron acceptor (cf. next section) under most environmental conditions, the only processes consuming phosphate are 1) the biological uptake for the formation of new biomass, 2) the adsorbtion onto particle surfaces or co-precipitation with in-situ-formed minerals and 3) the formation of authigenic carbonate fluorapatite (CFA). To examine the different processes of the P-cycle and to quantify the transfer and flux rates, Ruttenberg (1992) developed a selective leaching procedure which allows the speciation of solid-phase P into five reservoirs on the basis of their chemical reactivity. This established analytical scheme, known as the SEDEX method, was slightly modified several times to get more detailed information about specific phosphorus-containing phases (e.g. Slomp et al. 1996; Eijsink et al. 1997; Schenau and de Lange 2000). However, the following solid phases could be identified as the quantitatively most important P-reservoirs in marine sediments: 1) exchangeable or loosely sorbed P, 2) P bound to ferric oxides and oxyhydroxides, 3) fish debris, 4) CFA + biogenic hydroxyapatite + $CaCO_3$-bound P, 5) detrital apatite of igneous or metamorphic origin, and 6) organic P. Figure 6.12 shows an example for the distribution of these reservoirs in surface sediments from the upwelling area off Namibia. The main release of both ferrous iron and phosphate occurs in the suboxic zone at about 2-3 cm (oxygen penetration depth was determined at 1.5 cm). This indicates that the main source of both constituents are iron oxyhydroxides being reduced at this depth. Based on the pore water profiles, diffusive fluxes are directed upward into the oxidized surface layer and downward into the anoxic zone. At the redoxinterface ferrous iron is oxidized back to oxyhydroxides and phosphate is adsorbed. The internal P-cycle is closed by downward bioturbation of mainly in-situ-formed Fe-bound phosphorus. These processes are clearly reflected particularly by the distribution of the exchangeable and reducible P-fractions. In this specific example the P-sink in the deeper part of the sediment could not be identified, but may be related to the authigenic formation of CFA. Assuming steady state conditions, quantitative budgeting of flux rates with reservoirs gives indication that P has to run several hundred times through this internal cycle across the redox interface until it is buried. A very impressive hypothetical calculation showing the important role of iron oxyhydroxides for the P-cycle is given in Section 7.4.3.3. By studying the single P-reservoirs in marine sediments, Ruttenberg (1993) calculated the total burial flux of reactive/bio-available P in the range of $8 - 18.5 \cdot 10^{10}$ mol yr^{-1}, which results in a reassessment of the average residence time of oceanic phos-phate as 16-38 kyr.

Finally, it is obvious that the processes descri-bed above have a strong impact on the sedimen-tary ratio between organic carbon and phospho-rus (either as organic P also or as "reactive" P, which means the sum of authigenic, oxide-asso-ciated and organic P). While the initial organic matter can be approximated with a ratio close to Redfield (106:1), analysis on CFA gave values as low as 4:1 (Ruttenberg 1993). Therefore, several studies have discussed variations in the different types of C/P ratios as indicating temporal changes in the preservation efficiency of deposited organic matter (e.g. Ingall and Van Cappellen 1990; cf. Section 12.3.3), in the intensity of phosphorus regeneration relative to carbon (e.g. Slomp et al. 2004), or as a generally useful approach to describe the geochemical behavior of sedimentary P (Anderson et al. 2001).

6.3.2.2 Authigenic Formation of Phosphorites

There is no consistent definition for the use of the term phosphorites. Suggestions reach from a limi-ting P content of 6 wt.% (van Cappellen and Berner 1988) to a threshold value of 18 wt.% P_2O_5, as representative for authigenic and biogenic phosphate minerals (Jarvis et al. 1994). However, the formation of secondary P phases from initially more labile-P in marine sediments (see above) is one major sink for phosphorus on Earth, because this general process results in sequestration of P from the nutrient cycle in the water column (e.g. Compton et al. 2000).

Research on the benthic phosphorus cycle and phosphorites formation in particular was inten-sified especially in the seventies to early nineties of the last century (e.g. Burnet 1977; Burnett and Froelich 1988; Burnett and Riggs 1990; Nolton and Jarvis 1990; Föllmi 1996; Glenn et al. 2000). All present result clearly substantiate that the pro-cess of phosphogenesis (= the authigenic forma-tion of carbonate fluorapatite - CFA) is highly complex and not completely understood. However, hydroxy-apatite may be the primary authigenic P-

mineral phase. This initial apatite is forming inter-mediately under reducing conditions within the uppermost few centimeters of sediment. Following a recent study, the metabolic cycle of large sulfur bacteria may play a key role for this process (Schulz and Schulz 2005; cf. next Section). Subse-quent dissolution-re-precipitation processes are thought to be responsible for the transfer into the more stable and complex carbonate-fluoride varie-ty francolite, which is most common in phosphate-rich sediments and rocks (e.g. McClellan 1980; Kolodny and Luz 1992). The chemical composition of francolite is very complex and variable (Jarvis et al. 1994). Based on radiocarbon dating of phosphatic pellets from the Peru shelf, their forma-tion can be very quick, on time scales of only a few years (Burnett and Froelich 1988). To investi-gate the different stages of phosphogenesis, many interdisciplinary studies have been perfor-med, which include the sequential extractions mentioned before, the formation kinetics of special apatite crystals (e.g. Van Cappellen and Berner 1991), or the stable isotopic composition of sedimentary apatite reflecting the variability of environmental conditions (e.g. summarized in Kolodny and Luz 1992). Nevertheless, control factors of phosphogenesis in different environ-ments are still under debate, particularly with regard to the overall benthic C-cycle.

Because CFA is forming at the expense of organic P, high productive areas or at least organic-rich sediments favor phosphogenesis. Therefore, it is not surprising that sites of present-day phosphorite formation are found along continental margins where the organic detrital input is sufficient for intense microbial activity and suboxic to anoxic conditions close to the sediment surface. This is particularly the case in regions of intense coastal upwelling and below permanent oxygen minimum zones (Fig. 6.13).

6.3.2.3 Release of Phosphate by Bacterial Activity

In contrast to the different sulfur and nitrogen species, phosphorus in marine environments occurs almost exclusively in the oxidation state +5. Thus, with few exceptions (Schink and Friedrich 2000), bacteria cannot gain energy by the oxida-tion of reduced phosphorus species and cannot use phosphate as an electron acceptor. Neverthe-less, phosphorylation, the addition of a phos-phate group to another compound, plays a major

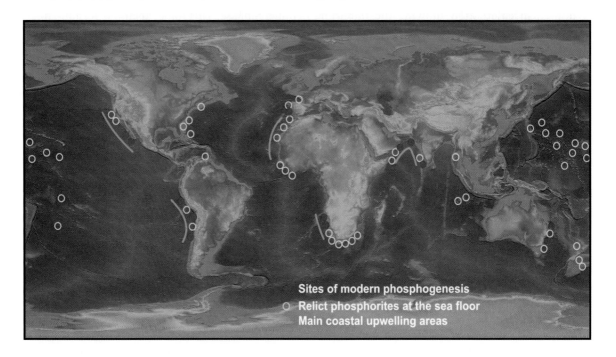

Sites of modern phosphogenesis
○ Relict phosphorites at the sea floor
Main coastal upwelling areas

Fig. 6.13 Locations of present-day phosphorite formation, relic phosphorites at the sea-floor, and zones of coastal upwelling (modified after Baturin (1982) and Föllmi (1996)).

role in the biological energy transfer (e.g. in the formation of ATP by the phosphorylation of ADP). Alternatively, energy can be conserved in linear polymers of phosphate called polyphosphates. The build-up of polyphosphates requires energy, whereas ATP is formed in the breakdown. Although all living organisms contain polyphosphates, only some microorganisms accumulate polyphosphates in larger amounts, visible as compact inclusions (e.g. Kornberg 1995).

Bacterial phosphate accumulation and release has been studied most thoroughly in wastewater treatment plants, where polyphosphate accumulating bacteria are used since decades to remove phosphate. To induce polyphosphate accumulation in bacteria an anoxic phase has to be introduced followed by an oxic phase. During the anoxic phase end products of fermentation such as acetate accumulate in the wastewater. The bacteria take up acetate and store it, e.g. as PHA (polyhydroxyalkanoate) inclusions. The energy for the storage of acetate is gained by the break down of polyphosphate, which is accompanied by a release of phosphate. In the following oxic phase the bacteria have a suitable electron acceptor, oxygen, to oxidize PHA and can gain large quantities of energy, which is partly conserved in the build-up of polyphos-

phate. Consequently, phosphate disappears from the water and the sludge containing polyphosphate can be removed (e.g. Mino 2000).

Analogous to wastewater it had been proposed that polyphosphate accumulating bacteria could also play a role in the phosphorus cycle of the ocean (Nathan 1993). Particularly, large sulfur bacteria (Fig. 6.7) have been suspected to play a role in the formation of phosphorite, as they occur in the same areas where recent and active phosphorite formation is observed and have been found as fossils in phosphorite deposits (Williams and Reimers 1983). Lately, these bacteria have been shown to accumulate polyphosphate. *Thiomargarita*, the largest sulfur bacterium thriving off the coast of Namibia, was observed to release phosphate into anoxic sediments leading to an over-saturation of the pore water with phosphate and rapid precipitation of hydroxyapatite. Laboratory experiments showed that the release of phosphate by *Thiomargarita* cells could be induced by the addition of acetate to the medium (Schulz and Schulz 2005). This might indicate that the processes of bacterial phosphate accumulation and release in eutrophic marine environments are similar to those observed in wastewater.

6.4   Determination of Consumption Rates and Benthic Fluxes

6.4.1   Fluxes and Concentration Profiles Determined by *In Situ* Devices

One reason for determining changes of oxidant concentrations or consumption / production rates in the pore water fraction of a sediment is to quantify the underlying respiration processes and to define the reactive horizons. Until today, however, there is no method to determine oxic respiration directly. Total oxic respiration has to be calculated from the difference between the oxygen demand of the sediment and the amount of oxygen consumed by oxidation of reduced species (see above). There are two main methods to determine diffusive or total oxygen uptake rates in deep-sea sediments. These are (1) the application of microelectrodes and optodes to obtain one- or even two-dimensional (planar optodes) depth profiles and (2) benthic chambers to reveal total areal uptake rates (cf. Section 12.2). Clark-type microelectrodes as they are commonly in use since more than two decades (e.g. Revsbech et al. 1980; Revsbech and Jørgensen 1986; Revsbech 1989; Gundersen and Jørgensen 1990) and the more recently invented optodes (Klimant et al. 1995) have become a driving force in performing measurements of oxygen consumption and penetration depths in any kind of soft sediment. Since a couple of years the two-dimensional oxygen distribution can be determined by so-called planar optodes (Glud et al. 1996) showing

an excellent correlation with measurements performed with microelectrodes. For the general principles of microelectrodes and optodes and their application in the deep sea we refer to the description in Section 3.5 and the references above.

Within the past two decades benthic lander systems have been increasingly applied in the deep-sea (e.g. Berelson et al. 1987; Jahnke et al. 1997; Reimers et al. 1992; Wenzhöfer et al., 2001a) to avoid artefacts resulting from sediment recovery (see discussion below). Some results obtained by these devices have already been shown in Chapter 3. The microelectrodes or optodes provide information on the oxygen penetration depth and its depth-dependent distribution, whereas the benthic chambers measure total fluxes across the sediment water interface. There is generally good agreement of fluxes obtained by both methods (Jahnke et al. 1990; Fig. 12.10), but total fluxes might exceed calculated diffusive fluxes from oxygen profiles. While diffusive transport of pore water is the dominant process in the large area of the oligotrophic oceans where the input of degradable organic matter to the sea-floor is low (Sayles and Martin 1995), this is not the case adjacent to continental margins. Glud et al. (1994) found that the total oxygen uptake was always larger than the diffusive uptake in continental slope sediments off Southwest Africa. The results shown in Figure 6.14a indicate a good correlation between the dry weight of macrofauna and the total oxygen uptake. Subtracting the diffusive oxygen uptake - measured with microelectrodes -

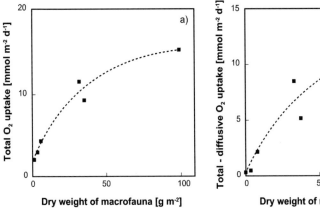
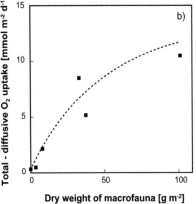

Fig. 6.14 Correlation of (a) total oxygen fluxes and (b) total-diffusive oxygen fluxes with the dry weight of organic macrofauna indicating the effect of macrobenthic activity for benthic respiration processes (adapted from Glud et al. 1994). Broken lines are fitted by eye.

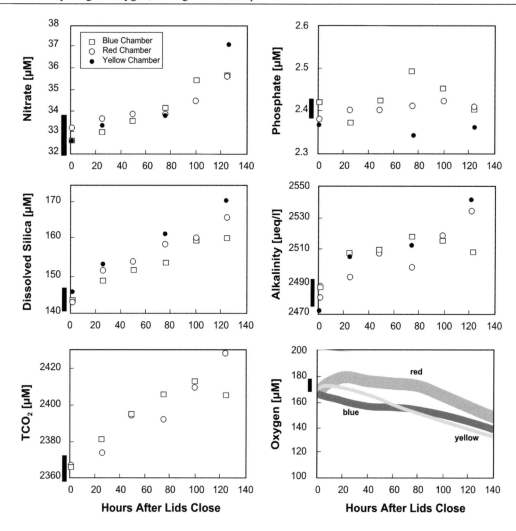

Fig. 6.15 Concentration versus time plots of different solutes measured with benthic chambers in the central equatorial Pacific (from Hammond et al. 1996). Black vertical bars indicate bottom water values.

from the total uptake rates reveals values close to zero for stations with a low dry weight of macrofauna obviously increasing with increasing population density (Fig. 6.14b). On one hand, the difference between both fluxes provides a measure of bioirrigation meaning that there is additional transport across the sediment-water interface due to active pumping of organisms and a higher oxygen demand due to an increased surface area at additional sites where oxygen is consumed (worm burrows etc.). On the other hand, the respiration by the macro- or meiofauna itself increases the oxygen consumption (Glud et al. 1994; Heip et al. 1995; Soetaert et al. 1997). Since the chamber system provides information on total mineralization rates and fluxes, and the profiling lander on the depth distribution of solutes and redox

processes, the application of both lander systems is required to properly investigate the oxygen demand and the pathways of oxygen in the sediment.

The determination of *in-situ* fluxes of nitrate is comparatively limited. Although there are electrodes for the determination of micro-concentration profiles in form of biosensors (Larsen et al. 1996), they are not yet suitable to be used on lander systems in the deep sea. The measurement of total nitrate fluxes by benthic chambers in continental slope sediments off California is shown in Figure 3.24 (Jahnke and Christiansen 1989). At this site, nitrate fluxes are directed into the sediment indicating strong denitrification supported by very low oxygen concentrations in the overlying bottom water.

Under normal deep-sea oxygen conditions, the nitrate flux is generally directed out of the sediment due to nitrification and lower denitrification as shown by results reported by Hammond et al. (1996) from measurements in the central equatorial Pacific (Fig. 6.15). It is further indicated that fluxes of oxygen, silicate, or ΣCO_2 are easier to determine than nitrate or, particularly, phosphate, because of the magnitude of concentration change over time. Phosphate in Figure 6.15 shows large scatter and does not allow a reliable flux calculation. The differences in the order of magnitude result from the reaction stoichiometry of organic matter decomposition (Eq. 6.1). Additional processes, like the dissolution of biogenic opal and calcium carbonate in the sediments, is of further significance for the parameters silicate and alkalinity. Whereas silicate fluxes depend on the amount and the surface area of soluble opal, the degree of silica undersaturation in pore waters, and the content of terrigenous components (cf. Section 12.3.3), variations of alkalinity and ΣCO_2 fluxes are controlled by respiratory production of CO_2 and dissolution of calcium carbonate (cf. Section 9.3.2).

The result of the phosphate measurement shown in Figure 6.15 indicates that flux chamber measurements are restricted to a certain number of measurable parameters. Even more, this is true for profiling lander systems. Apart from oxygen, only pH-, pCO_2-, H_2S and Ca-electrodes have been successfully employed on *in situ* lander systems (Cai et al. 1995; Hales and Emerson 1997; Wenzhöfer et al. 2001a; de Beer et al. 2005). There is, however, another microelectrode technique

described by Brendel and Luther (1995) which has, however, not yet been tested *in situ*. This voltammetric microelectrode technique allows the simultaneous determination of the most characteristic redox species: oxygen, manganese, iron and sulfide. Results from continental slope sediments of northeast Canada revealed clear and undisturbed vertical redox sequences, which is usually not the case when different methods are applied (Luther et al. 1997, 1998).

6.4.2 *Ex-Situ* Pore Water Data from Deep-Sea Sediments

Mostly, solutes are still determined *ex-situ* by extraction of pore water from multicorer or box corer samples (see Chapter 3). Additionally, numerous *ex-situ* measurements of oxygen and nitrate exist which were carried out by onboard core incubations or microelectrode measurements. Generally, the determination of a dissolved species in water samples does not pose a problem, if sampling is handled carefully and, in specific cases, contact with atmospheric oxygen is avoided. Manifold problems arise, however, when a sediment sample is retrieved from some thousand meters below the sea surface and subsequently during the extraction of pore water onboard a ship which leads to changes in pore water concentrations compared to *in situ* conditions. Such effects can be caused either by decompression and/or transient heating of a sample during its transport through the water column. These problems arise because of the large temperature difference between deep water and surface water in

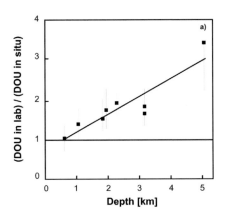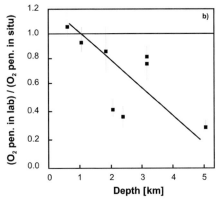

Fig. 6.16 Plots of (a) the ratio between diffusive oxygen uptake rates (DOU) *ex-situ* and *in-situ* and (b) oxygen penetration depth *ex-situ* and *in-situ* versus water depth from stations off the continental slope off Southwest Africa (from Glud et al. 1994). Solid lines indicate linear regressions. With increasing water depth fluxes appear to be overestimated and oxygen penetration underestimated when measured *ex-situ*.

most areas of the global ocean. With regard to oxygen a possible intensification of organic matter decay, triggered by the temperature increase or the lysis of cells, might result in a reduction of the oxygen penetration depth due to higher oxygen consumption. This effect has intensely been studied by Glud et al. (1994) and Wenzhöfer and Glud (2002) who could show that there might be significant discrepancies between *in-situ* and *ex-situ* measured fluxes and oxygen penetration depths. This effect is illustrated in Figure 6.16 and shows that the difference obviously increases with increasing water depth. At or above 1,000 m the sampling effect seems to be more or less negligible.

In analogy to oxygen, such artifacts are believed to exist also for nitrate profiles determined *ex-situ*. The occurrence of increased (compared to Redfield stoichiometry in Eq. 6.1) subsurface nitrate concentrations has been described in a number of studies (e.g. Hammond et al. 1996; Martin and Sayles 1996), but is mostly attributed to the centrifugation method in pore water extraction (see Section 3.3.2). Artificially increased subsurface nitrate concentration would consequently lead to an overestimation of benthic fluxes of nitrate. Berelson et al. (1990) and Hammond et al. (1996) found evidence for increased nitrate fluxes after pore water centrifugation compared to lander measurements, which were up to a factor of 3, but also agreement between both methods for quite a number of stations was found. However, the possibility of low C/N organic matter or oxidation of reduced nitrogen species (diffusing upwards from deeper layers) can be important natural factors increasing nitrate concentrations deviating from the general expected stoichiometry. Following results of Luther et al. (1997) the amount of subsurface nitrate production due to nitrification can also be regulated by the manganese oxide content of the sediment. As discussed in Section 6.3.2.3 high MnO_2 concentrations favor a catalytic reduction of nitrate to N_2 already in the oxic zone of the sediment so that nitrate peaks only occur when the solid-phase manganese content is low.

A further aspect of decompression that should be kept in mind is the degassing of CO_2 and the resulting precipitation of $CaCO_3$, which might affect pore water concentrations of phosphate by adsorption or co-precipitation (Jahnke et al. 1982). This effect can even imply negative fluxes as discussed below (see Section 6.5.1).

In general, fluxes, which are calculated on the basis of *ex-situ* data should be interpreted with caution; the overall result, however, in most cases reveals a reasonable approximation to real conditions (cf. Section 12.2).

6.4.3   Determination of Denitrification Rates

The downward flux of nitrate (e.g Fig. 6.5) indicates the depth of active denitrification, but is no measure for the total rate of denitrification. As mentioned above, denitrifying bacteria are facultative anaerobic and denitrification is generally located directly below the oxic zone (Christensen et al. 1989). The shape of a nitrate profile depends on the total rate of nitrification and the denitrification rate. Considering the example shown in Figure 6.5, the fit of the measured nitrate profile was achieved by determining an indirect nitrification rate (coupled oxidation and nitrification as described by Equation 6.1 and the C/N ratio) and denitrification occurring with a distinct rate at a depth of about 3 cm. A reduction of the denitrification rate would result in a greater nitrate penetration depth, i.e. not all the nitrate can be consumed in this depth zone. On the other hand, an increase of the denitrification rate would lead to a depression of the nitrate maximum and therefore reduce upward and downward fluxes, and finally induce a total nitrate flux from the bottom water into the sediment. To clarify these interactions we plotted three nitrate profiles from different regions of the South Atlantic in Figure 6.17. Two profiles indicate high respiration rates with a nitrate penetration depth of approximately 3 cm, but a distinct peak is visible only at one station. The station with the nearly linear gradient into the sediment is indicative for strong denitrification. The third profile shows a low gradient into the bottom water which is due to nitrification and remains nearly constant with depth indicating that denitrification does not occur close to the sediment surface.

Depletion of nitrate and the formation of dinitrogen are, however, not exclusively coupled to denitrification, since the microbially mediated reduction utilizing reduced species like Mn^{2+} or Fe^{2+} might occur. Based on field observations, a number of studies invoke the reduction of nitrate by Mn^{2+} to form N_2 instead of organic matter respiration (Aller 1990; Schulz et al. 1994; Luther et al. 1997,1998).

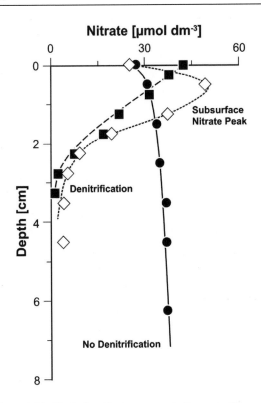

Fig. 6.17 Typical nitrate concentration profiles of surface sediments from different productivity regions in the South Atlantic Ocean. The profiles from the continental slope of the Argentine and the Cape Basin indicate denitrification at about 3 cm. The profile from an oligotrophic equatorial site shows no denitrification.

The direct determination of denitrification rates has been carried out by a number of different methods which all have certain advantages and restrictions. It is beyond the scope of this chapter to describe all of these methods in detail so that we will only give a short summary of the general principles. A more detailed overview of this subject can be found, for example, in Koike and Sørensen (1988); Seitzinger et al. (1993); and Kana et al. (1998).

The most important methods of measuring denitrification are (1) the detection of totally produced N_2 gas by incubation, (2) isotope-labeling methods with ^{15}N and ^{13}N, and (3) the acetylene (C_2H_2) inhibition technique.

(1) The first method aims at measuring the total production of N_2 as equal to the rate of denitrification (Eq. 6.6-6.8) by sediment incubation (Seitzinger et al. 1984; Devol 1991). At present, this is thought to be the best method to accurately determine rates of overall denitrification, although the contamination with atmospheric N_2 is possible

during long incubation periods and thought to be the main restriction (Koike and Sørensen 1988; Seitzinger 1988). More recently, however, the technique of membrane inlet mass spectrometry, which is able to detect small changes of dissolved nitrogen with high temporal resolution and without perturbation of the sediment has been developed (Kana et al. 1994, 1998; Hartnett and Seitzinger 2003). Perturbations by atmospheric nitrogen are avoided by detecting changes of N_2/Ar ratio relative to seawater standards.

(2) The intention of the isotope methods is to add $^{15}NO_3$ or $^{13}NO_3$ to the nitrate pool of the supernatant water during incubation and to measure the ^{15}N and ^{13}N content of the total amount of produced N_2. As the half-life of ^{13}N is 10 minutes, this method is only of limited use. The ^{15}N method already applied by Goering and Pamatmat (1970) to marine sediments off Peru is, in contrast, widely accepted and was used in a number of recent incubation studies of shallow marine and freshwater environments (e.g. Nielsen 1992; Rysgaard et al. 1994). This so-called ion-pairing method (Nielsen 1992) allows the determination of the total rate of denitrification and its dependence on the nitrate concentration in bottom water. Additionally, a number of authors (e.g. Nielsen 1992; Rysgaard et al. 1994; Sloth et al. 1995) believe that it also enables to distinguish between the source of nitrate, either as coming directly from the bottom water or from nitrification. This, however, has caused an intense discussion regarding the potential of the method and the benefits of its performance (Middelburg et al. 1996bc; Nielsen et al. 1996).

(3) At last, the C_2H_2 inhibition technique takes advantage of the property of acetylene to block the reduction of N_2O to N_2 after it is injected into the sediment. The total amount of N_2O produced is then the measure for the denitrification rate as it is easy to determine by gas chromatography (Andersen et al. 1984) or by microsensors (Christensen et al. 1989). The advantage of this method is that analyses can be carried out rapidly and sensitively. Problems are: (a) N_2O reduction is sometimes incomplete, (b) a homogenous distribution of C_2H_2 in the pore water is difficult to maintain, (c) C_2H_2 inhibits nitrification in the sediment meaning that the coupled system (nitrification / denitrification) might be seriously affected due to the applied method, and (d) might lose its inhibitory properties in the presence of hydrogensulfide (Sørensen et al. 1987;

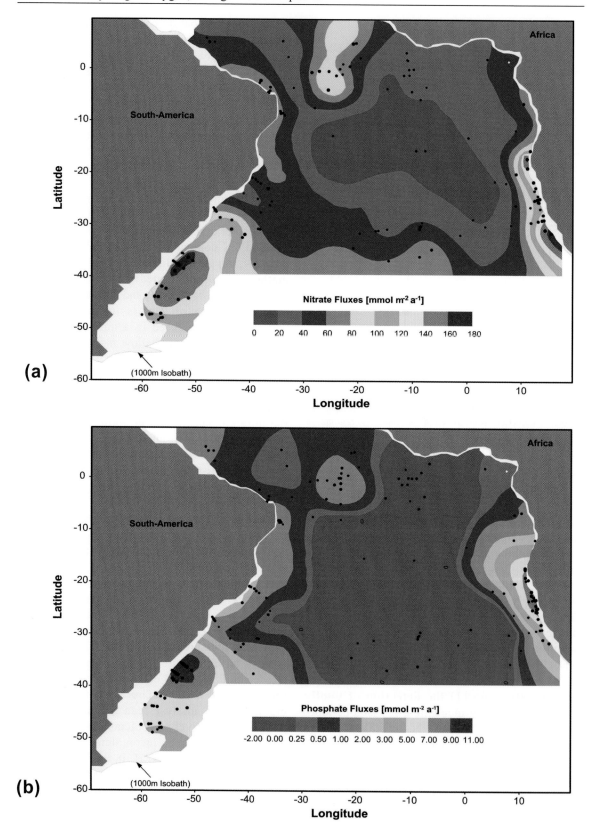

Fig. 6.18 Regional distribution of benthic nitrate (a) and phosphate (b) release rates in the South Atlantic below 1,000 m water depth (from Hensen et al. 1998).

Koike and Sørensen 1988; Hynes and Knowles 1978; Seitzinger 1988).

The application of acetylene and nitrogen labeling methods usually results in the determination of lower denitrification rates when compared to total nitrogen fluxes (Seitzinger et al. 1993; cf. Section 6.5). Apart from the above restrictions inherent to the methods themselves, the recently published concept of Luther et al. (1997) provides an additional explanation for the observed discrepancy. They tested the thermodynamic properties of several redox reactions and found evidence for a catalytic short circuit for the coupled process of nitrification-denitrification within the oxic zone (Eq. 6.12.). Organic nitrogen and ammonia released during oxic respiration are therefore oxidized to N_2 by MnO_2, instead of being further oxidized to NO_3.

$$2\,NH_3 + 3\,MnO_2 + 6\,H^+ \rightarrow 3\,Mn^{2+} + N_2 + 6\,H_2O$$

$$(6.12)$$

This process may outweigh nitrification in manganese-rich surface sediments and circumvents denitrification. Elevated N_2 fluxes without increasing denitrification rates and even N_2 production in oxidized sediments as observed by Seitzinger (1988) may be explained by this process.

6.5   Significance and Quantitative Approaches

After the description of the general biogeochemical processes controlling the distribution of oxygen and nitrate in marine sediments, including the possibilities and limitations of determining these inorganic compounds in the deep-sea, the following section will give an overview of the dimensions of their fluxes and their distribution in different marine environments.

6.5.1   Estimation of Global Rates and Fluxes

The basic mechanism inducing microbial activity is the supply of organic matter to the seafloor and this is generally coupled to surface water productivity. Most of the highly productive areas in the global ocean are adjacent to the continents, so

that we can expect a decrease of respiration intensity from the coastal marine environments over the continental shelves and slopes into the deep-sea. This becomes evident when we look at the data compiled by Middelburg et al. (1993) which indicate that 83% mineralization and 87% burial in marine sediments occurs in the coastal zone occupying only ~9% of the total ocean area. This means that the sediments with the highest respiration rates also have the highest burial efficiency in marine environments. For a more detailed discussion of this subject see also reviews by (Henrichs and Reeburgh 1987; Henrichs 1992; Canfield 1993). As shown in Figure 6.3 the organic matter supply is not only a function of productivity, but also of water depth generally amplifying this gradient between shallow and deep water environments.

Fluxes of oxygen and nitrate, therefore, vary over several orders of magnitude between oligotrophic open ocean areas and continental shelf and slope areas. This is about 50 to 6,000 mmol m^{-2} yr^{-1} for oxygen and -600 to 380 mmol m^{-2} yr^{-1} for nitrate (e.g. Devol and Christensen 1993; Glud et al. 1994; Berelson et al. 1994; Hammond et al. 1996; Luther et al. 1997; Hensen et al. 1998; Wenzhöfer and Glud 2002) where negative nitrate fluxes indicate fluxes into the sediment. The above minimum and maximum values do not permit differentiation between total and diffusive or *in situ* and *ex situ* fluxes. Figure 6.18 reveals the distribution of nitrate and phosphate fluxes released from sediments below 1,000 m water depth in the South Atlantic (Hensen et al. 1998) based on about 180 *ex-situ* concentration profiles. Averaged fluxes vary between 10 – 180 mmol m^{-2} yr^{-1} for nitrate and (-2)-11 mmol m^{-2} yr^{-1} for phosphate (where negative values are considered to be artifacts due to decompression and warming, cf. 6.3.2.2). Based on this compilation the total annual release for the whole area (about one tenth of the global deep ocean) is about 1.6 10^{12} mol NO_3 yr^{-1} and 3.5 10^{10} mol PO_4 yr^{-1}.

Global denitrification in marine sediments has been estimated by Middelburg et al. (1996a) to be about 1.64 – 2.03·10^{13} mol N yr^{-1} with a contribution of 0.71·10^{13} mol N yr^{-1} of shelf sediments. This model-based re-estimation produced values which are up to 3 to 20 times higher than those previously estimated by a number of authors in the mid-1980s. The predicted denitrification rates, however, are in the range of those derived from literature, and the total contribution to organic matter mineralization (7-11%) is well within the range of current estimates (see compilation in Middelburg et al. 1996a). Even higher rates of $3.2 \cdot 10^{13}$ mol N yr^{-1} have been estimated more recently by Codispoti et al. (2001) who explained this upward correction mainly by the need of balancing the isotopic nitrate pool at $\delta^{15}N$ of about 5‰ (Sigman et al. 1999, 2000).

Total denitrification rates vary between 0.4 mmol m^{-2}yr^{-1} in the deep-sea (Bender and Heggie 1984) and 1,200 mmol m^{-2} yr^{-1} (measured by the N$_2$ method) in continental margin sediments (Devol 1991) which is up to a factor of two higher than otherwise indicated by the highest fluxes of nitrate into the sediments. For estuarine and coastal areas Seitzinger et al. (1988) have summarized average rates between 440-2,200 mmol m^{-2} yr^{-1} with highest rates of up to 9,000 mmol m^{-2} yr^{-1}. However, considering the suggestions of Luther et al. (1997) a high amount of N$_2$ fluxes may be due to ammonia oxidation by MnO$_2$ in the oxic zone of the sediment bypassing denitrification and thus organic matter decay. If their estimate is correct that this process could contribute to up to 90% of N$_2$ production in continental margin sediments a careful evaluation and possibly a re-estimation of published denitrification rates for this environment is required.

6.5.2   Variation in Different Marine Environments: Case Studies

We already emphasized the importance of oxic respiration over other pathways in the deep-sea. Denitrification (or related processes) only account for a few percent of the carbon oxidation rate by

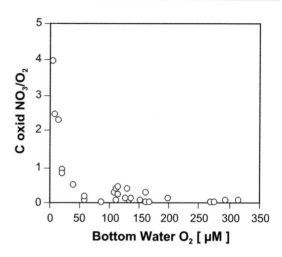

Fig. 6.19 Ratio of carbon oxidation by denitrification and oxic respiration as a function of bottom water oxygen content (after Canfield 1993).

oxic respiration, provided that oxygen is sufficiently available. In oxygen-depleted waters, the proportions can be dramatically shifted and denitrification might become an important pathway. Figure 6.19 represents data from different oceanic regions as compiled by Canfield (1993) where the ratio of carbon oxidation by oxygen and nitrate is plotted as a function of the oxygen concentration in bottom water. It clearly shows that denitrification becomes more important than oxic respiration below oxygen concentrations of about 20 µmol l^{-1}.

To illustrate the general trend of decreasing respiration processes from the continental margin to the deep-sea, Figure 6.20 shows the results of *in situ* oxygen microelectrode measurements for

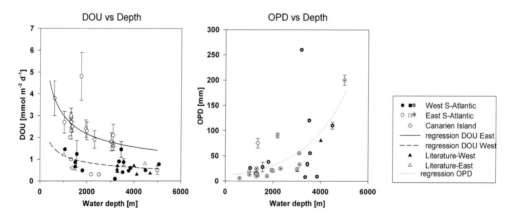

Fig. 6.20 Diffusive oxygen fluxes (a) and oxygen penetration depths (b) for a number of sites in the South Atlantic and the Canaries (from Wenzhöfer and Glud 2002). Oxic respiration decreases with increasing water depth resulting in higher oxygen penetration into the sediment.

the total South Atlantic as compiled by Wenzhöfer and Glud (2002). Both, the diffusive oxygen flux and the oxygen penetration depth at these stations can be clearly described as a function of water depth, even though scattering data points indicate some regional variability. The evident close correlation of diffusive oxygen flux and oxygen penetration depth is depicted in Figure 6.21.

Since microelectrode measurements are limited to a few centimeters of sediment depth, oxygen penetration depths have been difficult to obtain in strongly oligotrophic areas until the late 1990s. The invention of optode techniques, however, allows measurements up to several decimeters into the sediment (Fig. 6.22). The example is from a station located in the oligotrophic western equatorial Atlantic.

There is a number of studies that confirm the trend indicated in Figure 6.20, but on a regional scale there is much more variability, so that a simple relation between oxygen or nitrate fluxes and water depth cannot be found. The problem of regional flux variability will be dealt with further in Chapter 12, but generally, there is a high degree of small-scale variability related to sediment surface topography and the inhomogeneous distribution of easily degradable organic matter. Such variability can be observed when several oxygen microprofiles are recorded on a given surface area of about 10 cm² (the area is determined by the con-

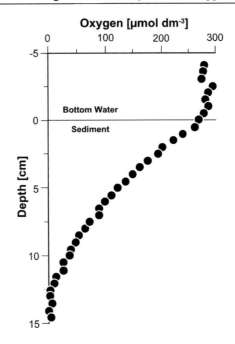

Fig. 6.22 Oxygen concentration profile measured with an *in situ* optode technique (after Wenzhöfer et. al. 2001b).

struction of the device; cf. Fig. 3.24) during lander deployments. This often reveals conspicuous differences in the shape of the profiles as well as the oxygen penetration depth. A more sophisticated method used to determine vertical and lateral oxygen distribution in sediments on a millimeter scale is provided by planar optodes (Glud et al. 1996; Wenzhöfer and Glud 2004). Their data show an excellent resolution of oxygen distribution within the sediment and the diffusive boundary layer. Furthermore, it shows that variations are due to differences in the surface topography. Another reason might be that other pathways, like denitrification or sulfate reduction, become more important in areas characterized by high sediment accumulation rates which are associated with a large input of degradable organic matter.

A general relation of the benthic oxygen flux to the availability of oxygen in the bottom water (as the limiting oxidant) and to the organic carbon content in the surface sediments (as limiting phase for respiration processes has been established by Cai and Reimers (1995). The highest oxygen fluxes across the continental margin of the Northeast Pacific were measured on the lower continental slope where the conditions for oxic respiration were optimal, because of the quanti-

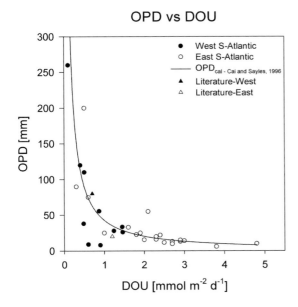

Fig. 6.21 Diffusive oxygen uptake vs. oxygen penetration depths for the same sites as shown in Fig. 6.20. The correlation follows the function found by Cai and Sayles (1996).

tative ratio between oxidant and organic matter availability (Fig. 6.23). On the upper slope, organic matter is available in excess, but oxygen is the limiting phase and reduces the total oxygen uptake in this area. The opposite situation can be observed for the deep ocean. Cai and Reimers (1995) also developed an empirical equation representing this obvious relationship between oxygen flux on the one hand and oxygen bottom water concentration and surface organic carbon content on the other hand with:

$$FO_2 = \frac{\pi \cdot [TOC] \cdot [O_2]_{BW}}{(126 + [O_2]_{BW})} \qquad (6.13)$$

where FO_2 is the oxygen flux in mmol m^{-2} yr^{-1}, $[TOC]$ is the concentration of organic matter in wt.% (dry sediment) and $[O_2]_{BW}$ is the oxygen concentration in bottom water (in µM).

More recently Seiter et al (2005) refined this approach based on a much larger data set:

$$FO_2 = \frac{((\ln([TOC] + k_3)) \cdot k_1 + k_2)[O_2]_{BW}}{k_{ox} + [O_2]_{BW}} \qquad (6.14)$$

$$with\, k_{ox} > 0,\ k_1 > 0, [TOC] > [TOC]_{lim}$$

where k_1, k_2 (both in mmol m^{-2} yr^{-1}), and k_3 (in wt%) are rate constants for the decay of organic matter and k_{ox} (in µM) is the saturation constant of oxygen in the bottom water.

Similarly, their approach considers depleted oxygen levels in the bottom water and the accumulation of organic matter above a certain threshold value $[TOC]_{lim}$, which has a global mean of about 0.6 wt.% (Seiter et al. 2005).

In regions without limitation by bottom water oxygen depletion, as prevailing in large portions of the Southern and Northern Atlantic, Equation 6.14 can be simplified to

$$\qquad \qquad \qquad \qquad \qquad (6.15)$$

$$with\ k > 0, [TOC] < [TOC]_{lim}$$

Both relations given above have been successfully applied to data sets in the Northeast Pacific and the Atlantic (cf. Section 12.5.2 and Fig. 12.18).

Looking at the above situation, we need to emphasize that any confusion with the total rate of carbon oxidation – which is probably higher in the upper slope sediments – must be avoided. As

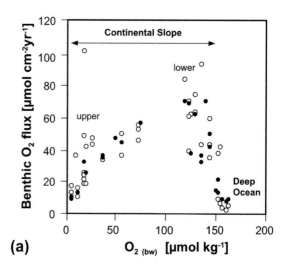

(a)

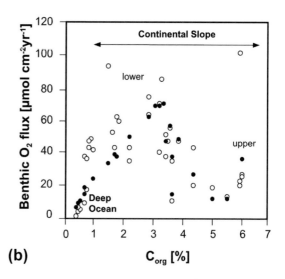

(b)

Fig. 6.23 Distribution of benthic oxygen fluxes across the continental slope in the Northeast Pacific related to (a) oxygen bottom water concentration and (b) organic carbon content in the surface sediments. Highest oxygen respiration occurs at the lower continental slope. Solid circles were calculated by applying Equation 6.12 (after Cai and Reimers, 1995).

stated previously, the mineralization rate and the burial rate are correlated to the input of organic carbon input, so that areas with the highest deposition of organic matter consequently have highest mineralization rates, but also the highest burial potential (Fig. 6.6). A further constraint for the standardization of empirical relations as given by Equation 6.12 or 6.13 is that temporal constancy, namely steady-state conditions, are required. The time dependent variability of early diagenetic processes, however, has been identified even in deep-sea sediments (e.g. Smith and Baldwin 1984). The important question in this regard is: How fast does oxic respiration react to the input of labile organic matter? If the reaction constant is high, organic matter will be quickly recycled at the sediment surface. If furthermore, the input of organic matter is episodic or seasonal, a highly variable oxygen flux at a given time interval might occur. Since organic particles are subject to burial, mixing, and other respiratory processes the surface content will always result in a more or less time-integrated value that does not necessarily reflect the oxygen flux at the time of a single measurement.

More recently, Soetaert et al. (1996) demonstrated the dependence of degradation rates of different mineralization pathways, as well as oxygen, nitrate, and other fluxes on seasonal variations in organic matter deposition and its reaction rate. Some of the model results compared to the measured carbon flux to the sediment and the oxygen uptake rates are plotted in Figure 6.24. The study was based on data derived from box corer and benthic chamber deployments in the abyssal Pacific and covered a time span of more than two years. The carbon flux function was derived from sediment trap data and sedimentation rates, whereas oxygen fluxes were obtained by the adaptation of total mineralization rates of organic material arriving at the sediment surface (and a number of other input parameters). A higher rate would account for

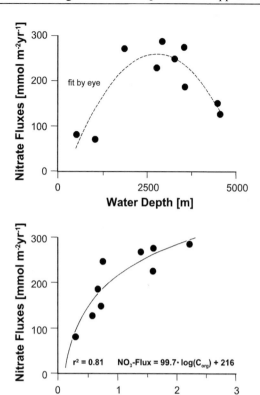

Fig. 6.25 Plot of diffusive benthic nitrate fluxes against (a) water depth and (b) organic carbon content in surface sediments off the Rio de la Plata mouth (Argentine Basin).

variation as reflected by the carbon flux curve, whereas a low rate would continually reduce the seasonal variability.

A situation different from that in the Northeast Pacific which does not comply to a general relationship can be found in the Argentine Basin. As shown in Figures 6.18 and 12.14, distribution maps of nutrient release from deep-sea sediments in the South Atlantic indicate high mineralization rates in this area. Figure 6.23a shows diffusive nitrate fluxes on several transects across the continental slope in front of the Rio de la Plata mouth. The highest release rates of nitrate were detected at intermediate and low depths of the slope. In contrast to the situation in the North-west Pacific, there is no oxygen limitation in the bottom water of the Argentine Basin, suggesting that other processes must be responsible for the observed flux distribution. In this case, it is assumed that intense downslope transport processes deliver large amounts of sediments and organic matter to the lower slope where most of the degradable material is deposited, whereas the

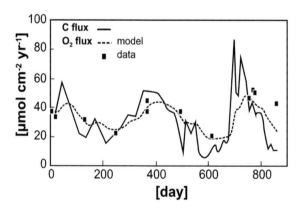

Fig. 6.24 Model results of Soetaert et al. (1996) for a site in the abyssal Pacific. The curve of mineralization rates (oxygen fluxes) is smoother and shows a slight shift compared to the sedimentary carbon flux caused by the effective reaction kinetics. Squares indicate oxygen fluxes determined on the basis of benthic chamber and box corer data.

upper slope surface sediments are partly depleted in organic carbon (Hensen et al. 2000; cf. section 12.3.2). This is indicated by the good correlation of nitrate fluxes with organic carbon content in the surface sediments (Fig. 6.25b).

The above examples have shown that there are large discrepancies between benthic biogeochemical processes in areas of high and low productivity, but there are no simple relationships or master variables to correlate benthic fluxes and mineralization rates with primary productivity or sediment parameters. However, the quantitative coupling between these processes will be a main objective of future research (cf. Chapter 12).

6.6   Summary

In this chapter we have summarized the general availability of oxygen, nitrate and phosphate in the oceans and aspects of their significance in the biogeochemical cycles of marine sediments. Oxic respiration is by far the most important pathway of organic carbon in deep-sea sediments and significantly determines the recycling of organic matter introduced onto the sediment surface. Even if some uncertainty remains, benthic oxygen fluxes reveal probably a reasonable approximation to the total carbon oxidation rate under deep-marine conditions. However, benthic fluxes vary considerably in the various ocean basins and the interactions between all parameters controlling benthic biogeochemical activities are not yet completely understood. Above, we showed some examples in which benthic activity deviates from the generally expected situation as suggested by export fluxes from the surface waters. We further elaborate these concepts in Chapter 12 where we summarize existent approaches for the regionalization of benthic fluxes and present methods how to access mineralization processes in surface sediments of the deep-sea globally and arrive at a definition of benthic biogeochemical provinces.

Acknowledgements

This is contribution No 0332 of the Research Center Ocean Margins (RCOM) which is financed by the Deutsche Forschungsgemeinschaft (DFG) at Bremen University, Germany.

6.7   Problems

Problem 1

Explain why oxygen consumption can be used as a measure for the sum of mineralization processes occurring in the sediment. Why is this - strictly speaking - not correct, but provides feasible results? In which areas of the seafloor would this method fail?

Problem 2

Explain the terms carbon limited and oxidant limited and give examples, where you would expect carbon- and oxidant-limited diagenesis.

Problem 3

Denitrification is a major pathway of carbon degradation in sediments of shallow-marine continental margin areas. Why is it so important for biogeochemical element cycles in the ocean, particularly in terms of regulation of nutrient levels and primary productivity?

Problem 4

Would you expect a higher (than Redfield) $C_{org.}/P_{org.}$ ratio in marine sediments under oxic or anoxic bottom water conditions?

Problem 5

Under which environmental conditions would you expect anammox (anaerobic ammonia oxidation) to be an important process?

References

Aller, R.C., 1990. Bioturbation and manganese cycling in hemipelagic sediments. Phil. Trans. Roy. Soc. London, A331, 51-68.

Altabet, M.A., and François, R., 1994. Sedimentary nitrogen isotopic ratio as a recorder for surface ocean nitrate utilization. Glob. Biogeochem. Cycles, 8(1), 103-116.

Altabet, M.A., François, R., Murray, D.W., and Prell W.L.,1995. Climate related variations in denitrification in the Arabian Sea from sediment 15N/14N ratios. Nature, 373, 506-509.

Altabet, M.A., Murray, D.W., Prell and W.L., 1999. Climatically linked oscillations in Arabian Sea denitrification over the 1 m.y.: implications for the marine N cycle. Paleoceanogr., 14(6), 732-743.

Andersen, T.K., Jensen, M.H. and Sørensen, J., 1984. Diurnal variation of nitrogen cycling in coastal marine sediments: 1. Denitrification. Mar. Biol., 83, 171-176.

Anderson, L.A. and Sarmiento, J.L., 1994. Redfield ratios of remineralization determined by nutrient data analysis. Glob. Biogeochem. Cycles, 8, 65-80.

Anderson, L.A., Delaney, M.L. and Faul, K.L., 2001. Carbon to phosphorus ratios in sediments: Implication for nutrient cycling. Glob. Biogeochem. Cycles, 15, 65-79.

Antoine, D., André, J.-M., and Morel, A., 1996. Oceanic primary production; 2. Estimation at global scale from satellite (coastal zone color scanner) chlorophyll. Glob. Biogeochem. Cycles, 10(1), 57-69.

Baden, S.P., Loo, L.-O., Pihl, L. and Rosenberg, R., 1990) Effects of eutrophication on benthic communities including fish: Swedish west coast.- Ambio., 19, 113-122.

Bartlett, K.B., Bartlett, D.S., Harriss, R.C. and Sebacher, D.I., 1987. Methane emissions along a salt marsh salinity gradient.- Biogeochemistry, 4, 183-202.

Baturin, G.N., 1982. Phosorites on the sea floor.- Developments in Sedimentology, Elsevier, Amsterdam, 33, 343 pp.

Behrenfeld, M.J. and Falkowski, P.G., 1997. Photosynthetic rates derived from satellite-based chlorophyll concentration. Limnol. Oceanogr., 42(1), 1-20.

Bender, M.L.H. and Heggie, D.T., 1984. Fate of organic Carbon reaching the deep-sea floor: a status report. Geochim. Cosmochim. Acta, 48, 977-986.

Berelson, W.M., Hammond, D.E. and Johnson, K.S., 1987. Benthic fluxes and the cycling of biogenic silica and carbon in two southern California borderland basins. Geochim. Cosmochim. Acta, 51, 1345-1363.

Berelson, W.M.; Hammond, D.E., O'Neill, D., Xu, X.M., Chin, C. and Zukin, J., 1990. Benthic fluxes and pore water studies from sediments of central equa-torial north Pacific: nutrient diagenesis. Geochim. Cosmochim. Acta, 54, 3001-3012.

Berelson, W.M., Hammond, D.E., McManus, J. and Kilgore, T.E., 1994. Dissolution kinetics of calcium carbonate in equatorial Pacific sediments. Glob. Biogeochem. Cycles, 8, 219 - 235.

Berger, W.H., Fischer, K., Lai, C. and Wu, G., 1987. Ocean productivity and organic carbon flux. I. Overview and maps of primary production and export production. Univ. California, San Diego, SIO Reference 87-30, 67 pp.

Betzer, P.R., Showers, W.J., Laws, E.A., Winn, C.D., DiTullio, G.R. and Kroopnick, P.M., 1984. Primary productivity and particle fluxes on a transect of the equator at 153°W in the Pacific Ocean. Deep-Sea Res., 31, 1-11.

Bock, E. and Wagner M., 2001. Oxidation of inorganic nitrogen compounds as an energy source. In: M. Dworkin et al. (eds): The prokaryotes: An evolving electronic resource for the microbiological community, 3rd edition, Springer-Verlag, New York.

Brandes, J.A., Devol, A.H., Yoshinari, T. and Jayakumar, D.A., 1998. Isotopic composition of nitrate in the central Arabian Sea and eastern tropical North Pacific: a tracer for mixing and nitrogen cycles. Limnol. Oceanogr., 43(7): 1680-1689.

Brendel, P.J. and Luther, G.W.III., 1995. Development of a gold amalgam voltammetric microelectrode for the determination of dissolved iron, manganese, O_2, and S(-II) in pore waters of marine and freshwater sediments. Environ. Sci. Technol., 29, 751-761.

Broda, E., 1977. Two kinds of lithotrophs missing in nature. Z. Allg. Mikrobiol., 17, 491-493.

Broecker, W.S. and Peng, T.-H., 1982. Tracers in the Sea. Palisades, New York.

Burnett, W.C., 1977. Geochemistry and origin of phosphorite deposits from off Peru and Chile. Geol. Soc. Am. Bull., 88, 813-823.

Burnett, W.C. and Froelich, P.N. (eds), 1988. The origin of marine phosphate. The results of the R.V. ROBERT D. CONRAD Cruise 23-06 to the Peru shelf. Mar. Geol., 80, 181-346.

Burnett, W.C. and Riggs, S.R. (eds), 1990. Phosphate deposits of the World – Vol. 3 Neogene to modern phosphorites. Cambridge Univ. Press, 464 pp.

Cai, W.-J. and Sayles, F.L., 1996. Oxygen penetration depths and fluxes in marine sediments. Mar. Chem., 52, 123–131.

Cai, W.-J., Reimers, C.E. and Shaw, T., 1995. Microelectrode studies of organic carbon degradation and calcite dissolution at a California Continental rise site. Geochim. Cosmochim. Acta, 59(3), 497-511.

Canfield, D.E., 1989. Sulfate reduction and oxic respiration in marine sediments: implications for organic carbon preservation in euxinic environments. Deep-Sea Res., 36, 121-138.

Canfield, D.E., 1993. Organic matter oxidation in marine sediments.- In: Wollast, R., Chou, L., Mackenzie, F. (eds.) Interactions of C, N, P, and S in biogeochemical cycles. NATO ASI Series, pp. 333-363, Springer Verlag.

Canfield, D.E., Jorgensen, B.B., Fossing, H., Glud, R.N., Gundersen, J.K., Ramsing, N.B., Thamdrup, B., Hansen, J.W. and Hall, P.O.J., 1993a. Pathways of organic carbon oxidation in three continental margin sediments. Mar. Geol., 113, 27-40.

Canfield, D.E., Thamdrup, B. and Hansen, J.W., 1993b. The anaerobic degradation of organic matter in Danish coastal sediments: Iron reduction, manganese reduction, and sulfate reduction. Geochim. Cosmochim. Acta, 57, 3867-3887.

Chapelle, F.H., 1993. Ground-Water Microbiology and Geochemistry. Wiley, New York, 424 pp.

Chester, R., 1990. Marine Geochemistry. Unwin Hyman, London, pp. 698.

Christensen, P.B., Nielsen, L.P., Revsbech, N.P. and Sørensen, J., 1989. Microzonation of denitrification activity in stream sediments as studied with a combined oxygen and nitrous oxide microsensor. Appl. Environ. Microbiol., 55, 1234-1241.

Codispoti, L.A., Brandes, J.A., Christensen, J.P., Devol, A.H., Naqvi, S.W.A., Paerl, H.W. and Yoshinara, T., 2001. The oceanic fixed nitrogen and nitrous oxide budgets: Moving targets as we enter the anthropocene? Scientia Marina, 65(2), 85-105.

Compton, J., Mallinson, D., Glenn, C.R., Filippelli, G., Föllmi, K., Shields, G. and Zanin, Y., 2000. Variations in the global phosphorus cycle. In: Glenn, Prévôt-Lucas and Lucas (eds) Marine Authigenesis: From Global to Microbial. SEPM special Publication No. 66, Tulsa Oklahoma US, 21-33.

Dalsgaard, T., Canfield, D.E., Petersen, J., Thamdrup, B. and Acuña-Gonzales, J., 2003. N_2 production by the anammox reaction in the anoxic water column of Golfo Dulce, Costa Rica. Nature, 422, 606-608.

Dauwe, B., Middelburg, J.J., Hermann, P.M.J. and Heip, C.H.R., 1999. Linking diagenetic alteration of amino acids and bulk organic matter reactivity. Limnol. Oceanogr., 44, 1809-1814.

de Beer, D., Wenzhöfer, F., Ferdelman, T., Boehme, S.E., Hüttel, M., van Beusekom, J., Böttcher, M.E., Musat, N. and Dubilier, N., 2005. Transport and mineralization rates in North Sea sandy intertidal sediments, Sylt-Rømø, Wadden Sea Limnol. Oceanogr., 50(1), 113-127.

Devol, A.H., 1991. Direct measurement of nitrogen gas fluxes from continental shelf sediments. Nature, 349, 319-321.

Devol, A.H. and Christensen, J.P., 1993. Benthic fluxes and nitrogen cycling in sediments of the continental margin of the eastern North Pacific. J. Mar. Res., 51, 345-372.

Eijsink, L.M., Krom, M.D. and de Lange, G.J., 1997. The use of sequential extraction techniques for sedimentary phosphorus in eastern Mediterranean sediments. Mar. Geol., 139, 147-155.

Falkowski, P.G., 1997. Evolution of the nitrogen cycle and its influence on the biological sequestration of CO_2 in the ocean. Nature, 387: 272-275.

Filippelli, G.M., 1997. Controls on phosphorus concentration and accumulation in oceanic sediments. Mar. Geol., 139, 231-240.

Filippelli, G.M., 2001. Carbon and phosphorus cycling in anoxic sediments of the Saanich inlet, British Columbia. Mar. Geol., 174, 307-321.

Föllmi, K.B., 1996. The phosphorus cycle, phosphogenesis and marine phosphat-rich deposites. Earth-Sci. Rev., 40, 55-124.

Fossing, H., Gallardo, V.A., Jørgensen, B.B., Hüttel, M., Nielsen, L.P., Schulz, H., Canfield, D.E., Forster, S., Glud, R.N., Gundersen, J.K., Küver, J., Ramsing, N.B., Teske, A., Thamdrup, B. and Ulloa, O., 1995. Concentration and transport of nitrate by the matforming sulphur bacterium Thioploca. Nature, 374, 713-715.

Froelich, P.N., 1988. Kinetic control of dissolved phosphate in natural rivers and estuaries: a primer on the phosphate buffer mechanism. Limnol. Oceanogr., 33, 649-668.

Froelich, P.N., Klinkhammer, G.P., Bender, M.L., Luedtke, N.A., Heath, G.R., Cullen, D. and Dauphin, P., 1979. Early oxidation of organic matter in pelagic sediments of the eastern equatorial Atlantic: suboxic diagenesis. Geochim. Cosmochim. Acta, 43, 1075-1090.

Froelich, P.N., Bender, M.L., Luedtke, N.A., Heath, G.R.

and De Vries, T., 1982. The marine phosphorus cycle. Am. J. Sci., 282, 474-511.

François R., Altabet M.A. and Burkle L.H., 1992. Glacial to interglacial changes in surface nitrate utilization in the Indian sector of the Southern Ocean as recorded by sediment $\delta^{15}N$. Paleoceanogr. 7, 589-606.

Faul, K.L., Paytan, A. and Delaney, M.L., 2005. Phosphorus distribution in sinking oceanic particulate matter. Mar. Chem., in press.

Gaye-Haake B., Lahajnar N., Emeis K.C., Unger D., Rixen T., Suthhof A., Ramaswamy V., Schulz H., Paropkari A. L., Guptha M. V. S. and Ittekkot V., 2005. Stable nitrogen isotopic ratios of sinking particles and sediments from the northern Indian Ocean. Mar. Chem. 96, 243-255.

Glenn, C.R., Prévôt-Lucas, L. and Lucas, J. (eds), 2000. Marine Authigenesis: From Global to Microbial. SEPM special Publication No. 66, Tulsa Oklahoma US, 536 pp.

Glud, R.N., Gundersen, J.K., Jørgensen, B.B., Revsbech, N.P. and Schulz, H.D., 1994. Diffusive and total oxygen uptake of deep-sea sediments in the eastern South Atlantic Ocean: in situ and laboratory measurements. Deep-Sea Res., 41, 1767-1788.

Glud, R.N., Ramsing, N.B., Gundersen, J.K. and Klimant, I., 1996. Planar optodes: A new tool for fine scale measurements of two dimensional O_2 distribution in benthic communities. Mar. Ecol. Prog. Ser., 140, 217-226.

Goering, G.G. and Pamatmat, M.M., 1970. Denitrification in sediments of the sea off Peru. Invest. Pesq., 35, 233-242.

Gundersen, J.K. and Jørgensen, B.B., 1990. Microstructure of diffusive boundary layers and the oxygen uptake of the sea floor. Nature, 345, 604-607.

Haese, R.R., 1997. Beschreibung und Quantifizierung frühdiagenetischer Reaktionen des Eisens in Sedimenten des Südatlantiks. Berichte, Fachbereich Geowissenschaften, Universität Bremen, 99, pp. 118.

Hales, B. and Emerson, S., 1997. Calcite dissolution in sediments of the Ceara Rise: In situ measurements of pore water O_2, pH, and $CO_2(aq)$. Geochim. Cosmochim. Acta, 61, 501-514.

Hammond, D.E., McManus, J., Berelson, W.M., Kilgore, T.E. and Pope, R.H., 1996. Early diagenesis of organic material in equatorial Pacific sediments: stoichiometry and kinetics. Deep-Sea Res. II, 43, 1365-1412.

Heip, C.H.R., Goosen, N.K., Herman, P.M.J., Kromkamp, J., Middelburg, J.J. and Soetaert, K., 1995. Production and consumption of biological particles in temperate tidal estuaries. Oceanogr. and Mar. Biol.: An annual review, 33, 1-149.

Henrichs, S.M., 1992. Early diagenesis of organic matter in marine sediments: progress and perplexity. Mar. Chem., 39, 119 - 149.

Henrichs, S.M. and Reeburgh W.S., 1987. Anaerobic Mineralization of Marine Sediment Organic Matter: Rates and the Role of Anaerobic Processes in the Oceanic Carbon Economy. Geomicrobiol. J., 5, 191-237.

Henriksen, K., Hansen, J.I. and Blackburn, T.H., 1981. Rates of nitrification, distribution of nitrifying bacteria and nitrate fluxes in different types of sediment from Danish waters. Mar. Biol., 61, 299-304.

Henriksen, K. and Kemp, W.M., 1988. Nitrification in estuarine and coastal marine sediments. In: Blackburn, T.H., Sørensen, J. (eds.) Nitrogen Cycling in Coastal Marine Environments. SCOPE Reports, 207-249, John Wiley and Sons, Chichester.

Hensen, C., Landenberger, H., Zabel, M., Gundersen, J.K., Glud, R.N. and Schulz, H.D., 1997. Simulation of early diagenetic processes in continental slope sediments off Southwest Africa: The Computer Model CoTAM tested. Mar. Geol., 144, 191-210.

Hensen, C., Landenberger, H., Zabel, M. and Schulz, H.D., 1998. Quantification of diffusive benthic fluxes of nitrate, phosphate and silicate in the Southern Atlantic Ocean. Glob. Biogeochem. Cycles, 12, 193-210.

Hensen, C., Zabel, M. and Schulz, H.D., 2000. A comparison of benthic nutrient fluxes from deep-sea sediments off Namibia and Argentina. Deep-Sea Res. II, 47, 2029-2050.

Hynes, R.K. and Knowles, R., 1978. Inhibition by acetylene of ammonia oxidation in Nitrosomonas europaea. FEMS Microbiol. Lett., 4, 319-321.

Ingall, E.D. and Van Cappellen, P., 1990. Relation between sedimentation rate and burial of organic phosphorus and organic carbon in marine sediments. Geochim. Cosmochim. Acta, 54, 373-386.

Jahnke, R.A. and Christiansen, M.B., 1989. A free-vehicle benthic chamber instrument for sea floor studies.- Deep-Sea Res., 36, 625-637.

Jahnke, R.A., Heggie, D., Emerson, S. and Grundmanis, V., 1982. Pore waters of the central Pacific Ocean: nutrient results.- Earth Planet. Sci. Letters, 61, 233-256.

Jahnke, R.A., Reimers, C.E. and Craven, D.B., 1990. Intensification of recycling of organic matter at the sea floor near ocean margins.- Nature, 348, 50-54.

Jahnke, R.A., Craven, D.B., McCorkle, D.C. and Reimers, C.E., 1997. $CaCO_3$ dissolution in California continental margin sediments: The influence of organic matter remineralization. Geochim. Cosmochim. Acta, 61(17), 3587-3604.

Jarvis, I., Burnett, W.C., Nathan., Y., Almbaydin, F., Attia, K.M., Castro, L.N., Flicoteaux, R., Hilmy, M.E., Husain, V., Qutawna, A.A., Serjani, A. and Zanin, Y.N., 1994. Phosphorite geochemistry: state-of-the-art and environmental concerns. In: Föllmi, K.B. (ed) Concepts and Controversies in Phosphogenesis. Eclogae Geol. Helv., 87, 643-700.

Johnson, K.S., Berelson, W.M., Coale, K.H., Coley, T.L., Elrod, V.A., Fairey, W.R., Iams, H.D., Kilgore, T.E. and Nowicki, J.L., 1992. Manganese flux from continental margin sediments in a transect through the oxygen minimum. Science, 257, 1242-1245.

Jørgensen, B.B., 1977. Bacterial sulfate reduction within reduced microniches of oxidized marine sediments. Mar. Biol., 41, 7-17.

Jørgensen, B.B. and Sørensen, J., 1985. Seasonal cycles of O_2, NO_3^-, and SO_4^{2-} reduction in estuarine sediments: The significance of an NO_3^- reduction maximum in spring. Mar. Ecol. Prog. Ser., 24, 65-74.

Jørgensen, K.S., Jensen, H.B. and Sørensen, J., 1984. Nitrous oxide production from nitrification and denitrification in marine sediment at low oxygen concentrations. Can. J. Microbiol., 30, 1073-1078.

Jørgensen, B.B. and Nelson, D.C., 2004. Sulfide oxidation in marine sediments: Geochemistry meets microbiology. In: Amen, J.P, Edwards, K.J., Lyons, T.W. (eds) Sulfur Biogeochemistry – Past and Present. Geological Society of America Special Paper, 379,63-81.

Kaplan, W.A., 1983. Nitrification. In: Carpenter, J.E.,

Capone, D.G. (eds.) Nitrogen in the marine environment, 139-190, Academic Press, New York.

Kennett, J., 1982. Marine Geology. Prentice Hall, New Jersey, pp. 813.

Klimant, I, Meyer, V. and Kühl., M., 1995. Fiber-optic oxygen microsensors, a new tool in aquatic biology. Limnol. Oceanogr., 40(6), 1159-1165.

Koike, I. and Sørensen, J., 1988. Nitrate Reduction and Denitrification in Marine Sediments. In: Blackburn, T.H., Sørensen, J. (eds.) Nitrogen Cycling in Coastal Marine Environments. SCOPE Reports, 251–273, John Wiley and Sons, Chichester.

Kolodny, Y. and Luz, B., 1992. Isotope signatures in phosphate deposits: Formation and diagenetic history. In: Clauer and Chaudhuri (eds) Isotopic signatures and sedimentary records, Lect. Notes in Earth Sci. 43, Springer, Heidelberg, 69-121.

Kornberg A., 1995. Inorganic Polyphosphate - Toward Making a Forgotten Polymer Unforgettable. J. Bacteriol., 177, 491-496.

Krom, M.D. and Berner, R.A., 1980. Adsorption of phosphate in anoxic sediments. Limnol. Oceanogr., 25, 797-806.

Kuypers, M.M.M., Sliekers, A.O., Lavik, G., Schmidt, M., Jørgensen, B.B., Kuenen, J.G., Sinninghe Damsté, J.S., Strous, M. and Jetten, M.S.M., 2003. Anaerobic ammonium oxidation by anammox bacteria in the Black Sea. Nature, 422, 608-611.

Kuypers, M.M.M., Lavik, G., Woebken, D., Schmidt, M., Fuchs, B.M., Amann, R., Jørgensen, B.B. and Jetten, M.S.M., 2005. Massive nitrogen loss from the Benguela upwelling system through anaerobic ammonium oxidation. Proc. Natl. Acad. Sci. USA, 102, 6478-6483.

Larsen, L.H., Revsbech, N.P. and Binnerup, S.J., 1996. A microsensor for nitrate based on immobilized denitrifying bacteria. Appl. Environm. Microbiol., 62 (4), 1248-1251.

Luther, G.W., III., Sundby, B., Lewis, B.L., Brendel, P.J. and Silverberg, N., 1997. Interactions of manganese with the nitrogen cycle: Alternative pathways to dinitrogen. Geochim. Cosmochim. Acta, 61(19), 4043-4052.

Luther, G.W., III., Brendel, P.J., Lewis, B.L., Sundby, B., Lefrançois, L., Silverberg, N. and Nuzzio, D.B., 1998. Simultaneous measurement of O_2, Mn, Fe, I⁻, and S(-II) in marine pore waters with a solid-state voltammetric microelectrode. Limnol. Oceanogr., 43(2), 325-333.

Martin, J.H., Knauer, G.A., Karl, D.M. and Broenkow, W.W., 1987. VERTEX: Carbon cycling in the Northeast Pacific. Deep-Sea Res., 34 (2), 267-285.

Martin, W.R. and Sayles, F.L., 1996. $CaCO_3$ dissolution in sediments of the Ceara Rise, western equatorial Atlantic. Geochim. Cosmochim. Acta, 60, 243-263.

McClellan, G.H., 1980. Mineralogy of carbonate fluorapatites. J. Geol. Soc. London, 137, 675-681.

McHatton, S.C., Barry, J.P.H., Jannasch, W. and Nelson, D.C., 1996. High nitrate concentrations in vacuolate, autotrophic marine Beggiatoa spp. Appl. Environ. Microbiol., 62, 954-958.

Middelburg, J.J., Vlug, T. and van der Nat, F.J.W.A., 1993. Organic matter mineralization in marine sediments. Glob. Planet. Change, 8, 47-58.

Middelburg, J.J., Soetaert, K., Herman, P.M.J. and Heip, C.H.R., 1996a. Denitrification in marine sediments: A model study. Glob. Biogeochem. Cycles, 10, 661-673.

Middelburg, J.J., Soetaert, K. and Herman, P.M.J., 1996b. Evaluation of the nitrogen isotope-pairing method for measuring benthic denitrification: A simu-lation analysis. Limnol. Oceanogr., 41, 1839-1844.

Middelburg, J.J., Soetaert, K. and Herman, P.M.J., 1996c. Reply to the comment by Nielsen et al. Limnol. Oceanogr., 41, 1846-1847.

Mills, M.M., Ridame, C., Davey, M., LaRoche, J. and Geider, R.J., 2004. Iron and phosphorus co-limit nitrogen fixation in the Eastern Tropical North Atlantic. Nature, 429, 292-294.

Mino, T., 2000. Microbial selection of polyphosphate-accumulating bacteria in activated sludge wastewater treatment processes for enhanced biological phos-phate removal. Biochemistry (Moscow), 65, 341-348.

Mulder, A., van de Graaf, A.A., Robertson, L.A. and Kuenen, J.G., 1995. Anaerobic ammonium oxidation discovered in a denitrifying fluidized bed reactor. FEMS Microbiol. Ecol., 16, 177-184.

Nathan Y., Bremner, J.M., Lowenthal, R.E. and Monteiro, P., 1993. Role of Bacteria in Phosphorite Genesis. Geomicrobiol., 11, 69-76.

Nielsen, L.P., 1992. Denitrification in sediment deter-mined from nitrogen isotope pairing. FEMS (Fed. Eur. Microbiol. Soc.), Microbiol. Ecol., 86, 357-362.

Nielsen, L.P., Risgaard-Petersen, N., Rysgaard, S. and Blackburn, T.H., 1996. Reply to the note by Middelburg et al. Limnol. Oceanogr., 41, 1845-1846.

Nolton, A.J.G. and Jarvis, I. (eds), 1990. Phosphorite research and development. Geol. Soc. London Special Publ. 52, Oxford, 326 pp.

Otte, S., Kuenen, J.G., Nielsen, L.P., Paerl, H.W., Zopfi, J., Schulz, H.N., Teske, A., Strotmann, B., Gallardo, V.A. and Jørgensen B.B., 1999. Nitrogen, carbon, and sulfur metabolism in natural Thioploca samples. Appl. Environ. Microbiol., 65, 3148-3157.

Painter, H.A., 1970. A review of literature on inorganic nitrogen metabolism in microorganisms. Water Res., 4, 393-450.

Pamatmat, M.M., 1971. Oxygen consumption in the seabed. IV. Shipboard and laboratory experiments. Limnol. Oceanogr., 16, 536-550.

Redfield, A.C., Ketchum, B.H. and Richards, F.A., 1963. The influence of organisms on the composition of sea-water. In: Hill, M.N. (ed.): The sea.- Interscience, New York, pp. 26-77.

Reid, J.L., 1994. On the total geostrophic circulation of the North Atlantic Ocean: Flow patterns, tracers, and transports. Prog. Oceanogr., 33(1), 1-92.

Reimers, C.E., Jahnke, R.H. and McCorkle, D.C., 1992. Carbon fluxes and burial rates over the continental slope and rise off central California with implications for the global carbon cycle. Glob. Biogeochem. Cycles, 6, 199-224.

Revsbech, N.P., 1989. An oxygen electrode with a guard cathode. Limnol. Oceanogr., 34, 474-478.

Revsbech, N.P. and Jørgensen, B.B., 1986. Micro-electrodes: Their use in microbial ecology. Adv. Microb. Ecol., 9, 293-352.

Revsbech, N.P., Jørgensen, B.B. and Blackburn, T.H., 1980. Oxygen in the sea bottom measured with a microelectrode. Science, 207, 1355-1356.

Rhoads, D.L. and Morse, J.W., 1971. Evolutionary and ecologic significance of oxygen-deficient marine basins. Lethaia, 4, 413-428.

Richards, F.A., Cline, J.D., Broenkow, W.W. and

Atkinson, L.P., 1965. Some consequences of the decomposition of organic matter in Lake Nitinat, an anoxic fjord. Limnol. Oceanogr., 10, 185-201.

Rowe, G.T., Boland, G.S., Phoel, W.C., Anderson, R.F. and Biscaye, P.E., 1994. Deep-sea floor respiration as an indication of lateral input of biogenic detritus from continental margins. Deep-Sea Res. II, 41, 657-668.

Ruttenberg, K.C., 1992. Development of a sequential extraction method for different forms of phosphorus in marine sediments. Limnol. Oceanogr., 37, 1460-1482.

Ruttenberg, K.C. and Berner, R.A., 1993. Authigenic apatite formation and burial in sediments from non-upwelling, continental margins. Geochim. Cosmo-chim. Acta, 57, 991-1007.

Rysgaard, S., Risgaard-Petersen, N., Sloth, N.P., Jensen, K. and Nielsen, L.P., 1994. Oxygen regulation of nitrification and denitrification in sediments. Limnol. Oceanogr., 39(7), 1643-1652.

Sayles, F.L. and Martin, W.R., 1995. In Situ tracer studies of solute transport across the sediment - water interface at the Bermuda Time Series site. Deep-Sea Res. I, 42(1), 31-52.

Schenau, S.J. and de Lange, G.J., 2000. A novel chemical method to quantify fish debris in marine sediments. Limnol. Oceanogr., 45, 963-971.

Schenau, S.J. and de Lange, G.J., 2001. Phosphorus regeneration vs. burial in sediments of the Arabian Sea. Mar. Chem., 75, 201-217.

Schenau, S.J., Slomp, C.P. and de Lange, G.J., 2000. Phosphogenesis and active phosphorite formation in sediments from the Arabian Sea oxygen minimum zone. Mar. Geol., 169, 1-20.

Schink, B. and Friedrich, M., 2000. Phosphite oxidation by sulphate reduction. Nature, 406, 37-37.

Schulz, H.D., Dahmke, A., Schinzel, U., Wallmann, K. and Zabel, M., 1994. Early diagenetic processes, fluxes, and reaction rates in sediments of the South Atlantic. Geochim. Cosmochim. Acta, 58, 2041-2060.

Schulz, H.N. and Schulz, H.D., 2005. Large sulfur bacteria and the formation of phosphorite. Science, 307, 416-418.

Schulz, H.N., Brinkhoff, T., Ferdelman, T.G., Marine, M.H., Teske, A. and Jørgensen, B.B., 1999. Dense populations of a giant sulfur bacterium in Namibian shelf sediments. Science, 284, 493-495.

Seiter, K., Hensen, C. and Zabel, M., 2005. Benthic carbon mineralization on a global scale. Global Biogeochemical Cycles, 19, GB1010, doi:10.1029/2004GB002225.

Seitzinger, S.P., 1998. Denitrification in freshwater and coastal marine environments: Ecological and geochemical significance. Limnol. Oceanogr., 33(4), 702-724.

Seitzinger, S.P., Nixon, S.W. and Pilson, M.E.Q., 1984. Denitrification and nitrous oxide production in a coastal marine ecosystem. Limnol. Oceanogr., 29, 73-83.

Seitzinger, S.P., Nielsen, L.P., Caffrey, J. and Christensen, P.B., 1993. Denitrification measurement in aquatic sediments: A comparison of three methods. Biogeo-chem., 23, 147-167.

Shapleigh, J.P., 2000. The denitrifying prokaryotes. In: Dworkin, M. et al. (eds), The Prokaryotes: An Evolving Electronic Resource for the Microbiological Community, 3rd edition, Springer-Verlag, New York.

Sigman, D.M., Altabet, M.A., François, R., McCorkle, D.C. and Gaillard, J.-F., 1999. The isotopic composition of diatom-bound nitrogen in Southern ocean sediments. Paleoceanogr., 14(2), 118-134.

Sigman, D.M., Altabet, M.A., McCorkle, D.C., François, R. and Fischer, G., 2000. The $\delta^{15}N$ of nitrate in the Southern Ocean: nitrogen cycling and circulation in the ocean interior. J. Geophys. Res., 105(C8): 19599-19614.

Sloth, N.P., Blackburn, T.H., Hansen, L.S., Risgaard-Petersen, N. and Lomstein, B.A., 1995. Nitrogen cycling in sediments with different organic loading. Mar. Ecol Prog. Ser., 116, 163-170.

Slomp, C.P., Epping, E.H.G., Helder, W. and van Raaphorst, W., 1996. A key role for iron bound phosphorus in authigenic apatite formation in North Atlantic continental platform sediments. J. Mar. Res., 75, 1179-1205.

Slomp, C.P., Thomson, J. and de Lange, G.J., 2002. Enhanced regeneration of phosphorus during formation of the most recent eastern Mediteranean sapropel (S1). Geochim. Cosmochi. Acta, 66, 1171-1184.

Slomp, C.P., Thomson, J., de Lange, G.J., 2004. Controls on phosphorus regeneration and burial during formation of eastern Mediterranean sapropels. Mar. Geol., 203, 141-159.

Smith, K.L. Jr. and Baldwin, R.J., 1984. Seasonal fluctuations in deep-sea sediment community oxygen consumption: central and eastern North Pacific. Nature, 307, 624-626.

Soetaert, K., Herman, P.M.J. and Middelburg, J.J., 1996. A model of early diagenetic processes from the shelf to abyssal depths. Geochim. Cosmochim. Acta, 60, 1019-1040.

Soetaert, K., Vanaverbeke, J., Heip, C., Herman, P.M.J., Middelburg, J.J., Sandee, A. and Duineveld, G., 1997. Nematode distribution in ocean margin sediments of the Goban Spur (northeast Atlantic) in relation to sediment geochemistry. Deep-Sea Res. I, 44, 1671-1683.

Straub, K.L., Benz, M., Schink, B. and Widdel, F., 1996. Anaerobic, nitrate-dependent microbial oxidation of ferrous iron. Appl. Environ. Microbiol., 62, 1458-1460.

Strous, M., Fuerst, J.A., Kramer, E.H.M., Logemann, S., Muyzer, G., Van de Pas-Scoonen, K.T., Webb, R., Kuenen, J.G. and Jette, M.S.M., 1999. Missing lithotroph identified as new planctomycete. Nature, 400, 446-449.

Stumm, W. and Morgan, J.J., 1996. Aquatic Chemistry. John Wiley & Sons, New York.

Sverdrup, H.U., Johnson, M.W. and Fleming, R.H., 1942. The oceans. Prentice Hall, New York.

Sweerts, J.-P.R.A., DeBeer, D., Nielsen, L.P., Verdouw, H., Vandenheuvel, J.C., Cohen, Y. and Cappenberg, T.E., 1990. Denitrification by sulfur oxidizing Beggiatoa spp. mats on freshwater sediments. Nature, 344, 762-763.

Sørensen, J., 1987. Nitrate reduction in marine sediment: pathways and interactions with iron and sulfur cycling. Geomicrobiol. J., 5, 401-421.

Sørensen, J., Rasmussen, L.K. and Koike, I., 1987. Micromolar sulfide concentrations alleviate blockage of nitrous oxide reduction by denitrifying Pseudomonas fluorescens. Can. J. Microbiol., 33, 1001-1005.

Takahashi, T., Broecker, W.S. and Langer, S., 1985.

Redfield ratio based on chemical data from isopycnal surfaces. J. Geophys. Res., 90, 6907-6924.

Tebo, B.M., Rosson, R.A. and Nealson, K.H., 1991. Potential for manganese(II) oxidation and manganese(IV) reduction to co-occur in the suboxic zone of the Black Sea. In: Izdar, E., Murray, W. (eds.): Black Sea Oceanography. Kluwer Academic Publishers, Dordrecht, 173-185.

Thamdrup, B. and Canfield, D.E., 1996. Pathways of organic carbon oxidation in continental margin sediments off Chile. Limnol. Oceanogr., 41(8), 1629-1650.

Thamdrup, B. and Dalsgaard, T., 2002. Production of N_2 through Anaerobic Ammonium Oxidation Coupled to Nitrate Reduction in Marine Sediments. Appl. Environ. Microbiol., 68, 1312-1318.

Thamdrup, B., Fossing, H. and Jørgensen, B.B., 1994. Manganese, iron, and sulfur cycling in a coastal marine sediment, Aarhus Bay, Denmark. Geochim. Cosmochim. Acta, 58(23), 5115-5129.

Thunell, R.C. and Kepple, B., 2004. Glacial-Holocene $\delta^{15}N$ record from the Gulf of Tehuantepec, Mexico: implications for denitrification in the eastern equatorial Pacific and changes in atmospheric N_2O. Glob. Biogeochem. Cycles 18(GB1001), doi:10.1029/2002GB002028.

Thunell, R.C., Sigman, D.M., Muller-Karger, F., Astor, Y. and Varela, R., 2004. Nitrogen isotope dynamics of the Cariaco Basin, Venezuela. Glob. Biogeochem. Cycles 18(GB3001), doi:10.1029/2003GB002185.

Tromp, T.K., van Cappellen, P. and Key, R.M., 1995. A global model for the early diagenesis of organic carbon and organic phosphorus in marine sediments. Geochim. Cosmochim. Acta, 59, 1259-1284.

Tyrrell, T., 1999. The relative influences of nitrogen and phosphorus on oceanic primary production. Nature, 400, 525-531.

van Cappellen, P. and Berner, R.A., 1988. A mathematical model for the early diagenesis of phosphorus and fluorine in marine sediments: Apatite precipitation. Am. J. Sci., 288, 289-333.

van Cappellen, P. and Berner, R.A., 1991. Fluorapatite crystal growth from modified seawater solutions. Geochim. Cosmochim. Acta, 55, 1219-1234.

van Cappellen, P., Gaillard, J.-P. and Rabouille, C., 1993. Biogeochemical transformations in sediments: Kinetic models of early diagenesis. In: Wollast, R., Mackenzie, F.T. and Chou, L. (eds) Interactions of C, N, P, and S biogeochemical cycles and global change. NATO ASI Series, 401-445, Springer Verlag.

van de Graaf, A.A., Mulder, A., De Bruijn, P., Jetten, M.S.M., Robertson, L.A. and Kuenen, J.G., 1995. Anaerobic oxidation of ammonium is a biologically mediated process. Appl. Environ. Microbiol., 61, 1246–1251.

van der Zee, C., Slomp, C.P. and van Raaphorst, W., 2002. Authigenic P formation and reactive P burial in sediments of the Nazaré canyon on the Iberian margin (NE Atlantic). Mar. Geol., 185, 379-392.

Wallmann, K., 2003. Feedbacks between oceanic redox states and marine productivity: a model perspective focused on benthic phosphorus cycling. Glob. Biogeochem. Cycles, 17, 1084, doi: 10.1029GB001968.

Wenzhöfer F. and Glud R. N., 2002. Benthic carbon mineralization in the Atlantic: a synthesis based on in situ data from the last decade. Deep-Sea Res. I, 49, 1255-1279.

Wenzhöfer F. and Glud R. N., 2004. Small-scale spatial

and temporal variability in coastal benthic O_2 dynamics: Effects of fauna activity. Limnol. Oceanogr., 49(5), 1471-1481.

Wenzhöfer, F., Adler, M., Kohls, O., Hensen, C., Strotmann, B., Boehme, S. and Schulz, H.D., 2001a. Calcite dissolution driven by benthic mineralization in the deep-sea: in situ measurements of Ca^{2+}, pH, pCO_2, O_2. Geochim. Cosmochim. Acta, 65(16), 2677-2690.

Wenzhöfer, F., Kohls, O. and Holby, O., 2001b. Deep penetrating benthic oxygen profiles measured *in situ* by oxygen optodes. Deep-Sea Res I, 48, 1741-1755.

Wenzhöfer, F., Greeff, O. and Riess, W., 2002. Benthic carbon mineralization in sediments of Gotland Basin, Baltic Sea, measured *in situ* with benthic landers. In: Taillefert, M., Rozan, T.F. (eds.) Environmental electrochemistry: analyses of trace element biogeochemistry. American Chemical Society Symposium Series 811, Washington DC, pp. 162-185.

Williams L.A. and Reimers, C., 1983. Role of bacterial mats in oxygen-deficient marine basins and coastal upwelling regimes: Preliminary report. Geology, 11, 267-269.

WOCE, 2002. World Ocean Circulation Experiment, Global Data, Version 3.0, WOCE International Project Office, WOCE Report, Southampton, UK; published by U.S. National Oceanographic Data Center, Silver Spring, 180/02, DVD-ROM.

7 The Biogeochemistry of Iron

RALF R. HAESE

7.1 Introduction

For our understanding of interactions between living organisms and the solid earth it is fascinating to investigate the reactivity of iron at the interface of the bio- and geosphere. Similar to manganese (chapter 11) iron occurs in two valence states as oxidized ferric iron, Fe(III), and reduced ferrous iron, Fe(II). Two principal biological processes are of importance: Microorganisms such as magnetotactic bacteria and phytoplankton (see chapter 2 and section 7.3) depend on the uptake of iron as a prerequisite for their cell growth (assimilation). Others conserve energy from the reduction of Fe(III) to maintain their metabolic activity (dissimilation). In this case ferric iron serves as an electron acceptor which is also termed oxidant. Apart from biotic reactions manifold abiotic reactions occur depending on thermodynamic and kinetic conditions. Due to redox-reactions dissolution and precipitation of iron-bearing minerals may result which has great influence on the sorption/desorption and co-precipitation/release behavior of various components such as phosphate and trace metals. From a geologic point of view it is striking to find discrete iron enriched layers such as black shales or strata of the banded iron formation, which challenge geochemists to reconstruct the environmental conditions of their formation.

7.2 Pathways of Iron Input to Marine Sediments

Within the continental crust iron is the fourth most abundant element with a concentration of 4.32 wt% (Wedepohl 1995). It is transported to marine sediments by four major regimes: fluvial, aeolian, submarine hydrothermal, and glacial input. For the investigation of iron reactivity it is important to differentiate regions of predominant input regimes since characteristic reactions occur at the interface of the transport regime and the marine environment. For the chemistry of hydrothermal fluids and reactions during mixing with seawater refer to chapter 13.

7.2.1 Fluvial Input

In Figure 7.1 averaged concentrations of dissolved and particulate iron, major fluxes and the respective predominant reactions are shown. Dissolved iron concentrations in averaged river and marine water clearly show less solubility in marine relative to river water. In contrast, the concentration of particulate iron does not change significantly and is similar to the average continental crust concentration. This is also reflected in the conservative behavior of iron under chemical weathering conditions. Along with Al, Ti and Mn, Fe belongs to the refractive elements (Canfield 1997). Note that in Fig. 7.1 the given value for particulate iron concentration in marine sediments is derived from pelagic clay sediments. Biogenic constituents such as carbonate and opal may significantly dilute the terrigineous fraction (chapter 1) and thus decrease the iron concentration.

The decrease of dissolved iron concentrations can be traced within estuarine mixing zones. Within river water, dissolved iron is mainly present as Fe(III)oxyhydroxide, which is stabilized in colloidal dispersion by high-molecular-weight humic acids (Hunter 1983). Due to increasing salinity and thus increasing ionic strength the colloidal dispersion becomes electrostatically and chemically destabilized which results in the coagulation of the fluvial colloids. This process is reflected in Fig. 7.2 showing dissolved iron along a transect off the mouth of the Congo (formerly: Zaire) river

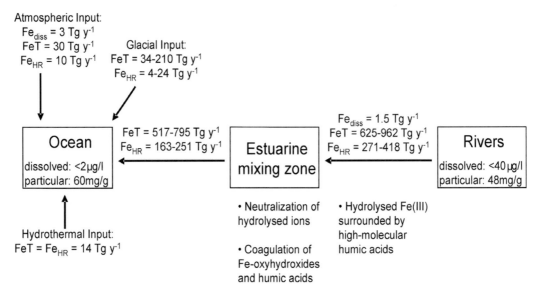

Fig. 7.1 Global fluxes of dissolved (Fe_{diss}), highly reactive iron (FeHR) and total iron (FeT) from riverine, glacial, atmospheric and hydrothermal sources. The estuarine mixing zone serves as a major sink for dissolved and highly reactive (dithionite-soluble) iron. Fluxes and concentrations are given according to Poulton and Raiswell (2002). The atmospheric dissolved iron flux was taken from Duce et al. (1991).

towards the open ocean. With increasing salinity the concentration of dissolved iron decreases exponentially. These results also point out the operatively defined differentiation of dissolved and particulate phase. The concentration of dissolved iron depends on the pore size of the used filters. Commonly, particles smaller than 0.4 µm are considered to be 'dissolved'.

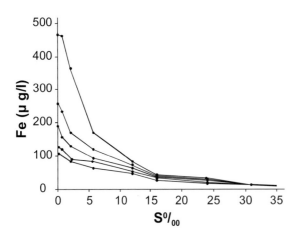

Fig. 7.2 Dissolved iron concentrations of surface water from a transect off the Congo River towards the open ocean. Pore sizes of 1.2, 0.45, 0.22, 0.05 and 0.025 µm (according to graphs from above downward) were used to separate particulate from dissolved phase (adopted from Figuères et al. 1978).

Apart from being a zone of coagulation the estuarine and coastal zone represents the most important sink for dissolved and particulate iron of fluvial origin. A recent study by Poulton and Raiswell (2002) provides quantitative evidence for the important role of estuaries for iron transformations at the Earth's surface (Fig. 7.1): The high degree of chemical weathering in the catchments of rivers discharging into the global ocean leads to a high proportion of highly reactive iron ($Fe_{HR,dithionite\ soluble}$) relative to total iron (FeT), with $Fe_{HR}/FeT = 0.43$. In contrast, glaciers discharge sediments which were formed predominantly under physical weathering conditions and reveal a mean Fe_{HR}/FeT of 0.11. Approximately 40 % of the riverine highly reactive iron fraction is retained in the estuarine and near-coastal zone, which, together with the low Fe_{HR} proportion in glacial sediments leads to a Fe_{HR}/FeT of 0.26 in marine sediments. The iron inputs from atmospheric and hydrothermal sources are relatively small, i.e. less than 10 % of Fe_{HR} and FeT as compared to the riverine input (Fig. 7.1).

7.2.2 Aeolian Input

In contrast to the fluvial transport regime the aeolian transport is highly efficient for the deposition of terrigenous matter in the deep sea.

Donaghay et al. (1991) published a global map of atmospheric iron flux to the ocean (Fig. 7.3). Among others the North African desert and Asia can be identified as major dust sources. Asian loess is transported across the northern Pacific making up to 100 % of the terrigenous fraction of pelagic sediments (Blank et al., 1985) and North African dust is spread out over the north equatorial Atlantic eventually reaching the Caribbean Sea (Carlson and Prospero 1972) and the northern coast of Brazil (Prospero et al., 1981). Iron settles out of wind-driven air layers by wet deposition which can be traced in equatorial areas (Murray and Leinen 1993) where a distinctively high humidity occurs within the Inner Tropical Convergence Zone (ITCZ). Therefore, a characteristic high deposition of iron occurs at the equatorial Atlantic (Fig. 7.3).

As iron is discussed as limiting nutrient for phytoplankton productivity of distinct oceanic regions (chapter 7.3) it is important to investigate the flux, degree of solubility and the bioavailability of the introduced iron into open ocean surface water. It has long been known that sorption onto suspended particles is highly efficient in the removal of trace metals from solution (Krauskopf 1956) resulting in a decrease of dissolved concentration relative to the thermodynamic saturated concentration. Chester (1990) listed the solubility of several trace metals in coastal and open ocean surface water and specified the solubility of aeolian transported iron with ≤ 7 %. This is in agreement with a short review of published results provided by Zhuang et al. (1990) revealing a range of iron solubility between 1 and 10 %. By contrast, Zhuang et al. (1990) themselves found a solubility of ~50 % of atmospheric iron suggesting that all the dissolved iron in North Pacific surface water is provided by atmospheric input. Zhuang and Duce (1993) could show that the concentration of suspended particles is the prime variable controlling the adsorbed fraction, yet, an increase in aeolian deposition would also result in an increase of net dissolved iron.

The reason for the dissolution of solid phase iron in the photic zone is the photochemical reduction of Fe(III), with UVB (280 – 315 nm) producing most of Fe(II), followed by UVA (315 – 400 nm) and visible light (400 – 700 nm) (Rijkenberg et al. 2005). As a consequence, about 10 % of atmospheric FeT reaches the ocean as dissolved iron (Duce et al. 1991; Fig. 7.1). In surface waters of the global ocean the produced Fe^{2+} is subsequently reoxidized, yet, the photoreduction

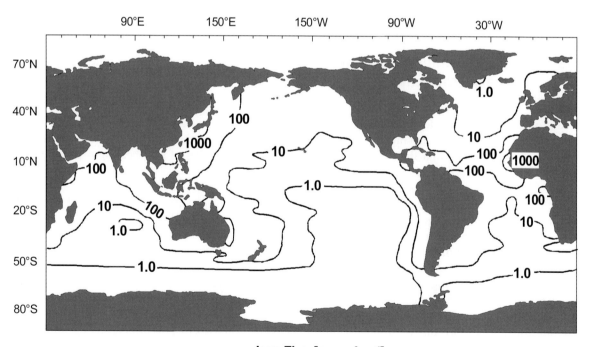

Iron Flux [mg m^{-2} yr^{-1}]

Fig. 7.3 Global map of atmospheric iron fluxes to the deep sea (adopted from Donaghay et al. 1991).

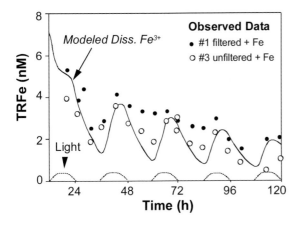

Fig. 7.4 Measured iron concentrations in incubations (dots) and model (solid line) results of dissolved iron within surface ocean water. The artificial light intensity (dashed line) drives the photochemical reduction. The gradual decrease of dissolved iron is caused by the uptake by phytoplankton (adopted from Johnson et al. 1994).

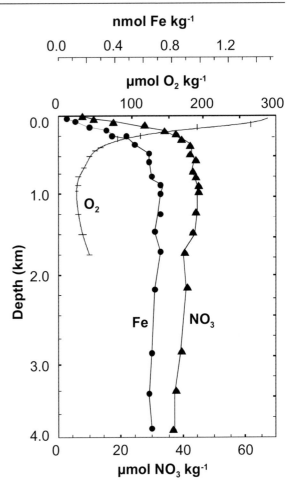

Fig. 7.5 Vertical distribution of NO_3^-, dissolved iron and oxygen at a station of the Gulf of Alaska (adopted from Martin et al. 1989).

causes a net increase of dissolved iron. Model and experimental results by Johnson et al. (1994) reflect the dependence of dissolved iron on irradiation (Fig. 7.4). During incubations diurnal cycles and gradually decreasing dissolved iron concentrations result from induced daily irradiation and a gradual uptake by phytoplankton.

7.3   Iron as a Limiting Nutrient for Primary Productivity

The growth of phytoplankton in the world ocean is undoubtedly one of the driving influences for the global carbon cycle and thus for the present and past climate (Berger et al. 1989; de Baar and Suess 1993). This primary productivity (PP) is limited by the availability of nutrients. Apart from the major nutrients nitrogen, phosphorous and silica so called micronutrients - especially iron - have long been speculated to have a limiting control on PP (Hart 1934). Yet, detailed investigations concerning their importance for the carbon cycle have only been possible for the past ten years due to analytical reasons. Virtually all microorganisms require iron for their respiratory pigments, proteins and many enzymes. Therefore, dissolved iron shows a similar vertical profile in the water column as nitrate being reduced to near zero within the

surface layer where PP takes place and being increased within the oxygen minimum zone due to the mineralization of iron bearing organic matter (Fig. 7.5).

Three major oceanic regions (20 % of the world's open ocean) are characterized by high-nitrate and low-chlorophyll (HNLC) concentrations. The PP of the Southern Ocean (Broecker et al., 1982), the equatorial Pacific (Chavez and Barber 1987) and the Gulf of Alaska (McAllister et al. 1960) are obviously not limited by nitrate. Alternatively, as atmospheric dust loads in the Antarctic and equatorial Pacific are the lowest in the world (Prospero 1981; Uematsu et al. 1983) the importance of iron as limiting micronutrient for PP became increasingly discussed.

The effect of added atmospheric dust to clean sea water from HNLC-regions were studied in

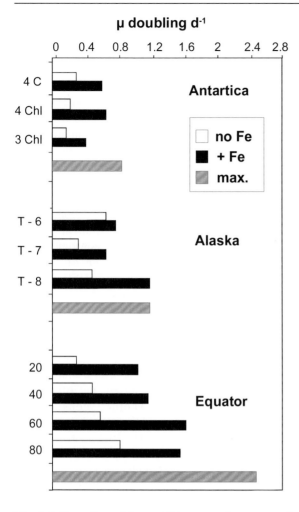

μ doubling d⁻¹

Fig. 7.6 The effect of iron addition to surface water of high-nutrient, low-chlorophyll (HNLC) regions. The doubling rate, μ, is an expression for the increase of primary productivity and maximum values, max., depend on light intensity and temperature. An increase of primary productivity by a factor of 2-4 is resulting due to the addition of atmospheric iron (adopted from Martin et al. 1991).

bottle experiments by Martin et al. (1991) demonstrating increased PP by a factor of 2-4 (Fig. 7.6). Since the role of large grazers living on phytoplankton was not considered by such experiments doubts and open questions to use these laboratory results for statements on large-scale environmental processes remained. Therefore, Martin and colleagues conducted a large-scale iron enrichment experiment south of the Galapagos Islands in the equatorial Pacific (Martin et al. 1994). A total volume of 15,600 l of iron solution (450 Kg Fe) were distributed by ship over an area of 64 km² which increased the original dissolved iron concentration of 0.06 nM to ~4 nM. Iron

concentrations, various parameters monitoring primary productivity and an inert tracer were continuously analyzed for 10 days. As a result they did find an increase of PP by a factor of 2-4 within the iron fertilized open ocean patch (Martin et al. 1994) which gave evidence for the importance of iron as limiting micronutrient within HNLC-areas.

Compelling evidence for the limitation of phytoplankton productivity by the availability of dissolved iron was also found in surface waters of the Peru shelf by comparing the nutrient inventories of the northern, chlorophyll-rich, 'brown' waters to the southern, chlorophyll-poor, 'blue' waters (Bruland et al. 2005). Surface waters of both areas receive large fluxes of macronutrients through upwelling, but only the bottom waters of the northern area are suboxic and rich in dissolved iron (> 50 nM). The constant replenishment of dissolved iron from iron-rich bottom waters leads to very high primary productivity in the northern Peru shelf region whereas the southern region is characterized by high (macro-) nutrient and low chlorophyll concentrations.

If iron is a limiting micronutrient for present-day PP, it may be an important link to explain glacial-interglacial climatic cycles of the past. Martin (1990) postulated the 'iron hypothesis' which explains decreased atmospheric CO_2 concentrations during glacial times with increased iron deposition by aeolian input resulting in increased PP and thus increased CO_2-

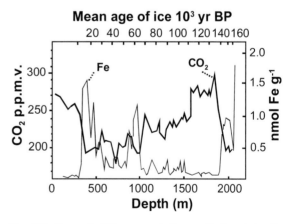

Fig. 7.7 Fe and CO_2 concentrations of the Antarctic Vostok ice core for the past 160,000 years (adopted from De Angelis et al. 1987). Measured Al concentrations were converted to Fe concentrations according to the average continental crust composition. The negative correlation of CO_2 and Fe supports the 'iron hypothesis' (see text).

uptake by the oceans. A comparison of iron and CO_2 concentrations of the past 160,000 years recorded within an Antarctic ice core revealed a strikingly negative correlation (Fig. 7.7) which supports the 'iron hypothesis'.

So far, dissolved iron has been discussed with respect to the assimilation by phytoplankton. The chemical state of bioavailable dissolved species is presently a matter of intensive studies and discussions. Due to thermodynamic reasons concentrations of free ions of dissolved iron are extremely low under oxic and pH-neutral conditions. A discussion paper by Johnson et al. (1997) reviews regional distributions and depth profiles of dissolved iron and points out that at greater depth the iron concentrations always remain constant of ~ 0.6 nM. Other elements with such short residence time (100 to 200 years) typically continuously decrease with depth and age. This suggests a substantial decrease in the iron removal rate below this concentration. As organic ligands with a binding capacity of 0.6 nM Fe have been found (Rue and Bruland 1995; Wu and Luther 1995), iron-organic complexes are regarded to be of great importance for the distribution of dissolved iron. Additional evidence for the interaction with dissolved organic molecules comes from the study of iron uptake mechanisms by organisms. To make dissolved iron more accessible, microorganisms have acquired the ability to synthesize chelators which complex ferric iron of solid phase. These chelators are commonly called siderophores and consist of a low-molecular-mass compound with a high affinity for ferric iron. Siderophores are secreted out of the microorganism where they form a complex with ferric iron. After transport into the cell, the chelated ferric iron is enzymatically reduced and released from the siderophore, which is secreted again for further complexation. For the open ocean environment ferric iron availability by the secretion of siderophores was shown for phytoplankton (Trick et al. 1983), as well as for bacteria (Trick 1989). Kuma et al. (1994) showed that total natural organic Fe^{3+}-chelators are abundant in open ocean regions of the eastern Indian Ocean and the western North Pacific Ocean where they control the dissolved iron concentration. Recent experiments have explored the mechanisms of the dissolution of iron oxyhydroxides under variable light and chelator conditions and it was concluded that the interplay of siderophores and light controls the overall process (Borer et al. 2005).

7.4   The Early Diagenesis of Iron in Sediments

The fundamental work by Froelich et al. (1979) established a conceptual model for the organic matter respiration in marine sediments which has been modified, verified and extended in numerous aspects since then. Froelich and colleagues found a succession of electron acceptors used by dissimilatory bacteria according to their energy gain. Consequently, a biogeochemical zonation of the sediment results where O_2, NO_3^-, bioavailable Mn(IV) and Fe(III) and SO_4^{2-} diminish successively with depth. Apart from the consumption of electron acceptors the production of reduced species such as NH_4^+, Mn^{2+}, Fe^{2+}, HS^- and CH_4 occurs. These components may be reoxidized abiotically under given thermodynamic conditions. As will be shown, these reoxidation reactions can also be microbiologically catalyzed. It is important to note that for the investigation of iron reactivity a differentiation of biotic and abiotic reactions is inherently important but often very difficult to achieve. Another considerable question with respect to the iron reactivity in sediments concerns the bioavailable fraction of iron bearing minerals. So far, it could be shown that ferric iron of iron oxides as well as certain sheet silicates can be used by dissimilatory iron reducing bacteria but their quantities and rates of reduction vary significantly. A final discussion of this section will compare different depositional environments with respect to the importance of dissimilatory iron reduction, chemical reduction and the availability of ferric iron.

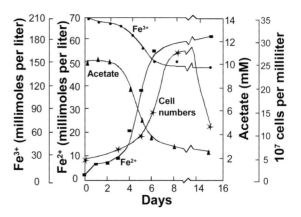

Fig. 7.8 Iron reduction by GS-15 with acetate as electron donor (adopted from Lovley and Philips 1988).

7.4.1   Dissimilatory Iron Reduction

Iron (and manganese) can be reduced enzymatically by various pathways, which are summarized in great detail by Lovley (1991). One of the prerequisites is the direct contact between the bacteria and solid phase ferric iron (Munch and Ottow 1982). The following given equations shall be considered as representative for a series of reactions within each group.

1. Fermentative Fe^{3+}-reduction:

$$C_6H_{12}O_6 + 24Fe^{3+} + 12H_2O$$
$$\Rightarrow 6HCO_3^- + 24Fe^{2+} + 30H^+ \qquad (7.1)$$

2. Sulfur-oxidizing Fe^{3+}-reduction:

$$S^0 + 6Fe^{3+} + 4H_2O$$
$$\Rightarrow HSO_4^- + 6Fe^{2+} + 7H^+ \qquad (7.2)$$

3. Hydrogen-oxidizing Fe^{3+}-reduction:

$$H_2 + 2Fe^{3+} \Rightarrow 2H^+ + 2Fe^{2+} \qquad (7.3)$$

4. Organic-acid-oxidizing Fe^{3+}-reduction:

$$acetate^- + 8Fe^{3+} + 4H_2O$$
$$\Rightarrow 2HCO_3^- + 8Fe^{2+} + 9H^+ \qquad (7.4)$$

5. Aromatic-compound-oxidizing Fe^{3+}-reduction:

$$phenol + 28Fe^{3+} + 17H_2O$$
$$\Rightarrow 6HCO_3^- + 28Fe^{2+} + 34H^+ \qquad (7.5)$$

For the case of fermentative Fe(III)-reduction it is important to note that during fermentation Fe(III) only serves as a minor sink for electrons and that organic substrates are primarily transformed to organic acids or alcohols.

A review of organisms reducing Fe(III), the respective electron donors and the applied forms of Fe(III) is given by Lovley (1987) and Lovley et al. (1997). In Figure 7.8 a typical result of ferric iron reduction mediated by a distinct ferric iron reducing bacteria (i.e. *Geobacter metallireducens*) and an appropriate electron donor (i.e. acetate) is shown. Note that the reproduction (cell numbers) ceases although acetate and ferric iron are still present. This is a general observation, which is explained by the limited bioavailability of the ferric iron fraction (section 7.4.2.2). Figure 7.9 represents a model for the degradation of natural organic matter with ferric iron as sole electron acceptor (Lovley 1991). The primary organic matter is hydrolyzed to smaller compounds such as sugars, amino and fatty acids, and aromatic

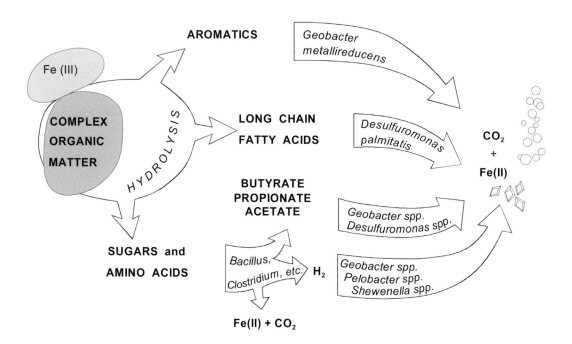

Fig. 7.9 Schematic model for the oxidation of complex organic matter with ferric iron as sole electron acceptor (adopted from Lovley 1997).

compounds through the activity of hydrolytic enzymes which are produced by a variety of microorganisms. Subsequently, these intermediate products can be used for the dissimilatory ferric iron reduction.

7.4.2   Solid Phase Ferric Iron and its Bioavailability

Solid phase ferric iron is present in various iron bearing minerals and amorphous phases of marine sediments. Two major groups can be distinguished: Iron oxyhydroxides (including iron oxides) and (sheet) silicates. As the iron bearing mineral assemblage varies considerably among different depositional environments the microbiological availability of specific Fe(III)-bearing compounds can be highly variable.

7.4.2.1   Properties of Iron Oxides

During terrestrial weathering a minor amount of the released iron from silicates (biotite, pyroxene, amphibole, olivine) is incorporated into clay minerals and a major fraction serves for the formation of iron oxides. Among iron oxides goethite (α-FeOOH) and hematite (α-Fe$_2$O$_3$) are the most abundant and are mostly associated with each other. Lepidocrocite (γ-FeOOH), maghemite (γ-Fe$_2$O$_3$) and magnetite (Fe^{2+}Fe$_2$$^{3+}O_4$) are generally quantitatively less abundant. Yet, with respect to the magnetization of the sediment magnetite and maghemite are of great importance

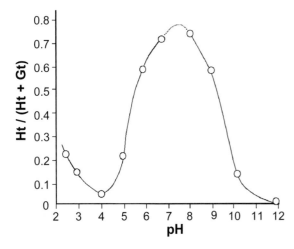

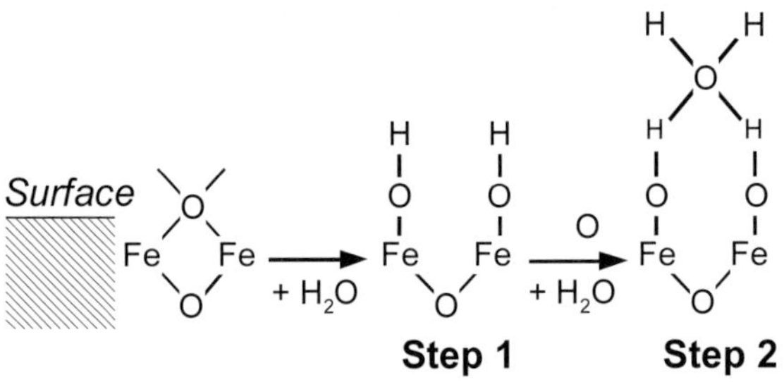

Fig. 7.11 In the presence of water the iron oxide surface is hydroxylated (step 1) and subsequently H$_2$O is adsorbed (step 2) (redrawn from Schwertmann and Taylor 1989).

due to their ferrimagnetic character (chapter 2). Iron phases precipitating from solution are collectively called iron oxyhydroxide. By means of mineralogical identification methods (X-ray diffraction, infrared and Mössbauer spectroscopy) the precipitate may be completely amorphous representing a ferric gel or it may be a poorly crystallized, water containing phase such as ferrihydrite (5Fe$_2$O$_3$·9H$_2$O). Formerly this fraction was called 'amorphous Fe(III)hydroxide' (Böhm 1925). Aging causes these earliest precipitates to increase their crystallinity, which means an increase in the ordering of the crystal lattice. Depending on the pH-value of ambient water the resulting proportion of goethite versus hematite varies. Hematite is favored under seawater conditions (Schwertmann and Murad 1983; Fig. 7.10). The effect of aging also explains the observation of a decrease in the highly reactive Fe-oxide fraction with water depth and distance to the coast although the total fraction of Fe-oxides and total iron concentration increases (Haese et al. 2000). With respect to the adsorption of anions and cations as well as to organic ligand formation (see below) it is important to know an approximate dimension of the specific surface area of iron oxides / oxyhydroxide. Crosby et al. (1983) determined specific surface areas for iron oxyhydroxide synthesized under natural conditions revealing 159-234 m^{2}g^{-1} for precipitates from Fe^{3+} solutions and 97-120 m^{2}g^{-1} for precipitates from Fe^{2+} solutions. Natural samples showed a range from 6.4-164 m^{2}g^{-1}.

Physical, chemical and mineralogical properties of the iron oxides can be variable. Aluminium (Al(III)) may substitute isomorphically for Fe(III) due to very similar ionic radii (Fe(III): 0.73 Å; Al(III): 0.61 Å) and the same valence. Norrish and

Fig. 7.10 Recrystallization products of a ferrihydrite suspension after 441 days. Aging causes a formation of hematite (Ht) and goethite (Gt) depending on the pH of ambient water (adopted from Schwertmann and Murad 1983).

Taylor (1961) found a maximum of one-third mole percent Al substitution for Fe within goethite and a maximum of one-sixth substitution was determined for hematite in soils (Schwertmann et al. 1979). As a product of terrestrial weathering magnetites usually contain minor amounts of Ti(IV) (ionic radius: 0.69 Å) as a substitution for Fe which is then called titano-magnetite. In contrast, bio-mineralized magnetite is a pure iron oxide. Different colors of iron oxides / oxyhydroxide are immediately apparent in nature. This holds true for different pure iron oxides, for different grain sizes of one oxide (e.g. goethite) as well as for distinct substitutions for Fe, e.g. by Mn or Cr (Schwertmann and Cornell 1991). The color of synthetic minerals and natural sediment can be quantitatively determined by reflectance spectroscopy (e.g. Morris et al. 1985).

In the presence of water the surface of an iron oxide is completely hydroxylated which can be understood as a two-step reaction which is shown schematically in Fig. 7.11.

As iron oxides / oxyhydroxide have a very high affinity for the adsorption of anions as well as cations under natural conditions the early diagenetic reactivity of iron is often of great significance for the behavior of compounds such as trace metals, phosphate and organic acids. The adsorption on iron oxides is caused by the hydroxylation of the mineral surface (S-OH). Depending on the pH of ambient water protonation or deprotonation occurs according to

$$S\text{-}OH + H^+ \quad \Leftrightarrow \quad S\text{-}OH_2^+ \qquad (7.6)$$

$$S\text{-}OH + OH^- \quad \Leftrightarrow \quad S\text{-}O^- + H_2O \qquad (7.7)$$

As a result, the surface charge and the surface potential vary depending on the concentration of H^+ ions in solution. Apart from the pH the surface charge is influenced by the concentration of the electrolyte and the valence of ions in solution. pH-values where the net surface charge is zero are called points of zero charge (pzc). For pure synthetic iron oxides these vary between a pH of 7 and 9. An excess of positive or negative charge is balanced by the equivalent amount of anions or cations. As representative for the adsorption of cations (Cu^{2+}), anions ($H_2PO_4^-$) and organic compounds (oxalate) the following reactions are given:

$$\begin{matrix} \text{-S-OH} \\ | \\ \text{-S-OH} \end{matrix} + Cu^{2+} \Leftrightarrow \begin{matrix} \text{-S-O} \\ | \quad Cu \\ \text{-S-O} \end{matrix} + 2H^+ \qquad (7.8)$$

$$\begin{matrix} \text{-S-OH} \\ | \\ \text{-S-OH} \end{matrix} + H_2PO_4^- \Leftrightarrow \begin{matrix} \text{-S-O}\quad O^- \\ | \quad P \\ \text{-S-O} \quad O \end{matrix} + 2H_2O \qquad (7.9)$$

$$\text{-S-OH} + H_2PO_4^- \Leftrightarrow \text{-S} \begin{matrix} O\text{-}C=O^- \\ | \\ O\text{-}C=O \end{matrix} + H_2O \qquad (7.10)$$

As the complexation of cations causes a release of H^+ and anions compete with surface bound hydroxyl groups ('ligand exchange') the adsorption is strongly pH-dependant (see above). As a result, anions are preferably adsorbed at lower pH-values whereas cations are primarily adsorbed at higher pH values (Fig. 7.12).

7.4.2.2 Bioavailability of Iron Oxides

As dissimilation is a biological process used to gain electrons for cell energetic functions, the energy gain from the induced reactions is noteworthy. The following reactions and their respective standard free energy, $\Delta G°$, values are given to provide an overview of the energy gain due to the reaction with various iron oxyhydroxide / oxides.

$$4Fe(OH)_3 + CH_2O + 7H^+$$
$$\Rightarrow 4Fe^{2+} + HCO_3^- + 10H_2O \qquad (7.11)$$

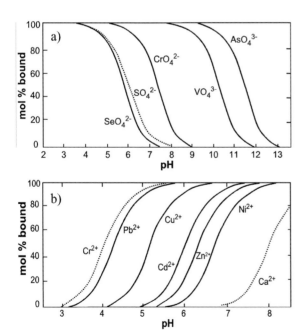

Fig. 7.12 The pH dependence of anion (a) and cation (b) sorption on hydrous ferric oxide (adopted from Stumm and Morgan 1996).

ΔG^0: (-376) - (-228) kJ mol^{-1}

$$4FeOOH + CH_2O + 7H^+$$
$$\Rightarrow 4Fe^{2+} + HCO_3^- + 6H_2O \qquad (7.12)$$

ΔG^0: (-387) - (-239) kJ mol^{-1}

$$2\alpha\text{-}Fe_2O_3 + CH_2O + 7H^+$$
$$\Rightarrow 4Fe^{2+} + HCO_3^- + 4H_2O \qquad (7.13)$$

ΔG^0: -236 kJ mol^{-1}

$$2Fe_3O_4 + CH_2O + 11H^+$$
$$\Rightarrow 6Fe^{2+} + HCO_3^- + 6H_2O \qquad (7.14)$$

ΔG^0: -328 kJ mol^{-1}

ΔG^0 values were adopted from Stumm and Morgan (1996). For Fe(OH)$_3$ and FeOOH a range of free energy is given as the solubility products $K_s = [Fe^{3+}] [OH^-]^3$ range from $10^{-37.3}$ to $10^{-43.7}$ depending on the mode of preparation, age and molar surface. For example, aged goethite (α-FeOOH) reveals a ΔG^0_f of -489 kJ mol^{-1} whereas for freshly precipitated amorphous FeOOH a value of -462 kJ mol^{-1} was determined (Stumm and Morgan 1996). The variability of FeOOH thermodynamic properties is a result of the metastability of freshly precipitated ferric iron phases (chapter 7.4.2.1, Fig. 7.10).

By means of microbial cultures growing on the various iron oxides as terminal electron acceptors it has been shown that they all can be used for the purpose of dissimilation (e.g. Lovley 1991; Kostka and Nealson 1995). Yet, it was generally found that the ferric iron phases were reduced at different rates and to varying degrees. Consequently, Ottow (1969) determined the following sequence of biological availability: FePO$_4$·4H$_2$O > Fe(OH)$_3$ > lepidocrocite (γ-FeOOH) > goethite (α-FeOOH) > hematite (Fe$_2$O$_3$). Note that the Fe-P phase may sequester significant amounts of P out of the nutrient loaded freshwater part of an estuary, but under brackish and marine conditions this phase is not stable as it rapidly transforms under the influence of free sulfide (Hyacinthe and Van Cappellen, 2004). Within natural soil samples a preferential reduction of amorphous to crystalline iron oxides was found (Munch and Ottow 1980). Later, Roden and Zachara (1996) discovered the importance of the solid phase specific surface area for the degree of reducibility. The greater the specific surface area was the more reduced ferric

iron was determined (Fig. 7.13). Additionally, they could show that after rinsing the solid phase, which could not be reduced any further, dissimilatory reduction of the original ferric iron was continued. They concluded that the reduction of iron oxides is limited by the adsorption of some components, possibly Fe^{2+}, which inhibits further microbial access. This finding also explains the results of Munch and Ottow (1980, see above) because amorphous and poorly crystalline phases are usually characterized by a higher specific surface area than well crystallized phases, and the greater the specific surface area the more can be adsorbed. This early finding agrees with the results of a recent study, which identified nano-phase goethite with diameters < 12 nm to be the predominant authigenic iron oxyhydroxide in freshwater and marine environments (van der Zee et al. 2003).

Once a framework for the availability of iron oxides is established, the kinetics of individual reactions provides insight into reaction rates and rate limiting steps for the overall reactivity of iron. Here, the kinetics of microbial iron oxide reduction is explored and in section 7.4.4.1 analog information are provided for the reduction by sulfide and ligands. Building on previous experimental results demonstrating the control of mineral surface area for the degree of iron reduction (Roden and Zachara 1996; Fig. 7.13), it was shown, that also the rate of microbial iron reduction in natural sediments is of first-order and controlled by the mineral surface area (Roden and Wetzel

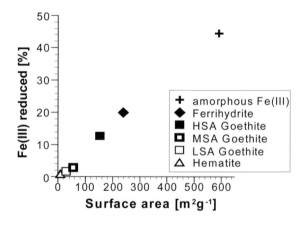

Fig. 7.13 The dependence of solid phase ferric iron reducibility from the specific surface area of iron oxyhydroxide / oxides (redrawn from Roden and Zachara 1996). HSA, MSA, and LSA represent high, medium, and low specific surface area.

2002). Kinetic experiments with synthetic iron oxyhydroxides have shown that the initial microbial reduction rate increases with increasing initial ferric iron concentration up to a given maximum reduction rate (Bonneville et al. 2004). This observation was explained by a saturation of active membrane sites with Fe(III) centers. The respective reaction was best described with a Michaelis-Menten rate expression with the maximum reduction rate per cell positively correlating with the solubility of the iron oxyhydroxides (Bonneville et al. 2004). Kinetic studies involving iron are not only inherently important to describe reaction pathways and to derive rate constants, which can be used in models. Kinetic studies also increasingly focus on iron isotopic fractionation to better understand the iron isotopic composition of ancient sediments, which may assist in the reconstruction of paleo-environments. Importantly, iron isotope fractionation occurs in abiotic and biotic processes; the degree of isotopic fractionation depends on individual reaction rates and the environmental conditions, e.g. whether reactions take place within an open or closed system (Johnson et al. 2004).

7.4.2.3 The Bioavailability of Sheet Silicate Bound Ferric Iron

As early as in 1972 Roth and Tullock published results on the chemical reduction of smectites. Subsequently, Rozenson and Heller-Kallai (1976a,b) studied the potential of reduction and reoxidation in various dioctahedral smectites with the aid of different reducing and oxidizing agents. Concurrent with the change of the intercrystalline redox state a change of the smectite color was observed. Oxidized smectites showed a white or yellowish color whereas the reduced smectites revealed a greenish-grey or black color. To balance the intercrystalline charge (de-) protonation was postulated. Additionally, as a consequence of smectite reduction a decrease of the specific surface area and the swellability in water, as well as an increase of nonexchangeable Na^+ was found (Lear and Stucki 1989). A potential importance of smectite redox reactivity in natural sediments was first pointed out by Lyle (1983) for sediments of the eastern equatorial Pacific where a distinctive color transition from tan (above) green-gray (below) was found in near-surface sediments. König et al. (1997) conducted a high resolution Mössbauer-spectroscopy study for

one core from the Peru Basin which revealed a present-day reduction of 42 % of total iron which can be differentiated into an immobile fraction (36 %) consisting of smectite bound iron and a mobile fraction (16 %) which diffuses back into the oxidized upper sediment layer. Evidence for the bioavailability of ferric iron bound to smectites was given by Kostka et al. (1996). During culturing of an iron reducing bacteria (*Shewanella putrefaciens* strain MR-1) on smectite as a sole electron acceptor a reduction of Fe^{3+} by 15 % within the first 4 hours and a total of 33 % after two weeks occurred.

7.4.3 Iron Reactivity towards S, O_2, Mn, NO_3^-, P, HCO_3^-, and Si-Al

The (microbial) dissimilatory iron reduction was shown in the previous section. In this section the reactions with major oxidants and reductants will be introduced. Additionally, the interactions between iron and phosphorus, as well as the formation of siderite and iron-bearing sheet silicates will be pointed out briefly to show to the variety of reactions in marine sediments coupled the reactivity of iron.

7.4.3.1 Iron Reduction by HS⁻ and Ligands

Apart from the dissimilatory iron reduction (section 7.4.1) iron oxides can be dissolved by protons, ligands and reductants. Dissolution reactions by protons and ligands are generally considered to be the rate-determining step for weathering processes. For iron bearing minerals in marine sediments proton-promoted dissolution is of no importance due to prevailing neutral or slightly alkaline conditions. Ligands (e.g. oxalates and citric acid) are by-products of biological decomposition and dissolve iron oxides by primary surface complexation onto the iron oxide surface resulting in a weakening of the Fe-O bond which is followed by a detachment of Fe^{3+}-ligand. Reductive dissolution (e.g. by HS⁻, ascorbate, and dithionite) is characterized by primary surface complexation followed by an electron transfer from the reductant to ferric iron and detachment of Fe^{2+}. The three pathways of Fe(III) (hydr)oxide dissolution are shown in Fig. 7.14.

Experimental determination of reduction rates (e.g. Pyzik and Sommer 1981; Dos Santos Afonso and Stumm 1992; Peiffer et al. 1992) reveal rates under well defined conditions and information on the reaction kinetics. Under natural conditions the

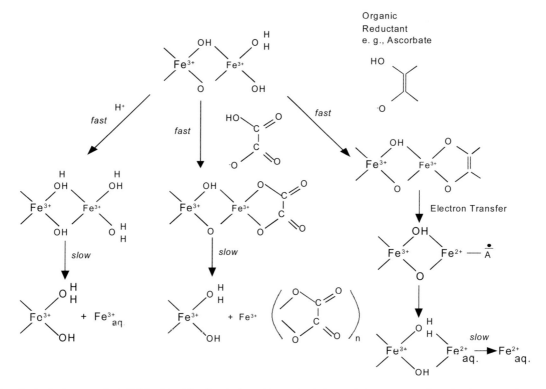

Fig. 7.14 Pathways of Fe(III)(hydr)oxide dissolution. From the left to right: Proton-, ligand- (here: oxalate), and reductant- (here: ascorbate) promoted dissolution is initiated by surface complexation. The subsequent step of detachment ($Fe^{3+}_{aq.}$, Fe^{3+}-ligand, $Fe^{2+}_{aq.}$) is rate determining. Note that the shown pathways of dissolution are fundamental for the described extractions (chapter 7.5) (adopted from Stumm and Morgan, 1996).

composition of dissolved compounds in solution is complex and may have significant influence on the dissolution rate. Biber et al. (1994) demonstrated the inhibition of reductive dissolution by H_2S and the ligand-promoted dissolution by EDTA due to the presence of oxoanions (e.g. phosphate, borate, arsenate) (Fig. 7.15 a,b).

A sequence of iron mineral reactivity towards sulfide was proposed more than 10 years ago (Canfield et al. 1992), which was later extended (Raiswell and Canfield 1996), and recently revised (Poulton et al. 2004). The revised sequence particularly accounts for the mineral surface area and can be grouped into two fractions excluding

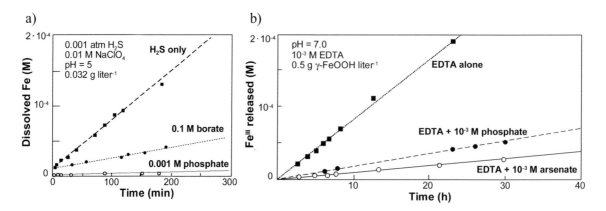

Fig. 7.15 Inhibition of reductive- and ligand-promoted dissolution of iron oxides by oxoanions. a: The dissolution of goethite by H_2S is inhibited by borate and phosphate. b: The dissolution of lepidocrocite by EDTA is inhibited by phosphate and arsenate (adopted from Stumm and Morgan 1996, original data from Biber et al., 1994).

Table 7.1 Reactivity of iron minerals towards sulfide (1000 iM ÓH$_2$S, pH 7.5, 25°C) according to 1: Poulton et al. (2004), 2: Canfield et al. (1992), and 3: Raiswell and Canfield (1996). The 'poorly-reactive silicate fraction' was determined operationally as (Fe$_{HCl, boiling}$ - Fe$_{Dithionite}$) / Fe$_{total}$

Iron Mineral / fraction	Half life, t$_{1/2}$
Hydrous Ferric Oxide[1]	5.0 minutes
2-line Ferrihydrite[1]	12.3 hours
Lepidocrocite[1]	10.9 hours
Goethite[1]	63 days
Magnetite [1]	72 days
Hematite [1]	182 years
Sheet silicates[2]	10 000 years
poorly-reactive silicate fraction[3]	2.4 x 10^6 years

iron bearing silicates: The poorly crystalline hydrous ferric oxide, 2-line ferrihydrite and lepidocrocite are reactive on a time scale of minutes to hours, whereas goethite, hematite and magnetite are reactive on a time scale of tens of days. In Table 7.1 the reactivity of iron oxyhydroxides and iron bearing silicate minerals towards sulfide is expressed as half-life (t$_{1/2}$).

The presence of iron minerals and their respective reactivity towards sulfide is of greatest importance for the pore water chemistry and the limitation for pyrite formation. In case of reactive iron rich sediments dissolved iron may build-up in pore water and dissolved sulfide is hardly present although sulfate reduction occurs. In contrast, in sediments characterized by a low content of reactive iron dissolved sulfide can build-up instead of dissolved iron (Canfield 1989). The degree of pyritisation (DOP) was originally defined by Berner (1970) and was later modified by Leventhal and Taylor (1990) and Raiswell et al. (1994). DOP is now defined as:

$$DOP = \frac{Fe_{pyrite}}{Fe_{pyrite} + Fe_{dithionite-soluble}} \quad (7.15)$$

A DOP-value of 1 means a complete pyritisation of reactive iron, which has been found in sediments overlain by anoxic-sulfidic bottom water (Raiswell et al., 1988). In sediments exposed to sulfide for more than one million years silicate-bound iron has only been partially turned into pyrite (Raiswell and Canfield 1996). This observation is explained by an overall slow rate of pyritisation of silicate-bound iron,

which is influenced by the mineral assemblage, degree of crystallinity and grain size. This clearly states the range of silicate iron reactivity towards sulfide, which is influenced by the mineral assemblage, degree of crystallinity and grain size.

7.4.3.2   Iron Oxidation by O$_2$, NO$_3^-$, and Mn^{4+}

The reaction of dissolved Fe^{2+} with oxygen is known to be fast and its rate was determined in sea-water by Millero et al. (1987):

$$\frac{-d[FeII]}{dt} = \frac{K_H[O_{2\,aq.}]}{[H^+]^2} \cdot [FeII] \quad (7.16)$$

at 20 °C, k$_H$=3x10^{-12} mol min^{-1}liter^{-1}. At a temperature of 5 °C the rate decreases by about a factor of 10. As the oxidation rate (-d[FeII]/dt) is inversely proportional to the power of the proton concentration ([H$^+$]^2) the importance of the pH becomes obvious. The lower the pH (= - log [H$^+$]), the lower is the rate of ferrous iron oxidation. Therefore, within a ferrous iron solution with a very low pH-value, e.g. acidified with HCl, the reaction is so slow that oxidation under air atmosphere is negligible over weeks. Under pH neutral conditions this reaction is so fast that dissolved iron may only escape from the sediment into the bottom water if the oxygen penetration depth is very little or even anoxic bottom water conditions are given. The effects on iron, manganese, phosphate and cobalt fluxes during a controlled decrease of oxygen bottom water concentration and the importance of the diffusive boundary layer within a benthic flux-chamber (see chapter 3) were studied by Sundby et al. (1986). They could demonstrate that due to a decrease of diffusive oxygen flux into the sediment manganese release increased prior to iron according to thermodynamic predictions (Balzer 1982). Stirring within a flux-chamber controls the thickness of the benthic boundary layer and thus the diffusive flux of oxygen into the sediment. A decrease or even an interruption of stirring results in a significant increase of benthic efflux of redox-sensitive constituents such as iron and manganese.

Buresh and Moraghan (1976) showed the thermodynamic potential of ferrous iron oxidation by nitrate, yet the reaction is not spontaneous. In the presence of solid phase Cu(II), Ag(I), Cd(II) , Ni(II), and Hg(II) serving as catalysts ferrous iron can reduce nitrate rapidly (Ottley et al. 1997). Similarly, the formation of a Fe(II)-lepidocrocite (γ-FeOOH)

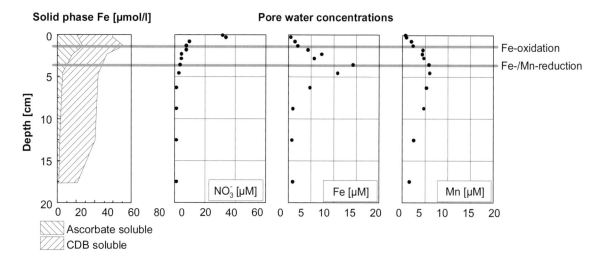

Fig. 7.16 Pore water and extraction results from hemipelagic sediments off Uruguay (redrawn from Haese et al. 2000). Dissolution and precipitation of Fe is reflected by the easy reducible iron oxyhydroxide fraction whereas less reducible iron oxides soluble by subsequent citrate/dithionite/bicarbonate (CDB) extraction remain constant. A concurrent liberation of Mn and Fe indicates dissimilatory iron reduction and subsequent iron reoxidation by manganese oxides, which results in the build-up of Mn^{2+}. Under these conditions the actual dissimilatory iron reduction rate is higher than deduced from iron pore water gradients.

surface complex was found to catalyze chemo-denitrification (Sørensen and Thorling 1991). The potential significance of microorganisms inducing ferrous iron oxidation was pointed out within the last years (Widdel et al. 1993; Straub et al. 1996). Ferrous iron was found to serve as electron donor in cultures of nitrate-reducing bacteria. Even in the presence of acetate as typical electron donor ferrous iron was additionally oxidized. This implies

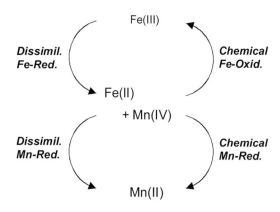

Fig. 7.17 Interaction of dissimilatory Fe / Mn reduction and abiotic reaction of Fe^{2+} with Mn(IV). Note that additional interactions with species and microbial processes typically occuring in surface sediments (e.g. sulfate reduction and subsequent reactions of HS⁻) are not considered and that Mn(IV) is not replenished.

that iron oxide formation typically occurring at the interface of nitrate and iron bearing pore water is at least in parts microbially mediated. Ehrenreich and Widdel (1994) have described a microbial mechanism of pure anaerobic oxidation by iron oxidizing photoautotrophs. This exciting observation challenges the conviction that the earliest iron oxidation on earth occurred during the build-up of free oxygen. One may now speculate that the accumulation of the Banded Iron Formation (Archaic age, ~ 3 billions years B.P.) was microbiologically induced under suboxic/anoxic conditions. Similarly, manganese oxidation rates in natural environments were determined to be considerably higher than determined in laboratory studies under abiotic conditions implying a microbially mediated manganese oxidation (Thamdrup et al. 1994; Wehrli et al. 1995).

Ferrous iron oxidation by manganese oxide was found to be especially fast as long as no iron oxyhydroxide precipitates, which presumably blocks reactive sites on the manganese oxide surface (Postma 1985). The oxidation of ferrous iron by manganese oxide has been proven to be important for the interpretation of pore water profiles and the precipitation of authigenic phases (Canfield et al. 1993a; Haese et al. 2000; van der Zee 2005). In Fig. 7.16 pore water profiles of iron and manganese reveal concurrent liberation of the two elements which was attributed to dissimilatory iron reduction and subsequent iron reoxidation by manganese oxides which in turn results in the production of $Mn^{2+}_{aq.}$.

Similar to the possibility of concurrent reduction of sulfate and ferric iron by a culture of a single bacteria (Coleman et al. 1993; see section 7.4.3.4) other iron reducing bacteria were found to additionally maintain dissimilation with more than one electron acceptors under suboxic conditions (Lovley and Phillips 1988) or even under oxic conditions (Myers and Nealson 1988a). In the presence of Fe(III) and Mn(IV) strain MR-1 was found to reduce both but additional manganese reduction occurred due to the immediate abiotic reaction with released Fe^{2+} (Myers and Nealson 1988b). The interactions of biotic and abiotic reactions are shown in Fig. 7.17.

7.4.3.3   Iron-bound Phosphorus

In section 7.4.2.1 the theoretical significance of phosphate adsorption onto iron oxides was illustrated. Numerous studies on natural sediments suggest that iron oxides control phosphate pore water and solid phase concentrations, as well as the overall sedimentary phosphate cycle (Krom and Berner 1980; Froelich et al. 1982; Sundby et al. 1992; Jensen et al. 1995; Slomp et al. 1996a,b). A generalized representation of the sedimentary phosphorus cycle is shown in Fig. 6.11. Apart from the Fe-bound P, organic P and authigenic carbonate fluorapatite are the principal carriers of solid phase P. HPO_4^{2-} is the predominant dissolved P species under sea water conditions (Kester and Pytkowicz 1967).

The following information and simple calculation allows the reader to assess and understand the important role of iron oxyhydroxides and their interactions with phosphorous: A maximum of 2.5 - 2.8 $\mu mol\ m^{-2}$ of adsorbed phosphate on iron oxides were found (Goldberg and Sposito 1984; Pena and Torrent 1984). In order to approximate a maximum adsorbed phosphate concentration in the sediment one can assume 1 cm^3 of sediment with a porosity of 75 %, a dry weight density of 2.65 g cm^{-3}, 50 $\mu mol/g_{Sediment}$ Fe bound to iron oxides and an iron oxide specific surface area of 120 $m^2\ g^{-1}$ (see section 7.4.2.1). For the wet sediment we can calculate an iron concentration of 33 $\mu mol\ cm^{-3}$ which is bound to iron oxides. This fraction has a specific surface area of ~ 0.22 m^2 within 1 cm^3 of wet sediment which may then adsorb up to ~ 0.57

μmol P (assuming an adsorption capacity of 2.6 μmol P per square meter of iron oxide). For a comparison of adsorbed and dissolved phosphate concentration we can furthermore assume a concentration of 5 μM phosphate within the interstitial water which is equivalent to 0.0037 $\mu mol\ cm^{-3}$ of wet sediment. Consequently, the adsorbed fraction of phosphate can be more than two orders of magnitude greater than the dissolved fraction due to the presence of iron oxides. In reality, the ratio of Fe bound to poorly crystalline iron oxides and P bound by these phases has been determined to be ~ 10Fe : 1P for coastal and shelf sediments (Slomp et al., 1996a). This differs significantly from the given theoretical sample in which we find a ratio of 58 (33 μmol Fe : 0.57 μmol P per 1 cm^3). Either P is additionally bound in the crystalline lattice of the iron oxides (i.e. Torrent et al., 1992) or the adsorption capacity for P in shelf sediments is much greater than derived from the calculated example. In the latter case, one must conclude that the specific surface area of Fe oxides is higher than assumed as another quantitatively important adsorbent of P in sediments is not likely. Instead of 120 $m^2\ g^{-1}$ iron oxide one needs to encounter a specific surface area of 650-700 $m^2\ g^{-1}$ to justify such high P adsorption.

7.4.3.4   The Formation of Siderite

In the marine environment siderite ($FeCO_3$) is hardly found relative to iron sulfides because it is thermodynamically not stable in the presence of even low dissolved sulfide activities. Postma (1982) proved the calculation of the solubility equilibrium between siderite and ambient pore water chemistry to be a reliable approach for the investigation of present-day siderite formation. In salt marsh sediments where the influence of salt and fresh water varies temporarily and spatially the formation of siderite and pyrite are closely interlinked. Mortimer and Coleman (1997) demonstrated that siderite precipitation is microbiologically induced. They could show that $\delta^{18}O$ values of siderite precipitated during the culturing of one specific iron-reducing microorganism, *Geobacter metallireducens*, were distinctively lower than expected according to equilibrium fractionation between siderite and water (Carothers et al. 1988) which is a contradiction to a pure thermodynamically induced reaction.

In fully marine systems siderite formation is probable to occur below the sulfate reduction zone where dissolved sulfide is absent, if reactive iron is still present and the Fe/Ca-ratio of pore water is high enough to stabilize siderite over calcite (Berner 1971). The coexistence of siderite and pyrite in anoxic marine sediments was shown by Ellwood et al. (1988) and Haese et al. (1997). Both studies attribute this observation to the presence of microenvironments resulting in different characteristic early diagenetic reactions next to each other within the same sediment depth. It appears that in one microenvironment sulfate reduction and the formation of pyrite is predominant, whereas at another site dissimilatory iron reduction and local supersaturation with respect to siderite occurs. Similarly, the importance of microenvironments has been pointed out for various other processes (Jørgensen 1977; Bell et al. 1987; Canfield 1989; Gingele 1992).

Apart from microenvironments, an explanation for the concurrent dissimilatory sulfate and iron reduction was provided by Postma and Jakobsen (1996). They demonstrated that the stabilities of iron oxides are decisive with respect to iron and/or sulfate reduction assuming that the fermentative step and not the overall energy yield is overall rate limiting. Additionally, it shall be noted that the typical sulfate reducing bacteria *Desulfovibrio desulfuricans* was found to reduce iron oxide enzymatically contemporarily or optionally (Coleman et al. 1993). When only very small concentrations of H_2 as sole electron donor were available iron oxide instead of sulfate was used as electron acceptor by *D. desulfuricans*.

7.4.3.5　The Formation of Iron Bearing Aluminosilicates

In 1966 the formation of aluminosilicates in marine environments was hypothesized by Mackenzie and Garrels (1966) who pointed out the potential significance of this process with respect to the oceanic chemistry and for global elemental cycles. As elements are transferred into solid phase and thus become insoluble this process is referred to 'reverse weathering'. Within the scope of this textbook only a brief overview of the major processes and conditions of formation is intended to be outlined.

Four major pathways for the formation of iron bearing aluminosilicates can be distinguished:

1. Formation from weathered basalt and volcanic ashes
2. Glauconite formation
3. Formation in the vicinity of hydrothermal vents
4. Formation under low temperature conditions

The first two pathways of formation will not be discussed here as they were found to be only of local/regional importance and are not considered to be of major importance for early diagenetic reactions. Iron bearing clay mineral formation under high-temperature conditions near a hydrothermal system of the Red Sea was studied by Bischoff (1972). A direct precipitation of an iron-rich smectite (nontronite) within the metalliferous sediments was found. This pathway of clay mineral formation was shown to occur at temperatures

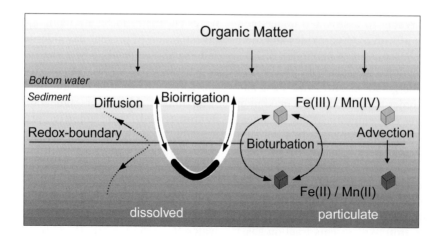

Fig. 7.18　Modes of transport in the sediment: molecular diffusion, bioirrigation, bioturbation, and advection.

typically ranging between 70 and 150 °C under oxic and anoxic conditions (Cole and Shaw 1983). Yet, lowest temperatures of formation were deduced to be ~ 20 °C (McMurtry et al 1983; Singer et al. 1984). Experimental studies by Decarreau et al. (1987) demonstrated the synthesis of dioctahedral smectite, containing Fe(III) within the octahedral sheet, only under strictly oxic conditions.

Experimental results by Harder (1976, 1978) gave evidence for the potential of iron bearing clay mineral formation under low temperature deep-sea floor conditions. Subsequent findings of sediments rich in montmorillonites in the north equatorial Pacific (Hein et al. 1979) and nontronite in the Bauer Deep of the eastern equatorial Pacific (Cole and Shaw 1983; Cole 1985) were attributed to authigenic aluminosilicate formation. For the formation of low-temperature iron-bearing aluminosilicates the deposition of skeletal opal (e.g. radiolarian) and iron oxyhydroxide (e.g. precipitation products of hydrothermal activity) as well as a low carbonate content are considered (Cole and Shaw 1983). Enhanced opal dissolution due to the presence of high iron oxide concentrations were reported (Mayer et al. 1991), yet kinetic reasoning of this observation remains unclear. As the skeletal opal closely associated with the iron oxyhydroxide becomes buried it dissolves and forms an amorphous Fe(III)-silica complex at the skeleton surface which subsequently recrystallizes to form nontronite on the surface of the partially dissolved skeletons (Cole 1985). By this analogy, Harder (1976,1978) also found an amorphous Fe(III)-silicate precipitate as a precursor which developed during aging under suboxic conditions into a crystalline iron-rich clay mineral. The presence of Fe^{2+} was a prerequisite for the synthesis of clay minerals under experimental conditions and therefore partial reduction of iron oxyhydroxide within the microenvironment of an opal skeleton must be assumed. The oxygen isotopic composition ($\delta^{18}O$) of the authigenic mineral can be used to reconstruct the prevailing temperature during formation by applying the geothermometric equation of Yeh and Savin (1977). For the aluminosilicates from the north equatorial Pacific and the Bauer Deep formation temperatures of ~ 3-4 °C were deduced representing authigenic formation under low-temperature conditions in deep-sea sediments. Similar to the above described deep-sea conditions, the Amazon delta represents an iron and silicate rich

depositional environment. Incubation experiments with sediments from the Amazon delta revealed substantial K-Fe-Mg-clay mineral formation within 1-3 years under low-temperature conditions (Michalopoulos and Aller 1995) implying significant elemental transfer into solid phase within the estuarine mixing zone.

7.4.4   Iron and Manganese Redox Cycles

Processes of early diagenesis can only be understood by integrating biogeochemical reactions and modes of transport in the sediment. With respect to the quantification of iron and manganese reactions molecular diffusion and bioirrigation need to be considered for the dissolved phase whereas bioturbation and advection are relevant for the particulate transport (Haese 2002). Bioirrigation is the term for solute exchange between the bottom water and tubes in which macro-benthic organisms actively pump water. For iron and manganese a recent study (Hüttel et al. 1998) points out the importance of solute transport in the sediment and across the sediment / bottom water interface due to pressure gradients induced by water flow over a rough sediment topography. Advection in the context of particulate transport describes the downward transport of particles relative to the sediment surface due to sedimentation. Strictly speaking, bioturbation (sometimes more generally termed mixing) also induces a vertical transport of dissolved phase. Yet, as molecular diffusive transport is usually much greater than dissolved transport by bioturbation the later is usually neglected.

The cycling of reduced and oxidized iron and manganese species are discussed together in this chapter since the driving processes are principally the same. In Fig. 7.18 the operating modes of transport are shown schematically along with a redox boundary. Above this boundary the reactive fraction of total solid phase iron or manganese is present as oxidized species whereas below the reduced species occur. Note that at this boundary a build-up in the pore water occurs if no immediate precipitation (e.g. FeS) or adsorption inhibits a release into ambient water. The change in the redox state implies oxidation above and reduction below by some electron donor / acceptor. In case of dissimilatory iron / manganese reduction the organic carbon serves as electron donor, the other most important oxidants and reductants are discussed in the sections 7.4.3.1 and 7.4.3.2 .

The intensity of the redox cycling and thus the importance for oxidation and reduction reactions in the sediment is terminated by either one of the following conditions: 1. In case of the absence of any efficient oxidant (e.g. O_2) in the upper-most layer or bottom water no oxidation will occur and the redox cycling cannot be maintained. 2. In case of the absence of a reactive fraction (bioavailable or 'rapidly' reducible by HS^-, see section 7.4.3.1) in the lower layer no reduction will occur and the redox cycle will cease. 3. A vertical transport mode must be maintained between the zone of oxidation and the zone of reduction. As advection is usually very much slower than the downward transport by bioturbation the intensity of bioturbation terminates the transport between the redox-zones.

For the most simple assumption of an homogeneously mixed layer the intensity of bioturbation is expressed by the biodiffusion (or mixing) coefficient, D_b, which can be deduced appropriately along with the sedimentation rate with the aid of natural radioactive isotopes. According to Nittrouer et al. (1983/1984) the general advection-diffusion equation can be rearranged to calculate the sedimentation rate, A:

$$A = \frac{\lambda x}{\ln \frac{C_0}{C_x}} - \frac{D_b}{x}\left(\ln \frac{C_0}{C_x}\right)$$

(7.17)

with λ: decay constant [y^{-1}], x: depth interval between two levels [cm], C_0, C_x: activity at an upper sediment level and at a lower level with the distance x below C_0 [decays per minute, dpm] D_b: biodiffusion coefficient [$cm^2\ y^{-1}$]. If mixing is negligible ($D_b = 0$) then the above equation can be simplified:

$$A = \frac{\lambda x}{\ln \frac{C_0}{C_x}}$$

(7.18)

In case of a very low sedimentation rate relative to mixing ($A^2 \ll \lambda \cdot D_b$) Eq. 7.17 can be rearranged to calculate the biodiffusion coefficient, D_b:

$$D_b = \lambda \left(\frac{x}{\ln \frac{C_0}{C_x}}\right)^2$$

(7.19)

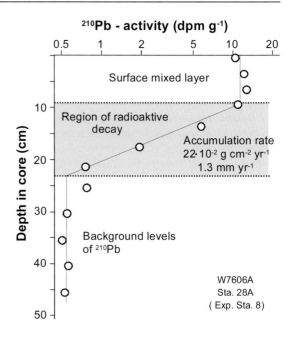

^{210}Pb - activity (dpm g^{-1})

Surface mixed layer

Region of radioaktive decay

Accumulation rate
$22 \cdot 10^{-2}$ g cm^{-2} yr^{-1}
1.3 mm yr^{-1}

Background levels of ^{210}Pb

W7606A
Sta. 28A
(Exp. Sta. 8)

Depth in core (cm)

Fig. 7.19 ^{210}Pb-activity depth profile from the Washington shelf (adopted from Nittrouer 1983/1984). The sediment surface layer is mixed as can be deduced from homogenous ^{210}Pb values. Below, constantly decreasing values imply no or hardly any mixing, this gradient can be used to calculate a sedimentation rate. The deepest part is characterized by a homogenous background activity resulting from the decay of ^{226}Ra in the sediment.

The above restrictions for the calculations of the sedimentation rate and the biodiffusion coefficient imply the use of radioactive isotopes with different half-lifes ($t_{1/2} = 0.693 \cdot \lambda^{-1}$) for different purposes and depositional environments. The higher the sedimentation rate, the shorter should be the half-life of the radioactive isotope. The more intense bioturbation in the surface layer, the shorter should be the half-life of the applied radioactive isotope be. For coastal and shelf sediments sedimentation rates of several decimeters to few meters per 1000 years are typical and can be determined by ^{210}Pb ($t_{1/2} = 22.3$ y). Shorter lived isotopes (e.g. $t_{1/2}$ of ^{234}Th = 24.1 d) are applicable for the determination of the mixing intensity. ^{230}Th ($t_{1/2} = 75,200$ y) is a commonly used radioactive isotope in oceanographic sciences to trace processes over longer periods of times. The above isotopes are rapidly scavenged by particles once they are formed from the decay of some parent isotopes and settle to the sea floor. Due to analytical reasons post-depositional processes can be traced for a time period of 4 to 5 times of the radioactive half-life which is approximately 100 years in case of ^{210}Pb.

In Figure 7.19 a typical ^{210}Pb-activity depth profile from the Washington shelf is shown. The uppermost 9 cm are characterized by constant ^{210}Pb activity implying intensive mixing in the surface layer. Below, ^{210}Pb activity decreases linearly (on a log-scale) indicating no mixing and continuous decay. The lowest part of the profile is characterized by constantly very low values resulting from the long-term decay of ^{226}Ra to ^{210}Pb in the sediment. This background value is subtracted from the above activities, which are then termed excess-^{210}Pb or unsupported-^{210}Pb. The depth interval showing a linear decrease on a log-activity scale is often used to calculate a sedimentation rate. Yet, a strong overestimation of the true sedimentation rate is possible as slight deep bioturbation in this part may not be seen exclusively by ^{210}Pb as it was shown by Aller and DeMaster (1984). The investigation of an additional, shorter-lived isotope within this depth interval will reveal a potential influence of bioturbation.

As mentioned in the beginning of this section bioturbation and advection by sedimentation cause particle transport in the sediment. In Fig. 7.20 three scenarios are schematically shown representing constant molecular diffusive and advective transport while bioturbation varies. As a result, the shape of the solid phase profile varies distinctively. In case of no bioturbation the molecular diffusive flux ($J_{diff.}$) from the zone of dissolution into the zone of precipitation causes a thin, sharp peak (enrichment) (Fig. 7.20a) whereas slight bioturbation and thus vertical up- and down-transport of particles (J_p) broadens the enrichment (Fig. 7.20b). In case of very intense particle transport relative to the molecular diffusive transport ($J_p \gg J_{diff.}$, Fig. 7.20c) hardly any or no enrichment will be formed although a distinctive depth of precipitation is still present. In summary, we can conclude that the solid phase profile is a result of the dissolution within the lower part of the enrichment, as well as of the sedimentation rate and of the bioturbation (mixing) intensity. This can be expressed mathematically by a one-dimensional transport-reaction model according to Aller (1980). If bioturbation and sedimentation with depth (no compaction) are constant, steady-state conditions apply (chapter 3), and solid phase decreases linearly over the interval of dissolution, then

$$P = -\frac{(C_1 - C_2)}{(x_2 - x_1)} \cdot D_b + A \cdot (C_1 - C_2)$$

$$(7.20)$$

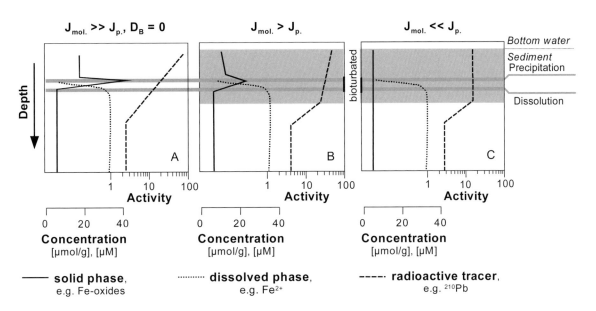

Fig. 7.20 To illustrate the influence of bioturbation (mixing) on the solid phase profile three schematic scenarios are drawn. For all scenarios the same molecular diffusive transport ($J_{mol.}$) and sedimentation is assumed, yet the particulate transport (J_p) by bioturbation is varied. A: As no bioturbation occurs a distinctive, thin solid phase enrichment is formed in the zone of precipitation. B: The enrichment broadens up- and downwards as slight bioturbation is present. C: In case of a much higher particulate transport relative to the diffusive transport ($J_p \gg J_{mol.}$) hardly any enrichment will be formed.

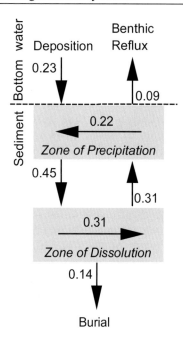

Fig. 7.21 Example of manganese cycling across the sediment/bottom water interface and within the sediment (modified after Sundby and Silverberg (1985). The applied depth-dependent flux model is described in the text. Depositional, burial, and molecular diffusive fluxes as well as the reduction rate within the zone of dissolution were calculated independently.

can be calculated according to Sundby and Silverberg (1985). P resembles the production (or dissolution) rate and the other variables are according to Eq. 7.17 except that C is given as concentration per volume [μmol cm^{-3}] as the depth distribution of solid phase strongly depends on the porosity.

In addition to the dissolution rate one can calculate the burial rate of non-reactive phase and the input rate to the sediment surface once the

sedimentation rate is known. Based on these independently calculated fluxes Sundby and Silverberg (1985) developed a depth-zonated flux model for manganese in the St. Lawrence estuary. Their depth-dependent reactive zones were surface water, bottom water, sediment depth of precipitation (oxidation), sediment depth of dissolution (reduction) and depth of eventually buried sediment. One example of their Mn-cycling results is given in Fig. 7.21.

The cycling of elements in bioturbated surface sediments can also be expressed in terms of turn-over times defined as period of time required for a complete oxidation - reduction cycle of the reactive fraction. Additional consideration of the bioturbation depth and the sedimentation rate then reveals the number of redox-cycles before ultimate burial. In Tab. 7.2 representative results for estuarine (coastal) and slope sediments are given.

7.4.5   Discussion: The Importance of Fe- and Mn-Reactivity in Various Environments

The above sections of this chapter have shown the high variability of iron-input modes, fluxes and reactivity towards oxidized and reduced species in marine sediments. Within this section the importance of iron and manganese reactivity with respect to the mineralization of organic matter as well as to the chemical oxidation (reduction) of reduced (oxidized) species within different depositional environments will be discussed and hopefully inspire further considerations.

In order to investigate the importance of iron and manganese reduction and oxidation processes one needs to determine their rates and compare

Table 7.2 Calculated turn-over times and times of redox cycling before burial in coastal and slope sediments. The dynamic of redox cycling becomes evident by envisaging a complete oxidation - reduction cycle on a 2 - 6 months time scale.
([1] Sundby and Silverberg 1985, [2] Aller 1980, [3] Canfield et al. 1993a, [4] Thamdrup and Canfield 1996)

Location	Fe/Mn	Turn-over time [d]	Times cycled before burial
St. Lawrence estuary [1]	Mn	43 - 207	
Long Island Sound estuary [2]	Mn	60 - 100	
Skagerrak [3]	Fe/Mn	70 - 250	130 - 300
Slope off Chile [4]	Fe	70	31 - 77

Table 7.3 Summary of results quantifying the relative contribution of dissimilatory iron and manganese reduction for the decomposition of organic matter.
([1] Wang and Van Cappellen 1996, [2] Canfield et al. 1993b, [3] Canfield et al. 1993a, [4] Aller 1990, [5] Thamdrup and Canfield 1996).

Location		Net-Fe/Mn deposition [$\mu mol\ cm^{-2}y^{-1}$]	Biodiff. coeff. [$cm^{-2}y^{-1}$]	Dissimilatory Fe/Mn reduction [%]	Total org. C decomposition [%]
Skagerrak, S4/S6	Fe	6 - 14 [3]	80 - 87 [2]		32 - 51 [2]
	Fe	13 - 24 [1]		71 - 84 [1]	
	Mn	1 - 15 [1]		3 - 24 [1]	0 [2]
Skagerrak, S9	Mn	5 - 10 [3]	19 [2]		90 [2]
	Mn	13 [1]		100 [1]	
	Fe	14 [1]		0 [1]	0 [2]
Panama Basin [4]	Mn	'high'	100	100	100
Cont. Slope Chile [5]	Fe	5.1	9; 29	46; 84	12; 29

these with other early diagenetic reaction pathways. For example, rates of organic carbon mineralization by each electron acceptor (O_2, NO_3^-, Mn(IV), Fe(III), SO_4^{2-}) are - strictly speaking - necessary to make a statement on the relative contribution of dissimilatory iron and manganese reduction. For the case of iron and manganese, the determination of such rates are problematic as both electron acceptors may be reduced by reduced species as well as by microbial respiration. Additionally, pore water fluxes usually strongly underestimate the true reduction rate as adsorption and precipitation of Fe^{2+}/Mn^{2+}-bearing minerals buffer the build-up within pore-water. Therefore, 'rates of dissimilatory Fe-oxide and Mn-oxide reduction are the least well quantified of the carbon oxidation pathways' (Canfield 1993).

An overview of methods to determine the various organic carbon oxidation pathways is provided by Canfield (1993). As these methods are technically highly demanding and time-consuming, only very few sediments have been investigated with respect to the contribution of the different organic carbon oxidation pathways. Among such studies different methods have been applied which may bear additional uncertainties. Thus, a present-day discussion on the importance of iron and manganese must be speculative to some degree. In Table 7.3 results of different studies concerning the relative contribution of dissimilatory iron and manganese reduction are summarized.

Notice that the results by Wang and Van Capellen (1996) are model results for which some results of Canfield et al. (1993 a,b) were used for the basic data set. A comparison of different locations reveals significant variabilities in the biodiffusion coefficient. For open ocean sediments one can expect even much lower deposition rates and biodiffusion coefficients. Similarly, the proportion of dissimilatory reduction (relative to dissimilatory plus chemical reduction) as well as the proportion of organic carbon mineralization by iron and manganese reduction (relative to total organic carbon oxidation) varies significantly.

The above sites of investigation are distinct by different depositional environments. Skagerrak sediments were retrieved from water depths of 200 (S4), 400 (S6), and 700 (S9) meters and site S9 is located in the central Norwegian Trough where solid phase manganese content made up to 3.5 - 4 wt%. The Panama Basin site is ~ 4000 m deep and is located in the equatorial upwelling region as well as in the vicinity of hydrothermal activity causing a delivery of large amounts of reactive organic matter and manganese to the sea floor. Sites from the continental slope off Chile were investigated during a period of intense upwelling. The intensity of redox-cycling is controlled by the intensity of bioturbation, the input and reduction of reactive Fe(III)/Mn(IV), as well as by the oxidation rate (section 7.4.4). Yet, Fe- and Mn-cycling may only become quantitatively significant if reduction is predominantly coupled to dissimilation and chemical reduction is unimportant. This is the case for the manganese dominated sites Skagerrak S9 and Panama Basin where neither the production of HS⁻ nor of Fe^{2+} was found. O_2 consumption was (almost) completely attributed to the reoxidation of Mn^{2+} in these cases. In contrast, Skagerrak sites S4 and S6 as well as sediments off Chile resemble situations where iron is reduced chemically (HS⁻) by up to ~ 50 % and only smaller amounts of ferric iron are available for the dissimilation. A third situation can be inferred for typical open ocean sediments where low Fe/Mn and organic matter deposition along with low bioturbation intensity occurs. Here organic matter decomposition is restricted to aerobic respiration and denitrification. Iron and manganese reduction rates are presumably negligible, yet over a long period of time a significant proportion of manganese is redistributed and the composition of iron oxide phases changes (Haese et al. 1998).

An extensive study in a shallow water, estuarine mixing zone elucidated ideal conditions for efficient manganese cycling (Aller 1994). Manganese turn-over was found to be most intense during warm periods with intensive bioturbation, well-oxygenated bottom water, and moderate organic matter input. Under such conditions manganese reoxidation consumed 30-50 % of the benthic oxygen flux and manganese reduction was mainly induced by HS⁻, FeS and FeS_2. As soon as bottom water became O_2-depleted the sedimentary Mn-cycle was reduced as dissolved Mn escaped out of the sediment. The influence of bioirrigation has not yet been explicitly investigated but modeling results by Wang and Van Capellen (1996) imply enhanced metal cycling efficiency with increasing irrigation due to more rapid Fe^{2+}- / Mn^{2+}-oxidation.

A conceptual model describing the importance of iron and manganese reactivity in different environments is only sketchy as results are scarce and manifold aspects deserve further investigations. A general complication concerns the differentiation of dissimilatory and chemical reduction of Mn-oxide when concurrent iron reduction is apparent and further investigations need to discriminate each reaction pathway. The reactivity of smectite-bound iron has only been shown qualitatively (Kostka et al. 1996; König et al. 1997), yet quantifications with respect to dissimilatory and/or chemical reactions are missing. An important role of adsorbed Fe^{2+} /

Mn^{2+} is indicated (Sørensen and Thorling 1991; Roden and Zachara 1996) but not proved by sediment studies. Investigations focusing on the importance of metal cycling and its influence on pathways of organic matter decomposition are scarce (Canfield et al. 1993a, Wang and Van Capellen 1996), yet necessary to understand an important link of the carbon cycle.

7.5   The Assay for Ferric and Ferrous Iron

In order to study iron reactivity qualitatively and quantitatively it is essential to quantify the ferrous and ferric iron fractions of the present minerals or mineral groups. With respect to the determination of iron speciation the principal problem is the rapid oxidation of ferrous iron. Atmospheric oxygen diffuses into pore water where it oxidizes dissolved ferrous iron 'immediately' and starts oxidizing FeS and FeS_2. Reduced smectites may also become oxidized under air atmosphere within hours. Therefore, if dissolved iron and iron speciation of solid phase are to be determined samples need to be conserved under inert gas atmosphere. No extra care is needed for the determination of total iron of solid phase.

Above a pH of 3 and in the absence of chelators dissolved iron is only present as ferrous iron under natural conditions. Therefore, the colored complex that results from the reaction between ferrous iron and a reagent can be analysed colorimetrically and correlates with the concentration of total dissolved iron. Most conveniently, one can mix a drop of Ferrozine® solution (Stookey 1970), one drop of H_2SO_4 (diluted 1:4) and 1 ml of pore water in the glove box, wait until complex formation is completed (20-30 minutes) and quantify the iron concentration by the intensity of the color at a wavelength of 562 nm. To avoid matrix effects standards should be prepared with artificial seawater.

The assay of solid phase ferrous and ferric iron has a long tradition due to the early interest in soil chemistry. Publications on extraction / leaching conditions and results from varying soils and sediments are extensive and thus, within the scope of a textbook, only important principals and a description of the (subjectively) most important extractions can be given. Since a great variety of the quantitatively most important iron bearing minerals / mineral groups is present in marine sediments a series of different extractions is necessary in order to achieve a complete distribution of ferrous and ferric iron. In general, the extraction conditions applied to natural sediments result from experiments conducted in advance proving the dissolution of individual minerals or mineral groups. Because the grain size, degree of crystallinity, ionic substitution within minerals and varying matrix constituents influence the dissolution kinetics during the extraction a clear-cut mineral specific determination is usually not possible with this approach. Yet, extractions have been successfully applied to show patterns of mineral (group) dissolution and precipitation and to deduce reaction rates. For a comparison with results of other studies exactly

Table 7.4 Solubility of iron bearing minerals derived under experimental conditions. + / - imply a solubility of $\geq 97\ \%$ / $\leq 3\ \%$, values indicate a percentage of release.

[1] Schwertmann (1964): 0.2 M NH_4^+-oxalate / 0.2 M oxalic acid; pH: 2.5, 2 h in darkness

[2] Ferdelman (1988): 10 g Na-citrate + 10 g Na-bicarbonate mixed in 200 ml distilled and deionized water, deaerated, before 4 g ascorbic acid are added; pH: 7.5, 24 h

[3] Lord (1980): 0.35 M acetate / 0.2 M Na-citrate + 1.0 g Na-dithionite for each sample (~ 1 g wet sediment in 20 ml solution); pH: 4.8, 4 hours.

[4] Chao and Zhou (1983): 1 M HCl, 30 min; Canfield (1988): 1 M HCl 20-23 h; Cornwell and Morse (1987): 1 M HCl, 45 min; Kostka and Luther (1994): 0.5 M, 1 h.

[5] Haese et al. (1997): 1 ml distilled and deionized water + 1 ml conc. H_2SO_4 + 2 ml HF were added to ~ 250 mg of wet sediment under inert gas atmosphere and constant stirring over few minutes.

[6] Canfield et al. (1986): 15 ml of O_2-free 1 M $CrCl_2$ in 0.5 M HCl + 10 ml of 12 M HCl under inert gas atmosphere.

[a] Chou and Zhou 1983, [b] Canfield 1988, [c] Kostka and Luther 1994, [d] Ruttenberg 1992, [e] Mehra and Jackson 1960, [f] Cornwell and Morse 1987, [g] Haese et al. 1997, [h] Canfield et al. 1986.

Mineral	Oxalate [1]	Ascorbate [2]	Dithionite [3]	HCl [4]	HF/H_2SO_4 [5]	Cr(II)/ HCl [6]
am. Fe(OH)₃	+ [a]			34 - 72 [a] + [b]		
Ferrihydrite	+ [b]	+ [c]	+ [b,c,d]	+ [b,c]		
Lepidocrocite	+ [b]		+ [b]	7 [b]		
Goethite	- [a,b]	- [c]	91 [c] + [b]	- [a,b,c]	+ [g]	
Hematite	- [a,b]	- [c]	63 [c] + [b,d,e]	- [a,b,c]	+ [g]	
Magnetite	60 [c]	- [c]	90 [c] - [b]	- [a,b,c]	+ [g]	
(am.) FeS	+ [c]			+ [f]		+ [h]
Pyrite (FeS₂)				- [b,f]		+ [h]
Chlorite	- [b,c]	- [c]	5 - 7 [c,b]	27 [c] 32 [b]	10 - 100 [g]	
Nontronite	- [b]		27 [b]	7 [b]		
Glauconite	- [b]		10 [b]	10 [b]		
Garnet	- [b]		- [b]	- [b]		

the same extraction conditions (reagent composition, sediment : solution ratio, contact time) must be applied.

Table 7.4 gives an overview of experimentally derived dissolution behavior of iron bearing minerals under some selected extraction conditions. The investigation of total-Fe from ascorbate and dithionite solution can be determined by ICP-AES or flame-AAS. Ferrous and ferric iron from non-reducing or non-oxidizing extractions can be determined by polarographic methods (Wallmann et al. 1993) or colorimetrically with and without the addition of a reducing agent (e.g. hydroxylamine hydrochloride, Kostka and Luther 1994). During the acidic extractions evolving sulfide can be trapped in a separate alkaline solution (e.g. Sulfur Antioxidant Buffer, SAOB, Cornwell and Morse 1987) where it can be determined polarographically, by precipitation titration with Pb or by a standard ion sensitive electrode. Sulfide evolving from HCl extraction is called Acid Volatile Sulfur (AVS).

The leaching with HF/H_2SO_4 as described above and the subsequent polarographic determination of ferrous and ferric iron is based on work by Beyer et al. (1975) and Stucki (1981) in order to quantify the silicate bound ferrous and ferric iron. This extraction has hardly been applied with respect to questions of early diagenesis so far, yet, the silicate bound iron fraction is quantitatively very important in marine sediments and even a small reactive fraction of this pool may be of overall significance for the iron reactivity. As a complementary method to the commonly applied extractions (Table 7.4) it renders the calculation of the total iron speciation in the sediment which may then be compared to Mössbauer-spectroscopic results (Haese et al. 1997; Haese et al. 2000).

One of the pitfalls in the interpretation of extraction results from natural sediments is caused by the fact that the presence of Fe^{2+} complexed by carboxylic acid catalyzes the reduction of crystalline iron oxides such as hematite (Sulzberger et al. 1989), magnetite (Blesa et al. 1989) and goethite (Kostka and Luther 1994). In order to avoid this catalytic dissolution of well-crystallized iron oxides by Fe^{2+} during the oxalate extraction Thamdrup and Canfield (1996) air-dried the sediment in advance, thereby oxidizing FeS and $FeCO_3$ to ferrihydrite. In addition, they applied the anoxic oxalate extraction and subtracted the released amount of Fe^{2+} from the amount of Fe^{3+} determined from the oxic extraction to calculate the poorly crystallized iron oxide fraction as intended according to Table 7.2.

Acknowledgements
I wish to thank Tim Ferdelman, Bo Thamdrup and Caroline Slomp for their critical reviews in the year 2000 leading to the first edition of this manuscript. The recent edition was improved by comments by Lynda Radke and Emmanuelle Grosjean, and it received permission for publication by the Chief Executive Officer of Geoscience Australia.

7.6   Problems

Problem 1

The Burdekin River is large river in northwestern Australia, which discharges about 3.4 million tons of sediment into the coastal sea.
a) Estimate how much total and highly reactive iron is discharged from the catchments. Assume that the sediment is of average continental crust composition. b) Estimate how much highly reactive iron remains in the estuary and in the near-coastal zone. c) Which other source of iron must be considered off-shore of northern Australia?

Problem 2

a) List 3 mineral properties of iron oxyhydroxides which influence the rate of microbial iron reduction. b) Explain why wet-chemical extractions are not mineral specific.

Problem 3

Explain why bioturbation is important for the rate of dissimilatory iron reduction in sediments.

Problem 4

Which are important variables affecting the importance of dissimilatory Fe- and Mn-reduction relative to other metabolic pathways?

Problem 5

a.) Define and explain the abbreviation DOP.
b.) Which extractions are used to determine DOP.

References

Aller, R.C., 1980. Diagenetic processes near the sediment-water interface of Long Island Sound. 2. Fe and Mn. Advances in Geophysics, 22: 351-415.

Aller, R.C. and DeMaster, D.J., 1984. Estimates of particle flux and reworking at the deep-sea floor using ^{234}Th/^{238}U disequilibrium. Earth and Planetary Science Letters, 67: 308-318.

Aller, R.C., 1990. Bioturbation and manganese cycling in hemipelagic sediments. Philosophical Transactions of the Royal Society of London A, 331: 51-68.

Aller, R.C., 1994. The sedimentary Mn cycle in Long Island Sound: Its role as intermediate oxidant and the influence of bioturbation, O_2, and C_{org} flux on diagenetic reaction balances. Journal of Marine Research, 52 (2): 259-295.

Balzer, W., 1982. On the distribution of iron and manganese at the sediment/water interface: thermodynamic vs. kinetic control. Geochimica Cosmochimica Acta, 46: 1153-1161.

Bell, P.E., Mills, A.L. and Herman, J.S., 1987. Biogeochemical conditions favoring magnetite formation during anaerobic iron reduction. Applied and Environmental Microbiology, 53: 2610-2616.

Berger, W.H., Smetacek. V.S. and Wefer, G., 1989. Ocean productivity and paleoproductivity - An overview. In: Berger, W.H., Smetacek, V.S. and Wefer, G. (eds), Productivity of the oceans: present and past. John Wiley and Sons, Chichester, 1-34.

Berner, R.A., 1970. Sedimentary pyrite formation. American Journal of Science, 268: 1-23.

Berner, R.A., 1971. Principles of chemical sedimentology. McGraw-Hill, New York., 240 pp.

Beyer, M.E., Bond, A.M. and McLaughlin, R.J.W., 1975. Simultaneous polarographic determination of ferrous, ferric and total iron in standard rocks. Analytical Chemistry, 47 (3): 479-482.

Biber, M.V., Dos Santos Afonso, M. and Stumm, W., 1994. The coordination chemistry of weathering: IV: Inhibition of the dissolution of oxides minerals. Geochimica Cosmochimica Acta, 58 (9): 1999-2010.

Bischoff, J.L., 1972. A ferroan nontronite from the Red Sea geothermal system. Clays and Clay Minerals, 20: 217-223.

Blank, M., Leinen, M. and Prospero J.M., 1985. Major Asian aeolian inputs indicated by the mineralogy of aerosols and sediments in the western North Pacific. Nature, 314: 84-86.

Blesa, M.A., Marinovich, H.A., Baumgartner, E.C., and Marota, A.J.G., 1987. Mechanism of dissolution of magnetite by oxalic acid - ferrous ion solution. Inorganic Chemistry, 26: 3713-3717.

Böhm, J., 1925. Über Aluminium- und Eisenoxide I. Zeitschrift der Anorganischen Chemie, 149: 203-218.

Bonneville, S., Van Cappellen, P., and Behrends, T., 2004. Microbial reduction of iron(III) oxyhydroxides: effects of mineral solubility and availability. Chemical Geology, 212: 255-268.

Borer, P.M., Sulzberger, B., Reichard, P., and Kraemer, S.M., 2005. Effect of siderophores on the light-induced dissolution of colloidal iron(III) (hydr)oxides. Marine Chemistry, 93: 179-193.

Broecker, W.S., Spencer, D.W., Craig, H., 1982. GEOSECS Pacific Expedition: Hydrographic Data. U.S. Goverment Printing Office, Washington, DC, 3: 137 pp.

Bruland, K.W., Rue, E.L., Smith, G.J., and DiTullio, G.R., 2005. Iron, macronutrients and diatom blooms in the Peru upwelling regime: brown and blue waters of Peru. Marine Chemistry, 93: 81-103.

Buresh, R.J. and Moraghan, J.T., 1976. Chemical reduction of nitrate by ferrous iron. Journal of Environmental Quality, 5: 320-325.

Canfield, D.E., Raiswell, R., Westrich, J.T., Reaves, C.M. and Berner, R.A., 1986. The use of chromium reduction in the analysis of reduced inorganic sulfur in sediments and shales. Chemical Geology, 54: 149-155.

Canfield, D.E., 1988. Sulfate reduction and the diagenesis of iron in anoxic marine sediments. Ph.D. thesis. Yale Univ., 248 pp.

Canfield, D.E., 1989. Reactive iron in marine sediments. Geochimica Cosmochimica Acta, 51: 619-632.

Canfield, D.E., Raiswell, R. and Botrell, S., 1992. The reactivity of sedimentary iron minerals towards sulfide. American Journal of Science, 292: 659-683.

Canfield, D.E., 1993. Organic matter oxidation in marine sediments. In: Wollast, R., Mackenzie, F.T. and Chou, L. (eds) Interactions of C, N, P and S biogeochemical cycles and global change. NATO ASI Series, 4, Springer, Berlin, Heidelberg, NY, pp. 333-363.

Canfield, D.E., Thamdrup, B. and Hansen, J.W., 1993a. The anaerobic degradation of organic matter in Danish coastal sediments: Iron reduction, manganese reduction, and sulfate reduction. Geochimica Cosmochimica Acta, 57: 3867-3883.

Canfield, D.E., Jørgensen, B.B., Fossing, H., Glud, R., Gundersen, J., Ramsing, N.B., Thamdrup, B., Hansen, J.W., Nielsen, L.P. and Hall, P.O.J., 1993b. Pathways of organic carbon oxidation in three continental margin sediments. Marine Geology, 113: 27-40.

Canfield, D.E., 1997. The geochemistry of river particles from the continental USA: Major elements. Geochimica Cosmochimica Acta, 61: 3349-3365.

Carlson, T.N. and Prospero, J.M., 1972. The large-scale movement of Saharan air outbreaks over the northern equatorial Atlantic. Journal of Applied Meteorology, 11: 283-297.

Carothers, W.W., Adami, L.H. and Rosenbauer, R.J., 1988. Experimental oxygen isotope fractionation between siderite-water and phosphoric acid liberated CO_2-siderite. Geochimica Cosmochimica Acta, 52: 2445-2450

Chavez, F.P. and Barber, R.T., 1987. An estimate of new production in the equatorial Pacific. Deep Sea Research, 34: 1229-1243.

Chester, R., 1990. Marine Geochemistry. Chapman & Hall, London, pp. 698.

Chou, T.T. and Zhou, L., 1983. Extraction techniques for selective dissolution of amorphous iron oxides from soils and sediments. Soil Science Society American Journal, 47: 225-232.

Cole, T.G. and Shaw, H.F., 1983. The nature and origin of authigenic smectites in some recent marine sediments. Clay Minerals, 18: 239-252.

Cole, T.G., 1985. Composition, oxygen isotope geochemistry, and the origin of smectite in the metalliferous sediments of the Bauer Deep, southeast Pacific. Geochimica Cosmochimica Acta, 49: 221-235.

Coleman, M.L., Hedrick, D.B., Lovley, D.R., White, D.C. and Pye, K., 1993. Reduction of Fe(III) in sediments by sulphate-reducing bacteria. Nature, 361: 436-438.

Cornwell, J.C. and Morse, J.W., 1987. The characterization of iron sulfide minerals in anoxic marine sediments. Marine Chemistry, 22: 193-206

Crosby, S.A., Glasson, D.R., Cuttler, A.H., Butler, I., Turner, D.R., Whitfield, M. and Millward, G.E., 1983. Surface areas and porosities of Fe(III)-Fe(II)-derived oxyhydroxides. Environmental Science and Technology, 17: 709-713.

De Angelis, M., Barkov, N.I., Petrov, V.N., 1987. Aerosol concentrations over the last climatic cycle (160 kyr) from Antarctic ice core. Nature, 325: 318-321.

De Baar, H.J.W. and Suess, E., 1993. Ocean carbon cycle and climate change - An introduction to the interdisciplinary union symposium. Global and Planetary Change, 8: VII-XI.

Decarreau, A., Bonnin, D., Badauth-Trauth, D., Couty, R. and Kaiser, P., 1987. Synthesis and crystallogenesis of ferric smectite by evolution of Si-Fe coprecipitates in oxidizing conditions. Clay Minerals, 22: 207-223.

Donaghay, P.L., 1991. The role of episodic atmospheric nutrient inputs in chemical and biological dynamics of oceanic ecosystems. Oceanography, 4: 62-70.

Dos Santos Afonso, M. and Stumm, W. 1992, Reductive dissolution of iron(III)(hydr)oxides by hydrogen sulfide. Langmuir, 8: 1671-1676.

Duce, R.A., Liss, P.S., Merill, J.T., Atlas, E.L., Buat-Menard, P., Hicks, B.B., Miller, J.M., Prospero, J.M., Arimoto, T., Church, T.M., Eillis, W., Galloway, J.N., Hansen, L., Jickells, T.M., Knap, A.H., Reinhardt, K.H., Schneider, B., Soudine, A., Tokos, J.J., Tsunogai, S., Wollast, R. and Zhou, M., 1991. The atmospheric input of trace species to the world ocean. Global Biogeochemical Cycles 5: 193-259.

Ehrenreich, A. and Widdel, F., 1994, Anaerobic oxidation of ferrous iron by purple bacteria, a new type of phototrophic metabolism. Applied and Environmental Microbiology, 60: 4517-4526.

Ellwood, B.B., Chrzanowski, T.H., Hrouda, F., Long, G.J. and Buhl, M.L., 1988. Siderite formation in anoxic deep-sea sediments: A synergetic bacterially controlled process with important implications in paleomagnetism. Geology, 16: 980-982

Ferdelman, T.G., 1980. The distribution of sulfur, iron, manganese, copper, and uranium in a salt marsh sediment core as determined by a sequential extraction method. Masters thesis, University Delaware.

Figuères, G., Martin, J.M. and Meybeck, M., 1978. Iron behaviour in the Zaire estuary. Netherlands Journal of Sea Research, 12 (3/4): 329-337.

Froelich, P.N., Klinkhammer, G.P., Bender, M.L., Luedtke, N.A., Heath, G.R., Cullen, D., Dauphin, P., Hammond, D. and Hartman, B., 1979. Early oxidation of organic matter in pelagic sediments of eastern equatorial Atlantic: suboxic diagenesis. Geochimica Cosmochimica Acta, 43: 1075-1090.

Froelich, P.N., Bender, M.L., Luedtke, N.A., Heath, G.R. and DeVries, T., 1982. The marine phosphorus cycle. American Journal of Science, 282: 474-511.

GESAMP (Group of Experts on the Scientific Aspects of Marine Pollution), 1987. Land/sea boundary flux of contaminants: Contributions from rivers. GESAMP Rep. Stud., 32: 172 pp.

Gingele, F., 1992. Zur klimaabhängigen Bildung biogener und terrigener Sedimente und ihre Veränderung durch die Frühdiagenese im zentralen und östlichen Südatlantik (in German). Berichte, 26, Fachbereich Geowissenschaften, Universität Bremen, 202 pp.

Goldberg, S. and Sposito, G., 1984. A chemical model of phosphate adsorption by soils. I. Reference oxide minerals. Soil Scienc Society American Journal, 48: 772-778.

Haese, R.R., Wallmann, K., Kretzmann, U., Müller, P.J. and Schulz, H.D., 1997. Iron species determination to investigate the early diagenetic reactivity in marine sediments. Geochimica Cosmochimica Acta, 61 (1): 63-72.

Haese, R.R., Petermann, P., Dittert, L. and Schulz, H.D., 1998. The early diagenesis of iron in pelagic sediments - a multidisciplinary approach. Earth and Planetary Science Letters, 157: 233-248.

Haese, R.R., Schramm, J., Rutgers van der Loeff, M.M. and Schulz, H.D., 2000. A comparative study of iron and manganese diagenesis in continental slope and deep sea basin sediments off Uruguay (SW Atlantic). International Journal of Earth Sciences, 88: 619-629.

Haese, R.R., 2002. Macrobenthic activity and its effects on biogeochemical reactions and fluxes, In: Wefer, G., Billet, D., Hebbeln, D., Jørgensen, B.B., Schlüter, M. and van Weering, T.C.E. (eds), Ocean margin systems, Springer-Verlag, Heidelberg-Berlin, 219-234.

Harder, H., 1976. Nontronite synthesis at low temperatures. Chemical Geology, 18: 169-180

Harder, H., 1978. Synthesis of iron layer silicate minerals under natural conditions. Clays and Clay Minerals, 26: 65-72.

Hart, T.J., 1934. On the phytoplankton of the south-west Atlantic and the Bellinghausen Sea, 1929-31. Discovery Reports VIII.

Hein, J.R., Yeh, H.-W. and Alexander, E., 1979. Origin of iron rich montmorillonite from the manganese nodule belt of the north equatorial Pacific. Clays and Clay Mineralogy, 27: 185-194.

Hunter, K.A., 1983. On the estuarine mixing of dissolved substances in relation to colloid stability and surface properties. Geochimica Cosmochimica Acta, 47: 467-473.

Hüttel, M., Ziebis, W., Forster, S. and Luther, G.W. III, 1998. Advective transport affecting metal and nutrient distributions and interfacial fluxes in permeable sediments. Geochimica Cosmochimica Acta, 62: 613-631.

Hyacinthe, C., and Van Cappellen, P., 2004. An authigenic iron phosphate phase in estuarine sediments: composition, formation and chemical reactivity. Marine Chemistry, 91: 227-251.

Jensen, H.S., Mortensen, B.P., Andersen, F.Ø., Rasmussen, E. and Jensen, A., 1995. Phosphorus cycling in a coastal marine sediment, Aarhus Bay, Denmark. Limnology and Oceanography, 40: 908-917.

Johnson, K.S., Coale, K.H., Elrod, V.A. and Tindale, N.W., 1994. Iron photochemistry in seawater from the equatorial Pacific. Marine Chemistry, 46: 319-334.

Johnson, K.S., Gordon, R.M. and Coale, K.H., 1997. What controls dissolved iron concentrations in the world ocean? Marine Chemistry, 57: 137-161.

Johnson, C.M., Beard, B.L., Roden, E.E., Newman, D.K., and Nealson, K.H., 2004. Isotopic constraints on biogeochemical cycling of Fe. In: Geochemistry of non-traditional stable isotopes, Eds: Johnson, C.M., Beard, B.L., and Albarède, F., Reviews in Mineralogy & Geochemistry, 55: 359-408.

Jørgensen, B.B., 1977. Bacterial sulfate reduction within reduced microniches of oxidized marine sediments. Marine Biology, 41: 7-17.

Kester, D.R. and Pytkowicz, R.M., 1967. Determination of apparent dissociation constants of phosphoric acid in sea water. Limnology and Oceanography, 12: 243-252.

Kostka, J.E. and Luther, G.W. III, 1994. Partitioning and speciation of solid phase iron in saltmarsh sediments. Geochimica Cosmochimica Acta, 58: 1701-1710.

Kostka, J.E. and Nealson, K.H., 1995. Dissolution and reduction of magnetite by bacteria. Environmental Science and Technology, 29: 2535-2540.

Kostka, J.E., Stucki, J.W., Nealson, K.H. and Wu, J., 1996. Reduction of structural Fe(III) in smectite by a pure culture of Shewanella putrefaciens Strain MR-1. Clays and Clay Minerals, 44: 522-529.

König, I., Drodt, M., Suess, E., Trautwein, A.X., 1997. Iron reduction through the tan-green color transition in deep-sea sediments. Geochimica Cosmochimica Acta, 61: 1679-1683

Krauskopf, K.B., 1956. Factors controlling the concentration of thirteen trace metals in seawater. Geochimica Cosmochimica Acta, 12: 331-334.

Krom, M.D., Berner, R.A., 1980. Adsorption of phosphate in anoxic marine sediments. Limnology and Oceanography, 25: 797-806.

Kuma, K., Nishioka, J., Matsunaga, K., 1994. Controls of iron(III) hydroxide solubility in seawater: The influence of pH and natural organic chelators. Limnology and Oceanography, 41: 396-407.

Lear, P.R. and Stucki, J.W., 1989. Effects of iron oxidation state on the specific surface area of nontronite. Clays and Clay Minerals, 37: 547-552.

Leventhal, J. and Taylor, C., 1990. Comparison of methods to determine the degree of pyritisation. Geochimica Cosmochimica Acta, 54: 2621-2625.

Lord, C.J. III., 1980. The chemistry and cycling of iron, manganese, and sulfur in salt marsh sediments. Ph.D. thesis, University Delaware, 177 pp.

Lovley, D.R., 1987. Organic matter mineralization with the reduction of ferric iron: A review. Geomicrobiology Journal, 5: 375-399.

Lovley, D.R. and Phillips, E.J.P., 1988. Novel mode of microbial energy metabolism: Organic carbon oxidition coupled to dissimilatory reduction of iron and manganese. Applied and Environmental Microbiology, 54: 1472-1480

Lovley, D.R., 1991. Dissimilatory Fe(III) and Mn(IV) reduction. Microbiological Reviews 55: 259-287.

Lovley, D.R., 1997. Microbial Fe(III) reduction in subsurface environments. FEMS Microbiological Reviews, 20: 305-313.

Lovley, D.R., Coates, J.D., Saffarini, D. and Lonergan, D.J., 1997. Diversity of dissimilatory Fe(III)-reducing bacteria. In: Winkelman, G. and Carrano, C.J. (eds) Iron and Related Transition Metals in Microbial Metabolism, Harwood Academic Publishers, Switzerland, pp. 187-215.

Lyle, M., 1983, The brown-green color transition in marine sediments: A marker of the Fe(III)-Fe(II) redox boundary. Limnology and Oceanography, 28: 1026-1033.

Mackenzie, F.T. and Garrels, R.M., 1966, Chemical mass balance between rivers and oceans. American Journal of Science, 264: 507-525.

Martin, J.H., Gordon, R.M., Fitzwater, S.E., Broenkow, W.W., 1989. VERTEX: phytoplankton/iron studies in the Gulf of Alaska. Deep-Sea Research, 36: 649-680.

Martin, J.H., 1990. Glacial-interglacial CO_2 change: The iron hypothesis. Paleoceanography, 5: 1-13.

Martin, J.H., Gordon, R.M. and Fitzwater, S.E., 1991. The case for iron. In: Chisholm, S.W. and Morel, F.M.M. (eds). What controls phytoplankton production in nutrient-rich areas of the open sea?, ASLO Symposium, Lake San Marcos, California, Feb. 22-24, 1991, Allen Press, Lawrence.

Martin, J.H., Coale, K.H., Johnson, K.S. and Fitzwater, S.E., 1994. Testing the iron hypothesis in ecosystems of the equatorial Pacific Ocean. Nature, 371: 123-129.

Mayer, L.M., Jorgensen, J., Schnitker, D., 1991. Enhencement of diatom frustule dissolution by iron oxides. Marine Geology, 99: 263-266.

McAllister, C.D., Parsons, T.R. and Strickland, J.D.H., 1960. Primary productivity and fertility at station "P" in the north-east Pacific Ocean. Journal du Conseil, 25: 240-259.

McMurtry, G.M., Chung-Ho, W. and Hsueh-Wen, Y., 1983. Chemical and isotopic investigation into the origin of clay minerals from the Galapagos hydrothermal mound field. Geochimica Cosmochimica Acta, 47: 291-300.

Mehra, O.P. and Jackson, M.L., 1960. Iron oxide removal from soils and clays by a dithionite-citrate system buffered with sodium carbonate. Proceedings of the National Conference on Clays and Clay Mineralogy, 7: 317-327.

Michalopoulos, P. and Aller, R.C., 1995. Rapid clay mineral formation in Amazon delta sediments: Reverse weathering and oceanic cycles. Science, 270: 614-617.

Millero, F.J., Sotolongo, S. and Izaguirre, M., 1987. The oxidation kinetic of Fe(II) in seawater. Geochimica Cosmochimica Acta, 51: 793-801.

Morris, R.V., Lauer, H.V. Jr., Lawson, C.A., Gibson, E.K. Jr., Nace, G.A. and Stewart C., 1985. Spectral and other physicochemical properties of submicron powders of hematite (α-Fe$_2$O$_3$), maghemite (γ-Fe$_2$O$_3$), magnetite (Fe$_3$O$_4$), goethite (α-FeOOH), and lepidocrocite (γ-FeOOH). Journal of Geophysical Research, 90: 3126-3144.

Mortimer, R.J.G., Coleman, M.L., 1997. Microbial influence on the oxygen isotopic composition of diagenetic siderite. Geochimica Cosmochimica Acta, 61: 1705-1711.

Munch, J.C. and Ottow, J.C.G., 1980. Preferential reduction of amorphous to crystalline iron oxides by bacterial activity. Journal of Soil Science,129: 15-21.

Munch, J.C. and Ottow, J.C.G., 1982. Einfluß von Zellkontakt und Eisen(III)oxidform auf die bakterielle Eisenreduktion. Zeitschrift der Pflanzenernährung und Bodenkunde, 145: 66-77.

Murray, R.W. and Leinen, M., 1993, Chemical transport to the seafloor of the equatorial Pacific across a latitudinal transect at 135°W: Tracking sedimentary major, minor, and rare earth element fluxes at the equator and the inner tropical convergence zone. Geochimica Cosmochimica Acta, 57: 4141-4163.

Myers, C.R. and Nealson, K.H., 1988a. Bacterial manganese reduction and growth with manganese oxide as the sole electron acceptor. Science, 240: 1319-1321.

Myers, C.R. and Nealson, K.H., 1988b. Micorbial reduction of manganese oxides: Interactions with iron and sulfur. Geochimica Cosmochimica Acta, 52: 2727-2732.

Nittrouer, C.A., DeMaster, D.J., McKee, B.A., Cutshall, N.H., Larsen, I.L., 1983/1984. The effect of sediment mixing on Pb-210 accumulation rates for the Washington continental shelf. Marine Geology, 54: 201-221.

Norrish, K., Taylor, R.M., 1961. The isomorphous replacement of iron by aluminium in soil goethites. Journal of Soil Sciences, 12: 294-306.

Ottley, C.J., Davison, W., Edmunds, W.M., 1997. Chemical catalysis of nitrate reduction by iron(II). Geochimica Cosmochimica Acta, 61: 1819-1828.

Ottow, J.C.G., 1969. Der Einfluss von Nitrat, Chlorat, Sulfat, Eisenoxidform und Wachstumsbedingungen auf das Ausmass der bakteriellen Eisenreduktion. Zeitschrift der Pflanzenernährung und Bodenkunde, 124: 238-253.

Peiffer, S., Dos Santos Afonso, M., Wehrli, B. and Gächter, R., 1992. Kinetics and mechanism of the reaction of H$_2$S with lepidocrocite. Environmental Science and Technology, 26: 2408-2412.

Pena, F. and Torrent, J., 1984. Relationships between phosphate sorption and iron oxides in alfisols from a river terrace sequence of Mediterranean Spain. Geoderma, 33: 283-296.

Postma, D., 1982, Pyrite and siderite formation in brackish and freshwater swamp sediments. American Journal of Science, 282: 1151-1183.

Postma, D., 1985. Concentration of Mn and separation from Fe in sediments. Kinetics and stoichiometry of the reaction between birnessite and dissolved Fe(II) at 10°C. Geochimica Cosmochimica Acta, 49: 1023-1033.

Postma, D., Jakobsen, R., 1996, Redox zonation: Equilibrium constraints on the Fe(II)/SO4-reduction interface. Geochimica Cosmochimica Acta, 60: 3169-3175.

Poulton, S.W. and Raiswell, R., 2002, The low-temperature geochemical cycle of iron: from continental fluxes to marine sediment deposition. American Journal of Science, 302: 774-805.

Poulton, S.W., Krom, M.D., and Raiswell, R., 2004. A revised scheme for the reactivity of iron (oxyhydr)oxide minerals towards dissolved sulfide. Geochimica Cosmochimica Acta, 68: 3703-3715.

Prospero, J.M., 1981. Eolian transport to the world ocean. In: Emiliani, C. (ed), The sea, 7, Wiley, New York, pp. 801-874.

Prospero, J.M., Glaccum, R.A. and Nees, R.T., 1981. Atmospheric transport of soil dust from Africa to South America. Nature, 289: 570-572.

Pyzik, A.J. and Sommer, S.E., 1981, Sedimentary iron monosulfides: kinetics and mechanisms of formation. Geochimica Cosmochimica Acta, 45: 687-698.

Raiswell, R., Buckley, F., Berner, R.A. and Anderson, T.F., 1988. Degree of pyritisation as a paleoenvironmental indicator of bottom water oxygenation. Journal of Sedimentary Petrology, 58: 812-819.

Raiswell, R., Canfield, D.E. and Berner, R.A., 1994. A comparison of iron extraction methods for the determination of degree of pyritisation and the recognition of iron-limited pyrite formation. Chemical Geology, 111: 101-110.

Raiswell, R. and Canfield, D.E., 1996. Rates of reaction between silicate iron and dissolved sulfide in Peru Margin sediments. Geochimica Cosmochimica Acta, 60: 2777-2787.

Rijkenberg, M.J.A., Fischer, A.C., Kroon, J.J., Geringa, L.J.A., Timmermans, K.R., Wolterbeek, H.Th., and de Baar, H.J.W., 2005. The influence of UV irradiation on the photoreduction of iron in the Southern Ocean. Marine Chemistry, 93: 119-129.

Roden, E.E. and Zachara, J.M., 1996. Microbial reduction of crystalline iron(III) oxides: Influence of oxides surface area and potential for cell growth. Environmental Science and Technology, 30: 1618-1628.

Roden, E.E., and Wetzel, R.G., 2002. Kinetics of microbial Fe(III) oxide reduction in freshwater wetland sediments. Limnology Oceanography, 47: 198-211.

Roth, C.B. and Tullock, R.J., 1972. Deprotonation of nontronite resulting from chemical reduction of structural ferric iron. Proceedings of the International Clay Conference, Madrid, 89-98.

Rozenson, I. and Heller-Kallai, L., 1976a. Reduction and oxidation of Fe^{3+} in dioctahedral smectites - 1: Reduction with hydrazine and dithionite. Clays and Clay Minerals, 24: 271-282

Rozenson, I. and Heller-Kallai, L., 1976b. Reduction and oxidation of Fe^{3+} in dioctahedral smectites - 2: Reduction with sodium sulphide solutions. Clays and Clay Minerals, 24: 283-288.

Rue, E.L. and Bruland, K.W., 1995. Complexation of Fe(III) by natural organic ligands in the Central North Pacific as determined by a new competitive ligand equilibration/ adsorptive cathodic stripping voltammetric method. Marine Chemistry, 50: 117-138.

Ruttenberg, K.C., 1992. Development of a sequential extraction method for different forms of phosphorous in marine sediments. Limnology and Oceanography, 37: 1460-1482.

Schwertmann, U., 1964. Differenzierung der Eisenoxide des Bodens durch photochemische Extraktion mit saurer Ammoniumoxalat-Lösung. Zeitschrift zur Pflanzenernährung und Bodenkunde, 195: 194-202.

Schwertmann, U. Fitzpatrick, R.W., Taylor, R.M. and Lewis, D.G., 1979. The influence of aluminium on iron oxides. Part II. Preparation and properties of Al substituted hematites. Clays and Clay Minerals, 11: 189-200.

Schwertmann, U. and Murad, E., 1983. Effect of pH on the formation of goethite and hematite from ferrihydrite. Clays and Clay Minerals, 31:277-284.

Schwertmann, U. and Taylor, R.M., 1989. Iron oxides. In: Dinauer, R.C. (ed), Minerals in soil environments, Soil Science Society of America Book Series, 1, Madison, WI, pp. 379-438.

Schwertmann, U. and Cornell, R.M., 1991. Iron oxides in the laboratory. VCH Verlagsgesellschaft mbH, Weinheim, 137 pp.

Singer, A., Stoffers, P., Heller-Kallai, L. and Szafranek, D., 1984. Nontronite in a deep-sea ore from the south Pacific. Clays and Clay Minerals, 32: 375-383

Slomp, C.P., Van der Gaast, S.J., Van Raaphorst, W., 1996a. Phosphorus binding by poorly crystalline iron oxides in North Sea sediments. Marine Chemistry, 52: 55-73.

Slomp, C.P., Epping, E.H.G., Helder, W. and Van Raaphorst, W., 1996b, A key role for iron-bound phosphorus in authigenic apatite formation in North Atlantic continental platform sediments. Journal of Marine Research, 54: 1179-1205.

Sørensen, J. and Thorling, L., 1991, Stimulation by lepidocrocite (γ-FeOOH) of Fe(II)-dependent nitrite reduction. Geochimica Cosmochimica Acta, 55: 1289-1294.

Stookey, L.L., 1970, Ferrozine - A new spectrophotometric reagent for iron. Analytical Chemistry, 42: 779-781.

Stucki, J.W., 1981, The quantitative assay of minerals for Fe^{2+} and Fe^{3+} using 1,10-phenanthroline: II. A photochemical method. Soil Science Society of America Journal, 45: 638-641.

Stumm, W. and Morgan, J.J., 1996. Aquatic chemistry, 3rd edition, Wiley & Sons, London, 1022 pp.

Straub, K.L., Benz, M., Schinck, B., Widdel, F., 1996, Anaerobic, nitrate-dependent microbial oxidation of ferrous iron. Applied and Environmental Microbiology, 62: 1458-1460.

Sulzberger, B., Suter, D., Siffert, C., Banwart, S., Stumm, W., 1989. Dissolution of Fe(III) hydroxides in natural waters: Laboratory assessment on the kinetics controlled by surface coordination. Marine Chemistry, 28: 127-144.

Sundby, B., Silverberg, N., 1985. Manganese fluxes in the benthic boundary layer. Limnology and Oceanography, 30: 372-381.

Sundby, B., Anderson, L.G., Hall, P.O.J., Iverfeldt, Å, Rutgers van der Loeff, M., Westerlund, S.F.G., 1986. The effect of oxygen on release and uptake of cobalt, manganese, iron, and phosphate at the sediment-water interface. Geochimica Cosmochimica Acta, 50: 1281-1288.

Sundby, B., Gobeil, C., Silcerberg, N., Mucci, A., 1992. The phosphorus cycle in coastal marine sediments. Limnology and Oceanography, 37: 1129-1145.

Thamdrup, B., Glud, R.N. and Hansen, J.W., 1994. Manganese oxidation and in situ manganese fluxes from a coastal sediment. Geochimica Cosmochimica Acta, 58: 2563-2570.

Thamdrup, B. and Canfield, D.E., 1996. Pathways of carbon oxidation in continental margin sediments off central Chile. Limnology and Oceanography, 41: 1629-1650.

Torrent, J., Barrón, V. and Schwertmann, U., 1992, Fast and slow phosphate sorption by goethite-rich natural materials. Clays and Clay Mineralogy, 40: 14-21.

Trick, C.G., Andersen, R.J., Gillam, A. and Harrison, P.J., 1983. Prorocentrin: An extracellular siderophore produced by the marine dinoflagellate *Prorocentrum minimum*. Science, 219: 306-308.

Trick, C.G., 1989. Hydroxomate-siderophore production and utilization by marine eubacteria. Current Microbiology, 18: 375-378.

Uematsu, M., Duce, R.A., Prospero, J.M., Chen, L., Merrill, J.T., McDonald, R.L., 1983. Transport of mineral aerosol from Asia over the North Pacific

Ocean. Journal of Geophysical Research, 88: 5343-5352.

Van der Zee, C., Roberts, D.R., Rancourt, D.G., and Slomp, C.P., 2003. Nanogoethite is the dominant reactive oxyhydroxide phase in lake and marine sediments. Geology, 31: 993-996.

Van der Zee, C., Slomp, C.P., Rancourt, D.G., De Lange, G.J. and van Raaphorst, W., 2005. A Mössbauer spectroscopic study of the iron redox transition in eastern Mediterranean sediments. Geochimica Cosmochimica Acta, 69: 441-453.

Wallmann, K., Hennies, K., König, I., Petersen, W. and Knauth, H.-D., 1993. A new procedure for the determination of 'reactive' ferric iron and ferrous iron minerals in sediments. Limnology and Oceanography, 38: 1803-1812.

Wang, Y. and Van Capellen, P., 1996. A multicomponent reactive transport model of early diagenesis: Application to redox cycling in coastal sediments. Geochimica Cosmochimica Acta, 60: 2993-3014.

Wedepohl, K.H., 1995. The composition of the continental crust. Geochimica Cosmochimica Acta, 59: 1217-1232.

Wehrli, B., Friedel, G. and Manceau, A., 1995. Reaction rates and products of manganese oxidation at the sediment-water interface. In: Huang, C.P., O'Melia, C.R. and Morgan, J.J. (eds), Aquatic chemistry: Interfacial and interspecies processes, ACS Advances in Chemistry, 244, pp. 111-134.

Widdel, F., Schnell, S., Heising, S., Ehrenreich, A., Assmus, B. and Schink, B., 1993. Ferrous iron oxidation by anoxygenic phototrophic bacteria. Nature, 362: 834-835.

Wu, J. and Luther, G.W. III, 1995. Complexation of Fe(III) by natural organic ligands in the Northwest Atlantic Ocean by competitive ligand equilibration method and kinetic approach. Marine Chemistry, 50: 159-177.

Yeh, H.-W. and Savin, S.M., 1977. Mechanism of burial metamorphism of argillaceaous sediments: 3. O-isotope evidence. Bulletin of the Geological Society of America, 88: 1321-1330.

Zhuang, G., Duce, R.A. and Kester, D.A., 1990. The dissolution of atmospheric iron in surface seawater of the open ocean. Journal of Geophysical Research, 59: 16,207-16,216.

Zhuang, G. and Duce, R.A., 1993. The adsorption of dissolved iron on marine aerosol particles in surface waters of the open ocean. Deep Sea Research 40: 1413-1429.

8 Sulfur Cycling and Methane Oxidation

Bo Barker Jørgensen and Sabine Kasten

This chapter deals with the biogeochemical transformations of sulfur and methane in marine sediments during early diagenesis. The term 'early diagenesis' refers to the whole range of post-depositional processes that take place in aquatic sediments and are coupled either directly or indirectly to the degradation of organic matter. We focus on the processes that drive sulfate reduction together with the manifold associated biotic and abiotic reactions that make up the sedimentary sulfur cycle – including the various pathways of sulfide oxidation. Furthermore, we give an overview of the quantitative significance of microbial sulfate reduction for the mineralization of organic matter and oxidation of methane in different depositional environments and discuss the different approaches to quantify sulfate reduction through radiotracer measurements or modeling of pore-water concentration profiles. As sedimentary pyrite represents the most important sink for seawater sulfate, the mechanisms of pyrite formation are discussed. The sulfidization of sediment organic matter is another sink for sulfur in the modern ocean, but will not be discussed here. For an overview of the processes and pathways involved in the incorporation of sulfur into organic matter, we refer to Orr and White (1990), Krein and Aizenshtat (1995), Schouten et al. (1995), and Werne et al. (2004).

Due to the profound alteration of the primary sediment geochemistry across and below the sulfate/methane transition (SMT, also called the sulfate/methane interface), where the process of anaerobic oxidation of methane (AOM) takes place, we also dedicate a part of this chapter to the processes of mineral dissolution and precipitation occurring at and below the SMT. We will demonstrate that sulfate reduction driven by AOM significantly alters primary mineral associations accompanied by strong perturbations of mineralogical, isotopic and rock magnetic signatures. These diagenetic processes

have a profound impact on the preservation of numerous paleoceanographic proxy variables and are therefore relevant for the interpretation of the geological record.

Reviews of the sulfur cycle from a biogeochemical or microbiological perspective have recently been presented, e.g. by Amend et al. (2004) and Canfield et al. (2005). The latter authors describe the relationship between the different microorganisms metabolizing sulfur compounds and their environment and emphasize the organisms as well as the biogeochemical processes.

8.1 Introduction

With $1.3 \cdot 10^9$ Tg (teragram = megaton = 10^{12} g) of sulfur present as sulfate, the oceans represent one of the largest sulfur pools on earth (Vairavamurthy et al. 1995). The main influx of sulfur to the oceans occurs via river water carrying the products of mechanical and chemical weathering of continental rocks. Relative to this fluvial input, the atmospheric transport of sulfur is of minor importance. It mainly consists of recycled oceanic sulfate from seaspray, volcanic sulfur gases, H_2S released by sulfate-reducing bacteria, organic S-bearing compounds released into seawater and subsequently into the atmosphere by phytoplankton, and anthropogenic emissions of sulfur dioxide. Due to the oxic conditions that prevail in the world's oceans, the dominant sulfur species in seawater is by far the sulfate ion (SO_4^{2-}). Sulfate is the second most abundant anion next to chloride and has a concentration of 29 mM (2.71 g/kg) in ocean water.

Marine sediments are the main sink for seawater sulfate which demonstrates that the sedimentary sulfur cycle is a major component of the global sulfur cycle. The most important mechanisms for removing sulfate from the oceans to the sediments are (1) the bacterial reduction of sulfate to hydrogen sulfide, which subsequently reacts with iron to form sulfide minerals, particularly pyrite (FeS_2), (2) the formation of organic sulfur, i.e. the incorporation of sulfur into sedimentary organic matter during early diagenesis, and (3) the precipitation of calcium sulfate minerals in evaporites (Vairavamurthy et al. 1995). With respect to the relative importance of each pathway, Vairavamurthy et al. (1995) point out that although marine evaporites were important sinks for sulfate from the Late Precambrian to the late Tertiary, their rate of formation in today's oceans, over the last few million years, is quantitatively insignificant. Thus, they conclude that the burial of sulfide minerals and, to a less extent, of organic sulfur represents the major sink for oceanic sulfur in the modern ocean.

8.2   Sulfate Reduction and the Degradation of Organic Matter

Throughout the water column of the ocean, ten billion tons of organic particles derived from plankton production in the photic surface layer steadily sink towards the sea floor. The annual deposition of organic material on the sea floor is also about ten billion tons, i.e. of the same magnitude as the total organic particle pool in the ocean. As the particles sink through the water column, the organic material is gradually degraded and respired back to CO_2 and nutrients by zooplankton and by microorganisms. Thus, with increasing distance from the coast and with increasing water depth, a gradually decreasing fraction of the sinking particles remains to ultimately land on the sea floor. On the shallow continental shelves this fraction is generally in the range of 20-50% of the phytoplankton productivity in the overlying water (Jørgensen 1983). In the deep sea the fraction is only 1-2% (Jahnke 1996). Deep sea sediments, therefore, play only a minor quantitative role in the oceanic carbon and nutrient cycles, as the following discussion will show.

As the particulate organic matter is deposited on the sea floor it is immediately attacked by a broad range of organisms that all contribute to its degradation and gradual mineralization. At the sediment surface, macrofauna plays a particular role by mechanically disintegrating the detritus and by mixing of the upper sediment layer through their burrowing and feeding activity (bioturbation), whereby the organic material becomes repeatedly exposed to oxygen. The benthic fauna also irrigates the burrows in order to transport oxygen down for

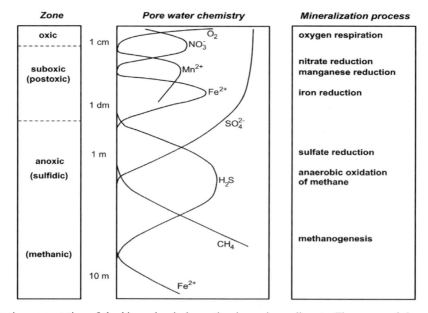

Fig. 8.1 Schematic representation of the biogeochemical zonation in marine sediments. The names of the main zones were proposed by Froelich et al. (1979) and Berner (1981, in parenthesis). The depth scale is quasi-logarithmic; the exact depths, however, vary strongly and increase from the shelf to the deep sea. The pore water chemistry shows relevant dissolved species. Peak heights and concentration scales are arbitrary. The chemical profiles reflect the depth sequence of the dominant mineralization processes through which organic matter is oxidized to CO_2 (Modified from Froelich et al. 1979).

respiration, and possibly food particles, and thereby extend the oxic zone (i.e. the zone containing O_2) deeper down into the sediment. Due to this macrofaunal activity, an extended surface layer of sediment remains oxidized (i.e. with positive redox potential and with many chemical species such as iron or manganese in their oxidized state). The macrofauna thus has a critical influence on the early diagenesis in sediments and affects both the pathways and the rates of many processes.

8.2.1   Geochemical Zonation

Apart from the heterogeneous chemical structure caused by faunal activity, marine sediments have a distinct biogeochemical zonation of the main aerobic and anaerobic mineralization processes. Fig. 8.1 shows how the dominant oxidants for mineralization change with depth, regulated partly by their concentration in sea water and partly by the energy yield of the process by which they are consumed (cf. Chapter 5). Many microorganisms gain energy from the oxidation of organic matter with an external oxidant (electron acceptor). Thermodynamically, oxygen is the most favorable electron acceptor. The supply of oxygen from seawater to the sediment is, however, highly transport-limited. In coastal marine sediments or in ocean areas of high productivity, e.g. upwelling regions, a high organic matter flux or low oxygen content of the bottom water reduces the thickness of the oxic surface layer of the sediment to only a few mm or cm. In deep sea sediments the depth of oxygen penetration may be a dm or a m or even more (Chapter 6).

Where oxygen has been consumed by aerobic respiration, the sediment is anoxic, i.e. O_2-free, and microorganisms utilize other terminal electron acceptors for the mineralization of organic matter. Listed in an order of decreasing energy gain these are: nitrate (NO_3^-), manganese oxides (represented by Mn(IV)), iron oxides (represented by Fe(III)), and sulfate (SO_4^{2-}) (Fig. 8.1). The sediment layer where NO_3^-, Mn(IV) and Fe(III) reduction predominate has been termed the suboxic zone (Froelich et al. 1979). Although these electron acceptors are energetically more favorable than sulfate, they are usually less important biogeochemically because of their limited supply to the sediments. The processes of O_2, NO_3^- and Fe(III) reduction are discussed in detail in Chapter 6 and 7.

Below the suboxic zone, sulfate reduction is the main pathway of organic carbon oxidation and

it generally extends many meters down into the sediment. The high concentration of sulfate in seawater makes it a dominant electron acceptor (Henrichs and Reeburgh 1987). With an average concentration of about 29 mmol/l (Vairavamurthy et al. 1995), sulfate concentrations in seawater are more than two orders of magnitude higher than in freshwater (about 0.1 mmol/l) (Bowen 1979). The mineralization of organic material by sulfate in marine sediments is, therefore, much more important than in freshwater.

In the even deeper subsurface sediment, where also sulfate is exhausted, there is little net oxidation of the organic carbon but rather a degradation to CH_4 and CO_2. As a stable end product of carbon degradation in the absence of oxidants, methane tends to accumulate in deep sub-surface sediments from where it slowly diffuses up towards the sulfate zone. Upon entry into the lower sulfate zone, however, also methane becomes oxidized completely to CO_2. The organic material initially deposited on the sediment surface thus undergoes an efficient progressing degradation as it becomes buried deeper and deeper down through this sequence of diagenetic processes. Only a small fraction escapes mineralization and remains after thousands or millions of years to contribute to the great pool of fossil organic carbon in marine sediments.

The main degradation of organic material takes place as it gradually becomes buried down through the oxic and suboxic zones. The fraction that still remains once it reaches the sulfate reduction zone therefore depends on how deep oxygen, nitrate, manganese and iron reduction predominate. In coastal sediments with high organic sedimentation these oxidants are rapidly depleted and the main sulfate reduction zone starts already a few cm below the sediment surface. Although it is not so apparent from the chemical zonation, sulfate reduction also occurs in suboxic and even in oxic sediment where the produced sulfide is rapidly reoxidized (Jørgensen and Bak 1991). In the deep sea, the metal oxides may be exhausted only several meters below the surface and little organic material is available once it becomes buried down into the sulfate reduction zone. The overall sulfate reduction in sediments is therefore very sensitive to the organic deposition rate and is geographically strongly shifted towards the continental margins and shelf sediments. In fact, a large part of the deep sea floor lies under low-productivity ocean regions where the organic sedimentation is so scanty that sulfate reduction is hardly detectable throughout the entire sediment column. Fig. 8.2 shows such an example, with data taken from an expedition of the Ocean Drilling Program in the eastern tropical Pacific Ocean.

Fig. 8.2 shows some of the main chemical changes with depth in the 320 m thick deposit overlying old ocean crust of Miocene age. The mineralization of organic carbon and organic nitrogen is apparent from the increase in pore water concentration of total CO_2 and ammonium. It is striking that these concentrations drop back towards sea water values at the bottom of the sediment column. This is due to a slow advection of sea water through the porous ocean crust which removes these products of mineralization and thereby influences the entire chemistry of the sea bed and its exchange with sea water in this part of the Pacific Ocean. Surprisingly, the microbial activity at the bottom of the sediment column is so low that even nitrate (and possibly oxygen) remains in the crustal fluid and provides an additional source of oxidant from below (Fig. 8.2, middle frame). The Mn^{2+} gradient in the upper 100 m and its slight increase at 200-300 mbsf show the importance of manganese reduction throughout this deep sea sediment. Consequently, sulfate reduction is inhibited by electron acceptors with higher energy yield, and even a sediment layer of 320 m thickness generates only a marginal drop in sulfate concentration (Fig. 8.2, right frame).

Methane producing microorganisms are even more strongly inhibited and, although present throughout the entire sediment deposit, methane does not exceed a trace concentration of 0.2 µM.

8.2.2   Dissimilatory Sulfate Reduction

Two types of sulfate reduction can be distinguished: (1) assimilatory sulfate reduction which serves the biosynthesis of sulfur-containing organic compounds that are part of the cell biomass, and (2) dissimilatory sulfate reduction from which microorganisms conserve energy and release H_2S to the environment. With respect to the mineralization of organic matter and the main source of H_2S in sediments, only dissimilatory sulfate reduction plays a role and is the process referred to when we discuss microbial sulfate reduction. The sulfate reducing bacteria use sulfate as the terminal electron acceptor for their anaerobic respiration (see Chapter 5). As energy and carbon source they use mostly short-chain fatty acids (acetate, formate, propionate, butyrate) and other small organic molecules that are produced from the degradation and fermentation of sediment organic matter. H_2 is also a product of fermentation and serves as an important energy source for sulfate reduction in marine sediments.

Most of the known dissimilatory sulfate reducers are bacteria, but also some thermophilic archaea belong to this group (Rabus et al. 2004; Stetter et al. 1993). For an overview of the most

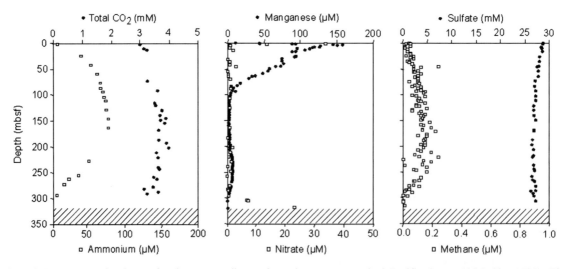

Fig. 8.2 Pore water chemistry of a deep sea sediment from the eastern tropical Pacific Ocean (ODP Site 1225). The core was obtained during Leg 201 of the Ocean Drilling Program and spans the entire sediment deposit, from the sediment surface at 3760 m water depth down to the 11 million years old basaltic crust (hatched) at 320 mbsf (meter below sea floor). Data from D'Hondt et al. (2003).

Table 8.1 Mineralization of deposited organic carbon (C_{org}) in marine sediments and the role of sulfate reduction. Data range from the shelf to the deep sea and are listed according to decreasing significance of sulfate reduction. The Black Sea is included at the end as an example of an anoxic basin totally dominated by sulfate reduction. The data are based on radiotracer measurements of sulfate reduction rates (* represents a modeled rate).

Location	Water depth	Rate of C_{org} degradation	C_{org} mineralized by sulfate reduction	Reference
	[m]	[mmol m^{-2} d^{-1}]		
Cape Lookout Bight	9	114	72%	Crill and Martens (1987)
Chilean shelf	34-122	10	56-79%	Thamdrup and Canfield (1996)
Baltic Sea	16	9.8	44%	Jørgensen (1996)
Gulf of Maine	50-300	10.7	43%	Christensen (1989)
St. Lawrence Estuary	335	10	26%	Edenborn et al. (1987)
E. South Atlantic	850-3000	0.28-3.1	9-40%	Ferdelman et al. (1999)
Eastern tropical Pacific	3760	0.5	0.05%*	D'Hondt et al. (2004) Jahnke (1996)
Black Sea, anoxic	130-1176	1.30-2.86	~100%	Jørgensen et al. (2004)

common sulfate-reducing bacteria in marine sediments we refer to Widdel (1988). Besides sulfate there are also other oxidized sulfur compounds, e.g. thiosulfate ($S_2O_3^{2-}$) and elemental sulfur (S^0), that can serve as the terminal electron acceptor (Ehrlich 1996). They are, however, not of similar quantitative importance as sulfate.

Dissimilatory sulfate reduction can be described by the following net equation:

$$2 \, [CH_2O] + SO_4^{2-} \rightarrow 2 \, HCO_3^- + H_2S \qquad (8.1)$$

where [CH_2O] simply represents organic material. When the sulfate reducing bacteria are growing, a part of the [CH_2O] will be assimilated to produce cell material. Hydrogen, in contrast, is used only as an energy source and the cell assimilates CO_2 or an organic subtrate:

$$4 \, H_2 + SO_4^{2-} + 2 \, H^+ \rightarrow 4 \, H_2O + H_2S \qquad (8.2)$$

8.2.3 Sulfate Reduction and Organic Carbon Mineralization

Continental margin sediments play a major role for the overall budget of the global carbon cycle in the modern ocean. Most burial of organic carbon takes place on the continental shelf, in particular in deltaic and other coastal sediments (Berner 1982; Hedges and Keil 1995). Thus, 82% of the buried organic carbon is stored in shelf sediments

and 16% in continental slope sediments (Wollast 1998). Only 2% of the marine organic carbon burial takes place in deep sea sediments. It is important to note, however, that during glaciations the sea level has in the past dropped by more than 100 m and thereby exposed vast shelf areas to erosion and transport of stored material into deeper water. The current burial of organic material in shelf sediments is, therefore, in a geological perspective a transient state that may undergo major reallocation during a future glaciation. On the other hand, in case of a long-term global warming the sea level would rise and the area and sediment accumulation on the continental shelves would increase.

The aerobic and anaerobic mineralization of organic material is also strongly focused towards the ocean margins. Table 8.1 presents examples of the areal rates of organic carbon degradation in a number of representative sediments, ranging from coastal embayments to the deep sea. The table also shows the fraction of the overall organic carbon mineralization that takes place by sulfate reduction. Sulfate reduction predominates especially in sediments underlying highly productive and/or oxygen-depleted coastal waters, e.g. in the Black Sea, Cape Lookout Bight, or the upwelling areas of the Chilean shelf. In other shelf sediments sulfate reduction generally accounts for 25-50% of the mineralization (Jørgensen 1982). The relative importance of sulfate reduction drops to <1% as we move down the continental slope and it

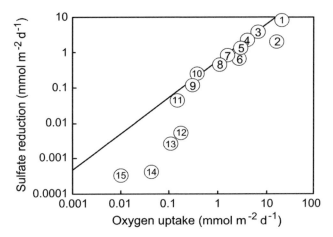

Fig. 8.3 Sulfate reduction versus oxygen uptake rates in marine sediments grouped according to ocean margin type and water depth: (1) Salt marsh; (2) mangrove; (3) shallow, high deposition; (4) seagras beds; (5) intertidal; (6) estuaries and embayments; (7) upwelling; (8) shelf - depositional; (9) shelf - non depositional; (10) upper slope (200-1000 m); (11) lower slope (1000-2000 m); (12) rise (2000-3000 m); (13) abyss (3000-4000 m); (14) abyss (4000-5000 m); (15) abyss (>5000 m). The line indicates equimolar mineralization rates of organic carbon by oxygen consumption and by sulfate reduction. The double-logarithmic plot is based on data compiled by Canfield (1993) and Canfield et al. (2005).

is insignificant in vast regions of the deep sea. The additional source of H$_2$S from the anaerobic degradation of organic sulfur plays a minor role relative to this dissimilatory sulfate reduction.

Radiotracer measurements of sulfate reduction rates have mainly been carried out in shallower waters, so that the data base for sediments below 500 m water depth is very limited. A detailed study was done by Ferdelman et al. (1999) between 855 and 4766 m water depth on the continental margin of southwest Africa. This area forms part of the Benguela upwelling system. The authors demonstrated that the depth-integrated sulfate reduction rates over the upper 20 cm of the sediment strongly correlated with the concentrations of organic carbon in the surface sediments. They further estimated that sulfate reduction at for example 3700 m water depth accounts for 3 to 16 % (Table 8.1) of total oxygen consumption. Thus, even on the lower continental slope in this region, a small but significant fraction of organic matter is degraded anaerobically through sulfate reduction.

Canfield (1993) and Canfield et al. (2005) have compiled published data on oxygen uptake and sulfate reduction in marine sediments and grouped these into different coastal types and into different depth regions of the ocean. Based on their data, Fig. 8.3 presents the relationship between aerobic mineralization and sulfate-based anaerobic mineralization in sediment types comprising the entire ocean floor. The sulfate reduction rates

were determined experimentally by the radiotracer method in shelf and slope sediments whereas they were modeled from pore water sulfate profiles in deep sea sediments (see Section 8.6). It is striking that shelf and slope sediments generally fall along or slightly below a line of equimolar carbon mineralization by oxygen and by sulfate, thus confirming that sulfate reduction in ocean margin sediments accounts for 25-50% of the entire organic carbon mineralization. Mangrove sediments provide an exception as they have predominant aerobic mineralization, perhaps due to an efficient oxygen transport by the mangrove aerial roots down into the root zone where most degradation of organic material takes place. Below 2000 m depth in the ocean sulfate reduction clearly looses importance relative to oxygen respiration and to suboxic processes such as nitrate, manganese or iron reduction.

This trend is illustrated in Fig. 8.4 where the 15 sediment types presented in Fig. 8.3 have been grouped into five depth zones of the global ocean. The graph shows how strongly the mineralization of organic matter in the sea bed is shifted towards ocean margin and shelf sediments. The continental shelf out to 200 m water depth comprises only 8% of the global ocean area of 3.6 · 10^8 km^2. Yet, 61% of the entire benthic oxygen uptake and 68% of the sulfate reduction take place here. If we include also the continental slope down to 1000 m water depth, this upper ocean margin includes even 70% of the oxygen uptake and 96% of the sulfate reduction. In other words, according to the data compiled by Canfield et al. (2005) only 30% of the global oxygen uptake of the sea floor and 4% of the sulfate reduction take place at depths below 1000 m, covering 87% of the ocean. Although a larger fraction of oxygen respiration than sulfate reduction occurs in deep sea sediments below 1000 m, still ca. 40% of the aerobic mineralization below 1000 m ocean depth occurs along the continental margins (Jahnke 1996).

The accuracy of such global estimates is clearly limited by the available data from which they were calculated. Other estimates of the total global mineralization rates in the sea bed vary by at least two-fold. Thus, the estimate of $23 \cdot 10^{13}$ mol yr^{-1} by Canfield et al. (2005) is slightly higher than the $19 \cdot 10^{13}$ mol yr^{-1} estimated by Jørgensen (1983) and close to the $22 \cdot 10^{13}$ mol yr^{-1} estimated by Smith and Hollibaugh (1993). Middelburg et al. (1997) calculated a range of $15\text{-}26 \cdot 10^{13}$ mol yr^{-1}, depending on whether the geometric or the arithmetic mean value of each depth interval was used. The low contribution of deep sea sediments to the global benthic oxygen respiration is consistent with the earlier budget calculations of Jørgensen (1983). Middelburg et al. (1997) estimated that 28% of the global sea floor mineralization and 7% of the sulfate reduction take place below 2000 m water depth. It should be noted that the low contribution of sulfate reduction in deep sea sediments may be affected by the method of sulfate reduction rate determination. In

sediments of the shelf and upper slope this is primarily by radiotracer experiments whereas in the deep sea rates were modeled. The modeling approach provides net rates rather than gross rates as obtained by the radiotracer method and thereby tends to underestimate the total rates, in particular in the near-surface sediment layer (see Section 8.6).

Since methane formation is shifted at least as strongly as sulfate reduction towards sediments with high organic deposition, an equally small fraction of the global biogenic methane production presumably takes place in the deep sea below 1000 m.

The global mineralization of sediment organic matter by oxygen was estimated in Fig. 8.4 to be $23 \cdot 10^{13}$ mol O_2 yr^{-1} while the global sulfate reduction was $7.5 \cdot 10^{13}$ mol SO_4^{2-} yr^{-1}. According to simple stoichiometries of oxygen respiration and sulfate reduction, one mol of oxygen oxidizes one mol of organic carbon to CO_2 whereas one mol of sulfate oxidizes two mol of organic carbon (Chapter 5):

$$[CH_2O] + O_2 \rightarrow CO_2 + H_2O \qquad (8.3)$$

$$2\,[CH_2O] + SO_4^{2-} + 2\,H^+ \rightarrow 2\,CO_2 + H_2S + 2\,H_2O$$

$$(8.4)$$

This means that the global estimate of sulfate reduction should be multiplied by two in order to convert it to carbon equivalents. Thus, sulfate

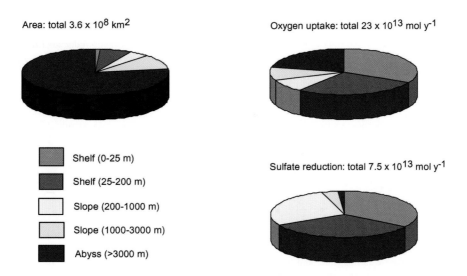

Area: total 3.6×10^8 km^2

Oxygen uptake: total 23×10^{13} mol y^{-1}

Sulfate reduction: total 7.5×10^{13} mol y^{-1}

Shelf (0-25 m)
Shelf (25-200 m)
Slope (200-1000 m)
Slope (1000-3000 m)
Abyss (>3000 m)

Fig. 8.4 Relative distributions of total ocean area and of global ocean oxygen uptake or sulfate reduction, grouped into five depth zones of the ocean. Based on data compiled by Canfield (1993) and Canfield et al. (2005).

oxidizes $15 \cdot 10^{13}$ mol organic carbon per year. This is equivalent to 65% of the global sediment oxygen uptake. The major part of the sulfide produced from sulfate reduction, around 90% in many shelf sediments, does not become buried in the sediment but is directly or indirectly reoxidized back to sulfate (see Section 8.5). The complex process of sulfide oxidation will thereby ultimately consume oxygen corresponding to the net equation:

$$H_2S + 2\,O_2 \rightarrow SO_4^{2-} + 2\,H^+ \tag{8.5}$$

It is a striking consequence of these simple calculations that about half of the entire oxygen uptake in shelf sediments, where most sulfate reduction takes place, is used for the reoxidation of sulfide or of reduced metals originally reduced by reaction with sulfide (cf. Jørgensen 1977). Thus, if 90 % of $15 \cdot 10^{13}$ mol O_2 yr^{-1} is consumed in this process, it means that 13 out of 23 mol O_2 yr^{-1} or 56% of the oxygen is somehow involved in the reoxidation of sulfide (see Section 8.5). Only the remaining half of the oxygen is therefore available for all the animals and heterotrophic microorganisms that oxidize organic matter directly through their aerobic respiration.

8.3   Anaerobic Oxidation of Methane (AOM)

Below the sulfate zone, methanogenesis is the main terminal pathway of organic carbon mineralization. Methane is produced exclusively by anaerobic archaea that utilize a narrow spectrum of substrates for the process (Whitman et al. 1999). Methanogenic bacteria or eukaryotes are not known to exist. The primary sources of methane formation in marine sediments are from the splitting of acetate (CH_3COO^-) and from the reduction of CO_2 by hydrogen (Eq. 8.6 and 8.7):

$$CH_3COO^- + H^+ + \rightarrow CH_4 + CO_2 \tag{8.6}$$

$$CO_2 + 4\,H_2 \rightarrow CH_4 + 2\,H_2O \tag{8.7}$$

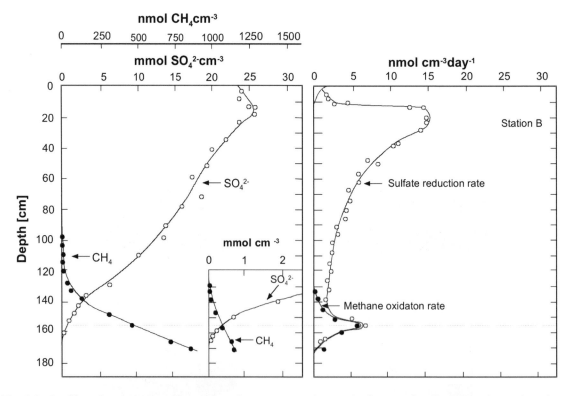

Fig. 8.5 Profiles of pore-water sulfate and methane concentrations and of rates of sulfate reduction and methane oxidation for a sediment core recovered from the Kattegat (Station B; 65 m water depth). The broken horizontal line denotes the depth where sulfate and methane were at equimolar concentrations - indicating the peak of the sulfate/methane transition. From Iversen and Jørgensen (1985).

As minor sources, also methanol, formalde-hyde, and some methylated compounds such as methyl amines and methyl mercaptans may be used for methanogenesis. The energy yields by these terminal steps in organic degradation are lower than by anaerobic respiration, e.g. with nitrate, metal oxides or sulfate. Sulfate reducing bacteria therefore compete effectively with the methanogenic archaea for substrates which both groups of organisms can utilize, primarily acetate and H_2. As a result, methane is not produced as the main end product in marine sediments until sulfate has been largely depleted by sulfate reducing bacteria and electron acceptors for a respiratory oxidation of the organic matter are therefore no longer available (Martens and Berner 1974). Only small amounts of methane may still be generated in the sulfate zone from non-competi-tive methylated substrates. This is a major diffe-rence from the degradation of organic matter in freshwater sediments where sulfate is highly limited and methane is the main end product of anaerobic mineralization.

Methane is a chemically unreactive molecule that requires microbiological catalysis for the activation and oxidation to CO_2 at environmental temperatures. Aerobic bacteria that oxidize methane with O_2 are well known from diverse environments, including sea water and marine sediments, and have been studied in pure culture in the laboratory (Bowman 2000). These earlier physiological and biochemical studies indicated that biological methane oxidation requires molecular oxygen. Anaerobic oxidation of methane (AOM) was recognized by geochemists already in the 1970'ies, however, as a key process in marine sediments (Reeburgh 1969, 1976; Martens and Berner 1974; Barnes and Goldberg 1976). A large number of studies based on reaction-transport modeling, radiotracer experiments, inhibition techniques, and stable isotope data have since then firmly established that methane is oxidized biologically in the absence of O_2 at the transition between sulfate and methane. The microorganisms responsible for the process remained elusive, however, and even today it has not been possible to bring anaerobic methane oxidizers into pure culture for detailed laboratory study. In spite of this, the identity of the organisms has now been discovered and research on the biogeochemistry of anaerobic oxidation of methane has made significant progress in recent years (see reviews by Valentine and Reeburgh 2000; Hinrichs and Boetius 2002).

8.3.1 The AOM Zone in Marine Sediments

Most of the methane produced in the sulfate depleted sub-surface sediment diffuses upwards along a steep methane gradient until it reaches the lower sulfate zone and becomes oxidized. Within this reaction zone, referred to as the "sulfate-methane transition" (SMT), pore-water methane and sulfate are both consumed to depletion. This distinct zone is typically located one to several meters below the sediment surface in continental margin sediments and plays a key role in the biogeochemistry of the sea bed. In this zone most of the energy and reducing power of organic carbon mineralized to CO_2 and CH_4 below the sulfate zone comes into contact with an efficient electron acceptor, sulfate, and is finally oxidized to CO_2. Since this methane flux comprises most of the subsurface methanogenesis, the anaerobic oxidation of methane integrates the degradation of buried organic carbon from the lower boundary of the sulfate zone to very deep sediment layers which were deposited many thousands to millions of years ago. The coupled sulfate-methane reac-tion has been proposed to proceed according to the following net equation, assuming a one to one stoichiometry between methane and sulfate (e.g. Murray et al. 1978; Devol and Ahmed 1981):

$$CH_4 + SO_4^{2-} \rightarrow HCO_3^- + HS^- + H_2O \qquad (8.8)$$

Figure 8.5 shows an example of the sulfate-methane transition zone in a continental shelf sediment from Kattegat (Denmark) at the transi-tion between the Baltic Sea and the North Sea. The sediment was characterized by a high sedi-mentation rate (0.16 cm yr^{-1}) and consisted of fine-grained silt and clay with an average organic matter content of 12% (Iversen and Jørgensen 1985). Sulfate diffused downwards to become depleted by sulfate reduction at 160 cm sediment depth while methane diffused upwards and penetrated ca. 40 cm into the lower sulfate zone to ca. 120 cm depth. In order to determine anaerobic methane oxidation rates in the sulfate-methane transition zone, Iversen and Jørgensen (1985) carried out radiotracer measurements of sulfate reduction using $^{35}SO_4^{2-}$ and of methane oxidation using $^{14}CH_4$. Fig. 8.5 (right frame) shows that the rates of sulfate reduction were high, starting from below the 10-15 cm deep suboxic sediment, and gradually dropped with increasing depth and age in the sediment. In the SMT, the rates increased

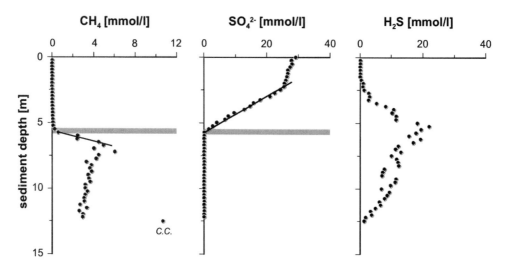

Fig. 8.6 Pore-water concentration profiles from gravity core GeoB 3714-9 from the Benguela upwelling area (2060 m water depth), South Atlantic. The shaded bar marks the sulfate-methane transition zone. The methane sample labeled C.C. was taken from the core catcher immediately after core recovery. From Niewöhner et al. (1998).

again with a peak just where sulfate and methane reached the same molar concentrations (small insert in Fig. 8.5). At that depth also the anaerobic oxidation of methane peaked with similar rates as the sulfate reduction. The integrated rates of methane oxidation in the transition zone accounted for 89 % of sulfate reduction at this depth. This example illustrates how bacterial sulfate reduction changes from organiclastic (based on the degradation of organic material) to methanotrophic (based on the oxidation of methane) down through the sediment column.

The pore water gradients of sulfate and methane and the depth of the AOM zone may vary strongly among sediments, depending on water depth, organic flux and other factors. Fig. 8.6 shows an example from the continental slope of the Benguela upwelling area in the southern Atlantic. The sulfate profile here shows little depletion in the top 0-2 m below which sulfate dropped almost linearly down to the sulfate-methane transition located at 5-6 m depth below sea floor. The concentration of H_2S peaked right at the SMT from where the H_2S concentration decreased both upwards towards the sediment surface and downwards (discussed below in Section 8.5). Based on pore-water concentration profiles, Niewöhner et al. (1998) calculated the diffusive flux of sulfate and methane into the transition zone in the sediment. Their calculations showed that anaerobic methane oxidation accounted for 100% of the deep sulfate reduction

within the sulfate-methane transition zone, i.e. it could consume the total diffusive sulfate flux. These findings demonstrate that methane can be the primary electron donor for sulfate reduction in the sulfate-methane transition zone.

The methane profile in Fig. 8.6 shows that below the steep gradient at the top of the methane zone the methane concentrations gradually drop again. This is an artifact due to degassing of methane immediately upon core retrieval. Due to the relatively low solubility of methane and the large pressure difference between *in situ* depth and sea surface, methane is highly supersaturated in the pore water upon recovery of sediment cores and escapes as bubbles before sampling on board ship. For this reason, true methane concentration gradients above atmospheric pressure are difficult to determine and only recently have pressurized core barrels been introduced (Dickens et al. 2003).

8.3.2   Energy Constraints and Pathway of AOM

Microorganisms able to perform anaerobic oxidation of methane were only recently discovered through intensive studies, not of the "normal" subsurface SMT, but of unique marine environments in which AOM dominates the carbon and sulfur cycles, such as sediments associated with gas hydrates and methane seeps. (Gas hydrates are ice-like solids - generally composed of water and methane - which occur naturally in sediments under conditions of high pressure, low temperature and high methane concentrations; see also Chapter 14). In such sediments from the Hydrate Ridge at 700 m water depth off the Oregon coast, conspicuous microbial aggregates, apparently responsible for the anaerobic oxidation of methane, were first observed (Boetius et al. 2000). The 3-5 μm large aggregates were stained with fluorescent molecular probes and studied under the fluorescence microscope. They were all found to consist of a central colony of archaea overgrown by sulfate reducing bacteria. In the AOM zone of the sediment, the biomass of these aggregates exceeded the biomass of all other microorganisms by an order of magnitude, thus indicating that they were indeed responsible for the methane oxidation with sulfate. Since then, similar aggregates have been found world-wide in diverse methane enriched surface sediments (Orphan et al. 2001, 2002; Michaelis et al. 2002; Knittel et al. 2005).

Further evidence for the identity of the anaerobic methane oxidizers had come from isotope analyses of specific lipid biomarkers in sediments with high rates of AOM. Relative to marine organic carbon or bicarbonate, methane is highly enriched in the lighter carbon isotope, ^{12}C, over the heavier carbon isotope, ^{13}C, with $\delta^{13}C$ values ranging from -50 to -90‰ (Whiticar 1999; see also Chapter 10). The archaeal biomarkers in these sediments were strongly enriched in ^{12}C, which indicated that methane served as the main carbon source for the microorganisms (Hinrichs et al. 1999; Elvert et al. 2000; Thiel et al. 2001). Also biomarkers derived from bacteria showed extreme

^{12}C enrichment and indicated that the associated sulfate reducers were also involved in the AOM process (Hinrichs et al. 2000; Elvert et al. 2000; Pancost et al. 2000). Final evidence that the aggregates carried out anaerobic oxidation of methane was provided by a recently developed SIMS technique (secondary ion mass spectrometry) by which carbon isotopic analysis could be done on individual microscopic AOM aggregates and thereby confirm that their cell material contained the ^{12}C depleted isotope signal of methane (Orphan et al. 2001).

Equation 8.8 above proposed a net equation for anaerobic oxidation of methane by sulfate that is based on modeled pore water profiles and laboratory experiments with sediment enrichments. No single microorganism is yet known, however, that is able to carry out such a complete oxidation of methane to bicarbonate and it remains an open question whether such organisms exist. Instead, the process may proceed in two steps involving an intermediate electron carrier that is transferred from the archaea to the sulfate reducers within the AOM aggregates. Hoehler et al. (1994) first suggested that AOM is carried out by a consortium of archaea and sulfate reducing bacteria and that hydrogen might serve as such an intermediate that is transferred from the methane oxidizers to the sulfate reducers in a two-step reaction (Fig. 8.7):

$$\text{I.} \quad CH_4 + 2\,H_2O \rightarrow CO_2 + 4\,H_2 \tag{8.9}$$

$$\text{II.} \quad SO_4^{2-} + 4\,H_2 + H^+ \rightarrow HS^- + 4\,H_2O \tag{8.10}$$

Sum of Eq. 8.9 and 8.10 = Eq. 8.8

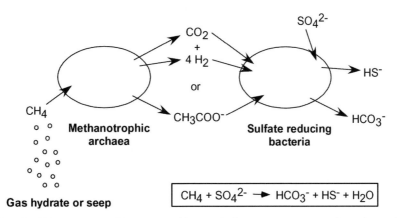

Fig. 8.7 Schematic of AOM aggregate with a syntrophic metabolism among two members that only in combination can catalyze the complete oxidization of methane with sulfate. A transfer of H_2 or acetate (CH_3COO^-) from methanotrophic archaea to sulfate reducing bacteria has been proposed as an intermediate in the net process. Such an interspecies transfer of hydrogen or organic carbon is still hypothetical and has not been directly demonstrated.

The first step in this reaction (Eq. 8.9) is obviously a reversal of methanogenesis from CO_2 and H_2 shown in Eq. 8.7. How can a metabolic process run in two opposite directions and yet yield energy for both types of archaea? Hoehler et al. (1994, 1998) showed that the energy yield is highly dependent on the H_2 concentration which, in turn, is regulated by the predominant process of mineralization. In the deep methanogenic zone the H_2 concentration is relatively high, about 10 nM, so that methanogenesis (Eq. 8.7) is an exergonic, energy yielding process. In the sulfate zone, the sulfate reducing bacteria maintain such a low H_2 concentration, about, 1 nM, that it falls below the concentration required for an exergonic methanogenesis. On the contrary, the reversal of methanogenesis (Eq. 8.9) becomes exergonic and AOM may proceed. This strong dependence on H_2 concentration could explain why the methane oxidizing archaea grow in aggregates together with the sulfate reducing bacteria. Only when the sulfate reducers efficiently scavenge H_2 as it is produced by the archaea are these able to continue oxidizing methane.

Although this idea of a syntrophic association based on inter-species hydrogen transfer is very appealing, it explains only a part of the observations on AOM. Since both the methanotrophic archaea and the sulfate reducing bacteria carry the light carbon isotopic signal of methane, it is a question how also carbon is transferred in this syntrophic association. An alternative pathway could therefore be the conversion of methane to acetate and the subsequent oxidation of acetate (CH_3COO^-) by the sulfate reducers with a concurrent incorporation of part of the acetate into their cell biomass:

$$CH_4 + HCO_3^- \rightarrow CH_3COO^- + H_2O \qquad (8.11)$$

$$SO_4^{2-} + CH_3COO^- \rightarrow HS^- + 2\,HCO_3^- \qquad (8.12)$$

Further possibilities exist, such as a combined transfer of acetate and hydrogen or a transfer of formate, both of which may explain the $\delta^{13}C$ observations and be more compatible with the energetic constraints on AOM (Valentine and Reeburgh 2000; Sørensen et al. 2001). Further research is required to understand the basic mechanism and regulation of this key process in the methane cycle.

Although AOM provides an almost complete barrier to methane escape from the sea floor in sediments with diffusive methane flux, the process is surprisingly sluggish and takes place over a sediment horizon and a time scale that highly exceed that of most other microbial processes in gradient environments. Thus, the life-time of methane upon diffusion up into the sulfate zone is generally on the order of months to years (Jørgensen et al. 2004). This is a reason why we prefer the term "sulfate-methane transition" rather than "sulfate-methane interface". A reason for the slow turnover may be that the methane oxidizing community is operating near the theoretical minimum in energy yield required for microbial growth. The estimated energy yield of AOM in a number of sedimentary environments is -20 to -25 kJ per mol of methane consumed (Hoehler et al. 1994) which is less than the energy required for the formation of ATP (see Chapter 5). If the process does indeed involve a two-step reaction such as suggested in Eq. 8.9-8.10 or 8.11-8.12, then this energy even has to be shared among two partners. This may explain why the growth of AOM aggregates is exceedingly slow and requires many months for a doubling of the biomass in laboratory incubations (Nauhaus et al. 2002). It may also explain why anaerobic methane oxidizing microorganisms have not yet been isolated in pure culture in the laboratory. They simply grow so slow that isolation may take many years. These observations demonstrate, however, that processes catalyzed by microbial communities in the environment can be highly efficient in energy conservation and operate close to the theoretical thermodynamic limits (Schink 1997; Hoehler et al. 2001; Jackson and McInerney 2002).

By the use of molecular methods based on sequence information from 16S rRNA genes, a broad diversity of archaea that all belong to the *Euryarchaeota* (Orphan et al. 2002; Knittel et al. 2005) and sulfate reducing bacteria mostly belonging to the *Desulfosarcina-Desulfococcus* branch of the *Delta-proteobacteria* (Boetius et al. 2000; Orphan et al. 2001) has been consistently found to be associated with anaerobic oxidation of methane. Although the genetic identity of AOM microorganisms is now revealed, the physiology and biochemistry of AOM are still incompletely understood. A recent clue came from the discovery that the terminal key enzyme of methanogenesis, methyl coenzyme M reductase, and its encoding gene seem also to be involved in the process of AOM in a slightly modified form, thus indicating that AOM may partly be a reversal of

Table 8.2 Role of methane as a carbon source for sulfate reduction in marine sediments. The compiled data show cumulative sulfate reduction rates measured by radiotracer technique, either over the entire sulfate zone, or in the upper 0-15 cm combined with modeling below that depth. The contribution of methane was calculated from the diffusion flux of methane up into the lower sulfate zone. In other data sets where sulfate reduction rates are determined only by modeling, or where also methane oxidation was measured by radiotracer technique, the calculated % of SRR from CH_4 is higher than shown here. (SRR = sulfate reduction rate).

Location	Water depth	SRR	CH_4 flux,	References
	[m]	[mmol m^{-2} d^{-1}]	% of SRR	
Skan Bay, Alaska	65	19	8%	Alperin and Reeburgh (1988)
Århus Bay, Baltic Sea	18	2.90	3%	Thode-Andersen and Jørgensen (1989)
				Knab and Fossing (unpubl.)
Kattegat-Skagerrak	65-200	4.8-5.6	4-6%	Iversen and Jørgensen (1985)
Saanich Inlet	225	17	9%	Devol et al. (1984)
Namibian slope	1372-2060	1.25-2.22	10-12%	Fossing et al. (2000)
Chilean slope	799-2744	0.53-0.82	8-24%	Treude et al. (2005)
Cape Lookout Bight	9	24-27	40%	Crill and Martens (1983, 1986)
				Chanton (1985)
Black Sea, anoxic zone	130-1176	0.65-1.43	7-18%	Jørgensen et al. (2001)

methanogenesis in certain archaea (Krüger et al. 2003; Hallam et al. 2003). In this respect, microbiological and biogeochemical research is currently providing mutually supporting evidence for the nature of AOM.

8.3.3   Quantitative Role of AOM

Since methane formation is largely restricted to sediment layers below the depth of sulfate penetration, there is relatively little organic carbon with low reactivity left to fuel methanogenesis. When integrated through the sediment column, the amount of organic carbon mineralized with the formation of methane is generally only 5-20% of that mineralized by sulfate reduction (Canfield 1993; Canfield et al. 2005; Table 8.2), with 10% possibly representing a mean value. Since sulfate reduction typically accounts for half of the organic mineralization in ocean margin sediments, it should only be about 5% of the total mineralized organic carbon that is ultimately degraded to methane. In deep sea sediments the fraction is presumably lower but relevant data to confirm this are lacking. In some coastal environments with very high organic loading, such as Cape Lookout Bight on the Atlantic coast of North America, much more of the organic carbon may be buried below the sulfate zone and be degraded to methane (Crill and Martens 1986). Also in ocean

margin sediments where the methane flux is enhanced due to subduction or is focused in association with methane seeps or surficial gas hydrate accumulations, methane may provide the main carbon source for the entire sulfate reduction (e.g. Boetius and Suess 2004). Table 8.2 provides selected examples of the contribution of methane as a carbon and energy source for sulfate reduction in ocean margin sediments.

As discussed in Section 8.6, the sulfate reduction rates determined from modeling of sulfate profiles may underestimate the rates in the most active layers of near-surface sediment, and model comparisons of sulfate and methane fluxes therefore tend to indicate a greater role of methane for the entire sulfate reduction than data based on experimental rate measurements. From interstitial flux calculations, Reeburgh (1976, 1982) thus estimated that approximately 50 % of the net downward sulfate flux at a Cariaco Trench station - an anoxic basin - could be accounted for by methane oxidation. For Saanich Inlet sediments, 75 % of the downward sulfate flux was attributed to anaerobic methane oxidation according to simple box model calculations (Murray et al. 1978). Devol et al. (1984) obtained lower percentages of 23 to 40 % of the downward sulfate flux consumed by methane oxidation for these same sediments using a coupled reaction diffusion model. Iversen and Jørgensen (1985) reported that in Kattegat and Skagerrak sediments methane oxidation accounted for only 10 % of the electron donor requirement for the depth integrated sulfate reduction. In this case sulfate reduction rates were measured by radiotracer technique throughout the entire sulfate zone.

The proportion of sulfate used for organic matter degradation in comparison to sulfate used for anaerobic methane oxidation may be calculated under molecular diffusion conditions from the shape of sulfate profiles. Borowski et al. (1996) inferred that the linear sulfate pore water concentration profiles found in Carolina Rise and Blake Ridge sediments imply that anaerobic methane oxidation is the dominant sulfate consuming process. They calculated the upward methane flux from measured sulfate pore water profiles assuming that the downward sulfate flux is stoichiometrically balanced by upward methane flux. In this way they used the sulfate concentration profiles to calculate the methane flux from the underlying methane gas hydrates which had been documented by seismic profiling. Niewöhner et al. (1998) found similar linear sulfate concentration profiles in sediments of the Benguela upwelling area (see Fig. 8.6) and concluded that anaerobic methane oxidation could account for the entire deep sulfate flux.

These studies demonstrate that the sulfate-consuming process occurring in the deeper sediments, i.e. AOM, is the dominant factor determining the shape of the sulfate pore water profiles, although oxidation of organic matter demonstrably occurs throughout the entire sulfate zone. Linear sulfate profiles can, therefore, be used to calculate the upward methane flux (e.g. Devol and Ahmed 1981; Borowski et al. 1996; Niewöhner et al. 1998) but they do not provide accurate sulfate reduction rates occurring in near-surface sediments. Concave-down pore water profiles may develop when methane dependent sulfate reduction in deeper sediment layers is less important as is schematically illustrated in Figure 8.8 or if transient conditions prevail in the pore water system (Hensen et al. 2003; Kasten et al. 2003).

Reeburgh et al. (1993) estimated the global methane flux in marine sediments to a mean value of 70 Tg yr^{-1} or $0.5 \cdot 10^{13}$ mol CH$_4$ yr^{-1}. This is equivalent to 2% of the entire organic carbon mineralization via oxygen respiration and about 7% of the global sulfate reduction. A more recent estimate by Hinrichs and Boetius (2002) of AOM

in marine sediments yielded a four-fold higher number, 300 Tg yr^{-1} or about $2 \cdot 10^{13}$ mol CH$_4$ yr^{-1}. These calculations do not include hot-spots of methane fluxes from cold seeps, hot vents or surficial gas hydrates, since reliable estimates of the total areal distribution of such sites are not yet available. Methane seeps occur both at active and passive margins but are highly focused and often variable over time. Judd et al. (2002) made an estimate of the global flux of methane from the sea bed of 16-40 Tg yr^{-1}, i.e. much less than the estimated subsurface flux. Data compiled by Hinrichs and Boetius (2002) indicate that, even if the area affected by methane seepage at continental margins is below 1%, this might have

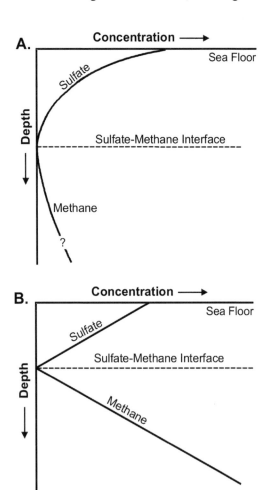

Fig. 8.8 Schematic profiles of sulfate and methane in marine sediments. A) In sediments with low methane flux, sulfate reduction based on oxidation of sediment organic matter predominates throughout the sulfate zone. B) In sediments with high methane flux, sulfate reduction based on anaerobic oxidation of methane tends to straighten out the sulfate profile.

Table 8.3 Reactivity of iron (oxyhydr)oxides towards sulfide. Half-lives ($t_{1/2}$) are given for reductive dissolution in seawater at pH 7.5 and at a sulfide concentration of 1000 µM. [a]data from Poulton et al. (2004); [b]data from Canfield et al. (1992) and Raiswell et al. (1994). (Adopted from Nüster, 2005).

Mineral	$t_{1/2}$[a]	$t_{1/2}$[b]
Freshly precipitated hydous ferric oxide	5 min.	
2-line ferrihydrite	12.3 hours	2.8 hours
Lepidocrocite	10.9 hours	< 3 days
Goethite	63 days	11.5 days
Magnetite	72 days	105 years
Hematite	182 days	< 31 days

a significant impact on the total methane budget. Therefore, a more extensive mapping of seepage areas, as well as a broader data base on diffusive methane fluxes, is needed for more accurate calculations of the methane cycling in the global sea bed (see also Chapter 14).

8.4 Effects of Sulfate Reduction on Sedimentary Solid Phases

Sulfate reduction, occurring either due to the oxidation of methane or the mineralization of organic material, can lead to a pronounced overprint or modification of the primary sediment composition by dissolution/reduction of minerals and precipitation of authigenic mineral phases. Since most of the minerals affected are also commonly used for paleoceanographic and paleoclimatologic reconstructions it is crucial to consider and assess the extent of diagenetic alteration of the sedimentary record driven by sulfate reduction.

8.4.1 Reactions with iron

Iron sulfides represent the most important minerals that form in association with both organoclastic and methanotrophic sulfate reduction, or - more precisely - as a result of the hydrogen sulfide produced by these processes. The different pathways of pyrite formation via intermediate iron sulfides will be described in more detail in Section 8.4.2. The first step in all pyrite forming sequences involves a reaction of hydrogen sulfide with either dissolved Fe^{2+} or solid-state iron (oxyhydr)oxides. The reactivity of oxidized iron minerals towards sulfide varies

significantly as shown in Table 8.3. In their recent study, Poulton et al. (2004) demonstrated that minerals with a lower degree of crystal order react within minutes to hours, while more ordered minerals react on time scales of days to years (c.f. Chapter 7). The largest discrepancy in reactivity between the studies of Canfield et al. (1992) and Raiswell et al. (1994) on the one hand and Poulton et al. (2004) on the other hand exists for the mineral magnetite (105 years versus 72 days; Table 8.3). This difference was explained by the surface area of magnetite, which was taken into consideration in the study of Poulton et al. (2004).

Within or around the zone of anaerobic oxidation of methane, other important mineral precipitates comprise authigenic carbonates – such as Mg-calcite, aragonite and dolomite – which precipitate due to the high concentrations of HCO_3^- ions generated by AOM (c.f., Eq. 8.8; e.g., Greinert et al. 2002; Moore et al. 2004), as well as diagenetic barite (e.g., Brumsack 1986; Torres et al. 1996). Due to the high rates of AOM, the formation of carbonate precipitates is particularly pronounced in sedimentary settings influenced by venting and seepage of methane-rich fluids and/or the presence of gas hydrates. The formation of carbonates in such methane dominated environments, which are mostly driven by advective transport processes, is discussed in Chapters 13 and 14. The factors and conditions determining the dissolution and preservation of the different carbonate phases are the subject of Chapter 9.

8.4.2 Pyrite Formation

Iron-sulfide minerals are important sinks for iron and sulfur as well as for trace metals and play an important role in the global cycles of these elements. Over the past 30 years extensive studies –

both in the field and in the laboratory – have been performed to elucidate the mechanisms of pyrite formation and in particular to understand the main factors which control the formation of pyrite in natural environments. Although these studies have provided significant new insights, fundamental questions remain with respect to (a) the different reaction rates obtained on the basis of field studies and laboratory experiments, (b) the role of FeS surface reactions and the electron acceptor involved in the conversion to pyrite, and (c) the role of sulfate-reducing bacteria – in particular their cell walls - in directing and promoting the pyrite precipitation process. A comprehensive review of pyrite formation in sedimentary environments by Rickard et al. (1995) was recently updated by Schoonen (2004). Besides the detailed description of the different pathways and mechanisms of sedimentary pyrite formation, the review by Schoonen (2004) also discusses the synthesis of nanoscale pyrite, trace element incorporation, and the formation of electronic defects during the formation process.

All pyrite forming pathways identified so far involve several reaction steps. First, hydrogen sulfide, produced during sulfate reduction (Eq. 8.1, 8.2 and 8.8), reacts with dissolved iron or reactive (towards sulfide) iron minerals to form amorphous iron sulfides, such as mackinawite (FeS). The precipitated amorphous iron sulfide is highly unstable and rapidly transforms to metastable iron sulfide phases such as pyrrhotite ($Fe_{x-1}S$) or greigite (Fe_3S_4), both of which represent intermediates in the reaction pathways to pyrite (FeS_2). The conversion of amorphous FeS to pyrite requires an electron acceptor and a change in the molar Fe:S ratio from about 1:1 to 1:2. The electron acceptor is needed to oxidize the S(-II) component in FeS to the mean oxidation state of –I in FeS_2. Concurrently, the Fe:S ratio has to decrease either via the addition of sulfur or the loss of iron. Three

general pathways have been reported to convert FeS to pyrite:

(1) Addition of sulfur with the sulfur species acting as electron acceptor (Berner 1970; Berner 1984; Luther 1991). This pathway has been termed the "polysulfide pathway".

$$FeS + S^0 \rightarrow FeS_2 \qquad (8.13)$$

(2) Reaction with hydrogen sulfide, i.e. addition of sulfur with a non-sulfur electron acceptor (Rickard and Luther 1997). This conversion mechanism is known as the "H_2S pathway".

$$FeS + H_2S \rightarrow FeS_2 + H_2 \qquad (8.14)$$

(3) Loss of iron, combined with an (additional) electron acceptor (Wilkin and Barnes 1996), known as the "iron-loss pathway".

$$2FeS + 2H^+ \rightarrow FeS_2 + Fe^{2+} + H_2 \qquad (8.15)$$

As reviewed in detail by Schoonen (2004) these different conversion mechanisms – and in particular the H_2S pathway – have received controversial discussion. However, field studies have shown that hydrogen sulfide can indeed sulfidize amorphous FeS and form pyrite. Rickard (1997) found that the H_2S process is by far the most rapid of the pyrite-forming reactions hitherto identified and suggested that it represents the dominant pyrite forming pathway in strictly anoxic systems. In addition, Morse (2002) discussed that the oxidation of FeS by hydrogen sulfide is the faster process compared with the oxidation by elemental sulfur. Berner (1970) suggested that, in the presence of zero-valent sulfur, a complete transformation of FeS to pyrite should be possible on a time scale of years. An incomplete conversion of FeS to pyrite, as often observed, e.g. in

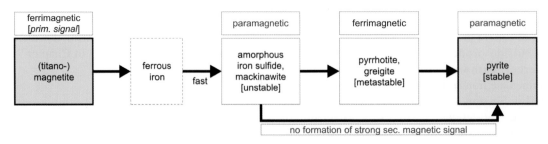

Fig. 8.9 Schematic representation of the major pathways of the transformation of iron oxides to iron sulfides in anoxic marine environments, in relation to the alteration of the magnetic record. From Riedinger (2005).

Table 8.4 Authigenic iron sulfides formed in marine sediments. Modified from Schinzel (1993).

Mineral	Composition	Crystal class	Magnetic properties
mackinawite	FeS	tetragonal	paramagnetic
pyrrhotite	$Fe_{1-x}S$	hexagonal or orthorhombic	ferrimagnetic, antiferrimagnetic
pyrite	FeS_2	cubic	paramagnetic
markasite	FeS_2	orthorhombic	diamagnetic
greigite	Fe_3S_4	cubic	ferrimagnetic

the anoxic sediments of Kau Bay, Indonesia, (Middelburg 1990) or the Amazon Fan (Kasten et al. 1998), can be due to a shortage of zero-valent sulfur during pyrite formation. Zero-valent sulfur may form as a result of the incomplete oxidation of H_2S or FeS by oxidants such as O_2, NO_3^-, MnO_2 or FeOOH. A limited transformation of FeS to pyrite can therefore be due to either (a) a limited amount of sulfide available to be oxidized to zero-valent sulfur or (b) a shortage in the amount of oxidants supplied by diffusion, bioturbation or burial.

Direct precipitation of pyrite without intermediate iron sulfide precursors was reported for salt marsh sediments, where pore waters were undersaturated with respect to amorphous FeS but oversaturated with respect to pyrite (Howarth 1979; Giblin and Howarth 1984). In these sediments, the oxidizing activity of the roots favored the formation of elemental sulfur and polysulfides which were thought to react directly with Fe^{2+}. The direct reaction pathway may proceed within hours, resulting in the formation of single small, euhedral pyrite crystals (Rickard 1975; Luther et al. 1982). Framboidal pyrite, apart from that formed by the mechanism presented by Rickard (1997), is formed slowly (over years) via intermediate iron sulfides (Sweeney and Kaplan 1973; Raiswell 1982).

Irrespective of the pyrite forming pathway the conversion of FeS into pyrite via intermediate iron sulfides goes along with a pronounced change in the rock magnetic properties of the particular iron sulfide phases (Fig. 8.9, Table 8.4). Amorphous FeS is paramagnetic and therefore has a low magnetic susceptibility and does not contribute to the remanent magnetization of sediments. In

contrast, the metastable iron sulfides, pyrrhotite and greigite, which represent precursor phases of pyrite, are ferrimagnetic and thus have a significant magnetic potential. This becomes obvious from maxima in magnetic susceptibility located slightly above enrichments of amorphous FeS in the form of black bands in sediments of the Amazon Fan (see Fig. 8.11 below; Kasten et al. 1998) and in the Black Sea (Neretin et al. 2004). The presence of greigite and pyrrhotite – responsible for the peaks in magnetic susceptibility – document the progress in the sequence of conversion of FeS to pyrite. With a further progression in the conversion pathway and the formation of the stable pyrite the magnetic signal is lost again as pyrite is paramagnetic (Fig. 8.9, Table 8.4).

It is generally assumed and found that there is a gradual decrease in the amount of amorphous iron sulfides and a corresponding increase in the amount of pyrite – thus a drop in the AVS to pyrite ratio – upon burial with increasing sediment depth. The discussion and examples presented in the following sections will, however, show that this is not always the case. In particular in sedimentary sequences affected by sulfidization fronts the sequence of conversion of FeS into pyrite can be reversed with depth and the most immature and unstable Fe sulfides be found at greater depth coinciding with the current depth of the diffusional H_2S/Fe^{2+} boundary. Furthermore, iron sulfide formation often occurs at or along fixed geochemical boundaries or reaction fronts. Thus, the sequence of iron sulfide formation is often restricted to or even bound to certain depth horizons rather than occurring gradually with increasing sediment depth. Mineral precipitation at sulfidization fronts can also occur cyclically within the sedimentary sequence – superimposed by episodic changes in depositional or other environmental conditions.

8.4.3   Magnetite and Barite

Besides the precipitation of mineral phases, dissolution of numerous minerals initially supplied to the seafloor can occur in sediments that experience sulfate reduction and/or sulfate-depleted conditions. Iron (oxyhydr)oxides and barite ($BaSO_4$) are two examples with particular relevance for paleoceanographic research.

Magnetic iron (oxyhydr)oxides, especially magnetite (Fe_3O_4), are the main carriers of remanent magnetization in sediments. Magnetostratigraphy of deep sea sediment cores is a valuable method of dating and comparing sedimentary records. Dissolution of the magnetic minerals under sulfate- and iron-reducing conditions and/or subsequent precipitation of authigenic iron minerals may, however, alter the initial remanent magnetization and seriously compromise the interpretation of the sedimentary geomagnetic record (e.g. Karlin and Levi 1983, 1985; Funk et al. 2003a, 2003b; Reitz et al. 2004). The effects of AOM-driven sulfate reduction on the transformation of iron minerals and the associated modification of rock magnetic properties will be demonstrated below using examples from the Amazon Fan and the western Argentine Basin.

A second sedimentary component important for paleoceanographic reconstructions, which is prone to dissolution under conditions of sulfate reduction – or more precisely in sulfate-depleted sediments - is the barium sulfate mineral, barite ($BaSO_4$). Since a correlation has been detected between barite deposition and the flux of organic matter through the water column, the concentration of barite in sediments has been proposed and applied as a geochemical tracer to reconstruct past changes in ocean productivity (e.g., Bishop

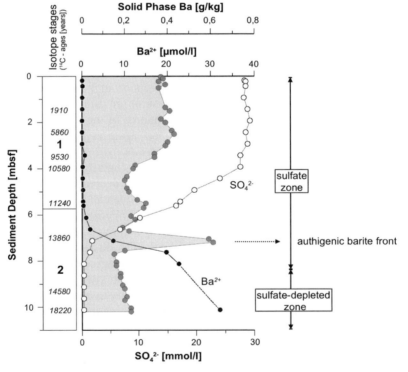

Fig. 8.10 Geochemical data for core GeoB 1023-4 recovered off north Angola (17°09.6'S, 10°59.9'E, 2047 m water depth). Barium and sulfate pore-water concentration profiles as well as the distribution of solid-phase barium indicate the precipitation of authigenic barite at a front slightly above the depth of complete sulfate consumption. Below the sulfate/methane transition barite becomes undersaturated and is thus subject to dissolution due to the total depletion of pore-water sulfate. Dissolved barium diffuses upwards into the sulfate zone where the mineral barite becomes supersaturated and so-called authigenic or diagenetic barite precipitates at a front at the base of the sulfate zone. Modified from Gingele et al. (1999), after Kölling (1991).

1988; Dehairs et al. 1991; Dymond et al. 1992; Gingele and Dahmke 1994; Paytan et al. 1996a). Sedimentary barite has also been used to reconstruct the strontium isotope composition of seawater over time (e.g., Paytan et al. 1993; Mearon et al. 2003), to determine the sulfur isotope ratio of marine sulfate (Cecile et al. 1983; Goodfellow and Jonasson 1984; Paytan et al. 1998), and to characterize Holocene sedimentation rates by using excess ^{226}Ra decay (Paytan et al. 1996b; van Beek and Reyss 2001; van Beek et al. 2002).

Although it is frequently stated that the process of sulfate reduction promotes barite dissolution it is, in fact, the *complete depletion* of interstitial sulfate which leads to an undersaturation of the pore water with respect to barite and to its subsequent dissolution (e.g., von Breymann et al. 1992; Torres et al. 1996; Gingele et al. 1999). This process is obvious from pore water Ba^{2+} concentrations which typically increase to micromolar concentrations with depth below the SMT. Dissolution of barite below the SMT and reprecipitation of diagenetic barite slightly above the sulfate penetration depth in so-called authigenic or diagenetic barite fronts can drastically obscure the primary barite record and thus lead to wrong paleoceanographic interpretations. The effect of barite dissolution in the zone of total sulfate depletion, and the subsequent reprecipitation of diagenetic barite at higher sediment levels is illustrated in Figure 8.10 for a sediment core recovered from the continental margin off Angola. The use of barite as a geochemical tracer or archive for paleoceanographic reconstructions is therefore limited – not to say precluded - in sediment intervals that are either currently sulfate-free or have been strongly sulfate depleted in the past (e.g. von Breymann et al. 1992; Gingele and Dahmke 1994; Torres et al. 1996; Gingele et al. 1999; Dickens 2001). While the current geochemical zonation of a sediment column can be determined from pore water sulfate data, a past migration of the SMT – and particularly a downward strike over time as a result of a transient decrease in methane flux from below – has also to be considered a possible cause that may alter the solid-phase Ba contents.

In a novel approach, Dickens (2001) used sedimentary Ba records to assess temporal changes during the Late Pleistocene in the upward flux of methane within sediments of the Blake Ridge that are rich in sub-surface gas hydrates. Due to a lack of Ba enrichment above the present depth of the barite front the author concluded that the upward methane flux from the underlying gas hydrate reservoir has not varied significantly across major changes in sea level and hydrostatic pressure. A possible means to identify the origin/source of barite enrichments is the analysis of the stable sulfur isotopic composition, δ^{34}S, of the barite particles. As the sulfur isotopic composition of sulfate in the water column, where the productivity-related biogenic barite is assumed to be formed, is significantly lighter (around +21 ‰; Paytan et al. 2002) than that of pore water sulfate at the SMT, where diagenetic barite precipitates (around + 40 ‰; Torres et al. 1996), it is generally assumed that the δ^{34}S of barite should give insight into its formation mechanism or origin (Paytan et al. 2004).

8.4.4   Non-Steady State Diagenesis

A particularly pronounced alteration of the sedimentary solid phase by mineral dissolution and authigenic mineral precipitation can take place in connection with sulfate reduction during non-steady-state diagenesis, i.e. during phases of deposition by which sedimentary conditions are not constant over time and sediment geochemistry adjusts to the new situation. Non-steady-state diagenesis can be initiated by any changes in the fluxes of electron donors and acceptors and environmental conditions, for example by changes in type of depositing sediment, oxygen content of the bottom water, sedimentation rate, flux of organic matter to the seafloor, and upward flux of methane (e.g., Kasten et al. 2003). The non-steady-state processes that occur during the transition from one depositional situation to another, or at the interface between two sediment types, typically comprise the development of geochemical reaction fronts. These fronts can either be fixed at particular sediment horizons for a prolonged period of time or move downwards or upwards within the sediment column. During a fixation or a slow migration of reaction fronts or geochemical boundaries the diagenetic processes acting at these fronts can produce higher concentrations of the authigenic minerals formed at specific sediment levels than under steady-state conditions. On the other hand, a fast upward migration of a geochemical front may support the burial and thus the preservation of substantial amounts of metastable minerals (Riedinger et al. 2005, see below).

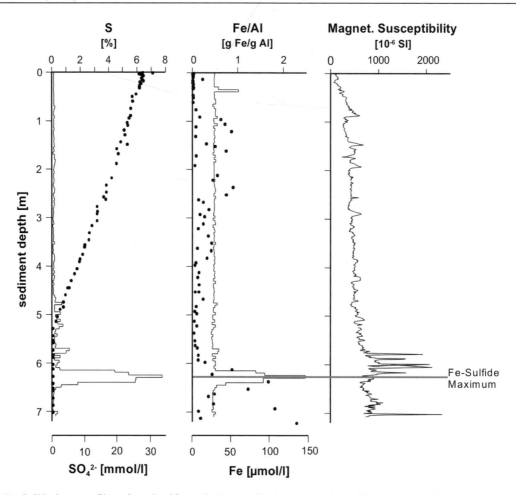

Fig. 8.11 Solid phase profiles of total sulfur and Al-normalized concentration of Fe from total digestion (solid lines), magnetic susceptibility (Funk, unpublished data), as well as pore-water concentration profiles of SO_4^{2-} and Fe^{2+} (solid circles) for core GeoB 1514-6 recovered from the Amazon deep-sea fan (3509 m water depth). The horizontal line demonstrates that the iron peak, which represents mainly iron sulfides, does not directly coincide with the peak in magnetic susceptibility. Modified from Kasten et al. (1998), after Schulz et al. (1994).

In the glacial sediments of the Amazon Fan, which were deposited during the Pleistocene sea level low-stand, Kasten et al. (1998) reported a pronounced enrichment of iron sulfides located several meters below the seafloor, near the present-day SMT (Fig. 8.11). They concluded that this iron sulfide enrichment formed during a period of non-steady state diagenesis, when the zone of anaerobic methane oxidation became fixed at this particular sediment level for an extended period. The condition that is likely to have initiated the non-steady state diagenetic situation was a pronounced decrease in sedimentation and organic carbon accumulation rate during the transition from the Pleistocene to the Holocene. Similar distinct enrichments of authigenic minerals due to a fixation of the SMT during non-steady state

depositional conditions induced by decreases in sedimentation rate or even pauses in sedimentation have also been reported by Raiswell (1988), Torres et al. (1996) and Bréhéret and Brumsack (2000).

A second striking feature of the Amazon Fan sediments (gravity core GeoB 1514-6) is the occurrence of a pronounced maximum in magnetic susceptibility, located slightly above the zone of strong enrichment of iron sulfide minerals (Fig. 8.11). As the total iron sulfides did not display a local maximum at the depth of the susceptibility peak, Kasten et al. (1998) suggested the presence of a specific iron sulfide with a high magnetic potential. X-ray diffraction analyses performed on sediment samples from this depth revealed the presence of the magnetic iron sulfide mineral, greigite (Fe_3S_4). These findings demonstrate that non-steady-state diagenesis does not only lead to a modification of the bulk sediment composition, but can also generate distinct magnetic signals of post-depositional origin within the sedimentary record.

The continental margin off Argentina and Uruguay is a highly dynamic sedimentary setting characterized by gravity driven mass-flow deposits and is therefore ideally suited to study diagenetic processes under shifting depositional conditions. As a typical feature of the deposits in this area, distinct minima in magnetic susceptibility are found a few meters below the sediment surface (c.f. Fig. 8.12). In order to reveal the origin of these susceptibility "gaps", Riedinger et al. (2005) carried out extensive geochemical and rock-magnetic investigations as well as numerical

transport-reaction modeling using the program CoTReM (Chapter 16). Pore water data revealed that the conspicuous minima in susceptibility coincide with the current depth of the SMT (Fig. 8.12). The hydrogen sulfide generated by this process reacts with the abundant iron (oxyhydr)-oxides resulting in the precipitation of iron sulfides accompanied by a nearly complete loss of the magnetic signal. Below the sulfidic sediment interval, where the magnetic susceptibility had not significantly suffered from diagenetic overprint, high amounts of iron oxides were present. Numerical modeling of geochemical data suggests that these high amounts of preserved Fe(III) as well as the distinct and spatially restricted loss in susceptibility can only be produced by extremely high glacial sedimentation rates ($\geq$100 cm/kyr) - shielding the Fe(III) minerals from complete

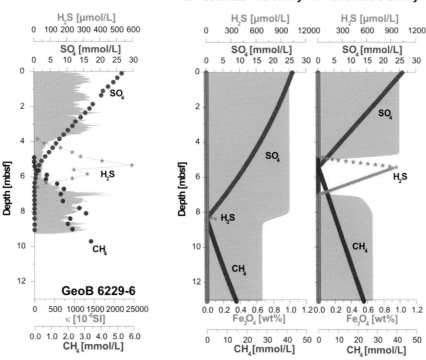

Fig. 8.12 *Left frame:* Sulfate (red circles), methane (blue circles), and sulfide (stars) pore water profiles for core GeoB 6229-6 (3446 m water depth) from the western Argentine Basin off the Rio de la Plata (sulfate data are from Hensen et al. 2003). The magnetic susceptibility is shown in grey. *Middle and right frame:* Results of numerical modeling of diagenetic alteration of magnetite to iron monosulfide with a major change of mean sedimentation rate (SR) for a sediment porosity of 75%. (a) A mean sedimentation rate of 100 cm kyr^{-1} leads to reduction of only about one third of the magnetite. (b) If the mean sedimentation rate is decreased to 5 cm kyr^{-1}, a time interval of ~9000 years is needed to reduce the total amount of magnetite initially contained within an interval of 2 m thickness. Modified from Riedinger et al. (2005).

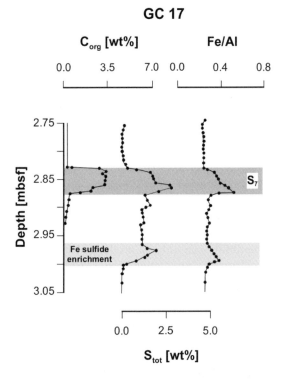

GC 17

Fig. 8.13 Concentration versus depth profiles of organic carbon (C_{org}), total sulfur (S_{tot}) and Fe/Al ratio (mol/g Fe divided by mol/g Al) through sapropel S_7 in gravity core GC17 from the eastern Mediterranean. The upper dark-grey bar marks the stratigraphical position of the sapropel visible in the core. The lower light-grey bar marks the coincident peaks of S_{tot} and Fe/Al below the sapropel which represent a Fe-sulfide band formed by Liesegang phenomena. Modified from Passier et al. (1996).

reduction - and a subsequent drastic drop during the glacial/Holocene transition (Fig. 8.12). Similar to the conditions on the Amazon Fan the strong decrease in sedimentation rate encountered during the last climatic transition induced a fixation of the SMT and an enhanced overprint of rock magnetic and mineralogical properties at this particular sediment layer. To obtain the observed geochemical and magnetic patterns, the SMT must have remained at a fixed position for about 9000 years – a time span which closely corresponds to the time since the Pleistocene/Holocene transition.

Sulfate reduction can also occur within discrete, organic-rich layers which can lead to a distinct overprint of the primary sediment composition within the organic-rich layers or in the sediment above and below. The non-steady state diagenetic processes occurring in and below the organic-rich layers (sapropels) of the Eastern Mediterranean have been studied by Passier et al.

(1996). They presented a model for the formation of distinct iron sulfide enrichments below the sapropels (Fig. 8.13) by the development of a downward moving sulfidization front, similar to Liesegang phenomena (formation of distinct iron sulfide bands) described by Berner (1969). A Liesegang situation exists in depositional systems that are characterized by intermediate contents of reactive (towards sulfide) iron, i.e. in systems that are neither iron nor sulfide dominated. Passier et al. (1996) concluded that excess hydrogen sulfide generated by dissimilatory sulfate reduction within the sapropel was able to migrate downwards (downward sulfidization). This resulted in the formation of pyrite below the sapropel by the reaction of hydrogen sulfide with solid-phase ferric iron and Fe^{2+} diffusing upwards from underlying sediments as schematically illustrated in Figure 8.14.

Downward progressing sulfidization fronts have also been reported to be initiated by transitions from limnic to brackish/marine conditions in the Baltic Sea (Böttcher and Lepland 2000; Neumann et al. 2005) and the Black Sea (Jørgensen et al. 2004; Neretin et al. 2004). In contrast to the example from the Eastern Mediterranean presented above, the sulfide driving the downward sulfidization in these sedimentary settings is derived primarily from AOM and from the increase in sulfate concentration in the water column during the Holocene. At the sites on the western continental slope of the Black Sea investigated by Jørgensen et al.

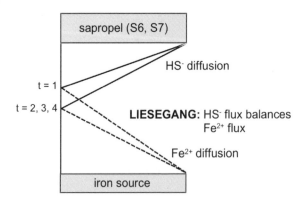

Fig. 8.14 Model for formation of iron sulfide bands below sapropels. The schematic depicted here represents a Liesegang situation (Berner 1969). The front at which downward diffusing hydrogen sulfide and upward diffusing iron react to form iron sulfides is fixed at particular levels below the sapropel for a prolonged period of time. Modified from Passier et al. (1996).

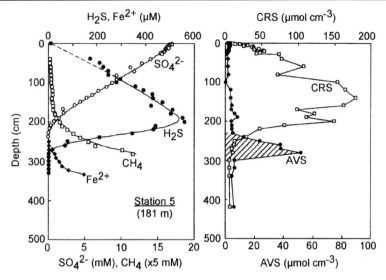

Fig. 8.15 Sulfur geochemistry of a 4-m deep sediment core from the upper slope of the western Black Sea. Left frame: SO_4^{2-}, H_2S, CH_4 and Fe^{2+} (notice scales) in the pore water. The smooth curves are model fits to the data based on the PROFILE model (Berg et al. 1998). Right frame: Chromium reducible sulfur (CRS) and acid volatile sulfide (AVS), the latter showing the black band of iron sulfide at 250-300 cm depth due to the downward progressing sulfidization front. From Jørgensen et al. (2004).

(2004) and Neretin et al. (2004) the sulfate-methane transition is typically located around 2 m sediment depth. Sulfide liberated into the pore water at this depth is diffusing up into the anoxic water column and is also drawn downward to a sulfidization front where it reacts with iron (oxyhydr)oxides and with Fe^{2+} diffusing up from the deeper iron-rich limnic deposits (Fig. 8.15). The current depth position of this HS^-/Fe^{2+} diffusion front is marked by a black band which mostly consists of amorphous iron sulfides termed AVS (acid volatile sulfide) in Fig. 8.15. These mineral phases are also responsible for the distinct black coloration of the sediment. Above (i.e., behind) this downward progressing sulfidization front the amorphous iron sulfides have been and are still converted into pyrite as seen from a distinct horizon of greigite and pyrite formation (c.f., CRS – chromium reducible sulfur representing pyrite; Fig. 8.15). Due to the unusual progression in the reaction sequence towards pyrite, the degree of pyritization (see Section 8.4.2) in this case decreases with increasing age of the deposits, i.e. the least mature or stable iron sulfides are found at greatest sediment depth.

The examples given above illustrate that sulfide produced by dissimilatory sulfate reduction, in particular in organic-rich layers or within the zone of AOM, can produce a profound diagenetic alteration of the sediment up to thousands or hun-

dreds of thousands of years after initial deposition and thereby cause a delayed chemical, mineralogical, isotopic, and magnetic lock-in, i.e. a formation of a relatively stable sedimentary signal at a defined depth. As a consequence, the age of the particular authigenic mineral does not correspond to the age of the sediment layer, in which it is formed, but is much younger. Counter-intuitive as it may seem, in the case of downward moving sulfidization fronts the age of the mineral precipitate becomes younger with increasing sediment depth. From these considerations it becomes obvious that the post-depositional alterations of mineral phases and element associations generated in this way complicate or even prevent interpretations of the geochemical environment during the time of original sediment deposition.

8.5   Pathways of Sulfide Oxidation

Vast amounts of sulfide, corresponding to 7 megaton of H_2S daily, are generated in marine sediments as the product of bacterial sulfate reduction. A small fraction of this sulfide is trapped within the sediment, mainly by reaction and precipitation with iron to form pyrite, or by the sulfidization of organic matter, and it thereby becomes buried in the sea bed (Brüchert and Pratt 1996). In continental slope and deep sea sediments this provides a net burial of reducing power over geological time that contributes to maintain an oxidized surface of the earth and thus, together with the burial of organic carbon, to regulate the atmospheric oxygen level (Berner 1982, 1989). Berner and Raiswell (1983) found that organic carbon and reduced sulfur were buried in the present ocean bed at a mean ratio of ca. 2.8 ($\pm$1.5) on a weight basis. This corresponds to a C_{org}:S_{red} burial ratio of 7.4 on a molar basis. Since the complete oxidation of organic carbon to CO_2 represents a change of 4 oxidation steps while the complete oxidation of reduced sulfur (pyrite) to sulfate represents 7 oxidation steps, the burial ratio is 4.3 in terms of reducing equivalents. The latter ratio corresponds to the amount of oxygen that can potentially accumulate in the atmosphere due to the burial of organic carbon versus reduced sulfur. Thus, the reduced sulfur is equivalent to ca. 20% of the buried reducing equivalents in the sea bed.

In deltaic and shelf sediments, where most sulfide production takes place, the intensive sulfide burial during interglacial periods is interrupted during glaciations that bind ocean water in polar ice caps and thereby may lower the sea water level by 100 m. The glacial sea water low-stand causes erosion of accumulated shelf sediments and partly reoxidation of the exposed pyrite, thereby returning much of the accumulated sulfide into the oceanic sulfate pool. Berner (1982) estimated the current burial of pyrite in marine sediments to be 39 Mt S yr^{-1} or $0.12 \cdot 10^{13}$ mol S yr^{-1}. This is equivalent to the sulfide production over 5-6 days per year or to 1.6% of the total sulfate reduction in the global ocean floor (cf. Fig. 8.4). Thus, 98.4% of the produced sulfide is reoxidized back to sea water sulfate on a geological time scale (million years).

The short-term (years to thousand years) burial of sulfide in the form of pyrite in ocean margin sediments is more efficient and generally accounts for 5-20% of the entire sulfide production (Jørgensen et al. 1990; Lin and Morse 1991; Canfield and Teske 1996). This burial provides a sink in the dynamic cycling of sulfur that is limited by the availability of reactive iron to bind the large excess of sulfide. Raiswell and Canfield (1998) found that, of the total iron in marine sediments, on average only 25-28% is highly reactive, 23-31% is poorly reactive, and 41-42% is unreactive. In spite of the small fraction of net

pyrite burial, a much larger fraction of the sulfur cycle, however, does go through pyrite and more labile iron sulfides, in particular in the surficial sediment. These metal-bound sulfides are recycled within the sediment together with the free sulfide so that 80-95% of the entire sulfide production is gradually oxidized back to sulfate. The reoxidation takes place at all depths of the sediment, most rapidly in the upper oxidized zone but also in the deeper and sulfidic part where the process is more difficult to detect.

The oxidation of organic material by sulfate reduction yields only a fraction of the energy that is available by aerobic respiration of the same organic compounds. Consequently, a large part of the potential chemical energy is still conserved in the product, H_2S, from sulfate reduction and this energy may become available to other microorganisms, provided a useful oxidant such as O_2 or NO_3^- is present. In coastal sediments where the organic deposition, and therefore the sulfate reduction, is particularly high, the reactive metal oxides may become completely reduced by sulfide. In this extreme case, H_2S may diffuse freely up to the sediment surface and reach the thin oxic skin of the surface sediment. A H_2S-O_2 interface thereby develops within the uppermost few mm of the sediment where the gradient-type of colorless sulfur bacteria may flourish on the chemical energy from H_2S. Such hotspots of sulfide oxidation may be recognizable from the dark coloration of the sediment surface due to black iron sulfide ("black spots"; Rusch et al. 1998). The sediments may also develop a distinct coating of filamentous sulfur bacteria, such as *Beggiatoa*, that store light refracting sulfur globules in their cells and thus provide the sediment with a distinct whitish appearence. Such white *Beggiatoa* mats are typical of the sediments around hydrothermal vents and cold seeps that bring H_2S from the subsurface in direct contact with oxygenated sea water (Jannasch et al. 1989). In extreme cases where the water column overlying the sediment is anoxic, e.g. in the permanently stratified Black Sea or during summer in some eutrophic coastal environments, sulfide is not retained at the sediment surface but penetrates directly up into the sea water (e.g., Roden and Tuttle 1992).

In oxic marine sediments, a brown layer rich in iron and manganese oxides generally separates O_2 and H_2S and thereby prevents a direct sulfide oxidation by oxygen (e.g. Thamdrup et al. 1994a). In this suboxic zone, neither O_2 nor H_2S is present in detectable concentrations. The metal oxides constitute an efficient sulfide barrier by oxidizing and binding H_2S that diffuses up from the sulfidic sediment below. In this case, H_2S oxidation by oxygen is the exception and requires that the metal oxide layer is penetrated by advective transport, for example by bioirrigation by burrowing animals that pump oxic water for respiration directly down into the sulfidic sediment. Another advective mechanism may be current-induced advective pore water transport in porous sandy sediments (Huettel et al. 1998), or oxygen transport down into the root zone of sea grass beds (Ballbjerg et al. 1998). Through such oxygen penetration also pyrite may be oxidized with oxygen, a process that has been extensively studied (e.g. Lowson 1982; Luther 1987; Moses and Herman 1991; Morse 1991). Pyrite oxidation with O_2 is a rather fast process that may be purely abiotic, catalyzed by an electron shuttle between adsorbed Fe(II) and Fe(III) ions transferring electrons from pyrite to O_2. Sulfate is the end product of the sulfur oxidation and iron oxides often coat the surface of the oxidized pyrite grains.

Most sulfide oxidation in marine sediments is anoxic (i.e. takes place in the absence of oxygen) and generally involves the precipitation of iron sulfide and the subsequent oxidation of iron-sulfur minerals back to sulfate. Evidence for anoxic sulfide oxidation comes from studies of pore water chemistry, solid phase distributions of metal oxides and sulfides, mass balance calculations, and direct experiments. By the use of radio-labeled H_2S added to anoxic sediment cores or slurries, a rapid transfer of the label could be observed into sulfur fractions defined as acid volatile sulfide (mostly FeS), chromium reducible sulfide (mostly FeS_2), elemental sulfur (S^0), or sulfate (Fossing and Jørgensen 1990). The radiolabeled FeS and S^0 were readily oxidized to sulfate in the anoxic sediment. Pyrite, in contrast, is more stable and is oxidized only over longer incubations. Yet, pyrite comprises the main sulfur pool in marine sediments and undergoes slow transport and oxidation of critical importance for the sulfur cycle.

These processes are illustrated in Fig. 8.16. Sulfate that penetrates down into the sediment from the overlying sea water is reduced to H_2S by sulfate reducing bacteria that use the deposited organic material as their energy source. Also methane diffusing up from below feeds sulfate reduction in the lower sulfate zone. At depth in

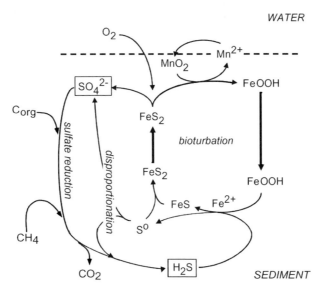

Fig. 8.16 The sulfur cycle in marine sediments. The cycle is energetically driven by deposited organic material and methane, both of which are used by sulfate reducing bacteria to produce H_2S. Much of the H_2S reacts chemically with iron (oxyhydr)oxides to form FeS and a range of intermediate oxidation states including S^0 and FeS_2. The further oxidation of these species back to sulfate is mediated by the vertical conveyer belt of bioturbation caused by burrowing macrofauna. Reoxidation of the solid phase sulfur species to sulfate at the sediment surface may be by oxygen, nitrate or manganese oxide. The same conveyer belt brings the oxidized iron back down towards the sulfide production zone where it reacts with further H_2S. A highly efficient recycling of sulfur is thereby achieved. (From Jørgensen and Nelson 2004).

marine sediments, the production of H_2S may exceed the availability of reactive metal oxides and H_2S accumulates in the pore water (see Fig. 8.6, 8.12, 8.13, and 8.17). The H_2S diffuses upwards along a concentration gradient that generally reaches zero at the bottom of the suboxic zone. Concurrently, the H_2S reacts with buried iron oxides to form FeS, FeS_2, S^0 and a number of other solid or dissolved intermediate products. Once the reduced sulfur is bound in the solid phase, e.g. as pyrite, its further oxidation depends on a slow reaction with further Fe(III) species or its transport up to near-surface layers with oxidants of higher redox potential.

Pyrite transport in near-surface sediments generally takes place through the conveyer belt of bioturbation whereby burrowing macrofauna mix the sediment or directly move sediment particles as part of their deposit feeding behavior (Fig. 8.16). As the pyrite reaches up into the suboxic

zone it may react with oxidants such as oxygen or manganese oxide and be converted into sulfate and iron oxides (or iron (oxyhydr)oxides). The iron oxides are in turn transported downwards through the same conveyer belt and thereby become available for further binding of sulfide and pyrite formation. The pyrite oxidation by manganese oxide has been implied from chemical profiles (Canfield et al. 1993) and demonstrated directly through experiments (Schippers and Jørgensen 2001, 2002). The process is interesting in that it involves the reaction between two mineral phases in the sediment that must be in close proximity for the oxidation to proceed. The initial reaction is purely chemical and was proposed to occur by a Fe(II)/Fe(III)-shuttle in the pore fluid between the mineral surfaces of FeS_2 and MnO_2. The immediate products of the oxidation are thiosulfate and polythionates. These can be further oxidized to sulfate by manganese reducing bacteria, thus

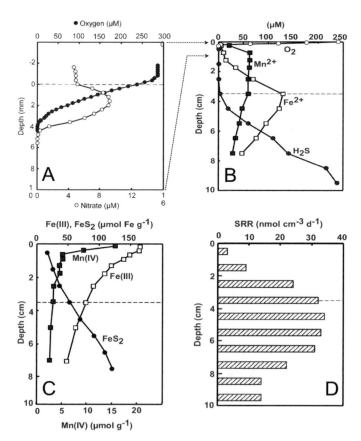

Fig. 8.17 Biogeochemical profiles of sulfur, manganese and iron species in a coastal marine sediment (Aarhus Bay, Denmark, 16 m water depth). A) Oxygen and nitrate profiles measured with O_2 and NO_3^- microsensors. B) Pore water profiles of dissolved manganese, iron and H_2S. C) Profiles of solid phase oxidized manganese and iron and of pyrite. D) Distribution of sulfate reduction rates (SRR) measured by ^{35}S-technique. The broken line at 4 cm depth indicates the transition between the suboxic zone and the sulfidic zone. Data in A) were measured at the same site but a different year than data in B)-D). (Data from Kjær 2000 and Thamdrup et al. 1994a; reproduced from Jørgensen and Nelson 2004).

making the complete pyrite oxidation to sulfate dependent on microbial catalysis (Schippers and Jørgensen 2001):

$$FeS_2 + 7.5\ MnO_2 + 11\ H^+ \rightarrow$$
$$Fe(OH)_3 + 2\ SO_4^{2-} + 7.5\ Mn^{2+} + 4\ H_2O \quad (8.16)$$

Pyrite oxidation by nitrate could not be demonstrated through short-term sediment experiments but FeS is readily oxidized, both by nitrate and MnO_2 (Aller and Rude 1988; Schippers 2004). The immediate product by FeS oxidation is not thiosulfate but polysulfide which subsequently degrades into elemental sulfur:

$$FeS + 1.5\ MnO_2 + 3\ H^+ \rightarrow$$
$$Fe(OH)_3 + S^0 + 1.5\ Mn^{2+} \quad (8.17)$$

The processes of sulfide oxidation described here are reflected in the chemical zonations of pore water chemistry, solid phase chemistry and bacterial activity. As an example, Fig. 8.17 shows data from organic-rich coastal sediment at the Baltic Sea - North Sea transition. Oxygen penetrated less than 0.4 cm into the sediment and showed no contact with the zone of detectable H_2S which started only from ca. 4 cm depth. Nitrate penetrated only slightly deeper than oxygen, with a peak at 0.2 cm depth due to aerobic oxidation of ammonium (nitrification) diffusing up from the sediment below. The upper 0.4-4 cm of the sediment comprised the suboxic zone where maxima in dissolved reduced manganese and iron showed the zones where the reduction of these metals was most intensive (peaks at 1- 2 cm depth for manganese and 3-4 cm depth for iron). The concentration of solid phase manganese oxide dropped steeply with depth from the sediment surface to 1 cm depth where its reduction was most intensive. Oxidized iron decreased more gradually as it was most intensively reduced in the lower part of the suboxic zone. Measurements of sulfate reduction showed that H_2S was produced throughout the sediment with maximum rates at the top of the sulfidic zone. Sulfate reduction also took place in the suboxic zone but the produced H_2S was here rapidly reoxidized and did not reach detectable concentrations. Only by the short-term experimental measurements using $^{35}SO_4^{2-}$ could the SRR activity therefore be demonstrated. Due to this sulfate reduction, a part of the manganese and iron reduction was driven by sulfide oxidation while another part was due to the direct oxidation of organic matter by heterotrophic, metal-reducing bacteria (Thamdrup et al. 1994a).

The intermediate sulfur species in sulfide oxidation, such as thiosulfate and elemental sulfur, are not stable in the sediment but are further transformed by microorganisms. Thiosulfate is turned over within hours or days and generally occurs only in sub-micromolar concentration in the sediment pore water (Thamdrup et al. 1994b). Elemental sulfur accumulates to much higher concentration in the solid phase but may also turn over on a seasonal or longer time scale in near-surface sediments (Troelsen and Jørgensen 1982). The preferred pathway of bacterial thiosulfate or sulfur transformation depends strongly on the chemical environment in the sediment. In the near-surface sediment with suitable oxidants such as oxygen, nitrate or metal oxides, these sulfur species may be used as energy sources by chemoautotrophic bacteria and be oxidized completely to sulfate (Fig. 8.17, see also Chapter 5). When formed below the suboxic zone, they may be used as oxidants (electron acceptors) in bacterial respiration to oxidize organic material. Through such a bacterial thiosulfate or sulfur respiration these intermediates are reduced back to sulfide. When the availability of oxidants and organic material are both limited, the two sulfur species may be disproportionated.

The ability of certain anaerobic bacteria to disproportionate intermediate sulfur species such as thiosulfate was discovered only in the late 1980'ies (Bak and Pfennig 1987; Krämer and Cypionka 1989; Finster et al. 1998) but has since been shown to play an important role in the sulfur cycle of aquatic sediments (Jørgensen 1990; Jørgensen and Bak 1991; Thamdrup et al. 1993). It is characteristic for the disproportionation that the sulfur species are concurrently reduced to sulfide and oxidized to sulfate. This is an energy yielding reaction under appropriate sediment conditions and is independent of external reductants or oxidants. The process can be considered a type of inorganic fermentation and it provides sufficient energy for bacteria to live on. By thiosulfate disproportionation, the inner (sulfonate) sulfur atom changes oxidation step from +5 in $S_2O_3^{2-}$ to +6 in SO_4^{2-}, while the outer (sulfane) atom changes from -1 in $S_2O_3^{2-}$ to -2 in H_2S (Vairavamurthy at al. 1993; Eq. 8.18). By elemental sulfur with an oxidation state of 0, some of the atoms are reduced to H_2S and some oxidized to SO_4^{2-} in a proportion of 3:1 that maintains electron balance (Eq. 8.19):

$$S_2O_3^{2-} + H_2O \rightarrow H_2S + SO_4^{2-} \qquad (8.18)$$

$$4\,S^0 + 4\,H_2O \rightarrow 3\,H_2S + SO_4^{2-} + 2\,H^+ \qquad (8.19)$$

Disproportionation reactions do not cause a net oxidation of the sulfur species, yet they have a key function in sulfide oxidation. Disproportionation provides a shunt in the sulfur cycle whereby the H_2S formed by this reaction may be oxidized again to the same sulfur intermediate by metal oxides. Manganese oxide, for example, rapidly oxidizes H_2S to S^0 without participation of bacteria, but does not oxidize the S^0 further to sulfate (Burdige 1993). The elemental sulfur may, however, be disproportionated (Eq. 8.19) whereby a fourth of it is oxidized completely to sulfate while the remaining three fourths return to the sulfide pool. Through repeated partial oxidation of sulfide to elemental sulfur with manganese oxide and subsequent disproportionation of the elemental sulfur to sulfate and sulfide a complete oxidation of sulfide to sulfate by manganese oxide may be achieved (Fig. 8.16; Thamdrup et al. 1993; Böttcher and Thamdrup 2001):

$$4\,H_2S + 4\,MnO_2 \rightarrow 4\,S^0 + 4\,Mn^{2+} + 8\,OH^- \qquad (8.20)$$

Sum of Eq. 8.20 and 8.19:
$$H_2S + 4\,MnO_2 + 2\,H_2O \rightarrow$$
$$SO_4^{2-} + 4\,Mn^{2+} + 6\,OH^- \qquad (8.21)$$

An interesting biological mechanism of sulfide oxidation in coastal upwelling regions and other high-productivity coastal ecosystems was discovered in the mid 1990'ies (Fossing et al. 1995). The sediment underlying some of the most intensive upwelling systems, e.g. off the Pacific coast of South and Central America or the Atlantic coast of southwest Africa, is densely populated with sulfur bacteria (species of *Thioploca*, *Thiomargarita*, and *Beggiatoa*) that have a peculiar mode of life and an unusually large cell size (Schulz et al. 1999). Within a liquid vacuole inside each cell they accumulate nitrate from the ambient sea water and later use this nitrate down in the sediment as an electron acceptor for sulfide oxidation from which they gain energy. The sulfide is oxidized first to elemental sulfur, which is stored transiently in the cells, and then to sulfate. This adaptation is unique in that it enables the motile types of the bacteria to commute up and down between the nitrate source in the sea water and the sulfide source in the sediment without having access to both at the same time (Jørgensen and Gallardo 1999; Schulz and Jørgensen 2001; see Chapter 6).

8.6   Determination of Process Rates

The rates of biogeochemical processes such as sulfate reduction in marine sediments can be determined by different approaches that all have strengths and weaknesses and may demonstrate fundamentally different aspects of the process. It is important for each application to consider carefully the properties of the sediment to be analyzed and the limitations of the method applied. The study of shallow and highly dynamic surface sediments requires a different approach than the study of deep sediments that have undergone stable diagenesis over a long time period. By the shallow sediment an experimental measurement of the process rate may be optimal whereas deep in the sediments generally a modeling approach is preferred. Dependent on the method, however, either the gross or the net rate of sulfate reduction is measured. The extent to which these differ has been determined only in a few cases.

Gross sulfate reduction rates (gross SRR) can be measured by experimental incubation of recovered sediment samples on board ship or in the laboratory. Today, this is mostly done by the use of radioactively labeled sulfate, $^{35}SO_4^{2-}$, as a tracer in order to obtain sufficiently high sensitivity of the method (see Chapter 5). Originally developed by Sorokin (1962), Ivanov (1968), Jørgensen (1978), this method has been modified and refined over the years (e.g., Howarth 1979; Canfield et al. 1986; Fossing and Jørgensen 1989; Kallmeyer et al. 2004). By adaptation of the radiotracer method it has been possible to directly measure sulfate reduction rates that vary over more than 7 orders of magnitude, for example on the Peruvian shelf, from >1000 nmol SO_4^{2-} cm^{-3} day^{-1} at the sediment surface to <0.001 nmol SO_4^{2-} cm^{-3} day^{-1} at 100 m subsurface (Parkes et al. 2005; J. Kallmeyer et al. in prep.).

The *gross SRR* as measured by $^{35}SO_4^{2-}$ technique comes closest to the overall sulfate reduction that takes place in the sediment. The radiotracer experiment generally lasts for only a few hours to a day and little reoxidation of produced ^{35}S-sulfide takes place during this time interval, unless the sediment is rich in reactive oxidized iron which may be the case in the suboxic zone near the sediment surface. The measurement should be kept short in order to minimize changes in the chemistry or microbial activity of the sediment under the laboratory conditions used. This is an important criterion for the use of radiotracer instead of just following the gradual disappearance of pore water sulfate over time as has been used earlier and is still applied when safety considerations prohibit the use of radio-tracer. Tracking the sulfate concentration over time is, however, a rather insensitive method that mostly requires incubations over many days or weeks during which significant changes in mea-sured rates occur.

Another approach is based on the experiment that nature has already done, namely by genera-ting decreasing sulfate concentrations down through the sediment column due to sulfate reduction acting over many years. The resulting pore water profile of sulfate reflects the rate of *net SRR*, which represents the gross SRR minus the (long-term) rate of reoxidation of reduced sulfur species back to sulfate. Such profiles, combined with solid-phase data on the sulfur chemistry and organic carbon content, have been used to calculate the distribution of sulfate reduction over longer periods by 1-dimensional reaction-transport modeling (e.g., Berner 1980; Canfield 1991; Schulz et al. 1994). Modeling of the net SRR generally assumes steady state conditions of reduction rate, transport and sedimentation and requires qualified information on the transport coefficients of pore water species down through the sediment column. Since disturbing factors such as burrow irrigation by the benthic macro-fauna or current-induced advective pore water flow may strongly enhance transport in addition to molecular diffusion, it is mostly difficult to provide accurate transport coefficients in the near-surface sediment. At depth in the sediment, however, molecular diffusion and steady state are more likely to prevail. Here the modeling approach has its obvious strength relative to the radiotracer measurements that are increasingly prone to disturbance artifacts the deeper in the sediment the samples are taken.

In the following, we will use data of Jørgensen et al. (2004) from the Black Sea as an example to illustrate the differences between the two approaches. Fig. 8.18 shows results from the deep sulfidic part of the Black Sea where macrofauna are unable to live and where bioirrigation is there-fore absent (although current-induced advective pore water transport may take place near the sedi-ment surface). Sulfate reduction rates were mea-sured experimentally by the radiotracer method down to 20 cm depth in the sediment (Fig. 8.18 A).

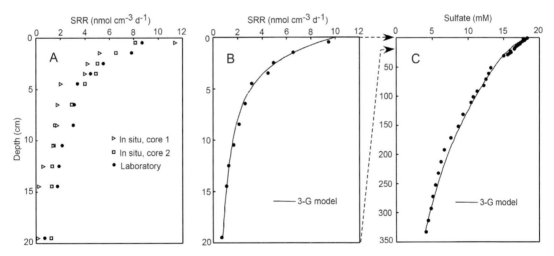

Fig. 8.18 Determination of sulfate reduction rates in sediment cored by multicorer (A and B) and by gravity corer (C) from 1176 m water depth in the sulfidic part of the western Black Sea. A) Sulfate reduction rates (SRR) measured experimentally using [35]SO_4^{2-}, either directly *in situ* on the sea floor using a benthic lander or shipboard in the laboratory. B) Mean experimental sulfate reduction rates (data points) and a model curve fitted to these rates. C) Sulfate concentrations measured in the pore water (data points) and modeled sulfate profile (smooth curve) that matches the fitted SRR in the uppermost 20 cm. After Jørgensen et al. (2001).

A multi-G model of Westrich and Berner (1984) was applied to fit a smooth curve to these rate data. The model assumes that the sediment organic matter that feeds sulfate reduction consists of three pools, each of which is degraded exponentially over depth and time, but each having its own pool size and exponential decay constant. Fig. 8.18 B and C show that this model fits well with both the measured sulfate reduction rates in the upper sediment layer and the sulfate curve in the deeper sediment down to 330 cm depth.

A point to observe in Fig. 8.18 C is that the measured and modeled profiles of pore water sulfate in the uppermost 0-20 cm hardly show any curvature and, thus, indicate zero sulfate reduction in this layer. The radiotracer measurements in Fig. 8.18 B, however, show that 66% of the sulfate reduction in the entire sulfate zone takes place in the top 15 cm where, in the deep sulfidic part of the Black Sea, sulfate reduction is the predominant pathway of organic carbon mineralization (Jørgensen et al. 2001). One possible reason for this discrepancy is that modeling has low sensitivity over narrow depth intervals because diffusion is fast over short distances and thereby reduces changes in concentration in spite of high reaction rates. In contrast, over the entire sulfate zone even a deep zone of relatively low sulfate reduction rates may affect the entire sulfate profile. This is why enhanced reduction rates in the sulfate-methane transition tend to shape the entire sulfate distribution in the sediment as seen in Fig. 8.18 C. Thus, experimental measurements are required to determine the sulfate reduction in the near-surface sediment. Such a discrepancy between measured and modeled rates is characteristic also of other sediments and is important to consider in relation to the global SRR estimates made in Fig. 8.3 and 8.4. Data from ocean margin sediments were there mostly determined by the radiotracer method (gross SRR) whereas those from the deep sea were modeled (net SRR).

8.7   The Sulfur Cycle

The sea bed functions as a giant anaerobic reactor where element cycles are coupled in a way that differs fundamentally from the cycles in the oxic ocean. The microbiological and geochemical processes of sulfur transformation thereby play key functions that control the mineralization of deposited organic matter and thereby the chemistry of the ocean. In the following we briefly summarize some main aspects of the sulfur cycle in marine sediments and discuss its role for the dynamics of sulfate in the ocean.

Rather than being a simple cycle, composed of anaerobic bacterial reduction of sulfate to hydrogen sulfide and aerobic reoxidation of H_2S to SO_4^{2-}, the transformations of sulfur in aquatic sediments include a combination of intermediate cycles or shunts and interactions with other element cycles (e.g., Jørgensen 1990; Luther and Church 1991; Thamdrup et al. 1993; van Cappellen and Wang 1996) schematically illustrated in Figure 8.16. Within this complex cycle, sulfur compounds occur in oxidation states ranging from -2 (H_2S) to +6 (SO_4^{2-})

By bacterial sulfate reduction H_2S is produced as the extracellular end-product (Widdel and Hansen 1991). During the oxidation of H_2S, oxic or anoxic, chemical or biological, compounds such as zero-valent sulfur (in elemental sulfur, polysulfides, or polythionates), thiosulfate ($S_2O_3^{2-}$), and sulfite (SO_3^{2-}) are produced (Cline and Richards 1969; Pyzik and Sommer 1981; Kelly 1988; Dos Santos Afonso and Stumm 1992). These intermediates may then be further transformed by one or several of the following processes:

- Respiratory bacterial reduction to H_2S,
- bacterial or chemical oxidation,
- chemical precipitation (e.g. FeS formation), or
- bacterial disproportionation to H_2S and SO_4^{2-}.

By bacterial disproportionation H_2S and SO_4^{2-} are produced concurrently without participation of an external electron acceptor or donor (Bak and Pfennig 1987; Thamdrup et al. 1993). The biogeochemical transformations of sulfur in marine sediments are closely coupled to the cycles of iron and manganese. Sulfate, iron oxides, and manganese oxides all serve as electron acceptors in the respiratory degradation of organic matter. As there are also non-enzymatic reactions between iron, manganese and H_2S within the sediment, the quantification of dissimilatory, heterotrophic Fe and Mn reduction is particularly difficult.

The precipitation of (authigenic) iron sulfides resulting from the reaction between H_2S and Fe phases exerts an important control on the distribution of H_2S in marine pore waters (Goldhaber and Kaplan 1974; Canfield 1989; Canfield et al.

1992). Depending on the abundance and reactivity of the Fe(III) minerals present as well as on the rate of sulfate reduction, H_2S may be undetectable despite its production at high rates. As the H_2S accumulates in the pore water, it diffuses up towards the sediment surface where most of it is ultimately oxidized back to sulfate. Only a small fraction, 5-20%, of the produced H_2S is trapped as solid phase iron-sulfur minerals or organic sulfur within the sediment. Over longer periods of thousands to millions of years, most of this trapped sulfur becomes exposed to oxidants and again returns to the great sulfate reservoir of the ocean.

Over geological time, ocean sulfate is thus continuously recycled through sulfate reduction in the sea bed. During this recycling, the sulfur atom in sulfate undergoes a major redox transformation of 8 oxidation steps between sulfate (+6) and sulfide (-2) and is thereby an effective oxidant for the mineralization of marine organic matter. Importantly, the sulfur atom in sulfate is separated from the oxygen atoms during the reduction process ($SO_4^{2-} \rightarrow H_2S$), and is subsequently combined with new oxygen atoms from sediment pore water upon reoxidation of the sulfide. How fast is this recombination of the elements and how long is the residence time of sulfate in sea water? We have calculated above that the magnitude of sulfate reduction in the global sea bed is $7.5 \cdot 10^{13}$ mol SO_4^{2-} yr^{-1}. The mean sulfate concentration of sea water is 29 mM and the total volume of ocean water is 1.37 billion km^3 (Garrison 1997). Thus, the ocean contains $1.37 \cdot 10^{21}$ liter of sea water with a total pool of $4.0 \cdot 10^{19}$ mol SO_4^{2-}. The turnover time of this global ocean sulfate pool through bacterial sulfate reduction is $(4.0 \cdot 10^{19})/(7.5 \cdot 10^{13})$ = ca. 0.5 million years.

Sulfate is reduced in the sea bed not only through bacterial catalysis but also through thermochemical catalysis at high temperature. At mid-oceanic ridges sea water is slowly convected through the hot magmatic crust where a part of it is heated to >350°C (see Chapter 13). At such high temperatures sulfate reacts as an oxidant for ferrous iron and other reduced minerals and the sulfate is reduced to H_2S without the participation of microorganisms which are excluded by temperatures above ca. 100°C (Weber and Jørgensen 2002). The global sea water flux that undergoes heating to >350°C has been estimated to be $3-6 \cdot 10^{13}$ liter yr^{-1} (Elderfield and Schultz 1996). By a complete reduction of the 29 mM of sea water sulfate, this corresponds to the reduction of $0.08-0.16 \cdot 10^{13}$ mol SO_4^{2-} yr^{-1} or 1-2% of the global microbiological sulfate reduction. The turnover time of the entire volume of ocean water at >350°C, and thus of the oceanic sulfate pool through thermochemical reduction, in hot mid-oceanic crust is thus 20-40 million years (cf. Elderfield and Schultz 1996).

The hydrothermal systems associated with the mid-oceanic ridges harbor some of the richest biological communities on the ocean floor and include giant clams, mussels, tube worms and shrimps. These animals are nourished by symbiotic bacteria that oxidize sulfide and methane from the vent water and use the energy for chemoautotrophic biomass production that in turn feeds the hosts. It is an interesting perspective that their energy source is independent of light and photosynthesis but is instead of geothermal origin and based on chemical reactions at high temperature. (It should be remembered, however, that the oxygen used for the oxidation of sulfide and methane is derived from photosynthesis in the surface ocean, so the chemosynthetic processes would still not run without sunlight.) Although the hydrothermal vent communities are rich oases of life on the sea floor, their total contribution to the oceanic carbon cycle is marginal. Bach and Edwards (2003) estimated that the global chemosynthetic biomass production from oxidation of new ocean crust may be up to 10^{12} g C yr^{-1}. The total potential for chemosynthetic primary production at the deep sea hydrothermal vents is globally estimated to be about 10^{13} g biomass per year (McCollom and Shock 1997). This represents only about 0.02% of the global primary production by photosynthesis in the oceans. As the chemosynthetic production takes place in the otherwise nutrient-poor deep sea, however, it makes a much more important contribution to the local carbon supply to the deep sea floor. Based on the oxygen uptake data presented in Fig. 8.4, the global organic carbon mineralization by oxygen in the deep sea bed at >3000 m water depth is only $5 \cdot 10^{13}$ mol yr^{-1}, relative to which the chemoautotrophic production at mid-oceanic ridges corresponds to 20%.

Acknowledgements

This is contribution No 0333 of the Research Center Ocean Margins (RCOM) which is financed by the Deutsche Forschungsgemeinschaft (DFG) at Bremen University, Germany. BBJ was also supported by the Max Planck Society and the Fonds der Chemischen Industrie and SK by the Helmholtz Society.

8.8   Problems

Problem 1

Organic matter is oxidized in the suboxic zone through iron or manganese reduction. This may be a direct oxidation by metal reducing bacteria or an indirect oxidation via sulfate reduction and sulfide oxidation. Does it matter for the end products which pathway dominates?

Problem 2

H_2S can be oxidized completely to sulfate by manganese oxide in marine sediments although the chemical reaction oxidizes H_2S only to elemental sulfur, S^0. How is the oxidation to sulfate then possible?

Problem 3

Which factors determine how large a fraction of the deposited organic material in the sea bed is mineralized through sulfate reduction relative to other mineralization pathways?

Problem 4

Only about 5% of the total mineralized organic carbon in marine sediments is ultimately degraded to methane. Why, then, does the anaerobic oxidation of methane with sulfate often generate quasi-linear sulfate profiles that indicate methane to be the dominant energy source for sulfate reduction?

Problem 5

Which factors may cause a lock-in of a diagenetic front in the sea bed and thereby lead to the diagenetic overprinting of paleoceanographic signals in a specific sediment horizon?

References

Aller, R.C., and Rude, P.D., 1988. Complete oxidation of solid phase sulfides by manganese and bacteria in anoxic marine sediments. Geochimica et Cosmochimica Acta, 52: 751-765.

Amend, J.P., Edwards, K.J., and Lyons, T.W. (eds), 2004. Sulfur Biogeochemistry - Past and Present. Geological Society of America Special Paper 379, Boulder, Colorado, 205 p.

Alperin, M.J., and Reeburgh, W.S., 1988. Carbon and hydrogen isotope fractionation resulting from anaerobic methane oxidation. Global Biogeoche-mical Cycles, 2: 279-288.

Bach W., and Edwards, K.J., 2003. Iron and sulfide oxidation within the basaltic ocean crust: Implications for chemolithoautotrophic microbial biomass production. Geochimica et Cosmochimica Acta, 67: 3871-3887.

Bak, F., and Pfennig, N., 1987. Chemolithotrophic growth of *Desulfovibrio sulfodismutans* sp. nov. by disproportionation of inorganic sulfur compounds. Archives of Microbiology, 147: 184-189.

Ballbjerg, V., Mouritsen, K.N., and Finster, K., 1998. Diel cycles of sulfate reduction rates in sediments of a *Zostera marina* bed (Denmark). Aquatic Microbial Ecology, 15: 97-102.

Barnes, R.O., and Goldberg, E.D., 1976. Methane production and consumption in anoxic marine sediments. Geology, 4: 297-300.

Berg, P., Rysgaard-Petersen, N. and Rysgaard, S., 1998. Interpretation of measured concentration profiles in the sediment porewater. Limnology Oceanography, 43: 1500-1510.

Berner, R.A., 1969. Migration of iron and sulfur within anaerobic sediments during early diagenesis. American Journal of Science, 267: 19-42.

Berner, R.A., 1970. Sedimentary pyrite formation. American Journal of Science, 268: 1-23.

Berner, R.A., 1980. Early diagenesis: A theoretical approach. Princeton University Press, Princeton, New Jersey.

Berner, R.A., 1981. A new geochemical classification of sedimentary environments. Journal of Sedimentary Petrology, 51: 359-365.

Berner, R.A., 1982. Burial of organic carbon and pyrite sulfur in the modern ocean: Its geochemical and environmental significance. American Journal of Science, 282: 451-473.

Berner, R.A., 1984. Sedimentary pyrite formation: An update. Geochimica et Cosmochimica Acta, 48: 605-615.

Berner, R.A., 1989. Biogeochemical cycles of carbon and sulfur and their effect on atmospheric oxygen over phanerozoic time. Global and Planetary Change 75: 97-122.

Bishop, J.K.B., 1988. The barite-opal-organic carbon association in oceanic particulate matter. Nature, 332: 341-343.

Boetius, A., and Suess, E., 2004. Hydrate Ridge: a natural laboratory for the study of microbial life fueled by methane from near-surface gas hydrates. Chemical Geology, 205: 291-310.

Boetius, A., Ravenschlag, K., Schubert, C., Rickert, D., Widdel, F., Gieseke, A., Amann, R., Jørgensen, B.B.,

Witte, U., and Pfannkuche, O., 2000. A marine microbial consortium apparently mediating anaerobic oxidation of methane. Nature, 407: 623-626.

Böttcher, M.E. and Lepland, A., 2000. Biogeochemistry of sulfur in a sediment core from the west-central Baltic Sea: Evidence from stable isotopes and pyrite textures. Journal of Marine Systems, 25: 299-312.

Böttcher, M.E., and Thamdrup, B., 2001. Anaerobic sulfide oxidation and stable isotope fractionation associated with bacterial sulfur disproportionation in the presence of MnO_2. Geochimica et Cosmochimica Acta 65: 1573-1581.

Borowski, W.S., Paull, C.K., and Ussler III, W., 1996. Marine pore-water sulfate profiles indicate in-situ methane flux from underlying gas hydrate. Geology, 24: 655-658.

Bowen, H.J.M., 1979. Environmental chemistry of the elements. Academic Press, London.

Bowman, J., 2000. The Methanotrophs - The families Methylococcaceae and Methylocystaceae. In: Dworkin, M., Balows, A., Trüper, H.G., Harder, W., and Schleifer, K.-H. (eds), The Prokaryotes, 3rd Ed. Springer, New York.

Bréhéret, J.-G. and Brumsack, H.-J., 2000. Barite concretions as evidence of pauses in sedimentation in the Marnes Bleues Formation of the Vocontian Basin (SE France). Sedimentary Geology, 130: 205-228.

Brüchert, V., and Pratt, L.M., 1996. Contemporaneous early diagenetic formation of organic and inorganic sulfur in estuarine sediments from St. Andrew Bay, Florida, USA. Geochimica et Cosmochimica Acta 60: 2325-2332.

Brumsack, H.-J., 1986. The inorganic geochemistry of Cretaceous black shales (DSDP Leg 41) in comparison to modern upwelling sediments from the Gulf of California. In: Shackelton, N.J. and Summerhayes, C.P. (eds) North Atlantic paleoceanography. Geological Society Special Publication 21, London, pp. 447-462.

Burdige, D.J., 1993. The biogeochemistry of manganese and iron reduction in marine sediments. Earth Science Reviews, 35: 249-284.

Canfield, D.E., 1989. Sulfate reduction and oxic respiration in marine sediments: implications for organic carbon preservation in euxinic environments. Deep-Sea Research, 36: 121-138.

Canfield, D.E., 1991. Sulfate reduction in deep-sea sediments. American Journal of Science, 291: 177-188.

Canfield, D.E., 1993. Organic matter oxidation in marine sediments. In: Wollast, R., Mackenzie, F.T., and Chou, L. (eds), Biogeochemical cycles and global change. Springer, Berlin, pp. 333-363.

Canfield, D.E., and Teske, A., 1996. Late Proterozoic rise in atmospheric oxygen concentration inferred from phylogenetic and sulphur-isotope studies. Nature, 382: 127-132.

Canfield, D.E., Raiswell, R., Westrich, J.T., Reaves, C.M., and Berner, R.A., 1986. The use of chromium reduction in the analyses of reduced inorganic sulfur in sediments and shales. Chemical Geology, 54: 149-155.

Canfield, D.E., Raiswell, R. and Bottrell, S., 1992. The reactivity of sedimentary iron minerals toward sulfide. Am. J. Sci., 292: 659-683.

Canfield, D.E., Thamdrup, B. and Hansen, J.W., 1993. The anaerobic degradation of organic matter in Danish coastal sediments: Iron reduction, manganese reduction, and sulfate reduction. Geochimica et Cosmochimica Acta, 57: 3867-3883.

Canfield, D.E., Kristensen, E., and Thamdrup, B., 2005. Aquatic Geomicrobiology. Elsevier, San Diego, California.

Cecile, M.P., Shakur, M.A. and Krouse, H.R., 1983. The isotopic composition of western Canadian barites and the possible derivation of oceanic sulfate $\delta^{34}S$ and $\delta^{18}O$ age curves. Canadian Journal of Earth Sciences, 20: 1528-1535.

Chanton, J.P., 1985. Sulfur mass balance and isotopic fractionation in an anoxic marine sediment. Ph.D. Thesis, University of North Carolina, Chapel Hill, USA.

Christensen, J.P., 1989. Sulphate reduction and carbon oxidation rates in continental shelf sediments, an examination of offshelf carbon transport. Continental Shelf Research, 9: 223-246.

Cline, J.D. and Richards, F.A., 1969. Oxygenation of hydrogen sulfide in seawater at constant salinity, temperature, and pH. Environment Sci. Technology, 3: 838-843.

Crill, P.M., and Martens, C.S., 1983. Spatial and temporal fluctuations of methane production in anoxic coastal marine sediments. Limnology and Oceanography, 28: 1117-1130.

Crill, P.M., and Martens, C.S., 1986. Methane production from bicarbonate and acetate in an anoxic marine sediment. Geochimica et Cosmochimica Acta, 50: 2089-2097.

Crill, P.M., and Martens, C.S., 1987. Biogeochemical cycling in an organic-rich coastal marine basin. 6. Temporal and spatial variations in sulfate reduction rates. Geochimica et Cosmochimica Acta, 51: 1175-1186.

Dehairs, F., Stoobants, N. and Goeyens, L., 1991. Suspended barite as a tracer of biological activity in the Southern Ocean. Marine Chemistry, 35: 399-410.

Devol, A.H. and Ahmend, S.I., 1981. Are high rates of sulphate reduction associated with anaerobic oxidation of methane? Nature, 291: 407-408.

Devol, A.H., Anderson, J.J., Kuivila, K., and Murray, J.W., 1984. A model for coupled sulfate reduction and methane oxidation in the sediments of Saanich Inlet. Geochimica et Cosmochimica Acta, 48: 993-1004.

D'Hondt, S., Jørgensen, B.B., Miller, J., et al., 2003. Proceedings of the ODP, Initial Reports, 201 [CD-ROM]. Available from: Ocean Drilling Program, Texas A&M University, College Station, Texas 77845-9547, USA.

D'Hondt, S., Jørgensen, B.B., Miller, D.J., et al., 2004. Distributions of microbial activities in deep subseafloor sediments. Science, 306: 2216-2221.

Dickens, G.R., 2001. Sulfate profiles and barium fronts in sediment on the Blake Ridge: Present and past methane fluxes through a large gas hydrate reservoir. Geochimica et Cosmochimica Acta, 65: 529-543.

Dickens, G.R., Schroeder, D., Hinrichs, K.-U., and the Leg 201 Scientific Party, 2003. The pressure core sampler (PCS) on ODP Leg 201: general operations and gas release. In: D'Hondt, S.L., Jørgensen, B.B., Miller, D.J., et al. (eds), Proceedings of the ODP, Initial Reports, 201, 1-22 [CD-ROM]. Available from: Ocean Drilling Program, Texas A&M University, College Station TX 77845-9547, USA.

Dos Santos Afonso, M. and Stumm, W., 1992. The reductive dissolution of iron (III) (hydr) oxides by hydrogen sulfide. Langmuir, 8: 1671-1676.

Dymond, J., Suess, E. and Lyle, M., 1992. Barium in deep-sea sediment: A geochemical proxy for paleo-productivity. Paleoceanography, 7: 163-181.

Edenborn, H.M., Silverberg, N., Mucci, A., and Sundby, B., 1987. Sulfate reduction in deep coastal marine sediments. Marine Chemistry, 21: 329-345.

Ehrlich, H.L., 1996. Geomicrobiology. Marcel Dekker, NY, 719 pp.

Elderfield, H., and Schultz, A., 1996. Mid-ocean ridge hydrothermal fluxes and the chemical composition of the ocean. Annual Review of Earth and Planetary Science, 24: 191-224.

Elvert, M., Suess, E., Greinert, J., and Whiticar, M.J., 2000. Archaea mediating anaerobic methane oxidation in deep-sea sediments at cold seeps of the eastern Aleutian subduction zone. Organic Geochemistry, 31: 1175-1187.

Ferdelman, T. G., Fossing, H., Neumann, K., and Schulz, H.D., 1999. Sulfate reduction in surface sediments of the southeast Atlantic continental margin between 15 degrees 38′S and 27 degrees 57′S (Angola and Namibia). Limnology and Oceanography, 44: 650-661.

Finster, K., Liesack, W., and Thamdrup, B., 1998. Elemental sulfur and thiosulfate disproportionation by *Desulfocapsa sulfoexigens* sp. nov., a new anaerobic bacterium isolated from marine surface sediment. Applied and Environmental Microbiology, 564: 119-125.

Fossing, H., and Jørgensen, B.B., 1989. Measurement of bacterial sulfate reduction in sediments: Evaluation of a single-step chromium reduction method. Bio-geochemistry, 8: 205-222.

Fossing, H., and Jørgensen, B.B., 1990. Oxidation and reduction of radiolabeled inorganic sulfur compounds in an estuarine sediment, Kysing Fjord, Denmark. Geochimica et Cosmochimica Acta, 54: 2731-2742.

Fossing, H., Ferdelman, T.G., and Berg, P., 2000. Sulfate reduction and methane oxidation in continental margin sediments influenced by irrigation (South-East Atlantic off Namibia). Geochimica et Cosmo-chimica Acta, 64: 897-910.

Fossing, H., Gallardo, V.A., Jørgensen, B.B., Hüttel, M., Nielsen, L.P., Schulz, H., Canfield, D.E., Forster, S., Glud, R.N., Gundersen, J.K., Küver, J., Ramsing, N.B., Teske, A., Thamdrup, B., and Ulloa, O., 1995. Concentration and transport of nitrate by the mat-forming sulphur bacterium *Thioploca*. Nature, 374: 713-715.

Froelich, P.N., Klinkhammer, G.P., Bender, M.L., Luedtke, N.A., Heath, G.R., Cullen, D., Dauphin, P., Hammond, D., Hartman, B., and Maynard, V., 1979. Early oxidation of organic matter in pelagic sediments of the eastern equatorial Atlantic: Suboxic diagenesis. Geochimica et Cosmochimica Acta, 43: 1075-1090.

Funk, J.A., von Dobeneck, T. and Reitz, A., 2003a. Integrated rock magnetic and geochemical quantification of redoxomorphic iron mineral diagenesis in Late Quaternary sediments from the Equatorial Atlantic. In: Wefer, G., Mulitza, S. and Ratmeyer, V. (eds) The South Atlantic in the Late Quaternary: Reconstruction of Material Budget and Current Systems. Springer, Berlin, pp. 237-260.

Funk, J.A., von Dobeneck, T., Wagner, T. and Kasten, S., 2003b. Late Quaternary sedimentation and early diagenesis in the equatorial Atlantic Ocean: Patterns, trends and processes deduced from rock magnetic and geochemical records. In: Wefer, G., Mulitza, S. and Ratmeyer, V. (eds) The South Atlantic in the Late Quaternary: Reconstruction of Material Budgets and Current Systems. Springer, Berlin, pp. 461-497.

Garrison, T., 1997. Oceanography. Wadsworth Publishing Company, Belmont, California.

Giblin, A.E. and Howarth, R.W., 1984. Porewater evidence for a dynamic sedimentary iron cycle in salt marshes. Limnology and Oceanography, 29: 47-63.

Gingele, F.X. and Dahmke, A., 1994. Discrete barite particles and barium as tracers of paleoproductivity in South Atlantic sediments. Paleoceanography, 9: 151-168.

Gingele, F.X., Zabel, M., Kasten, S., Bonn, W.J. and Nürnberg, C.C., 1999. Biogenic barium - methods and constraints of application as a proxy for paleoproductivity. In: Fischer G. and Wefer, G. (eds), Use of proxies in paleoceanography: examples from the South Atlantic. Springer, Berlin, pp. 345-364.

Goldhaber, M.B. and Kaplan, I.R., 1974. The sulfur cycle. In: Goldberg, E.D. (ed), The Sea, 5, Wiley, pp 569-655.

Goodfellow, W.D. and Jonasson, I.R., 1984. Ocean stagnation and ventilation defined by ^{34}S secular trends in pyrite and barite, Selwyn Basin, Yukon. Geology, 12: 583-586.

Greinert, J., Bollwerk, S.M., Derkachev, A., Bohrmann, G. and Suess, E., 2002. Massive barite deposits and carbonate mineralization in the Derugin Basin, Sea of Okhotsk: precipitation processes at cold seeps. Earth and Planetary Science Letters, 203: 165-180.

Hallam, S.J., Girguis, P.R., Preston, C.M., Richardson, P.M., and DeLong, E.F., 2003. Identification of methyl coenzyme M reductase A (*mcrA*) genes asso-ciated with methane oxidizing archaea. Applied and Environmental Microbiology, 69: 5483-5491.

Hedges, J.I., and Keil, R.G., 1995. Sedimentary organic matter preservation: An assessment and speculative synthesis. Marine Chemistry, 49: 81-115.

Henrichs, S.M. and Reeburgh, W.S., 1987. Anaerobic mineralization of marine sediment organic matter: rates and the role of anaerobic processes in the oceanic carbon economy. Geomicrobiology Journal, 5: 191-237.

Hensen, C., Zabel, M., Pfeifer, K., Schwenk, T., Kasten, S., Riedinger, N., Schulz, H.D. and Boetius, A., 2003. Control of sulfate pore-water profiles by sedimen-tary events and the significance of anaerobic oxida-tion of methane for burial of sulfur in marine sedi-ments. Geochimica et Cosmochimica Acta, 67: 2631-2647.

Hinrichs, K.-U., and Boetius, A., 2002. The anaerobic oxidation of methane: New insights in microbial ecology and biogeochemistry. In: Wefer, G., Billett, D., Hebbeln, D., Jørgensen, B.B., Schlüter, M., and van Weering, T. (eds), Ocean margin Systems. Springer, Berlin, pp. 457-477.

Hinrichs, K.-U., Hayes, J.M., Sylva, S.P., Brewer, P.G., and DeLong, E.F., 1999. Methane-consuming archaebacteria in marine sediments. Nature, 398: 802-805.

Hinrichs, K.-U., Summons, R.E., Orphan, V., Sylva, S.P., and Hayes, J.M., 2000. Molecular and isotopic analyses of anaerobic methane-oxidizing communi-ties in marine sediments. Organic Geochemistry, 31: 1685-1701.

Hoehler, T.M., Alperin, M.J., Albert, D.B., and Martens, C.S., 1994. Field and laboratory studies of methane oxidation in an anoxic marine sediment: Evidence for a methanogen-sulfate reducer consortium. Global Biogeochemical Cycles, 8: 451-463.

Hoehler, T.M., Alperin, M.J., Albert, D.B., and Martens, C.S., 1998. Thermodynamic control on hydrogen concentration in anoxic sediments. Geochimica et Cosmochimica Acta, 62: 1745-1756.

Hoehler, T.M., Alperin, M.J., Albert, D.B., and Martens, C.S., 2001. Apparent minimum free energy requirements for methanogenic Archaea and sulfate-reducing bacteria in an anoxic marine sediment. FEMS Microbiology Ecology, 38: 33-41.

Howarth, R.W., 1979. Pyrite: Its rapid formation in a salt marsh and its importance in ecosystem metabolism. Science, 203: 49-51.

Huettel, M., Ziebis, W., Forster, S., and Luther III, G.W., 1998. Advective transport affecting metal and nutrient distributions and interfacial fluxes in permeable sediments. Geochimica et Cosmochimica Acta, 62: 613-631.

Ivanov, M.V., 1968. Microbiological processes in the formation of sulfur deposits. Israel Program for Scientific Translations, Jerusalem.

Iversen, N., and Jørgensen, B.B., 1985. Anaerobic methane oxidation rates at the sulfate-methane transition in marine sediments from Kattegat and Skagerrak (Denmark). Limnology and Oceanography, 30: 944-955.

Jackson, B.E., and McInerney, M.J., 2002. Anaerobic microbial metabolism can proceed close to thermodynamic limits. Nature, 415: 454-456.

Jahnke, R.A., 1996. The global ocean flux of particulate organic carbon: Areal distribution and magnitude. Global Biogeochemical Cycles, 10: 71-88.

Jannasch, H.W., Nelson, D.C., and Wirsen, C.O., 1989. Massive natural occurrence of unusually large bacteria (Beggiatoa sp.) at a hydrothermal deep-sea vent site. Nature, 342: 834-836.

Jørgensen, B.B., 1977. The sulfur cycle of a coastal marine sediment (Limfjorden, Denmark). Limnology and Oceanography, 22: 814-832.

Jørgensen, B.B., 1978. A comparison of methods for the quantification of bacterial sulfate reduction in coastal marine sediments. I. Measurements with radiotracer techniques. Geomicrobiology Journal, 1: 11-27.

Jørgensen, B.B., 1982. Mineralization of organic matter in the sea bed - The role of sulfate reduction. Nature, 296: 643-645.

Jørgensen, B.B., 1983. Processes at the sediment-water interface. In: Bolin, B. and Cook, R.C. (eds), The major biogeochemical cycles and their interactions. SCOPE, pp. 477-509.

Jørgensen, B.B., 1990. A thiosulfate shunt in the sulfur cycle of marine sediments. Science, 249: 152-154.

Jørgensen, B.B., 1996. Case Study: Aarhus Bay. In: Jørgensen, B.B., and Richardson, K. (eds), Eutrophication in a coastal marine environment. Coastal and Estuarine Studies, American Geophysical Union, Washington, DC, pp. 137-154

Jørgensen, B.B., and Bak, F., 1991. Pathways and microbiology of thiosulfate transformations and sulfate reduction in a marine sediment (Kattegat, Denmark). Applied and Environmental Microbiology, 57: 847-856.

Jørgensen, B.B., and Gallardo, V.A., 1999. Thioploca spp.: filamentous sulfur bacteria with nitrate vacuoles. FEMS Microbiology Ecology, 28: 301-313.

Jørgensen, B.B., and Nelson D.C., 2004. Sulfide oxidation in marine sediments: Geochemistry meets microbiology. In.: Amend, J.P., Edwards, K.J., and Lyons, T.W. (eds), Sulfur Biogeochemistry - Past and Present. Geological Society of America Special Paper 379, Boulder, Colorado, pp. 63-81.

Jørgensen, B.B., Bang, M., and Blackburn, T.H., 1990. Anaerobic mineralization in marine sediments from the Baltic Sea - North Sea transition. Marine Ecology Progress Series, 59: 39-54.

Jørgensen, B.B., Weber, A., and Zopfi, J., 2001. Sulfate reduction and anaerobic methane oxidation in Black Sea sediments. Deep-Sea Research, 48: 2097-2120.

Jørgensen, B.B., Böttcher, M.E., Lüschen, H., Neretin, L., and Volkov, I., 2004. Anaerobic methane oxidation and a deep H_2S sink generate isotopically heavy sulfides in Black Sea sediments. Geochimica et Cosmochimica Acta, 68: 2095-2118.

Judd, A.G., Hovland, M., Dimitrov, I.I., Garcia Gil, S., and Jukes, V., 2002. The geological methane budget at Continental Margins and its influence on climate change. Geofluids, 2: 109-126.

Kallmeyer, J., Ferdelman, T.G., Weber, A., Fossing, H., and Jørgensen, B. B., 2004. A cold chromium distillation procedure for radiolabeled sulfide applied to sulfate reduction measurements. Limnology and Oceanography Methods, 2: 171-180.

Karlin, R. and Levi, S., 1983. Diagenesis of magnetic minerals in recent hemipelagic sediments. Nature, 303: 327-330.

Karlin, R. and Levi, S., 1985. Geochemical and sedimentological control of the magnetic properties of hemipelagic sediments. Journal of Geophysical Research, 90: 10373-10392.

Kasten, S., Zabel, M., Heuer, V. and Hensen, C., 2003. Processes and signals of nonsteady-state diagenesis in deep-sea sediments and their pore waters. In: Wefer, G., Mulitza, S. and Ratmeyer, V. (eds), The South Atlantic in the Late Quaternary: Reconstruction of Material Budget and Current Systems. Springer, Berlin, pp. 431-459.

Kasten, S., Freudenthal, T., Gingele, F.X., von Dobeneck, T. and Schulz, H.D., 1998. Simultaneous formation of iron-rich layers at different redox boundaries in sediments of the Amazon Deep-Sea Fan. Geochimica et Cosmochimica Acta 62: 2253-2264.

Kelly, D.P., 1988. Oxidation of sulfur compounds. In: Cole, A.S. and Ferguson, S.J. (eds), The Nitrogen and Sulfur Cycles. Soc. Gen. Microbiol., 42, pp. 65-98.

Kjær, T., 2000. Development and application of new biosensors for microbial ecology. Ph.D. Thesis, University of Aarhus, Denmark, 324 p.

Knittel, K., Lösekann, T., Boetius, A., Kort, R., and Amann, R., 2005. Diversity and distribution of methanotrophic archaea at cold seeps. Applied and Environmental Microbiology, 71: 467-479.

Kölling, A., 1991. Frühdiagenetische Prozesse und Stoff-Flüsse in marinen und ästuarinen Sedimenten. Berichte, 15, Fachbereich Geowissenschaften, Universität Bremen, 140 pp.

Krämer, M., and Cypionka, H., 1989. Sulfate formation via ATP sulfurylase in thiosulfate- and sulfite-disproportionating bacteria. Archives of Microbiology, 122: 183-188.

Krein, E.B. and Aizenshtat, Z., 1995. Proposed thermal pathways for sulfur transformations in organic macromolecules: Laboratory simulation experiments. In: Vairavamurthy, M.A. and Schoonen, M.A.A. (eds), Geochemical Transformations of Sedimentary Sulfur, ACS symposium series, 612, Washington, DC, pp. 110-137.

Krüger, M., Meyerdierks, A., Gloeckner, F.O., Amann, R., Widdel, F., Kube, M., Reinhardt, R., Kahnt, J., Boecher, R., Thauer, R.K., and Shima, S., 2003. A conspicuous nickel protein in microbial mats that oxidize methane anaerobically. Nature, 426: 878-881.

Lin, S., and Morse, J.W., 1991. Sulfate reduction and iron sulfide mineral formation in Gulf of Mexico anoxic sediments. American Journal of Science, 291: 55-89.

Lowson, R.T., 1982. Aqueous oxidation of pyrite by molecular oxygen. Chemical Reviews, 82: 461-497.

Luther, G.W., III, 1987. Pyrite oxidation and reduction: Molecular orbital theory considerations. Geochimica et Cosmochimica Acta, 51: 3193-3199.

Luther III, G.W., 1991. Pyrite synthesis via polysulfide compounds. Geochimica et Cosmochimia Acta, 55: 2839-2849.

Luther III, G.W. and Church, T.M., 1991. An overview of the environment chemistry of sulfur in wetland systems. In: Howarth, R.W. et al (eds), Sulfur cycling on the continents. John Wiley, pp: 125-144.

Luther III, G.W., Giblin, A., Howarth, R.W. and Ryans, R.A., 1982. Pyrite and oxidized iron mineral phases formed from pyrite oxidation in salt marsh and estuarine sediments. Geochimica et Cosmochimica Acta, 46: 2665-2669.

Martens, C.S., and Berner, R.A., 1974. Methane production in the interstitial waters of sulfate-depleted marine sediments. Science, 185: 1167-1169.

McCollom, T.M., and Shock, E.L., 1997. Geochemical constraints on chemolithoautotrophic metabolism by microorganisms in seafloor hydrothermal systems. Geochimica et Cosmochimica Acta, 61: 4375-4391.

Mearon, S., Paytan, A. and Bralower, T.J., 2003. Cretaceous strontium isotope stratigraphy using marine barite. Geology, 31: 15-18.

Michaelis, W., Seifert, R., Nauhaus, K., Treude, T., Thiel, V., Blumenberg, M., Knittel, K., Gieseke, A., Peterknecht, K., Pape, T., Boetius, A., Amann, R., Jørgensen, B.B., Widdel, F., Peckmann, J., Pimenov, N.V., and Gulin, M.B., 2002. Microbial reefs in the Black Sea fueled by anaerobic oxidation of methane. Science, 297: 1013-1015.

Middelburg, J.B.M., 1990. Early diagenesis and authigenic mineral formation in anoxic sediments of Kau Bay, Indonesia. PhD Thesis, Universitiy of Utrecht, Utrecht, 177 pp.

Middelburg, J.J., Soetaert, K., and Herman, P.M.J., 1997. Empirical relationships for use in global diagenetic models. Deep-Sea Research I, 44: 327-344.

Moore, T.S., Murray, R.W., Kurtz, A.C. and Schrag, D.P., 2004. Anaerobic methane oxidation and the formation of dolomite. Earth and Planetary Science Letters, 229: 141-154.

Morse, J.W., 1991. Oxidation kinetics of sedimentary pyrite in seawater. Geochimica et Cosmochimica Acta, 55: 3665-3667.

Morse, J.W., 2002. Sedimentary geochemistry of the carbonate and sulphide systems and their potential influence on toxic metal bioavailability. In: Gianguzza, A., Pelizzetti, E. and Sammartano, S. (eds), Chemistry of Marine Water and Sediments. Springer, pp. 165-189.

Moses, C.O., and Herman, J.S., 1991. Pyrite oxidation at circumneutral pH. Geochimica et Cosmochimica Acta, 55: 471-482.

Murray, J.W., Grundmanis, V. and Smethie, W.M. Jr, 1978. Interstitial water chemistry in sediments of Saanich Inlet. Geochimica et Cosmochimica Acta, 42: 1011-1026.

Nauhaus, K., Boetius, A., Krüger, M., and Widdel, F., 2002. In vitro demonstration of anaerobic oxidation of methane coupled to sulphate reduction in sediment from a marine gas hydrate area. Environmental Microbiology, 4: 296-305.

Neretin, L., Böttcher, M.E., Jorgensen, B.B., Volkov, I.I., Lüschen, H. and Hilgenfeldt, K., 2004. Pyritization processes and greigite formation in the advancing sulfidization front in the upper Pleistocene sediments of the Black Sea. Geochimica et Cosmochimica Acta, 68: 2081-2093.

Neumann, T., Rausch, N., Leipe, T., Dellwig, O., Berner, Z. and Böttcher, M.E., 2005. Intense pyrite formation under low-sulfate conditions in the Achterwasser lagoon, SW Baltic Sea. Geochimica et Cosmochimica Acta, 69: 3619-3630.

Niewöhner, C., Hensen, C., Kasten, S., Zabel, M. and Schulz, H.D., 1998. Deep sulfate reduction completely mediated by anaerobic methane oxidation in sediments of the upwelling area off Namibia. Geochimica et Cosmochimica Acta, 62: 455-464.

Nüster, J., 2005. New methods in biogeochemistry: The development of electrochemical tools for the measurement of dissolved and solid state compounds in natural systems. PhD Thesis, University of Bremen, 216 pp.

Orphan, V.J., House, C.H. Hinrichs, K.-U., McKeegan, K.D., DeLong, E.F., 2001. Methane-consuming archaea revealed by directly coupled isotopic and phylogenetic analysis. Science, 293: 484-487.

Orphan, V.J., House, C.H., Hinrichs, K.-U., McKeegan, K.D., DeLong, E.F., 2002. Multiple archaeal groups mediate methane oxidation in anoxic cold seep sediments. Proceedings of the National Academy of Sciences USA, 99: 7663-7668.

Orr, W.L. and White, C.M. (eds), 1990. Geochemistry of sulfur in fossil fuels. ACS Symposium Series, 429, Washington, DC.

Pancost, R.D., Sinninghe Damsté, J.S., Lint, S.D., van der Marel, M.J.E.C., Gottschal, J.C., and Shipboard Scientific Party, 2000. Biomarker evidence for widespread anaerobic methane oxidation in Mediterranean sediments by a consortium of methanogenic archaea and bacteria. Applied and Environmental Microbiology, 66: 1126-1132.

Parkes, R.J., Webster, G., Cragg, B.A., Weightman, A.J., Newberry, C.J., Ferdelman, T.G., Kallmeyer, J., Jørgensen, B.B., Aiello, I.W., and Fry, J.C., 2005. Deep sub-seafloor prokaryotes stimulated at interfaces over geological time. Nature, 436: 390-394.

Passier, H.F., Middelburg, J.J., Os, B.J.H.v. and Lange, G.J.de, 1996. Diagenetic pyritisation under eastern Mediterranean sapropels caused by downward sulphide diffusion. Geochimica et Cosmochimica Acta, 60: 751-763.

Paytan, A., Kastner, M., Martin, E.E., MacDougall, J.D., and Herbert, T., 1993. Marine barite as a monitor of seawater strontium isotope composition. Nature, 366: 445-449.

Paytan, A., Kastner, M. and Chavez, F.P., 1996a. Glacial to interglacial fluctuations in productivity in the equatorial Pacific as indicated by marine barite. Science, 274: 1355-1357.

Paytan, A., Moore, W.S. and Kastner, M., 1996b. Sedimentation rate as determined by ^{226}Ra activity in marine barite. Geochimica et Cosmochi-mica Acta, 60: 4313-4319.

Paytan, A., Kastner, M., Campbell, D. and Thiemens, M.H., 1998. Sulfur isotope composition of Cenozoic seawater sulfate. Science, 282: 1459-1462.

Paytan, A., Mearon, S., Cobb, K. and Kastner, M., 2002. Origin of marine barite deposits: Sr and S isotope characterization. Geology, 30: 747-750.

Paytan, A., Martinez-Ruiz, F., Eagle, M., Ivy, A. and Wankel, S.D., 2004. Using sulfur isotopes to elucidate the origin of barite associated with high organic matter accumulation events in marine sediments. In: Amend, J.P., Edwards, K.J. and Lyons, T.W. (eds), Sulfur Biogeochemistry – Past and Present. Geological Society of America Special Paper 379, pp. 151-160.

Poulton, S.W., Krom, M.D. and Raiswell, R., 2004. A revised scheme for the reactivity of iron (oxyhydr)oxides towards dissolved sulfide. Geochimica et Cosmochimica Acta, 68: 3703-3715.

Pyzik, A.J. and Sommer, S.E., 1981. Sedimentary iron monosulfides: kinetics and mechanism of formation. Geochimica et Cosmochimica Acta, 45: 687-698.

Rabus, R., Hansen, T., and Widdel, F., 2004. Dissimilatory Sulfate- and Sulfur Reducing Prokaryotes. In: Dworkin, M. et al. (eds), The Prokaryotes: An Evolving Electronic Resource for the Microbiological Community, 3rd edition, Springer-Verlag, New York, (http://link.springer-ny.com/link/service/books/10125/).

Raiswell, R., 1982. Pyrite texture, isotopic composition and the availability of iron. American Journal of Science, 282: 1244-1265.

Raiswell, R., 1988. Chemical model for the origin of minor limestone-shale cycles by anaerobic methane oxidation. Geology, 16: 641-644.

Raiswell, R., and Canfield, D.E., 1998. Sources of iron for pyrite formation in marine sediments. American Journal of Science, 298: 219-245.

Raiswell, R., Canfield, D.E. and Berner, R.A., 1994. A comparison of iron extraction methods for the determination of degree of pyritization and the recognition of iron-limited pyrite formation. Chemical Geology, 111: 101-110.

Reeburgh, W.S., 1969. Observations of gases in Chesapeake Bay sediments. Limnology and Oceanography, 14: 368-375.

Reeburgh, W.S., 1976. Methane consumption in Cariaco Trench waters and sediments. Earth and Planetary Science Letters, 47: 345-352.

Reeburgh, W.S., 1982. A major sink and flux control for methane in sediments: Anaerobic consumption. In: Fanning, K.A. and Manheim, F.T. (eds), The dynamic environment. Heath, Lexington, MA, pp. 203-217.

Reeburgh, W.S., Whalen, S.C., and Alperin, M.J., 1993. The role of methylotrophy in the global methane budget. In: Murrell, J.C., and Kelly, D.P. (eds),

Microbial growth on C_1 compounds. Intercept, Andover, UK, pp. 1-14.

Reitz, A., Hensen, C., Kasten, S., Funk, J. and de Lange, G., 2004. A combined geochemical and rock-magnetic investigation of a redox horizon at the last glacial/interglacial transition. Physics and Chemistry of the Earth, 29: 921-931.

Rickard, D.T., 1975. Kinetics and mechanisms of pyrite formation at low temperatures. American Journal of Science, 275: 636-652.

Rickard, D. and Luther III, G.W., 1997. Kinetics of pyrite formation by the H_2S oxidation of iron(II) monosulfide in aqueous solutions between 25 and 125°C: The rate equation. Geochimica et Cosmochimica Acta, 61: 115-134.

Rickard, D., Schoonen, M.A.A., Luther, G.W., 1995. Chemistry of iron sulfides in sedimentary environments. In: Vairavamurthy, M.A. and Schoonen, M.A.A. (eds), Geochemical Transformations of Sedimentary Sulfur. ACS Symposium Series 612, Washington DC, pp. 168-193.

Riedinger, N., 2005. Preservation and diagenetic overprint of geochemical and geophysical signals in ocean margin sediments related to depositional dynamics. Berichte, 242, Fachbereich Geowissenschaften, Universität Bremen, 91 pp.

Riedinger, N., Pfeifer, K., Kasten, S., Garming, J.F.L., Vogt, C. and Hensen, C., 2005. Diagenetic alteration of magnetic signals by anaerobic oxidation of methane related to a change in sedimentation rate. Geochimica et Cosmochimica Acta, 69: 4117-4126.

Roden, E.E. and Tuttle, J.H., 1992. Sulfide release from estuarine sediments underlying anoxic bottom water. Limnology and Oceanograpy 37: 725-738.

Rusch, A., Töpken, H., Böttcher, M.E., and Höpner, T., 1998. Recovery from black spots: results of a loading experiment in the Wadden Sea. Journal of Sea Research, 40: 205-219.

Schink, B., 1997. Energetics of syntrophic cooperation in methanogenic degradation. Microbiological and Molecular Biological Reviews, 61: 262-280.

Schinzel, U., 1993. Laboratory experiments on early diagenetic reactions of iron(III) oxyhydroxides in marine sediments (in German). Berichte, 36, Fachbereich Geowissenschaften, Universität Bremen, 189 pp.

Schippers, A., 2004. Biogeochemistry of metal sulfide oxidation in mining environments, sediments, and soils. In: Amend, J.P., Edwards, K.J., and Lyons, T.W. (eds), Sulfur biogeochemistry - Past and Present. Geological Society of America Special Paper 379, Boulder, Colorado, pp. 49-62

Schippers, A., and Jørgensen, B.B., 2001. Oxidation of pyrite and iron sulfide by manganese in marine sediments. Geochimica et Cosmochimica Acta, 65: 915-922.

Schippers, A., and Jørgensen, B.B., 2002. Biogeochemistry of pyrite and iron sulfide oxidation in marine sediments. Geochimica et Cosmochimica Acta, 66: 85-92.

Schoonen, M.A.A., 2004. Mechanisms of sedimentary pyrite formation. In: Amend, J.P., Edwards, K.J. and Lyons, T.W. (eds), Sulfur Biogeochemistry – Past and Present. Geological Society of America Special Paper 379, pp. 117-134.

Schouten, S., Eglington, T.I., Sinninghe Damsté, J.S. and de Leeuw, J.W., 1995. Influence of sulfur cross-linking on the molecular size distribution of sulfur-rich macromolecules in bitumen. In: Vairavamurthy,

M.A. and Schoonen, M.A.A. (eds), Geochemical transformation of sedimentary sulfur. ACS Symposium 612, Washington, DC, pp. 80-92.

Schulz H.D., Dahmke A., Schinzel U., Wallmann K., and Zabel M., 1994. Early diagenetic processes, fluxes and reaction rates in sediments of the South-Atlantic. Geochimica et Cosmochimica Acta, 58: 2041-2060.

Schulz, H.N., and Jørgensen, B.B., 2001. Big bacteria. Annual Reviews in Microbiology, 55: 105-137.

Schulz, H.N., Brinkhoff, T., Ferdelman, T.G., Hernandez Marine, M., Teske, A., and Jørgensen, B.B., 1999. Dense populations of a giant sulfur bacterium in Namibian shelf sediments. Science, 284: 493-495.

Smith, S.V., and Hollibaugh, J.T., 1993. Coastal metabolism and the oceanic organic carbon balance. Reviews in Geophysics, 31: 75-89.

Sørensen, K.B., Finster, K., and Ramsing, N.B., 2001. Thermodynamic and kinetic requirements in anaerobic methane oxidizing consortia exclude hydrogen, acetate, and methanol as possible electron shuttles. Microbial Ecology, 42: 1-10.

Sorokin, Yu.L., 1962. Experimental investigation of bacterial sulfate reduction in the Black Sea using S35. Microbiology, 31: 329-335.

Stetter, K.O., Huber, R., Blöchl, E., Kurr, M., Eden, R.D., Fielder, M., Cash, H., and Vance, I., 1993. Hyperthermophilic archaea are thriving in deep North Sea and Alaskan oil reservoirs. Nature, 365: 743-745.

Sweeney, R.E. and Kaplan, I.R., 1973. Pyrite Framboid Formation: Laboratory Synthesis and Marine Sediments. Economic Geology, 68: 618-634.

Thamdrup, B., and Canfield, D.E., 1996. Pathways of carbon oxidation in continental margin sediments off central Chile. Limnology and Oceanography, 41: 1629-1650.

Thamdrup, B., Finster, K, Hansen, J.W., and Bak, F., 1993. Bacterial disproportionation of elemental sulfur coupled to chemical reduction of iron or manganese. Applied and Environmental Microbiology, 59: 101-108.

Thamdrup, B., Fossing, H., and Jørgensen, B.B., 1994a. Manganese, iron, and sulfur cycling in a coastal marine sediment, Aarhus Bay, Denmark. Geochimica et Cosmochimica Acta, 58: 5115-5129.

Thamdrup, B., Finster, K., Fossing, H., Hansen, J.W., and Jørgensen, B.B., 1994b. Thiosulfate and sulfite distributions in porewater of marine sediments related to manganese, iron and sulfur geochemistry. Geochimica et Cosmochimica Acta, 58: 67-73.

Thiel, V., Peckmann, J., Richnow, H.H., Luth, U., Reitner, J., and Michaelis, W., 2001. Molecular signals for anaerobic methane oxidation in Black Sea seep carbonates and a microbial mat. Marine Chemistry, 73: 97-112.

Thode-Andersen, S., and Jørgensen, B.B., 1989. Sulfate reduction and the formation of ^{35}S-labeled FeS, FeS$_2$, and S° in coastal marine sediments. Limnology and Oceanography, 34: 793-806.

Torres, M.E., Brumsack, H.J., Bohrmann, G. and Emeis, K.C., 1996. Barite fronts in continental margin sediments: A new look at barium remobilization in the zone of sulfate reduction and formation of heavy barites in diagenetic fronts. Chemical Geology, 127: 125-139.

Treude, T., Niggemann, J., Kallmeyer, J., Wintersteller, P., Schubert, C.J., Boetius, A., and Jørgensen, B.B., 2005. Anaerobic oxidation of methane and sulfate reduction along the Chilean continental margin.

Geochimica et Cosmochimica Acta, 69: 2767-2779.

Troelsen, H., and Jørgensen, B.B., 1982. Seasonal dynamics of elemental sulfur in two coastal sediments. Estuarine and Coastal Shelf Science, 15: 255-266.

Vairavamurthy, A., Manowitz, B., Luther III, G.W., Jeon, Y., 1993. Oxidation state of sulfur in thiosulfate and implications for anaerobic energy metabolism. Geochimica et Cosmochimica Acta, 57: 1619-1623.

Vairavamurthy, M.A., Orr, W.L. and Manowitz, B., 1995. Geochemical transformation of sedimentary sulfur: an introduction. In: Vairavamurthy, M.A. and Schoonen, M.A.A. (eds), Geochemical tranformation of sedimentary sulfur. ACS Symposium, 612, Washing-ton, DC, pp. 1-17.

Valentine, D.L., and Reeburgh, W.S., 2000. New perspectives on anaerobic methane oxidation - Mini-review. Environmental Microbiology, 2: 477-484.

Van Beek, P. and Reyss, J.-L., 2001. ^{226}Ra in marine barite: New constraints on supported ^{226}Ra. Earth Planetary Science Letters, 187: 147-161.

Van Beek, P., Reyss, J.-L., Paterne, M., Gersonde, R., Rutgers van der Loeff, M. and Kuhn, G., 2002. ^{226}Ra in barite: Absolute dating of Holocene Southern Ocean sediments and reconstruction of sea-surface reservoir ages. Geology, 30: 731-734.

van Cappellen, P., and Wang, Y., 1996. Cycling of iron and manganese in surface sediments: A general theory for the coupled transport and reaction of carbon, oxygen, nitrogen, sulfur, iron, and manganese. American Journal of Science, 296: 197-243.

Von Breymann, M.T.K., Emeis, K.C. and Suess, E., 1992. Water depth and diagenetic constraints on the use of barium as a paleoproductivity indicator. In: Summerhayes, C.P. (ed) Upwelling Systems: Evolution since the Early Miocene. Geological Society Special Publication 64, pp 273-284.

Weber, A., and Jørgensen, B.B., 2002. Bacterial sulfate reduction in hydrothermal sediments of the Guaymas Basin, Gulf of California, Mexico. Deep-Sea Research I, 49: 827-841.

Werne, J.P., Hollander, D.J., Lyons, T.W. and Sinninghe Damsté, J.S., 2004. Organic sulfur biogeochemistry: Recent advances and future research directions. In: Amend, J.P., Edwards, K.J. and Lyons, T.W. (eds), Sulfur Biogeochemistry – Past and Present. Geological Society of America Special Paper 379, pp. 135-150.

Westrich, J.T., and Berner, R.A., 1984. The role of sedimentary organic matter in bacterial sulfate reduction: The G model tested. Limnology and Oceanography, 29: 236-249.

Whiticar, M.J., 1999. Carbon and hydrogen isotope systematics of bacterial formation and oxidation of methane. Chemical Geology, 161: 291-314.

Whitman, W.B., Bowen, T.L., and Boone, D.R., 1999. The methanogenic bacteria. In: Dworkin, M., Balows, A., Trüper, H.G., Harder, W., and Schleifer, K.-H. (eds), The Prokaryotes, 3rd. Ed. Springer, New York.

Widdel, F., 1988. Microbiology and ecology of sulfate- and sulfur-reduction bacteria. In: Zehnder, A.J.B. (ed), Biology of anaerobic microorganisms. Wiley & Sons, NY, pp. 469-585.

Widdel, F. and Hansen, T.A., 1991. The dissimilatory sulfate- and sulfur-reducing bacteria. In: Balows, H. et al. (eds), The Procaryotes. Springer, pp. 583-624.

Wilkin, R.T. and Barnes, H.L., 1996. Pyrite formation by reactions of iron monosulfides with dissolved inorganic and organic sulfur species. Geochimica et Cosmochimica Acta, 60: 4167-4179.

Wollast, R., 1998. Evaluation and comparison of the global carbon cycle in the coastal zone and in the open ocean. In: Brink, K.H., and Robinson, A.R. (eds), The Sea, Vol. 10, pp. 213-252.

9 Marine Carbonates: Their Formation and Destruction

Ralph R. Schneider, Horst D. Schulz
and Christian Hensen

9.1 Introduction

For marine carbonates, an overwhelming amount of information exists in a variety of specialized journals addressing marine geochemistry and carbon cycling, as well as in many books summarizing the state of knowledge on this topic. Therefore it would be far beyond the scope of this chapter to try to completely review what is known about marine calcareous sediments and their diagenesis. On the other hand, although intensively investigated since the previous century (e.g. Murray 1897), marine carbonates have gained increasing attention by marine biologists, geochemists, paleoceanographers and paleoclimatologists over the last three decades. Among other reasons this is because marine carbonates, together with the oceanic CO_2-carbonic acid-system, act as a sink or source of carbon within the global carbon cycle which has become a key topic of investigation and modelling related to the role of the greenhouse gas CO_2 in future global climate change.

Descriptions of sedimentary carbonates in different oceanic environments always have dealt with the formation of the calcium carbonate and the biologically, physically and chemically mediated processes governing the observed distribution of sedimentary carbonates in the marine realm. In this context special emphasis was often given to the complex pattern of inorganic and organic carbon exchange between the atmosphere, the world ocean and the continents. These are additional factors determining the distribution of inorganic carbon dissolved in seawater and the accumulation or destruction of calcium carbonate in marine sediments (Berger 1976; Andersen and Malahoff 1977; Broecker and Peng 1982; Sundquist and Broecker 1985; Morse and Mackenzie 1990). The intention of this chapter is

furthermore to review the knowledge on the rate of calcium carbonate production in the ocean, the fluxes through the water column, as well as the rates of inorganic carbon accumulation and destruction in different marine environments. For this purpose this chapter summarizes studies dealing with the estimation of global carbonate reservoirs and fluxes in the state of ongoing production, accumulation and dissolution or physical erosion (Milliman 1993; Milliman and Droxler 1996; Wollast 1994). In addition, this chapter addresses the principles and presents examples for calculation of carbonate saturation conditions under variable boundary conditions in the oceanic CO_2-carbonic acid-calcite system, i.e. temperature, pressure, salinity, and CO_2 exchange with the atmosphere.

9.2 Marine Environments of Carbonate Production and Accumulation

This section primarily focuses on the description of the deposition and accumulation of carbonates in shallow waters and in the deep ocean. The main depocenters for calcium carbonates are the continental shelf areas, as well as island arcs or atolls, which are the typical shallow water environments for massive carbonate formation, and the pelagic deep-sea sediments above the calcite compensation depth catching the rain of small calcareous tests formed by marine plankton in the surface waters.

9.2.1 Shallow-Water Carbonates

Shallow-water carbonates are formed either by sediments made up of a variety of carbonate particles which have their origin in biotic and abiotic processes, or by massive reefs and platforms built up by skeleton-forming organisms. On a global scale, these shallow-water carbonates in the modern environment are mainly constituted by particles of skeletal origin. However, aside from the corals, the understanding of the physico-chemical and vital factors affecting the biomineral composition of shallow platform calcareous sediments in warm waters is still incomplete. Shallow environment precipitates form ooids and aragonitic needle muds, whereby the former involve primarily abiotic processes and the latter have both an abiotic and biotic source.

For long time, the classical picture of shallow water carbonates was suggesting that most of their formation was restricted to tropical and subtropical regions within the 22°C isotherm of annual mean surface water temperatures (e.g. Berger and Seibold 1993), but it now has become evident that a significant amount of carbonate can also be formed as so-called 'cool-water' carbonate banks and reefs in temperate and cold latitudes (review by James 1997; Freiwald 2002). In the temperate and cold-water zones, particularly shelf and upper slope areas with only very low inputs of terrigenous sediments are covered by cool-water carbonate bioherms (Fig. 9.1a). Different from the warm-water environment, where the major portion of skeletal carbonate is predominantly formed by an association of hermatypic corals and green algae referred to as 'Photozoan Association' (James 1997), the cool-water carbonates can be composed of molluscs, foraminifers, echinoderms, bryozoans, barnacles, ostracods, sponges, worms, ahermatypic corals and coralline algae. For differentiation from the warm-water Photozoan Association, the group of organisms forming cool-water carbonates in shallow waters that are colder than 20°C is defined by James 1997 as Heterozoan Association (see also Fig. 9.1). According to Freiwald (2002) the largest coral reef provinces occur at greater water depths under cold and dark conditions usually below the storm wave base from the high to low latitudes of both hemispheres. In the North Atlantic the dominant reef-forming corals belong to scleractinian species. A first attempt to calculate the amount of $CaCO_3$ produced by cold-water reefs on global carbonate production was provided by Lindberg & Mienert (2005). Their estimate is in the order of 4 to 12 % of that of tropical reefs which results in a tentative estimate > 1 % of total marine carbonate production. Since

these first estimates are relying on data from the Norwegian shelf only, it is difficult to come up with really reliable global carbonate production esti-mates for cold-water reefs at the moment. Therefore the estimates of Wollast (1994); Milliman & Droxler (1996), and Vecsei (2004) for worldwide shelf areas are considered here. For budget considerations it seems feasible to separate shal-low-water carbonates according to Milliman (1993) into coral reefs, carbonate platforms which consist of non-reef habitats, such as banks and embayments dominated by the sedimentation of biogenic and abiogenic calcareous particles, and shelves which can be further subdivided into car-bonate-rich and car-bonate-poor shelves (Table 9.1 adopted from Milliman (1993) and Milliman and Droxler (1996), taking into account Vescei (2004) estimates.

Reefs

Hermatypic coralalgal reefs and their fore-reef sections occupy an area of about $0.35 \cdot 10^6$ km² and are considered as the most productive carbonate environments in modern times. Carbonate accretion is the result of corals and green algae and, to a minor extent also of benthic foraminifera. Measured on a global scale, the mean calcium carbonate accumulation in coral reefs is in the order of 900-2700 g $CaCO_3$ m⁻²yr⁻¹. The total present-day global $CaCO_3$ production by coral reefs then is $6.5-8.3 \cdot 10^{12}$ mol yr⁻¹ from which about $7 \cdot 10^{12}$ mol yr⁻¹ accumulate, while the rest (up to $1.5 \cdot 10^{12}$ mol yr⁻¹) undergoes physical erosion and offshore transport, as well as biological destruction (Milliman 1993).

Carbonate platforms

These platforms are the second important tropical to subtropical environment where high amounts of carbonate are produced and accumulated at water depths shallower than 50 m. The areal extension is about $0.8 \cdot 10^6$ km². In contrast to reefs, on carbonate platforms production is mainly carried out by benthic red/green algae, mollusks and benthic foraminifera. Estimates of biotic and, to a much lesser extent, abiotic carbonate production on platforms range between 300-500 g $CaCO_3$ m⁻² yr⁻¹, which amounts to $4 \cdot 10^{12}$ mol yr⁻¹ on a global scale. Accumulation of platform carbonate is difficult to assess because a lot of it is dissolved or can be found as exported material in several 10 to 100 m

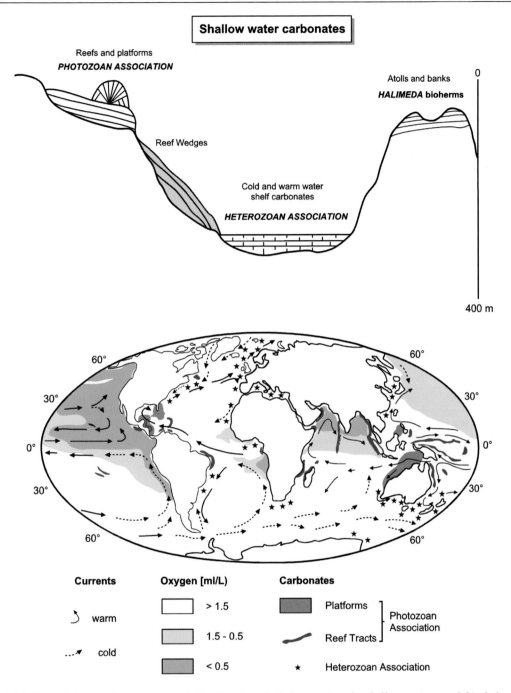

Fig. 9.1 Marine calcium carbonate accumulation in a) typical depocenters in shallow waters and b) their global distribution (James and Clarke 1997).

thick sediment wedges of Holocene age at the fringes of the platforms. Milliman (1993) estimates that only one half of the carbonate accumulates on shallow platforms where it is produced ($2 \cdot 10^{12}$ mol yr^{-1}), while the other half leaves the shallow water environment in the form of sediment lobes or wedges downslope.

Continental shelves

Only very little knowledge exists concerning the quantitative carbonate production, export and its accumulation on continental shelves. Two shelf types, carbonate-rich and carbonate-poor shelves, are distinguished by Milliman and Droxler (1996).

Their areas amount to 15 and $10 \cdot 10^{12}$ km^2, respectively (Hay and Southam 1977). However for the two shelf types well-constrained estimates of how much carbonate is produced are missing. In the context of shelves it may be important to separate two other specific bioherms which could have a great potential in shallow-water carbonate production. These are sedimentary carbonates exclusively built up by the calcareous green algae *Halimeda* in tropical latitudes (e.g. Roberts and Macintyre 1988) and extensive biotic cold-water carbonate reefs or banks as described above for mid to high latitudes. For *Halimeda* bioherms total carbonate production and accumulation is about $1.5 \cdot 10^{12}$ mol yr^{-1} (Table 9.1) while the estimates for open shelves given by Wollast (1994) and Milliman and Droxler (1996) do not differentiate a

budget for cold-water shelf areas on its own. Mean carbonate production on lower-latitude shelf areas may range between 50 and 100 g m^{-2}yr^{-1} (Table 9.1); in total $6 \cdot 10^{12}$ mol yr^{-1}. How much of this annual production actually accumulates, is dissolved or exported to the deep-sea remains questionable, because shelf sediments are commonly a mixture of modern and relictic components. Moreover, accumulation rates on broad sandy shelves like that off Argentina are not well-known, because they have not been investigated in such detail yet as other shelf environments. According to Milliman and Droxler (1996) half of the production on carbonate-rich shelves is accumulating ($3 \cdot 10^{12}$ mol yr^{-1}; Table 9.1) while the other half is either transported downslope to the deep sea or dissolved. For carbonate-poor shelves carbonate

Table 9.1 Maximum estimates of present-day carbonate production and accumulation. The difference between production and accumulation is taken as an estimate for dissolution (after Milliman 1993, Milliman and Droxler 1996, Vescei 2004). All numbers labeled with ? are supposed to have an possible error of greater than 100 %.

Habitat	Area ($\cdot 10^6$ km^2)	CaCO$_3$ Production (g m^2 yr^{-1})	CaCO$_3$ Production (10^{12} mol yr^{-1})	Accumulation (10^{12} mol yr^{-1})	Dissolution/Erosion (10^{12} mol yr^{-1})
Shallow waters:					
Coral Reefs	0.35	1800	9	7	2
Platforms	0.8	500	4	2	2
Shelves:					
Carbonate-poor	15	25 ?	4 ?	1 ?	3 ?
Carbonate-rich	10	20-100 ?	6 ?	3 ?	3 ?
Halimeda bioh.	0.05 ?	3000 ?	1.5 ?	1.5 ?	
Total Shallow-water carbonates			**24.5**	**14.5**	**10**
Deep Sea:					
Slopes	32	15	5	4	1
(improved from shelfes)			3.5 ?	2	1.5
Pelagic Surface Waters	290	23	60-90		
(Flux at 1000m		8		24	36 ?)*
Sea floor				11-19	5-13 (in the deep waters & surficial sediments)
Total Deep-sea carbonates			**68.5-90 ?**	**17-25**	**51.5-67**

* Problem of very low fluxes at 1000 m water depth estimated from sediment trap data. Implies very high dissolution rates above the lysocline.

production should at least exceed 20 g m^{-2}yr^{-1} which is the average value of planktonic production in coastal waters. While global carbonate production on carbonate-poor shelves may be relatively high (4·10^{12} mol yr^{-1}; Milliman and Droxler 1996), carbonate accumulation in this environment is negligible.

The rate of total neritic carbonate production in the modern ocean is roughly 25·10^{12} mol yr^{-1} from which 60 % (15·10^{12} mol yr^{-1}) accumulate as shallow-water carbonates. The difference, 10·10^{12} mol yr^{-1}, is the contribution of the neritic environment to the pelagic environment, either in the form of flux of total dissolved inorganic carbon or particulate accumulation on continental slopes and in the deep sea.

9.2.2   Pelagic Calcareous Sediments

Calcium carbonate is the most important biogenic component in pelagic marine sediments which cover an area of about 320·10^6 km^2. Carbonate-rich sediments (>30 % CaCO$_3$) form about 55 % of the deposits on the continental slopes and the deep-sea floor (Lisitzin 1996; Milliman 1993). A new compilation on the distribution of carbonate-rich pelagic sediments (Fig. 9.2b) has been recently carried out by Archer (1996a).

The distribution of pelagic marine carbonates is a result of a variety of influences. The most important components of the pelagic carbonate system affecting their formation and destruction are shown in Figure 9.2b. These can be divided in external contributions such as riverine input and CO$_2$-exchange with the atmosphere and the lithosphere (hydrothermal venting), and internal processes. The main internal processes are those by which CaCO$_3$ is produced and dissolved in the ocean. In the pelagic realm carbonate is mainly produced by planktonic organisms in ocean surface waters supersaturated with calcite and aragonite. Production of biogenic calcite or aragonite comes from planktonic organisms like coccolithophorids, foraminifers and pteropods, as well as calcareous dinoflagellates. The amount of the total production is determined by temperature, light, and nutrient-conditions in the surface ocean. On upper slopes, where food supply is high, part of the carbonate production also originates from benthic foraminifera and small mollusks. Estimates of the global calcium carbonate production by pelagic organisms are difficult to assess, because the amount of production per unit of water volume

or area cannot be directly measured. In addition, according to sparse data available from plankton hauls and shallow floating trap catchments they vary significantly between oligotrophic and eutrophic, and particularly between silicon-poor and silicon-rich surface waters (Fischer et al. 2004). Mean global estimates are either derived from sediment trap fluxes at about 1000 m water depth (Milliman 1993) or from theoretical approaches in which the production required to explain the observed water-column profiles of alkalinity and total dissolved inorganic carbon (Fig. 9.3) is calculated, taking the modern ocean circulation pattern and residence times of the different water masses into account. However, these estimates differ by a factor of two to three, whereby the sediment trap data and surface-water properties suggest a mean global calcium carbonate production of 23 or 60 to 90·10^{12} mol yr^{-1}, respectively (see comments and discussion in Morse and Mackenzie 1990; Milliman and Droxler 1996; Wollast 1994).

How much of the carbonate produced in the surface ocean finally accumulates on the sea floor and is buried there, depends on the calcite dissolution in the water column during the settling of particles and on the calcite dissolution within the sediment. For slope sediments also dilution with sedimentation of terrigeneous material affects the carbonate content. Dissolution of pelagic carbonates is driven by the aragonite and carbonate saturation levels in the deep-sea, controlled by temperature and pressure conditions, as well as by pH and alkalinity of circulating deep-water masses (Berger 1976; Broecker and Peng 1982). On its way to the deep sea the settling biogenic carbonate reaches waters which are increasingly undersaturated with respect to aragonite and later calcite, due to the decrease in temperature, increases of pressure, and the latter CO$_2$ resulting from the reoxidation of organic matter. Increase of CO$_2$ due to benthic respiration also affects CaCO$_3$ which reaches the sea floor. Early diagenesis forces calcite dissolution in the upper sediment column even at supralysoclinal and lysoclinal water depths due to excess pore water CO$_2$ originating from remineralization of sedimentary organic matter (Archer 1991, 1994; Jahnke et al. 1994; Martin and Sayles 1996). The latter will be discussed in more detail in Section 9.4. Moreover, the ocean's circulation also changes the depth distribution of water masses with different saturation levels of aragonite and calcite. As a consequence, only a small fraction

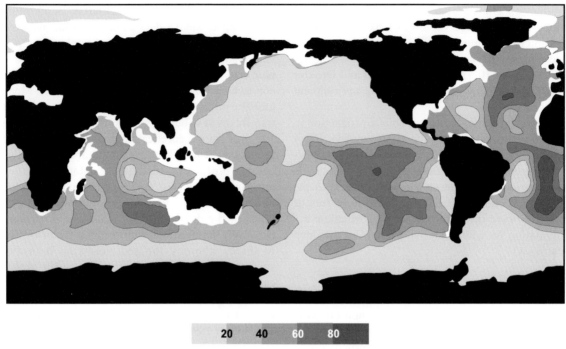

CaCO$_3$ in surface sediments [%]

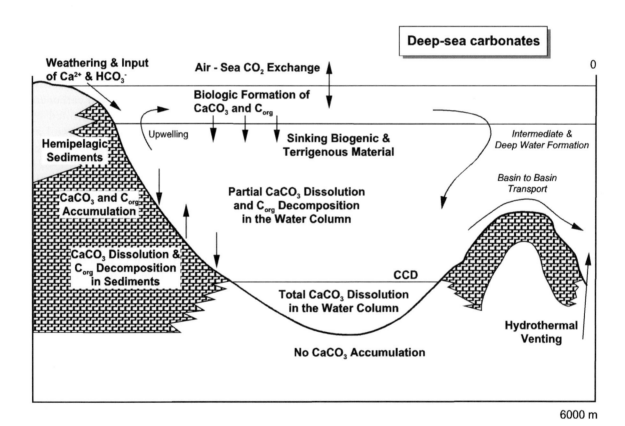

Fig. 9.2 Environments of pelagic carbonate deposition: a) deep-sea distribution of carbonate-rich sediments (new compilation from Archer 1996a) and b) factors controlling pelagic carbonate production and dissolution (modified from Morse and Mackenzie 1990).

(15 to 20 %; see Table 9.1, or Archer 1996b, Wollast 1994) of the total amount of pelagic carbonates produced in the upper ocean is buried in deep-sea sediments. Different estimates for mean global calcium carbonate accumulation rates from various authors (Milliman 1993; Morse and Mackenzie 1990; Wollast and Mackenzie 1987) are in good agreement for the deep sea, all ranging between $10 \cdot 10^{12}$ mol yr^{-1} and $12 \cdot 10^{12}$ mol yr^{-1} (1 to 1.2 bt CaCO$_3$ yr^{-1}).

Essentially all the above mentioned factors have the potential to change the calcite-carbonate equilibrium in the ocean over time and thus exert major control on the distribution and amount of calcareous sediments in the deep sea. On the other hand, given the 60 times greater carbon reservoir of the ocean compared to that of the atmosphere, changes in the oceanic calcite-carbonate equilibrium can modify the oceanic CO_2 uptake/release balance with respect to the atmosphere (Archer and Maier-Reimer 1994; Berger 1982; Broecker and Peng 1987; Maier-Reimer and Bacastow 1990; Opdyke and Walker 1992; Siegenthaler and Wenk

1984; Wolf-Gladrow 1994; see also review in Dittert et al. 1999). Therefore, the principles of the calcite-carbonate system will be re-examined in the following Section 9.3 and examples will be given for modeling this system under specific boundary conditions typical for the modern ocean. As we know from the geological record of calcareous sediment distribution, this system has changed dramatically in the past and definitely will change in the future, leaving us with the questions of how much, how fast and in which direction the ocean carbonate system will respond to anthropogenic disturbances of the carbon cycle in the future, which may result in a more acid ocean (Feely et al. 2004, Orr et al. 2005).

9.3 The Calcite-Carbonate-Equilibrium in Marine Aquatic Systems

The calcite-carbonate-equilibrium is of particular importance in aquatic systems wherever carbon dioxide is released into water by various processes, and wherever a concurrent contact with calcite (or other carbonate minerals) buffers the system by processes of dissolution/precipitation. Carbon dioxide may either originate from the gaseous exchange with the atmosphere, or is formed during the oxidation of organic matter. In such a system this equilibrium controls the pH-value - an essential system parameter which, directly or indirectly, influences a number of secondary processes, e.g. on iron (cf. Chap. 7) and on manganese (cf. Chap. 11). In the following, the calcite-carbonate equilibrium will be described in more detail and with special attention given to its relevant primary reactions.

The description of Sections 9.3.1 and 9.3.2 will assume in principle a solution at infinite dilution in order to reduce the system to really important reactions. The Sections 9.3.3 and 9.3.4 provide examples calculated for seawater, including all constituents of quantitative importance. In such calculations, two different approaches are possible:

The use of measured concentrations (calcium, carbonate, pH) together with so-called 'apparent' equilibrium constants. These apparent constants (e.g. Goyet and Poisson 1989; Roy et al. 1993; Millero 1995) take the difference between activities and concentrations into account, as well as

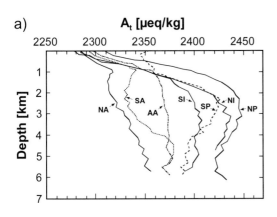

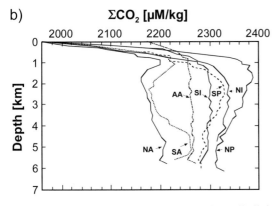

Fig. 9.3 The mean vertical distribution of a) alkalinity and b) total CO_2 concentration normalized to the mean world ocean salinity value of 34.78 (after Takahashi et al. 1980, adapted from Morse and Mackenzie 1990)

the fact, that for instance only part of the measured calcium is present in the form of Ca^{2+}-ions, whereas the rest forms complexes or ion pairs ($CaSO_4{}^0{}_{aq}$, $CaHCO_3{}^+{}_{aq}$, or others). This is true for seawater, since the activity coefficients of these complexes or ion pairs with sufficient accuracy exert a constant influence.

The other approach employs a geochemical computer model, such as PHREEQC (Parkhurst 1995; also Chap. 15) with an input of a complete seawater analysis. Such a model will then calculate the activity coefficients and the species distribution of the solution according to the complete analysis and the constants of the thermodynamic database used. These constants are well known with an accuracy which is usually better than the accuracy of most of our analyses at least for the major aquatic species. Together with the 'real' constant of the solubility product a reliable saturation index (SI = log Ω) is then calculated. The constants of solubility products are not accurately known for some minerals, but for calcite, and also for most other carbonates, these constants and their dependence on temperature and pressure are very well documented.

Both approaches used in the analysis of seawater lead to identical results, thus leaving it undetermined which method should be preferred for seawater analysis. However, for marine pore water the first approach is *not valid*, since quite different concentrations of complexes or ion pairs are likely to occur, for instance, sulfate reduction and/or an increase of alkalinity. Therefore we rely on the second approach in our examples and strongly recommend it for the marine geochemistry.

9.3.1   Primary Reactions of the Calcite-Carbonate-Equilibrium with Atmospheric Contact in Infinitely Diluted Solutions

Upon making contact with the atmosphere, the partial pressure (pCO_2) of carbon dioxide ($CO_{2,g}$) determines the concentration of carbonic acid ($H_2CO^0{}_{3,aq}$) in solution. The equilibrium reaction is written as:

$$CO_{2,g} + H_2O \Leftrightarrow H_2CO_3{}^0{}_{aq} \qquad (9.1)$$

The equilibrium is determined by the equation:

$$K_{CO2} = [H_2CO_3{}^0{}_{aq}] / \{pCO_2 \cdot [H_2O]\} = 3.38E\text{-}2$$
$$(25°C, 1 \text{ atm}) \qquad (9.2)$$

In this form the reaction combines two different steps: One leads to the formation of $CO_{2,aq}$ in aqueous solution, and a second leads from $CO_{2,aq}$ to the formation of $H_2CO_3{}^0{}_{aq}$. The brackets [...] denote activities, whereas parentheses embrace molar concentrations. The activity of water [H_2O] is equal to 1.0 in infinitely diluted solutions, whereas it is slightly less in seawater (0.981). The value of the constant K_{CO2} – as for all other constants in the following – is determined at a temperature 25°C and under a pressure of 1 atm. The temperature dependency of the constants was empirically investigated by Plummer et al. (1978) for the range between 5°C and 60°C. The computer program PHREEQC (Parkhurst 1995) also accounts for this temperature dependence. The pressure dependence, which is important in great water depths, is considered only in the model SOLMINEQ (Kharaka et al. 1988). The constants belonging to this program, corrected for specific water depths, can be extracted and explicitly entered into the database of the generally more versatile model PHREEQC.

The aqueous complex $H_2CO_3{}^0{}_{aq}$ dissociates into protons and bicarbonate ions:

$$H_2CO_3{}^0{}_{aq} \Leftrightarrow H^+ + HCO_3{}^- \qquad (9.3)$$

This equilibrium is described by the equation:

$$K_{H2CO3} = \{[H^+] \cdot [HCO_3{}^-]\} / [H_2CO_3{}^0{}_{aq}] = 4.48E\text{-}7$$
$$(25°C, 1 \text{ atm}) \qquad (9.4)$$

The second step of dissociation is determined by:

$$HCO_3{}^- \Leftrightarrow H^+ + CO_3{}^{2-} \qquad (9.5)$$

This equilibrium is described by the equation:

$$K_{HCO3} = \{[H^+] \cdot [CO_3{}^{2-}]\} / [HCO_3{}^-] = 4.68E\text{-}11$$
$$(25°C, 1 \text{ atm}) \qquad (9.6)$$

Furthermore, the dissociation of water needs to be included:

$$H_2O \Leftrightarrow H^+ + OH^- \qquad (9.7)$$

with the constant K_{H2O}:

$$K_{H2O} = [H^+] \cdot [OH^-] = 1.00E\text{-}14$$
$$(25°C, 1 \text{ atm}) \qquad (9.8)$$

The contact made by processes of dissolution and precipitation with the solid phase of calcite is described by the reaction:

$$CaCO_{3,calcite} \Leftrightarrow Ca^{2+} + CO_3^{2-} \qquad (9.9)$$

The equilibrium is described by the equation:

$$K_{calcite} = [Ca^{2+}] \cdot [CO_3^{2-}] = 3.31E\text{-}9$$
$$(25°C, 1 \text{ atm}) \qquad 9.10$$

Another essential condition for describing the calcite-carbonate-equilibrium consists in the fulfillment of the neutral charge rule. This holds that charges may be neither lost nor gained on reaching the equilibrium state.

The balance of the charges is done indirectly on the basis of the ions that are the carriers of these charges. Therefore, calculations must consider divalent ions twice:

$$2 (Ca^{2+}) + (H^+) = (OH^-) + (HCO_3^-) + 2 (CO_3^{2-})$$
$$(9.11)$$

It should be noted that the parentheses (...) in this charge balance represent molar concentrations, since the chemical activities are not of importance here, instead the really existent amounts of the substances and their charges. Equation 9.11 describes in a simplified form the conditions in which the equilibrium state is reached starting from pure water. Only in this case all final concentrations produce charges equal to zero in the balance.

In an impure system, any initial concentration $(...)_i$ or initial activity $[...]_i$ may be present. In this context, the index 'i' stands for 'initial'. On the basis of these concentrations, those reached on attaining the final equilibrium state $(...)_f$ or $[...]_f$ will then have to be sought for. The index 'f' in this case stands for 'final'. The statement as to the charge balance made in Equation 9.11 is not accomplished by observing the total concentration of each reactant, but by the difference between 'initial' to 'final'. The generalized case described in Equation 9.12 resumes the condition of Equation 9.11 in which all initial concentrations are set to zero (pure water).

$$2 (Ca^{2+})_f - 2 (Ca^{2+})_i + (H^+)_f - (H^+)_i =$$
$$(OH^-)_f - (OH^-)_i + (HCO_3^-)_f - (HCO_3^-)_i +$$
$$2 (CO_3^{2-})_f - 2 (CO_3^{2-})_i \qquad (9.12)$$

In a system following this description, the solution for all six variables (Ca^{2+}), (H^+), (OH^-), $(H_2CO_3^0)$, (HCO_3^-), (CO_3^{2-}) is sought for the condition of a final equilibrium state determined by the Equations 9.2, 9.4, 9.6, 9.8, 9.10, and 9.12. It should be noted that in Equations 9.2, 9.4, 9.6, 9.8, and 9.10 the activities written in brackets [...] are to be understood as final activities $[...]_f$, which are products of an activity coefficient and the single concentrations in equilibrium $(...)_f$ (cf. Eq. 15.2 in Sect. 15.1.1). For a system at infinite dilution, however, these activity coefficients are 1.0, and thus concentrations are (...) equal to activities [...].

The system of six non-linear equations with six unknown variables can be solved mathematically for $(H^+)_f$ by inserting the equations into each other until only constants, initial concentrations $(...)_i$, and pCO_2 describe the complete equation. This way of solving the problem leads to a cubic equation for which an analytical solution exists. In practice, however, this strategy is irrelevant, since such a set of equations cannot be analytically solved for other boundary conditions (e.g. a system which is closed with respect to CO_2, having no contact with the atmosphere) or if more constituents are included (e.g. complex species like $CaHCO_3^+{}_{aq}$ or $CaCO_3^0{}_{aq}$). In such a case, it must be solved by an iteration process, as will be described in the succeeding section. The application of spreadsheet calculation programs make such an iterative solution easy to find by pursuing the procedures below:

First, the concentration of $(H_2CO_3^0)_f$ is calculated on the basis of the given pCO_2 using Equation 9.2

Then, the pH-value of the equilibrium is estimated. For a first approximation, the pH-value derived from the 'initial' concentration of $(H^+)_i$ may be used. But any other pH-value, for instance pH 7, is permitted as well.

With this pH-value, the concentrations of $(H^+)_f$ and $(OH^-)_f$ are calculated using Equation 9.8.

Successively, $(HCO_3^-)_f$ and $(CO_3^-)_f$ are calculated using the Equations 9.4 and 9.6 and the values so far determined. The concentration of $(Ca^{2+})_f$ is obtained from Equation 9.12.

Now the saturation index $(SI_{calcite})$ can be calculated with Equation 9.10 (cf. Eq. 15.1 in Sect. 15.1.1). If the saturation index equals zero, then the estimation of the pH-value in equilibrium, as mentioned in step 2, has been correct, as well as all other equilibrium concentrations $(...)_f$ calculated in steps 3 and 4. If the saturation index $(SI_{calcite})$ is not equal to zero with sufficient accuracy, the estimation for the pH in step 2 will have to be improved until the correct value is found.

Irrespective whether a solution to this system is found in the iterative manner described, or whether it is solved by finding an analytical solution of the cubic equation, the pH-value of the equilibrium is at all times determined by the partial pressure pCO_2 in the atmosphere. As it remains in contact with the solution, such a system is referred to as open to CO_2. It is characteristic of such systems that the reactions can cause the release of CO_2 into the atmosphere at any time and that CO_2 can be drawn from the atmosphere without directly affecting total atmospheric pCO_2.

9.3.2   Primary Reactions of the Calcite-Carbonate-Equilibrium without Atmospheric Contact

A system which is closed to CO_2 exists wherever the final calcite-carbonate-equilibrium is reached without any concurrent uptake or release of atmospheric CO_2 in its process. This implies that the Equations 9.1 and 9.2 are no longer valid. In their stead, a balance of various C-species is related to calcium. This balance maintains that the sum of C-species must equal the calcium concentration in solution, since both can enter the solution only by dissolution of calcite or aragonite:

$$(Ca^{2+}) = (H_2CO_3^0) + (HCO_3^-) + (CO_3^{2-}) \qquad (9.13)$$

Analogous to Equations 9.11 and 9.12, Equation 9.13 is only valid if the initial solution consists of pure water. In the normal case, in which the solution will already contain dissolved carbonate- and/or calcium ions, the difference between equilibrium concentrations $(...)_f$ and the initial concentrations $(...)_i$ needs to be regarded:

$$(Ca^{2+})_f - (Ca^{2+})_i =$$
$$(H_2CO_3^0)_f - (H_2CO_3^0)_i + (HCO_3^-)_f -$$
$$(HCO_3^-)_i + (CO_3^{2-})_f - (CO_3^{2-})_i \qquad (9.14)$$

The six Equations 9.4, 9.6, 9.8, 9.10, 9.12, and 9.14 must now be solved for the six variables (Ca^{2+}), (H^+), (OH^-), $(H_2CO_3^0)$, (HCO_3^-), (CO_3^{2-}). There is no analytical solution for this non-linear equation system. If there were one, it would be irrelevant because an iterative solution, similar to the one described in the previous section, would always be easier to handle and besides be more reliable. However, still independent of the method of mathematical solution is the fact that the equilibrium is exclusively determined by the carbonate concentration which the system previously contained and which is defined by the concentrations $(...)_i$. At this particular point, the amount of CO_2 would also have to be considered which is liberated into such a system, e.g. from the oxidation of organic matter. This amount would be simply added to the 'initial' concentrations and would influence the equilibrium in the same way as the pre-existing carbonate.

9.3.3   Secondary Reactions of the Calcite-Carbonate-Equilibrium in Seawater

In seawater, the differences between activities and concentrations must always be considered (cf. Sect. 15.1.1). The activity coefficients for monovalent ions in seawater assume a value around 0.75, for divalent ions this value usually lies around 0.2. In most cases of practical importance, the activity coefficients can be regarded with sufficient exactness as constants, since they are, over the whole range of ionic strengths in solution, predominately bound to the concentrations of sodium, chloride, and sulfate which are not directly involved in the calcite-carbonate-equilibrium. The proportion of ionic complexes in the overall calcium or carbonate content can mostly be considered with sufficient exactness as constant in the free water column of the ocean. Yet, this cannot be applied to pore water which frequently contains totally different concentrations and distributions of complex species due to diagenetic reactions.

Figure 9.4 shows the distribution of carbonate and calcium species in ocean water and in an anoxic pore water, calculated with the program PHREEQC (Parkhurst 1995). It is evident that about 10 % of total calcium is prevalent in the form of ionic complexes and 25 – 30 % of the total dissolved carbonate in different ionic complexes other than bicarbonate. These ionic complexes are not included in the equations of Sections 9.3.1 and 9.3.2. Accordingly, the omission of these complexes would lead to an erraneous calculation of the equilibrium. The inclusion of each complex shown in Figure 9.4 implies further additions to the system of equations, consisting in another concentration variable (the concentration of the complex) and a further equation (equilibrium of the complex concentration relative to the non-

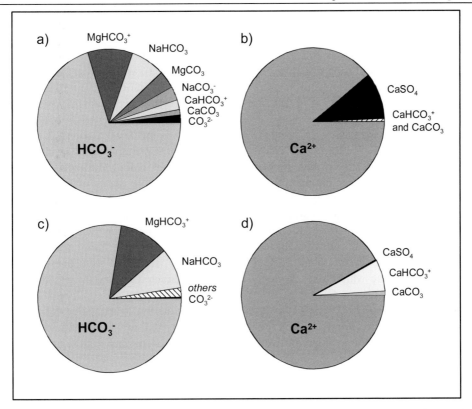

Fig. 9.4 Distribution of carbonate species (a and c) and calcium species (b and d) in seawater after Nordstrom et al. 1979 (a and b) and in an anoxic pore water (c and d). The pore water sample was extracted from the core previously shown in Figure 3.1 and was taken from a depth of 14.8 m below the sediment surface. The calculation of species distributions was performed with the program PHREEQC (Parkhurst 1995)

complexed ions). If only the ion complexes shown in Figure 9.4 are added, as there are ($MgHCO_3^+$), ($NaHCO_3^0$), ($MgCO_3^0$), ($NaCO_3^-$), ($CaHCO_3^+$), ($CaCO_3^0$), and ($CaSO_4^0$), the system would consequently be extended to 6+7=13 variable concentrations and just as many non-linear equations. Such a system of equations is solvable with the aid of an appropriate spread-sheet calculation program, however, that would be certainly not reasonable and would also be more laborious than using an equilibrium model of the PHREEQC type (Parkhurst 1995). Function and application of PHREEQC are described in Section 15.1 in more detail. In the following section, some calculation examples with regard to carbonate will be presented.

9.3.4 Examples for Calculation of the Calcite-Carbonate-Equilibrium in Ocean Waters

Here, exemplary results will be introduced which were obtained from model calculations of the cal-

cite-carbonate-equilibrium under applied boundary conditions close to reality. To this end, the model PHREEQC (Parkhurst 1995) has been used. The chemical composition of the solution studied in these examples is based on analytical data published by Nordstrom et al. 1979 (cf. Table 15.1).

In the first example shown in Table 9.2, a warm (25°C) water from the ocean's surface (1 atm) has been modeled. In the zone near the equator, where this water sample had been taken, carbon dioxide partial pressures of 400 μatm (equivalent to a pCO_2 of 0.0004 atm or a log pCO_2 of -3.40) have been measured which is somewhat higher than the corresponding atmospheric value. This sample of ocean water thus displays a CO_2-gradient directed towards the atmosphere and therefore continually releases CO_2 into the atmosphere. This situation is accounted for in the model by pre-setting pCO_2 to 0.0004 atm as an open boundary condition with regard to CO_2. Accordingly, a state of supersaturation ensues equivalent to a $SI_{calcite}$ value of 0.77 or an $\Omega_{calcite}$ value of 5.9. Such a supersaturation state is, according to the

321

analytical evidence, obviously permanent by its nature, due to the processes in the surface-near ocean water and the fluxes across the interface ocean/atmosphere. The model results for a theoretical precipitation of calcite describing the state of calcite-carbonate-equilibrium ($SI_{calcite}$ = 0.0) are shown in the lower part of Table 9.2.

In the model calculation presented in Table 9.3, the constant of the solubility product of calcite was adjusted to temperature (6°C) and pressure (100 atm) relative to a depth of about 1000 m. Therefore, the pressure-corrected constant derived from the model SOLMINEQ (Kharaka et al. 1988) was entered into the database of the model PHREEQC, and the correction procedure for temperature integral to this model was used. Furthermore, an organic substance which also contained nitrogen and phosphorus in a C:N:P ratio of 106:16:1 (Redfield 1958), was decomposed down to a low residual concentration of dissolved oxygen. The calculation was then run under conditions

closed to CO_2, i.e. concentrations of the previous example from Table 9.2 were regarded as 'initial'. The partial pressure pCO_2 in this case no longer represents a fixed boundary condition, but has now become a model output. The calcite-carbonate-equilibrium was kept at a constant state of supersaturation ($SI_{calcite}$ = 0.26) throughout all steps of the model calculation.

In the next example outlined in Table 9.4, a cold (2°C), surface-near (1 atm) ocean water sample has been modeled. In zones of high latitudes, where this water type occurs, a partial pressure of 280 µatm has been measured (equivalent to a pCO_2 of 0.00028 atm or a log pCO_2 of –3.55) which is somewhat lower than determined in the atmosphere. Here, the atmosphere displays a CO_2-gradient directed toward the ocean into which CO_2 is continually taken up.

This situation is accounted for in the model by pre-setting pCO_2 to 0.00028 atm, the boundary condition of a system open with regard to CO_2. Accord-

Table 9.2 Model calculation using the PHREEQC program (Parkhurst 1995) on a sample of warm surface water from the ocean near the equator. The lower part of the table shows the results for a theoretical ($SI_{calcite}$ = 0.0) precipitation of calcite.

Model of warm surface seawater

input concentrations:

dissolved constituents from Nordstrom et al. 1979, cf. Tabel 15.1

boundary conditions:

temperature	25 °C	
pCO_2	400 µatm	(i.e. log pCO_2 = -3.40)
log k calcite	-8,48	(at 25 °C and 1 atm pressure)

input situation without calcite-carbonate-equilibrium:

pH	8,24	
sum of carbonate species (TIC)	2.17 mmol/l	
sum of calcium species	10.65 mmol/l	
$SI_{calcite}$	0,77	(i.e. $\Omega_{calcite}$ = 5.9)

reactions:

calcite-carbonate-equilibrium constant at SI = 0.0

PHREEQC model results:

pH	7,50	
sum of carbonate species (TIC)	1.92 mmol/l	
sum of calcium species	10.40 mmol/l	
$SI_{calcite}$	0,00	(i.e. $\Omega_{calcite}$ = 1.0)

Table 9.3 Model calculation applying the computer program PHREEQC (Parkhurst 1995) to a sample of ocean water in a water depth of approximately 1000 m, near the equator. The constant of the solubility product for calcite is corrected for temperature and pressure. The decomposition of organic substance due to the presence of free oxygen in the water column has been included.

Model of low latitude seawater at 1000 m depth

---

input concentrations:

model calculation of warm surface seawater

boundary conditions:

temperature	6 °C	
log k calcite	-8.26	(at 6 °C and 100 atm pressure)

input situation without calcite-carbonate-equilibrium:

pH	8.08	
sum of carbonate species (TIC)	2.26 mmol/l	
sum of calcium species	10.61 mmol/l	
$SI_{calcite}$	0.26	(i.e. $\Omega_{calcite}$ = 1.8)

reactions:

$(CH_2O)_{106}(NH_3)_{16}(H_3PO_4)$ reacts with dissolved O_2
calcite supersaturation constant at SI = 0.26

PHREEQC model results:

pH	8.01	
sum of carbonate species (TIC)	2.60 mmol/l	
sum of calcium species	10.79 mmol/l	
$SI_{calcite}$	0.26	(i.e. $\Omega_{calcite}$ = 1.8)

---

ingly, a state of supersaturation ensues equivalent to a $SI_{calcite}$ value of 0.43 or an $\Omega_{calcite}$ value of 2.7. As a theoretical example (lower part of Table 9.4) for the modeling in this case a constant state of supersaturation according to a $SI_{calcite}$ value of 0.50 or an $\Omega_{calcite}$ value of 3.2 was calculated.[1]

In the calculated model summarized in Table 9.5, the equilibrium constants for calcite was adjusted to temperature (2°C) and pressure (600 atm) of a water depth of about 6000 m. In this example as well, the pressure-corrected constant derived from the model SOLMINEQ (Kharaka et

al. 1988) was entered into the database of the model PHREEQC, and the correction procedure for temperature of this model was used. Calculation was run under the boundary conditions of a system closed with regard to CO_2, i.e. the concentrations of the previous example in Table 9.4 were regarded as 'initial'. The partial pressure pCO_2 is not a fixed boundary condition of an open system, and it did not change as in the example shown in Table 9.3, since no reactions were determined. A comparable decomposition of organic matter as in Table 9.3 was thus excluded from this example. After pressure-correction, the saturation index of calcite documents an undersaturation displaying a value of –0.16 (equivalent to a $\Omega_{calcite}$ value of 0.69). In the case of an additional decomposition of organic matter, this undersaturation of calcite is likely to increase further. Such a reaction could be in fact applicable to deep-sea waters of pelagic deep ocean areas.

---

[1] $\Omega_{calcite}$ is still in use in chemical oceanography for describing the state of saturation. It corresponds to the saturation index without the logarithm, hence: log $\Omega_{calcite}$ = $SI_{calcite}$. However, the SI value is more useful, since the same amounts of undersaturation and supersaturation respectively display the same SI values, only distinguished by different signs (+) for supersaturation and (-) for undersaturation. A SI value of zero describes the state of saturation.

Table 9.4 Model calculation applying the computer program PHREEQC (Parkhurst 1995) to a sample of cold surface water of the ocean from higher latitudes. The constant of the solubility product for calcite is accordingly corrected for temperature.

Model of cold surface seawater

---

input concentrations:

dissolved constituents from Nordstrom et al. 1979, cf. Tabel 15.1

boundary conditions:

temperature	2 °C	
pCO_2	280 µatm	(i.e. log pCO_2 = -3.55)
log k calcite	-8,34	(at 2 °C and 1 atm pressure)

input situation without calcite-carbonate-equilibrium:

pH	8,23	
sum of carbonate species (TIC)	2.28 mmol/l	
sum of calcium species	10.63 mmol/l	
$SI_{calcite}$	0,43	(i.e. $\Omega_{calcite}$ = 2.7)

reactions:

calcite supersaturation constant at SI = 0.50

PHREEQC model results:

pH	8,30	
sum of carbonate species (TIC)	2.30 mmol/l	
sum of calcium species	10.65 mmol/l	
$SI_{calcite}$	0,50	(i.e. $\Omega_{calcite}$ = 3.2)

9.4   Carbonate Reservoir Sizes and Fluxes between Particulate and Dissolved Reservoirs

Present-day production of calcium carbonate in the pelagic ocean is calculated to be in the order of 6 to 9 billion tons (bt) per year ($60\text{-}90 \cdot 10^{12}$ mol yr^{-1}), from which about 1.1 to 2 bt (11 to $20 \cdot 10^{12}$ mol yr^{-1}) accumulate in sediments (Tables 9.1 and 9.6). Together with the accumulation in shallow waters of $14.5 \cdot 10^{12}$ mol yr^{-1} (Table 9.1, Fig. 9.5), the total carbonate accumulation in the world ocean amounts to 3.5 bt per year. This latter number is twice as much calcium as is brought into the ocean by rivers and hydrothermal activity (1.6 bt, Wollast 1994). Wollast (1994) and Milliman and Droxler (1996)

therefore consider the carbonate system of the modern ocean to be in non-steady state conditions, because production is not equal to the input of Ca^{2+} and HCO_3^- by rivers and hydrothermal vents (Fig. 9.5). Consequently, the marine carbonate system is at a stage of imbalance, or the output controlled by calcite sedimentation or input conveiled by dissolution has been overestimated or underestimated, respectively. On the other hand, one or more input sources may have been not detected so far. Published calcium carbonate budgets vary strongly, because the various authors have used different data sets and made different assumptions with respect to the production and accumulation of carbonates (e.g. Table 9.6), as well as for the sources of dissolved calcium and carbonate in marine waters (see summaries in Milliman 1993; Milliman and Droxler 1996; Wollast 1994).

Table 9.5 Model calculation using the computer program PHREEQC (Parkhurst 1995) on deep-sea waters of the ocean. The constant of the solubility product for calcite is accordingly corrected for temperature and pressure. A comparable decomposition of organic matter as contained in Table 9.3 was excluded in this example.

Model of seawater at 6000 m depth

input concentrations:

model calculation of cold surface seawater

boundary conditions:

temperature	2 °C	
pCO$_2$	280 µatm	(i.e. log pCO$_2$ = -3.55)
log k calcite	-7.75	(at 2 °C and 600 atm pressure)

input situation without calcite-carbonate-equilibrium:

pH	8.23	
sum of carbonate species (TIC)	2.28 mmol/l	
sum of calcium species	10.63 mmol/l	
SI$_{calcite}$	0.43	(i.e. $\Omega_{calcite}$ = 2.7)

no reactions, but pressure changed to 600 atm

PHREEQC model results:

pH	8.23	
sum of carbonate species (TIC)	2.28 mmol/l	
sum of calcium species	10.63 mmol/l	
SI$_{calcite}$	-0.16	(i.e. $\Omega_{calcite}$ = 0.7)

9.4.1 Production *Versus* Dissolution of Pelagic Carbonates

This chapter summarizes the most recent compilations of carbonate reservoir size in the ocean and sediments, as well as the particulate and dissolved fluxes (Fig. 9.5) provided by the above mentioned authors. Coral reefs are probably the best documented shallow-water carbonate environment. Carbonate production on reef flats range as high as 10.000 g CaCO$_3$ m^{-2}yr^{-1}, with a global mean of about 1800 g CaCO$_3$ m^{-2}yr^{-1}. Totally this amounts to 24.5·10^{12} mol yr^{-1} (Table 9.1) from which 14.5·10^{12} mol yr^{-1} accumulate and 10·10^{12} mol yr^{-1} are transported to the deep-sea either by particulate or dissolved export. One of the most uncertain numbers in all these budget calculations are the estimates of the global carbonate production in the open ocean. Milliman's (1993) estimate was only about 24·10^{12} mol yr^{-1}, based on carbonate flux rates at about 1000 m water depth, measured by long-term time series of sediment trap moorings, which is approximately 8 g CaCO$_3$ m^{-2}yr^{-1} accounting for a global flux rate of particulate pelagic carbonate to be in the range of 24·10^{12} mol yr^{-1}. However, as discussed in Wollast (1994), in order to produce the measured water column profiles of total inorganic carbon and carbonate alkalinity (Fig. 9.3) a much higher surface ocean carbonate production is required. Thus estimates reported more recently are in the order of 60 to 90·10^{12} mol yr^{-1} (Table 9.6). The discrepancy between the very high global production rates and measured fluxes obtained in sediment traps then can only be explained if one accepts that a substantial portion (30 to 50 %) of carbonate produced in the pelagic

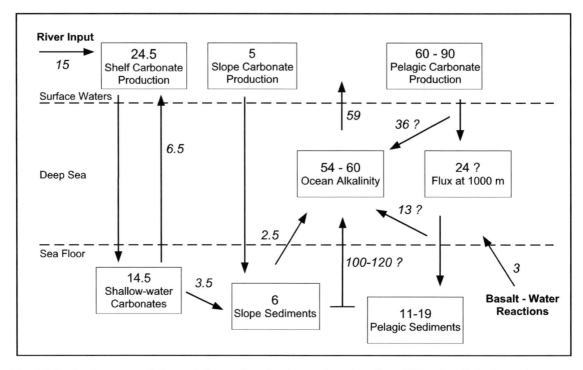

Fig. 9.5 Production, accumulation and fluxes of marine inorganic carbon (in x 10^{12} mol yr^{-1}) in the modern ocean, summarizing production, accumulation, and fluxes of particulate inorganic carbon, as well as carbonate dissolution rates as given in Table 9.1 (modified after Milliman and Droxler 1996, Wollast 1994).

ocean is already dissolved in the upper 1000 m of the ocean. On the other hand, if calcite dissolution above lysocline depths due to benthic respiration of organic matter (e.g. Emerson and Bender 1981; see also discussion and references below) is much higher as previously thought, then the fraction of pelagic carbonate production which reaches the sea floor prior to benthic respiration maybe in the order of 50 to 70 % and not only about 20 % as given in Table 9.6 for the

more recent budget estimates. According to Archer (1996b) 20-30 % of the carbonate flux to the sea floor finally escapes dissolution. If this is correct, it would require a much less total carbonate production in the surface ocean (in the order of the estimate of Milliman (1993); see Table 9.6), to maintain measured water column profiles of alkalinity and total dissolved inorganic carbon, as well as $11 \cdot 10^{12}$ mol yr^{-1} of pelagic carbonate sedimentation. To solve the problem whether high

Table 9.6   Comparison of carbonate rain, accumulation and dissolution estimates for the pelagic ocean (all values in 10^{12} mol yr^{-1}).

Author	Pelagic Production	Burial	Dissolution	Burial fraction of Production
Milliman (1993)	24	11	12	47%
Milliman & Droxler (1996)	60	11	13	18%
			36 already at 1000 m, Table 9.1, with about 6 from sediments above the hydrogr. lysocline (Hensen et al. 2003).	
Wollast (1994)	65	11	54	18%
Archer (1994) (as cited in Archer 1996b, Fig. 12)	86	19	67	22%
Hensen et al. (2003)			22-81	

values of alkalinity and total dissolved inorganic carbon in intermediate and deep waters are the result of dissolving calcite particles settling through the water column, or of sedimentary dissolution above the lysocline, would require better knowledge of the total amount of calcite production in the surface ocean and of the dissolution rates in the sediment. While the first is very difficult to measure, the second may be reached by improving data sets of inorganic carbon flux rates from the sediment to ocean at intermediate and deep water levels. The next Section 9.4.2. describes the approach of estimating the fluxes of total carbon and inorganic carbon resulting from calcite dissolution in more detail.

9.4.2 Inorganic and Organic Carbon Release from Deep Sea Sediments

Although only a small fraction of carbon arriving at the seafloor is finally buried, over geological time scales deep sea sediments have formed the largest reservoir of carbon on earth mainly consisting of biogenic carbonates (~50 Mio. Gt) and organic detritus (~12 Mio. Gt; i.e. de Baar & Suess, 1993). This huge storage potential of marine carbonates is a major factor for maintaining reasonably low atmospheric CO_2-levels in earth history. However, the ratio of accumulation versus recycling of carbon at the seafloor reflects a dynamic equilibrium depending on various external forcing factors. Hence, the question for the "how much is being recycled" is an important issue, but quite difficult to explore.

Diagenetic mineralization and dissolution processes in deep sea sediments have been recognized as important control factors and numerous efforts have been made to estimate the contribution by benthic reflux to the ocean budget. The diagenetic reactions are most intense at the sediment-water interface where the most labile components become rapidly mineralized. Hence, this is the place where the determination between burial and recycling is made. The driving forces of carbon release from the sediments are the degradation of organic matter and the dissolution of calcium carbonate, which are in turn dependent on the supply by the remnants of biological production in the surface waters. However, the dissolution of calcium carbonate in deep-sea sediments is controlled by two major factors, the degree of undersaturation of the deep ocean waters with respect to calcite and aragonite,

and the reaction with carbon dioxide from respiration processes. Two of the key parameters are thus, the "rain ratio" (C_{CaCO_3}/C_{POC}) of the sinking material and the water depth of final deposition, above or below the CO_3^{2-} saturation horizon (lysocline). Because the solubility of calcium carbonate increases with increasing pressure, the ocean is usually super-saturated at shallow to intermediate depths and undersaturated in the deep basins.

Another important effect, which has been outlined in Chapter 6.2, is the ageing, or better the respiratory enrichment in CO_2 of the deep waters along the flow path from the North Atlantic to the North Pacific causing large differences in saturation state above the seafloor in the Atlantic and the Pacific (Fig. 9.6).

While calcite dissolution below saturation horizons in the water column can be described by the equations reported in section 9.3.2., dissolution due to CO_2 release from organic carbon respiration is expressed by the following equations:

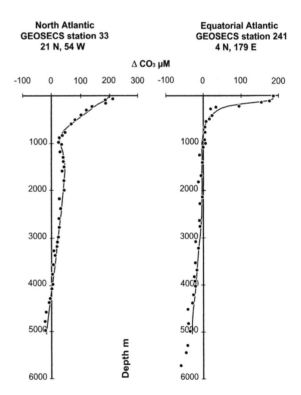

Fig. 9.6 ΔCO_3^{2-} (giving the difference between the saturation concentration of CO_3^{2-} and the ambient concentration of CO_3^{2-} at a specific depth) based on GEOSECS-data (Takahashi et al., 1980) of two locations in the Atlantic and the Pacific ocean. The intersection of the data points with the 0 µM line denotes the position of the lysocline (from Archer, 1996).

$$(CH_2O)_{106}(NH_3)_{16}(H_3PO_4) + 138\ O_2 \rightarrow$$

$$106\ HCO_3^- + 16\ NO_3^- + HPO_4^{2-} + 16\ H_2O + \mathbf{124\ H^+}$$

$$(9.15)$$

$$124\ CaCO_3 + \mathbf{124\ H^+} \rightarrow 124\ HCO_3^- + 124\ Ca^{2+}$$

$$(9.16)$$

where organic matter with Redfield C:N:P ratio is oxidized and the produced acid is neutralized by sedimentary calcium carbonate.

Carbonate dissolution induced by metabolic processes in deep-sea sediments has been neglected for a long time. Emerson and Bender (1981) were about the first who explicitly stated that the degradation of organic matter may significantly drive calcite dissolution, and hence, affect the preservation of calcium carbonate in deep-sea sediments even above the lysocline. A number of subsequent studies has identified this problem and generally focused on the differentiation between calcium carbonate dissolution by undersaturation of bottom waters and organic matter remineralization. These studies specifically considered the dissolution kinetics of calcium carbonate in deep-sea sediments (i.e. Berelson et al. 1990; Berelson et al. 1994; Hales and Emerson 1996 and 1997a; Jahnke et al. 1994 and 1997; Martin and Sayles 1996; Wenzhöfer et al. 2001). It is very important to understand whether the dissolution of calcium carbonate is driven by one or the other process in order to correctly interpret the accumulation of calcium carbonate in sediments over time. For example, calcium carbonate preservation at a given site may be reduced by stronger bottom water undersaturation or the decrease of the rain ratio by accelerating the metabolic CO_2 release in the sediments. Archer and Maier-Reimer (1994) demonstrated that a shift to higher rain ratios could explain both, reduced pCO_2-levels during the last glacial and sedimentary calcium carbonate concentrations in deep-sea sediments. The calcite dissolution by oxic respiration of organic matter might therefore be able to mask effects of changes in carbonate productivity and deep-water chemistry in the sedimentary carbonate record (Martin and Sayles, 1996).

For a long time it was not possible to calculate the benthic total carbon dioxide or alkalinity flux because of artifacts introduced by decompression processes during core recovery. Moreover there

was no established method for predicting a true concentration profile or benthic flux (Murray et al. 1980; Emerson and Bender 1981; Emerson et al. 1980; Emerson et al. 1982). What happens during recovery of cores from some thousand meters of water depth is that the solubility of CO_2 in the pore water is increasingly reduced due to decompression and warming. Probably, dependent on the calcium carbonate content of the sediment providing nucleation sites, calcium carbonate is then precipitated from the pore water on the way through the water column. Calculation of the diffusive alkalinity flux across the sediment-water interface from such cores may thus, largely underestimate the real flux or even suggest a flux directed into the sediments (cf. chapter 6).

In the last decade, however, in-situ techniques have been developed to overcome these problems. Profiling lander systems were deployed to record the pore water microprofiles of oxygen, pH and pCO_2, and Ca whereas benthic chambers were deployed to measure solute fluxes across the sediment-water interface directly. Very often, reactive-transport models are used to explain the interrelation between measured microprofiles, to predict overall calcite dissolution rates by defining the dissolution rate constants, and to distinguish between dissolution driven by organic matter oxidation and by the undersaturation of the bottom water.

In most of recently published studies, the calcium carbonate dissolution in seawater and in pore water of surface sediments is assumed to follow a kinetic process that can be described by the equation (Morse 1978; Keir 1980):

$$R_d = k_d\ (1 - \Omega)^n \qquad (9.17)$$

$$\Omega = \frac{[Ca^{2+}][CO_3^{2-}]}{k} \qquad (9.18a)$$

$$SI = \log\frac{[Ca^{2+}][CO_3^{2-}]}{k} \qquad (9.18b)$$

where R_d is the calcite dissolution rate, k_d is the calcite dissolution rate constant, and Ω or SI describe the degree of saturation (ion activity product divided by k), and k the solubility constant of the calcium carbonate species in question. Mostly, k′ is used instead of k, which is defined as the apparent solubility constant and is related to concentrations instead of activities. Compilations of apparent solubility constants are available e.g. by Mehrbach et al. (1973). In several studies, a further dependence of R_d on the calcium carbonate content (respectively the surface area) in the sediment is considered. (for a more detailed overview of this subject see Zeebe and Wolf-Gladrow 2000).

The most extensively used reaction order for modeling calcite dissolution is 4.5 and 4.2 for aragonite (as suggested by Keir 1980). A more extensive re-evaluation of this topic has been provided by Cai et al. (1995). However, the discussion concerning the "correct" reaction order still continues. The values of k_d reported so far range over several orders of magnitude from 0.005-0.16 % d^{-1} (Berelson et al. 1994; Hales and Emerson 1996, 1997a) up to laboratory values of 10-1000 % d^{-1} (Keir 1980, 1983)[2]. The reason for this huge discrepancy is not clearly known. Important and regionally variable factors, however, may be the grain size and thus the surface area of calcium carbonate crystals in the sediments or adsorbed coatings like phosphate ions protecting calcium carbonate grains from corrosive pore waters (Jahnke et al. 1994; Hales and Emerson 1997a). In contrast, Hales and Emerson (1997b) found evidence that *in-situ* pH measurements in pore waters of calcite rich deep-sea sediments are more consistent with a first-order instead of 4.5[th] order dependence. Applied to their data, they rewrote Equation 9.17 to

$$R_d = 38\ (1-\Omega)^1 \qquad (9.19)$$

In contrast, the study of calcite dissolution kinetics in $CaCO_3$-poor sedi-ments of the equatorial Atlantic, Adler et al. (2001) again favored higher reaction orders. In this sense, the observed dissolution rate constants are highly variable, which seems to be mainly dependent on differences in the physical (e.g. surface area) and chemical properties (high/low Mg-calcite) of the calcite mineral phase.

It is also important to consider where in the sediment dissolution occurs. Metabolically produced CO_2 released immediately at the sediment-water interface is probably much less effective for carbonate dissolution than in deeper sediment strata, because neutralization with bottom water CO_3^{2-} might occur instead of dissolution. If the particulate organic matter is more rapidly mixed down, i.e. by bioturbation, and oxidized in deeper sediment strata, the CO_2 released into the pore waters can probably more effectively dissolve carbonates (Martin and Sayles 1996).

It is generally reported that dissolution at and above the saturation horizon is solely attributed to the oxidation of organic matter. The importance of oxidation-related dissolution decreases with increasing undersaturation of the bottom water, however, the efficiency by which C_{org} oxidation drives $CaCO_3$ dissolution increases with increasing undersaturation of the bottom water (Martin and Sayles 1996). In the same sense, Berelson et al. (1994) suggested that the undersaturation of the bottom water is more important for better soluble calcite phases (high k-values), whereas for more resistant calcite phases (low k-values), the rain of C_{org} to the seafloor, and thus mineralization, becomes the driving force for calcite dissolution (Fig. 9.7). This can be understood as an alteration process, where dissolution starts at the sediment-water interface, right after deposition of the calcite phase, and becomes more resistant to dissolution during burial. In the zone of oxic respiration the removal of organic coatings around calcite grains might also play an important role since they help to expose larger surface areas to the pore water, thus triggering dissolution processes.

In summary, a closer evaluation of Fig. 9.7 reveals:

(1) It is obvious that for low oxygen fluxes the relative importance of bottom water undersaturation is the driving force for $CaCO_3$ dissolution, meaning that the y-axis intercept is exceptionally determined by k_d and Ω.

(2) It is implicit from Eq. 9.17 that a higher degree of undersaturation increases the differences of dissolution fluxes between the chosen values of k_d (distance between hatched and solid lines).

(3) Higher oxygen fluxes (higher amount of metabolically released CO_2) are obviously more efficient in carbonate dissolution when bottom waters are stronger undersaturated (lower Ω).

---

[2] The unit % d^{-1} originates from experimental studies (e.g. Morse 1978; Keir 1980) and is used in most studies dealing with carbonate dissolution in marine sediments. The use is, however, not always consistent regarding the units of the dissolution rate and the parameters used in the equation and should, therefore, generally be evaluated with caution.

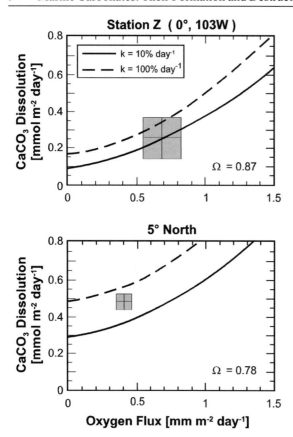

Fig. 9.7 Model results of calcium carbonate dissolution rates as a function of oxygen uptake rates and degree of saturation with different values of k_d from Berelson et al. (1994). The boxes represent averaged benthic lander fluxes for each station (see text for explanation).

(4) For high oxygen fluxes the model predicts a ratio of calcite dissolution and oxygen flux of 0.85 which is close to the stoichiometrical ratio in Eq. 9.15.

More recently *in situ* microsensor measurements of O_2, pH, pCO_2, and Ca could be obtained from the upper continental slope off Gabon (Wenzhöfer et al. 2001; Adler et al. 2001; Pfeifer et al. 2002), which, to date, provide one of the most complete sets of non-corrupted pore water data from deep-sea sediments. The bottom water at a water depth of about 1300 m is slightly oversaturated with respect to calcite ($\Omega = 1.07$ and 1.10, Adler et al. 2001), thus, dissolution must exclusively be mediated by metabolically produced CO_2. For the first time, *in situ* measurements of all parameters describing the carbon dioxide system, combined with O_2 microprofiles, permitted the quantification of the amount of C_{org} mineralization relative to $CaCO_3$ dissolution. These were completed by a number of *ex situ* parameters.

The numerical model CoTReM was applied to investigate the depth dependent effects of respiration and redox processes related to $CaCO_3$ dissolution (Pfeifer et al. 2002; cf. Fig. 15.16 in chapter 15). Interestingly, if calculated until a steady-state situation is reached, the model-derived calcite dissolution and precipitation rates produce an almost perfect fit to the measured $CaCO_3$ profile in the sediment (Fig. 9.8), which suggests that ~90 % of the $CaCO_3$ flux to the sea floor is redissolved in the sediment.

It is important to note that calcite dissolution is exceptionally driven by oxic respiration (Eq. 9.15, 9.16) and re-oxidation of reduced species like HS⁻. Subsequent degradation processes like the reduction of iron oxides (Eq. 9.20) may even have the opposite effect leading to the precipitation of calcium carbonate. Sulfate reduction (Eq. 9.21) is less efficient than oxic respiration, which is related to at least three important facts, which are (1) lower overall reaction rates, (2) a lower potential of acid production, and (3) a higher buffer capacity due to rising alkalinity levels in the pore water with increasing sediment depth. Because manganese and iron reduction are only of minor importance in terms of total mineralization (cf. Chapter 6), significant amounts of carbonate precipitation are usually not observed (Raiswell and Fisher 2004). A major process driving carbonate precipitation in sediments is the anaerobic oxidation of methane (AOM; see Chapter 8). However, the formation of visible or massive authigenic carbonates (crusts and chimneys) is restricted to cold seep environments, where methane-enriched fluids are advected at sufficient rates (Luff and Wallmann 2003; Luff et al. 2004).

$$(CH_2O)_{106}(NH_3)_{16}H_3PO_4 + 424Fe(OH)_3 \rightarrow$$
$$106HCO_3^- + 16NH_4^+ + HPO_4^{2-} + 424Fe^{2+} + 304H_2O + \mathbf{756OH^-}$$

$$(9.20)$$

$$(CH_2O)_{106}(NH_3)_{16}H_3PO_4 + 53SO_4^{2-} \rightarrow$$
$$106HCO_3^- + 16NH_4^+ + HPO_4^{2-} + 53HS^- + \mathbf{39H^+}$$

$$(9.21)$$

In the study of Pfeifer et al. (2002) calcite dissolution fluxes have been quantified in two different ways: (1) by calculating the Ca^{2+}-gradient from the sediment into the bottom water of the Ca-microprofile (Wenzhöfer et al. 2001b) and (2) by the model derived dissolution rates (Fig. 9.8). Following the model approach, Ca-fluxes resulted

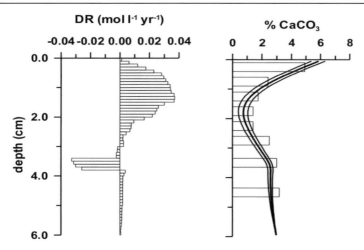

Fig. 9.8 Measured and simulated pore water profiles at station GeoB 4906 and corresponding rates of primary and secondary redox processes. (b) Corresponding calcite dissolution and precipitation rates and the resulting steady-state distribution of sedimentary $CaCO_3$ for input fluxes of 40, 42, and 44 g $m^{-2}yr^{-1}$, respectively, compared to measured concentrations (bars) (after Pfeifer et al. 2002).

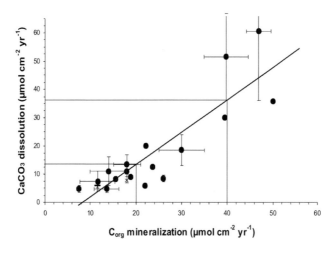

Fig. 9.9 Compilation of literature-derived $CaCO_3$ dissolution fluxes vs. C_{org} mineralization rates from deep-sea sediments (after Pfeifer et al. 2002). Equation 9.22 implies that there exists a considerable threshold of mineralization, which has to be exceeded before $CaCO_3$ dissolution is initiated. The increasing proportion of $CaCO_3$ dissolution with increasing mineralization is indicated by the vertical and horizontal lines.

in 35.8 µmol cm^{-2} yr^{-1} at site GeoB 4906 and 33 µmol cm^{-2} yr^{-1} at site GeoB 4909, whereas calcite dissolution fluxes calculated directly from Ca-microprofiles are distinctively lower with 20.1 and 21.2 µmol cm^{-2} yr^{-1}, respectively (attributed to scattering data and inconsistencies in the measured profiles).

Combining these results with existing data on calcite dissolution fluxes and C_{org} mineralization rates from deep-sea sediments located above the saturation horizon and slightly below ($\Omega \sim 0.8$; Fig. 9.9) results in a good correlation, which has been used to derive a general empirical formu-

lation relating calcite dissolution and C_{org} mineralization (Pfeifer et al. 2002):

$$DR_{CaCO_3} = 1.1 \cdot MR_{C_{org}} - 9.3 \qquad (9.22)$$

where DR_{CaCO3} is the calcite dissolution flux and MR_{Corg} is the C_{org} mineralization rate (both in µmol cm^{-2} yr^{-1}).

Using Eq. 9.19, Hensen et al. (2003) calculated an estimate of the global calcite dissolution flux above the hydrographic lysocline by application of a GIS-system (Fig. 9.10). To define the position of the hydrographic lysocline, the gridded global

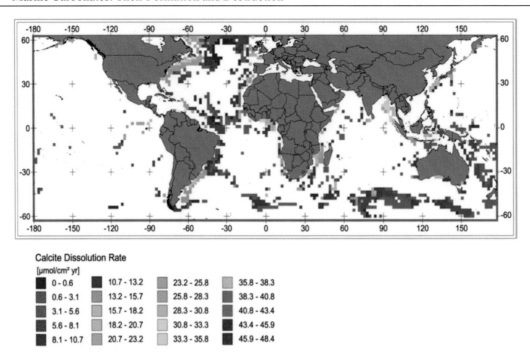

Calcite Dissolution Rate
[µmol/cm² yr]

■ 0 - 0.6	■ 10.7 - 13.2	23.2 - 25.8	35.8 - 38.3
■ 0.6 - 3.1	13.2 - 15.7	25.8 - 28.3	38.3 - 40.8
■ 3.1 - 5.6	15.7 - 18.2	28.3 - 30.8	■ 40.8 - 43.4
■ 5.6 - 8.1	18.2 - 20.7	30.8 - 33.3	■ 43.4 - 45.9
■ 8.1 - 10.7	20.7 - 23.2	33.3 - 35.8	■ 45.9 - 48.4

Fig. 9.10 Map showing the global distribution of supralysoclinal calcite dissolution by applying Equation 9.19, based on global grids of bottom water "CO_3^{2-} (Archer 1996a) and oxygen consumption rates (Jahnke 1996) (from Hensen et al. 2003).

field of bottom water ΔCO_3^{2-} data (Archer 1996a) has been used and C_{org}-mineralization rates from the grid of oxygen consumption rates after Jahnke (1996) have been calculated, assuming a Redfield ratio of decomposed organic material. The global sea floor area above the lysocline was estimated to about $7.4 \cdot 10^7$ km² (corresponding to ~20 % of the total ocean area) and the dissolution flux sums up to about $5.7 \cdot 10^{12}$ mol $CaCO_3$.

Taking into account more recent estimates for the global oxygen demand of the seafloor (i.e. Wenzhöfer and Glud 2002) respiratory driven calcite dissolution can be expected to be 50 % higher in total. Compared to Archer's study in 1996 (see Table 9.7), this would add ~10-20 % to the total estimate.

In comparison with the study of Archer (1996b), the estimate of Milliman et al. (1999) clearly underestimates the total benthic $CaCO_3$-dissolution. Actually, simply using the Archer-flux would help to resolve a long lasting debate and balances – more or less – the overall budget shown in Fig 9.5.

Milliman et al. (1999) postulated an enormous loss of $CaCO_3$ in the upper water column for, to date, unknown reasons. Jansen et al. (2002) used a numerical model in order to investigate and predict, whether such a loss could be attributed to

respiration-driven dissolution of skeletal material in microenvironments of sinking detritus. Although the efficiency of this process is largely dependent on the sinking velocity and the size of the sinking spheres or particles, these authors could show that this pathway of carbonate dissolution does not account significantly to the overall loss as observed by Milliman et al. (1999).

Applying Equation 9.22 for the total sea floor area should give a reasonable estimate for the metabolically induced $CaCO_3$-dissolution on a global scale, even though this must be less than the total dissolution flux from respiration and bottom water undersaturation. The application C_{org}-mineralization rates derived from various estimates of the global oxygen consumption in deep-sea sediments (Jahnke 1996; Christensen 2000; Wenzhöfer and Glud 2002; Seiter et al. 2005) results in global fluxes of $CaCO_3$-dissolution between $14.8 \cdot 10^{12}$ and $47.2 \cdot 10^{12}$ mol yr⁻¹. These comparatively low efficiencies of $CaCO_3$ dissolution in combination with a wide range of estimates are the result of the significant threshold value of about 8 µmol cm⁻² yr⁻¹ before $CaCO_3$ dissolution is initiated (Fig. 9.9). More precisely, the low average mineralization rates of 12.9 µmol cm⁻² yr⁻¹ as derived from the lowest estimate by Jahnke (1996) would proportionally be much less efficient in terms of $CaCO_3$ dissolution than higher estimates of up to 31.6 cm^{-2} yr^{-1} (derived from Wenzhöfer & Glud 2002). This is in contradiction with studies of Reimers et al. (1992) and Hammond et al. (1996) who predict that mineralization and carbonate dissolution contribute about one half each to the total dissolved carbon (DIC) or alkalinity fluxes. This 1:1 relationship obviously proves true only for higher mineralization rates (Fig. 9.9).

Nevertheless, in combination with Archer's results from the non-respiratory dissolution runs (7-34 · 10^{12} mol yr^{-1}) the global $CaCO_3$-dissolution flux may range between 22 to 81 · 10^{12} mol yr^{-1} (Hensen et al. 2003). On the one hand, it confirms Archer's results based on a number of newly obtained *in situ* measurements from different deep-sea locations, and, on the other hand, it suggests that calcite dissolution fluxes to the seafloor that even Archer's maximum estimation of 54 · 10^{12} mol yr^{-1} might be even an underestimation, if higher benthic mineralization rates as derived from Christensen (2000) and Wenzhöfer and Glud (2002) prove true (compare with Milliman's estimate).

A compilation of recent data on carbon re-flux from deep-sea sediments is provided in Table 9.7.

Four flux categories are given in Table 9.7: The CO_2 produced due to oxic respiration, calcite dissolution, the alkalinity as a sum parameter for calcium carbonate dissolution and CO_2 from oxic respiration, and, hitherto neglected in the discussion, the dissolved organic carbon (DOC).

It is very interesting to note that many of the more recently obtained calculations agree very much with those of Berelson et al. (1994) who calculated the benthic alkalinity input to the deep ocean for the Pacific and the Indo-Pacific (Table 9.7). They suggested that most of the carbonate dissolution in the deep ocean (Fig. 9.5) occurs within the sediments (85 %). The extension of their results from Pacific and Indian Ocean to the Atlantic Ocean leading to 120 · 10^{12} mol yr^{-1} of global dissolved carbon fluxes from sediments may, however, be critical because of the completely different deep-water conditions in the Indo-Pacific and the Atlantic. Deep ocean waters in the Indian and Pacific Oceans are known to be much older and depleted in CO_3^{2-} implying that a much higher proportion of calcite dissolution contributes to the total alkalinity input there. However, despite this problem of different bottom-

Table 9.7 Carbon fluxes from deep-sea sediments (below 1000 m water depth) in 10^{12} mol yr^{-1} estimated by using different parameters. Global estimations of regional data compilations are made by multiplication with surface area factors.

Parameter	Area	Flux	Source
Respiratory CO_2[1]	Global	40	after Jahnke (1996)
		61	after Christensen (2000)
		60-75	Wenzhöfer and Glud (2002)
		43	Seiter et al. (2005)
Calcite Dissolution	Global	27-54	Archer (1996b)
	Global	22-81	Hensen et al. (2003)
	Global	13	Milliman (1999)
Alkalinity / TCO_2	Pacific	55	Berelson et al (1994)
	Indo-Pacific	91	Berelson et al (1994) [2]
	Global	100	after Berelson et al (1994)
		120	after Berelson et al (1994) [2]
		120	Mackenzie et al. (1993)
		100	Summary, this study
DOC	Atlantic [3]	4	Otto (1996)
	Global [3]	18	Otto (1996)
Total C	Global	120	Summary, this study

[1] Estimated as oxic respiration of organic matter.
[2] Using data of Broecker and Peng (1987).
[3] Including water depth above 1000 m.

water saturation conditions the global estimate extending the Berelson et al. (1994) approach is in agreement to that of Mackenzie et al. (1993). Based on Eq. 9.22 and assuming an total mineralization of $55 \cdot 10^{12}$ mol C yr^{-1} as well as an average contribution of $15 \cdot 10^{12}$ mol yr^{-1} from non-respiratory dissolution (Archer 1996b) we may come up with a total alkalinity flux of about $100 \cdot 10^{12}$ mol yr^{-1} as a best current estimate.

In combination with the flux of dissolved organic carbon (DOC), which may add another $20 \cdot 10^{12}$ mol yr^{-1} (Otto 1996; Table 9.7), a total carbon release from deep-sea sediments of about $120 \cdot 10^{12}$ mol yr^{-1} seems to be the best recent approximation regarding all sources of uncertainty.

Summary

The main subjects addressed in Chapter 9 are listed below:

- The major site of marine carbonate accumulation is the neritic environment, including coral reefs, banks and continental shelves, and pelagic calcite-rich sediments. In total, about $35 \cdot 10^{12}$ mol $CaCO_3$ accumulate per annum in the marine realm.

- Based on budget calculations of calcium carbonate, reservoir sizes in the world ocean and exchange fluxes between reservoirs the carbonate system is not in steady state.

- However, calcium carbonate budget calculations are strongly biased by inexact estimations of calcite production in the surface ocean and of the dissolution of pelagic biogenic calcite in the water column and in sediments above the calcite lysocline. In addition, the uncertainty is enhanced by the difficulty to estimate dissolved inorganic carbon release from sediments.

- The total carbon release from deep-sea sediments is estimated to be about $120 \cdot 10^{12}$ mol yr^{-1}, but is subject to great uncertainty due to the complexity of processes controlling carbon remobilization.

- Both bottom water undersaturation and organic matter decay are responsible for calcium

carbonate dissolution in the sediments at more or less equal levels.

- The efficiency of calcium carbonate dissolution by metabolic CO_2 strongly depends on the organic carbon / calcium carbonate rain ratio at the sediment surface, the oxidation rate of organic matter (and the depth horizon, where oxidation occurs), as well as the saturation state of bottom water (Ω) and the dissolution rate constant k_d.

9.5   Problems

Problem 1

Where would you expect more $CaCO_3$ production: On the continental shelves, on the slopes, or in the deep sea?

Problem 2

Explain the difference between $CaCO_3$ production and $CaCO_3$ accumulation. Discuss the difference between both in connection with the processes of $CaCO_3$ dissolution in the water column and in sediment.

Problem 3

At which water depth would you expect the calcite compensation depth (CCD) in waters of high latitudes and in which water depth in waters of low latitudes? Explain your answer.

Problem 4

Which carbonate species has the highest concentrations in sea-water? How much of the total calcium concentration in normal sea-water exists as $CaSO_4$ complex? Under which conditions is this complex rather insignificant in anoxic pore-water?

Problem 5

$CaCO_3$ dissolution fluxes from sediment to ocean bottom water can be estimated from C_{org} mineralization rates. What are the geochemical processes behind this correlation? Which calcite dissolution rate would you expect for sediments in an upwelling area, and which for deep-sea sediments?

References

Adler, M., Hensen, C., Kasten, S. and Schulz, H.D., 2000. Computer simulation of deep-sulfate reduction in sediments off the Amazon Fan. International Journal of Earth Sciences (Geol. Rdsch.), 88: 619-629.

Adler, M., Hensen, C., Wenzhöfer, F., Pfeifer, K. and Schulz, H.D., 2001. Modeling of calcite dissolution by oxic respiration in supralysoclinal deep-sea sediments. Marine Geology, 177: 167-189.

Andersen, N.R. and Malahoff, A., 1977. The fate of fossil fuel CO2 in the Oceans. Plenum Press, NY, 749 pp.

Archer, D.E., 1991. Modeling the calcite lysocline. Journal of Geological Research, 96: 17037-17050.

Archer, D. and Maier-Reimer, E., 1994. Effect of deep-sea sedimentary calcite preservation on atmospheric CO2 concentration. Nature, 367: 260-263.

Archer, D.E., 1996a. An atlas of the distribution of calcium carbonate in sediments of the deep sea. Global Biochemical Cycles, 10: 159-174.

Archer, D.E., 1996b. A data-driven model of the global calcite lysocline. Global Biochemical Cycles, 10: 511-526.

Berelson, W.M., Hammond, D.E. and Cutter, G.A., 1990. In situ measurements of calcium carbonate dissolution rates in deep-sea sediments. Geochimica et Cosmochimica Acta, 54: 3013-3020.

Berelson, W.M., Hammond, D.E., McManus, J. and Kilgore, T.E., 1994. Dissolution kinetics of calcium carbonate in equatorial Pacific sediments. Global Biogeochemical Cycles, 8: 219-235.

Berger, W.H., 1976. Biogenic deep-sea sediments: Production, preservation and interpretation. In: Riley, J.P. and Chester, R. (eds) Chemical Oceanography, 5, Academic Press, London, pp. 266-388.

Berger, W.H., 1982. Increase of carbon dioxide in the atmosphere during deglaciation: The coral reef hypothesis. Naturwissenschaften, 69, 87.

Broecker, W.S. and Peng, T.-H., 1982. Tracers in the Sea. Lamont-Doherty Geol. Observation, Eldigo Press, Palisades, NY, 690 pp.

Broecker, W.S. and Peng, T.-H., 1987. The Role of CaCO3 compensation in the glacial to interglacial atmospheric CO2 change. Global Biogeochemical Cycles, 1: 15-29.

DeBaar, H.J.W. and Suess, E., 1993. Ocean carbon cycle and climate change - An introduction to the interdisciplinary Union Symposium. Global and Planetery Change, 8: VII-XI.

Dittert, N., Baumann, K.H., Bickert, T., Henrich, R., Huber, R., Kinkel, H., Meggers, H. and Müller, P., 1999. Carbonate dissolution in the Deep-Sea: Methods, quantification and paleoceanography application. In: Fischer, G. and Wefer, G. (eds) Use of proxies in paleoceanography: examples from the South Atlantic, Springer, Berlin, Heidelberg, 255-284.

Emerson, S.R., Jahnke, R., Bender, M., Froelich, P., Klinkhammer, G., Bowser, C. and Setlock, G., 1980. Early diagenesis in sediments from the eastern equatorial Pacific. 1. Pore water nutrient and carbonate results. Earth Planet Science Letters, 49: 57-80.

Emerson, S. and Bender, M., 1981. Carbon fluxes at the sediment-water interface of the deep-sea: calcium carbonate preservation. Journal of Marine Research, 39: 139-162.

Emerson, S., Grundmanis, V. and Graham, D., 1982. Carbonate chemistry in marine pore waters: MANOP sites C and S. Earth and Planetary Science Letters, 61: 220-232.

Feely, R.A., Sabine, C.L., Lee, K., Berelson, W., Kleypas, J., Fabry, V.J. and Millero, F.J., 2004. Impact of Anthropogenic CO2 on the CaCO3 System in the Oceans. Science, 305: 362-366.

Fischer, G., Wefer, G., Romero, O., Dittert, N., Ratmeyer, V. and Donner, B., 2004. Transfer of Particles into the Deep Atlantic and the Global Ocean. In: Wefer, G., Mulitza, S. and Ratmeyer, V. (eds) The South Atlantic in the Late Quaternary. Springer Berlin, Heidelberg, New York, pp. 21-46.

Freiwald, A., 2002. Reef-forming cold-water corals. In: Wefer, G., Billet, D., Hebbeln, D., Jørgensen, B.B., Schlüter, M., and Van Weering, T. (eds) Ocean margin systems, Springer, Berlin, Heidelberg, 365-385.

Goyet, C. and Poisson, A., 1989. New determination of carbonic acid dissociation constants in seawater as a function of temperature and salinity. Deep-Sea-Research, 36: 1635-1654.

Hales, B. and Emerson, S., 1996. Calcite dissolution in sediments of the Ontong-Java Plateau: In situ measurements of pore water O2 and pH. Global Biogeochemical Cycles, 10: 527-541.

Hales, B. and Emerson, S., 1997a. Calcite dissolution in sediments of the Ceara Rise: In situ measurements of porewater O2, pH, and CO2(aq). Geochimica et Cosmochimica Acta, 61: 501-514.

Hales, B. and Emerson, S., 1997b. Evidence in support of first-order dissolution kinetics of calcite in seawater. Earth and Planetary Science Letters, 148: 317-327.

Hammond, D.E., McManus, J., Berelson, W.M., Kilgore, T.E. and Pope, R.H., 1996. Early diagenesis of organic material in equatorial Pacific sediments: stoichiometry and kinetics. Deep-Sea Research, 43: 1365-1412.

Hay, W.W. and Southam, J.R., 1977. Modulation of marine sedimentation by continental shelves. In: Andersen, N.R. and Malahoff, A. (eds) The fate of fossil fuel CO2 in the Oceans. Plenum Press, NY, pp. 564-604.

Hensen, C., Landenberger, H., Zabel, M., Gundersen, J.K., Glud, R.N. and Schulz, H.D., 1997. Simulation of early diagenetic processes in continental slope sediments in Southwest Africa: The computer model CoTAM tested. Marine Geology, 144: 191-210.

Hensen, C., Landenberger, H., Zabel, M. and Schulz, H.D., 1998. Quantification of diffusive benthic fluxes of nitrate, phosphate and silicate in the Southern Atlantic Ocean. Global Biogeochemical Cycles, 12: 193-210.

Jahnke, R.A., Craven, D.B. and Gaillard, J.-F., 1994. The influence of organic matter diagenesis on CaCO3 dissolution at the deep-sea floor. Geochimica et Cosmochimica Acta, 58: 2799-2809.

Jahnke, R.A., 1996. The global ocean flux of particulate

organic carbon: Areal distribution and magnitude. Global Biogeochemical Cycles, 10: 71-88.

Jahnke, R.A., Craven, D.B., McCorkle, D.C. and Reimers, C.E., 1997. CaCO3 dissolution in California continental margin sediments: The influence of organic matter remineralization. Geochimica et Cosmochimica Acta, 61: 3587-3604.

James, N.P. and Clarke, J.A.D., 1997. Cool-water carbonates, SEPM Spec. Publ., 56, Tulsa, Oklahoma, 440 pp.

Jansen, H., R.E., Z. and Wolf-Gladrow, D.A., 2002. Modeling the dissolution of settling CaCO3 in the Ocean. Glob. Biogeochem. Cycles, 16(2): 11, 1-16.Keir, R.S., 1983. Variation in the carbonate reactivity of deep-sea sediments: determination from flux experiments. Deep-Sea Res., 30(3A): 279-296.

Keir, R.S., 1980. The dissolution kinetics of biogenic calcium carbonates in seawater. Geochimica et Cosmochimica Acta, 44: 241-252.

Kharaka, Y.K., Gunter, W.D., Aggarwal, P.K., Perkins, E.H. and DeBraal, J.D., 1988. SOLMINEQ88: a computer program for geochemical modeling of water-rock-interactions. Water-Resources Invest. Report, 88-4227, US Geol. Surv., 207 pp.

Lindberg, B. and Mienert, J., 2005. Postglacial carbonate production by cold-water corals on the Norwegian Shelf and their role in the global carbonate budget. Geology, 33(7): 537-540, doi: 10.1130/G21577.1.

Lisitzin, A.P., 1996. Oceanic sedimentation: Lithology and Geochemistry (English Translation edited by Kennett, J.P.). Amer. Geophys. Union, Washington, D.C., 400 pp.

Luff, R., Wallmann, K., 2003. Fluid flow, methane fluxes, carbonate precipitation and biogeochemical turnover in gas hydrate-bearing sediments at Hydrate Ridge, Cascadia Margin: Numerical modeling and mass balances. Geochim. Cosmochim. Acta 67 (18), 3403-3421.

Luff, R., Wallmann, K., Aloisi, G., 2004. Numerical modeling of carbonate crust formation at cold vent sites: significance for fluid and methane budgets and chemosynthetic biological communities. Earth Planet. Sci. Lett. 221, 337-353.

Mackenzie, F.T., Ver, L.M., Sabine, C., Lane, M. and Lerman, A., 1993. C, N, P, S global biogeochemical cycles and modeling of global change. In : Wollast, R., Mackenzie, F.T. and Chou, L. (eds), Interactions of C, N, P and S biogeochemical cycles and global change. NATO ASI Series, 14, Springer Verlag, pp 1-61.

Maier-Reimer, E. and Bacastow, R., 1990. Modelling of geochemical tracers in the ocean. Climate-Ocean Interaction. In: Schlesinger, M.E. (ed), Climate-ocean interactions, Kluwer, pp. 233-267.

Martin, W.R. and Sayles, F.L., 1996. CaCO3 dissolution in sediments of the Ceara Rise, western equatorial Atlantic. Geochimica et Cosmochimica Acta, 60: 243-263.

Mehrbach, C., Culberson, C., Hawley, J.E. and Pytkowicz, R.M., 1973. Measurement of the apparent dissociation constants of carbonic acid in seawater at atmospheric pressure. Limnology and Oceanography, 18: 897-907.

Millero, F.J., 1995. Thermodynamics of the carbon dioxide systems in the oceans. Geochimica et Cosmochimica Acta, 59: 661-677.

Milliman, J.D., 1993. Production and accumulation of calcium carbonate in the ocean: budget of a nonsteady state. Global Biogeochemical Cycles, 7: 927-957.

Milliman, J.D. and Droxler, A.W., 1996. Neritic and pelagic carbonate sedimentation in the marine environment: ignorance is not a bliss. Geologische Rundschau, 85: 496-504.

Milliman J. D., Troy P. J., Balch W. M., Adams A. K., Li Y.-H., and Mackenzie F. T., 1999. Biologically mediated dissolution of calcium carbonate above the chemical lysocline? Deep-Sea Res. I 46, 1653–1669.

Morse, J.W., 1978. Dissolution kinetics of calcium carbonate in sea water: VI. The near-equilibrium dissolution kinetics of calcium carbonate-rich deep-sea sediments. American Journal of Science, 278: 344-353.

Morse, J.W. and Berner, R.A., 1979. Chemistry of calcium carbonate in the deep ocean. In: Jenne, E.A. (ed), Chemical modelling in aqueous systems. Am. Chem. Soc., Symp. Ser., 93, pp. 499-535.

Morse, J.W. and Mackenzie, F.T., 1990. Geochemistry of sedimentary carbonates. Elsevier, Amsterdam, 707 pp.

Mucci, A., Sundby, B., Gehlen, M., Arakaki, T., Zhong, S., Silverberg, N., 2000. The fate of carbon in continental shelf sediments of eastern Canada: a case study. Deep-Sea Res. II 47, 733-760.

Murray, J.W., 1897. On the distribution of the pelagic foraminifera at the surface and on the sea floor of the ocean. Nat. Sci., 11: 17-27.

Murray, J.W., Emerson, S. and Jahnke, R.A., 1980. Carbonate saturation and the effect of pressure on the alkalinity of interstitial waters from the Guatemala Basin. Geochimica et Cosmochimica Acta, 44: 963-972.

Nordstrom, D.K., Plummer, L.N., Wigley, T.M.L., Wolery, T.J., Ball, J.W., Jenne, E.A., Basset, R.L., Crerar, D.A., Florence, T.M., Fritz, B., Hoffman, M., Holdren, G.R. Jr., Lafon, G.M., Mattigod, S.V. McDuff, R.E., Morel, F., Reddy, M.M., Sposito, G. and Thrailkill, J., 1979. A comparison of computerized chemical models for equilibrium calculations in aqueous systems. In: Jenne, E.A. (ed), Chemical modeling in aqueous systems, speciation, sorption, solubility, and kinetics, 93, American Chemical Society, pp. 857-892.

Opdyke, B.D. and Walker, J.C.G., 1992. Return of the coral reef hypothesis: basin to shelf partitioning of CaCO3 and its effects on atmospheric CO2. Geology, 20: 733-736.

Orr, J.C., Fabry, V.J., Aumont, O., Bopp, L., Doney, S.C., Feely, R.A., Gnanadesikan, A., Gruber, N., Ishida, A., Joos, F., Key, R.M., Lindsay, K., Maier-Reimer, E., Matear, R., Monfray, P., Mouchet, A., Najjar, R.G., Plattner G.-K., Rodgers, K.B., Sabine, C.L., Sarmiento, J.L., Schlitzer, R., Slater, R.D., Totterdell, I.J., Weirig, M.-F., Yamanaka, Y. and Yool, A., 2005. Anthropogenic ocean acidification over the twenty-first century and its impact on calcifying organisms. Nature 437: 681-686, doi:10.1038/nature04095.

Otto, S., 1996. Die Bedeutung von gelöstem organischen Kohlenstoff (DOC) für den Kohlenstofffluß im Ozean. Berichte, 87, Fachbereich Geowissenschaften, Universität Bremen, 150 pp.

Palmer, A.N., 1991. The origin and morphology of limestone caves. Geological Society American Bulletin, 103: 1-21.

Parkhurst, D.L., Thorstensen, D.C. and Plummer, L.N., 1980. PHREEQE - a computer program for geochemical calculations. Water-Resources Invest. Report, 80-96, US Geol. Surv., 219 pp.

Parkhurst, D.L., 1995. User's guide to PHREEQC: a computer model for speciation, reaction-path, advective-transport, and inverse geochemical calculation. Water-Resources Invest. Report, 95-4227, US Geol. Surv., 143 pp.

Pfeifer, K., Hensen, C., Adler, M., Wenzhöfer, F., Weber, B. and Schulz, H.D., 2002. Modeling of subsurface calcite dissolution, including the respiration and reoxidation processes of marine sediments in the region of equatorial upwelling off Gabon. Geochim. Cosmochim. Acta, 66(24): 4247-4259.

Plummer, L.N., Wigley, T.M.L. and Parkhurst, D.L., 1978. The kinetics of calcite dissolution in CO2-water systems at 5°C to 60°C and 0.0 to 1.0 atm CO2. Am. J. Sci., 278: 179-216.

Plummer, L.N., Wigley, T.M.L. and Parkhurst, D.L., 1979. Critical review of the kinetics of calcite dissolution and precipitation. In: Jenne, E.A. (ed), Chemical modelling in aqueous systems. Am. Chem. Soc., Symp. Ser., 93, pp. 537-572.

Raiswell, R., Fisher, Q.J. (2004) Rates of carbonate cementation associated with sulphate reduction in DSDP/ODP sediments: implications for the formation of concretions. Chem. Geol. 211, 71-85.

Redfield, A.C., 1958. The biological control of chemical factors in the environment. Am. Scientist, 46: 206-2226.

Reimers, C.E., Jahnke, R.H. and McCorkle, D.C., 1992. Carbon fluxes and burial rates over the continental slope and rise off central California with implications for the global carbon cycle. Global Biogeochemical Cycles, 6: 199-224.

Roberts, H.H. and Macintyre, I.G. (eds), 1988. Special issue: Halimeda. Coral Reefs, 6(3/4), 121-280.

Roy, R.N., Roy, L.N., Vogel, K.M., Moore, C.P., Pearson, T., Good, C.E., Millero, F.J. and Campbell, D.M., 1993. Determination of the ionization constance of carbonic acid in seawater. Marine Chemistry, 44: 249-268.

Seiter, K., Hensen, C. and Zabel, M., 2005. Benthic carbon mineralization on a global scale. Glob. Biogeochem. Cycles, 19: GB1010, doi:10.1029/2004GB002225.

Siegenthaler, H.H. and Wenk, T., 1984. Rapid atmospheric CO2 variations and ocean circulation. Nature, 308: 624-626.

Sundquist, E.T. and Broeker, W.S., 1985. The carbon cycles and atmospheric CO2: natural variations archean to present. American Geophysical Union, Washington, D.C., 627 pp.

Svensson, U. and Dreybrodt, W., 1992. Dissolution kinetics of natural calcite minerals in CO2-water systems approaching calcite equilibrium. Chemical Geology, 100: 129-145.

Takahashi, T., Broecker, W.S., Bainbridge, A.E. and

Weiss, R.F., 1980. Carbonate chemistry of the Atlantic, Pacific and Indian Oceans: The results of the Geosecs expeditions, 1972-1978, Natl. Sci. Found., Washington D. C.

Vecsei, A., 2004. A new estimate of global reefal carbonate production including fore-reefs. Global and Planetary Change, 43: 1-18.

Wenzhöfer, F., Adler, M., Kohls, O., Hensen, C., Strotmann, B., Boehme, S. and Schulz, H.D., 2001. Calcite dissolution driven by benthic mineralization in the deep-sea: in $situ$ measurements of Ca^{2+}, pH, pCO_2, O_2. Geochim. Cosmochim. Acta, 65(16): 2677-2690.

Wolf-Gladrow, D., 1994. The ocean as part of the global carbon cycle. Environ. Sci. & Pollut. Res., 1: 99-106.

Wollast, R., 1994. The relativ importance of biomineralisation and dissolution of CaCO3 in the global carbon cycle. In: Doumenge, F., Allemand, D. and Toulemont, A. (eds), Past and present biomineralisation processes: Considerations about the carbonate cycle. Bull. de l'Institute océanographique, 13, Monaco, pp. 13-35.

Zeebe, R.E. and Wolf-Gladrow D.A., 2001. CO_2 in Seawater: Equilibrium, Kinetics, Isotopes. Elsevier Oceanography Series, 65, 346 pp.

10 Influence of Geochemical Processes on Stable Isotope Distribution in Marine Sediments

TORSTEN BICKERT

10.1 Introduction

Stable isotope geochemistry has become an essential part of marine geochemistry and has contributed considerably to the understanding of the ocean's changing environment and the processes therein. In some fields, such as biogeochemistry and paleoceanography, the application of stable isotopes is still growing due to new microanalytical techniques, permitting a relatively precise analysis of very small samples or single compounds, which allows the investigation of a new generation of problems. Stable isotopes are useful tracers for reconstructing past temperatures, salinities, productivity, pCO_2, nutrients, etc. However, limitations exist on the application of these tracers because of diagenetic processes which may considerably alter the primary signals, due to preferential preservation, decomposition or relocation of particular tracers. On the other hand, stable isotope geochemistry offers the chance to iden-tify and better understand such diagenetic processes, which is essential for the interpreta-tion of isotope variability in marine sediments of the past.

Within this chapter, we focus on four elements (C, O, N, S), which participate in most marine geochemical reactions and which are important elements in the biological system. Some recent developments in the use of B isotopes are added to this chapter. We summarize the influence of geochemical processes on the stable isotope distribution of those elements in ocean water and marine sediments. After a short review on the fundamentals of stable isotope fractionation and mass spectrometry, the most important fractionation mechanisms for each el-ement within the marine environment are described, followed by the identification of geochemical processes responsible for syn- or post-sedimentary changes in isotope signals.

10.2 Fundamentals

Excellent summaries of the use of stable isotopes in the study of sediments and the environment in which they are deposited are given by Hoefs (2004), and by Faure and Mensing (2005). The fundamentals of stable isotope fractionation for selected elements (esp. H, C, O, S) are also presented in a special volume edited by Valley and Cole (2001).

10.2.1 Principles of Isotopic Fractionation

The stable isotopic compositions of elements having low atomic numbers (e. g. H, C, N, O, S) vary considerably in nature as a consequence of the fact that certain thermodynamic properties of molecules depend on the masses of the atoms of which they are composed. The partitioning of isotopes between two substances or two phases of the same substance with different isotope ratios is called isotopic fractionation. In general, isotopic fractionation occurs during several kinds of physical processes and chemical reactions:

- Isotope exchange reactions involving the redistribution of isotopes of an element among different molecules containing that element.

- Kinetic effects, which are associated with uni-directional and incomplete processes such as condensation or evaporation, crystallization or melting, adsorption or desorption, biologically mediated reactions, and diffusion.

In general, light isotopes are more mobile and more affected by such processes than heavy isotopes. The isotope fractionation that occurs during these processes is indicated by the fractionation factor α which is defined as the ratio R_A of the heavy to the light isotopes in one compound or phase A divided by the corresponding isotope ratio R_B for the compound or phase B:

$$\alpha_{A\text{-}B} = R_A / R_B \qquad (10.1)$$

For example, the fractionation factor for the exchange of ^{18}O and ^{16}O between water and calcium carbonate is expressed as:

$$H_2^{18}O + 1/3\ CaC^{16}O_3 \Leftrightarrow$$
$$H_2^{16}O + 1/3\ CaC^{18}O_3 \qquad (10.2)$$

with the fractionation factor $\alpha_{CaCO3\,-\,H2O}$ defined as:

$$\alpha_{CaCO3\,-\,H2O} =$$
$$(^{18}O\,/\,^{16}O)_{CaCO3}\,/\,(^{18}O\,/\,^{16}O)_{H2O} =$$
$$1.031\ \text{at}\ 25°C \qquad (10.3)$$

Because isotopic fractionation factors are close to 1, they can be expressed in ‰ with the introduction of the ε-value defined as

$$\varepsilon_{A\text{-}B} = (\alpha_{A\text{-}B} - 1) \cdot 1000 \qquad (10.4)$$

For geochemical purposes, the dependence of isotope fractionation factors on temperature is the most important property. In principle, fractionation factors for isotope exchange reactions are also slightly pressure-dependent, but experimental studies have shown the pressure dependence to be of no importance within the outer earth environments (Hoefs 2004). Occasionally, the fractionation factors can be calculated by means of partition functions derivable from statistical mechanics. However, the interpretation of observed variations of the isotope distribution in nature is largely empirical and relies on observations in natural environments or experimental results obtained in laboratory studies. A brief summary of the theory of isotope exchange reactions is given by Hoefs (2004).

10.2.2   Analytical Procedures

Stable isotope measurements on light elements are made on gases, i.e. H_2 for hydrogen, CO_2 for carbon and oxygen, N_2 for nitrogen, and SO_2 or SF_6 for sulfur isotopes. A variety of techniques is used to convert samples to a compound suitable for analysis. The most important aspect of sample preparation is to avoid isotopic fractionation. Since molecules with different isotopic masses have different reaction rates, procedures with less than a quantitative (i.e. 100%) output may produce a reaction product that does not have the same isotopic composition as the original sample. Furthermore, a pure gas is necessary to avoid interference by contaminants in the mass spectrometer. Contamination may result from incomplete evacuation of the vacuum preparation system or degassing of the sample, as well as from unwanted side reactions in the preparation procedures. In general, the error attributable to sample preparation is greater than the instrumental analysis of the product gas.

Isotopic abundance measurements for geochemical research are determined using mass spectrometry. A mass spectrometer separates and detects ions based on their motions in magnetic or electrical fields. For detailed information on mass spectrometry and the according analytical techniques we refer the reader to the comprehensive review by Hoefs (2004). This volume also gives an introduction in new microanalytical techniques, such as laser-assisted ablation, which allows the on-line transfer of submilligram quantities of mineral into a standard gas-source spectrometer (Kyser 1995), and the gas chromatography combined with mass spectrometry, which allows the determination of the isotopic composition of single compounds previously separated by means of a gas chromatograph (Hayes et al. 1990; Merrit and Hayes 1994).

Boron isotope ratios are determined by positive or negative thermal ionization mass spectrometry (TIMS). Whereas the positive ionisation technique requires microgram quantities of boron producing $Na_2BO_2^+$ or $Cs_2BO_2^+$ ions (Spivack and Edmond 1986), the negative ionisation technique using BO_2^- ions allows for analyzing nanogram quantities of boron (Hemming and Hanson 1994). Recently, Lecuyer et al. (2002) described the use of multiple collector ICP-MS technique for boron isotopic measurements of waters, carbonates,

Table 10.1 Absolute isotope ratios of international standards and laboratory standards (after Hoefs 2004)

Standard	Ratio	Accepted value $\cdot 10^6$ within 95% confid. interval	Lab standard	δ-value $[^o/_{oo}]$
SMOW Standard Mean Ocean Water	D/H	155.8 ± 0.1	VSMOW	0.00
			SLAP	-428.00
	$^{18}O/^{16}O$	2005.2 ± 0.4	VSMOW	0.00
			SLAP	-55.50
PDB Pee Dee Belemnite	$^{13}C/^{12}C$	11237.2 ± 2.9	NBS 19 (calcite)	+1.95
	$^{18}O/^{16}O$	2067.1 ± 2.1	NBS 19 (calcite)	-2.20
			NBS 19 (carbonatite)	-23.01
N$_2$ (atm.) Air nitrogen	$^{15}N/^{14}N$	3676.5 ± 8.1	Air nitrogen	0.00
CDT Canyon Diablo Troilite	$^{34}S/^{32}S$	45004.5 ± 9.3	CDT (FeS)	0.04
NIST 951 Searles Lake Borax	$^{11}B/^{10}B$	4.04558 ± 0.00033	NIST 951 (boric acid)	0.25

phosphates, and silicates with an external reproducibility of ±0.3‰.

In Earth sciences, the relative differences in isotopic ratios between a sample and a standard are mostly used for reporting stable isotope abundances and variations. The reason is that the absolute value of an isotopic ratio is difficult to determine with sufficient accuracy for geochemical applications. The reporting notation employed is the δ-value, defined as

$$\delta \text{ in } \permil = (R_{sample} - R_{standard}) / R_{standard} \cdot 1000$$

(10.5)

where R_{sample} is the isotopic ratio of the sample ($^{13}C/^{12}C$, $^{18}O/^{16}O$, $^{15}N/^{14}N$, $^{34}S/^{32}S$, etc.) and $R_{standard}$ is the corresponding rate in a standard. Nevertheless, the determination of absolute isotope ratios (Table 10.1) is essential, since these numbers form the basis for the calculation of the relative differences.

Isotope laboratories use different reference gases or working standards for the measurement of relative isotope ratios by mass spectrometry. However, all results are reported relative to an internationally accepted standard (Table 10.1). The selection of standards is an important procedure in isotope geochemistry because their definition and availability controls the extent to which results from different laboratories can be compared.

Since the supply of PDB, the working standard introduced by H. C. Urey's laboratory at the University of Chicago, as well as of SMOW, a water sample prepared by H. Craig for distribution by the IAEA, have been exhausted for years, some confusion and irregularities occurred in the past regarding standards, particularly oxygen isotope standards. These problems may be resolved following the recommendations of the Commission on Atomic weights and isotopic abundances of the International Union of Pure and Applied Chemistry, published in 1995 (see Appendix in Coplen 1996). Isotope reference materials may be obtained from the National Institute of Standards and Technology (NIST), Gaithersburg, MD, or the International Atomic Energy Agency (IAEA), Vienna, Austria.

10.3 Geochemical Influences on $^{18}O / ^{16}O$ Ratios

10.3.1 $\delta^{18}O$ of Seawater

Principles of Fractionation

The oxygen isotopic composition of seawater ($\delta^{18}O_w$) is controlled by fractionation effects due to evaporation and precipitation at the sea surface, freezing of ice in polar regions, the admixing of water masses containing different $^{18}O/^{16}O$ ratios such as melt water, river runoff, etc., and the global isotope content of the oceans (Craig and Gordon 1965; Broecker 1974). Since the salinity of seawater is similarly affected by these processes, Craig and Gordon (1965) and later Fairbanks et al. (1992) defined a set of regression relationships between salinity and $\delta^{18}O_w$ with different slopes for several modern water masses, varying between 0.1 for humid tropical and 1.0 for arid polar surface water masses with a global mean of 0.49 (Fig. 10.1). Higher slopes represent areas where evaporation exceeds precipitation, and vice versa. However, the extent of oxygen isotope enrichments due to evaporation is limited due to the recycling of atmospheric moisture by the different pathways of precipitation. The slope of the global trend extrapolates to a $\delta^{18}O$-value of $-17‰$ at zero salinity, reflecting the influx of high-latitude precipitation and glacial meltwater. For the Antarctic continental ice, even $\delta^{18}O$-values as low as $-54‰$ have been determined (Weiss et al. 1979; Jacobs et al. 1985). However, the slope of the $\delta^{18}O_w$-salinity relationship may have changed through geological time (see discussion below).

For water masses deeper than 2000 m, Zahn and Mix (1991) obtained a slope as high as 1.53. This gradient is explained with the formation of sea-ice in the source areas especially for southern component water masses. Since the freezing of polar surface waters raises the salinity, but does not fractionate oxygen isotopes, southern source deep water masses, like the Antarctic Bottom Water, exhibit relatively low $\delta^{18}O_w$ values, and so do other water masses, which are derived from the admixture of south polar water masses (Mackensen 2001).

Modern Range of Values and Historical Variability

The modern $\delta^{18}O_w$ values of seawater are close to 0‰ (V-SMOW) and vary only within narrow limits. From the GEOSECS $\delta^{18}O$ sections for the today's world oceans, compiled by Birchfield

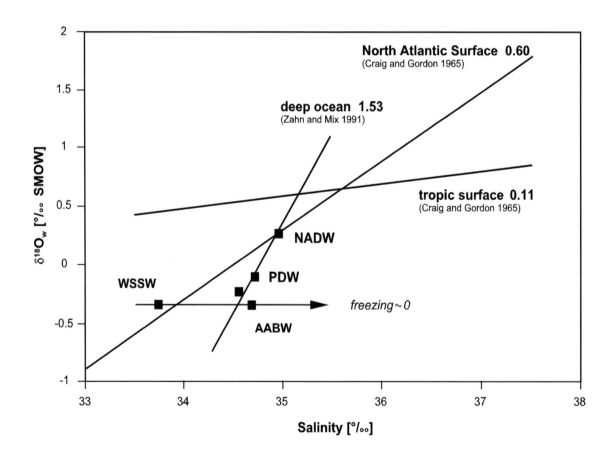

Fig. 10.1 Relationship between salinity and $\delta^{18}O_w$ of major water masses (after Craig and Gordon 1965; deep water line according to Zahn and Mix 1991).

(1987), a range of +1.0‰ in the mid-latitude surface Atlantic to –0.6‰ in the northern surface Pacific is present. Deep water mass $\delta^{18}O_w$ ranges from +0.3‰ in the core of North Atlantic Deep Water to –0.1‰ in the Circumpolar Deep Water (for a recent compilation search the Global Seawater Oxygen-18 Database, Schmidt et al. 1999). Since $\delta^{18}O_w$ exhibits only a narrow range within open ocean conditions, this proxy is an excellent tracer for indicating the influence of freshwater input to ocean water masses, since river discharge or meltwater release is always depleted in ^{18}O (Craig and Gordon 1965). For example, Mackensen et al. (1996, 2001) showed isotopically light meltwaters of the Antarctic Peninsula shelves to cascade down slopes the Weddell Sea Basin and thereby contributing to bottom water formation.

In geologic history, the $\delta^{18}O_W$ has been shown to vary considerably. For the sea-level low stand of -120 m during the last glacial maximum, Fairbanks (1989) showed Barbados coral oxygen isotopic composition to be enriched by 1.2‰, which coincides with an isotopic change of 0.10‰ corresponding to a 10 m sea-level change earlier estimated by Shackleton and Opdyke (1973). A slightly lower glacial interglacial range of 0.7 to 1.1‰ has been reported from porewater analyses of different drill holes in the Atlantic (Adkins et al. 2002). In an ice-free world, like in the Cretaceous, the global mean should have been depleted by another -0.8‰. On longer time scales, additional processes affecting the $\delta^{18}O_W$ have to be considered. A compilation of Phanerozoic $\delta^{18}O_W$ values showed an 5‰ increase in $\delta^{18}O_W$ since the Cambrian (Veizer et al. 1999 with a recent update of the database in 2004). By mass balance calculations, Wallmann (2001) demonstrated that this increase in $\delta^{18}O_W$ is in agreement with the ocean's 6-10% loss of seawater over the last 600 mio years because the subduction of water structurally bound in altered oceanic crust exceeds the water emission by mantle degassing. However, the large isotopic shifts observed episodically in the Paleozoic, as recorded in fossils (Veizer et al. 1999) and abiotic marine calcites (Lohmann and Walker 1989; Carpenter et al. 1991), exceed by far the variability in isotopes observed in Neogene and Quaternary times and are too fast to be explained with ocean crust-seawater interactions. Those changes must be attributed to changes in earth

climate systems, like ice volume and ocean circulation (e.g. Railsback 1990; Bickert et al. 1997).

10.3.2 $\delta^{18}O$ in Marine Carbonates

Principles of Fractionation

The oxygen isotope ratios in carbonates are a function of both temperature and the $\delta^{18}O_w$ of the surrounding seawater. Since the early equation given by Epstein et al. (1953), many equations have been published, which substantiated the potential of oxygen isotope paleothermometry for biogenically precipitated calcite. The first equation based on laboratory experiments with planktonic foraminifera was generated by Erez and Luz (1983). Their measurements on the cultured symbiotic species *G. sacculifer* were approximated by the second order polynom

$$T = 17.0 - 4.52 \cdot (\delta^{18}O_c - \delta^{18}O_w) + 0.03 \cdot (\delta^{18}O_c - \delta^{18}O_w)^2 \qquad (10.6)$$

with T standing for the *in-situ* temperature during calcite precipitation (°C), $\delta^{18}O_c$ representing the oxygen isotopic composition of the calcite (as ‰ PDB), and $\delta^{18}O_w$ representing the $\delta^{18}O$ value (‰ PDB) of the seawater from which the calcite has been precipitated. Since oxygen isotope analyses of waters are commonly reported relative to SMOW, the conversion of $\delta^{18}O_w$ to the PDB scale can be calculated according to Hoefs (2004)

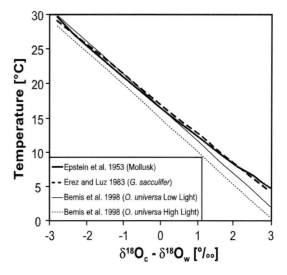

Fig. 10.2 Comparison of the results of different paleo-temperature equations (see Bemis et al. 1998 for review).

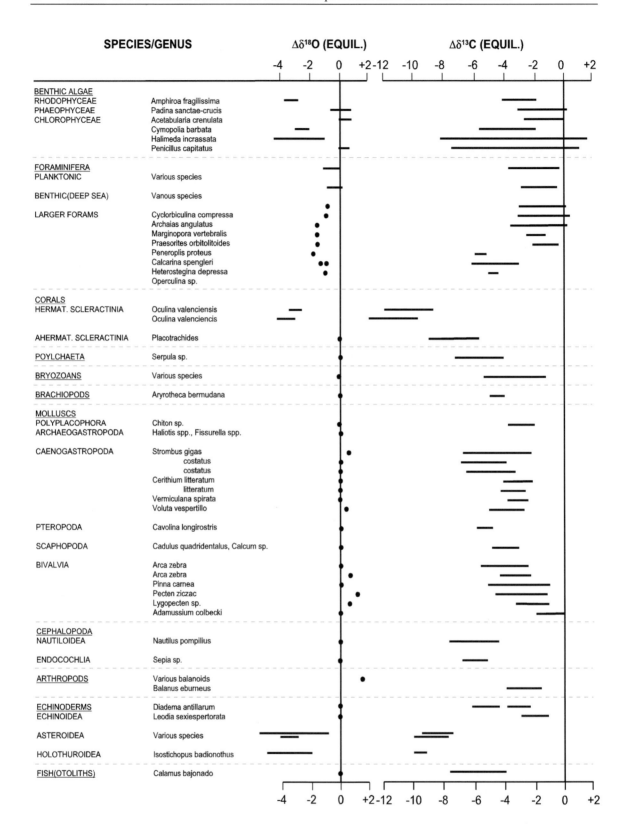

Fig. 10.3 $\delta^{18}O$ and $\delta^{13}C$ deviations from equilibrium isotope composition of selected calcareous species (adopted from Wefer and Berger 1991).

$$\delta^{18}O_{PDB} = 0.97002 \cdot \delta^{18}O_{SMOW} - 29.98 \qquad (10.7)$$

Diagenesis

A $\delta^{18}O_w$ correction of $-0.27‰$ is therefore necessary to compare $\delta^{18}O$ values measured in CO_2 produced by the reaction of calcite with H_3PO_4 with those measured in CO_2 equilibrated with water.

Most of the paleotemperature equations appear to be similar to the one stated above (Eq. 10.6), but temperature reconstructions can differ as much as $2°C$ when ambient temperature varies between $5°C$ and $25°C$ (Fig. 10.2; Bemis et al. 1998; for a recent review see Mulitza et al. 2003). The reason is that, in addition to *in-situ* temperature and water isotopic composition, the shell $\delta^{18}O$ may be affected by the photosynthetic activity of algal symbionts and by the carbonate ion concentration in seawater. Wefer and Berger (1991) summarized the importance of such effects, previously called 'vital effects', on a broad spectrum of organisms (Fig. 10.3). For oxygen isotopes, most organisms are shown to precipitate $CaCO_3$ close to equilibrium with the water in which they live. However, some organisms, such as planktonic foraminifera and hermatypic corals, exhibit significant differences from isotopic equilibrium (see Waelbroek et al. 2005 for a review). Laboratory experiments with live planktonic foraminifera demonstrated that an increase in the symbiont photosynthetic activity results in a decrease in shell $\delta^{18}O$ values (Spero 1992; Spero and Lea 1993). The mechanisms driving the effects of symbiont photosynthesis on shell $\delta^{18}O$ are not well understood, but appear to be linked to the carbonate ion concentration. During photosynthetic activity, CO_2 uptake by the symbionts increases the pH in the microenvironment around the shell (Jørgensen et al. 1985; Rink 1998; Zeebe 1999). Consequently, more alkaline conditions correspond to locally elevated $[CO_3^{2-}]$. The ^{18}O depletion of shells due to higher symbiont photosynthetic activity is consistent with the effect of higher ambient $[CO_3^{2-}]$. Spero et al. (1997) obtained a $\delta^{18}O/[CO_3^{2-}]$ slope of $-0.002‰$ $\mu mol^{-1}kg^{-1}$ from experiments with symbiotic (*O. universa*) and non-symbiotic (*G. bulloides*) foraminifer species. This previously undocumented carbonate isotope effect may help to solve inconsistencies in temperature reconstructions by applying oxygen isotope paleothermometry relative to other marine or terrestrial temperature proxies (see discussions in Spero et al. 1997; Zeebe 1999).

The isotopic composition of a carbonatic shell will remain unchanged until the shell material dissolves and recrystallizes during diagenesis. Diagenetic modification, however, could begin immediately after deposition or even in the water column due to corrosive deep ocean or pore waters. Such waters are generally enriched in CO_2 due to the respiration of organic matter mediated by specific bacteria. In shells of the planktonic foraminifera *P. obliquiloculata* sampled in a depth profile in the western equatorial Pacific Wu and Berger (1989) showed that below the depth of the modern lysocline the $\delta^{18}O$ of this species increased with water depth due to increasing calcite dissolution. Close to the modern depth of calcium carbonate compensation, the observed deviation reached a maximum value of $+0.9‰$. This effect of differential dissolution, i.e. the preferential removal of the light isotope ^{16}O, has recently helped to unravel the so-called cool tropics paradox. Using exceptionally well preserved planktonic foraminifer shells extracted from clay-rich late Cretaceous to Eocene sediments, Pearson et al. (2001) could show that the previously reported cooler tropical sea surface temperatures are the result of up to 3 ‰ increased $\delta^{18}O$ values caused by a subtle diagenetic recrystallization on a micrometer scale, without obliterating surface ornaments or internal layering features. Pearson et al. (2001) emphasized that this kind of diagenesis is hard to detect compared to more obvious modes of alteration like the precipitation of euhedral inorganic calcite resulting in substantial infillings or overgrowth.

The conversion of sediment into limestone within deep-sea sediments results from pressure and temperature rises which both increase with burial depth. Within carbonate sediments, diagenesis generally transforms the less stable aragonite and Mg-calcite into a low-Mg calcitic cement by means of a dissolution-reprecipitation process. Theoretically, the oxygen isotope composition of carbonates should not change significantly with burial, since the ^{18}O in pore waters originates from seawater. However, in many cores recovered during DSDP/ODP drilling, deep-sea carbonates and often pore waters as well, exhibit ^{18}O depletions by several permil (e.g. Lawrence 1989). Mass balance calculations by Matsumoto (1992) indicated that the ^{18}O shift in pore water towards lower values in sediments of the Japan Sea is controlled by a low-temperature alteration of basement basalts (see also discussion in Sect. 10.3.1), which is slightly compensated by the transformation of biogenic opal to quartz. Furthermore, detailed measurements on different generations of carbonate cements suggest that late cements exhibit lower $\delta^{18}O$ values compared to early precipitates. This $\delta^{18}O$ trend may be due to the increasing temperatures with increasing burial depth or to the isotopic evolution of pore waters during precipitation (Hoefs 2004).

10.4   Geochemical Influences on $^{13}C / {}^{12}C$ Ratios

10.4.1 $\delta^{13}C_{\Sigma CO2}$ of Seawater

Principles of Fractionation

The carbon isotopic composition of ΣCO_2 in seawater is mainly controlled by two processes, the biochemical fractionation due to the formation and decay of organic matter, and the physical fractionation during gas exchange at the air-sea boundary (Broecker and Maier-Reimer 1992). Surface water is enriched in ^{13}C, because photosynthesis preferentially removes ^{12}C from the CO_2. Deeper water masses have lower $\delta^{13}C$ values, since nearly all of the organic matter that is produced by photosynthesis is subsequently remineralized in the water column. Broecker and Maier-Reimer (1992) showed that if there were no air-sea gas exchange, the relationship between $\delta^{13}C$ and PO_4 in the ocean would be

$$\delta^{13}C - \delta^{13}C_{m.o.} =$$

$$\varepsilon_p / \Sigma CO_{2\,m.o.} \cdot C / P_{org} \cdot (PO_4 - PO_{4\,m.o.})$$

$$(10.8)$$

where ε_p (‰) is the isotopic effect associated with the photosynthetic fixation of carbon, C/P_{org} is the Redfield Ratio, and the subscript m.o. stands for mean ocean values. When reasonable values are substituted ($\delta^{13}C_{m.o.} = 0.3$‰, $\varepsilon_p = -19$‰, $\Sigma CO_{2\,m.o.} = 2200$ µmol kg^{-1}, $C/P_{org} = 128$, $PO_{4\,m.o.} = 2.2$ µmol kg^{-1}), the predicted relationship closely matches the relationship for waters in the deep Indian and Pacific Oceans ($\delta^{13}C = 2.7 - 1.1 \cdot PO_4$). This is to be expected, as the effect of air-sea exchange should be constant for these deep water masses due to the

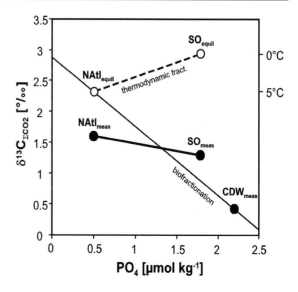

Fig. 10.4   Schematic diagram showing the effect of biofractionation and thermodynamic effects on the carbon isotopic composition of total dissolved inorganic carbon (according to Zahn and Keir 1994).

homogeneity of source waters for the deep Indian and Pacific Oceans.

On the other hand, carbon isotope fractionation during air-sea gas exchange is also an important factor in determining the isotopic composition of carbon in surface water (Charles and Fairbanks 1990; Broecker and Peng 1992; Lynch-Stieglitz et al. 1995). If the CO_2 in the atmosphere were in isotopic equilibrium with the dissolved inorganic carbon in the ocean, the dissolved inorganic carbon would be enriched in ^{13}C by about 8‰ at 20°C relative to the atmosphere CO_2 (Zhang et al. 1995). This thermodynamic fractionation depends on the temperature of equilibration, with ΣCO_2 becoming more enriched relative to the atmospheric value by about 1‰ per 10°C cooling (Fig. 10.4; Mook et al. 1974). If the surface ocean were in complete isotopic equilibrium with atmospheric CO_2, one expects a 3‰ range in oceanic $\delta^{13}C$ for the 30°C range in ocean temperatures, similar to the magnitude of $\delta^{13}C$ change induced by biological processes. In fact, CO_2 exchange rates between the ocean and the atmosphere are slow enough (relative to mass transport of ΣCO_2 within the ocean) that the range of $\delta^{13}C_{\Sigma CO2}$ in surface waters is less than 3‰. Moreover, Zahn and Keir (1994) showed in their ocean-atmosphere box-model that even in the absence of the temperature effect, the air-sea gas exchange modifies the $\delta^{13}C$ distribution of the upper ocean and the Atlantic deep water. The observed deviations occur because net biological production in the Southern Ocean is low relative to the upwelling fluxes of nutrients and dissolved carbon, allowing low $^{13}C/^{12}C$ ratios in the dissolved carbon to outcrop and exchange with the atmosphere. As a consequence, a dynamic balance is achieved where CO_2 evading from Antarctic waters to the atmosphere has a lower $\delta^{13}C$ than the invading CO_2, while the isotope ratio of the evading CO_2 from the warm ocean is greater than that of the return flux of CO_2.

Modern Range of Values and Historical Variability

The modern $\delta^{13}C_{\Sigma CO2}$ values of seawater are close to 0‰ (PDB) and vary only within a small range. From the GEOSECS $\delta^{13}C_{\Sigma CO2}$ sections for today's world oceans, compiled by Kroopnick (1985), a range of +2.5‰ in the mid-latitude surface Atlantic to +0.7‰ in the northern surface Pacific has been determined. Deep water mass $\delta^{13}C_{\Sigma CO2}$ ranges from +1.2‰ in the core of North Atlantic Deep Water and +0.4‰ in the Circumpolar Deep Water to −1.0‰ in the northern Pacific Deep Water. As stated above, $\delta^{13}C_{\Sigma CO2}$ of a deep water mass behaves like a conservative tracer for phosphate, since the water mass left its sea surface source area. On the other hand, paired $\delta^{13}C_{\Sigma CO2}$ and PO_4 measurements on two sections in the Southern Ocean clearly support the thermodynamic imprint during Weddell Sea Bottom Water formation by a significant deviation from the Redfield $\delta^{13}C_{\Sigma CO2}/PO_4$ relationship (Mackensen et al. 1996; Mackensen 2001). A recent matter of concern is the so-called oceanic Suess effect, which describes the uptake of ^{13}C-depleted anthropogenic CO_2 by the oceans and the consequent decrease in the oceanic $\delta^{13}C$. Comparisons between modern sediment trap and core top foraminiferal $\delta^{13}C$ reveal a modern depletion in $\delta^{13}C$ of the upper water column of up to 0.65‰ in the Subantarctic Zone and up to 0.9‰ in the Arctic realm (e.g., King and Howard 2004). This effect has to be considered before using surface sediment values for carbon cycle proxy calibrations.

The $\delta^{13}C_{\Sigma CO2}$ varied considerably in geologic history. Generally, three explanations are given for changes of $\delta^{13}C_{\Sigma CO2}$ distribution in the ocean: (1) changes in the surface-ocean productivity which cause variable fractionation between surface and deep water carbon isotopic composition, (2)

changes in the gas exchange rates between ocean and atmosphere due to changes in surface temperatures and ocean circulation, and (3) changes in the marine carbon budget by variations in the reservoirs of the atmosphere, the ocean, or the lithosphere. Of course, these different processes act on different time scales. Variations on a scale of 10^1 to 10^5 years might be attributed to changes in the effectivity of the biologic pump (Berger and Vincent 1986), to rapid changes between land and ocean reservoirs (Shackleton 1977; Broecker 1982), or to changes in ocean circulation responding to climate variability (e.g. Sarnthein et al. 1994; Bickert and Mackensen 2003).

On scales of 10^5 years and longer, changes in the influx and in the burial of sedimentary inorganic and organic carbon exert a primary control on the $\delta^{13}C$ of the ocean and the atmosphere (Holser 1997). Main sources for carbon comprise the erosional flux of sedimentary carbon (as C_{org} and carbonate) and the degassing of volcanic CO_2; main sinks are the burial of organic matter and the deposition of carbonate. Changes in the oceanic $\delta^{13}C$ are assumed to be the result of variations in the ratio of inorganic to organic carbon contributed to sediments (e.g. Derry and France-Lanord 1996). An increase in the burial of organic carbon would preferentially remove ^{12}C from seawater, so that the ocean reservoir would become isotopically heavier, and vice versa.

10.4.2 $\delta^{13}C$ in Marine Organic Matter

Principles of Fractionation

The carbon isotopic composition of marine organic matter produced in the photic zone depends on the isotope ratio of the total dissolved inorganic carbon and the degree to which this inorganic pool is utilized (Degens et al. 1968). Although more than 90% of the inorganic carbon pool is in the form of HCO_3^-, phytoplankton utilizes mostly carbon of the very small reservoir (1%) of dissolved $CO_{2(aq)}$. While the equilibrium exchange fractionation is only small between atmospheric CO_2 and dissolved $CO_{2(aq)}$ (about 1‰), and moderate between $CO_{2(aq)}$ and HCO_3^- (with a range of 10‰ to 8‰ between 5°C and 25°C; Mook et al. 1974), a large kinetic fractionation accompanies the biological carbon fixation during photosynthesis. The main isotope-discriminating steps are (1) the uptake and intracellular diffusion of CO_2, and (2) the enzymatic carbon fixation (Park

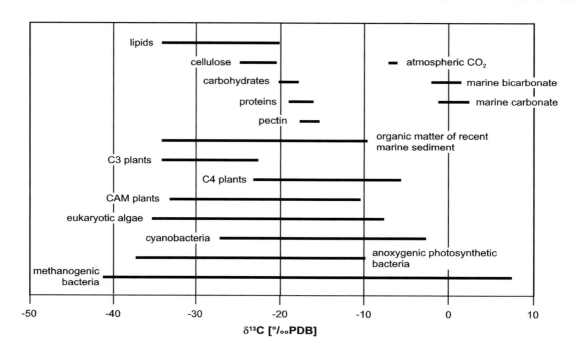

Fig. 10.5 Carbon isotopic composition of autotrophic marine and terrestrial organisms in comparison to the range observed in anorganic compounds (after Schidlowski 1988) and organic constituents (after Degens 1969).

and Epstein 1960). The first, reversible step is associated with a moderate fractionation of about −4‰, the second, irreversible step causes a kinetic fractionation of up to −40‰ (O'Leary 1981).

The $\delta^{13}C_{org}$ values of marine phytoplankton range between −10‰ and −31‰, but most of the warm-water plankton exhibits values between −17‰ to −22‰ (Fig. 10.5; Degens et al. 1968, 1969). The two-step model of carbon fixation clearly suggests that isotope fractionation depends on the concentration of $CO_{2(aq)}$. The fractionation decreases with decreasing CO_2 availability, such that in warm marine surface waters one would expect higher $\delta^{13}C_{org}$ values than in colder waters, since the solubility of CO_2 increases with decreasing temperature. However, other factors, such as species composition, light intensity, growth rate and cell geometry may also influence organic $\delta^{13}C$ values of particulate organic matter (e.g. Hayes 1993; Bidigare et al. 1997; Rau et al. 1997; Popp et al. 1998).

An important application of carbon stable isotope ratios lies in the reconstruction of the carbon dioxide concentration in surface waters and subsequently in the atmosphere at the time the organic matter was produced. Experimental and field studies have shown that the fractionation of stable carbon isotopes during photosynthesis by plankton depends on the concentration of ambient dissolved molecular carbon dioxide ($CO_{2(aq)}$), which led to the suggestion that sedimentary $\delta^{13}C_{org}$ may serve as a proxy for surface water $CO_{2(aq)}$ (Popp et al. 1989; Rau et al. 1991). The applicability of this tool in paleoceanographic studies has been shown repeatedly (e.g. Jasper and Hayes 1990, 1994; Fontugne and Calvert 1992; Müller et al. 1994), but the reconstruction of paleo-pCO_2 has become more and more complex due to the growing knowledge of factors influencing carbon isotopic fractionation during organic matter construction. Recent laboratory and field experiments as well as theoretical considerations indicate that the isotopic fractionation of carbon during photosynthesis (ε_p), and consequently also the sedimentary $\delta^{13}C$ of organic matter, could be influenced by physiological processes and environmental factors such as growth rates, temperature, nutrient supply, irradiance (Laws et al. 1995; Bidigare et al. 1997; Burkhardt et al. 1999; Rost et al. 2002), cell geometries and membrane permeability (Popp et al. 1998; Burkhardt et al. 1999), and active carbon uptake (for a review see Schulte et al. 2003). To better follow the discussion of this complex matter, the consecuting paragraphs introduce the use of $\delta^{13}Corg$ in pCO_2 reconstruction, knowing the fact that there is an ongoing debate on the applicability of this proxy.

For algae where CO_2 is thought to reach the photosynthetic site only by passive diffusion, ε_p can be calculated by the following equation (Bidigare et al. 1997):

$$\varepsilon_p = \varepsilon_f - b / [CO_{2(aq)}] \qquad (10.9)$$

where ε_f (‰) is the maximum isotopic effect associated with the photosynthetic fixation of carbon mediated by the enzyms Rubisco and β-carboxylase ($\approx$25 ‰; Bidigare et al. 1997), and the variable b reflects the intracellular carbon demand. Because the b-value combines a suite of factors, like growth rate, membrane permeability, cell geometry, and boundary layer thickness, significant variations in b (–109 ‰μM to –164 ‰μM) have been observed in empirical fits to field and experimental data (Laws et al. 1995). For example, higher growth rates lead to higher b-values (Bidigare et

al. 1997; Rau et al. 1997). Based on chemostat experiments with the coccolithophorid *E. huxleyi*, Bidigare et al. (1997) found a close correlation between ε_p and the growth rate μ:

$$\varepsilon_p = 24.6 - 137.9 \cdot \mu / [CO_{2(aq)}] \quad (10.10)$$

To obtain $CO_{2(aq)}$, Equation 10.9 has to be rearranged, and values for ε_p are determined from the carbon isotopic compositions of the primary photosynthate $\delta^{13}C_p$ (‰) and of the ambient dissolved molecular carbon dioxide $\delta^{13}C_d$ (‰) (Fig. 10.6):

$$\varepsilon_p = ((\delta^{13}C_p + 1000) / (\delta^{13}C_d + 1000) - 1) \cdot 1000$$

$$(10.11)$$

Assuming that diagenetic isotopic alterations can be neglected (see below), $\delta^{13}C_p$ may be directly substituted by the measured $\delta^{13}C_{org}$ value.

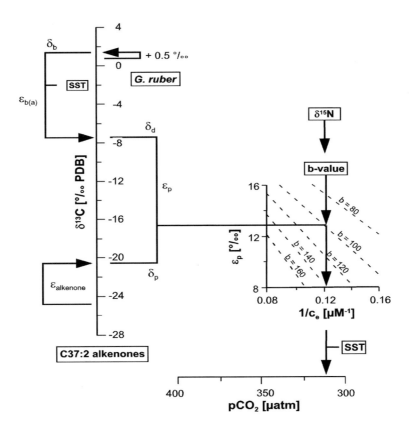

Fig. 10.6 Diagram outlining the estimation of surface ocean $CO_{2(aq)}$ and pCO_2, when $\delta^{13}C_{alkenones}$, $\delta^{13}C_{calcite}$, $\delta^{15}N$ as a proxy for carbon demand b, and surface ocean temperature are provided (Andersen et al. 1998; modified from Jasper and Hayes 1994).

Values for $\delta^{13}C_d$ are estimated on the basis of the $\delta^{13}C$ of calcite tests assuming that the calcite was precipitated in equilibrium with the ΣCO_2, correcting for the temperature-dependent fractionation $\varepsilon_{b(a)}$ between $CO_{2(aq)}$ and dissolved bicarbonate according to Mook et al. (1974):

$$\begin{aligned}\delta^{13}C_d &= \delta^{13}C_{\Sigma CO2} - \varepsilon_{b(a)} \\ &= \delta^{13}C_{\Sigma CO2} - 24.12 - 9866/T \end{aligned} \quad (10.12)$$

T is the temperature of surface waters measured in K. The $CO_{2(aq)}$ concentrations are then converted to CO_2 partial pressure values (pCO_2 in µatm) using Henry's Law:

$$pCO_2 = CO_{2(aq)}/\alpha \quad (10.13)$$

where the solubility coefficient α is mainly a function of temperature and, to a minor extent, of salinity (Rau et al. 1991). α may be calculated according to Weiss (1974). The success of the equations in determining surface $CO_{2(aq)}$ is dependent on the estimation of the sea-surface temperatures and on the determination of the complex variable b for the past. While several proxies exist for reconstructing past SST, the determination of ancient b-values is difficult. A promising approach is the use of bulk sediment $\delta^{15}N$ as proxy for b (Andersen et al. 1999). However, the use of $\delta^{15}N$ as a proxy for carbon demand introduces a number of potential problems associated with the complex fractionation mechanisms of nitrogen isotopes (see Sect. 10.5).

A topic of much concern regarding the use of bulk $\delta^{13}C_{org}$ in paleoceanographic investigations is the masking of the marine $\delta^{13}C_{org}$ signal by terrestrially-derived organic matter. Since terrigenous material is usually substantially depleted in ^{13}C (estimated mean value of –27‰, since 90% of land plants are C_3 plants) compared to marine-derived matter (mean value of –19‰), the carbon isotopic composition of ocean sediments have been used to trace the origin of sedimentary organic carbon (Newman et al. 1973; Rühlemann et al. 1996). However, the underlying assumption regarding the mixing of a marine and a terrestrial $\delta^{13}C_{org}$ end member is not always supported because of the possible variability of $\delta^{13}C_{org}$ values produced in terrestrial systems (e.g. change in the relative contribution of C_3 and C_4 plants). Also, marine values may change by processes occurring in the water column (e.g. changing phytoplankton growth conditions, see above).

To circumvent the problem of influencing carbon isotope signals by terrigenous organic mat-

ter contamination or by marine photosynthesizers with varying carbon fixation pathways and cell geometries, recent studies concentrated on certain organic molecules (biomarker) from known marine sources unique to a particular class of organism. In this context, C_{37} alkenones have been shown to serve as excellent biomarkers (e.g., Andersen et al. 1999), because they derive exclusively from some haptophyte algae, which have a narrow range of cell geometries and diameters (Bidigare et al. 1997; Popp et al. 1998), thereby minimizing these variables as a significant control on isotopic compositions. As an application, Pagani et al. (1999) concluded on the basis of $\varepsilon_{p37:2}$ records that early to late Miocene (25-9 Ma) pCO_2 levels were similar to those recorded for Pleistocene glacial-interglacial intervals. This pattern of low pCO_2 levels throughout the Neogene has been confirmed independently by Pearson and Palmer (2000) using boron isotopes (see chapter 10.7)

Diagenesis

Early diagenesis begins in the photic zone of the oceans, continues during the sinking of particles, and is intense in the bioturbated surface layer of sediments. Therefore, only a few percent of the initially produced organic matter becomes buried in the sediments (Berger et al. 1989). However, despite the extensive loss of organic matter due to remineralization, the carbon isotopic composition of particulate organic matter appears to undergo only little change (see review in Popp et al. 1997). On the other hand, large and nonsystematic differences have been observed between the isotopic composition of total organic carbon and single organic compounds, e.g. phytoplankton biomarkers, in the same deposit (Fig. 10.5; Degens et al. 1969). Therefore, the assumption that the carbon isotopic composition of marine sedimentary organic matter directly or indirectly reflects the carbon isotopic composition of phytoplanktonic matter must be taken cautiously (Popp et al. 1997). Furthermore, since marine organic matter is easily digestible, whereas terrestrial organic matter is rather resistant, such preferential decomposition of organic matter has great impact on the estimation of the average marine and terrestrial $\delta^{13}C_{org}$ component (de Lange et al. 1994).

The carbon isotopic composition of pore waters reflects both the decomposition of organic matter, which releases ^{13}C-depleted CO_2 to the pore water, and the dissolution of $CaCO_3$, which adds CO_2 relatively enriched in ^{13}C. Since the depletion of organic matter δ^{13}C exceeds by far the range of δ^{13}C exhibited in sedimentary carbonate, the net result of these two processes is to make dissolved CO_2 in pore water isotopically lighter compared to the overlying bottom water. Below the sediment/water interface, where pore water has a δ^{13}C-value near that of seawater, a δ^{13}C-gradient exists in the uppermost few centimeters. McCorkle et al. (1985) and McCorkle and Emerson (1988) showed that the gradient of δ^{13}C-profiles observed in several box cores vary systematically with the rain of organic matter to the seafloor. In slowly deposited oxic sediments containing little amounts of organic carbon, only a small decrease in δ^{13}C relative to seawater δ^{13}C occurs. In organic-carbon rich sediments, the rates of organic matter decomposition and the production of reduced nitrogen compounds would rapidly decrease the pore water δ^{13}C. In these anoxic sediments, the situation is even more complex due to bacterial methanogenesis, which follows the sulfate reduction during early diagenesis. Since methane-producing bacteria metabolize methane highly enriched in ^{12}C ($-50‰$ to $-100‰$; Deines 1981), the pore water becomes significantly enriched in ^{13}C. Such trends in δ^{13}C of total dissolved CO_2 from pore waters within anoxic sediments have been recovered in various Deep Sea Drilling sites (Arthur et al. 1983). However, because pore waters are not truly closed systems, other factors like carbon loss due to upward diffusion of methane and other dissolved carbon species as well as precipitation of authigenic carbonates may also affect pore water δ^{13}C. For example, if significant amounts of methane are utilized in sulfate reduction, the rate of δ^{13}C decrease associated with the ΣCO_2 increase would be anomalously high (Arthur et al. 1983).

10.4.3 δ^{13}C in Marine Carbonates

Principles of Fractionation

Precipitation of carbonate, largely from total dissolved carbon which is present primarily as HCO_3^-, involves a much smaller fractionation of carbon isotopes compared to the photosynthetic fixation of carbon. In fact, the δ^{13}C of calcite is relatively insensitive to changes in temperature (e.g. $\approx 0.035‰$ $°C^{-1}$; Emrich et al. 1970) such that carbonate minerals can be used generally to monitor

changes in the $\delta^{13}C_{\Sigma CO2}$ in the waters from which they precipitate. However, the possible effects of biologically mediated carbonate precipitation have been considered already for oxygen isotopes (see Sect. 10.3.2). Observations summarized by Wefer and Berger (1991) reveal that most organisms exhibit δ^{13}C-values which are in disequilibrium with the $\delta^{13}C_{\Sigma CO2}$ of the waters, from which shell carbonate is precipitated (Fig. 10.3). As discussed for the influences on oxygen isotopic composition in carbonate, there may be several reasons for these deviations. McConnaughy et al. (1997) distinguish metabolic and kinetic effects on the carbon isotopic disequilibria. Kinetic isotope effects result from the discrimination against ^{13}C during hydration and hydroxylation of CO_2. Such effects appear to be associated with rapid calcification, as observed in corals. Metabolic effects apparently result from changes in the $\delta^{13}C_{\Sigma CO2}$ of the microenvironment due to photosynthesis and respiration. For example, infaunal benthic foraminifera precipitate their calcite in equilibrium with ambient pore water $\delta^{13}C_{\Sigma CO2}$, and therefore monitor the vertical carbon isotope gradient due to organic matter remineralization rather than bottom water $\delta^{13}C_{\Sigma CO2}$ (Fig. 10.7; McCorkle et al. 1990, 1997). Mackensen et al. (1993) discuss several ways, how the δ^{13}C of even epibenthic living foraminifer species could be influenced by the decay of organic matter. An example for the consequences of such a 'Mackensen' effect is discussed by Bickert and Wefer (1999) for the late Quaternary reconstruction of South Atlantic deep water circulation. An extreme depletion in δ^{13}C is recorded in endobenthic foraminifers sampled from ODP Site 680. Wefer et al. (1994) attributed these excursions to the release of methane in the continental margin off Peru sediments, which lowers substantially the carbon isotopic composition of CO_2 in the porewater. Furthermore, the laboratory experiments with live planktonic foraminifera which demonstrated symbiont photosynthetic effect on shell δ^{18}O values (Spero 1992; Spero and Lea 1993) revealed a similar effect on shell δ^{13}C values. Spero et al. (1997) obtained from experiments with symbiotic (*O. universa*) and non-symbiotic (*G. bulloides*) foraminifer species a δ^{13}C/[CO_3^{2-}] slope of $-0.007‰$ $\mu mol^{-1}kg^{-1}$. Furthermore, Bemis et al. (2000) showed a clear disequilibrium effect related to the calcification temperature, which has recently been approved in a field study by King and Howard (2004). These effects might partly be responsible for the shifts observed in foraminiferal

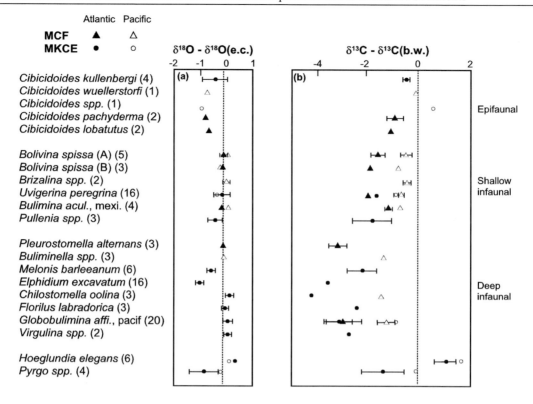

Fig. 10.7 $\delta^{18}O$ and $\delta^{13}C$ values of benthic foraminifera species in comparison to equilibrium calcite $\delta^{18}O$ values and to the pore water $\delta^{13}C_{\Sigma CO2}$ gradient, respectively (McCorkle et al. 1990).

calcite previously attributed to global changes in $\delta^{13}C_{\Sigma CO2}$ of the ocean (see discussion in Spero et al. 1997).

Carbonate Dissolution and Precipitation

Some aspects of the influence of diagenesis on the isotopic composition of carbonate have already been discussed for oxygen isotopes (see Sect. 10.3.2). The isotopic composition of a carbonatic shell will remain unchanged until the shell material dissolves and recrystallizes during diagenesis. As stated above, diagenetic modification could begin immediately after deposition or even in the water column due to corrosive deep ocean or pore waters. McCorkle et al. (1995) showed in shells of the benthic foraminifera C. *wuellerstorfi* sampled in a depth profile in the western equatorial Pacific that below the depth of the modern lysocline the $\delta^{13}C$ of this species decrease with water depth. Close to the modern depth of calcium carbonate compensation, the observed deviation reached a minimum value of −0.35‰. Since the effect of differential dissolution would cause a preferential removal of the

light isotope ^{12}C, an increase would be expected to occur with increasing dissolution. The opposite is observed and requires therefore other explanations, beside the effect of carbonate ion concentration (Spero et al. 1997).

Under oxic porewater conditions, no further change in the carbon isotopic composition of shell material is expected, until dissolution occurs under deep burial. But while a temperature-effect is negligible (Emrich et al. 1970), an alteration of the primary $\delta^{13}C$ value may arise during mineral transformations (aragonite $\rightarrow$ low Mg-calcite), depending on the amount of carbon present in the postdepositional solutions. Under suboxic conditions in sediments with high organic carbon contents, the carbon isotope signal may be affected by early diagenesis (Nissenbaum et al. 1972; Irwin et al. 1977). The degradation of organic matter will cause an intense degree of carbonate dissolution and the reprecipitation of authigenic calcite standing in equilibrium with porewater $\delta^{13}C_{\Sigma CO2}$. Schneider et al. (1992) estimated for continental slope sediments off Angola that a precipitation of 5% authigenic calcite on foraminiferal tests would produce a decrease of −1‰ in the

$\delta^{13}C$ of these tests. Again, the fractionation of $\delta^{13}C$ during precipitation has only a minor effect on the carbon isotope composition of the solid phase, the major effect is determined by the $\delta^{13}C_{\Sigma CO2}$ of the pore water which itself is controlled by the degradation of organic matter.

10.5 Geochemical Influences on ^{15}N / ^{14}N Ratios

10.5.1 $\delta^{15}N$ in Marine Ecosystems

Principles of Fractionation

The use of nitrogen isotopes in marine sediments is a relatively new tool in the field of paleoceanography. The usefulness of this proxy lies in its recording changes of nutrient dynamics in the water column. Major biological transformations of nitrogen in marine systems include the utilization of the dissolved forms of inorganic nitrogen (NO_2^-, NO_3^-, NH_4^+) by phytoplankton, the consumption of phytoplankton by grazers, and the remineralization of organic nitrogen by animals and bacteria. In addition, various marine prokaryotes are able to carry out a suite of reactions that move nitrogen in and out of the major inorganic pools. These include nitrogen fixation and dissimilatory transformations of nitrogen, such as nitrification, denitrification, and anaerobic ammonium oxidation (anammox) which has recently been recognized as a major sink for fixed inorganic nitrogen in oxygen minimum zones (e.g. Kuypers et al. 2003, 2005). The marine nitrogen cycle involves multiple inorganic and organic pools which are coupled by rapid biological transformations of nitrogen (cf. Chap. 6).

Since most marine autotrophs require combined nitrogen as a substrate for growth, the isotopic composition of dissolved inorganic nitrogen acts as a sort of master variable in setting the baseline isotopic composition of marine plankton. A number of biological processes may alter the isotopic composition of the marine pool of inorganic nitrogen. The most important are nitrogen fixation, nitrification, denitrification, and anaerobic ammonium oxidation, all of which may lead to a net transfer between the oceanic pool of combined nitrogen and the atmosphere. The basic biochemical reactions are the following:

Nitrogen fixation:
$$N_2 \text{ atm.} \Rightarrow \text{organic N} \tag{10.14}$$

It is well known that symbiontic bacteria in the roots of plants, but also bacteria and algae in the ocean can fix nitrogen. The high energy needed to break the triple bond of the nitrogen gas molecule makes natural nitrogen fixation a process feasable by only few organisms.

Ammonification:
$$N_{org} \Rightarrow NH_4^+ \tag{10.15}$$

Nitrification:
$$NH_4^+ + 3/2\,O_2 \Rightarrow NO_2^- + H_2O + 2\,H^+ \tag{10.16}$$

$$NO_2^- + 1/2\,O_2 \Rightarrow NO_3^- \tag{10.17}$$

Although ammonium is the direct product of decomposition of nitrogenous organic compounds (deamination), most inorganic nitrogen in the ocean occurs in the form of nitrate. Conversion of ammonia into nitrate is carried out by nitrifying organism.

Denitrification:
$$5\,(CH_2O) + 4\,NO_3^- + 4\,H^+ \Rightarrow$$
$$5\,CO_2 + 7\,H_2O + 2\,N_2 \tag{10.18}$$

or Anammox:
$$NH_4^+ + NO_2^- \Rightarrow N_2 + 2\,H_2O \tag{10.19}$$

After oxygen is exhausted, nitrate can serve as electron acceptor in the degradation of organic matter. Denitrification results in the conversion of nitrate to nitrogen gas and therefore balances the biological fixation of nitrogen. If there were no denitrification or anammox, atmospheric nitrogen would be exhausted in less than 100 million years. Denitrification and anaerobic ammonium oxidation occur in stratified or stagnant water masses of the ocean (e.g. within oxygen minimum layers) and in the sea bed.

The main processes involved in the biological utilization of nitrogen are all associated with kinetic fractionation effects, nevertheless they exhibit different fractionation factors. The isotope effect of nitrogen fixation ($\alpha = 1.000$ to 1.004) is small relative to the effects of bacterial nitrification, denitrification, or anammox ($\alpha = 1.02$ to 1.04) (see Montoya et al. 1994: Table 1). The results by Miyake and Wada (1971) indicate that little overall isotope fractionation occurs in the bacterial mineralization of organic nitrogen. Equilibrium isotope exchange reactions, which commonly occur in nature, are the ammonia volatilization and the solution of nitrogen gas. The former process has a significant isotopic effect (α= 1.034), whereas the latter process induces only a small fractionation (α = 1.0008).

Modern Range of Values and Historical Variability

The isotopic composition of some important nitrogen compounds is summarized in Fig. 10.8. NO_3^- dominates the oceanic pool of dissolved inorganic nitrogen, and the available data indicate that the $\delta^{15}N$ of NO_3^- from oxygenated deep waters ranges between 3‰ and 7‰ with a mean value around 6‰ (Liu and Kaplan 1989). The $\delta^{15}N$ of NO_3^- is significantly higher than this mean in and around anoxic water masses where denitrification occurs (Liu and Kaplan 1989). For the eastern tropical North Pacific Ocean values as high as 18.8‰, for the western Caribbean Sea values up to 12‰ have been reported to occur within the active denitrification zone (Cline and Kaplan 1975).

On geological time-scales, the entire nitrogen cycle has to be considered to describe the factors controlling the nitrogen isotopic balance in the ocean. The major input of nitrogen compounds into the ocean results from precipitation, river discharge and the fixation of molecular nitrogen. Nitrogen is removed from the ocean mainly by burial in the sediment and denitrification. Although it is generally assumed that the nitrogen balance in the ocean is in a steady state, it is diffcult to estimate the budget for mass balance as well as for isotopic balance (for a review of estimates see Gruber and Sarmiento 1997). However, denitrification in the water column and in the sediment is the dominant process maintaining mass balance. It is also the only process that can produce a large fractionation. All other processes add or remove nitrogen compounds with $\delta^{15}N$ values mostly in the range between –2‰ and +8‰. Changes in the global rate of denitrification could therefore lead to significant changes in the $\delta^{15}N$ of marine NO_3^-, which in turn would affect the $\delta^{15}N$ of marine plankton, since the isotopic composition of the phytoplankton at the base of the food web depends at least in part on the $\delta^{15}N$ of the dissolved inorganic nitrogen available (Altabet and Curry 1989). Today, it appears that denitrification (and maybe also anammox; Kuypers et al. 2005) is the principal mechanism that keeps the marine nitrogen compounds at a higher $\delta^{15}N$ value than atmospheric nitrogen.

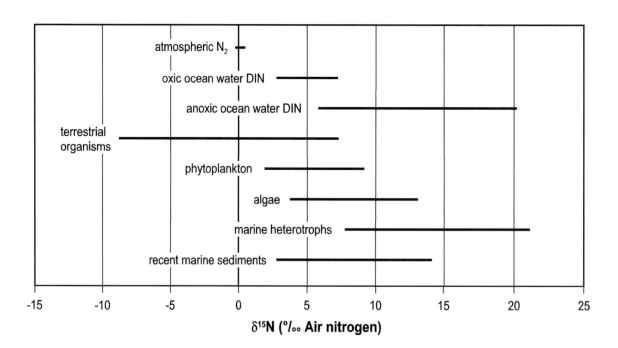

Fig. 10.8 $\delta^{15}N$ ranges of some important nitrogen compounds (according to Arthur et al. 1983).

10.5.2 δ¹⁵N in Marine Organic Matter

Principles of Fractionation

The nitrogen isotopic composition of marine organic matter produced in the photic zone depends on the isotope ratio of nitrate and the degree to which this inorganic pool is utilized (Wada 1980; Altabet et al. 1991; Voss et al. 1996). The preferential uptake of $^{14}NO_3^-$ leads to a depletion of ^{15}N in organic matter relative to the dissolved inorganic nitrogen used as a substrate for growth (Montoya 1994). Subsequently, the remaining nitrogen pool becomes progressively enriched in ^{15}N according to Rayleigh fractionation kinetics (Cifuentes et al. 1988). The degree of isotopic fractionation associated with primary productivity varies between taxa and with growth conditions. Culture experiments suggest that diatoms may discriminate more strongly than flagellates (Wada and Hattori 1978). The ensuing transfer of nitrogen through trophic levels is associated with a systematic increase in $\delta^{15}N$, with each trophic step resulting in an enrichment of 3.5‰ (Montoya 1994). This effect, however, should not affect the bulk sedimentary $\delta^{15}N$ values because of mass balance considerations.

An important use of nitrogen stable isotope ratios is the reconstruction of the degree of nitrate utilization in surface waters at the time the organic matter was produced. The applicability of this tool in paleoceanographic studies has been shown repeatedly (e.g. Altabet and Francois 1994; Holmes et al. 1997; Freudenthal et al. 2001). To determine changes in the fraction of unutilized nitrate in surface waters, Altabet and Francois (1994) defined the following equations:

$$\delta^{15}NO_{3\ (f)}^- = \delta^{15}NO_{3\ (f=1)}^- - \epsilon_u \cdot \ln(f)$$

$$(10.20)$$

$$\delta^{15}N\text{-}PN_{(f)} = \delta^{15}NO_{3\ (f)}^- - \epsilon_u$$

(instantaneous product) (10.21)

$$\delta^{15}N\text{-}PN_{(f)} =$$

$$\delta^{15}NO_{3\ (f=1)}^- + f/(1-f) \cdot \epsilon_u \cdot \ln(f)$$

(accumulated product) (10.22)

where f is the fraction of unutilized NO_3^- remaining (i.e. $[NO_3^-]/[NO_3^-]_{initial}$), ϵ_u is the fractionation factor associated with the NO_3^- uptake, and $\delta^{15}N\text{-}PN$

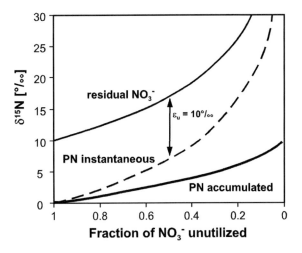

Fig. 10.9 Effects of nitrate utilization (assuming an ϵ_u of 10 ‰) on the $\delta^{15}N$ of the residual nitrate (solide line; Eq. 10.19), the instantaneous product (broken line; Eq. 10.20) and the accumulated product (heavy line; Eq. 10.21) of a reaction (according to Altabet and Francois 1994).

and $\delta^{15}NO_3^-$ refers to the measured $\delta^{15}N$ of the particulate nitrogen and of NO_3^-, respectively. The success of these equations in determining surface nitrate utilization depends on the estimation of the fractionation factor ϵ_u exhibited by phytoplankton during photosynthesis. The magnitude of ϵ_u sets an upper limit to the amplitude in $\delta^{15}N$ observed. ϵ_u may be obtained from the slope of the regression line between the photosynthate $\delta^{15}N$ (routinely, sedimentary $\delta^{15}N$ is used) and $\ln[NO_3^-]_{surface}$. A limited number of culture experiments with marine phytoplankton have shown significant variations in ϵ_u between phylogenetic groups (1‰ to 9‰) and growth conditions (0‰ to 16‰; Montoya 1994). Field estimates of ϵ_u in regions of moderate to high nitrate concentrations in the open ocean have fallen within a narrower range of 5‰ to 9‰ (Altabet and Francois 1994). At 90% utilization of NO_3^-, this latter range in ϵ_u corresponds to an increase of 12‰ to 21‰ in $\delta^{15}N$.

$[NO_3^-]_{initial}$ is the nitrate concentration of newly upwelled water or surface waters before the onset of springtime productivity. Equation 10.20 underscores that changes in $\delta^{15}N$ are in reality a function of nitrate utilization and not simply concentration per se. Equation 10.21 refers to particulate nitrogen produced at any one point in the course of reaction (instantaneous product) (Fig.10.9). Observations would fit Equation 10.21, only if particulate nitrogen is rapidly removed from the system. On the other hand, if there is no removal of particulate nitrogen from the system, the change of $\delta^{15}N$ within this accumulating pool is described by Equation 10.22, the integral of Equations 10.20 and 10.21. However, this latter case almost never occurs (Altabet and Francois 1994).

A topic of much concern regarding the application of bulk $\delta^{15}N$ in paleoceanographic investigations is the effect of terrestrially derived organic matter on the marine $\delta^{15}N$ signal. Since terrigenous material is mostly substantially depleted in ^{15}N relative to marine-derived matter, the nitrogen isotopic composition of ocean sediments have been used to trace the origin of sedimentary nitrogen (Sweeny and Kaplan 1980; Wada et al. 1987). However, the underlying assumption regarding the mixing of a marine and terrestrial $\delta^{15}N$ end members is not always conclusive because of the wide range of $\delta^{15}N$ values produced in terrestrial systems (Sweeney et al. 1978). Furthermore, marine values may be subject to changes produced by processes in the water column (as discussed above).

Diagenesis

Reconstructing past variations of nutrient conditions in surface waters requires the preservation of the nitrogen isotope surface-water signal in the sedimentary organic matter. Diagenesis, however, may significantly alter the nitrogen isotopic composition of particles even before they are buried at the seafloor. Altabet and McCarthy (1985) suggested that bacterial remineralization of sinking particles resulted in a ^{15}N enrichment in the residual organic matter. Francois et al. (1992) reported a ^{15}N enrichment of 5‰ to 7‰ in core top values of Southern Ocean sediments compared to measurements of organic matter in the photic zone. However, Altabet and Francois (1994) demonstrated that there was only little offset between sinking particles at 150 m water depth in the equatorial Pacific and the respective core tops. Sigman et al. (1999) suggested that the difference in preservation of the isotopic signal existent in the two regions may be explained with the high opal content in the Southern Ocean sediments, in which organic matrices with high $\delta^{15}N$ may be captured. In any case, the findings of these authors suggest that where differences between the surface-generated signal and sediments exist, the offset appears to be relatively constant within geographic regions.

Further decomposition of organic matter occurs at the sediment/water interface and in the upper sections of the sediment. However, observations

from equatorial Pacific sediments indicate that the loss of organic carbon as well as of nitrogen in the uppermost few millimeters is not accompanied by a corresponding shift in $\delta^{15}N$ over this depth interval (Altabet and Francois 1994). Freudenthal et al. (2001) found that early diagenesis in sediments of the eastern tropical Atlantic led to variability in $\delta^{15}N$ of even less than 1‰. With further diagenesis of organic matter, nitrogen may occur as ammonium incorporated in the lattice of clay minerals, where it replaces potassium. But even this complex substitution is not associated with an isotopic effect, since this nitrogen exhibits an isotopic composition very similar to the organic matter from which it is derived (Williams 1995). Therefore, beside the possible fractionation mechanisms, which may alter the nitrogen isotopic composition of organic matter in the water column, there seems to be no significant diagenetic alteration of the isotope signal within the sediments. The nitrogen isotopic composition of sediments is primarily determined by the source organic matter.

10.6   Geochemical Influences on ^{34}S / ^{32}S Ratios

10.6.1 $\delta^{34}S$ of Seawater and Pore Waters

Principles of Fractionation

In the marine environment, sulfur occurs most commonly in its oxidized form as dissolved sulfate in seawater or as precipitated sulfate in evaporites and in its reduced form as sedimentary pyrite. The ratio of $^{34}S/^{32}S$ is a sensitive indicator for the

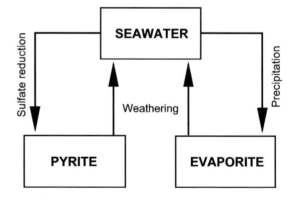

Fig. 10.10 Main reservoirs of the sedimentary sulfur cycle (adopted from Strauss 1997).

transfer of sulfur between the different reservoirs, which is often associated with a change in the oxidation state of the element. The main processes to be considered in the sulfur cycle are the riverine input of sulfate as a weathering product of sulfur-bearing rocks, the precipitation of evaporites from seawater and the biological reduction of seawater sulfate and subsequent formation of sedimentary pyrite (Fig. 10.10). The associated fractionation mechanisms are therefore:

- isotope exchange reactions between sulfate and sulfides

- kinetic isotope effects in the bacterial reduction of sulfate or in the oxidation of sulfides

Only a small fractionation is associated with the precipitation of sulfates in seawater. Theoretical considerations as well as sulfur-isotope measurements in natural environments revealed an isotopic difference between seawater sulfate and

crystallized gypsum of 0 to +2.4‰ (Raab and Spiro 1991). Therefore, within the generally observed variability in an evaporitic deposit, calcium sulfate is assumed to record the isotopic composition of seawater sulfate at the time of deposition (Claypool et al. 1980; Strauss 1997).

By far the largest fractionation is associated with the bacterial sulfate reduction (see a recent review by Canfield 2001). Due to the activity of sulfate-reducing bacteria, such as *Desulfovibrio desulfuricans*, organic matter is oxidized according to the following equation:

$$2\ (CH_2O) + SO_4^{2-} \Rightarrow H_2S + 2\ (HCO_3^-) \qquad (10.23)$$

The resulting hydrogen sulfide reacts with sedimentary iron, which is available in the reactive non-silicate bond form (oxy-hydroxides), and is fixed as iron sulfide (e.g. pyrite). In general, a substantial depletion of $\delta^{34}S$ occurs due to preferential utilization of the ^{32}S-isotope by the sulfate reducing bacteria. Whereas the fractionation

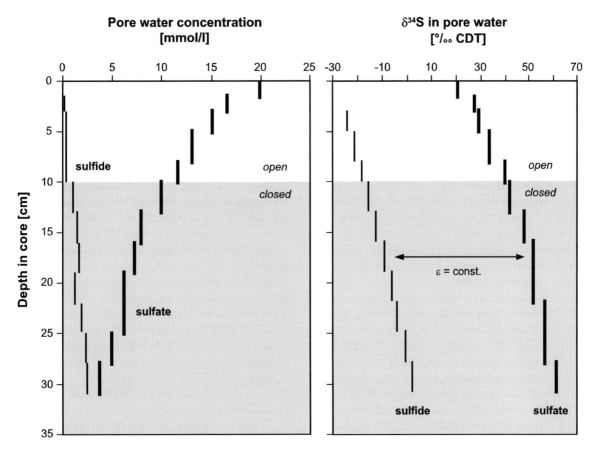

Fig. 10.11 Variation of sulfate and sulfide concentration and δ^{34}S values of pore water in core 2092 from the western Baltic Sea (Hartmann and Nielsen 1969).

is small during the entrance of sulfate into the cell (-3 to 0‰), the subsequent breaking of the S-O bonds is causing a large change   in the isotope composition of the reactants (SO_4^{2-} => SO_3^{2-} : 22‰, SO_3^{2-} => H_2S: up to 18‰; see Canfield 2001). As a result, sedimentary sulfide is depleted in ^{34}S relative to ocean water sulfate (Kaplan and Rittenberg 1964). The depletion is usually in the order of 20‰ to 70‰ (Hartmann and Nielsen 1969; Ohmoto 1990), although cultural experiments yielded only depletions between 10‰ and 30‰, with a maximum reported value of 49‰ (Kaplan and Rittenberg 1964; Bolliger 2001). The much larger fractionation ocurring under natural conditions   may be generated by a concurrent recycling of H_2S via an intermediate oxidation to elemental sulfur or thiosulfate by buried Fe(III) and a disproportionation of these intermediates to SO_4^{2-} and H_2S (Jørgensen 1990; Habicht and Canfield 1997). The disproportionation shunts produce ^{34}S-enriched SO_4^{2-} and ^{34}S-depleted H_2S and thereby increase the resulting isotopic difference between SO_4^{2-} and H_2S in the sediment pore water (for a recent discussion see Canfield 2001).

The extent of depletion during bacterial sulfate reduction is affected by the magnitude of reduction rates and by the availability of sulfate. Generally, an increasing rate of sulfate reduction leads to a decrease in fractionation. Furthermore, if the availability of sulfate is reduced, e.g. through limited diffusion in non-bioturbated or impermeable sediments, the isotopic composition of both reactant and product shifts towards higher values. Such a Rayleigh fractionation process has been shown by Hartmann and Nielsen (1969) for dissolved sulfate and sulfide in the pore water of a core retrieved from the western Baltic Sea (Fig. 10.11). The $\delta^{34}S$-values of both reactant and product increase with depth, because the two isotopes in sulfate or in sulfide do not diffuse in the same proportion as they occur in the pore water (Jørgensen 1979).

Fractionation also occurs during processes in the oxidative part of the sulfur cycle. Sulfide oxidation, which is pervasive in marine environments (for example, 90% or more of the sulfide produced during sulfate reduction in coastal sediments are reoxidized) includes various processes, which are all associated with inverse isotope fractionation effects. These processes include the phototrophic oxidation by a variety of anoxygenic bacteria (-2 to 0‰), the non-phototrophic oxydation (up to -18‰), and the already mentioned disproportionation of sulfur compounds with intermediate oxidation states as

elemental sulfur or thiosulfate (up to -33 ‰) (Canfield 2001).

Modern Range of Values and Historical Variability

The isotopic composition of sulfur in modern ocean sulfate is constant within very narrow limits and is represented by a $\delta^{34}S$-value of +20‰ with a standard deviation of ±0.12‰ (Longinelli 1989). The small isotopic range over a great variety of localities and at various depths is explained with the residence time of sulfate in the ocean which is well in excess of the ocean mixing time (Holland 1978). This sulfate homogeneity should be considered for the evaluation of isotope data analyzed in ancient deposits.

The sulfur isotopic composition in oceanic sulfate through Phanerozoic time shows a Cambrian maximum of about +30‰, a decrease to a Permian minimum of about +10‰ and an increase towards the modern ocean value of +20‰ (Claypool 1980; Strauss 1997). Superimposed over this long-term secular variation is a pronounced short-term variability exhibiting about the same isotopic range as documented for the entire Phanerozoic. The observed variations are assumed to reflect changes in the sulfur redox cycle within the ocean. Since bacterial sulfate reduction and the subsequent formation of sedimentary pyrite is associated with a substantial fractionation, any changes in the isotopic mass balance are generally interpreted as changes in the burial of reduced sulfur. Furthermore, both the rate of sulfur input and its isotopic composition may vary as a function of time depending on the weathering rates of different sedimentary rocks, or on the intensity of volcanic activity. For example, increased erosion of black shales containing ^{34}S-depleted sulfides may lower the $\delta^{34}S$ value of oceanic sulfate. On the other hand, the increased reduction of sulfate by bacteria and removal of ^{34}S-depleted sulfide from the ocean may increase its $\delta^{34}S$ value. However, the fact that the deposition of evaporites in the Phanerozoic history occurred in rather short episodes raises the question about the validity of the sulfate isotope curve with respect to representativeness and homogeneity. The specific causes for the variation of $\delta^{34}S$ of marine sulfate at a particular geologic time interval are still largely speculative (for further discussion see Strauss 1997).

A special case in the modern ocean sulfur isotope distribution is the isotopic composition of hydrogen sulfide in anoxic basins. Fry et al. (1991) compared depth profiles of the Black Sea and the

Cariaco Basin. Sulfide isotopic compositions in deep waters of the Black Sea were roughly constant near −41.5‰, while those of the Cariaco Trench averaged −31‰. In contrast, in the uppermost 50 m of the Black Sea's sulfidic waters (that is in water depths between 100 and 150m) the sulfide isotopic compositions change significantly in a region of sulfide consumption, increasing up to 5‰ vs. the deep-water background. These increases may be due to sulfide oxidation mediated by MnO_2 or oxygen, but are not consistent with sulfide oxidation by photosynthetic bacteria (Fry et al. 1991). Growth experiments with sulfate-reducing bacteria suggested that part of the increase in sulfide isotopic compositions could be attributed to rapid rates of sulfate reduction in the oxic/anoxic interface region.

10.6.2 $\delta^{34}S$ in Marine Sediments

Evaporitic Sulfate

The general aspects of isotopic fractionation during the evaporation of sulfates have been addressed in the previous section. Due to a minor fractionation during precipitation, marine evaporitic sulfates are assumed to reflect the original seawater signature. However, exceptions might occur in sulfates precipitated during late stages of evaporation. Sulfur isotope measurements from the Permian Zechstein in Germany (e.g. Nielsen and Ricke 1964) and experiments by Raab and Spiro (1991) indicate that sulfate deposited within the halite or even the potash-magnesia facies is depleted in ^{34}S in comparison to sulfate precipitated within the calcium sulfate facies. Progressive crystallization of ^{34}S-enriched sulfate minerals results in a residual brine depleted in ^{34}S, by possibly as much as 4‰. Therefore, to obtain a true record of seawater sulfate isotopic composition in the past no late-stage sulfates should be included.

Sedimentary Pyrite

Modern sedimentary pyrites are generally depleted in ^{34}S relative to ocean water sulfate with average $\delta^{34}S$-values around −20‰ and a range between +20‰ and −50‰. The majority of isotopic compositions of pyrites in modern sediments indicate the sedimentary sulfides having formed through bacterial sulfate reduction under open system conditions with respect to sulfate availability. These measurements agree well with ex-

perimental data as discussed above. However, higher values have been reported for pyrites formed under anoxic bottom-water conditions like in the Black Sea (Vinograd et al. 1962; Lyons and Berner 1992). They are interpreted to result from an open-system diagenesis combined with the effect of differential diffusion (Jørgensen et al. 2004).

During Earth history, the isotope fractionation between contemporaneous seawater sulfate and sedimentary reduced sulfur compounds like pyrite increases (Shen et al. 2001). With a few exceptions, sulfur isotope fractionations seem to have been negligible during the Archean but increased during the Proteozoic, presumably as a consequence of increased oceanic sulfate concentrations (Habicht et al. 2002). In sediment deposits younger than about 800 Myr, as in modern anoxic systems like the Black Sea, sulfur fractionations often exceed what appears to be the maximum reached during sulfate reduction in surface sediments as well as in cultures of sulfate reducers, as has been mentioned above. The isotope record for sedimentary sulfides of Phanerozoic age shows a much higher variability than observed for marine sulfates (Strauss 1997). Beside a long-term trend with a Cambrium maximum of about +2 ‰, a Permian to Cretaceous minimum around −30 ‰ and an increase to about −10 ‰ to −20 ‰ in the modern ocean sediments, there is a considerable variation as high as 80 ‰ within a single sedimentary unit. The observed range clearly reflects that in addition to the factors which control the sulfate isotopic composition in ancient seawater, changes in the geochemical conditions during sulfate reduction and pyrite formation may affect the $d^{34}S$ of sulfides. For example, changing bottom-water conditions (oxic to anoxic and vice versa) during the time of deposition may alter the conditions for pyrite formation. The amount of pyrite formed in sediments is known to be limited by (1) the amount of metabolizable organic matter (2) the amount of sulfate and (3) the amount of reactive iron. Under normal marine conditions, the amount of organic matter limits the reduction process (Raiswell and Berner 1986). This is documented in a positive correlation of organic carbon and pyrite sulfur through geologic history (Berner and Raiswell 1983). In anoxic environments, reactive iron appears to be the limiting factor for pyrite formation, since the amount of H_2S exceeds its removal by the reaction with iron (Raiswell and Berner 1985).

In that case, some H_2S diffuses back to the sediment surface, where it is reoxidized, whereas some H_2S undergoes further reactions. All these processes are associated with different fractionation effects on the sulfur isotopic composition of sulfides.

In addition to these primary, syngenetic effects, the isotope variability may reflect the diagenetic history of the sediments with a progressive evolution from early to late diagenetic sulfides (i.e. framboids to concretions or overgrowth). In summary, due to the large number of processes that might affect the isotopic composition of sulfides, their potential for interpreting features of the Phanerozoic sulfur cycle appears to be rather limited (Strauss 1997).

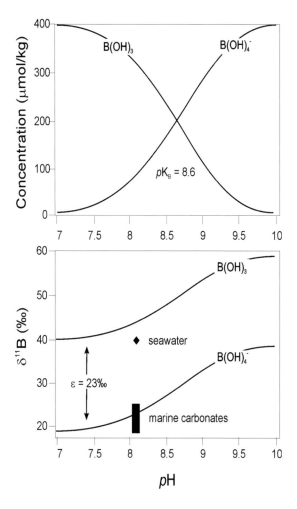

Fig. 10.12 Relative abundance of boron acid and borate in seawater, and their isotopic composition versus pH (according to Hemming and Hanson 1992).

10.7   Geochemical Influences on $^{11}B / ^{10}B$ Ratios

10.7.1 $\delta^{11}B$ of Seawater and Pore Waters

Principles of Fractionation

In ocean water, boron occurs in two dominant species, boric acid $B(OH)_3$ and borate $B(OH)_4^-$. The relative concentration of these two species is pH dependent such that the boric acid $B(OH)_3$ is dominant at pH < 9.0, whereas the tetrahedral complex of $B(OH)_4^-$ dominates at pH >9.0 (Fig.10.12; Hemming and Hanson 1992). The principal process that causes fractionation of boron isotopes in aqueous solutions is the following exchange reaction:

$$^{10}B(OH)_3 + {}^{11}B(OH)_4^- \Leftrightarrow {}^{11}B(OH)_3 + {}^{10}B(OH)_4^-$$

(10.24)

In seawater, $B(OH)_3$ is enriched in ^{11}B compared to $B(OH)_4^-$ by 23‰ at 25°C (Kakihana et al. 1977). Consequently, as the relative concentration of the dissolved species changes with pH, also their isotopic composition changes, but with a constant offset (Fig. 10.12).

Modern Range of Values and Historical Variability

The $\delta^{11}B$ value of modern ocean water is constant at about 40‰, relative to average continental material like the borax at Searles Lake, California, which serves as a standard reference material for boron isotope analyses (see Tab. 10.1).The ^{11}B enrichment of seawater is explained with the preferential absorption of $^{10}B(OH)_4^-$ by hydrothermally altered basalts of the oceanic crust, by clay-rich marine sediments, and by the precipitation of carbonate minerals (e.g., Hemming and Hanson 1992). Typical values for geologically important reservoirs summarized by Hoefs (2004) range between -2 to +10‰ for (altered) oceanic basalts, -18 to +20‰ for clay-rich sediments, and +18 to +35‰ for marine evaporites. Varying $\delta^{11}B$ values in geologic history might reflect changes in the relative contribution of boron from different reservoirs, especially due to selective erosion of rocks or enhanced production and alteration of fresh marine basalts, or by changes in the oceanic pH value in the past (see separate paragraph below).

10.7.2   δ^{11}B in Marine Carbonates

Principles of Fractionation

Diverse biogenic and inorganic marine carbonates record boron isotopic compositions close to the boron isotopic composition of B(OH)$_4^-$ at modern seawater pH. Therefore, Hemming and Hanson (1992) suggested that it is the charged borate species that is preferably incorporated into marine carbonates. This interpretation has been confirmed by a number of empirical calibration studies on inorganic calcium carbonates (Hemming et al. 1995; Sanyal et al. 2000), corals (Hönisch et al. 2004; Reynaud et al. 2004), and planktonic foraminifers (Sanyal et al. 1996; Sanyal et al. 2001) over a wide range of culture water pH. Although biogenic carbonates and even inorganic-precipitated calcite show δ^{11}B offsets of up to 4‰ relative to the theoretical borate curve, all empirical calibration curves follow the shape of the theoretical curve and therefore support the validity of the calculated isotope fractionation factor by Kakihana et al. (1977) (cf. Fig. 10.13).

However, the causes for these offsets are yet not completely understood. Culture experiments using living planktonic foraminifers (Hönisch et al. 2003) and numerical modeling experiments (Zeebe et al. 2003) identified physiological processes, such as symbiont photosynthetic activity, respiration and calcification, to be important. These processes influence the pH in the microenvironment of foraminifers and corals, and lead to specific offsets relative to seawater pH. Because the observed offsets between different carbonate producers are constant across the pH-range investigated, it is likely that the boron isotope fractionation related to these physiological processes is not influenced by the varying pH itself. Nevertheless, an important consequence resulting from these observations is the effect on shell precipitation during the life cycle of foraminifers. δ^{11}B data measured in different shell size classes of the symbiont-bearing planktonic foraminifer species *Globigerinoides sacculifer* showed a significant increase of δ^{11}B with increasing shell size (Hönisch and Hemming 2004), which could only be explained by a deeper growth habitat of smaller shells, where symbiont photosynthetic activity is reduced because of lower light levels. In addition to the original shell size effect, Hönisch and Hemming (2004) discovered that partial shell dissolution significantly lowers δ^{11}B. Because of the greater dissolution susceptibility of smaller shells the diagenetic effect on δ^{11}B was even more pronounced in shells of decreasing shell size (Fig. 10.14).

Application as a paleo-pH proxy

One of the first studies that the boron isotopic composition in planktonic foraminifera shells reflects climate driven changes in marine carbonate chemistry was provided by Sanyal et al. (1995), who demonstrated an approximately 0.2 unit higher pH in the Last Glacial Maximum surface water compared to today. Other surface ocean boron isotope studies estimated the strength of modern and past upwelling areas in the Holocene and late Pleistocene. In a comparative study of the northwest African and eastern equatorial Pacific upwelling zones Sanyal and Bijma (1999) proposed that the eastern equatorial Pacific upwelling system was a significantly larger source of CO$_2$ to the atmosphere during the last glacial period compared to the eastern Atlantic upwelling zone. Similarly, Palmer and Pearson (2003) reconstructed surface water pH and aqueous pCO$_2$ in the western equatorial Pacific over the past 23 ky and suggested that this area was a strong source of CO$_2$ to the atmosphere during the last deglaciation.

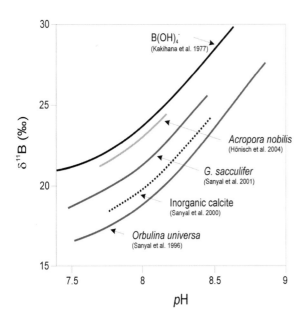

Fig. 10.13 Empirical δ^{11}B calibrations of laboratory precipitated inorganic calcite and cultured biogenic carbonates vs. pH.

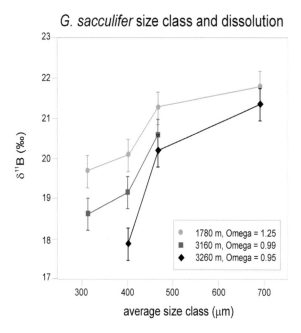

Fig. 10.14 $\delta^{11}B$ of symbiont-bearing planktonic foraminifer species from core-top samples at the different water depth on the Ontong Java Plateau, western equatorial Pacific (Hoenisch and Hemming 2004). Ω values indicate the carbonate saturation with respect to calcite at different sites.

Although still limited in their spatial coverage, the studies cited above provide strong support for using the boron isotope pH proxy to evaluate the role of different ocean parts to estimate past sources and sinks of CO_2 to the atmosphere. However, the application of the boron isotope proxy in mixed benthic foraminifera species for deep ocean pH reconstruction led to some doubt, because it suggested a 0.3 units higher glacial pH (i.e., +2.5‰ in $\delta^{11}B$) compared to modern deep water pH (Sanyal et al. 1995), which is not corroborated by sedimentary records of glacial calcite preservation suggesting no significant change in deep ocean carbonate ion concentration and pH. But this too large increase in pH was most probably due to the use of mixed benthic species instead of a single species, only (B. Hoenisch, pers. comm.).

Pearson and Palmer (2000) extended the $\delta^{11}B$ application to the Paleogene and estimated a pH of 7.4 and corresponding pCO_2 values higher than 2000 ppmv for the early Eocene time period. However, several uncertainties may limit the confidence in their interpretation, because of the residence time of boron in the ocean of only 3-5 million years (Lemarchand et al. 2000) and the use of extinct

foraminifer species without a significant crosscalibration against modern species. So far, such records need to be interpreted with caution, and considerable efforts will have to be made to improve confidence in pre-Quaternary reconstructions.

Acknowledgements

This is contribution No 0335 of the Research Center Ocean Margins (RCOM) which is financed by the Deutsche Forschungsgemeinschaft (DFG) at Bremen University, Germany.

10.8   Problems

Problem 1

Antarctic and Greenland ice sheets store today about 2 % of the ocean water mass with an average $\delta^{18}O$ value of -45‰. What would be the $\delta^{18}O_w$ value of an ice-free world?

Problem 2

A number of authors have proposed that the 0.3‰ lower $\delta^{13}C$ mean ocean value during the last glacial maximum (about 21000 years ago) resulted from an increased erosion of organic carbon from terrigenous soils and shelf sediments during a drop in sea-level as a result of continental glaciation. Assuming an oceanic carbon mass of 38,000 Gt and an average $\delta^{13}C$ value of terrestrial organic carbon of -25‰, calculate the transfer of organic carbon from the terrestrial to the oceanic reservoir necessary to explain the carbon isotope shift. Is the result realistic? What other explanations would be reasonable?

Problem 3

Given the range of b-values (equation 10.9) between 80 and 160 ‰µM as derived from South Atlantic core tops (Schulte et al. 2003), what would be its effect on atmospheric pCO_2 assuming an ε_p of 12‰, an enzymatic fractionation ε_f of 25‰, and a solubility coefficient $\alpha = 0.032$ at 20°C?

Problem 4

Seasonal $\delta^{15}N$ values of particulate organic matter collected in a sediment trap off Namibia varied between 3‰ at the end of spring and 8‰ during mid fall. Applying the principles of Rayleigh fractionation to the nitrate utilization in ocean surface water (eq. 10.22), what is the range of nitrate utilization assuming an initial $\delta^{15}NO_3$ value of 10‰ and an ε_u of 10‰? How to interpret those data in terms of seasonal changes in productivity?

Problem 5

Last glacial boron isotope composition of planktonic foraminifers has been shown to be 2‰ higher compared to the modern value of 23‰. What is the result for surface ocean pH during the last glacial? What happened to the carbonate ion concentration and consequently to atmospheric pCO_2 assuming the ΣCO_2 to be the same as today (see chapter 9)?

References

Adkins, J.F., McIntyre, K., Schrag, D.P., 2002. The salinity, temperature, and $\delta^{18}O$ of the glacial deep ocean. Science 298: 1769-1773.

Altabet, M., 1988. Variations in nitrogen isotopic composition between sinking and suspended particles: implications for nitrogen cycling and particle transformation in the open ocean. Deep-Sea Research 35: 535-554.

Altabet, M., Francois, R., 1994. Sedimentary nitrogen isotopic ratio as recorder for surface ocean nitrate utilization. Global Biogeochemical Cycles 8: 103-116.

Altabet, M., McCarthy, J., 1985. Temporal and spatial variation in the natural abundance of ^{15}N in PON from a warm-core ring. Deep-Sea Research 32: 755-722.

Altabet, M., Curry, W.B., 1989. Testing models of past ocean chemistry using foraminifera $^{15}N/^{14}N$. Global Biogeochemical Cycles 3: 107-119.

Altabet, M., Deuser, W.G., Honjo, S., Stienen, C., 1991. Seasonal and depth-related changes in the source of sinking particles in the North Atlantic. Nature 354: 136-139.

Altenbach, A.v., Sarnthein, M., 1989. Productivity record in benthic foraminifera. In: Berger, W.H., Smetacek, V.S., Wefer, G., (eds), Productivity of the ocean: Present and Past. Wiley & Sons, New York, pp 255-269.

Andersen, N., Müller, P.J., Kirst, G., Schneider, R.R., 1999. Late Quaternary PCO_2 variations in the Angola Current inferred from alkenone $\delta^{13}C$ and carbon demand estimated by $\delta^{15}N$. In: Fischer, G., Wefer, G., (eds), Proxies in Paleoceanography: Examples from the South Atlantic. Springer, Berlin, pp 469-488.

Arthur, M.A., Anderson, T.F., Kaplan, I.R., Veizer, J., Land, L.S., 1983. Stable isotopes in sedimentary geology. SEPM Short Course No. 10. SEPM, Tulsa, OK, 432 pp.

Bemis, B.E., Spero, H.J., Bijma, J., Lea, D.W., 1998. Reevaluation of the oxygen isotopic composition of planktonic foraminifera: Experimental results and revised paleotemperature equations. Paleoceanography 13: 150-160.

Berger, W.H., Smetacek, V.S., Wefer, G., 1989. Productivity of the ocean: Present and Past. Wiley & Sons, New York, 471 pp.

Berger, W.H., Vincent, E., 1986. Deep-sea carbonates: reading the carbon-isotope signal. Geol Rundsch 75: 249-269.

Berner, R.A., Raiswell, R., 1983. Burial of organic carbon and pyrite sulfur in sediments over Phanerozoic time: a new theory. Geochimica et Cosmochimica Acta 47: 885-892.

Bickert, T., Pätzold, J., Samtleben, C., Munnecke, A., 1997. Paleoenvironmental changes in the Silurian indicated by stable isotopes in brachiopod shells from Gotland, Sweden. Geochimica et Cosmochimica Acta 61: 2717-2739.

Bickert, T., Mackensen, A., 2003. Last Glacial to Holocene changes in South Atlantic deep water circulation. In: Wefer, G., Mulitza, S., Rathmeyer, V., (eds), The South Atlantic during the Late Quaternary. Springer, Berlin, pp. 671-695.

Bickert, T., Wefer, G., 1999. South Atlantic and benthic foraminifer δ^{13}C-deviations: Implications for reconstructing the Late Quaternary deep-water circulation. Deep-Sea Research, 46: 437-452.

Bidigare, R.R., Fluegge, A., Freeman, K.H., Hanson, K.L., Hayes, J.M., Hollander, D., Jasper, J., King, L.L., Laws, E.A., Milder, J., Millero, F.J., Pancost, R., Popp, B.N., Steinberg, P.A., Wakeham, S.G., 1997. Consistent fractionation of ^{13}C in nature and in the laboratory: Growth rate effects in some haptophyte algae. Global Biogeochem Cycles 11: 279-292.

Birchfield, G.E., 1987. Changes in deep-ocean water δ^{18}O and temperature from last glacial maximum to the present. Paleoceanography 2: 431-442.

Bolliger, C., Schroth, M.H., Bernasconi, S.M., Kleikemper, J., and Zeyer, J., 2001. Sulfur isotope fractionation during microbial sulfate reduction by toluene-degrading bacteria. Geochimica et Cosmochimica Acta 65: 3289–3298.

Broecker, W.S., 1982. Ocean chemistry during glacial time. Geochimica et Cosmochimica Acta 46: 1689-1705.

Broecker, W.S., Maier-Reimer, E., 1992. The influence of air and sea exchange on the carbon isotope distribution in the sea. Global Biogeochemical Cycles 6: 315-320.

Burkhardt, S., Riebesell, U., Zondervan, I., 1999. Stable carbon isotope fractionation by marine phytoplankton in response to daylength, growth rate, and CO_2 availability. Marine Ecology Progress Series 184: 31-41.

Canfield, D.E., 2001. Biogeochemistry of sulfur isotopes. Reviews in Mineralogy and Geochemistry 43: 607-636.

Carpenter, S.J., Lohmann, K.C., Holden, P., Walter, L.M., Huston, T.J., Halliday, A.N., 1991. δ^{18}O values, ^{87}Sr/^{86}Sr and Sr/Mg ratios of Late Devonian abiotic marine calcite: Implications for the composition of ancient seawater. Geochimica et Cosmochimica Acta 55: 1991-2010.

Charles, C.D., Fairbanks, R.G., 1990. Glacial to interglacial changes in the isotopic gradients of the Southern Ocean surface water. In: Bleil, U., Thiede, J., (eds), Geological history of the Polar Oceans: Artic versus Antarctic. Kluwer, Dordrecht, pp 519-538.

Cifuentes, L.A., Sharp, J.H., Fogel, M.L., 1988. Stable carbon and nitrogen isotope biogeochemistry in the Delaware estuary. Limnology and Oceanography 33: 1102-1115.

Claypool, G.E., Holser, W.T., Kaplan, I.R., Sakai, H., Zak, I., 1980. The age curves of sulfur and oxygen isotopes in marine sulfate and their mutual interpretation. Chemical Geology 28: 190-260.

Cline, J.D., Kaplan, I.R., 1975. Isotopic fractionation of dissolved nitrate during denitrification in the easter tropical North Pacific Ocean. Marine Chemistry 3: 271-299.

Coplen, T.B., 1996. More uncertainty than necessary. Paleoceanography 11: 369-370.

Craig, H., Gordon, L.I., 1965. Deuterium and oxygen-18 variations in the ocean and marine atmosphere. In: Tongiorgi, E., (eds), Stable isotopes in oceanic studies and paleotemperatures. Consiglio Nazionale Delle Ricerche, Laboratorio di Geologia Nucleare, Pisa, pp 9-130.

Curry, W.B., Oppo, D.W., 2005. Glacial water mass geometry and the distribution of δ^{13}C of ΣCO_2 in the western Atlantic Ocean. Paleoceanography 20: PA1017.

de Lange, G.J., van Os, B., Pruysers, P.A., Middelburg, J.J., Castradori, D., van Santvoort, P., Müller, P.J., Eggenkamp, H., Prahl, F.G., 1994. Possible early diagenetic alteration of palaeo proxies. In: Zahn, R., Pedersen, T.F., Kaminski, M.A., Labeyrie, L., (eds), Carbon cycling in the glacial ocean: Constraints on the ocean´s role in global climate. Springer, Berlin, pp 225-258.

Degens, E.T., 1969. Biogeochemistry of stable carbon isotopes. In: Eglington, G., Murphy, M.T.J., (eds), Organic geochemistry. Methods and results. Springer, Berlin, pp 304-329.

Degens, E.T., Guillard, R.R.L., Sackett, W.M., Hellebust, J.A., 1968. Metabolic fractionation of carbon isotopes in marine plankton. I. Temperature and respiration experiments. Deep Sea Research 15: 1-9.

Deines, P., 1981. The isotopic composition of reduced organic carbon. In: Fritz, P., Fontes, J.C., (eds), Handbook of environmental geochemistry, vol. 1., Elsevier, New York, pp 239-406.

Derry, L.A., France-Lanord, C., 1996. Neogene growth of the sedimentary organic carbon. Paleoceanography 11: 267-276.

Emrich, K., Ehhalt, D.H., Vogel, J.C., 1970. Carbon isotope fractionation during the precipitation of calcium carbonate. Earth and Planetary Science Letters 8: 363-371.

Epstein, S., Buchsbaum, R., Lowenstam, H.A., Urey, H.C., 1953. Revised carbonate-water isotopic temperature scale. Bull Geol Soc Am 64: 1315-1325.

Erez, J., Luz, B., 1983. Experimental paleotemperature equation for planktonic foraminifera. Geochimica et Cosmochimica Acta 47: 1025-1031.

Fairbanks, R.G., 1989. A 17,000-year glacio-eustatic sea level record: Influence of glacial melting rates on the Younger Dryas event and deep-ocean circulation. Nature 342: 637-642.

Fairbanks, R.G., Charles, C.D., Wright, J.D., 1992. Origin of global meltwater pulses. In: Taylor, R.E., (eds), Radiocarbon after four decades. Springer, New York, pp 473-500.

Faure, G., Mensing, T.M., 2005. Isotopes: Principles and applications, 3rd edition. John Wiley & Sons, New York, 928 pp.

Fontugne, M.R., Calvert, S.E., 1992. Late Pleistocene variability of the carbon isotopic composition of organic matter in the eastern Mediterranean: Monitor of changes in carbon sources and atmospheric CO_2 concentrations. Paleoceanography 7: 1-20.

Francois, R., Altabet, M.A., Burckle, L.H., 1992. Glacial to interglacial changes in surface nitrate utilization in the Indian sector of the Southern Ocean as recorded by sediment $\delta^{15}N$. Paleoceanography 7: 589-606.

Freudenthal, T., Neuer, S., Meggers, H., Davenport, R., Wefer, G., 2001. Influence of lateral particle advection and organic matter degradation on sediment accumulation and stable nitrogen isotope ratios along a productivity gradient in the Canary Islands region. Marine Geology, 177: 93-109.

Fry, B., Jannasch, H.W., Molyneaux, S.J., Wirsen, C.O., Muramato, Ja., King, S., 1991. Stable isotopes of the carbon, nitrogen and sulfur cycles in the Black Sea and the Cariaco Trench. Deep-Sea Research 38, Suppl. 2: S1003-S1019.

Gruber, N., Sarmiento, J.L., 1997. Global patterns of marine nitrogen fixation and denitrification. Global Biogeochem Cycles 11: 235-266.

Habicht, K.S., Canfield, D.E., 1997. Sulfur isotope fractionation during bacterial sulfate reduction in organic-rich sediments. Geochimica et Cosmochimica Acta 61: 5351-5361.

Hartmann, M., Nielsen, H., 1969. ^{34}S Werte in rezenten Meeressedimenten und ihre Deutung am Beispiel einiger Sedimentprofile aus der westlichen Ostsee. Geologische Rundschau 58: 621-655.

Hayes, J.M., 1993. Factors controlling the ^{13}C contents of sedimentary organic compounds: principals and evidence. Marine Geology 113: 111-125.

Hayes, J.M., Freeman, K.H., Popp, B.N., Hoham, C.H., 1990. Compound-specific analyses, a novel tool for reconstruction of ancient biogeochemical processes. Organic Geochemistry 16: 1115-1128.

Hemming, N.G., Hanson, G.N., 1992. Boron isotopic composition and concentration in modern marine carbonates. Geochimica et Cosmochimica Acta 56: 537-543.

Hemming, N.G., Hanson, G.N. 1994. A procedure for the isotopic analysis of boron by negative thermal ionization mass spectrometry. Chemical Geology 114: 147–156.

Hemming, N.G., Reeder, R.J., Hanson, G.N., 1995. Mineral-fluid partitioning and isotopic fractionation of boron in synthetic calcium carbonate, Geochimica et Cosmochimica Acta 59: 371-379.

Hoefs, J., 2004. Stable isotope geochemistry. 5th ed. Springer, Berlin, 244 pp.

Holland, H.D., 1978. The chemistry of the atmosphere and oceans. Wiley, New York, 351 pp

Holmes, M.E., Müller, P.J., Schneider, R.R., Segl, M., Wefer, G., 1997. Reconstruction of past nutrient utilization in the eastern Angola Basin based on sedimentary $^{15}N/^{14}N$ ratios. Paleoceanography 12: 604-614.

Holser, W.T., 1997. Geochemical events documented in inorganic carbon isotopes. Palaeogeography, Palaeoclimatology, Palaeoecology 132: 173-182.

Hönisch, B., Bijma, J., Russell, A.D., Spero, H.J.,

Palmer, M.R., Zeebe, R.E., Eisenhauer, A., 2003. The influence of symbiont photosynthesis on the boron isotopic composition of foraminifera shells. Mar. Micropaleontol. 49: 87-96.

Hönisch B, Hemming NG (2004) Ground-truthing the boron isotope paleo-pH proxy in planktonic foraminifera shells: Partial dissolution and shell size effects. Paleoceanography 19: PA4010.

Hönisch, B., Hemming, N.G., Grottoli, A.G., Amat, A., Hanson, G.N., Bijma, J., 2004. Assessing scleractinian corals as recorders for paleo-pH: Empirical calibration and vital effects. Geochimica et Cosmochimica Acta 68: 3675-3685.

Irwin, H., Curtis, C., Coleman, M., 1977. Isotopic evidence for source of diagenetic carbonates formed during burial of organic-rich sediments. Nature 269: 209-213.

Jacobs, S.S., Fairbanks, R.G., Horibe, Y., 1985. Origin and evolution of water masses near the antarctic continental margin: Evidence from $H_2^{18}O/H_2^{16}O$ ratios in seawater. Antarctic Research Series 43: 59-85.

Jasper, J.P., Hayes, J.M., 1990. A carbon isotope record of CO_2 levels during the late Quaternary. Nature 347: 462-464.

Jasper, J.P., Hayes, J.M., 1994. Reconstruction of Paleoceanic PCO_2 levels from carbon isotopic compositions of sedimentary biogenic components. In: Zahn, R., Pedersen, T.F., Kaminski, M.A., Labeyrie, L., (eds), Carbon cycling in the glacial ocean: Constraints on the ocean's role in global climate. Springer, Berlin, pp 323-342.

Jørgensen, B.B., 1979. A theoretical model of the stable sulfur isotope distribution in marine sediments. Geochimica et Cosmochimica Acta 43: 363-374.

Jørgensen, B.B., 1990. A thiosulfate shunt in the sulfur cycle of marine sediments. Science 249: 152-154.

Jørgensen, B.B., Erez, J., Revsbech, N.P., Cohen, Y., 1985. Symbiontic photosynthesis in a planktonic foraminifera, *Globigerinoides sacculifer* (Brady), studied with microelectrodes. Limnol Oceanogr 30: 1253-1267.

Jørgensen, B.B., Böttcher, M.E., Lüschen, H., Neretin, L.N., Volkov, I.I., 2004. Anaerobic methane oxidation and a deep H_2S sink generate isotopically heavy sulfides in Black Sea sediments. Geochimica et Cosmochimica Acta 68: 2095-2118.

Kakihana, H., Kotaka, M., Satoh, S., Nomura, M., Okamoto, M., 1977. Fundamental studies on the ion-exchange of boron isotopes. Bull. Chem. Soc. Japan 50: 158-163.

Kaplan, I.R., Rittenberg, S.C., 1964. Microbiological fractionation of sulfur isotopes. J Gen Microbiol 34: 195-212.

King, A.L., Howard, W.R., 2004. Planktonic foraminiferal $\delta^{13}C$ records from Southern Ocean sediment traps: New estimates of the oceanic Suess effect. Global Biogeochemical Cycles 18, GB2007.

Kroopnick, P., 1985. The distribution of ^{13}C of ΣCO_2 in the world oceans. Deep-Sea Research 32: 57-84.

Kuypers, M.M.M., Lavik, G., Woebken, D., Schmid, M.,

Fuchs, B.M., Amann, R., Jørgensen B.B., Jetten, M.S.M., 2005. Massive nitrogen loss from the Benguela upwelling system through anaerobic ammonium oxidation. Proceedings of the National Academy of Science 102: 6478-6483.

Kuypers, M.M.M., Sliekers, A.O., Lavik, G., Schmid, M., Jørgensen, B.B. Kuenen, J.G., Sinninghe Damsté, J.S., Strous, M., Jetten, M.S.M, 2003. Anaerobic ammonium oxidation by anammox bacteria in the Black Sea. Nature 422: 608-611.

Kyser, T.K., 1995. Micro-analytical techniques in stable isotope geochemistry. Can Mineral 33: 261-278.

Lawrence, J.R., 1989. The stable isotope geochemistry of deep-sea pore water. In: Fritz, P., Fontes, J.C., (eds), Handbook of environmental isotope geochemistry, vol. 3. Elsevier, Amsterdam, pp 317-356.

Laws, E.A., Popp, B.N., Bidigare, R.R., Kennicutt, M.C., Macko, S.A., 1995. Dependence of phytoplankton carbon isotopic composition on growth rate and $(CO_2)_{aq}$: Theoretical considerations and experimental results. Geochimica et Cosmochimica Acta 59: 1131-1138.

Lecuyer, C., Grandjean, P., Reynard, B., Albarede, F., Telouk, P., 2002. $^{11}B/^{10}B$ analysis of geological materials by ICP-MS PLasma 54: application to boron fractionation between brachiopod calcite and seawater. Chem Geol 186: 45-55.

Lemarchand, D., Gaillardet, J., Lewin, E., Allègre, C., 2000. The influence of rivers on marine boron isotopes and implications for reconstructing past ocean pH. Nature 408: 951-954.

Liu, K.K., Kaplan, I.R., 1989. The eastern tropical Pacific as a source of ^{15}N-enriched nitrate in seawater off southern California. Limnol Oceanography 34: 820-830.

Longinelli, A., 1989. Oxygen-18 and sulphur-34 in dissolved oceanic sulphate and phosphate. In: Fritz, P., Fontes, J.C., (eds), Handbook of environmental isotope geochemistry, vol. 3. Elsevier, Amsterdam, pp 219-255.

Lynch-Stieglitz. J., Stocker, T.F., Broecker, W.S., Fairbanks, R.G., 1995. The influence of air-sea exchange on the isotopic composition of oceanic carbon: observations and modeling. Global Biogeochemical Cycles 9: 653-665.

Lyons, T.W., Berner, R.A., 1992. Carbon-sulfur-iron systematics of the uppermost deep-water sediments of the Black Sea. Chemical Geology 99: 1-27.

Mackensen, A., 2001. Oxygen and carbon stable isotope tracers of Weddell Sea water masses: New data and some paleoceanographic implications. Deep-Sea Research 48: 1401-1422.

Mackensen, A., Hubberten, H.W., Bickert, T., Fischer, G., Fütterer, D.K. 1993. $\delta^{13}C$ in benthic foraminiferal tests of *Fontbotia wuellerstorfi* (SCHWAGER) relative to $\delta^{13}C$ of dissolved inorganic carbon in Southern Ocean deep water: implications for glacial ocean circulation models. Paleoceanography 8: 587-610.

Mackensen, A., Hubberten, H.W., Scheele, N., Schlitzer, R., 1996. Decoupling of $\delta^{13}C_{\Sigma CO2}$ and phosphate in recent Weddell Sea Deep and Bottom Water:

Implications for glacial Southern Ocean paleoceanography. Paleoceanography 11: 203-215.

Matsumoto, R., 1992. Causes of the oxygen isotopic depletion of interstitial waters from sites 798 and 799, Japan Sea, Leg 128. Proc ODP, Sci Res 127/128: 697-703.

McConnaughey, T.A., Burdett, J., Whelan, J.F., Paull, C.K., 1997. Carbon isotopes in biological carbonates: Respiration and photosynthesis. Geochimica et Cosmochimica Acta 61: 611-622.

McCorkle, D.C., Corliss, B.H., Farnham, C.A., 1997. Vertical distributions and stable isotopic compositions of live (stained) benthic foraminifera from the North Carolina and California continental margins. Deep-Sea Research 44: 983-1024.

McCorkle, D.C., Emerson, S.R., 1988. The relationship between pore water carbon isotopic composition and bottom water oxygen concentration. Geochimica et Cosmochimica Acta 52: 1169-1178.

McCorkle, D.C., Emerson, S.R., Quay, P.D., 1985. Stable carbon isotopes in marine porewaters. Earth and Planetary Science Letters 74: 13-26.

McCorkle, D.C., Keigwin, L.D., Corliss, B.H., Emerson, S.R., 1990. The influence of microhabitats on the carbon isotopic composition of deep-sea benthic foraminifera. Paleoceanography 5: 161-185.

McCorkle, D.C., Martin, P.A., Lea D.W., Klinkhammer, G.P., 1995. Evidence of a dissolution effect on benthic foraminiferal shell chemistry: $\delta^{13}C$, Cd/Ca, Ba/Ca, and Sr/Ca results from the Ontong Java Plateau. Paleoceanography 10: 699-714.

Merrit, D.A., Hayes, J.M., 1994. Factors controlling precision and accuracy in isotope-ratio-monitoring mass spectrometry. Anal Chem 66: 2336-2347.

Miyake, Y., Wada, E., 1971. The isotope effect on the nitrogen in biochemical oxidation-reduction reactions. Rec Oceanogr Works Japan 11: 1-6.

Montoya, J.P., 1994. Nitrogen fractionation in the modern ocean: Implications for the sedimentary record. In: Zahn, R., Pedersen, T.F., Kaminski, M.A., Labeyrie, L., (eds), Carbon cycling in the glacial ocean: Constraints on the ocean's role in global change. Springer, Berlin, pp 259-279.

Mook, W.G., Bommerson, J.C., Staverman, W.H., 1974. Carbon isotope fractionation between dissolved bicarbonate and gaseous carbon dioxide. Earth and Planetary Science Letters 22: 169-176.

Mulitza, S., Donner, B., Fischer, G., Paul, A., Pätzold, J., Rühlemann, C. Segl, M., 2003. The South Atlantic oxygen isotope record of planktic foraminifera. In: The South Atlantic in the Late Quaternary: Reconstruction of Material Budgets and Current Systems. Wefer, G., Mulitza, S., and Ratmeyer, V., editors. Springer, Berlin, pp. 121-142.

Müller, P.J., Schneider, R., Ruhland, G., 1994. Late Quaternary pCO2 variations in the Angola Current: Evidence from organic carbon $\delta^{13}C$ and alkenone temperatures. In: Zahn, R., Pedersen, T.F., Kaminski, M.A., Labeyrie, L., (eds), Carbon cycling in the glacial ocean: Constraints on the ocean's role in global change. Springer, Berlin, pp 343-366.

Newman, J.W., Parker, P.L., Behrens, E.W., 1973. Organic carbon ratios in Quaternary cores from the Gulf of Mexico. Geochimica et Cosmochimica Acta 37: 225-238.

Nielsen, H., Ricke, W., 1964. Schwefel-Isotopen-Verhältnisse von Evaporiten aus Deutschland: ein Beitrag zur Kenntnis von ^{34}S im Meerwasser-Sulfat. Geochimica et Cosmochimica Acta 28: 577-591.

Nissenbaum, A., Presley, B.J., Kaplan, I.R., 1972. Early diagenesis in a reducing fjord, Saanich Inlet, British Columbia - I. Chemical and isotopic changes in major components of interstitial water. Geochimica et Cosmochimica Acta 36: 1007-1027.

Ohmoto, H., Kaiser, C.J., Geer, K.A., 1990. Systematics of sulphur isotopes in recent marine sediments and ancient sediment-hosted basemetal deposits. Univ Western Australia Publ 23: 70-120.

O'Leary, M.H., 1981. Carbon isotope fractionation in plants. Phytochemistry 20: 553-567.

Palmer, M.R., Pearson, P.N., 2003. A 23,000-year record of surface water pH and PCO$_2$ in the Western Equatorial Pacific Ocean. Science 300: 480-482.

Park, R., Epstein, S., 1960. Carbon isotope fractionation during photosynthesis. Geochimica et Cosmochimica Acta 21: 110-126.

Pearson, P.N., Palmer, M.R., 2000. Atmospheric carbon dioxide concentrations over the past 60 million years. Nature 406: 695-699.

Pearson, P.N., Ditchfield, P.W., Singano, J., Harcourt-Brown, K.G., Nicholas, C.J., Olsson, R.K., Shackleton, N.J. and Hall, M.A., 2001. Warm tropical sea surface temperatures in the Late Cretaceous and Eocene epochs. Nature 413: 481-487.

Popp, B.N., Anderson, T.F., Sandberg, P.A., 1986. Brachiopods as indicators of original isotopic compositions in some Paleozoic limestones. Geological Society America Bulletin 97: 1262-1269.

Popp, B.N., Laws, E.A., Bidigare, R.R., Dore, J.E., Hanson, K.L., Wakeham, S.G., 1998. Effect of phytoplankton cell geometry on carbon isotopic fractionation. Geochimica et Cosmochimica Acta 62: 69-77.

Popp, B.N., Parekh, P., Tilbrook, B., Bidigare, R.R., Laws, E.A., 1997. Organic carbon δ^{13}C variations in sedimentary rocks as chemostratigraphic and paleoenvironmental tools. Palaeogeography, Palaeoclimatology, Palaeoecology 132: 119-132.

Popp, B.N., Tagiku, R., Hayes, J.M., Louda, J.W., Baker, E.W., 1989. The post-paleozoic chronology and mechanism of δ^{13}C depletion in primary marine organic matter. American Journal of Sciences 289: 436-454.

Raab, M., Spiro, B., 1991. Sulfur isotopic variations during seawater evaporation with fractional crystallization. Chemical Geology 86: 323-333.

Railsback, L.B., 1990. Influence of changing deep ocean circulation on the Phanerozoic oxygen isotopic record. Geochimica et Cosmochimica Acta 54: 1501-1509.

Raiswell, R., 1997. A geochemical framework for the application of stable sulphur isotopes to fossil pyritization. Journal of the Geological Society, London 154: 343-356.

Raiswell, R., Berner, R.A., 1985. Pyrite formation in euxinic and semi-euxinic sediments. American Journal of Sciences 285: 710-724.

Raiswell, R., Berner, R.A., 1986. Pyrite and organic matter in Phanerozoic normal marine shales. Geochimica et Cosmochimica Acta 50: 1967-1976.

Rau, G.H., Froehlich, P.N., Takahashi, T., Des Marais, D.J., 1991. Does sedimentary organic δ^{13}C record variations in Quaternary ocean [CO$_{2(aq)}$]? Paleoceanography 6: 335-347.

Rau, G.H., Riebesell, U., Wolf-Gladrow, D., 1997. CO$_{2aq}$-dependent photosynthetic ^{13}C fractionation in the ocean: A model versus measurements. Global Biogeochemical Cycles 11: 267-278.

Reynaud, S., Hemming, N.G., Juillet-Leclerc. A., Gattuso. J.P., 2004. Effect of pCO$_2$ and temperature on the boron isotopic composition of a zooxanthellate coral: Acropora sp. Coral Reefs 23: 539-546.

Rink, S., Kühl, M., Bijma, J., and Spero, H. J., 1998. Microsensor studies of photosynthesis and respiration in the symbiontic foraminifer O. universa.. Marine Biology, 131: 583-596.

Rost, B., Zondervan, I., Riebesell, U., 2002. Light-dependent carbon isotope fractionation in the coccolithophorid Emiliania huxleyi. Limnol. Oceanogr. 47: 120-128.

Rühlemann, C., Frank, M., Hale, W., Mangini, A., Mulitza, S., Müller, P.J., Wefer, G., 1996. Late Quaternary productivity changes in the western equatorial Atlantic: Evidence from ^{230}Th-normalized carbonate and organic carbon accumulation rates. Marine Geology 135: 127-152.

Sanyal, A., and Bijma, J., 1999. A comparative study of northwest Africa and eastern equatorial Pacific upwelling zones as sources of CO$_2$ during glacial periods based on boron isotope paleo-pH estimation, Paleoceanography 14: 753-759.

Sanyal, A., Hemming, N.G., Broecker, W.S., Lea, D.W., Spero, H.J., Hanson, G.N., 1996. Oceanic pH control on the boron isotopic composition of foraminifera: Evidence from culture experiments. Paleoceanography 11: 513-517.

Sanyal, A., Nugent, M., Reeder, R.J., Bijma, J., 2000. Seawater pH control on the boron isotopic composition of calcite: Evidence from inorganic calcite precipitation experiments. Geochim. Cosmochim. Acta. 64: 1551-1555.

Sanyal, A., Bijma, J., Spero, H.J., Lea, D.W., 2001. Empirical relationship between pH and the boron isotopic composition of G. sacculifer: Implications for the boron isotope paleo-pH proxy. Paleoceanography 16: 515-519.

Sanyal, A., Hemming, N.G., Hanson, G.N., Broecker, W.S., 1995. Evidence for a higher pH in the glacial ocean from boron isotopes in foraminifera. Nature 373: 234-236.

Sarnthein, M., Winn, K., Jung, S.J.A., Duplessy, J.C., Labeyrie, L., Erlenkeuser, H., Ganssen G., 1994. Changes in east Atlantic deepwater circulation over the last 30,000 years: Eight time slice reconstructions. Paleoceanography 9: 209-268.

Schmidt, G.A., Bigg, G.R., and Rohling, E.J., 1999. Global seawater oxygen-18 database. http://data.giss.nasa.gov/o18data/.

Schneider, R., Dahmke, A., Kölling, A., Müller, P.J., Schulz, H.D., Wefer, G., 1992. Strong deglacial minimum in the $\delta^{13}C$ record from planktonic foraminifera in the Benguela Upwelling Region: paleoceanographic signal or early diagenetic imprint. In: Summerhayes, C.P., Prell, W.L., Emeis, K.C., (eds), Upwelling systems: Evolution since the early Miocene. Geological Society Special Publication 63, pp 285-297.

Schrag, D.P., and DePaolo, D.J., 1993. Determination of $\delta^{18}O$ of seawater in the deep ocean during the last glacial maximum: Paleoceanography, 8: 1-6.

Schrag, D.P., DePaolo, D.J., and Richter, F.M., 1995. Reconstructing past sea surface temperatures correcting for diagenesis of bulk marine carbonate: Geochimica et Cosmochimica Acta 59: 2265-2278.

Schulte, S., Benthien, A., Andersen, N., Müller, P.J., Rühlemann, C., Schneider, R.R., 2003. Stable carbon isotopic composition of the C37:2 alkenone: A proxy for $CO_{2(aq)}$ concentration in oceanic surface waters? In: The South Atlantic in the Late Quaternary: Reconstruction of Material Budgets and Current Systems. G. Wefer, S. Mulitza and V. Ratmeyer, (eds), Springer, Berlin, pp. 195-211.

Shackleton, N.J., 1977. Tropical rainforest history and the equatorial Pacific carbonate dissolution cycles. In: Anderson, N.R., Malahoff, A., (eds), Fate in fossil fuel CO_2 in the oceans. Plenum, New York, pp 401-427.

Shackleton, N.J., Opdyke, N.D., 1973. Oxygen isotope and paleomagnetic stratigraphy of equatorial Pacific core V 28-238: Oxygen isotope temperatures and ice volumes on a 10^5 year scale. Quartenary Research 3: 39-55.

Sigman, D.M., Altabet, M., Francois, R., McCorkle, D.C., Gaillard, J.F., 1999. The isotopic composition of diatom - bound nitrogen in Southern Ocean sediments. Paleoceanography, 14, 118 – 134.

Spero, H.J., 1992. Do planktonic foraminifera accurately record shifts in the carbon isotopic composition of ΣCO_2? Marine Micropaleontology 19: 275-285.

Spero, H.J., Bijma, J., Lea, D.W., Bemis, B.E., 1997. Effect of seawater carbonate chemistry on planktonic foraminiferal carbon and oxygen isotope values. Nature 390: 497-500.

Spero, H.J., Lea, D.W., 1993. Intraspecific stable isotope variability in the planktic foraminifera *Globigerinoides sacculifer*: Results from laboratory experiments. Marine Micropaleontology 22: 221-234.

Spivack, A.J., Edmond, J.M., 1986. Determination of boron isotope ratios by thermal ionization mass spectrometry of the dicesium metaborate cation. Anal Chem 58: 31-35.

Strauss, H., 1997. The isotopic composition of sedimentary sulfur through time. Palaeogeography, Palaeoclimatology, Palaeoecology 132: 97-118.

Sweeney, R.E., Kaplan, I.R., 1980. Natural abundances of ^{15}N as a source indicator for near-shore marine

sedimentary and dissolved nitrogen. Marine Chemistry 9: 81-94.

Sweeney, R.E., Liu, K.K., Kaplan, I.R., 1978. Oceanic nitrogen isotopes and their uses in determining the source of sedimentary nitrogen. In: Robinson, B.W., (eds), Stable isotopes in earth sciences. Dept. Scientific and Industrial Research, Wellington, pp 9-26.

Valley, J.W., and Cole, D.R., (eds), 2001. Stable Isotope Geochemistry. Reviews in Mineralogy and Geochemistry, The Mineralogical Society of America, Blacksburg, vol. 43, pp. 662.

Veizer, J., Ala, D., Azmy, K., Bruckschen, P., Buhl, D., Bruhn, F., Carden, G.A.F., Diener, A., Ebneth, S., Godderis, Y., Jasper, T., Korte, C., Pawellek, F., Podlaha, O., Strauss, H., 1999. $^{87}Sr/^{86}Sr$, $\delta^{13}C$ and $\delta^{18}O$ evolution of Phanerozoic seawater. Chemical Geology 161, 59-88. With data update 2004: http://www.science.uottawa.ca/geology/isotope_data/.

Veizer, J., Bruckschen, P., Pawellek, F., Diener, A., Podlaha, O.G., Carden, G.A.F., Jasper, T., Korte, C., Strauss, H., Azmy, K., Ala, D., 1997. Oxygen isotope evolution of Phanerozoic seawater. Palaeogeography, Palaeoclimatology, Palaeoecology 132: 159-172.

Vinogradov, A.P., Grinenko, V.A., Ustinov, V.I., 1962. Isotopic composition of sulfur compounds in the Black Sea. Geokhimiya 10: 973-997.

Voss, M., Altabet, M.A., von Bodungen, B., 1996. $\delta^{15}N$ in sedimenting particles as indicator of euphotic-zone processes. Deep-Sea Research 43: 33-47.

Wada, E., Hattori, A., 1978. Nitrogen assimilation effects in the assimilation of inorganic nitrogenous compounds by marine diatoms. Geomicrobiol J 1: 85-101.

Wada, E., Minagawa, M., Mizutani, H., Tsuji, T., Imaizumi, R., Karasawa, K., 1987. Biogeochemical studies on the transport of organic matter along the Otsuchi River watershed, Japan. Estuarine, Coastal and Shelf Sciences 25: 321-336.

Waelbroeck, C., Mulitza, S., Spero, H., Dokken, T., Kiefer, T., Cortijo, E., 2004. A global compilation of Holocene planktonic foraminiferal $\delta^{18}O$: relationship between surface water temperature and $\delta^{18}O$, Quaternary Science Reviews 24: 853-868.

Wallmann, K., 2001. The geological water cycle and the evolution of marine $\delta^{18}O$ values. Geochimica et Cosmochimica Acta 65: 2469-2485.

Wefer, G., Berger, W.H., 1991. Isotope paleontology: growth and composition of extant calcareous species. Marine Geology 100: 207-248.

Wefer, G., Heinze, P.M., Berger, W.H., 1994. Clues to ancient methane release. Nature, 369: 282.

Weiss, R.F., 1974. Carbon dioxide in water and seawater: The solubility of a non-ideal gas. Marine Chemistry 2: 203-215.

Weiss, R.F., Östlund, H.G., Craig, H., 1979. Geochemical studies of the Weddell Sea. Deep-Sea Research 26: 1093-1120.

Williams, L.B., Ferell, R.E., Hutcheon, I., Bakel, A.J., Walsh, M.M., Krouse, H.R., 1995. Nitrogen isotope geochemsitry of organic matter and minerals during diagenesis and hydrocarbon migration. Geochimica et Cosmochimica Acta 59: 765-779.

Wu, G., Berger, W.H., 1989. Planktonic foraminifera: differential dissolution and the Quaternary stable isotope record in the west Equatorial Pacific. Paleoceanography 4: 181-198.

Zahn, R., Keir, R., 1994. Tracer-nutrient correlations in the upper ocean: observational and box model constraints on the use of benthic foraminiferal $\delta^{13}C$ and Cd/Ca as paleo-proxies for the intermediate-depth ocean. In: Zahn, R., Pedersen, T.F., Kaminski, M.A., Labeyrie, L., (eds), Carbon cycling in the glacial ocean: Constraints on the ocean's role in global change. Springer, Berlin, pp 195-223.

Zahn, R., Mix, A.C., 1991. Benthic foraminiferal $\delta^{18}O$ in the ocean's temperature-salinity-density field: Constraints on ice age thermohaline circulation. Paleoceanography 6: 1-20.

Zeebe, R.E., Wolf-Gladrow, D.A., Bijma, J., Hšnisch, B., 2003. Vital effects in foraminifera do not compromise the use of $d^{11}B$ as a paleo-pH indicator: Evidence from modeling. Paleoceanography 18: PA 1043.

Zeebe, R.E., and Sanyal, A., 2002. Comparison of two potential strategies of planktonic foraminifera for house building: Mg^{2+} or H^+ removal? Geochim. Cosmochim. Acta, 66: 1159-1169.

Zeebe, R.E., 1999. An explanation of the effect of seawater carbonate concentration on foraminiferal oxygen isotopes. Geochimica et Cosmochimica Acta 63: 2001-2007.

Zeebe, R.E., 2001. Seawater pH and isotopic paleo-temperatures of Cretaceous oceans. Palaeogeography, Palaeoclimatology, Palaeoecology, 170:49-57.

11 Manganese: Predominant Role of Nodules and Crusts

Geoffrey P. Glasby

11.1 Introduction

The importance of manganese in the marine environment can be deduced from the fact that it is the tenth most abundant element in the Earth's crust (av. conc. 0.093%) and is available in two valency states whose stability boundary lies within the range of the natural environment (Glasby 1984). Manganese oxides also have a high adsorption capacity. δMnO_2, for example, has a surface area of about 260 $m^2\,g^{-1}$ and a pH_{zpc} of 2.25 and can therefore adsorb cations such as Ni^{2+}, Cu^{2+} and Zn^{2+} from natural waters. By comparison, iron is the fourth most abundant element in the Earth's crust (av. conc. 5.17%) giving an average Mn/Fe ratio of 0.02. It also occurs in two valency states whose stability boundary lies within the range of the natural environment. Fe oxyhydroxides have a high adsorption capacity and large surface area. Goethite has a pH_{zpc} of 7.1 and can adsorb both cations (REE^{3+}) and anions (e.g. PO_4^{3-}, MoO_4^{2-} and WO_4^{2-}).

Both elements can therefore migrate under the influence of redox gradients. They can also fractionate from each other, particularly under acid or reducing conditions such as in lakes and shallow seas or in marine sediments. FeS_2 plays a major role in separating Mn and Fe in anoxic marine environments such as in Baltic Sea sediments where SO_4^{2-} ions are present. The reduced form of manganese, Ca-rhodochrosite (Ca-$MnCO_3$), is much rarer than FeS_2 but it is found in environments such as Gotland Deep sediments. Mn is more mobile than Fe. Mn therefore migrates more readily than Fe but Fe deposits more readily than Mn. This leads to the fractionation of Mn from Fe as, for example, in submarine hydrothermal systems or in anoxic environments such as Baltic Sea deeps.

The formation of huge quantities of manganese nodules and crusts on the deep-sea floor is a function of the fact that manganese and iron are relatively abundant in the Earth's crust and migrate from less oxidizing to more oxidizing environments. The increased glaciation of Antarctica about 12 Ma led to the initiation of Antarctic Bottom Water (AABW) flow and the increased ventilation of the deep ocean. The modern, well-oxygenated deep-sea has therefore become the ultimate repository for manganese. Deep-sea Mn nodules formed since the lower Miocene unconformity (12 Ma) hold about 10^{11} t of Mn (about 16 times the total Mn in terrestrial deposits) reflecting the importance of manganese nodules in the global cycle of manganese (Glasby 1988). The high contents of Co, Ni, Cu and Zn in manganese nodules and crusts make these deposits an important potential economic resource for these elements and are a function of the sorption characteristics of the manganese and iron oxyhydroxides.

11.2 Manganese, Iron and Trace Elements in Seawater

Manganese occurs in seawater mainly as Mn^{2+} or $MnCl^+$ (Bruland 1983). The dissolved Mn concentration in the open ocean is in range 0.2-3 nmol kg^{-1} which is above the equilibrium concentration with respect to MnO_2 or MnOOH. This situation reflects the slow rate of oxidation of Mn^{2+} in solution.

Figure 11.1 shows the vertical distribution of Mn in seawater at the VERTEX-IV site situated in the centre of the central North Pacific subtropical gyre, a region of low biological productivity (Bruland et al. 1994). The high content of Mn in the surface waters reflects the input of eolian material into this region from Asia. Photoreduction of particulate MnO_2 to Mn(II) takes place in the surface waters resulting in 99% of the Mn in the surface waters (0-100 m) being in the dissolved form. By contrast, only 80% of the Mn in the deep water (500-4,000 m) is in the dissolved form. Mn therefore

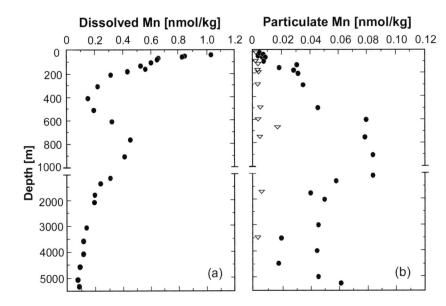

Fig. 11.1 Vertical profiles of (a) dissolved manganese (nmol kg^{-1}) and (b) particulate manganese (nmol kg^{-1}) at the VERTEX-IV site (after Bruland et al. 1994). For particulate manganese, filled circles represent the acetic acid leachable fraction and open triangles represent refractory manganese.

behaves in seawater as a scavenged-type metal. Bacterial mediation plays a key role in the scavenging and oxidation of dissolved Mn in intermediate and deep water. The higher concentration of particulate Mn in North Atlantic deep water (0.15 nmol kg^{-1}) than in central North Pacific deep water (0.05 nmol kg^{-1}) reflects the higher input of eolian material into the Atlantic compared to the Pacific Ocean.

Mn is also influenced by redox processes in the water column. Both dissolved and particulate Mn display maxima at the oxygen minimum zone which occurs at a depth of 500-1,000 m in the central North Pacific (Fig. 11.1). In this case, the maximum dissolved Mn concentration in the oxygen minimum zone is about 0.4 nmol kg^{-1}. This high concentration was considered to be the result of lateral transport of Mn from the continental margins to the open ocean along the oxygen minimum zone (Martin et al. 1985) and to be particularly important in the highly oxygen-deficient waters of the oxygen minimum zone in parts of the eastern North Pacific Ocean (Burton and Statham 1988). However, Johnson et al. (1996) have subsequently argued that a decrease in the oxidation rate of Mn (II) within the oxygen minimum zone is a more likely mechanism for the enrichment of manganese there, thus obviating the need for the lateral transport of Mn from the continental margins.

About 90% of the Mn introduced to the oceans has a hydrothermal origin (Glasby 1988). Hydrothermal Mn anomalies in seawater can be detected over 1,000 km from the source in the Pacific Ocean (Burton and Statham 1988). When hydrothermal fluids are discharged at the sea floor, a buoyant hydrothermal plume is formed on mixing of the hydrothermal fluid with seawater (Lilley et al. 1995; Lupton 1998; German and Van Damm 2004). The plume can rise tens to hundreds of meters above the sea floor to a level of neutral buoyancy where it forms a distinct hydrographic layer with a distribution extending tens to thousands of kilometers from the vent. The dilution factor of the vent fluid with respect to seawater is of the order of 10^4-10^5. In the first centimeters to meters above the vent, up to 50% of the iron is precipitated as sulfides. The chalcophile elements (Cu, Zn, Cd and Pb) tend to be incorporated in the sulfide minerals at this stage. The remaining Fe is precipitated over a longer time period as fine-grained iron oxyhydroxide particles. The half life for Fe (II) precipitation is 2-3 minutes. The iron oxyhydroxide particles scavenge anionic species such as HPO_4^{2-}, CrO_4^{2-}, VO_4^{2-} and $HAsO_4^{2-}$ as well as the rare earth elements (REE). Precipitation of particulate Mn oxides takes place much more slowly, mainly in the neutrally buoyant plume where the oxidation is bacterially mediated. Because of the slow precipitation rate, particulate Mn concentrations increase in the plume to a maximum 80-150 km from the vent. 80% of the hydrothermal Mn is deposited on the sea floor within several hundred km of the vent field but the remaining Mn still raises the background concentration of Mn in seawater severalfold (Lavelle et al. 1992). The residence time of hydrothermal Mn in seawater is several years. German and Angel (1995) have estimated that the total hydrothermal Mn flux to the oceans is 6.85·10^9 kg yr^{-1}. This compares with the flux of Mn from the rivers of $0.27 \cdot 10^9$ kg yr^{-1} (Elderfield and Schulz 1996) confirming the original calculation of Glasby (1988).

The predominance field of Mn in seawater (Glasby and Schulz 1999) is best illustrated by the use of an E_H, pH diagram (Fig. 11.2). At the conditions prevalent in seawater (E_H +0.4 V, pH 8), the stable form of Mn is seen to be the aqueous species, Mn^{2+}, and not any of the solid phases of Mn. The fact that Mn oxides are abundant on the seafloor can be explained on the basis that the Mn oxyhydroxides initially formed in seawater (β-manganite) are not pure mineral phases but have significant concentrations of transition elements and are fine grained, both of which help to stabilize them (Glasby 1974). In fact, 10 Å manganate and δMnO$_2$ are the principal manganese oxide minerals found in deep-sea manganese nodules (see section 11.4.8) but the free energies of formation of these fine-grained minerals have not been determined. However, the fact that the aqueous species of Mn appear more stable than any of the solid phases of Mn under deep-sea conditions probably explains the slow rate of oxidation of Mn in sea-water.

The kinetics of oxidation of Mn^{2+} in seawater have been discussed by Murray and Brewer (1977) and the following reaction sequence has been proposed.

$$Mn(II) + O_2 \rightarrow MnO_{2\,s} \qquad\qquad slow$$

$$Mn(II) + MnO_{2\,s} \rightarrow (Mn(II).MnO_2)_s \quad fast$$

$$(Mn(II).MnO_2)_s + O_2 \rightarrow 2MnO_{2\,s} \qquad slow$$

$$-d(Mn(II))/dt \rightarrow k_0(Mn(II)) + \\ k_1(Mn(II))(MnO_{2\,s})(P_{O2})(OH^-)^2$$

This rate law demonstrates that the oxidation of Mn (II) in seawater is autocatalytic. From this equation, it was calculated that it would take about 1,000 years to oxidize 90% of the Mn present in seawater. Surface catalysis on MnO$_2$ or FeOOH or bacterial oxidation is therefore required to increase the rate of deposition of Mn^{2+} from solution (Cowen and Bruland 1985; Mandernack et al. 1985; Ehrlich 1996; Hastings and Emerson 1986; Tebo et al. 1997, 2004). Giovanoli and Arrhenius (1988) also proposed that the surface catalyzed oxidation of Mn^{2+} by FeOOH is a rate-controlling step in the formation of marine manganese

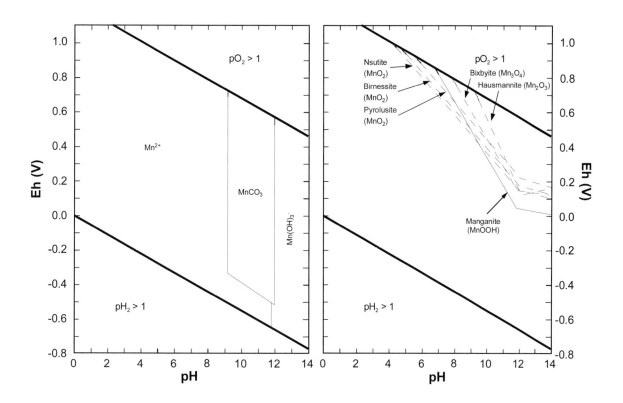

Fig. 11.2 E_H, pH diagram for Mn calculated for the chemical conditions prevailing in the deep sea (after Glasby and Schulz 1999). Note that, under seawater conditions and at an E_H, of +0.4 V and a pH of 8, the stable form of Mn is the aqueous species, Mn^{2+}, and not any of the solid phases of Mn.

deposits. Rate equations and rate constants for the oxidation of manganese in seawater based on field data have been determined by Yeats and Strain (1990), von Langen et al. (1997) and Morgan (2005).

By contrast, the form and concentration of iron in seawater remain poorly known because of the pronounced tendency of Fe(III) species to hydrolyze in aqueous solution (Bruland 1983) (see Chap. 7). Iron occurs in seawater mainly in the hydrolyzed forms $Fe(OH)_3^{\circ}$ and $Fe(OH)_2^{+}$ and as $FeCl^{+}$. The concentration range of Fe in seawater is 0.1-2.5 nmol kg^{-1} giving a Mn/Fe ratio of about unity which is much greater than that in the Earth's crust.

Figure 11.3 shows the vertical distribution of Fe in seawater at the VERTEX-IV site (Bruland et al. 1994). Dissolved Fe shows a maximum concentration in the surface mixed layer (0.35 nmol kg^{-1}) but this declines to a minimum in the subsurface stratified layer (70-100 m) (0.02 nmol kg^{-1}). The high concentration of Fe in the surface waters reflects the strong eolian input into this region. The Fe in the surface layer is rapidly consumed by plankton but is recycled within days. Below 100 m, dissolved Fe displays a nutrient-type distribution. In deep water, the dissolved Fe concentration reaches a value of 0.38 nmol kg^{-1} Fe. Iron is a limiting nutrient in open-ocean surface waters characterized by high nitrate and low chlorophyll contents. Proposals have been made to seed the oceans with Fe in order to stimulate phyto-plankton growth which would hopefully reduce the levels of atmospheric CO_2 (de Baar et al. 1995; Fitzwater et al. 1996). Johnson et al. (1997) have recently presented a model describing the factors controlling the dissolved Fe content in seawater.

Particulate Fe does not show an eolian influence in the surface layer and its distribution is more constant with depth. 48% of Fe in the surface water (0-100 m) is in the dissolved form and 55% in deep water (500 -4,000 m). Most of the particulate Fe in the intermediate and deep water is in the form of refractory alumino-silicate minerals of eolian origin. The higher concentration of particulate Fe in North Atlantic deep water (1.2 nmol kg^{-1}) than in central North Pacific deep water (0.3 nmol kg^{-1}) again reflects the higher input of eolian material into the Atlantic compared to the Pacific Ocean.

Fe behaves in seawater in part as a scavenged-type element and in part as a nutrient-type element as shown by the strong correlation of dissolved Fe with nitrate and phosphate at depths below 100 m.

The stability field of Fe in seawater is best illustrated by the use of an E_H, pH diagram (Fig. 11.4). Solid $Fe(OH)_3$ is shown as the metastable form of iron at the conditions prevalent in sea-water (E_H +0.4 V, pH 8). Actually, akagenéite (β-FeOOH) is the more stable form found in deep-sea manganese nodules (see Sect. 11.4.8) but its free energy of formation has not been determined.

Of the trace elements that are of most interest in the formation of manganese nodules, Co is present in deep ocean water at concentrations of <0.1 nmol kg^{-1}, mainly as Co^{2+}, $CoCO_3^{\circ}$ and $CoCl^{+}$ (Bruland 1983; Burton and Statham 1988; Nozaki 1997). Its extremely low concentration suggests that it is rapidly removed from seawater, probably scavenged by manganese oxides. Ni is present in deep ocean water at concentrations of about 10 nmol kg^{-1}, mainly as Ni^{2+}, $NiCO_3^{\circ}$

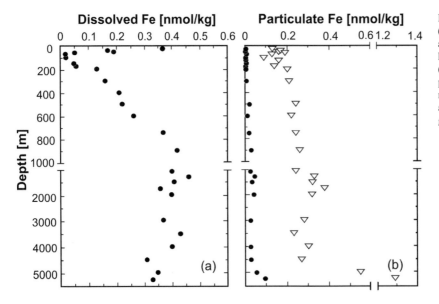

Fig. 11.3 Vertical profiles of (a) dissolved iron (nmol kg^{-1}) and (b) particulate iron (nmol kg^{-1}) at the VERTEX-IV site (after Bruland et al. 1994). For particulate iron, filled circles represent the acetic acid leachable fraction and open triangles represent refractory iron.

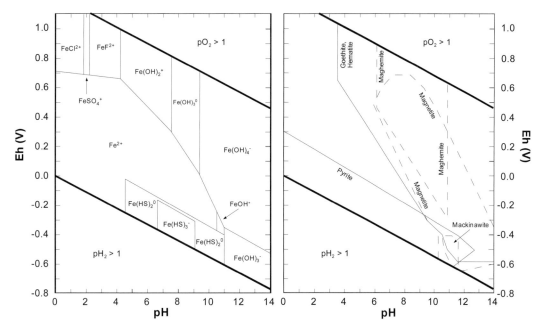

Fig. 11.4 E_H, pH diagram for Fe calculated for the chemical conditions prevailing in the deep sea (after Glasby and Schulz 1999). Note that, under seawater conditions and at an E_H, of +0.4 V and a pH of 8, the metastable form of Fe is $Fe(OH)_3$ and the stable forms magnetite ((Fe, Mg) Fe_2O_4) and maghemite (Fe_2O_3).

and $NiCl^+$, and displays a nutrient-type behavior in seawater. Cu is present in deep ocean water at concentrations of about 6 nmol kg^{-1}, mainly as $CuCO_3°$, $CuOH^+$ and Cu^{2+}, and has a distribution intermediate between that of nutrient-type elements and Mn. Zn is present in deep ocean water at concentrations of about 8 nmol kg^{-1}, mainly as Zn^{2+}, $ZnOH^+$, $ZnCO_3°$, and $ZnCl^{2+}$, and displays a nutrient-type behavior in seawater.

Trace metals in seawater can be characterized as scavenged-type or nutrient-type elements (Bruland et al. 1994; Bruland and Lohan, 2004). On this basis, Co is mainly a scavenged-type element, Ni and Zn are nutrient-type elements, Fe displays an intermediate behavior and Mn is a scavenged-type element strongly influenced by redox processes. Nutrient-type elements have much longer deep-sea residence times than scavenged-type elements. For Zn, this has been estimated to be 22,000 - 45,000 years, for Fe 70 - 140 years and for Mn 20 - 40 years. The residence times for Fe and Mn are short compared to the general oceanic turnover time of about 1,500 years (Bender et al. 1977). The rapid removal of Mn from seawater accounts for its significant fractionation between oceanic basins as well as the widespread occurrence of manganese deposits on the deep-sea floor.

11.3   Sediments

11.3.1   Manganese, Iron and Trace Elements in Deep-Sea Sediments

The distribution of elements in deep-sea sediments has been discussed by Bischoff et al. (1979), Stoffers et al. (1981, 1985), Meylan et al. (1982), Aplin and Cronan (1985), Baturin (1988), Chester (1990), Glasby (1991) and Miller and Cronan (1994).

Deep-sea sediments cover more than 50% of the earth's surface and consist of carbonates, red clay and siliceous ooze (cf. Chap. 1). On average, red clay covers about 31% of the world's ocean basins but its abundance is much higher in the Pacific (49%) than in the Atlantic (26%) and Indian (25%) Oceans (Glasby 1991). Carbonates act as a diluent for the transition elements in deep-sea sediments because of the low contents of these elements in them and the composition of deep-sea sediments is therefore often presented on a carbonate-free basis.

Red clays are mainly allogenic in origin (Glasby 1991). In the Pacific, this allogenic component is dominantly eolian dust. The high input of dust from the deserts of central Asia explains the higher sedimentation rates of the noncarbonate fraction of sediments from the North Pacific (0.5-6 mm ka^{-1}) than those of sediments from the South Pacific (0.4 mm ka^{-1}). Because of its composite origin, red clay has a similar composition to that of average shale. However, a number of elements are enriched in red clay relative to average shale. These include the transition elements (Mn, Co, Ni, Cu) and Ba (Table 11.1). The transition elements constitute the authigenic fraction of the sediments whereas Ba is biogenically introduced into the sediments as barite. The Fe contents of red clay and average shale, on the other hand, are similar. Red clays are enriched in Mn relative to average shale by a factor of 7, in Co by a factor of 4, in Ni by a factor of 3, in Cu by a factor of 5 and in Fe by a factor of 1.4. By contrast, mildly reducing or seasonally oxidizing nearshore sediments do not incorporate an authigenic fraction and the composition of these sediments is similar to that of average shale. Nonetheless, elevated concentrations of manganese in surficial coastal sediments are well documented (Overnell et al. 1996; Overnell 2002).

The occurrence of an authigenic fraction in red clays reflects the fact that Mn occurs in two valencies and can migrate to regions where more oxidizing conditions prevail. Within red clays, Mn is dominantly in the tetravalent state with an average O:Mn ratio of 1.89 ± 0.05 (Murray et al. 1984). The presence of Mn oxides in the sediments leads to scavenging of transition elements such as Co, Ni and Cu.

The hemipelagic/pelagic transition is important in defining the nature of red clays. This boundary is a redox transition zone. Red clays form in regions of low sedimentation rate where the rate of organic matter deposition is very low. In these sediments, diffusion of oxygen into the sediments exceeds its rate of consumption in oxidizing organic matter. The sediments therefore remain brown throughout their length and Mn, Co, Ni and Cu are enriched in the authigenic fraction of the sediments. Oxic sediments are characterized by sedimentation rates of <40 mm ka^{-1} in regions of moderate productivity. In these sediments, iron is present dominantly in the trivalent state and reduction of nitrate in the pore waters by organic carbon does not occur.

An inverse relationship has been observed between the transition metal contents (Mn, Co, Ni and Cu) of red clays and the sedimentation rate (Krishnaswamy 1976). This relationship suggests that these sediments are characterized by a uniform rate of deposition of the authigenic elements superimposed on a variable deposition rate of detrital elements. This model explains the high concentration of authigenic elements in Pacific pelagic clays where sedimentation rates are low. The higher Mn content of red clays from the South Pacific compared to those from the North Pacific reflects the lower rates of sedimentation there. Chemical leaching techniques have been used to identify the forms of elements in red clays. About 90% Mn, 80% Co and Ni and 50% Cu are considered to be of authigenic origin whereas >90% of Fe is thought to be of detrital origin. Fe therefore occurs dominantly in the allogenic phase of these sediments. Table 11.2 suggests that most of the transition elements are delivered to the sediment/water interface associated with large organic aggregates.

Chemical analyses of red clays taken on a series of transects across the Southwestern Pacific Basin have shown that there is a systematic change in the composition of the red clays across the basin. On a transect from the base of the New Zealand continental slope to Rarotonga, Mn was shown to increase from 0.3-1.4%, Fe from 3-8 %, Co from 25-250 ppm, Ni from 50-250 ppm and Cu from 75-325 ppm (Meylan et al. 1982). However, relative to each other, the elements show an enrichment sequence along the transect of Co>Ni>Mn ≈ Cu>Fe. Mössbauer studies also showed a marked increase in the Fe^{3+}:Fe^{2+} ratio in the sediments with increasing distance from New Zealand which was attributed to the incorporation of Fe oxyhydroxides, probably ferrihydrite, into the sediments (Johnston and Glasby 1982). By contrast, Fe^{2+} is thought to occur in the sediments mainly in montmorillonite and chlorite. A decrease in sedimentation rate from 32 to 2 mm ka^{-1} was also observed along this transect (Schmitz et al. 1986). The sediments on this transect therefore show a decrease in grain size, an increased darkening of the sediments from pale yellowish brown to dusky

Table 11.1 Comparison of the transition metal and Ba contents of Pacific Pelagic Clay and average shale. Mn and Fe in per cent; Co, Ni, Cu and Ba in ppm (after Glasby 1991).

	Pacific Pelagic Clay	Average shale
Mn	0.43	0.05
Fe	5.4	5.2
Co	113	8
Ni	210	29
Cu	230	45
Ba	3900	250

Table 11.2 Rates of deposition ($\mu g\ cm^{-2}\ 10^3 yr^{-1}$) of transition elements from eolian dust and organic aggregates and into red clays (after Glasby 1991).

	Dust	Organic aggregates	Red clay
Mn	20-70	1565	1020-1230
Fe	3000-5800	16175	1560-3500
Co	2.7-4	12	4-14
Ni	14-20	15	20
Cu	20-30	490	28-34

brown, increasing transition metal contents, increasing oxidation state of iron, decreasing sedimentation rate and increased abundance of deep-sea manganese nodules with increasing distance from New Zealand. The dusky brown sediments are rich in phillipsite and manganese micronodules (Glasby et al. 1980). Several of the changes along the transect were thought to reflect an increase in the degree of oxidation of the sediments with decreasing sediment accumulation rate caused by a longer contact time of the sediment surface with well-oxygenated ocean bottom water. On a transect from the crest of the East Pacific Rise to New Zealand at 42°S, Stoffers et al. (1985) found similar increases in the contents of Mn, Fe, Co, Ni and Cu in sediments with increasing distance from New Zealand.

In a detailed comparison of sediments from the equatorial North Pacific high productivity zone (Area C) and the low productivity SW Pacific subtropical anticyclonic gyre (Area K), Stoffers et al. (1981) showed that the siliceous oozes from the equatorial North Pacific have much higher contents of Mn, Ni, Cu and Ba but lower contents of Fe and Co than the red clays from the SW Pacific (Table 11.3). Sedimentation rates on a carbonate-free basis for the two areas are of the same order (1-3 mm ka^{-1} for the equatorial North Pacific and 0.5-1 mm ka^{-1} for the SW Pacific). Differences in the transition metal contents of these sediments were therefore considered to be controlled by sediment type rather than sedimentation rate. Calculations based on the equations of Bischoff et al. (1979) confirmed that the hydrogenous (authigenic) component was much higher in Area C (7.9%) than in Area K (3.2%) sediments.

In addition to the authigenic component, there may also be a hydrothermal component in deep-sea sediments. During DSDP cruise 92, the variation in compo-

sition of sediments from three drill cores and several piston cores taken on a transect away from the crest of the East Pacific Rise was determined (Lyle et al. 1986; Marchig and Erzinger 1986). It was shown that the Mn accumulation rate in sediments falls off rapidly with increasing distance becoming relatively small 1,000 km from the ridge crest. At 19-20°S, for example, the Mn accumulation rate in the surface sediments declined from 36 mg cm^{-2} ka^{-1} at the ridge crest to 0.2 mg cm^{-2} ka^{-1} 1,130 km away. The corresponding decrease for Fe was from 120 mg cm^{-2} ka^{-1} to 0.68 mg cm^{-2} ka^{-1}. In a similar study at 42°S, the hydrothermal component was shown to decline from about 75% at the ridge crest to zero about 1,000 km away (Stoffers et al. 1985) (Fig. 11.5).

11.3.2 Diagenetic Processes in Deep-sea Sediments

The nature of the diagenetic changes occurring in pelagic sediments depends on the influx of decomposable organic matter to the sediment and the metabolic rate of oxidation (Müller et al. 1988). Three types of diagenetic processes can be distinguished: oxic diagenesis, suboxic diagenesis and anoxic diagenesis. Oxic diagenesis takes place when oxygen remains in the pore waters as in red clays. Mn concentrations in the pore waters remain extremely low (of the order of 2 $\mu g\ l^{-1}$) compared to a concentration of about 0.2 $\mu g\ l^{-1}$ in ocean bottom water. Suboxic diagenesis takes place when nitrate reduction occurs in the core and the oxygen content in the pore waters becomes very low. Dissolved Mn concentrations in the pore water can then increase by several orders of magnitude (>1,000 $\mu g\ l^{-1}$) compared to ocean bottom water. Anoxic diagenesis takes place in stratified anoxic basins

Table 11.3 Comparison of the transition metal and Ba contents of sediments (on a carbonate-free basis) from Areas C and K. Mn and Fe in per cent; Co, Ni, Cu and Ba in ppm (after Stoffers et al. 1981).

	Area C	Area K
Mn	1.97	0.8
Fe	4.8	8.39
Co	129	155
Ni	677	235
Cu	1044	275
Ba	3921	730

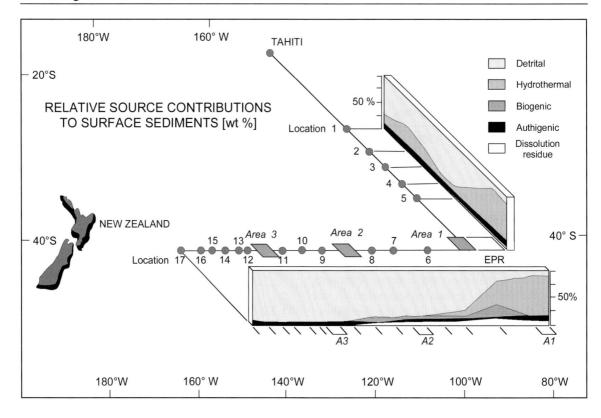

Fig. 11.5 Distribution of the normative sediment components; weight percent of the five individual components present in the sediments along the Tahiti - EPR - New Zealand transect (after Stoffers et al. 1985)

characterized by a well-developed halocline which prevents mixing of the anoxic basinal waters with the overlying sea-water as well as in the deeper layers of sediments from well-oxygenated coastal and upwelling environments. This leads to extensive diffusion of Mn and Fe into the water column from the underlying sediments. As an example, the dissolved Mn and Fe contents attain concentrations of <700 and 120 µg l⁻¹ respectively in the anoxic waters of the Gotland Deep, Baltic Sea (Glasby et al. 1997). The diagenetic pathway taken depends principally on the rate of organic carbon accumulation in the sediment.

Within red clays, the diffusive flux of Mn is small (23 µg cm⁻² ka⁻¹) and corresponds to about 7% of the total sedimentation flux of Mn (Glasby 1991). Mn is therefore largely immobilized in red clays. By contrast, 96% of the Cu in red clays is regenerated from the sediment into the bottom water as a result of the diagenetic flux across the sediment/water interface. Red clays therefore provide a relatively low flux of Mn and transition elements to the sediment surface and this is not an important source of metals for manganese nodule formation in red clay areas.

Müller et al. (1988) have distinguished between 'deep diagenesis' and 'surficial diagenesis'. In Pacific red clays, deep diagenesis results in no significant net upward flux of Mn, Fe, Ni or Cu. Surficial diagenesis is more significant. In siliceous ooze sediments, the regeneration rate of Mn in the surface sediments is of the same order as the accretion rate of Mn in the associated Mn nodules. Surficial diagenesis is therefore a significant source of metals to manganese nodules in siliceous ooze areas. 96% of the metals in the associated manganese nodules come from this source.

In general, the thickness of the oxidized layer in the sediment increases from near-shore and hemipelagic to pelagic environments. This is illustrated in Figure 11.6 which shows the trends for the eastern equatorial Pacific. This diagram confirms the inverse relationship between the thickness of the oxidized layer in the sediment and the biological productivity in the overlying surface waters. In general, there is a transition from tan to green within these sediments resulting from the *in-situ* reduction of Fe (III) to Fe (II) in smectites at the iron redox transition-zone (Lyle 1983; Köning et al. 1997).

In a comparison of pore water profiles of red clays and hemipelagic sediments, Sawlan and Murray (1983) showed that Mn and Fe are below the detection limit in the pore waters of red clays, Ni is present in the same concentration as in ocean bottom water and Cu shows a pronounced maximum at the sediment-water interface. In hemipelagic clays, denitrification becomes important and remobilization of Mn and Fe takes place. Ni correlates with Mn in the pore waters suggesting that it is associated with the Mn oxides in the solid phase. Cu is regenerated very rapidly at the sediment-water interface. The diffusive flux of Mn in hemipelagic sediments was determined to be in the range 2,200-33,000 $\mu g \, cm^{-2} \, ka^{-1}$. More detailed studies of pore water profiles in five different areas of the Californian Borderland confirmed the importance of Mn recycling in the surface sediments when the oxygen content of the bottom waters exceeds 0.1 ml l^{-1} (Shaw et al. 1990). Co and Ni appeared to be scavenged by Mn oxides and trapped in the surface sediments whereas the accumulation of Cu appeared to be more closely related to the flux of biogenic material to the sediment.

Based on a detailed study of pore water profiles in sediment cores from the eastern equatorial Atlantic, Froelich et al. (1979) were able to show that oxidants are consumed in the order of decreasing energy production per mole of organic carbon oxidized ($O_2 >$ Mn oxides $\approx$ nitrate $>$ Fe oxides $>$ sulfate). A schematic representation of the profiles is shown in Figure 11.7. From this diagram, it is seen that the reduction and remobilization of Mn takes place in zone 4. This is followed by the upward diffusion and reoxidation of Mn in zone 3. This process enables Mn to be stripped from the sediments as they accumulate and to be redeposited as a discrete layer within the sediment column. An example of this process is given in Figure 3.8. Of course, the depth at which these processes take place is controlled by the influx of organic matter to the sediments which governs the nature of the diagenetic process occurring there. In regions of extremely high productivity characterized by organic-rich hemi-pelagic sediments (such as found in the Panama Basin), burrowing by macrofauna can markedly increase the rate of recycling of Mn in the bioturbated zone of these sediments (Aller 1990). In this situation, Mn-oxide rich sediments and organic matter are mixed into the anoxic layers of the sediment by bioturbation, thereby permitting remobilization of Mn^{2+} in the sediment column (Thamdrup and Canfield 1996). However, reduction of Mn oxides was shown to play only a minor role in the oxidation of organic carbon in continental margin sediments taken off Chile (Thamdrup and Canfield 1996) (see Chapter 7). The main-

tenance of high Mn oxide reduction rates therefore depends on the continuous mixing of Mn oxides and fresh organic matter into the sediments through bioturbation or other mixing processes.

In addition to the above examples of steady state diagenesis, non-steady state diagenesis may occur in deep-sea sediments when turbidites are deposited in abyssal plains (Thomson et al. 1987). A color difference is often seen in the upper layers of the turbidite. This is a consequence of the 'oxidation front' in which oxygen from the overlying pelagic sediment, often carbonate ooze, diffuses down into the underlying turbidite sequence and oxidizes organic carbon there. As a result, oxygen and nitrate are reduced to almost zero below the front but Mn and Fe are mobilized in the sediment. Mn migrate upwards and is immobilized at the oxidation front. Ultimately, it may be fixed in the sediment as a manganese carbonate. Within a long sediment core, a number of turbidite sequences, and therefore fossil oxidation fronts, may be seen. Other examples of the non-steady state deposition of Mn in deep-sea sediments have been recorded at the glacial/interglacial boundary (Wallace et al. 1988; Gingele and Kasten 1994).

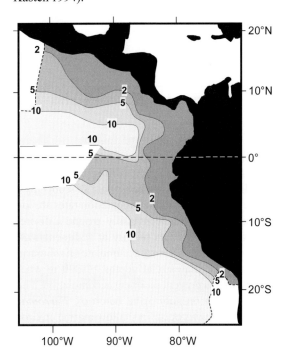

Fig. 11.6 Variation of the thickness (in cm) of the oxidized surface layer of sediments from the eastern equatorial Pacific (after Lyle 1993). In conjunction with the regional distribution of the biological productivity of the surface waters (Fig. 11.12), this pattern indicates that early diagenesis in the sediments is controlled on a regional scale by the input of biological detritus into the sediments.

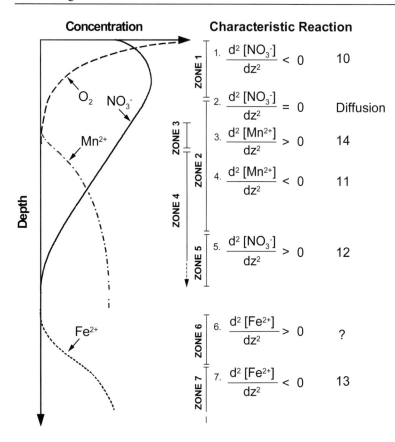

Fig. 11.7 Schematic representation of trends in pore water profiles for the principal oxidants in marine sediments. Depths and concentrations in arbitrary units (after Froelich et al. 1979). This pattern reflects the sequence of reduction of the principal oxidants in the sediment column (O_2>Mn oxides=nitrate>Fe oxides>sulfate).

11.4   Manganese Nodules and Crusts

There are three principal types of manganese deposits in the marine environment. *Manganese nodules* generally accumulate in deep water (>4,000 m) in oceanic basins where the sedimentation rates are low. They usually grow concentrically around a discrete nucleus. Growth occurs mainly at sediment-water interface. *Manganese crusts* accumulate on submarine seamounts and plateaux at depths >1,000 m where bottom currents prevent sediment accumulation. They generally form on submarine outcrops. *Ferromanganese concretions* occur in shallow marine environments (e.g. the Baltic and Black Seas) and in temperate-zone lakes. They grow much faster than deep-sea nodules and are quite different in shape, mineralogy and composition from them.

There are also three principal modes of formation of these deposits. *Hydrogenous deposits* form directly from seawater in an oxidizing environment. They are characterized by slow growth (about 2 mm Ma⁻¹). Manganese nodules tend to form on red clays and

Co-rich manganese crusts on rock substrates. The high Mn/Fe ratio in deep sea water compared to the earth's crust is mainly responsible for the formation of hydrogenous manganese deposits with Mn/Fe ratios of about unity. *Diagenetic deposits* result from diagenetic processes within the underlying sediments leading to upward supply of elements from the sediment column. These deposits are characterized by faster growth rates (10-100 mm Ma⁻¹) and are often found on siliceous oozes. *Hydrothermal deposits* precipitate directly from hydrothermal solutions in areas with high heat flow such as mid-ocean ridges, back-arc basins and hot spot volcanoes. They are characterized by high to extremely high growth rates (>1,000 mm Ma⁻¹) and low to very low trace element contents. They tend to be associated with hydrothermal sulfide deposits and iron oxyhydroxide crusts.

11.4.1   Deep-Sea Manganese Nodules

Deep-sea manganese nodules occur mainly in deep-ocean basins characterized by low sedimentation rates (i.e. <5 mm ka⁻¹) where inputs of calcareous ooze, turbidity flows and volcanic ash are low. They therefore occur in highest abundances on red clay and siliceous ooze far from land. They occur worldwide and are found in most major oceanic basins. Their distribution is also related to the patterns of oceanic bottom water flow and, to a lesser extent, to the availability of potential nuclei on which they grow such as weathered volcanic rock, pumice, whales' ear-bones, sharks' teeth, fragments of older nodules and indurated sediment. Antarctic Bottom Water (AABW) is the major oceanic bottom current in the Pacific. Its influence is seen in the lowered sedimentation rates, and therefore increased nodule abundance, along its flow path. A huge literature on deep-sea nodules has developed over the last 30 years, including a number of standard texts on the subject (Mero 1965; Horn 1972; Glasby 1977; Anon 1979; Bischoff and Piper 1979; Sorem and Fewkes 1979; Cronan 1980; Varentsov and Grasselly 1980; Roy 1981; Teleki et al. 1987; Baturin 1988; Halbach et al. 1988; Nicholson et al. 1997; Cronan 2000).

In the Pacific Ocean, the distribution of manganese nodules has been mapped as part of the Circum-Pacific Map Project using data from 2,500 bottom camera stations and from sediment cores (Piper et al. 1987) (Fig. 11.8). Although considerable variability in nodule abundance within individual stations was observed, high coverage of nodules was recorded in five main regions: between the Clarion and Clipperton Fracture Zones (C-C F.Z.) in the equatorial North Pacific and extending westwards into the northern sector of the Central Pacific Basin, in the abyssal plain area around the Musicians Seamounts in the Northeast Pacific Basin, in the central sector of the Southwestern Pacific Basin, in an E-W trending belt in the Southern Ocean coincident with the Antarctic Convergence, and in the northern sector of the Peru Basin. Andreev and Gramberg (2004) have recently published a more detailed map of the mineral resources of the world ocean including manganese nodules.

In discussing the composition of deep-sea manganese nodules, it should be born in mind that the composition of nodules varies within individual nodules as seen in the discrete micro-banding in the nodules, locally (on the scale of hundreds of meters) and regionally (over thousands of km) (e.g von Stackelberg and Marchig 1987). In spite of this, regional patterns in the composition of nodules are commonly observed such that we can reasonably compare and contrast the characteristics of nodules from different physiographic provinces of the world ocean as attempted here (cf. Cronan 1977; Piper and Williamson 1977; Sorem and Fewkes 1979).

In the following section, the distribution, mineralogy and composition of manganese nodules from three of these regions is considered in order to illustrate the different modes of formation of nodules in different settings. A detailed comparison of the characteristics of the nodules from these three regions has already been presented by Glasby et al. (1983).

Southwestern Pacific Basin

The Southwestern Pacific Basin has an area of $10 \cdot 10^6$ km^2. It is bounded by New Zealand-Tonga-Kermadec Arc, the East Pacific Rise and the Polynesian island chain and has a maximum depth 5,800 m. It lies beneath subtropical anticyclonic gyre which is a low productivity area. Two cruises of R.V. Tangaroa were undertaken in 1974 and 1976 to study the distribution and mode of formation of nodules in the Southwestern Pacific and Samoan Basins (Glasby et al. 1980). On a transect from New Zealand to Rarotonga in the Cook Islands, it was shown that the maximum abundance of nodules (>20 kg m^{-2}) occurs on the dusky brown clays in the region 220-745 km S.W. of Rarotonga. The western sector of the basin, particularly N.E. of New Zealand, was largely devoid of nodules. This was attributed to the influx of terrigenous sediments from the New Zealand landmass which raised the sedimentation rate above the threshold for nodule formation. The morphology of the nodules in the Southwestern Pacific Basin is somewhat variable but those taken S.W. of Rarotonga are 40% s[S]m, 17% m[S]m and 16% s[E]m (see box for explanation). 72% of these nodules are small (<30 mm), 26% medium (30-60 mm) and 2% large (>60 mm). The nodules are dominantly spheroidal with smooth surface texture when small but become more ellipsoidal and develop equatorial rims with increasing size. The larger nodules also tend to exhibit differences in the surface texture between the upper and lower surfaces. This reflects the fact that the larger nodules have been static at the seafloor for longer than is necessary to form the external layer of the nodule (i.e. their rate of rolling is slower than the rate of growth). Mineralogically, the nodules consist of δMnO_2, quartz and feldspar. Table 11.4 lists the average composition of nodules from S.W. Pacific Basin and the adjacent Samoan Basin. The average Ni+Cu+Co content of the nodules on the transect from New Zealand to Rarotonga is 1.00%. This is well below the level considered necessary for economic exploitation even though the Southwestern Pacific Basin is thought to contain $10 \cdot 10^9$ t of nodules. A subsequent E-W transect across the Southwestern Pacific Basin at 42°S undertaken during cruise SO-14 of R.V. Sonne showed

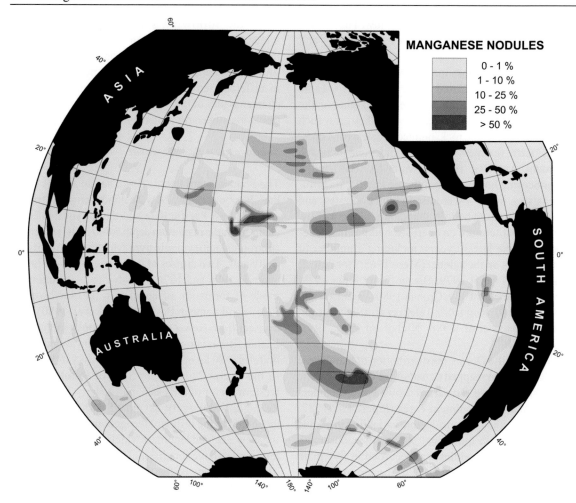

Fig. 11.8 Schematic map showing the distribution of manganese nodules in the Pacific Ocean compiled as part of the Circum-Pacific Map Project (Piper et al. 1985). The contours represent the percentage cover of the ocean floor by manganese nodules. Diagram prepared with the permission of D.Z. Piper, USGS.

that the highest abundances of nodules occur in the far eastern sector of the basin furthest from land (Plüger et al. 1985).

Descriptive classification of manganese nodule types (after Glasby et al. 1980).

Prefix: (Nodule size based on maximum diameter)

s =   small (<30 mm)
m =   medium (30-60 mm)
l =   large (>60 mm)

Bracketed: (Primary nodule shape)

S =   speroidal
E =   ellipsoidal
D =   discoidal
P =   polynucleate
T =   tabular
F =   faceted (polygonal)
V =   scoriaceous (volcanic)
B =   biological (shape determined by shark's tooth)

Suffix: (Nodule surface texture)

s =   smooth
m =   microbotryoidal
b =   botryoidal
r =   rough

Examples: m[S]m = medium-sized spheroidal nodule with microbotryoidal surface texture.

l[D]^b_r = large discoidal nodule with botryoidal upper surface texture and rough lower surface texture.

s[S-F]s = small spheroidal nodule with significant facing, smooth surface texture.

Southwestern Pacific Basin nodules are considered to be hydrogenous in origin based on their average Mn/Fe ratio of about unity, Ni+Cu contents of <1%, δMnO_2 as the principal Mn oxide phase and growth rates of 1-2 mm Ma^{-1} (Dymond et al. 1984).

Clarion-Clipperton Fracture Zone Belt

The C-C F.Z. belt is the area bounded by the Clarion F.Z. to north and the Clipperton F.Z. to the south. The water depth varies from 5,300 m in the west to 4,300 m in the east (Andreev and Gramberg 1998). The C-C F.Z. is a high productivity area characterized by the occurrence of siliceous sediments. The sedimentation is controlled largely by bottom current activity. The area is dominated by an abyssal hill topography. There is a great variability in nodule distribution and type which depends on the sediment type and accumulation rate, bottom current activity, benthic biological activity (bioturbation), bottom topography, availability of potential nodule nuclei and productivity of the surface waters (Friedrich et al. 1983; Halbach et al. 1988; von Stackelberg and Beiersdorf 1991; Skornyakova and Murdmaa 1992; Jeong et al. 1994, 1996; Knoop et al. 1998; Morgan 2000).

Within the C-C F.Z., there appear to be two discrete nodule types; larger 'mature' nodules and smaller 'immature' nodules. The larger 'mature' nodules make up 26% of the nodules by number but 92% by weight. The larger nodules are dominantly m-l[E,D]$_b$^{s}, the so-called hamburger-shaped nodules which have an equatorial rim corresponding to the sediment-water interface. The nodules become flatter with increasing size and have a maximum diameter of 140 mm. These nodules are characterized by hydrogenous growth on the upper surface and diagenetic growth on the lower surface. Figures 11.9 and 11.10 show the principal features of these hamburger nodules. By contrast, the morphology of the nodules from the slopes of seamounts is s-m[S,P]s-m. These nodules form at the sediment surface and are characterized by hydrogenous growth. The principal minerals present are todorokite and δMnO_2. Table 11.4 lists the average composition of nodules from the Clarion-Clipperton F.Z. belt. There is an inverse relation between nodule grade and abundance which vary from 3.4% Ni+Cu and 3 kg m^{-2} to 0.3% Ni+Cu and 16 kg m^{-2}. The high Ni+Cu nodules are dominantly diagenetic in origin and the low Ni+Cu nodules dominantly hydrogenous in origin. The high contents of Ni+Cu in the diagenetic

1 cm

Fig. 11.9 Hamburger-shaped nodule from the Clarion-Clipperton F.Z. region displaying a smooth surface texture on the upper surface and a rough surface texture on the underside (l [D] ^{s}$_r$). Sample collected at Stn 321 GBH, R.V Sonne cruise SO-25 (7°57.30'N, 143°28.92'W, 5153 m). Photograph courtesy of U. von Stackelberg, BGR.

Fig. 11.10 Vertical section of a hamburger-shaped nodule from the Clarion-Clipperton F.Z. region displaying a layered growth structure (1 D ^s_r). The nodule nucleus is a nodule fragment. Sample collected at Stn 178 GBH, R.V Valdivia cruise VA 13/2 (9°22.5'N, 145°53.1'W, 5231 m). Photograph courtesy of U. von Stackelberg, BGR.

nodules from this area ensure that the C-C F.Z. is the prime target area for the economic exploitation of manganese nodules (cf. Morgan 2000).

C-C F.Z. nodules may be considered to be diagenetic in origin based on their average Mn/Fe ratio of about 2.5, Ni+Cu contents of <2.5%, 10 Å manganate as the principal Mn oxide phase and growth rates of 10-50 mm Ma^{-1} (Dymond et al. 1984). The influence of oxic diagenesis on nodule formation is the result of low sedimentation rates and low inputs of organic carbon to sediments. The sediments are oxidizing with an E_H of +0.4 V at 8m depth in the sediment column

(Müller et al. 1988). Ni and Cu on the undersides of the nodules are ultimately derived from the dissolution of siliceous tests in the sediment (Glasby and Thijssen 1982). Dissolved silicate concentrations in the sediment pore waters may attain values of up to 900 μM. Ni^{2+} and Cu^{2+} substitute in the interlayer spacings of 10 Å manganate on the undersides of the nodules resulting in differences in the compositions of the upper and lower surfaces of the nodules (Dymond et al. 1984). In general, the tops of these nodules are enriched in Fe, Co and Pb and the bottoms in Mn, Cu, Zn and Mo (Raab and Meylan 1977).

Table 11. 4 Comparison of the average compositions of Mn nodules from S.W. Pacific Basin and Samoan Basin (after Glasby et al. 1980), the Clarion-Clipperton Fracture Zone region (after Friedrich et al. 1983) and the Peru Basin (after Thijssen et al. 1985).

	S.W. Pacific Basin	Samoan Basin	C-C F.Z.	Peru Basin
Mn (%)	16.6	17.3	29.1	33.1
Fe (%)	22.8	19.6	5.4	7.1
Co (%)	0.44	0.23	0.23	0.09
Ni (%)	0.35	0.23	1.29	1.4
Cu (%)	0.21	0.17	1.19	0.69
Mn/Fe	0.73	0.88	5.4	4.7

Peru Basin

The Peru Basin is a relatively shallow oceanic basin with depths typically in the range 3,900-4,300 m which is close to the depth of the Carbonate Compensation Depth (CCD) (4,250 m) and lies in a region of high biological productivity. Manganese nodules were initially recovered during cruises SO-04 and SO-11 of R.V. Sonne and investigated in part because of their economic potential (Thijssen et al. 1985; von Stackelberg 1997, 2000). Subsequently, the nodule field was investigated during cruises SO-61, SO-79 and SO-106 as part of the DISCOL program, a long-term, large-scale disturbance-recolonization experiment to assess the potential environmental impact of nodule mining (von Stackelberg 1997; Thiel 2001).

Peru Basin nodules are quite different from those in other regions of the ocean floor. They tend to be large and mamillated with an average Mn/Fe ratio of 4.7 and an average Ni+Cu content of about 2.1% (Thijssen et al. 1985). Table 11.4 lists the average composition of nodules from the Peru Basin. The nodules are friable and contain fragments of fish bones or broken nodules as nodule nuclei. Nodule density averages >10 kg m^{-2} but achieves a maximum value of 53 kg m^{-2}. The highest abundance of nodules is found

Figs 11.12 Vertical section of an asymmetric nodule from the Peru Basin with a cetacean earbone nucleus (l [S] ^r_s). Sample collected at Stn 63 KG, R.V Sonne cruise SO-79 (6°45.58'N, 90°41.48'W, 4257 m). Photograph courtesy of U. von Stackelberg, BGR.

near the CCD. Although only 3.8% of the nodules by number are in the >80 mm size class, these nodules make up 56% of the total weight of the nodules. The largest nodule recovered was 190 mm in diameter. The morphology of the nodules is a function of the size class. Nodules <20 mm in diameter are dominantly fragments of larger nodules and less commonly discrete ellipsoidal nodules with botryoidal surface texture. Nodules in the 20-60 mm size classes are dominantly ellipsoidal to discoidal with the position of the sediment-water interface clearly defined. The upper surface of these nodules is mamillated with micro-botryoidal surface texture whereas the lower surface is less mamillated and smoother. Nodules in the >60 mm size class are dominantly highly mamillated and spheroidal with botryoidal surface texture on the upper surface and smooth to micro-botryoidal on the lower surface. Figures. 11.11 and 11.12 show the principal features of these large nodules. For every nodule encountered at the sediment surface, 1.25 nodules occur to a depth of 2.5 m in the sediment column. Buried nodules have the same morphology as surface nodules but different surface texture indicating that material has been removed from the surface of the nodule by dissolution after burial. von Stackelberg (1997) has shown that the redox boundary separating soft oxic surface sediments from stiffer suboxic sediments occurs at a depth of about 100 mm in the sediment column. Highest growth rates occur on the underside of large nodules immediately above the

Figs 11.11 Sperical nodule with a cauliflower-shaped surface from the Peru Basin displaying a rough surface texture on the upper surface and a smooth surface texture on the underside (l [S] ^r_s). Sample collected at Stn 25 KG, R.V Sonne cruise SO-79 (6°51.99'N, 90°26.52'W, 4170 m). Photograph courtesy of U. von Stackelberg, BGR.

redox boundary. Strong bioturbation occurs throughout the region and helps maintain the nodules at the sediment surface.

Peru Basin nodules are considered to be diagenetic in origin with maximum Mn/Fe ratios of >50, corresponding to Ni+Cu in these samples of <1.4%. 10 Å manganate as the principal Mn oxide phase and growth rates of 100-200 mm Ma^{-1} (Dymond et al. 1984). Nodule compositional data for this assessment are taken from Halbach et al. (1980) and Thijssen et al. (1985) and nodule growth rates from Reyss et al. (1985) and Bollhöffer et al. (1996, 1999). The influence of suboxic diagenesis on nodule growth is the result of higher sedimentation rates and higher inputs of organic carbon into the sediments. Mn is strongly remobilized by dissolution of Mn micronodules within the sediment column. This leads to the high Mn/Fe ratios and low Ni+Cu contents in the nodules. The nodules are characterized by increasing substitution of Mn^{2+} in the interlayer spacings of 10 Å manganate.

11.4.2   Influence of Diagenesis on Nodule Growth

Perhaps the most comprehensive explanation of the role of hydrogenetic and diagenetic processes on manganese nodule accretion has been presented by Dymond et al (1984). On their classification, *hydrogenous deposition* involves the direct precipitation or accumulation of colloidal metals oxides from seawater. Strictly, this involves deposition of manganese oxides on surfaces in contact with seawater such as involved in the formation of manganese crusts. In practice, manganese nodules formed on red clays have charac-

teristics very similar to those of manganese crusts and are therefore considered to be hydrogenous in origin. However, Aplin and Cronan (1985) have argued that no nodules resting on or in marine sediments can be considered entirely free of diagenetic influences.

Oxic diagenesis refers to processes occurring within oxic sediments. Decomposition and oxidation of labile organic matter and the dissolution of labile biogenic components such as siliceous tests may release biologically-bound metals into the sediment pore waters which may ultimately be incorporated into the nodules. Dissolution of siliceous tests may also introduce silica into the sediment pore waters which may react with amorphous ferromanganese oxides to form nontronite (Dymond and Eklund 1978). This process fixes silica and Fe in the sediment column and releases transition elements such as Mn, Co, Ni, Cu and Zn into the sediment pore waters for incorporation into the nodules. Jung and Lee (1999) have suggested that formation of manganese nodules may be episodic under the influence of oxic diagenesis. Under these conditions, manganese and associated transition metals would be remobilised in oxygen-depleted environments which develop as a result of the decomposition of organic matter in burrows and supplied episodically to nodule surfaces during periods of intermittent bottom water flow strong enough to stir up the bioturbated sediment.

Suboxic diagenesis involves the reduction of Mn (IV) to Mn (II) within the sediment column and then reoxidation of Mn (II) to Mn (IV) during the formation of the nodules. These diagenetic processes are controlled by redox processes within the sediment column and are therefore dependent on depth in the

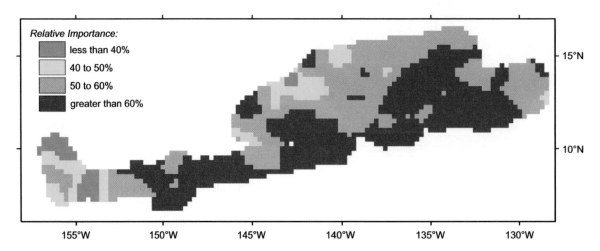

Fig. 11.13 Contour plot of the diagenetic accretion compositional end member of manganese nodules which shows that this end member increases in relative proportion as the equatorial zone of high productivity is approached as a consequence of the increase in both the primary productivity and sedimentation rate (after Knoop et al. 1998).

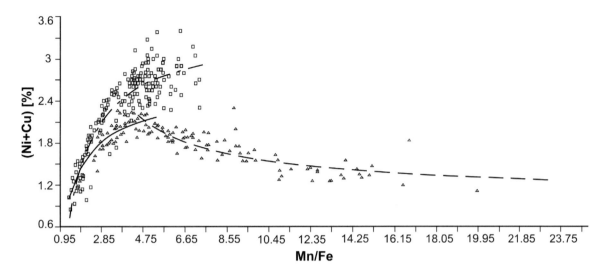

Fig. 11.14 Hyperbolic regression curves of Ni+Cu against Mn/Fe for nodules from the Clarion-Clipperton F.Z. region (upper curve) and Peru Basin (lower curve) (after Halbach et al. 1981).

sediment column. In most cases (as in the Peru Basin), these nodules are formed in oxic sediments overlying reducing sediments. The diagenetic supply of metals to the nodules normally takes place at or near the sediment surface. Because the metals are released within the sediment column and migrate upwards, they tend to be incorporated on the underside of nodules. This frequently results in differences in composition between the upper and lower surfaces of diagenetic nodules. Manganese micronodules may play a key role in retaining transition elements in the sediment column until they are released on further burial of the micronodules.

Diagenetic processes may therefore involve recycling of elements within the sediment column prior to their incorporation in nodules. The characteristics of each of these nodule types have been given in the previous sections. In addition to influencing the mineralogy and composition of the nodules, diagenetic processes also influence their surface texture; hydrogenous nodules tend to have smooth surface texture and diagenetic nodules botryoidal to rough surface texture.

From the above comments, it will be seen that the principal factor driving the diagenetic milieu in the sediments is the biological productivity of the oceanic surface waters. A map of biological productivity of Pacific ocean surface waters shows that the productivity of the northern sector of the Southwestern Pacific Basin is 50-100 gC m^{-2} yr^{-1}, in the C-C F.Z. is in the range 100-150 gC m^{-2} yr^{-1} and in the Peru Basin is 100-150 gC m^{-2} yr^{-1} (Fig. 12.5) (cf. Cronan 1987, 1997;

Müller et al. 1988). The importance of the productivity of the surface waters to the diagenetic component of manganese nodules in the C-C F.Z. is well illustrated in Figure 11.13 which shows that the diagenetic component of the nodules increases from west to east as the equatorial zone of high productivity is approached (cf. Skornyakova and Murdmaa 1992; Morgan 2000).

The influence of biological productivity on nodule composition is perhaps best illustrated by the hyperbolic regression curve of Cu and Ni against Mn/Fe for nodules from C-C F.Z. and Peru Basin (Fig. 11.14). This diagram can be divided into three parts. Hydrogenous nodules deposited entirely from seawater have Mn/Fe ratios of about unity and low Ni+Cu contents. Nodules influenced by oxic diagenesis have Mn/Fe ratios of up to 5 (but more typically about 2.5) with correspondingly high Ni+Cu contents. Nodules influenced by suboxic diagenesis have Mn/Fe ratios of up to 50 but lower Ni+Cu contents. The maximum Ni+Cu contents of nodules correspond to a Mn/Fe ratio of about 5 and is found at the so-called point of reversal.

For this purpose, we may therefore consider hydrogenous deep-sea nodules to be a baseline. When oxic diagenesis takes place, Mn, Ni and Cu are released into the sediment pore waters and are ultimately incorporated into the nodules. Ni^{2+} and Cu^{2+} substitute in the phyllomanganate lattice (see below). This explains the high Mn/Fe ratios and Ni+Cu contents of oxic nodules. When sub-oxic diagenesis takes place, Mn^{2+} is remobilized into the reducing sediments and migrates upwards towards the sediment surface where it is either oxidized to Mn (IV) at the nodule surface or substitutes in the phyllomanganate lattice as Mn^{2+}. This behavior explains both the very high Mn/Fe ratios of suboxic nodules and their correspondingly low Ni+Cu contents. The boundary between oxic and suboxic diagenesis is marked by a Mn/Fe ratio in the nodules of about 5. The nature of the diagenetic processes occurring in the sediments depends directly on the input of organic carbon and therefore on the productivity of the overlying surface waters.

Based on a study of manganese nodules taken on the Aitutaki-Jarvis Transect in the S.W. Pacific, Cronan and Hodkinson (1994) also demonstrated the influence of the CCD on nodule composition. Above the CCD, accumulation of carbonate tests was thought to dilute the organic carbon in the sediments, thus inhibiting diagenesis, whereas, below the CCD, the decay of organic material in the water column was thought to reduce its effectiveness in driving diagenetic reactions. Cronan and Hodkinson (1994) indeed found that Mn, Ni, Cu and Zn were most concentrated in nodules taken near the CCD at the north of the transect in an region of high productivity and least concentrated in nodules taken away the CCD at the south of the transect in a region of low productivity. They therefore inferred that diagenetic cycling of Mn, Ni, Cu and Zn to the nodules is enhanced as a result of the decay of organic material near the CCD.

On a more local scale, the influence of diagenetic processes on nodule morphology and composition has been related to variations in sedimentation rate in areas of hilly topography within the C-C F.Z. (von Stackelberg and Marchig 1987; von Stackelberg and Beiersdorf 1991). On the flanks of hills and in parts of basins, sedimentation rates are lower and dominantly diagenetic nodules are formed whereas, in areas of sediment drift, sedimentation rates are higher and dominantly hydrogenous nodules are formed. These processes are related to the degree of decomposition of organic matter within the sediment column. Where sedimentation rates are low, organic matter is rapidly consumed within the sediment releasing Mn^{2+} into the pore water and resulting in the formation of dominantly diagenetic nodules. In areas of sediment drift, on the other hand, the organic carbon is buried within the sediment stimulating bioturbation and the resultant biogenic lifting of the nodules to the sediment surface.

Bioturbation is an essential requirement for benthic lifting of manganese nodules and maintaining the nodules at the sediment surface. In both the C-C F.Z. and the Peru Basin, the biological productivity of the oceanic surface waters is high enough that the amount of organic matter reaching the sea floor is sufficient to stimulate the activity of sediment-feeding organisms. As a result, bioturbation appears to be the dominant factor in maintaining nodules at the sediment surface in the C-C F.Z. (von Stackelberg and Beiersdorf 1991; Skornyakova and Murdmaa 1992) and the Peru Basin (von Stackelberg 1997, 2000). Several mechanisms have been proposed by which nodules could be maintained at the sediment surface by bioturbation. These include biological pumping, burrowing by crack propagation as well as bioturbation by benthic fauna and megafauna (Sanderson 1985; McCave 1988; Banerjee 2000 and Dogan et al. 2005). The biological productivity of the surface waters in the Southwestern Pacific Basin, on the other hand, is much lower than that in the C-C F.Z. and the Peru Basin and the abundance of benthic fauna on the sea floor is therefore much reduced. This may account for the greater influence of ocean bottom currents in maintaining nodules at the sediment surface there as proposed by Glasby et al. (1983). The biological productivity of the oceanic surface waters may therefore also influence the way in which nodules are maintained at the sediment surface.

In addition to the above, Calvert and Piper (1984) have proposed a diagenetic source of metals derived from sediments far away for nodules occurring in an erosional area with thin sediment cover in the C-C F.Z. The metals were thought to be transported to the site of deposition by oceanic bottom water, probably AABW. Although this process is well known in shallow-water continental margin areas such as the Baltic Sea, only limited evidence to support this hypothesis has been presented for the deep-sea environment.

11.4.3   Rare Earth Elements (REE) as Redox Indicators

A key parameter in understanding nodules formation is the redox milieu of the environment at the time of deposition. This can influence the mineralogy and therefore composition of the nodules (see below). In this regard, Glasby (1973) used the Ce/La ratio of nodules as a redox indicator. He showed that deep-sea nodules from the NW Indian Ocean had much higher Ce/La ratios (4.4) and ΣREE contents (490 ppm) than those from shallow-water continental margin environments such as Loch Fyne, Scotland (1.9 and 29 ppm, respectively). The higher REE contents of the deep-sea nodules were taken to reflect the more oxidizing conditions in the deep sea which facilitated the oxidation of the trivalent Ce^{3+} in seawater to CeO_2 on the surface of the nodules. The lower ΣREE contents of the shallow-water continental margin concretions was taken to reflect the remobilization of Mn in the sediment column which resulted in the fractionation of Mn from Fe, Co, Ni, Cu and REE in the concretions. It was suggested that the REE were adsorbed from seawater onto colloidal FeOOH in the nodules. The high rate of manganese remobilization in the shallow-water continental margin sediments also resulted in growth rates of the associated concretions two orders of magnitude higher than those of the deep-sea nodules (Ku and Glasby 1972).

Based on these ideas, Glasby et al. (1987) attempted to use the Ce/La ratios of deep-sea manganese nodules as a means of tracing the flow path of the AABW in the S.W. and Central Pacific Ocean. On this basis, the Ce/La ratios in the deep-sea nodules should decrease along the flowpath of the AABW as the oxygen in the bottom waters is consumed. The data confirmed that the ΣREE contents of the nodules are strongly correlated with Fe resulting in higher REE contents in nodules from the Southwestern Pacific Basin compared to those from the C-C F.Z (La 169 v 93 ppm). They also showed a systematic decrease in the Ce/La ratios of the nodules from the Southwestern Pacific Basin to the C-C F.Z. and on to the Peru Basin as follows: Area III (9.6) > Area K (8.0) > Area C (3.7) > Peru Basin (2.0) > Area G (1.5) > Area F (1.4). The Peru Basin and Areas G and F are situated in the equatorial South Pacific in regions which do not lie directly under the flow path of the AABW and which are characterized by sluggish bottom water flow (Nemoto and Kroenke 1981; Davies 1985). This factor explains the low Ce/La ratios of these samples. The sequence of these ratios confirms that the Ce/La ratio of deep-sea manganese nodules can be used to trace the flow path of the AABW and therefore that the Ce/La ratios of manganese nodules can be used as a redox indicator in the deep-sea environment. Subsequent work on the REE contents of manganese nodules taken on an E-W transect across the Southwestern Pacific Basin at 42°S showed that the highest Ce/La ratios and ΣREE contents of the nodules occur in the central part of the basin which is exposed to the strongest flow of AABW (and therefore the most oxygenated conditions) (Kunzendorf et al. 1993).

More recently, Kasten et al. (1998) have determined the REE contents of a suite of manganese nodules from the South Atlantic to see if a similar relationship between the Ce/La ratios of the nodules and AABW flow could be observed there. In fact, the Ce/La ratios of the nodules decrease in the sequence Lazarev Sea, Weddell Sea (10.4 and 9.7) > East Georgia Basin (6.5-7.1) > Argentine Basin (5.0) but then increase in the

Brazil Basin (6.2) and Angola Basin (9.8 and 15.1). A further decrease was observed in the Cape Basin (7.6). In addition, an extremely high Ce/La ratio of 24.4 had already been determined for nodules sampled north of the Nares Abyssal Plain in the North Atlantic. These data reflect the more complicated pattern of bottom water flow in the Atlantic compared to the Pacific Ocean. It is believed that the penetration of more oxygenated North Atlantic Deep Water (NADW) into the South Atlantic accounts for the higher Ce/La ratios in the nodules from the Angola and Brazil Basins. The influence of AABW could therefore be traced only as far north as the Argentine Basin. These data confirm the potential importance of REE studies in deep-sea manganese nodules in tracing oceanic bottom water flow.

Of particular interest in this regard is the study of REE patterns of different types of nodules from DOMES Site A in the C-C F.Z. by Calvert et al. (1987) on nodules which had previously been investigated by Piper and Blueford (1982) and Calvert and Piper (1984) (see sections 11.4.2 and 11.4.10.2). These authors demonstrated that there were three end-member types of nodules at this site, granular nodules from the valley which formed on thin (<20 cm) Quaternary sedimentary sections and displayed the highest Mn/Fe, Cu/Ni, todorokite/δMnO$_2$ ratios and the lowest Ce/La ratios, smooth nodules from the uplands which formed on thick (>36 cm) Quaternary sections and displayed the lowest Mn/Fe, Cu/Ni, todorokite/δMnO$_2$ ratios and highest Ce/La ratios and an intermediate type of nodules. The authors interpreted these differences in Ce/La ratios of the nodules in terms of variations in the composition of seawater from which the nodules formed, either from geochemically more evolved AABW in the valley or from more local sources of seawater in the highlands. However, other authors have interpreted such local variations in nodule type in other areas of rugged bottom topography in the C-C F.Z. in terms of the formation of three genetic types of nodules, hydrogenous, diagenetic, and mixed hydrogenous/diagenetic (Heye 1978; Glasby et al. 1982; Skornyakova and Murdmaa 1992). At DOMES Site A, the extreme hydrogenous endmember is characterized by a Mn/Fe ratio of 1.1, Ni + Cu of 1.0% and a Ce/La ratio of 2.7 and the extreme diagenetic endmember by a Mn/Fe ratio of 5.2, Ni + Cu of 2.9% and a Ce/La ratio of 2.5. Although the Ce/La ratios for these two nodules are quite similar, the highest Ce/La ratio for the hydrogenous nodules was 3.7 and the lowest ratio for the diagenetic nodules 1.6 (see Glasby et al. 1987; Kunzendorf et al. 1993). This suggests that variations in the REE patterns of nodules at DOMES

Site A may be better interpreted in terms of variations in nodule type within the area rather than local variations in seawater chemistry as proposed by Calvert et al. (1987).

In addition, De Carlo and McMurtry (1992) have analyzed 32 samples of Co-rich Mn crusts from the Hawaiian Archipelago for REE (see section 11.4.4). These samples contained on average 287 ppm La and 1,277 ppm Ce giving an average Ce/La ratio of 4.5. This is intermediate between the ratios in nodules from the Aitutaki Passage (Area K) and the C.C. FZ (Area C) and is compatible with a hydrogenous origin for these crusts.

11.4.4   Co-Rich Mn Crusts

Co-rich Mn crusts may be defined as hydrogenous manganese crusts having Co contents >1% (Manheim 1986; Mangini et al. 1987; Hein et al. 2000). These crusts are typically 5-100 mm thick and occur on older seamounts (100-60 Ma) in many of the seamount chains in the equatorial Pacific such as the Mid-Pacific Mountains and Line Islands. The crusts are commonly found on exposed rock on seamount slopes or on the summits of oceanic plateaux at water depths of 3,000-1,100 m. Principal substrates include basalts, hyaloclastites, indurated phosphorite and claystone. A vertical section of a Co-rich Mn crust is shown in Fig. 11.15. On plateaux and flat terraces where fragments of rock or manganese crusts have accumulated, manganese nodules may be seen lying on the surface of calcareous ooze. Ripple marks indicating the presence of strong bottom currents are sometimes observed on this ooze. The average composition of crusts from the 1,500-1,100 m depth zone in the Mid-Pacific Mountains has been reported

as Mn 28.4%, Fe 14.3%, Co 1.18%, Ni 0.50%, Cu 0.03%, Pt 0.5 ppm, Mn/Fe 2.0. δMnO_2 is the principal manganese mineral present and they have a growth rate of 1-2 mm Ma^{-1}. The crusts have attracted economic interest as a potential source of Co and, to a lesser extent, Pt. The areas in which these crusts generally form lie well above the CCD. Crusts therefore tend to form in regions of strong bottom current activity which can prevent the deposition of calcareous ooze by erosion.

The variation in the composition of manganese crusts from the Mid-Pacific Mountains with water depth is presented in Table 11.5. These data show that these crusts are hydrogenous in origin (based on their Mn/Fe ratios) but they tend to have much higher Co and lower Cu contents than deep-sea nodules from the same region. The Mn/Fe ratios, Co and Ni contents are highest but the Cu contents lowest in crusts in the depth range 1,900-1,100 m. The positive correlation of Mn, Co and Ni reflects that association of these elements in δMnO_2. Overall, the Pt content of the crusts is very high, in the range 0.2-1.2 ppm with an average of 0.5 ppm.

Halbach and Puteanus (1984) showed that the dissolution of calcareous tests in the water column plays a key role in the incorporation of Fe into these crusts. The calcareous tests contain about 500 ppm Fe. The flux of Fe to the surface of the crusts derived from the release of colloidal Fe oxyhydroxide particles on dissolution of the calcareous tests was estimated to be about 15 µg cm^{-2} a^{-1} which is almost equivalent to the flux of Fe in the concretions of 22.4 - 44.8 µg cm^{-2} a^{-1}. The rate of incorporation of Fe into the crusts is therefore related to the position of the lysocline. Based on such considerations, these authors concluded that the metal supply from the water column to the crusts

Table 11.5 Variation of the composition of manganese crusts from the Mid-Pacific Mountains with water depth (after Mangini et al. 1987). Analyses expressed as wt. % of dried material.

water depth (m)	Mn	Fe	Co	Ni	Cu	Mn/Fe
1100 - 1500	28.4	14.3	1.18	0.50	0.03	1.99
1500 - 1900	24.7	15.3	0.90	0.42	0.06	1.61
1900 - 2400	25.5	16.1	0.88	0.41	0.07	1.58
2400 - 3000	20.5	19.5	0.69	0.18	0.09	1.05
3000 - 4000	20.5	18.0	0.63	0.35	0.13	1.41
4000 - 4400	19.7	16.7	0.67	0.24	0.10	1.17

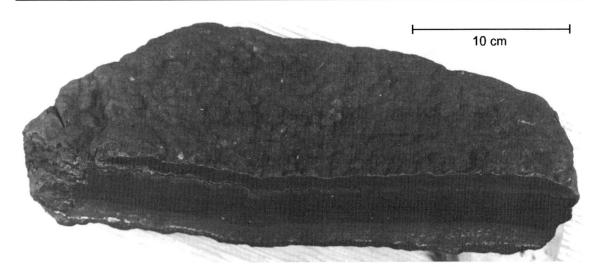

10 cm

Fig. 11.15 Vertical section of a Co-rich Mn crust from the flanks of a guyot on the Ogasawara Plateau; N.W. Pacific (25°18.9'N, 143°54.8'E; 1515 m) collected by dredge during a cruise of the GSJ with R.V. Hakurei-maru in 1986. The substrate (not clearly seen) is phosphatized limestone. The crust is more than 20 cm across, about 10 cm thick and has a knobby surface texture. Element contents are: Mn 21.7%, Fe 18.9%, Co 0.81%, Ni 0.31%, Cu 0.04%, Pb 0.26% and Pt 0.29 ppm. The upper layer of the crust displays the highest Co content and the bottom layer the highest Pt (0.78 ppm). δMnO_2 is the principal mineral present with minor quartz and plagioclase. Photograph courtesy of A. Usui, GSJ.

is controlled by a number of factors such as changes in oceanic productivity, depth of the lysocline, rate of $CaCO_3$ dissolution and strength of AABW flow.

The highest Co contents are found in crusts occurring in the depth range 1,900-1,100 m which corresponds to the depth of the oxygen minimum zone. In samples taken from summits and the upper parts of slopes at depths less than 1,500 m, the Co contents of the crusts sometimes exceed 2%. At this depth, Mn tends to remain in solution because the deposition of MnO_2 is not favored at these low oxygen contents. The deposition rate of MnO_2 is therefore at a minimum and the Mn content in seawater at a maximum of about 2 nmol kg^{-1}. The flux of Co to the surface of the crusts remains constant with water depth at about 2.9 μg cm^{-2} ka^{-1}. The Co content of crusts is therefore a maximum just below the oxygen minimum zone and no special source of Co is required to explain their formation (Puteanus and Halbach 1988).

Pt is also significantly enriched in Co-rich crusts (Hein et al. 1988, 1997; Halbach et al. 1989). However, the mechanism of enrichment in the crusts is not well understood. Hodge et al. (1985) argued that Pt is oxidized from the $PtCl_4^{2-}$ state in seawater to the tetravalent state in manganese nodules. This process was thought to be responsible for the anomalously high Pt/Pd ratios in nodules (50-1,000) compared to seawater (4.5). However, Halbach et al. (1989) considered that this process would not be possible

because the first formed tetravalent species, $PtCl_6^{2-}$, would be very stable in oxygenated seawater. Instead, they proposed that the $PtCl_4^{2-}$ would be reduced to Pt metal in the crusts with some minor amounts of Pt being introduced from cosmic spherules. Mn^{2+} was assumed to be the reducing agent for the reduction of the $PtCl_4^{2-}$.

However, Stüben et al. (1999) subsequently plotted stability field diagrams for Pt and Pd and showed that Pt is present in seawater dominantly as $Pt(OH)_2^{\circ}$ and is close to saturation under seawater conditions. From this, it was argued that Pt is enriched in Co-rich crusts mainly as a result of the preferential adsorption of $Pt(OH)_2^{\circ}$ onto the surface of Mn and Fe oxyhydroxide minerals. Pd, on the other hand, lies at the boundary of the stability fields of $PdCl_4^{2-}$ and $Pd(OH)_2^{\circ}$ and is probably undersaturated in seawater. Since $PdCl_4^{2-}$ can not be adsorbed on Mn oxyhydroxides which have a negative surface charge, this would explain the so-called negative Pd anomaly in marine manganese deposits.

For Cu, the EH-pH diagram shows that the boundary between Cu^{2+} and $CuCl_3^{2-}$ as the dominant species in seawater lies slightly above the normal range of redox conditions in seawater (E_H of +0.48 V) (Fig. 11.16). At an E_H of +0.4 V, the concentration of Cu^{2+} in seawater would still be sufficient for it to be incorporated into manganese nodules by sorption on the surface of negatively charged MnO_2 (Glasby 1974;

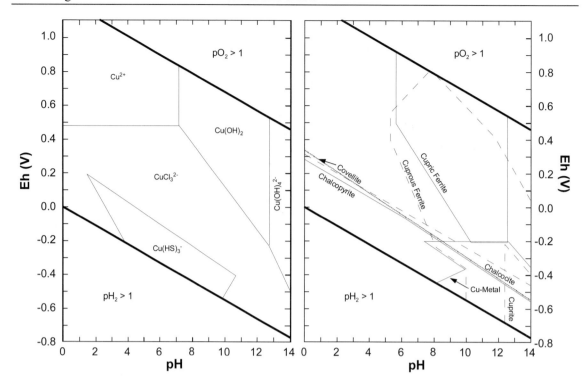

Fig. 11.16 E_H, pH diagram for Cu calculated for the chemical conditions prevailing in the deep sea (after Glasby and Schulz 1999). Note that, at pH 8, the metastable form of Cu is $Cu(OH)_2$ at an E_H, > +0.48 V. However, if the E_H, drops to +0.4 V corresponding to the E_H of seawater, the anionic species, $CuCl_3^{2-}$, becomes the more stable species. It is believed that the dominance of $CuCl_3^{2-}$ in seawater in the oxygen minimum zone accounts for the low Cu contents in Co-rich Mn crusts.

Glasby and Thijssen 1982). At somewhat lower E_H values, however, the concentration of Cu^{2+} in seawater would decline drastically and the anionic species $CuCl_3^{2-}$ would become the dominant species. Its sorption on MnO_2 would then be inhibited by charge considerations. This may well explain the high Ni/Cu ratios observed in cobalt-rich manganese crusts (max. 15) formed adjacent to the oxygen minimum zone where less oxidizing conditions prevail (Mangini et al. 1987; Meylan et al. 1990).

In many crusts, two generations of crustal growth can be observed, an older crust (18-12 Ma) and a younger crust (<12 Ma) which are separated by a thin phosphorite horizon (e.g. Halbach and Puteanus 1984; Mangini et al. 1987; Puteanus and Halbach 1988). However, McMurtry et al. (1994) and Jeong et al. (2000) subsequently obtained quite different ages for this boundary which are difficult to explain. For example, McMurtry et al. (1994) considered this boundary to have a minimum Oligocene age (28-33 Ma) and possibly a Late Paleocene age (55 Ma) based on dating of francolite in the Schumann Seamount crust. Jeong et al. (2000) also dated the age of this boundary in two crusts from seamounts in the Marshall Islands and Micronesia

at 25 Ma. Pulyaeva (1997), on the other hand, tentatively accepted the boundary age for the younger crust generation proposed by Halbach and Puteanus (1984) for custs from the Magellan Seamount chain.

However, these age differences may be a case of mistaken identity (see section 11.4.10.1). Hein et al. (1993) have reported that two major phosphogenic episodes took place in the oceans during the period of formation of older crust layers centred on the Eocene-Oligocene (~34 Ma) and Oligocene-Miocene (~24 Ma) boundaries with a minor event in the Middle Miocene (~15 Ma). It is therefore possible that older ages for phosphogenesis reported by both McMurtry et al. (1994) and Jeong et al. (2000) correspond to one of the older episodes of phosphogenesis described by Hein et al. (1993, 2000). This view is supported by the observation of Jeong et al. (2000) that phosphatization on the seamounts of the Marshall Islands lasted until the peak of the second major epoch of phosphogenesis (25 Ma) reported by Hein et al. (1993) and was much earlier than the middle Miocene phosphatization (18-12 Ma) which is generally believed to be the time of crust phosphatization in the Pacific.

Koschinsky et al. (1997) have also related differences in composition between the older and younger layers of Co-rich crusts to the expansion of the oxygen minimium zone during periods of phosphatization of the older crust layers which led to the diagenetic remobilization of certain elements within this layer of the crust.

In order to be of potential economic interest, crusts should have Co contents >0.8%, average crustal thicknesses >40 mm and be situated in an area of subdued small-scale topography. Based on their extensive experience of studying Co-rich Mn crusts, Hein et al. (1987) listed a number of other criteria for locating interesting crusts. The crusts should be situated on large volcanic edifices shallower than 1,500-1,000 m and older than 20 Ma occurring in areas of strong oceanic bottom currents. The volcanic structures should not be capped by large modern atolls or reefs and the seamount slopes should be stable. The area should not be influenced by the input of abundant fluvial or eolian debris and there should be no local active volcanism. Most importantly, the area should be characterized by a shallow and stable oxygen minimum zone. In particular, the importance of mass wasting of coral debris from atolls and guyots in tropical environments is stressed. In a detailed sampling programme around islands and seamounts in the area of the Manihiki Plateau in the equatorial South Pacific, Meylan et al. (1990) recovered only thin Mn crusts with a maximum thickness of 20 mm because of the extensive mass wasting of limestone debris in the region which gave insufficient time for thick crusts to develop.

11.4.5 Shallow-Marine Ferromanganese Concretions

Shallow-water continental margin ferromanganese concretions have been reported in a number of areas such as the Baltic Sea, Black Sea, Kara Sea, Loch Fyne, Scotland, and Jervis Inlet, British Columbia (Calvert and Price 1977). In fact, these were the first type of marine Mn deposits to be discovered, during the 1868 Sofia expedition to the Kara Sea led by A.E. Nordenskiøld (Earney 1990).

Ferromanganese concretions from the Baltic Sea have been described in detail by Glasby et al. (1997a). Three main types of concretions occur there (spheroidal, discoidal and crusts). The concretions frequently, although not always, form around a glacial erratic seed or nucleus and display alternate banding of Fe- and Mn-rich layers tens to hundreds of μm thick. Mineralogically, the concretions consist of 10 Å manganate with abundant quartz and lesser amounts of feldspar and montmorillonite. The composition of the concretions is highly variable reflecting, in part, the variable amount of erratic material in the concretion. The composition is controlled by the redox characteristics of the environment. In particular, the higher Mn/Fe ratios of concretions from Kiel Bay compared to those from other areas of the Baltic reflects the diagenetic remobilization of Mn from the adjacent muds into the concretions during the summer anoxia (Table 11.6). This may be considered to be an example of the role of *anoxic diagenesis* in the formation of shallow-marine ferromanganese concretions (Glasby et al. 1997a). Cu, Zn and Pb may be trapped as sulfides in the associated sediments and this may explain their low contents in

Table 11.6 Average composition of ferromanganese concretions from various basins of the Baltic Sea. Mn and Fe in per cent; Co, Ni, Cu and Zn in ppm (after Glasby et al. 1997a).

	Gulf of Bothnia	Gulf of Finland	Gulf of Riga	Gotland Region	Gdansk Bay	Kiel Bay
Mn	14.6	13.3	9.7	14	8.7	29.3
Fe	16.6	19.7	22.8	22.5	18.5	10.1
Co	140	96	64	160	91	77
Ni	260	35	47	750	148	97
Cu	80	9	17	48	42	21
Zn	200	113	135	80	137	340
Mn/Fe	0.88	0.68	0.43	0.62	0.47	2.9

the concretions compared to deep-sea nodules. The growth rates of the concretions have been estimated to be 3-4 orders of magnitude higher than those of deep-sea nodules. A 20 mm diameter concretion is therefore about 500-800 years old (Glasby et al. 1996). This means that these concretions are transient features on the sea floor. In active areas of the sea floor, concretions may be buried by sediment during storms and new concretions then begin to form on erratic material exposed at the sediment surface. Certain anthropogenic elements, notably Zn, are enriched in the outer layers of the concretions as a result of the pollution of Baltic seawater over the last 160 years or so. Zn profiles in the outer layers of these concretions can therefore be used to monitor heavy-metal pollution in Baltic (Hlawatsch 1999; Hlawatsch et al. 2002). This technique will be useful if the planned clean up of the Baltic Sea is to be achieved during the present century.

Baltic Sea concretions can be classified into three main types based on their abundance, morphology, composition and mode of formation: those from the Gulfs of Bothnia, Finland and Riga, from the Baltic Proper and from the western Belt Sea.

Concretions from the Gulf of Bothnia are most abundant in Bothnian Bay where the abundance reaches 15-40 kg m^{-2} in an area of about 200 km^2. This is equivalent to about 3 million tons of concretions and has led to these deposits being evaluated as a possible economic source of Mn. These concretions are mainly spheroidal up to 25-30 mm in diameter and are formed in the uppermost water-rich sediment layers at well-oxidized sites. They are most abundant where sedimentation rates are <0.4 mm a^{-1}.

Concretions from the Baltic Proper are found mainly around the margins of the deep basins in a depth range 48-103 m. The concretions are mainly discoidal 20-150 mm in diameter and crusts. Their abundance is mainly sporadic and more rarely common to abundant. Locally, abundances of 10-16 kg m^{-2} are attained. Their formation is the result of the build up of Mn and Fe in the anoxic waters of the deep basins of the Baltic Proper. During major inflows of North Sea water (>100 km^3) into the Baltic which occur on average once every 11 years, the anoxic waters are flushed out of the basins. Mn and Fe precipitate out as an unstable gel but are ultimately incorporated into the concretions as

Fig. 11.17 A photograph showing the upper surface of a hydrothermal Mn crust from the Tyrrhenian Sea (39°31.40'N, 14°43.67'E, 996 m). Part of the surface displays the characteristic black metallic sheen of hydrothermal crusts (after Eckardt et al. 1997).

oxyhydroxides. These concretions occur mainly on lag deposits in the vicinity of the halocline where strong bottom currents occur.

Concretions from Kiel Bay in the western Belt Sea occur in a narrow depth range of 20-28 m at the boundary between sands and mud in zones of active bottom currents. They occur as coatings on molluscs and as spheroidal and discoidal concretions. The formation of the concretions is influenced by the development of summer anoxia which leads to the diagenetic remobilization and lateral transport of Mn. This accounts for the high Mn/Fe ratios of these concretions.

Marcus et al. (2004) have studied the speciation of Mn, Fe, Zn and As in one of the Baltic Sea concretions described by Hlawatsch et al. (2002) by means of micro X-ray fluorescence, micro X-ray diffraction and micro X-ray spectroscopy. The concretion was shown to consist of thin, alternating Fe- and Mn-rich layers. The Fe-rich layers consisted of two-line ferrihydrite in which As was mostly pentavalent and the Mn-rich layers consisted of birnessite in which Zn is tetrahedrally coordinated and sorbed in the interlayers of birnessite. The occurrence of ~15% of As(III) in Fe-rich layers and of ~15% Mn^{3+} in Mn-rich layers was thought to reflect the stagnation of the bottom waters at the time of the deposition of the Fe- and Mn-rich layers.

11.4.6   Hydrothermal Manganese Crusts

At mid-ocean ridges, three types of submarine hydrothermal minerals are found, sulfide minerals associated with silicates and oxides, sharply fractionated oxides and silicates of localized extent and widely dispersed ferromanganese oxides. The ferromanganese oxides are generally considered to have precipitated last in this sequence and are thought to represent a late-stage, low-temperature hydrothermal phase with temperatures of deposition estimated to be in the range 20-5°C (Burgath and von Stackelberg 1995). The hydrothermal Mn deposits are characterized by high Mn/Fe ratios and low contents of Cu, Ni, Zn, Co, Pb and detrital silicate minerals. They have growth rates exceeding 1,000 mm Ma^{-1} in some cases, more than three orders of magnitude faster than that of hydrogenous deep-sea nodules and crusts.

Compared to hydrogenous Mn nodules and crusts, hydrothermal Mn crusts are relatively restricted in the marine environment and make up less than 1% of the total Mn deposits in the world ocean. These crusts occur in all types of active oceanic environments such

as at active mid-ocean spreading centers in the depth range 250-5,440 m, in back-arc basins in the depth range 50-3,900 m, in island arcs in the depth range 200-2,800 m (Eckhardt et al. 1997), in mid-plate submarine rift zones in the depth range 1,500-2,200 m (Hein et al. 1996) and at hot spot volcanoes in the depth range 638-1,260 m (Eckhardt et al. 1997). Fossil submarine hydrothermal manganese deposits have also been recovered from sediment cores in the Izu-Bonin Arc (Usui 1992) and the Central Pacific Basin (Usui et al. 1997).

Submarine hydrothermal Mn crusts have been reported from Enareta and Palinuro seamounts in the Tyrrhenian Sea (Eckhardt et al. 1977), along the Izu-Bonin-Mariana Arc (Usui et al. 1986, 1989; Usui and Nishimura 1992; Usui and Terashima 1997; Usui and Glasby 1998) and at the Pitcairn hotspot (Glasby et al. 1997b; Scholten et al. 2004). The Tyrrhenian Sea crusts consist of porous, black, layered Mn oxides up to 45 mm thick. In some cases, the surface has a black metallic sheen. The crusts overlie substrates such as calcareous sediment, siltstone and oyster shells. A photograph of a hydrothermal Mn crust from the Tyrrhenian Sea is shown in Figure 11.17. The crusts consist dominantly of 10Å manganate and 7Å manganate with minor quartz, illite, montmorillonite, plagioclase and goethite. The sample having the highest Mn content contained 54.2% Mn, 0.07% Fe, 33 ppm Ni, 200 ppm Cu, 20 ppm Zn, 11 ppm Pb and 910 ppm Ba with a Mn/Fe ratio of 774. It also had a low REE abundance and a negative Ce anomaly.

Submarine hydrothermal Mn crusts are also relatively common along the Izu-Bonin-Mariana Arc. Recent hydrothermal manganese crusts associated with active hydrothermal systems tend to occur on seamounts or rifts located about 5-40 km behind the volcanic front on the Shichito-Iwojima Ridge. Fossil hydrothermal Mn crusts associated with inactive hydrothermal systems occur on seamounts located on older ridges running parallel to the volcanic front in both forearc and backarc settings. Fossil hydrothermal Mn crusts are generally overlain by hydrogenetic Mn oxides. The thickness of the overlying hydrogenetic Mn crust depends on the length of time since hydrothermal activity ceased. Figure 11.18 shows the distribution of recent and fossil submarine hydrothermal Mn crusts across the Izu-Bonin Arc. Mineralogically, the hydrothermal Mn deposits consist of 10 Å manganate and/or 7 Å manganate. The Mn/Fe ratios of these deposits range from 10 to 4,670 and the contents of Cu, Ni, Zn, Co and Pb from 20 to 1,000, 1 to 1,403, 1 to 1,233, 6 to 209 and 0 to 93 ppm, respectively.

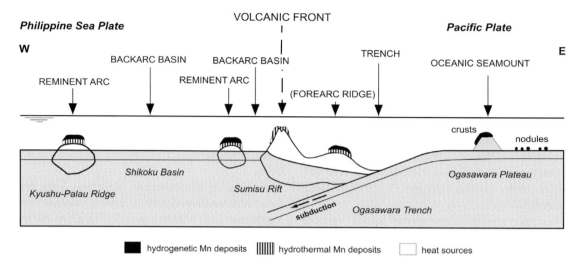

Fig. 11.18 A schematic diagram showing the distribution of hydrogenous and hydrothermal Mn deposits across an E-W section of the northwest Pacific island arc, south of Japan (after Usui and Terashima 1997). Four types of deposit are observed: i. hydrogenous Mn crusts and nodules occurring on Pacific seamounts and in deep-sea basins, ii. hydrogenous crusts on marginal seamounts of the remnant arcs, iii. modern hydrothermal Mn deposits associated with submarine volcanoes and backarc rifts in the active volcanic ridge, and iv. fossil hydrothermal Mn deposits usually overlain by younger hydrogenous Mn crusts from seamounts of the remnant arc. Each type of deposit has its own characteristic mineralogy, composition and growth rate. The growth of these deposits is closely related the evolution of the island arc system.

At the Pitcairn hotspot, massive hydrothermal Mn crusts display the highest Mn/Fe ratios (2,440), the lowest contents of Ni (18ppm) and Zn (21ppm) as well as the lowest aluminosilicate fraction (<1%) in individual horizons (sample 69-3 DS at a depth of 7-8 mm). This type of crust may therefore be considered to represent the most extreme hydrothermal endmember. The other types of hydrothermal Mn crust are probably formed as a result of the interpenetration and replacement of volcanoclastic sands or biogenic carbonates by hydrothermal Mn oxides. The low contents of Fe, Ni and Zn in the massive crusts were thought to reflect the rapid incorporation of these elements into sulphide minerals within the interior of the hotspot volcano such that only a small proportion of these elements are available for incorporation in the crusts. The small positive Eu anomaly in the crusts on a NASC-normalized basis indicated lower temperatures of the hydrothermal fluids within the hotspot volcano (<250°C) compared to those at mid-ocean ridges (c.350°C). A laser-ablation ICP-MS profile in one of the crusts revealed varying REE concentrations and patterns in the different layers of the crust. These data showed that the hydrothermal component was variable during the formation of the crust and was almost 100% in the upper layers of this crust but about 80% in the lower layers.

In addition to submarine hydrothermal manganese crusts formed near submarine hydrothermal vents,

submarine hydrothermal plumes can transport detectable amounts of manganese up to 1,000 km from the crests of mid-ocean ridges (Burton and Statham 1988) (see section 11.2). Under favourable circumstances, part of this manganese can then settle out over manganese nodule fields and contribute to nodule growth. As an example, Chen and Owen (1989) identified a hydrothermal component in deep-sea manganese nodules from the southeast Pacific based on Q-mode factor analysis of compositional data for 76 nodules. The hydrothermal component was shown to have a Mn/Fe ratio of 0.31 which is typical of metalliferous sediments and to be distributed in two areas, one west of the East Pacific Rise between 3 and 25°S and the other east of the East Pacific Rise south of 25°S. These observations are supported by the subsequent identification of hydrothermal helium plumes in the South Pacific by Lupton (1998) in which a pair of intense hydrothermal plumes was observed to extend westwards from the East Pacific Rise at 10°S and 15°S at a water depth of 2,500 m and a plume of lesser intensity to extend eastwards into the Chile Basin at 30°S. The study of Chen and Owen (1989) therefore showed that a hydrothermal component can be incorporated into deep-sea manganese nodules forming within 1,000 km or so of an active mid-ocean ridge and not just into metalliferous sediments as is commonly assumed (Kunzendorf et al. 1993; Marchig 2000; Dekov et al. 2003; van de Fliert et al. 2003a).

11.4.7   Micronodules

Manganese micronodules are an important reservoir for Mn and associated transition elements in oxic deep-sea sediments (Stoffers et al. 1984). They are generally <1 mm in diameter. By contrast, macronodules are generally >3-6 mm. There is therefore a size gap in which no Mn deposits are found. This is a consequence of the fact that macronodules require a discrete nucleus several mm in diameter on which to form (Heath 1981). The abundance of micronodules appears to be inversely related to the sedimentation rate with highest abundances occurring in dark brown clays where sedimentation rates are low. The shape and internal structure of the micronodules is dependent on the nature of the seed material around which they form, of which calcareous tests, siliceous tests and volcaniclastic material are the most important. Micronodules tend to predominate in the coarse fraction of the sediment. The mineralogy and composition of the micronodules tend to be similar to that of the associated nodules, although differences do exist. Micronodules tend to have higher Mn/Fe ratios than nodules from the same sediment core. These diffe-

rences are related to the depth of occurrence of the micronodules within the sediment column and reflect the importance of diagenetic processes in remobilizing elements within the sediment. In this sense, micronodules may be considered a transient feature which serves as a medium for recycling elements within the sediment and ultimately into manganese nodules. The study of micronodules has received relatively little attention because of the difficulty of hand picking them from the sediment.

In a detailed study of micronodules from various areas of the Pacific, Stoffers et al. (1984) showed that micronodules are characterized by regional variations in shape, texture, mineralogy and composition. Micronodules from the Southwestern Pacific Basin are most abundant in the dark brown clays of the central part of the basin. Mineralogically, these micronodules consist of 10 Å manganate and δMnO_2 with traces of quartz, phillipsite, quartz and feldspar. They have Mn/Fe ratios of 2.4 and Ni+Cu contents of 1.3%. These Mn/Fe ratios are somewhat higher than for the associated nodules but the Ni+Cu contents somewhat lower. The micronodules are botryoidal with smooth surface texture. Under the SEM, the surfaces of the micronodules display a honeycomb texture.

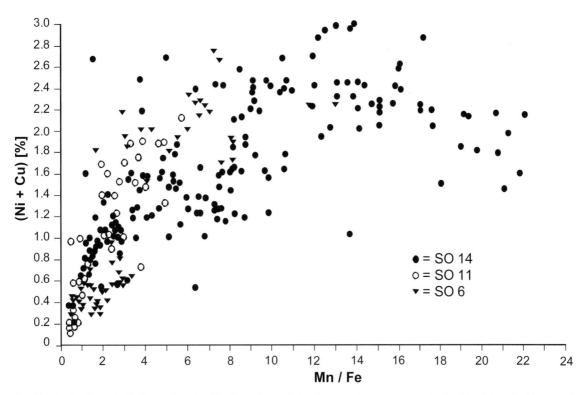

Fig. 11.19 A plot of Ni+Cu against Mn/Fe for micronodules from various areas in the Pacific (after Stoffers et al. 1984). SO-06 samples taken from the Clarion-Clipperton F.Z. region, SO-11 from the Peru Basin and SO-14 from the Southwestern Pacific Basin.

Micronodules from the C-C F.Z. occur mainly within siliceous ooze. They consist of 10 Å manganate with traces of quartz and sometimes phillipsite. They have Mn/Fe ratios of 4.7 and Ni+Cu contents of 1.7%. Again, the Mn/Fe ratios are somewhat higher than for the associated nodules but the Ni+Cu contents somewhat lower. The micronodules are dominantly spheroidal or have rod-like structures. Under the SEM, the surfaces of the micronodules consist of plates.

Micronodules from the Peru Basin occur in three stratigraphic horizons. The upper 50-100 mm of the core consists of a homogeneous brown calcareous-siliceous mud containing a high abundance of micronodules. The micronodules are larger (125-500 mm) than those taken deeper in the core (40-125mm) and have a mamillated botryoidal surface texture similar to that of the associated macronodules. They consist of 10 Å manganate with traces of quartz and feldspar. They have Mn/Fe ratios of 8.7 and Ni+Cu contents of 2.1%. These sediments are underlain by a highly bioturbated light yellowish brown mud containing varying amounts of calcareous and siliceous organisms. These micronodules have a smooth surface texture. The replacement of radiolaria and foraminifera can be clearly seen under the SEM. The micronodules are coarsely crystalline with well-defined rod-like structures. They consist of 10 Å manganate with traces of quartz and feldspar and have Mn/Fe ratios of 1.9 and Ni+Cu contents of 0.9%. The sharp drop in the Mn/Fe ratios of these micronodules compared with those from the sediment surface suggests that remobilization of Mn and associated transition elements has taken place within this sediment horizon. At the base lies a dark overconsolidated clay with a low content of siliceous tests but containing volcanic glass and fish teeth. These micronodules have Mn/Fe ratios of 5.0 and Ni+Cu contents of 1.3%. The high Mn/Fe ratios of these micronodules suggest a well-oxidized sedimentary environment from which no diagenetic remobilization has taken place. The importance of element remobilization in the yellowish brown muds is confirmed by the fact that micronodule abundance is high in the surface sediments, low in the yellowish brown muds and high again in the dark brown clays.

A plot of Ni+Cu against Mn/Fe for micro-nodules from various areas in the Pacific is presented in Figure 11.19. It is seen that, although Mn/Fe ratios in micro-nodules >20 can occur, particularly in the Peru Basin, there is no well-defined point of reversal such as observed for Mn nodules (Fig. 11.14), although there is a flattening of the curve at a Mn/Fe ratio of about 10. This is a consequence of the fact that the micro-nodules occur within the sediment column where

conditions are less oxidizing than at the sediment surface. Pore waters are not as enriched in Mn, Ni and Cu within the sediment as at the sediment surface and the Mn/Fe ratios and Ni+Cu contents of micronodules are never as high as in the associated macronodules. Diagenetic fractionation of elements in micronodules is therefore not as pronounced as in the associated macronodules.

11.4.8 Mineralogy

There are three principal phases in manganese nodules, Mn oxides which tend to incorporate cationic transition metal species such as Ni^{2+}, Cu^{2+} and Zn^{2+}, Fe oxyhydroxides which tend to incorporate anionic species such as HPO_4^{2-}, $HAsO_4^{2-}$, HVO_4^{2-}, MoO_4^{2-} and WO_4^{2-} as well as the REE and Co^{3+} and detrital aluminosilicates which consist of elements such as SiO_2, Al_2O_3, TiO_2 and Cr_2O_3. In addition, carbonate fluorapatite (francolite) also occurs in older generations of Co-rich manganese crusts and as a discrete layer between the older and younger crust generations. Concentrations of individual elements in deep-sea Mn nodules and crusts therefore tend to covary with the relative amounts of these four phases (cf. Koschinsky and Halbach 1995; Koschinsky and Hein, 2003).

Three principal Mn oxide minerals are found in Mn nodules and crusts. Their principal X-ray diffraction peaks are given below together with their alternative mineral names (in parenthesis):

10 Å manganate   9.7 Å   4.8 Å   2.4 Å   1.4 Å
(todorokite, buserite, 10 Å manganite)

7 Å manganate   7.3 Å   3.6 Å   2.4 Å   1.4 Å
(birnessite, 7 Å manganite)

δMnO_2 2.4 Å   1.4 Å
(vernadite)

Mn oxide minerals are characterized by fine grain size in the range 100-1,000 Å. These minerals are poorly crystalline and give diffuse X-ray diffraction patterns. FeOOH minerals also tend to be fine grained with sizes of the order of 100 Å and are generally X-ray amorphous (Johnson and Glasby 1969). In chloride-bearing solutions such as seawater, akagenéite (β-FeOOH) has been identified as the principal iron oxyhydroxide mineral present by Mössbauer spectroscopy (Johnston and Glasby 1978). However, there are no independent Fe-bearing minerals in Mn nodules.

Rather, δMnO_2 may be considered to be a randomly stacked Fe-Mn mineral. Some authors have suggested that there is usually an intimate association of δMnO_2 and $FeOOH \cdot xH_2O$ in deep-sea hydrogenous nodules due to the transport of minerals as colloidal particles and subsequent coagulation of these particles (Halbach et al. 1981). Detrital minerals such as quartz and feldspar normally give much stronger X-ray diffraction peaks than the Mn and Fe oxides. Within various types of nodules and crusts, the following Mn and Fe oxide minerals tend to be found.

Shallow-water concretions   10 Å manganate
(Baltic Sea, Loch Fyne, Scotland)

Hydrothermal Mn crusts   10 Å manganate +
                                 7 Å manganate

Deep-sea hydrogenous δMnO_2
Mn nodules and crusts

Deep-sea diagenetic Mn nodules 10 Å manganate

Within the Mn oxide minerals, there is a well-known dehydration sequence, 10 Å manganate → 7 Å manganate → δMnO_2 (Glasby 1972; Dymond et al. 1984). Samples collected on board ship must therefore be preserved moist in seawater to prevent mineralogical change on drying (Glasby et al. 1997a). A similar redox sequence 10 Å manganate → 7 Å manganate → $\delta MnO2$ is also observed in nodules. This means that nodules formed in less-oxidizing environments tend to contain 10 Å manganate whereas those formed in more-oxidizing environments tend to contain δMnO_2 (Glasby 1972). This trend is confirmed by the O/Mn ratios in nodules which vary from 1.60 in shallow-water concretions to 1.95 in deep-sea nodules. 98% of the Mn in deep-sea nodules is therefore in the Mn (IV) form (Murray et al. 1984; Piper et al. 1984).

Modern ideas on the structure of Mn oxides are based largely on the work of R. Giovanoli of the University of Berne. Much of this work was carried out on synthetic 10 Å (Na⁺)-manganate prepared by the rapid oxidation of $Mn(OH)_2$ with O_2 for 5 hrs. Two types of 10 Å manganate may be considered. Todorokite is a large tunnel-structure mineral based on MnO_6 octahedra which can not expand or contract on heating to 100°C (or even 400°C) (Burns and Burns 1977, 1980; Turner and Buseck 1981; Waychunas 1991; Mellin and Lei 1993; Lei 1996; Post 1999). Buserite, on the other hand, is a phyllomanganate mineral with an expandable or contractible sheet-like structure (Giovanoli and Bürki 1975; Giovanoli 1980, 1985; Giovanoli and Arrhenius 1988; Waychunas 1991; Kuma et al. 1994;

Usui and Mita 1995; Post 1999). It contains exchangeable interlayer cations (Ca^{2+}, Mg^{2+}, Cu^{2+}, Ni^{2+}, Co^{2+}, $2Na^+$) which occupy specific lattice sites. The ratio of these metals to Mn is 1:6 or 1:7. The uptake sequence of these metals into the buserite structure is $Cu^{2+} > Co^{2+} > Ni^{2+} > Zn^{2+} > Mn^{2+} > Ca^{2+} > Mg^{2+} > Na^+$. Fe can not enter the buserite lattice because any Fe^{2+} would be oxidized to insoluble FeOOH by Mn^{4+}. Instead, any excess FeOOH in the sediments reacts with the dissolving siliceous tests to form nontronite and therefore fix Fe in siliceous sediments. This explains the high Mn/Fe ratios in diagenetic nodules from oxic environments. Buserite may be considered to be an expandable or contractible sheet which can accommodate hydrated stabilizing interlayer cations. It can therefore be distinguished from todorokite by expanding the interlayer spacing to 25 Å on treatment with dodecylammonium hydrochloride or contracting it to 7 Å on heating in air to 100°C. Usui et al. (1989) had previously described the influence of divalent cations in stabilizing the structures of todorokite-like and buserite-like manganese oxide minerals in seawater, in air and dried in air at 110°C (cf. Usui 1979).

Mellin and Lei (1993) have explained the mineralogy of the principal marine Mn deposits in the following terms. Low-temperature hydrothermal 10 Å manganate has a less stable todorokite-like structure with tunnel walls composed of $Mn^{2+}O_{2x}^{2-}(OH)_{6-2x}$ octahedra. High-temperature hydrothermal 10 Å manganate has a more stable todorokite-like structure as a result of the oxidation of interlayer Mn^{2+}. In both cases, divalent cations such as Cu^{2+}, Ni^{2+} and Zn^{2+} are deposited as sulfides prior to the deposition of Mn minerals which explains the low contents of these metals in these deposits. Diagenetic 10 Å manganate, on the other hand, has an unstable buserite-like structure. Divalent cations such as Ni^{2+}, Cu^{2+}, Zn^{2+}, Mg^{2+} and Ca^{2+} can substitute for $2Na^+$ in the interlayer spacing. Diagenetic deep-sea nodules formed in oxic environments have high Cu^{2+} and Ni^{2+} contents (up to 2%) as a result of the release of these elements from siliceous tests. The resulting high contents of Ni+Cu in these nodules explain their commercial interest. Diagenetic shallow-water concretions have low Cu^{2+} and Ni^{2+} contents (<0.1%) as a result of trapping these elements in the sediment column as sulfides. The composition of marine manganese deposits therefore reflects both the mineralogy of the sample and the availability of the transition metal ions for nodule formation (Glasby and Thijssen 1982).

The mineralogical data reported above were obtained from X-ray diffraction analysis. However, this technique has distinct limitations when applied to the study of marine manganese nodules and crusts where the minerals making up these deposits are fine grained and the powder diffraction patterns for many of these minerals are similar (Post 1999). This makes the unambiguous identification of these minerals difficult (Manceau et al. 1992a). For this reason, other techniques have been used to study the mineralogy of these deposits at the micron (μm) level such as electron diffraction (Varentsov et al. 1991) and X-ray absorption spectroscopy (XAS). XAS includes both X-ray absorption near-edge structure (XANES) and extended X-ray absorption fine structure (EXAFS) (Manceau et al. 1992a,b, 2002; Drits et al. 1997; Silvester et al. 1997). Studies using these techniques were initiated by F.V. Chukhrov, Director of the Institute of Geology of Ore Deposits of the Russian Academy of Sciences in Moscow, in the 1980s (e.g. Chukhrov et al. 1989) and achieved world recognition following joint Franco-Russian work on this topic.

Using these methods, it was established that the 10 Å phase in manganese nodules actually belongs to one of at least seven minerals each with different structural characteristics: asbolane (a Co-Ni-bearing manganese oxide mineral), buserite-I, buserite-II, todorokite, mixed-layered phases: asbolane-buserite-I, buserite-I-buserite-II and buserite-I-"defective lithiophorite" (Manceau et al. 1987, 1992b; Drits et al. 1997). In the case of the manganese crust from Krylov Seamount located in the eastern North Atlantic on the Cape Verde plate, Varentsov et al. (1991a) showed that the lower layers of the crust consisted dominantly of Fe-vernadite (δMnO_2) and goethite with admixtures of mixed layered asbolane-buserite whereas the upper layers consisted of Fe-vernadite and feroxyhite ($\delta FeOOH$) with subordinate amounts of birnessite, mixed-layered asbolane-buserite and goethite (cf. Varentsov et al. 1991b). This sequence was thought to reflect a significant hydrothermal input into the crust when it first formed near the crest of the Mid-Atlantic Ridge which was followed by a dominantly hydrogenous origin. In addition, asbolane has been shown to occur as flakes and microflakes in manganese crusts and nodules from the Atlantis Fracture Zone and Krylov Seamount in the Atlantic Ocean, the Kurchatov Fracture Zone and the Chile Plate in the Pacific Ocean and in manganese micronodules in nannofossil ooze from the Indian Ocean (Chukhrov et al. 1982, 1983).

An important question is the anomalous position of Co in the geochemistry of manganese nodules. Glasby and Thijssen (1982) have interpreted this in terms of the crystal field characteristics of Co. On this basis, Co^{3+} (d^6) is stable in nodules in the low spin state with octahedral coordination and has an ionic radius of 0.53 Å which is almost identical to that of Fe^{3+} and Mn^{4+}. As such, Co may substitute in either MnO_2 or FeOOH as the trivalent ion but not in the interlayer spacing of 10 Å manganate as the divalent ion as is the case for Ni^{2+}, Cu^{2+}, Zn^{2+} and Mn^{2+}. The ability of Co to substitute in both Mn and Fe oxyhydroxides leads to its correlation with either Mn or Fe in nodules or crusts depending on the environment of formation (Burns and Burns 1977, 1980; Halbach et al. 1983; Giovanoli and Arrhenius 1988). Experimental evidence has confirmed that Co^{3+} is the dominant form of Co in Mn nodules (Dillard et al. 1984; Hem et al. 1985; Manceau et al. 1997).

11.4.9 Dating

Deep-sea Mn nodules and crusts are amongst the slowest growing minerals on Earth (Manheim 1986). They have a minimum growth rate of 0.8 mm Ma^{-1} (Puteanus and Halbach 1988) which is equivalent to the formation of about one unit cell every year. Shallow-water concretions grow about 10^4 times faster. Several methods have been used to date deep-sea manganese nodules and crusts (Ku 1977; Mangini 1988).

One of the earliest methods used to date nodules involved K-Ar dating of the volcanic nucleus of the nodule. This method is of limited value because submarine weathering of the core often invalidates the results. In addition, the method makes no allowance for any time gap between the formation of the nucleus and subsequent Mn accretion or for hiatuses in growth of the Mn deposit. Only a minimum average growth rate is therefore obtained. Although this method was used by early workers (e.g. Barnes and Dymond 1967), it is too crude to be of much value now.

In addition, Moore and Clague (2004) have recently demonstrated the progressive thickening of ferromanganese crusts along the Hawaiian Ridge with increasing distance from the Hawaiian hotspot (Loihi Seamount). From a comparision of the maximum thicknesses of the submarine crusts along the ridge with radiometric ages of adjacent subaerial features, these authors estimated the growth rates of the crusts to be about 2.5 mm Ma^{-1}. This growth rate was then used to deduce the ages of landslide deposits and volcanic fields occurring away from the axis of the ridge from the estimated ages of associated crusts. This may be considered to be an example of indirect dating of ferromanganese crusts.

Several attempts have been made to date Mn nodules and crusts by paleontological methods. Initial studies by Harada and Nishida (1976, 1978) established slow growth rates for the outer layers of some deep-sea nodules (1.0-6.7 mm Ma^{-1}) but higher rates for the inner layers of some nodules (39 mm Ma^{-1}) based on biostratigraphic analysis of calcareous microfossils. Cowen et al. (1993) subsequently determined the ages of different layers of a Mn crust (KK84-RD50 S1B) from Schumann Seamount in the North Pacific based on the identification of coccolith imprints at various depths in the crust (see section 11.4.10.1). In a detailed study, Pulyaeva (1997) also identified three main periods of growth of Co-rich Mn crusts from the Magellan Seamounts in the western Pacific based on a study of calcareous microfossils. On this basis, it was established that the crusts began forming inter-mittently in the Late Cretaceous in the presence of phosphatized nannoforaminifera. The crusts attained a composition similar to that of today in the Eocene but the Miocene-Pliocene was shown to be the most favourable period for growth of the crusts. Numerous interruptions in the growth of the crusts were observed with major regional hiatuses being recorded at the Cretaceous-Paleocene and Eocene-Oligocene boun-daries (see section 11.4.10.1). More recently, Shilov (2004) has reported on a major programme of dating manganese nodules from the C-C F.Z. based on age determinations of radiolaria in the nodules. From this, he was able to demonstrate that active growth of the manganese nodule fields started between the Late Oligocene and Early Miocene at about 23.7 Ma and that the principal layers which make up 50-80% of the total nodule volume in individual nodules were formed in the Middle Eocene (16.6-11.2 Ma) and Late Miocene-Pleistocene (11.2-1.8 Ma). He also established that the highest growth rates of the nodules (2.5-3.7 mm Ma^{-1}) occurred during the Pleistocene-Holocene (1.8-0 Ma). These ages are in agreement with the findings of Glasby (1978) who recorded the presence of buried manganese nodules in D.S.D.P. cores associated with sediments of Oligocene age and older (cf. Usui and Ito 1994; Ito et al. 1998) and by von Stackelberg and Beiersdorf (1991) who showed that most of the nodules in the C-C F.Z. southeast of Hawaii began their growth at Tertiary hiatuses. These ages are much older than the middle Miocene age (~12-15 Ma) obtained by Kadko and Burckle (1980) for two nodules from the C-C F.Z. based on the age of fossil diatoms in the nodules.

Nonetheless, most modern methods of dating Mn nodules and crusts rely on radiometric determinations. There are three principal methods available based on the following isotope ratios ^{230}Th/^{232}Th, ^{231}Pa/^{230}Th and ^{10}Be. The half lives of ^{230}Th, ^{231}Pa and ^{10}Be are

75,200 a, 32,000 a and 1.5 Ma, respectively (Turekian and Bacon 2004). The ^{230}Th/^{232}Th and ^{231}Pa/^{230}Th methods can therefore only date samples to ages of 300,000 and 125,000 a (corresponding to 4 half lives of the isotopes), respectively, which is equivalent to the outer 1-2 mm of a deep-sea nodule or crust. The ^{10}Be method, on the other hand, permits dating back to ~10 Ma which is equivalent to a depth of a few cm in some samples (Turekian and Bacon 2004). Hein et al. (2000) have compiled a list of all isotopically-deter-mined growth rates of hydrogenous Mn crusts to that date.

The ^{238}U and ^{235}U decay-series methods are based on the fact that ^{230}Th and ^{231}Pa are generated in seawater by the decay of U isotopes. The isotopes are then carried to the sea floor on particles and incor-porated into the nodules and crusts. The residence times for Th and Pa in seawater are short (less than 40 and 160 years respectively). The distribution of ^{230}Th and ^{231}Pa with depth in nodules or crusts can be measured by counting a particles after separation from the nodule material and plating on a planchet or more rapidly by α-track counting on nuclear emulsion plates. The growth rate of the nodule or crust can then be calculated from the formula

$$dC/dt \quad = \quad S\ dC/dx - \lambda C$$

$$\text{or} \qquad C_{(x)} \quad = \quad C_0 \exp(-x\lambda/S)$$

where $C_{(x)}$ is the concentration at depth x, ë is the decay constant of the nucleus (=ln $2/\tau_{1/2}$), $\tau_{1/2}$ the half life of the nuclide and S the nodule growth rate.

The ^{230}Th/^{232}Th and ^{231}Pa/^{230}Th methods were the first radiometric methods to be used for dating of manganese nodules but their application was limited because of the short half lives of ^{230}Th and ^{231}Pa. In spite of the limitation of working with samples from the uppermost 1-2 mm of the sample, high resolution studies of nodules have been achieved using this method. For example, Eisenhauer et al. (1992) analyzed 69 samples to a depth of 1.4 mm with a resolution of 0.02 mm in the VA 13/2 Mn crust sample (see section 11.4.10.1). The samples were analyzed by alpha counting after chromatographic separation of thorium from uranium and electroplating the elements on stainless steel holders. However, the development of the thermal-ionization mass spectrometric (TIMS) method for determining U and Th isotopes increased the precision of the measurements and extended the possibilities of this method (Chabaux et al. 1995, 1997).

In particular, Böllhofer et al. (1996, 1999) used this method to investigate the growth rates of diagenetic manganese nodules from the Peru Basin. These nodules were shown to have high growth rates of ~110 mm Ma^{-1} but with frequent oscillations in growth rate which were attributed to transitions in the glacial/interglacial cycle. These climatic variations were assumed to control the depth of the suboxic/oxic boundary in the sediment leading to changing growth rates and the development of distinct laminations within the nodules. However, von Stackelberg (1997, 2000) subsequently observed that the highest growth rates in these nodules occur on the underside of large nodules which repeatedly sink to a level immediately above the redox boundary where diagenetic recycling of Mn is at a maximum. These authors therefore suggested that layering of these nodules was the result of the lifting of these nodules by benthic organisms within the oxic surface sediments from a diagenetic to a hydrogenous environment rather than the result of climatic change as proposed by Böllhofer et al. (1996, 1999).

Han et al. (2003) also established that there are four orders of basic cyclic growth patterns in a deep-sea manganese nodule from the C-C F.Z. based on analysis of samples taken at intervals of 0.1 mm in the outer 1.3 mm of the nodule using the $^{230}Th_{excess}/^{232}Th$ method. These growth patterns were named laminae bonds, laminae zones, laminae groups and laminae pairs and had thicknesses of 402-454, 185-206, 58-67 and 15-18 μm, respectively. This rhythmic growth was considered to be related to Milankovitch cycles with the growth cycles of the laminae, bands and zones corresponding to the periods of eccentricity, obliquity and precession of the Earth's orbit, respectively. From the thicknesses of these laminae, it was was then possible to establish a net growth rate of the outer layers of the nodule of 4.5 mm Ma^{-1} based on the Earth's orbital cycle in close agreement with the rate of 4.6 mm Ma^{-1} determined by $^{230}Th_{excess}/^{232}Th$ dating of the nodule. This is an excellent example of the high resolution that can now be attained using $^{230}Th/^{232}Th$ dating.

^{10}Be dating involves a measurement of the depth profile of ^{10}Be in a deep-sea nodule or crust. The half life of ^{10}Be is 1.5 Ma which normally permits dating back to ~10 Ma, although earlier authors dated crusts back to ~14 Ma by analyzing larger amounts of material for the older samples (Segl et al. 1989). ^{10}Be is a cosmogenic nuclide which is deposited at the sea surface before being mixed into the oceans. Its residence time in the ocean is about 1,000 yrs. Because of its greater half life, ^{10}Be can be measured to much greater depths within the nodule or crust (> 40 mm) than the U decay-series isotopes (1-2 mm). Post-

depositional diffusion of ^{10}Be is not considered a problem in determining the growth rates of the nodule or crust (cf. Kusakabe and Ku 1984). The development of accelerator mass spectrometry (AMS) led to a reduction in sample sizes and counting times by several orders of magnitude compared to measurements based on radioactive decay. Much better resolution can therefore be obtained and dating to 15 Ma is now possible. This predates the age of the lower Miocene Antarctic glaciation. Discontinuities in the growth rates of nodules and crusts can also be determined from breaks in the ^{10}Be depth profile. As an example, Koschinsky et al. (1996) determined the growth rates of the surface layers of two NE Atlantic manganese crusts to be 3 and 4.5 mm Ma^{-1} using this method. Mangini et al. (1990) and Usui et al. (1993) also determined the growth rates of S.W. Pacific nodules to be in the range 1.1-6.0 mm Ma^{-1}.

A particularly interesting application of the ^{10}Be method involved the dating of entrapped manganese nodules in an ODP sediment core taken in the Campbell Nodule Field located at the foot of the Campbell Plateau in the Southwestern Pacific Basin (Graham et al. 2004). From the ages of the rims of the buried nodules, it was concluded that bottom water circulation was more intense, and therefore sedimentation rates were lower, during the periods 10-5.5 Ma and 1.5-0 Ma which facilitated the maintenance of the nodules at the sediment surface. However, a major limitation of the ^{10}Be method is that it does not permit the direct determination of the ages of the earliest sections of Co-rich Mn crusts. ^{10}Be dating must therefore be used in conjunction with other methods for this purpose.

Sr isotope stratigraphy has also been developed as a high resolution stratigraphic method for determining the growth rates of hydrogenous and hydrothermal Mn crusts based on a comparison of the Sr isotope ratios of individual layers in the crust with Sr isotope curves for seawater (Futa et al. 1988). In principal, this method can be used to identify hiatuses in the growth of the crusts to ages in excess of 20 Ma. However, von der Haar et al. (1995) subsequently demonstrated that Sr in the crusts is extracted from both the phosphatic and detrital phases of the crust during leaching and that the Sr in the Mn oxide phase exchanges with seawater throughout the history of the crust. The method therefore appears to be of little value in dating Mn crusts.

In addition, indirect methods of dating Mn crusts have been developed such as the one based on the inverse relationship between the Co content of the crusts and their rate of accumulation (Manheim and Lane-Bostwick 1988; Puteanus and Halbach 1988).

Based on the analysis of 520 samples of ferro-manganese crusts from ~250 locations, it was established that the flux of Co into ferromanganese crusts is constant over the entire period of growth of the crusts whereas the flux of Mn into the crusts is variable. As a consequence, the content of Co in crusts is inversely proportional to the rate of formation of the crust. It therefore became possible to derive an empirical equation relating the growth rate of each layer in a crust to the Co content in that layer by fitting data derived from the ^{10}Be dating of 20 ferromanganese crusts. This led to the formulation of the equation G (mm Ma^{-1}) = 1.28/[Co(%) – 0.24] where G is the growth rate of the layer of the crust under consideration from which the growth rate and therefore the age of the crust can be derived. However, modification of this equation is required in cases where phosphatization of the crust has occurred. The great advantage of this method is that it requires no further radiometric age determinations to date a crust of unknown age. It is therefore much easier to calculate the growth rates of manganese crusts using this method than by the other methods. Frank et al. (1999a) have confirmed the validity of the Co geo-chronometer and shown that it provides detailed information on the growth history of ferromanganese crusts prior to 10-12 Ma where the ^{10}Be method can not be applied.

Growth rates of shallow-water ferromanganese concretions from Mecklenburg Bay in the Baltic Sea have also been determined by ^{210}Pb dating (half life of ^{210}Pb is 22.3 years) and by comparison of the distribution of the Zn concentration (on a mm scale) in the outer layers of selected concretions with the distribution of Zn in dated sediment cores from the same area (Hlawatsch 1999). These data gave growth rates for the concretions in the range 0.018 - 0.21 mm a^{-1} which are 4 to 5 orders of magnitude higher than for the deep-sea manganese deposits. Liebetrau et al. (2002) also determined the growth rates of concretions from this area from the distribution of ^{226}Ra$_{excess}$/Ba ratios with depth in selected concretions (half life of ^{226}Ra is 1622 years) and obtained somewhat lower growth rates in the range 0.0075 - 0.021 mm a^{-1}. From these data, the maximum age of these concretions was calculated to be 4,300 ± 300 years which is close to the period when sea level stabilized at its present level in the Baltic about 5,500 to 4,500 years ago. By contrast, very little information is available on the growth rates of submarine hydrothermal manganese crusts at present (Lalou et al. 1983; Usui and Nishimura 1992; Usui and Terashima 1997).

11.4.10 Mn Crusts as Paleoceanographic Indicators

11.4.10.1 Recording Hiatuses in Mn Crusts

Mn crusts may be considered to be condensed stratigraphic sections that record variations in paleoceanographic conditions with time. High resolution dating of Mn crusts can therefore be used to record the occurrence of major paleoceanographic events. A number of such studies has been undertaken (Segl et al. 1989; McMurtry et al. 1994; Koschinsky et al. 1996; Frank et al. 1999a). In this section, the results of two detailed studies of dating deep-sea manganese crusts using various methods of radiometric dating are presented. In addition, the distribution of the long-lived radiogenic isotopes of Nd, Pb, Be, Hf and Os in Co-rich Mn crusts has been extensively used to study the patterns of deep ocean circulation and the provinciality of these crusts over the past 10-15 years. This topic has been excellently reviewed by Frank (2002).

One of the most thoroughly studied Mn crusts was collected on top of an abyssal seamount at 4,830 m water depth in the equatorial N. Pacific in 1976 during cruise VA 13/2 of R.V. Valdivia (sample 237 KD) (Friedrich and Schmitz-Wiechowski 1980). It is a large hemispherical Fe-Mn crust 500 mm in diameter and up to 250 mm thick partly covering a basalt substrate. It is quite different in character from the Co-rich Mn crusts described in section 11.4.4. The volcanic seamount on which it occurs was formed about 65 Ma near crest of the East Pacific Rise. Submarine weathering of glass on the surface of the submarine basalt led to formation of an 8 mm thick nontronite layer. The seamount migrated north across the equatorial high productivity belt during which time it was above the Carbonate Compensation Depth (CCD).

The mode of formation of this crust has been outlined by von Stackelberg et al. (1984). Below a depth of 40 mm, the crust was characterized by higher Fe contents which could be attributed to the dissolution of calcareous tests resulting in an increased supply of iron to the crust. Goethite was the main mineral formed and could be seen as yellowish-brown flecks. Calcareous tests were also observed within the crust. At 40 mm, there is an abrupt change in composition of the crust with higher Mn/Fe (1.6) and Ce/La (3.5) ratios and higher contents of Ni (0.45%) and Cu (0.25%) above that boundary. These changes were associated with the development of a strong AABW at 12 Ma. From 10-0 mm, the chemical compo-

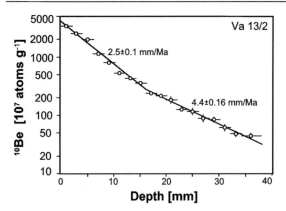

Fig. 11.20 Profile of ^{10}Be with depth in the Mn crust VA 13/2 from the Clarion-Clipperton F.Z. region showing the change in a growth rate in the crust at a depth of 16 mm (after Segl et al. 1989). Other time markers in the crust were marked by changes in the structure of the crust.

sition was again similar to that below 40 mm reflecting the weak influence of AABW at present.

^{10}Be dating of the crust was undertaken by Segl (1984, 1989). Figure 11.20 shows the variation in growth rates within the crust. An age of 12-13 Ma was determined at a depth of 36 mm in the crust. This was taken to correspond to the build up of East Antarctic ice sheet, the onset of high AABW velocities, increased ventilation of the deep ocean and a major erosional event in Pacific. At 6.2-6.7 Ma at a depth of 16 mm in the crust, there was a change of growth rate and a visible change in crustal structure. This was taken to correspond to a decrease in global δ^{13}C, a lowering of sea level, isolation and drying of Mediterranean, shoaling of the Panama Isthmus and an increase in bottom water circulation rates and oceanic fertility. At 3 Ma at a depth of 8 mm in the crust, there was a visible change in crustal structure. This was taken to correspond to closure of the Panama Isthmus and the onset of northern hemisphere glaciation. These events are recorded in many manganese crusts in the World Ocean. Subsequent determination of the distribution of Nd isotopes within this crust showed a decrease in the ε_{Nd} values from 3-5 Ma to the present (see later). These data suggest that the closure of the Panama Isthmus about 3-4 Ma may have played a role in reducing the inflow of Atlantic-derived Nd into the Pacific Ocean (Burton et al. 1997; Ling et al. 1997).

High resolution ^{230}Th profiles were also obtained by Eisenhauer et al. (1992) for this crust as part of a detailed study of growth rates in crusts during last 300 ka. 69 samples were taken in the upper 1.4 mm of crust at intervals of 0.02 mm. Fig. 11.21 shows the high resolution growth rates for the crust. From 0 - 0.88 mm, the growth rate was determined to be 6.6 ± 1.1 mm Ma^{-1},

from 0.9 - 1.28 mm 6.1 ± 1.7 mm Ma^{-1} and from 1.3-1.4 mm 5.8 ± 1.4 mm Ma^{-1}. Two breaks in the growth rate were observed. These correspond to standstills in growth at 284-244 ka and 138-128 ka which are equivalent to glacial stages 8 and 6. The growth rates of the crust were seen to be higher during interglacial (stages 1 and 5) and lower during glacial stages. These data imply a correlation between the growth rate of nodules and crusts and climate in the late Quaternary as previously proposed by Mangini et al. 1990a).

Most recently, supernova debris has been identified within the VA 13/2 crust based on a study of the distribution of ^{60}Fe in a well-resolved time profile within this crust (Knie et al. 2004). 28 samples between 1-2 mm thick representing a total period of about 13 Ma were dated by means of the ^{10}Be method. A ^{60}Fe anomaly was detected at a depth corresponding to an age of 2.8 Ma (the half life of ^{60}Fe is 1.49 Ma). The ^{60}Fe measured at this time horizon corresponds to an influx

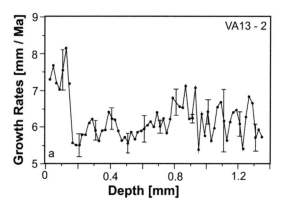

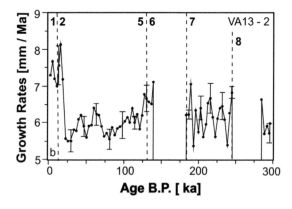

Fig. 11.21 High resolution growth rates for Mn crust VA 13/2 plotted as a function of depth in the crust (a) and of time (b) (after Eisenhauer et al. 1992). The major climatic changes are marked by dotted lines. High growth rates are mainly associated with interglacial stages (especially during the Holocene and stage 5) whereas low growth rates are associated with glacial stages. Periods of growth standstills are associated with glacial stages 6 and 8.

of about 2.9 x 10^6 atoms of ^{60}Fe cm^{-2} to the Earth's surface. This amount is compatible with the ^{60}Fe being derived from an exploding supernova located about 10 light years from Earth. This event began at ~3.0 Ma and lasted ~300 ka and therefore coincides with the onset of N. Hemisphere glaciation at 2.8 Ma. The explosion released as much energy as the sun will do in its entire lifetime. The authors speculated that the interstellar particles may have acted as nuclei for the condensation of water and therefore cloud formation in the atmosphere. This offers the possibility that the Northern Hemisphere glaciation may have been controlled by interstellar events and not by changes in CO_2 levels in atmosphere as is more commonly assumed. A weak ^{244}Pu signal from this explosion was also detected in the crust (Wallner et al. 2003).

The VA 13/2 crust has also been used for paleoceanographic studies by a number of authors (Abouchami et al. 1997; Ling et al. 1997; David et al. 2001; Frank et al. 1999b; van de Fliert et al. 2004a).

Detailed studies have also been carried out on the Co-rich Mn crust (KK84-RD50 S1-B), a 95 mm thick crust, which was sampled at a depth of 2,250-2,600 m from near the summit of Schumann Seamount, one of the Musicians Seamounts, located 700 km north of Kaui, Hawaii. The initial study was based on identification of coccolith imprints in the crust from which the ages of different layers of ages of the individual layers could be determined (Cowen et al. 1993). These ages were not well constrained and varied from 1-4 Ma to >10 Ma. However, the minimum age for the upper 27 mm of the crust was shown to be Eocene giving an average growth rate for the crust of about 0.5 mm Ma^{-1}. Extrapolating this growth rate through the entire 95 mm of the crust gave a very much older age for the crust than was accepted for Co-rich crusts at that time.

Subsequently, McMurtry et al. (1994) carried out a very detailed study of this crust using a variety of techniques including ^{10}Be profiling of the crust, determination of ^{87}Sr/^{86}Sr and δ^{18}O isotopic ratios of included phosphatized limestone debris and vein infillings and Co chronometry of the ferromanganese layers. Using the ^{10}Be method, it was possible to date only the upper 12 mm of this crust with the lower 82 mm of the crust lying beyond the limits of this method. However, by combining the ^{10}Be data with results obtained from microfossil dating (Cowen et al. 1993) and Co chronometry, it was possible to establish that formation of the crust began at least in the Eocene and possibly as far back as the Cretaceous. Within the crust, eight major disconformities were identified of which three of the upper disconformities in the crust were placed at the Plio-Pliocene, the Middle Miocene

and, tentatively, the Paleocene-Eocene boundaries. As previously described, McMurtry et al. (1994) considered the boundary between the older and younger crust layers had a minimum Oligocene age (28-33 Ma) but was possibly of Late Paleocene age (55 Ma) but this may reflect the fact that these boundaries correspond to different episodes of phosphogenesis from that described by Halbach and Puteanus (1984) (see section 11.4.4).

In order to assess the validity of the ^{87}Sr/^{86}Sr method of dating manganese crusts, von der Haar et al. (1995) carried out a series of leaching experiments on the Schumann Seamount crust in which it was established that Sr was leached from both phosphatic material and aluminosilicate detritus within the crust. It was also shown that Sr within the oxides exchanges with seawater strontium throughout the history of the deposit. These results cast severe doubt on the validity of the Sr isotopic method for determining the ages of these crusts (see section 11.4.9). In addition, De Carlo (1991) analyzed 11 samples at various depths in the Schumann Seamount crust for REE in an attempt to reconstruct its growth history (see section 11.4.3). The results are somewhat surprising in that they show a steady increase in the Ce/La ratio of the deposit from the base of the crust (4.5 at a depth of 78-95 mm) to a maximum of 7.0 at 63-68 mm and then a steady decline to 3.0 at a depth of 0-5 mm. It is not easy to understand why the most oxidizing conditions in the water column should prevail at a depth of 63-68 mm in the crust corresponding to a Paleocene age (McMurtry et al. 1994).

In addition, ^{10}Be dating of seven manganese nodules from the Southwestern Pacific Basin was carried out by Mangini et al. (1990b). In one nodule (143GB), extrapolation of the nodule growth rates indicated that the nodule began forming at about 14.5 Ma. At 6 Ma, there was a major change in structure of the nodule. Within the outer layer of the nodule, concentric layers filled with detritus corresponding to ages of 3.2 and 1.3 Ma. However, there was no corresponding change in growth rate at any of the intervals, the growth rate remaining constant with time. This study confirmed that these ages are important time markers which had previously been recorded in many Pacific pelagic sediments and manganese crusts and nodules.

The dates of the hiatuses reported above are consistent with the ages listed by Frank (2002) for major paleogeographic changes in the Pacific resulting from the opening and closing gateways which are thought to have been responsible for large-scale reorganizations of global ocean circulation and are linked to changes in global climate. In the Pacific, these include the rifting and northward drift of Australia from Antarctica which began at about 55 Ma (and is associated with an abrupt negative ~3 ppt global carbon isotope excursion which has been attributed to the dissociation of 2,000 - 6,000 Gt of methane hydrates; Norris and Röhling 1999; Zachos et al. 2001; Dickens 2004), the development of the deep-water passage through the Tasman Strait at about 36 Ma (Ling et al. 1997), the opening of the Drake Passage at about 23 Ma which is thought to have initiated ocean bottom water flow through the Samoan Passage (Lonsdale 1981), the development of Antarctic glaciation about 14 Ma ago, an increase in ocean bottom circulation rates at 6.2 Ma and closing of the Panama Isthmus about 3.5 Ma ago (see also Segl et al. 1989; Xu 1997). Of these, the most important for the formation of oceanic manganese nodules and crusts was the Middle Miocene climate transition which occurred between 14.3 and 13.8 Ma (Shevenell et al. 2004). This event resulted in a cooling of the surface waters in the southwest Pacific by 6 - 7°C at ~55°S and was responsible for the development of the modern pattern of deep ocean bottom water flow. In addition, there was a general cooling of the global climate since the Eocene climate maximum at about 50 Ma, although this general trend has been punctuated by a series of steps (Zachos et al. 2001). The observations presented above show that Co-rich manganese crusts record these changes in ocean circulation, although perhaps not with the precision that one might hope for.

It is well known that deep circulation of the oceans is much more pronounced during periods of „ice house Earth" (characterized by Polar ice sheets and a well-developed cryosphere) than during periods of „green-house Earth" (when the cryosphere is absent). The „ice house Earth" is generally taken to include the Oligocene, Late Miocene, Pliocene and Pleistocene and the „greenhouse Earth" the Cretaceous, Paleocene and the Eocene (Jenkyns and Wilson 1999; Zachos et al. 2001). Ventilation, and therefore the degree of oxygenation, of the deep ocean, is controlled by the density difference between surface seawater and deep ocean bottom water which induces deep circulation of the oceans and is much more pronounced during periods of „ice house Earth" than during periods of the „greenhouse Earth". This trend is illustrated by the dominance of deep-sea manganese nodules in D.S.D.P. cores since the Eocene (Glasby 1978).

Kaiko (1998) has confirmed this trend by demonstrating that the low oxygen events in the oceans during the past 120 Ma have coincided with warm episodes in the oceans based on his benthic foraminiferal dissolved-oxygen index. Pulyaeva (1997) has also shown that the average Co content of Co-rich crusts from the Magellan Seamounts has increased steadily from 0.16% in the Late Cretaceous to 0.57% in the Eocene and 0.72% in the Miocene-Pleistocene whereas P_2O_5 showed the opposite trend declining from 13.5% to 1.3% during this period (von Stackelberg et al. 1984; De Carlo 1991; Frank et al. 1999b; see section 11.4.9). This trend reflects the increased ventilation of the oceans through time and, in particular, the development of a more pronounced oxygen minimum zone in the ocean following the Middle Miocene climate transition.

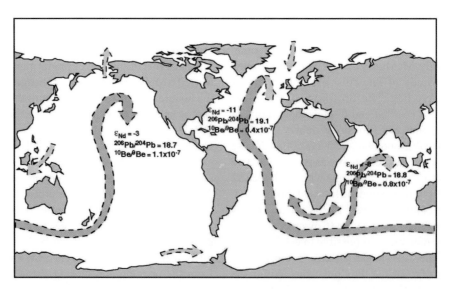

Fig. 11.22 Schematic representation of deep-water flow in the world ocean based on characteristic changes in the isotopic composition of Nd, Pb and Be in ferromanganese crusts with location. The isotopic data are taken from Albarède and Goldstein (1992) and von Blanckenburg (1996a,b) (after Hein et al. 2000, Fig. 9.10).

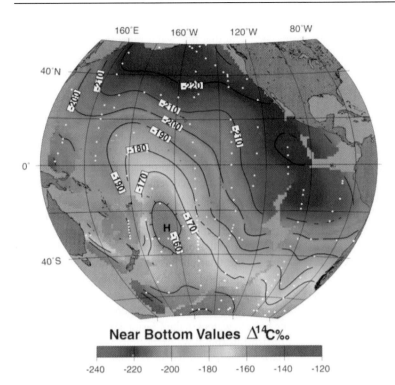

Near Bottom Values $\Delta^{14}C‰$

-240   -220   -200   -180   -160   -140   -120

Fig. 11.23 Schematic representation of the distribution of $\delta^{14}C$ in Circum Polar Deep Water at depths >3,500 m. The tongue of high $\delta^{14}C$ shows the present-day path of circum-polar water entering the Pacific. The water flows northwards along the western Pacific island arc system and then clockwise north of the equator. The oldest water is found in the NE sector of the Pacific (after Schlosser et al. 2001, plate 5.8.17).

11.4.10.2 Application of Isotopic Studies of Co-rich Mn Crusts to the Study of Present-day Deep-ocean Circulation

The release of trace metals with different isotopic ratios from the Earth's crust by weathering or hydrothermal activity is the basis for the observed variability of isotopic composition of these metals in the oceans (Frank 2002). Albarède and Goldstein (1992) pioneered the application of isotopic studies of ferromanganese deposits to the study of present-day deep-ocean circulation by mapping the Nd isotopic composition of 270 samples of ferromanganese crusts and nodules from the world ocean. The results showed a marked similarity between regional Nd isotopic variations of the crusts and nodules and the broad pattern of present-day deep-ocean circulation. This was somewhat surprising in view of the short residence time of Nd in seawater (~10 - 10^3 years) compared with the much longer time scale of ferromanganese formation which is measured in millions of years but reflects the stability of deep-sea circulation since the Pliocene integrated over time. Subsequently, von Blanckenburg et al. (1996a) measured Pb isotope ratios ($^{206}Pb/^{204}Pb$ and $^{208}Pb/^{204}Pb$) in ferromanganese crusts from the world ocean and showed that these isotopic ratios are remarkably uniform in the Pacific reflecting the fact that Pb is extremely well mixed as a result of its short residence time in ocean bottom waters (80-100 years).

The Pb isotope ratios also indicated that Pb is more juvenile in the Pacific than in the Atlantic Ocean. From this, it was deduced that the Pb in the Pacific Ocean is derived principally from the subduction of MORB and pelagic sediment at volcanic arcs and from continental margin volcanics whereas in the Atlantic Ocean it is derived principally from riverine particulate matter. Von Blanckenburg et al. (1996b) also measured beryllium isotope ratios ($^{10}Be/^9Be$) in ferromanganese crusts from the world ocean and showed that the $^{10}Be/^9Be$ ratios in the crusts are in close agreement with seawater values from the same area and that the $^{10}Be/^9Be$ ratios of the crusts decrease along the path of ocean bottom water flow and correlate with the $\Delta^{14}C$ age of the deep-ocean bottom water flow. The residence time of Be in the Pacific away from the margins was estimated to be 600 ± 100 years.

These observations enabled Hein et al. (2000) to prepare a schematic map of deep-ocean flow in the world ocean showing the characteristic changes in the Nd, Pb and Be isotopic composition of the ferromanganese crusts along the bottom water flow path (Fig. 11.22). Particular interest in this figure lies in comparing this schematic flow path with the actual flow path of Pacific deep water based on measurements of the $\Delta^{14}C$ age of the deep bottom waters as presented by Schlosser et al. (2001, Plate 5.8.17; Fig. 11.23). These data confirm that the ocean bottom water flow in the Pacific is counter-clockwise with the oldest waters being found in the

407

northeast Pacific. However, these figures give no indication as to whether these bottom waters are transported eastwards into the equatorial North Pacific. This is important because it determines whether manganese nodules from the C-C F.Z. were formed under the influence of the AABW (see section 11.4.3).

It is generally accepted that the Pacific Ocean bottom waters originate from the northward flow of the Circumpolar Deep Water (CDW) which flows east of the Campbell Plateau and Tonga-Kermadec Ridge and then onwards to the Samoan Passage (Reid and Lonsdale 1974; Taft et al. 1991; Schmitz 1995; Hogg 2001; Schlosser et al. 2001). Although the use of the term CDW to describe the bottom water flow in the western Pacific is correct (Clarke et al., 2001), most of the older literature cited in this chapter refers to this flow as AABW. Based on geostrophic calculations from the results of WOCE Leg 31 in 1994, it was estimated that the northward transport of bottom waters through the Samoan Passage at depths greater than 4,000 m is 7.8 Sv of which 4.8 Sv has a potential temperature below 0.8°C making it the coldest abyssal water in the South Pacific (Roemmich et al. 1996; Rudnick 1997). Previous studies had indicated that

~3 Sv of AABW flow northwards out of the Samoan Passage and that this water mass then bifurcates at a latitude of about 7° 50'S with the minor branch turning to the northeast (Edmond et al. 1971; Macdonald and Hollister 1973). This branch was then thought to flow southeastwards along the western side of the Line Islands Ridge before turning east and crossing the ridge at the Horizon and Clarion Passages (Gordon and Gerard 1970; Edmond et al. 1971; Johnson 1972). However, it should be emphasized that the flow paths of the ocean bottom currents in their approach to these two passages are not well constrained.

Edmond et al. (1971) estimated that the flow of water through the Horizon Passage was approximately 0.4 Sv and presumed that only a minor part of this flow passes through the Clarion Passage and into the C-C F.Z. However, Mantyla (1975) subsequently noted that the Clarion Passage is deeper and wider than the Horizon Passage and estimated the potential temperature of the bottom waters in the Clarion Passage to be 0.87°C confirming the flow of bottom water through the Clarion Passage. However, the magnitude of this flow is unknown and it is clear that the Line Islands Ridge constitutes a formidable barrier to the flow of deep

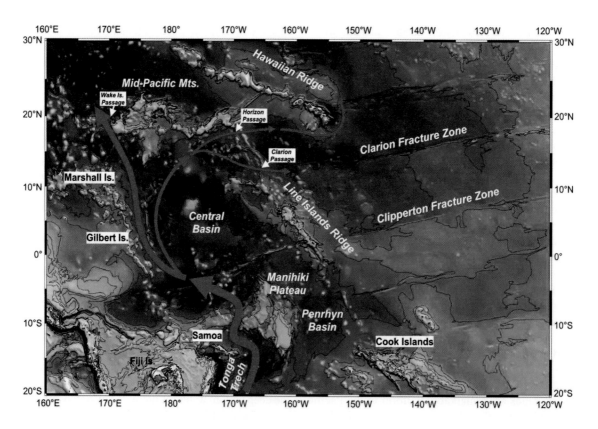

Fig. 11.24 Schematic map showing the inferred paths of oceanic bottom currents in the western Pacific (after Lonsdale 1981). Base map taken from the GEBCO bathymetric map of the Pacific Ocean.

water into the C-C F.Z. Fig. 11.24 illustrates the principal pathways of oceanic bottom water flow mentioned above.

In an early study of oceanic bottom water flow within the C-C F.Z., Johnson (1972) deployed free-fall bottom current meters in an area north of the Clipperton Fracture Zone where substantial sediment erosion was known to occur. The limited data showed that the bottom currents were generally slow (<10 cm sec⁻¹) but fluctuated markedly due to a strong semi-diurnal tidal component. It was also established that the currents flowed mainly to the east with minor variations due to topographic effects. In addition, data from a 14-day record of bottom current measurements taken at 210 m above the sea floor revealed an averge bottom water flow of 2.0 cm sec⁻¹ in an ENE direction with peak velocities of up to 16.5 cm sec⁻¹ at semidiurnal periods. These data showed that peak velocities of bottom water transport were strong enough to erode and transport sediment in the area (Amos et al. 1977).

More detailed studies in the C-C F.Z. were carried out during the period 1986 to 1989 in which a grid of bottom-moored instruments was deployed (Kontar and Sokov 1994). In general, the bottom currents were shown to be weak with velocities not greater than 4 cm sec⁻¹. However, a benthic storm lasting 10 days was recorded during these observations in which bottom currents with average velocities of 10 -15 cm sec⁻¹ were measured. It was thought that these storms were related to periods when the sea surface was elevated by more than 4 cm above the norm which would lead to the excess kinetic energy at the sea surface being transmitted to the seafloor at depths > 3,000 m (Kontar and Sokov 1994). Other measurements of oceanic bottom water flow in the DOMES areas of the C-C F.Z. have been reported by Hayes (1979, 1998).

Within the Samoan Passage, the most rapid drift deposition occurred in the Late Oligocene-Early Miocene and in the Late Miocene-Early Pliocene (Lonsdale 1981). Both these periods were characterized by intensified bottom circulation These times correspond roughly with periods of enhanced sediment erosion in the C-C F.Z. (van Andel et al. 1975) in accordance with the ages of the nodules in the C-C F.Z. as reported by Shilov (2004) (see section 11.4.9).

Mangini et al. (1982) also established that there had been a marked increase in deep circulation in the Pacific at ~70 ka corresponding to the end of interglacial cycle stage 5 based on ^{230}Th and ^{231}Pa dating in 9 out of 13 sediment cores from the C-C F.Z. and in all 4 cores from the Aitutaki Passage. These results supported the idea that the onset of glacial events increased ocean bottom water flow in the C-C F.Z.,

although this effect may have been masked by bioturbation in the sediment cores in some cases.

In a detailed study of the DOMES Site A in the C-C F.Z., Piper and Blueford (1982) described an apparently anomalous situation in which small, polynucleate nodules with smooth surface texture and δMnO_2 as the principal manganese oxide phase formed in Quaternary sediments with variable thickness >40 cm located on the side of an W-E-trending valley whereas large nodules with granular surface texture and todorokite as the principal manganese oxide phase formed on Tertiary sediments at sites of active erosion on the sea floor. This is contrary to the normal trend where the diagenetic influence on nodules growth is more pronounced in those areas with higher sedimentation rates (see section 11.4.2). However, this area is unusual in that the topographic fabric lies W-E. It was therefore thought that the AABW would be channelled through the valley creating an erosional area on its southern flanks. In other areas of the C-C F.Z., however, the alignment of the topographic fabric may lie normal to the flow path of the AABW.

The data presented above paint a picture of relatively weak bottom currents in the C-C F.Z. resulting from the limited inflow of bottom water through the Clarion Passage. However, physical oceanographic data suggest that there are at least three components to the bottom currents, namely AABW flow, tidal effects and benthic storms. At present, it is not possible to quantify the relative importance of these components, although it is believed that the intensity of bottom water flow has fluctuated through time, for example during glaciations. Nonetheless, there is indirect evidence that the C-C F.Z. has been a dynamic and fluctating environment during the entire period of formation of deep-sea manganese nodules. Of particular interest is the observation of Knoop et al. (1998) that the relative influence of oxic diagenesis declines and that of suboxic diagenesis increases along the flow path of the AABW through the C-C F.Z. This may explain the trend of decreasing Ce/La ratios of deep-sea manganese nodules along the flow path of the AABW in the equatorial North Pacific as noted in section 11.4.3. Skornyakova and Murdmaa (1992) have also suggested that benthic storms play a key role in the formation of diagenetic nodules in the C-C F.Z. because they resuspend the active surface layers of the radiolarian ooze resulting in the burial of the nodules under a thin layer of semi-fluid sediment. This then acts as a source of transition elements such as Mn, Ni, Cu and Zn to the nodules and leads to the development of the rhythmic structure characteristic of these nodules. By contrast, hydrogenous nodules in the area form under the influence of strong, stable bottom currents which ensure long-term contact of the nodule surface with seawater.

In the Peru Basin, two main flow paths into the basin have been recognized, both northwards from the South East Pacific Basin into the Chile Basin and thence into the Peru Basin (Lonsdale 1976). Inflow of oceanic bottom water into the Peru Basin from the northwest across the East Pacific Rise is blocked by the Galapagos Rise.

11.4.10.3 Application of Isotopic Studies of Co-rich Mn Crusts to the Study of Paleocean Circulation

The application of isotopic studies has also led to considerable advances in understanding the paleoceanographic conditions of formation of ferromanganese deposits in the Pacific Ocean and elsewhere (Frank 2002). Detailed studies on the isotopic composition of various elements have been carried out on a number of ferromanganese crusts throughout the Pacific. These include studies on Nd (Abouchami et al. 1997; Ling et al. 1997; O'Nions et al. 1998; Frank et al. 1999a,b; van de Fliert et al. 2004a,b,c), Pb (Abouchami et al. 1997; Christensen et al. 1997; Ling et al. 1997; O'Nions et al. 1998; Frank et al. 1999a,b; van de Fliert et al. 2003, 2004a,b,c), Hf (David et al. 1999; van de Fliert et al. 2004a,b,c), Os (Burton et al. 1999), Th (Rehkämper et al. 2004) and Fe (Beard et al. 2003; Lavasseur et al. 2004). Frank (2002) has plotted the time slice data for the Nd and Pb isotope distributions in the world ocean at 8 Ma, 6 Ma, 5 Ma, 1.5 Ma and the Present in which only five samples are shown for the entire Pacific Ocean reflecting the extremely sparse database on which to draw any conclusions.

As an example, we may consider the application of Nd isotopes of ferromanganese crusts to the study of paleoceanographic conditions in the Pacific. Nd is removed from the ocean water column by particle scavenging and has a global ocean residence time in deep waters of 600-2,000 years (Frank 2002). Its isotopic distribution in seawater is characteristic of individual water masses and is related to the weathering of Nd from different types continental crust with different ages. In the Pacific, the island arc rocks surrounding the ocean are composed of young mantle-derived material with high (radiogenic) ε_{Nd} values of up to + 20. The isotopic ratios are expressed as ε_{Nd} values because of the very small variations in the $^{143}Nd/^{144}Nd$ ratios in seawater which are mostly in the fourth or fifth decimal place (Frank 2002).

Ling et al. (1997) analyzed the Nd isotopic composition of three ferromanganese crusts taken from different water depths (4,800, 2,300 and 1,800 m) in the central Pacific. The ε_{Nd} pattern of the deepest crust over the past 20 Ma was shown to be similar to that of the two shallower crusts but about 1.0-1.5 units more negative. This was taken to indicate the long-term stratification of the Pacific over this period. O'Nions et al. (1998) subsequently confirmed the provinciality of Pacific Ocean water relative to Atlantic and Indian Ocean water based on an analysis of ε_{Nd} values of crusts from the Indian and Atlantic Oceans over this period. Ling et al. (1997) also demonstrated a steady increase in the ε_{Nd} values of these crusts from 10 to 3 Ma followed by a decrease to the present. This was interpreted to be the result of the progressively restricted access of Atlantic Ocean water to the Pacific via the Panama gateway which finally closed at 5-3 Ma leading to intensification of NADW (North Atlantic Deep Water) and greater production of AABW in the Pacific (Burton et al. 1997). However, this conclusion was disputed by Frank et al. (1999b) who argued that the contribution of Atlantic waters through the Panama gateway in the 3-4 Ma prior to its closure would have been very small. Instead, these authors proposed that this decrease in ε_{Nd} values during the last 3-4 Ma was related to an increased input of aeolian material into the Pacific (Frank 2002).

Van de Fliert et al. (2004b) also used Nd and Pb isotopic profiles in two carefully chosen ferromanganese crusts from the Lord Howe Rise in the S.W. Pacific and the Nova Trough in the equatorial Pacific to identify the relative influence of the equatorial Pacific and southwest Pacific bottom waters on the isotopic compositions of the crusts. The results were then used as a time series to reconstruct Cenozoic circulation patterns for the region. From this, it was possible to establish the influence of the progressive build up of Antarctic Circumpolar Current (ACC) from 38 Ma to 21/17 Ma as a result of the opening of the Tasmanian gateway and the Drake Passage and the development of the East Antarctic Ice Sheet from 17 Ma onwards. A major change in the Nd isotopic composition in the Lord Howe Rise crust from ~10 Ma onwards was also observed which was thought to reflect the increasing influence of equatorial Pacific deep water and was attributed to the closure of the Indonesian Seaway and reorganization of oceanic circulation in the S.W. Pacific. Abouchami et al. (1997) also showed that the Nd isotopic composition of the crusts has been more or less constant during the last glacial stages.

The above observations demonstrate not only the importance of isotopic studies on marine ferromanganese crusts in evaluating paleoceanographic conditions in the world ocean but also the relative paucity of data, especially in the Pacific Ocean.

11.4.11   Economic Prospects

In 1965, J.L. Mero proposed that manganese nodules could serve as an economic resource for Ni, Cu and Co. In his book, he painted a picture of an essentially limitless resource of over one trillion tonnes of manganese nodules on the Pacific deep-sea floor which were growing at a rate faster than could possibly be exploited (Mero 1965). Although he grossly overestimated the economic potential of the nodules, his work was hugely influential and set in train a number of major national programs to explore the resource potential of manganese nodules. In 1972-1982, the U.S. International Decade of Ocean Exploration (IDOE) program was initiated to facilitate scientific research on deep-sea Mn nodules in the equatorial North Pacific as it was then known. Strong German, French and Soviet programmes were also undertaken to evaluate nodules as an economic resource. The period 1972-1982 became the „Golden Age" for manganese nodule research. During this period, there were 30-40 U.S. cruises, 26 German cruises and 42 French cruises to study nodules focussed mainly on this region. In addition, seven consortia were formed to investigate

the possible commercial exploitation of the nodules. Mero (1977) subsequently estimated that the North Pacific high-grade area covering an area of about 6 million km^2 contained about 11,000 Mt of Mn, 115 Mt of Co, 650 Mt of Ni and 520 Mt of Cu. He also claimed that „within the next five to ten years, assuming no political and/or legal interference, the nodules should be in full-scale economic production as a valuable source of important industrial metals." More recently, Morgan (2000) has estimated that there are 34,000 Mt of nodules located in an area of 9 million km^2 in the C-C F.Z. and that these nodules contain approximately 7,500 Mt of Mn, 78 Mt of Co, 340 Mt of Ni and 265 Mt of Cu.

Several factors combined to end this optimism (Glasby 2000, 2002). For example, a major factor in this belief was the assumption of world metal prices would continue to rise in real terms based on the prediction of global mineral shortages by the Club of Rome. However, although world metal prices did indeed rise from about 1965 to about 1976, the long-term trend for Ni and Cu from then on was downwards (Fig. 11.25). In addition, the economic feasibility of deep-sea mining based on the assumption of very high capital investment and an enormous throughput of nodules (Medford 1969) never materialized. Indeed, the practicality of such an operation based on the deployment of a single mining ship at a given mine

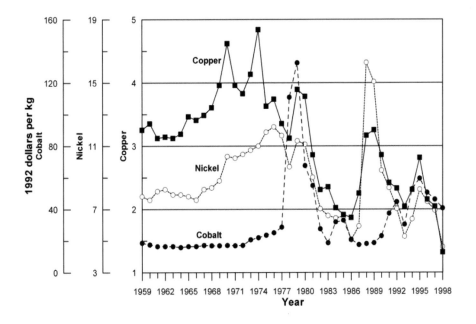

Fig. 11.25 Variation in the prices of Co, Ni and Cu in the United States from 1959-1998 expressed in constant 1992 dollars per kg (after Plunkert and Jones 1999). The prices of Co, Ni and Cu in 1999 in current dollar prices were Co $US 21, Ni $US 2 and Cu $US 0.75. Co is therefore easily the most valuable metal in deep-sea nodules and Co-rich crusts after Pt.

site was questioned by Glasby (1983). The feasibility of developing a reliable mining system was also cast into doubt when the testing of deep-sea nodule mining at the pilot-plant stage in 1978 ended abruptly with the loss of the entire mining system over the stern of the ship after about 800 t of nodules had been recovered from the seafloor. False expectations were also raised by the development of a deep-sea mining system by the Lockhead/OMCO consortium using the *Hughes Glomar Explorer* as the mother ship. The system developed was considered to be one of the greatest innovations in ship design and deep-sea technology in the twentieth century. Only later was it revealed that the *Hughes Glomar Explorer* had been built specifically to recover the K-129, a sunken Soviet nuclear submarine, and that the building of this ship had been funded entirely by the CIA at a cost of about $500 million in current dollars.

Nonetheless, the key problem was undoubtedly the provisions of the Third United Nations Convention on the Law of the Sea (UNCLOS III) which lasted from 1973 until 1982 and was the largest ever forum for international diplomacy (Anon 1982). These provisions were considered so onerous that several countries with an interest in deep-sea mining (including the U.S.A., U.K. and Germany) refused to sign the initial Law of the Sea Treaty in 1982. One provision that the United States, in particular, found unacceptable was the requirement to transfer technology to the Enterpise (representing Third World countries). Nonetheless, these countries eventually signed the convention in 1994 after a new agreement on deep-seabed mining was reached. This led to the International Seabed Authority (ISA) being set up in Kingston, Jamaica, to mangage the mineral resources of the International Seabed Area (http//:www.isa.org.jm). However, it was the marked drop in the price of Ni and Cu from 1979 to 1982 (Fig. 11.25) which essentially ended all commercial and almost all scientific interest in deep-sea manganese nodules in these countries in 1982.

Concurrently, there was a marked surge in the world price of Co in 1979 and 1980 (Fig. 11.25) following the invasion of the Shaba Province of Zaire by insurgents in 1978 (Manheim 1986). Cobalt is essential for the production of the superalloys used in jet aircraft engines. As a result, President Reagan created a strategic materials task force in 1980 to evaluate the availability of strategic minerals such as Co, Mn and the Pt metals. This stimulated interest in marine resources of cobalt, manganese and platinum. Although the cost of mining these deposits was considered to be too high to be practical, it was thought

that an increase in demand (and therefore in price) for these minerals or technological innovations which would reduce the cost of mining could make these deposits an economically viable source of these minerals. In 1983, President Reagan proclaimed a 200 n.m. Exclusive Economic Zone (EEZ) around the United States and the US-controlled islands in the Pacific and Caribbean. This paved the way for research on Co-rich manganese crusts around Johnson Island and the Hawaian Archipelago in the U.S. EEZ as well as within the EEZ's of the Federated States of Micronesia, Marshall Islands, Kiribati and from international waters in the Mid-Pacific Mountains. Hein (2004) has listed 36 cruises dedicated to the study of Co-rich crusts in the Pacific between 1981 and 1999 undertaken by groups from the USA, Japan, Korea, Germany, Russia, Australia, France and New Zealand. Since then, China has become a major player in Co-rich crust investigations. One of the advantages of mining Co-rich Mn crusts compared to deep-sea manganese nodules is that they mostly lie within the EEZ's of island nations and therefore outside the juridiction of the ISA. Mining rights could therefore be negotiated with the host government. However, the price of Co in 1998 was less than 30% of its peak value in 1979 in real terms (Fig. 11.25) and there is no serious talk of mining these deposits yet.

Following the adoption of the Law of the Sea Treaty in 1982, a new phase was initiated to assess the commercial potential of manganese nodules and a number of countries registered claims to deep-sea mining areas as „pioneer investors" through the United Nations on the basis of UNCLOS. These countries committed themselves to long-term programmes to establish the viability of nodule mining including nodule surveys, the development of mining systems, nodule processing and environmental surveys. So far, the following countries have registered, France, Russia, India and Japan (1987), People's Republic of China (1991), InterOceanMetal (1992) and Republic of Korea (1994). Fig 11.26 shows the locations of the mine site areas allocated to each of these countries within the C-C F.Z. India has registered its claim to an area within the Central Indian Ocean Basin (CIOB) south of India (Juahari and Pattan 2000; Mukhopadhyay et al. 2002). However, the commercial viability of nodules mining is by no means assured and no consortium is building a nodule mining system at present. This reflects the fact that world metal prices are too low for nodule mining to be profitable. As can be seen in Fig. 11.25, the price of both Ni and Cu in 1998 was less than 40% of their peak values in the 1970s in real terms. It would therefore appear that those countries still interested in the possibility of exploiting nodules are looking for a strategic source of metals rather than a resource that is economically cost competitive.

For deep-sea nodules, only those nodules with Cu+Ni+Co >2.5% and abundance >10 kg m^{-2} can be considered to be a potential economic resource. This represents only a small percentage of total quantity of nodules worldwide (i.e. diagenetic nodules from the C-C F.Z. and Central Indian Ocean Basin). To be economic, a 20 year-mine-site would be required to produce 3 Mt of nodules per year. This would cover an area >6,000 km^2. According to Lenoble (2004), the estimated total number of potential mine sites varies from 8 to 225 corresponding to a total of between 480-13,500 Mt of nodules. However, if the capacity of world metal markets to absorb production from the deep sea is taken into account, a more realistic assessment was taken to be 3-10 mine sites with a total tonnage of 100-600 Mt over this period. This reflects the marked differences in the ratios of the amounts of metals in nodules (Ni 7: Cu 6: Co 1) compared to the ratios of their consumption (Ni 27: Cu 267: Co 1) (Bernhard and Blissenbach 1988). As a result, if large tonnages of nodules were to be mined for Ni and Cu, it would lead to overproduction of Co with an associated drop in its world price. The markets would therefore have to be manipulated to avoid such disruptions. Based on an annual worldwide production rate of Ni of 0.9 Mt in

1998; Morgan (2000) considered Ni to be the primary metal of commercial interest in deep-sea nodules. Because deep-sea nodules occur at great depths in the oceans (greater than 4,000 m), they require sophisticated capital-intensive technology to mine and are in direct competition with land-based mineral resources. A decision to mine them must therefore ultimately be based on economic rather than technological considerations.

For Co-rich manganese crusts, the crusts should contain >0.8% Co and be >40 mm thick to be considered economically viable. The crusts occur in shallower water depths (1,000-2,500 m) than deep-sea nodules (>4,000 m) and could be mined within national Exclusive Economic Zones (EEZs) and therefore under national jurisdiction. Excellent accounts of the distribution and composition of Co-rich crusts in the world ocean have been presented by Andreev and Gramberg (1998) and Hein (2000) and several authors have attempted to assess of their economic potential (Wiltshire et al. 1999; Hein 2000, 2004; Wiltshire 2000). Andreev and Gramberg (2002) have estimated the total abundance of Co-rich crusts in the world ocean to be about 21·10^9 t.

The types of mining systems likely to be used for the recovery of deep-sea nodules and Co-rich Mn crusts are not well known. For nodules, Lenoble (2004) has described a system to mine the French area. This

Fig. 11.26 Schematic map showing the distribution of the allocated sectors of registered pioneer investors and applicants for pioneer investor status in the C-C F.Z. Most of the consortia interested in deep-sea mining have already made extensive studies of their future 'mine sites' in the C-C F.Z. The economic potential of this region is reflected in the large proportion of the area already under claim. Areas not subject to claims are often topographically unsuitable (too mountainous). The map is an updated version of the original version appearing in Kotlinski (1995).

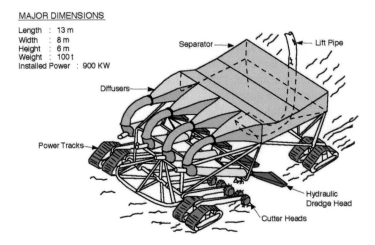

MAJOR DIMENSIONS

Length   : 13 m
Width   : 8 m
Height   : 6 m
Weight   : 100 t
Installed Power   : 900 KW

Fig. 11.27   Sketch of a proposed vehicle for mining Co-rich Mn crusts (after Hein et al. 2004, Fig. 9).

involves the deployment of a self-propelled dredge which crawls along the bottom, collects the nodules, crushes them and introduces them to a 600 m long flexible hose connected to a rigid pipe. The crushed nodules are then lifted to a semi-submersible surface platform in a 4,800 m rigid steel pipe by airlift or pumps. They are then transferred from the platform to the ore carrier through a flexible hose as thick slurry. The recovery rate using this method would be of the order of 1.5 Mt of nodules on a dry-weight basis per year.

Mining systems for the recovery of Co-rich Mn crusts have been described by Hein (2004). According to this author, mining crusts is technically much more difficult than mining nodules because the crusts are attached to a substrate and occur in undulating terrain. It is therefore necessary to separate the crust from the underlying substrate at the seafloor. One proposed mining system consists of a bottom crawling vehicle attached to the surface mining vessel by a hydraulic pipe lift system. The mining machine provides its own propulsion and travels at a speed of about 20 cm sec^{-1}. The miner has articulated cutters that would fragment the crusts and reduce the amount of substrate material collected. This system could possibly mine 1 Mt of crust with a 25% dilution by substrate on a dry-weight basis per year. Fig. 11.27 shows a possible design of a vehicle for mining Co-rich Mn crusts.

Both Hein (2004) and Lenoble (2004) have considered the profitability of mining deep-sea nodules and Co-rich Mn crusts in some detail and concluded that mining of these deposits is not commercially viable at current world metal prices. Mining these deposits therefore remains a matter for the future.

11.4.12   Future Prospects

The geochemistry of manganese in the marine environment is defined by the following characteristics: its relatively high abundance in the earth's crust, its availability in two valency states whose stability boundary lies within the range of the natural environment and the high adsorption capacity of its oxides for cationic species, especially Ni^{2+}, Co^{3+} and Cu^{2+}. As such, manganese naturally migrates to zones with the highest redox potential within the marine environment. The modern, well-oxygenated deep-sea is therefore the ultimate repository for manganese.

Manganese can also fractionate from iron and other transition metals to form shallow-marine ferromanganese concretions and submarine hydrothermal crusts under conditions in which iron and the associated transition metals are trapped as sulphides prior to the deposition of the manganese oxide minerals. Mn can also form deposits enriched in elements such as Co, Ni, Cu and Zn under conditions in which these elements can be adsorbed from seawater and sediment pore water onto the surfaces of manganese oxide minerals. The most favourable conditions for this to occur require extremely low growth rates of the manganese oxide minerals. These are usually to be found in low sedimentation regimes in the deep sea as with deep-sea manganese nodules or associated with the oxygen-minimum zone at mid-depths in the open ocean as with Co-rich manganese crusts. These factors are responsible for the formation of potential economic deposits of manganese in the deep oceans, namely deep-sea manganese nodules and Co-rich manganese crusts.

In addition, the rate of formation of marine manganese deposits in the deep ocean is extremely slow. The formation of these deposits therefore takes place over long periods of time. Deposition of manganese oxides in the deep oceans is also a function of the degree of oxygenation of the oceans. This in turn is controlled by the ventilation of the deep ocean which varies according to the climate. Deep-sea manganese nodules and Co-rich Mn crusts therefore record variations in the degree of oxygenation and pattern of circulation of the deep ocean. Mn crusts can therefore be considered to be condensed stratigraphic sections which can be used to elucidate variations in paleoceanographic conditions on timescales from Milankovitch cycles to geological era on the basis of their isotopic ratios. As a result, both deep-sea manganese nodules and Co-rich manganese crusts may be thought of as archives of paleoenvironmental data and therefore important targets for paleoceanographic studies. Because of their much longer period of formation, Co-rich manganese crusts are more useful than deep-sea manganese nodules in this respect.

Although the boom period of research into deep-sea Mn nodules and Co-rich Mn crusts associated with the drive to establish these deposits as an economic resource in the 1970s and 1980s is over, there remain many interesting topics which have been neglected over the last two decades. These include the study of shallow-marine concretions, lake concretions, hydrothermal Mn crusts and fossil manganese deposits. Each of these deposits has its own characteristic modes of formation. In addition, much of the work on deep-sea manganese nodules has focussed on areas of potential economic interest. There remains considerable scope for studying nodules from a much wider range of settings than has previously been attempted, especially for investigating their internal characteristics in much greater detail using modern instrumental techniques.

The economic potential of marine manganese deposits and the potential of these deposits as paleoenvironmental indicators are directly related to the chemistry of manganese, its associated transition metals and other key elements adsorbed on to the surface of these deposits. It is this complex interplay of these factors which makes the study of marine manganese deposits so intriguing. Unfortunately, research on deep-sea manganese nodules and crusts has been held hostage to the economic potential of these deposits. Time will tell if these deposits are to become a major source of metals for the world as was once thought.

11.5 Problems

Problem 1

Figure 11.28 is a simplified E_H-pH diagram for Mn at 25°C, 1 atmosphere pressure and in pure water for a limited number of aqueous species and solid phases of Mn taken from Garrels and Christ (1965 Fig 7.28a). The boundaries between the dissolved species and solid phases are taken to be at a total activity of the dissolved species of 10^{-6} and 10^{-2}. The total dissolved carbonate species is taken to be $10^{-1.4}$.

In your opinion, what are the advantages of plotting an E_H, pH diagram of Mn of the type shown in Fig. 11.2 compared with this older type of Eh, pH diagram pioneered by Garrels and Christ (1965)? How do you account for the fact that solid manganese oxide minerals do not appear to be thermodynamically stable under seawater conditions in the diagram of Glasby and Schulz (1999; Fig. 11.2).

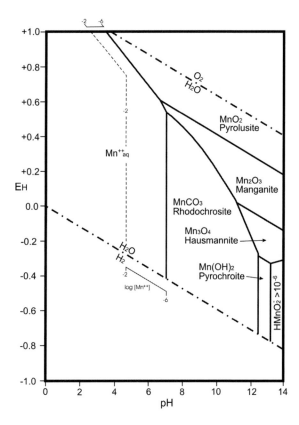

Figs 11.28 Simplified E_H-pH diagram for Mn at 25°C, 1 atmosphere pressure and in pure water for a limited number of aqueous species and solid phases of Mn (after Garrels and Christ 1965 Fig 7.28a).

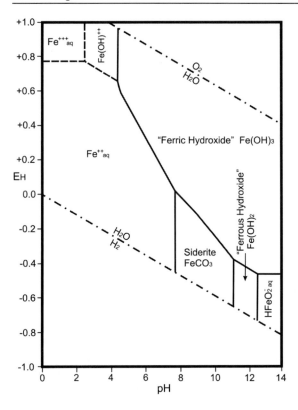

Figs 11.29 Simplified E_H-pH diagram for Fe at 25°C, 1 atmosphere pressure and in pure water for a limited number of aqueous species and solid phases of Fe (after Garrels and Christ 1965 Fig 7.14).

Problem 2

Figure 11.29 is a simplified E_H-pH diagram for Fe at 25°C and 1 atmosphere pressure showing the relations between the aqueous Fe species and metastable iron hydroxides and siderite. The boundaries between the dissolved species and solid phases are taken to be at a total activity of the dissolved species of 10^{-6}. The total dissolved carbonate species is taken to be 10^{-2} (Garrels and Christ 1965 Fig 7.14).

In your opinion, what are the advantages of plotting an E_H, pH diagram of the type shown in Fig. 11.4 compared with this older type of E_H, pH diagram pioneered by Garrels and Christ (1965)? What are the limitations of Fig. 11.4 in deducing the form of Fe in deep-sea manganese nodule?

Problem 3

Estimate the flux of Mn^{2+} from marine sediments from Loch Etive, Scotland. Some marine sediments, particularly in hypoxic fjord environments and in some hemipelagic environments, contain elevated concentrations of manganese. The sediments receive this manganese in the form of solid MnO_2 through the water column. In return, they release some Mn^{2+} in solution back to the water column. The extent of this exchange needs to be measured in order to determine the importance of Mn in oxidation-reduction processes in the sediment quantitatively. The flux of Mn^{2+} within the sediment and from the sediment to the water column can be calculated from the Mn^{2+} gradient using Fick's first law.

The following is a table (Tab. 11.7, Fig. 11.30) of measured pore water manganese concentrations in reducing sediment at the deepest station in the upper basin of Loch Etive, Scotland, where Mn^{2+} diffuses up to the surface of the sediment following the concentration gradient. The temperature of the sediment is 10°C and the porosity 0.84 (Overnell 2002; Overnell et al. 2002).

a) Calculate the maximum upward flux of Mn^{2+} towards the sediment surface (J_{Mn}) in the reducing sediments of the upper basin of Loch Etive. Use the depth interval with the steepest upward gradient for the calculation. The Mn^{2+} diffusion coefficient for sediment along with the in situ porosity and temperature are given in chapter 3 together with the necessary equations.

b) At what depth below the sediment surface are the Mn-oxides reduced and Mn^{2+} ions released to the pore water?

c) Are there indications that precipitation of Mn oxides is already taking place in the uppermost few millimeters of the sediment?

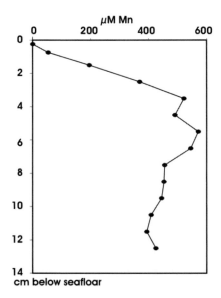

Figs 11.30 Measured pore water manganese concentrations in reducing sediment at the deepest station in the upper basin of Loch Etive, Scotland.

Table 11.7 Measured pore water manganese concentrations in reducing sediment at the deepest station in the upper basin of Loch Etive, Scotland.

Depth cm	Mn^{2+} μM	Depth cm	Mn^{2+} μM
0.25	1.5	6.5	544.9
0.75	54.2	7.5	455.1
1.5	195.6	8.5	453.1
2.5	369.1	9.5	444.7
3.5	521.4	10.5	408.4
4.5	490.6	11.5	393.8
5.5	571.2	12.5	425.0

Problem 4

Plot the NASC-normalized REE distribution patterns (Table 11.8) for the following types of submarine and subaerial Mn deposits. Remember to include all REE on the x-axis, even those elements that have not been analyzed. What do these REE patterns tell us about the modes of formation of these various manganese deposits?

Problem 5

Radiometric dating has revealed wide variations (by four orders of magnitude) for the growth rates of marine manganese deposits: Co-rich Mn crusts (0.8 mm Ma^{-1}; Puteanus and Halbach 1988), deep-sea manganese nodules on red clay substrates (1-2 mm Ma^{-1}; Hu and Ku 1984), deep-sea manganese nodules on siliceous ooze substrates (3-8 mm Ma^{-1}; Hu and Ku 1984), deep-sea manganese nodules on hemipelagic clay substrates (20-50 mm Ma^{-1}; Hu and Ku 1984), Baltic Sea concretions from 1,700-21,000 mm Ma^{-1}; Liebetrau et al. 2002) and submarine hydrothermal manganese crusts >1,000 mm Ma^{-1}. These growth rates vary by a four orders of magnitude. How do you account for such wide variations in growth rates?

Table 11.8 REE distribution patterns for submarine and subaerial Mn deposits.
Source of data: a Kunzendorf et al. (1993), b Glasby et al. (1987), c De Carlo and McMurtry (1992), d Szefer et al. (1998), e Glasby et al. (1997), f Glasby et al. (2004), g Gromet et al. (1984)   n.a. = not analyzed.

	Deep-sea nodule SW Pacific	Deep-sea nodule CC FZ	Deep-sea nodule Peru Basin	Co-rich Mn crust	Baltic Sea ferromanganese concretion	Submarine hydrothermal crust	Vani Mn deposit	NASC
source of data	a	b	b	c	d	e	f	g
La	167	93	55	287	24.1	38.1	19.2	31.1
Ce	952	344	112	1277	40.8	80.8	28	66.7
Nd	183	134	46	260	20.9	39.1	11.2	27.4
Sm	40	33.3	12.3	50.5	4.58	8.44	2.2	5.59
Eu	10.7	7.8	3.0	14.0	1.08	2.51	3.3	1.18
Gd	n.a.	n.a.	n.a.	60.8	4.46	7.54	2.9	5.5
Tb	6.5	4.0	1.9	n.a.	0.63	1.09	0.2	0.85
Dy	n.a.	n.a.	n.a.	48.7	3.42	6.38	2.4	5.54
Er	n.a.	n.a.	n.a.	26.5	1.92	3.15	1.6	3.27
Yb	19.5	12.9	8.7	25.3	2.1	2.64	1.6	3.06
Lu	2.9	1.8	1.4	3.57	0.29	0.36	0.2	0.456

Acknowledgements

The author would like to thank Professor G.M. McMurtry (University of Hawaii), Dr C.L. Morgan (Planning Solutions, Hawaii), Professor A. Usui (Koichi University) and Dr U. von Stackelberg (BGR) for their reviews of the manuscript. The author is also greatly indebted to Dr S.I. Andreev, Dr J. Fenner, Professor M. Frank, Dr J.R. Hein, Dr J. Overnell, Dr I. Pulyaeva, Dr V.V. Shilov, Dr I.M. Varentsov and Dr J. Wiltshire for their helpful contributions.

References

Abouchami, W., Goldstein, S.L., Galer, S.J.G., Eisenhauer, A. and Mangini, A., 1997. Secular changes in lead and neodymium in central Pacific seawater recorded in a Fe-Mn crust. Geochimica et Cosmochimica Acta, 61: 3957-3974.

Albarède, F. and Goldstein, S.L., 1992. World map of Nd isotopes in sea-floor ferromanganese deposits. Geology, 20: 761-763.

Aller, R.C., 1990. Bioturbation and manganese cycling in hemipelagic sediments. Philosophical Transactions of the Royal Society of London, A331: 51-68.

Amos, A.F., Roels, O.A., Garside, C., Malone, T.C. and Paul, A.Z., 1977. Environmental aspects of nodule mining. In: Glasby, G.P. (ed), Marine manganese deposits. Elsevier, Amsterdam, pp. 391-437.

Andreev, S.I. and Gramberg, I.S., 1998. The explanatory note to the metallogenic map of the world ocean. VNIIOkeanologia (St Petersburg) and InterOceanMetall (Szczecin), 212 pp. + illustrations (in English and Russian)

Andreev, S.I. and Gramberg, I.S., 2002. Cobalt-rich ores of the world ocean. Ministry of Natural Resources of the Russian Federation and All-Russian Research Institute for Geology and Mineral Resources of the World Ocean, St. Petersburg, 167 pp. + appendix (in Russian with English abstract)

Anon, 1979. La Genèse des nodules de manganese. Colloques Internationaux du Centre National de la Recherche Scientifique (CNRS), 287: 410 pp.

Anon 1982. United Nations Convention on the Law of the Sea of 10 Dcember 1982 (http://www.un.org/Depts/los/index.htm.Convention_agreements/texts/unclos/unclos_e.pdf)

Aplin, A.C. and Cronan, D.S., 1985. Ferromanganese deposits from the central Pacific Ocean, II. Nodules and asso-ciated sediments. Geochimica et Cosmochmica Acta, 49: 437-451.

Banerjee, R., 2000. A documentation of burrows in hard substrates of ferromanganese crusts and associated soft sediments from the Central Indian Ocean Basin. Current Science, 79: 517-521.

Barnes, S.S. and Dymond, J., 1967. Rates of accumulation of ferro-manganese nodules. Nature, 213: 1218-1219.

Baturin, G.N., 1988. The geochemistry of manganese and manganese nodules in the ocean. D. Reidel, Dordrecht, 342 pp.

Beard, D.L., Johnson, C.M., Von Damm, K.L. and Poulson, R.L., 2003. Iron isotope constraints on Fe cycling and mass balance in oxygenated Earth oceans. Geology, 31: 629-632.

Bender, M.L., Klinkhammer, G.P. and Spencer, D.W., 1977. Manganese in seawater and the marine manganese balance. Deep-Sea Research, 24: 799-812.

Berger, W.H., Fischer, K., Lai, C. and Wu, G., 1987. Ocean productivity and organic carbon flux. Scripps Institution of Oceanography Reference Series 87-30.

Bernhard, H.-H. and Blissenbach, E., 1988. Economic importance. In: Halbach, P., Friedrich, G. and von Stackelberg, U. (eds), The manganese nodule belt of the Pacific Ocean Geological environment, nodule formation, and mining aspects. Enke Verlag, Stuttgart, pp. 4-9.

Bischoff, J.L. and Piper, D.Z., 1979. Marine geology and oceanography of the Pacific manganese nodule province. Plenum Press, NY, 842 pp.

Bischoff, J.L., Heath, G.R. and Leinen, M.L., 1979. Geochemistry of deep-sea sediments from the Pacific manganese nodule province: DOMES Sites A, B, and C. In: Bischoff, J.L. and Piper, D.Z. (eds), Marine geology and oceanography of the Pacific manganese nodule province. Plenum Press, NY, pp. 397-436.

Böllhofer, A., Eisenhauer, A., Frank, N., Pech, D. and Mangini, A., 1996. Thorium and uranium isotopes in a manganese nodule from the Peru basin determined by alpha spectrometry and thermal ionization mass spectrometry (TIMS): Are manganese supply and growth related to climate? Geologische Rundschau, 85: 577-585.

Böllhofer, A., Frank, N., Rohloff, S., Mangini, A. and Scholten, J.C., 1999. A record of changing redox conditions in the northern Peru Basin during the Late Quaternary deduced from Mn/Fe and growth rate variations in two diagenetic manganese nodules. Earth and Planetary Science Letters, 170: 403-415.

Bruland, K.W., 1983. Trace elements in sea-water. In: Riley, J.P. and Chester, R. (eds), Chemical oceanography. Academic Press, London, pp. 157-220.

Bruland, K.W., Orians, K.J. and Cowen, J.P., 1994. Reactive trace metals in the stratified central North Pacific. Geochimica et Cosmochimica Acta, 58: 3171-3182.

Bruland, K.W. and Lohan, M.C., 2004. Controls of trace metals in seawater. In: Elderfield, H. (ed.), Treatise on geochemistry volume 6 The oceans and marine geochemistry. Elsevier, Amsterdam, pp. 23-47.

Burgarth, K.P. and von Stackelberg, U., 1995. Sulfide - impregnated volcanics and ferromanganese incrustations from the southern Lau Basin (Southwest Pacific). Marine Georesources and Geotechnology, 13: 263-308.

Burns, R.G. and Burns, V.M., 1977. Mineralogy. In: Glasby, G.P. (ed.), Marine manganese deposits. Elsevier, Amsterdam, pp. 185-248.

Burns, R.G. and Burns, V.M., 1980. Manganese oxides. Reviews in Mineralogy, 6: 1-46.

Burton, J.D. and Statham, P.J., 1988. Trace metals as tracers in the ocean. Philosophical Transactions of the Royal Society of London, A325: 127-145.

Burton, K.W., Ling, H.-F. and O'Nions, R.K., 1997. Closure of the central American Isthmus and its effect on deep-water foand lacustrine nodules: Distribution and geochemistry. In: Glasby, G.P. (ed.), Marine manga-nese deposits. Elsevier, Amsterdam, pp. 45-86.

Calvert, S.E. and Piper, D.Z., 1984. Geochemistry of ferromanganese nodules from DOMES Site A, Northern Equatorial Pacific: Multiple diagentic metal sources in the deep sea. Geochim. et Cosmochim. Acta, 48: 1913-1928.

Calvert, S.E.., Piper, D.Z. and Baedecker, P.A., 1987. Geochemistry of the rare earth elements of ferromanganese nodules from DOMES Site A, northern equatorial Pacific.

Geochimica et Cosmochimica Acta, 51: 2331-2338.

Chabaux, F., Cohen, A.S., O'Nions, R.K. and Hein, J.R., 1995. ^{238}U - ^{234}U - ^{230}Th chronometry of Fe-Mn crusts: Growth processes and recovery of thorium isotopic ratios of seawater. Geochimica et Cosmochimica Acta, 59: 633-638.

Chabaux, F., O'Nions, R.K., Cohen, A.S. and Hein, J.R., 1997. ^{238}U - ^{234}U - ^{230}Th disequilibrium in hydrogenous oceanic Fe-Mn crusts: Paleoceanographic record or diagenetic alteration? Geochimica et Cosmochimica Acta, 61: 3619-3632.

Chen, J.C. and Owen, R.M., 1989. The hydrothermal component in ferromanganese nodules from the southeast Pacific Ocean. Geochimica et Cosmochimica Acta, 53: 1299-1305.

Chester, R., 1990. Marine geochemistry. Chapman & Hall, London, 698 pp.

Christensen, J.N., Halliday, A,.N., Godfrey, L.V., Hein, J.R. and Rea, D.K., 1997. Climate and ocean dynamics of lead isotope records in Pacific ferromanganese crusts. Science, 277: 913-918.

Chukhrov, F.V., Gorshkov, A.I., Ermilova, L.P. et al., 1982. On mineral forms of manganese and iron occurrences in oceanic sediments. International Geological Review, 24: 466-480.

Chukhrov, F.V., Gorshkov, A.I., Drits, V.A., Shterenberg, L.Ye., Sivtsov, A.V. and Sakharov, B.A., 1983. Mixed-layered asbolite-buserite minerals and asbolite in oceanic iron-manganese nodules. International Geological Review, 25: 838-847.

Chukhrov, F.V., Gorshkov, A.I., and Drits, V.A., 1989. Supergene manganese oxides. Nauka, Moscow, 208 pp. (In Russian)

Clarke, A., Church, J. and Gould, J., 2001. Ocean processes and climate phenomena. In: Siedler, G., Church, J. and Gould, J.J. (eds), Ocean circulation & climate observation and modeling of the global ocean. Academic Press, San Diego, pp. 11-30.

Cowen, J.P. and Bruland, K.W., 1985. Metal deposits associated with bacteria: implications for Fe and Mn marine geochemistry. Deep-Sea Research, 32A: 253-272.

Cowen, J.P., DeCarlo, E.H. and McGee, D.L., 1993. Calcareous nannofossil biostratigraphic dating of a ferromanganese crust from Schumann Seamount. Marine Geology, 115: 289-306.

Cronan, D.S., 1977. Deep-sea nodule: distribution and chemistry. In: Glasby, G.P. (ed.), Marine manganese deposits. Elsevier, Amsterdam, pp. 11-44.

Cronan, D.S., 1980. Underwater minerals. Academic Press, London, 362 pp.

Cronan, D.S., 1987. Controls on the nature and distribution of manganese nodules in the western equatorial Pacific Ocean. In: Teleki, P.G., Dobson, M.R., Moore, J.R. and von Stackelberg, U. (eds), Marine minerals Advances in research and resource assessment. D. Reidel, Dordrecht, pp. 177-188.

Cronan, D.S., 1997. Some controls on the geochemical variability of manganese nodules with particular reference to the tropical South Pacific. In: Nicholson, K., Hein, J.R., Bühn, B. and Dasgupta, S. (eds), Manganese mineralization: Geochemistry and mineralogy of terrestrial and marine deposits. Geological Society Special Publication, 119: 139-151.

Cronan, D.S. (ed.), 2000. Handbook of marine minerals. CRC Press, Boca Raton, Florida, 406 pp.

Cronan, D.S. and Hodkinson, R.A., 1994. Element supply to surface manganese nodules along the Aitutaki-Jarvis Transect, South Pacific. Journal of the Geological Society of London, 151: 392-401.

David, K., Frank, M., O'Nions, R.K., Belshaw, N.S. and Arden, J.W., 2001. The Hf isotopic composition of global seawater and the evolution of Hf isotopes in deep Pacifc Ocean from Fe-Mn crusts. Chemical Geology, 178: 23-42.

Davies, T.A., 1985. Mesozoic and Cenozoic sedimentation in the Pacific Ocean Basin. In: Nairn, A.E.M., Stehli, F.G. and Uyeda, S. (eds), The ocean basins and margins 7A The Pacific Ocean. Plenum Press, NY, pp. 65-88.

de Baar, H.W., Bacon, M.P., Brewer, P.G. and Bruland, K.W., 1985. Rare earth elements in the Pacific and Atlantic Oceans. Geochimica et Cosmochimica Acta, 49: 1943-1959.

de Baar, H.J.W., de Jong, J.T.M., Bakker, D.C.E., Löscher, B.M., Veth, C., Bathmann, U. and Smetacek, V., 1995. Importance of iron for pankton blooms and carbon dioxide drawdown in the Southern Ocean. Nature, 373: 412-415.

De Carlo, E.H., 1991. Paleoceanographic implications of the rare earth element variabiliy within a Fe-Mn crust from the central Pacific ocean. Marine Geology, 98: 449-467.

De Carlo, E.H. and McMurtry, G.M., 1992. Rare-earth element geochemistry of ferromanganese crusts from the Hawaiian Archipelago, central Pacific. Chemical Geology, 95: 235-250.

Dekov, V.M., Marchig, V., Rajita, I. and Uzonyi, I., 2003. Fe-Mn micronodules born in the metalliferous sediments of two spreading centres: the East Pacific Rise and Mid-Atlantic Ridge. Marine Geology, 199: 101-121.

Dickens, G.R. 2004. Hydrocarbon-driven warming. Nature, 429: 513-515.

Dillard, J.G., Crowther, D.L. and Calvert, S.E., 1984. X - ray photoelectron spectroscopic study of ferromanganese nodules: Chemical speciation for selected transition metals. Geochimica et Cosmochimica Acta, 48: 1565-1569.

Dorgan, K.M., Jumars, P.A., Johnson, B., Boudreau, B.P. and Landis, E., 2005. Burrow extension by crack propagation. Nature, 433: 475.

Drits, V.A., Silvester, E., Gorshkov, A.I. and Manceau, A., 1997. Structure of synthetic monoclinic Na-rich birnessite and hexagonal birnessite: I. Results from X-ray diffraction and selected-area electron diffraction. American Mineralogist, 82: 946-961.

Dymond, J. and Eklund, W., 1978. A microprobe study of metalliferous sediment components. Earth and Planetary Science Letters, 40: 243-251.

Dymond, J., Lyle, M., Finney, B., Piper, D.Z., Murphy, K., Conard, R. and Pisias, N., 1984. Ferromanganese nodules from MANOP Sites H, S, and R - Control of mineralogical and chemical composition by multiple accretionary processes. Geochimica et Cosmochimica Acta, 48: 931-949.

Earney, F.F.C., 1990. Marine mineral resources. Routledge, London, 387 pp.

Eckhardt, J.D., Glasby, G.P., Puchelt, H. and Berner, Z., 1997. Hydrothermal manganese crusts from Enareta and Palinuro seamounts in the Tyrrhenian Sea. Marine Georesources and Geotechnology, 15: 175-209.

Edmond, J.M., Chung, Y. and Sclater, J.G., 1971. Pacific bottom water: Penetration east around Hawaii. Journal of Geophysical Research, 76: 8089-8097.

Ehrlich, H.L., 1996. Geomicrobiology. Marcel Dekker, NY, 719 pp.

Eisenhauer, A., Gögen, K., Pernicka, E. and Mangini, A., 1992. Climatic influences on the growth rates of Mn crusts during the Late Quaternary. Earth and Planetary Science Letters, 109: 25-36.

Elderfield, H. and Schulz, A., 1996. Mid-ocean ridge hydrothermal fluxes and the chemical composition of the ocean. Annual Review of Earth and Planetary Sciences, 24: 191-224.

Finney, B., Heath, G.R. and Lyle, M., 1984. Growth rates of manganese-rich nodules at MANOP Site H (Eastern North Pacific). Geochimica et Cosmochimica Acta, 48: 911-919.

Fitzwater, S.E., Coale, K.H., Gordon, M., Johnson, K.S. and Ondrusek, M.E., 1996. Iron deficiency and plankton growth in the equatorial Pacific. Deep-Sea Research, 43: 995-1015.

Fleet, A.J. 1983. Hydrothermal and hydrogenous ferro-manganese deposits: Do they form a continuum? In: Rona, P., Boström, K., Laubier, L. and Smith, K.L. (eds) Hydrothermal processes at seafloor spreading centers. NATO Conference Series, 12: 535-555.

Frank, M., 2002. Radiogenic isotopes: Tracers of past ocean circulation and erosional input. Reviews of Geophysics, 40(1): Article number 1001.

Frank, M., O'Nions, R.K., Hein, J.R. and Banakar, V.K., 1999a. 60 Myr records of major elements and Pd-Nd isotopes from hydrogenous ferromanganese crusts: reconstruction of seawater paleoceanography. Geochimica et Cosmochimica Acta, 63: 1689-1708.

Frank , M., Reynolds, B.C. and O'Nions, R.K., 1999b. Nd and Pb isotopes in Atlantic and Pacific water masses before and after closure of the Panama gateway. Geology, 27: 1147-1150.

Friedrich, G., Glasby, G.P., Thijssen, T. and Plüger, W.L., 1983. Morphological and geochemical characteristics of manganese nodules collected from three areas on an equatorial Pacific transect by R.V. Sonne. Marine Mining, 4: 167-253.

Friedrich, G. and Schmitz-Wiechowski, A., 1980. Mineralogy and chemistry of a ferromanganese crust from a deep-sea hill, central Pacific, „VALDIVIA" Cruise VA 13/2. Marine Geology, 37: 71-90.

Froelich, P.N., Klinkhammer, G.P., Bender, M.L., Luedtke, N.A., Heath, G.R., Cullen, D., Dauphin, P., Hammond, D., Hartmann, B. and Maynard, V., 1979. Early oxidation of organic matter in pelagic sediments of the eastern equatorial Atlantic: suboxic diagenesis. Geochimica et Cosmochimica Acta, 43: 1075-1090.

Futa, K., Peteman, Z.E. and Hein, J.R., 1988. Sr and Nd isotopic variations in ferromanganese crusts from the Central Pacific: Implications for age and source provenance. Geochimica et Cosmochimica Acta, 52: 2229-2233.

Garrels, R.M. and Christ, C.L., 1965. Solutions, minerals, and equilibria. Harper & Row, N.Y. 450 pp.

German, C.R. and Angel, M.V., 1995. Hydrothermal fluxes of metal to the oceans: a comparison with anthropogenic discharges. In: Parson, L.M., Walker, C.L. and Dixon, D.R. (eds), Hydrothermal vents and processes. Geological Society Special Publication, 87: 365-372.

German, C.R. and Von Damm, K.L., 2004. Hydrothermal processes. In: Elderfield, H. (ed.), Treatise on geochemistry volume 6 The oceans and marine geochemistry. Elsevier, Amsterdam, pp. 181-222.

Gingele, F.X. and Kasten, S., 1994. Solid-phase manganese in Southeast Atlantic sediments: Implications for the paleoenvironment. Marine Geology, 121: 317-332.

Giovanoli, R., 1980. On natural and synthentic manganese nodules. In: Varentsov, I.M. and Grasselly, Gy. (eds), Geology and geochemistry of manganese, volume 1. Hungarian Academy of Sciences, Budapest, pp. 159-202.

Giovanoli, R., 1985. A review of the todorokite-buserite problem; implications to the mineralogy of marine manganese nodules: discussion. American Mineralogist, 70: 202-204.

Giovanoli, R. and Bürki, P., 1975, Comparison of X-ray evidence of marine manganese nodules and non-marine manganese ore deposits. Chimia, 29: 266-269.

Giovanoli, R. and Arrhenius, G., 1988. Structural chemistry of marine manganese and iron minerals and synthetic model compounds. In: Halbach, P., Friedrich, G. and von Stackelberg, U. (eds), The Manganese Nodule Belt of the Pacific Ocean Geological environment, nodule formation and mining aspects. Enke Verlag, Stuttgart, pp. 20-37.

Glasby, G.P., 1972. The mineralogy of manganese nodules from a range of marine environments. Marine Geology, 13: 57-72.

Glasby, G.P., 1973. Mechanism of enrichment of the rarer elements in marine manganese nodules. Marine Chemistry, 1: 105-125.

Glasby, G.P., 1974. Mechanism of incorporation of manganese and associated trace elements in marine manganese nodules. Oceanography and Marine Biology: An Annual Review, 12: 11-40.

Glasby, G.P. (ed), 1977. Marine manganese deposits. Elsevier, Amsterdam, 523 pp.

Glasby, G.P., 1978. Deeep-sea manganese nodules in the stratigraphic record: Evidence from DSDP cores. Marine Geology, 28: 51-64.

Glasby, G.P., 1983. The Three-Million-Tons-Per-Year Manganese Nodule „Mine Site": An optimistic assumption? Marine Mining, 4: 73-77.

Glasby, G.P., 1984. Manganese in the marine environment. Oceanography and Marine Biology: An Annual Review, 22: 169-194.

Glasby, G.P., 1988. Manganese deposition through geological time: Dominance of the Post-Eocene environment. Ore Geology Reviews, 4: 135-144.

Glasby, G.P., 1991. Mineralogy and geochemistry of Pacific red clays. N.Z. Journal of Geology and Geophysics, 34: 167-176.

Glasby, G.P., 2000. Lessons learned from deep-sea mining. Science, 289: 551, 553.

Glasby, G.P., 2002. Deep-seabed mining: Past failures and future prospects. Marine Georesources and Geotechnology, 20: 161-176.

Glasby, G.P., Meylan, M.A., Margolis, S.V. and Bäcker, H., 1980. Manganese deposits of the Southwestern Pacific Basin. In: Varentsov, I.M. and Grasselly, Gy. (eds), Geology and geochemistry of manganese, volume 3. Hungarian Academy of Sciences, Budapest, pp. 137-183.

Glasby, G.P. and Thijssen, T., 1982. Control of the mineralogy and composition of the manganese nodules by the supply of divalent transition metal ions. Neues Jahrbuch für Mineralogie Abhandlung, 145: 291-307.

Glasby, G.P., Friedrich, G.; Thijssen, T.; Plüger, W.L., Kunzendorf, H., Ghosh, A.K. and Roonwal, G.S., 1982: Distribution, morphology, and geochemistry of manganese nodules from the *Valdivia* 13/2 area, equatorial North Pacific. Pacific Science, 36: 241-263.

Glasby, G.P., Stoffers, P., Sioulas, A., Thijssen, T. and Friedrich, G., 1983. Manganese nodule formation in the Pacific Ocean: a general theory. Geo-Marine Letters, 2: 47-53.

Glasby, G.P., Gwozdz, R., Kunzendorf, H., Friedrich, G. and Thijssen, T., 1987. The distribution of rare earth and minor elements in manganese nodules and sediments from the equatorial and S.W. Pacific. Lithos, 20: 97-113.

Glasby, G.P., Uscinowicz, S.Z. and Sochan, J.A., 1996. Marine ferromanganese concretions from the Polish exclusive economic zone: Influence of Major Inflows of North Sea Water. Marine Georesources and Geotechnology, 14: 335-352.

Glasby, G.P., Emelyanov, E.M., Zhamoida, V.A., Baturin, G.N., Leipe, T., Bahlo, R. and Bonacker, P., 1997a. Environments of formation of ferromanganese concretions in the Baltic Sea: a critical review. In: Nicholson, K., Hein, J.R., Bühn B. and Dasgupta, S. (eds), Manganese mineralization: Geochemistry and mineralogy of terrestrial and marine deposits. Geological Society Special Publication, 119: 213-237.

Glasby, G.P., Stüben, D., Jeschke, G., Stoffers, P. and Garbe-Schönberg, C.-D., 1997b. A model for the formation of hydrothermal manganese crusts from the Pitcairn Island hotspot. Geochim. et Cosmochim. Acta, 61: 4583-4597.

Glasby, G.P. and Schulz, H.D., 1999. E$_H$, pH diagrams for Mn, Fe, Co, Ni, Cu and As under seawater conditions: Application of two new types of E$_H$, pH diagrams to the study of specific problems in marine geochemistry. Aquatic Geochemistry, 5: 227-248.

Glasby, G.P., Papavassiliou, C.T., Mitsis, J., Valsami-Jones, E., Liakopoulos, A. and Renner, R.M., 2004. The Vani manganese deposit, Milos island, Greece: A fossil stratabound Mn-Ba-Pb-Zn-As-Sb-W-rich hydrothermal deposit. In: Fytikas, M. and Vougioukalakis, G. (eds), The South Aegean Volcanic Arc Present knowledge and future perspectives. Developments in Volcanology, Elsevier, Amsterdam (in press)

Gordon, A.L. and Gerard, R.D., 1970. North Pacific bottom potential temperatures. In: Hayes, J.D. (ed.), Geological investigations of the North Pacific. Geological Society of America Memoir, 126, pp. 23-39.

Graham, I.J., Carter, R.M., Ditchburn, R.G. and Zondervan, A., 2004. Chronostratigraphy of ODP 181, Site 1121 sediment core (Southwest Pacific Ocean), using ^{10}Be/^{9}Be dating of trapped ferromanganese nodules. Marine Geology, 205: 227-247.

Gromet, L.P., Dymek, R.F., Haskin, L.A., and Korotev, R.L., 1984. The „North American Shale Composite": its compilation, major and trace element characteristics. Geochimica et Cosmochimica Acta, 48: 2469-2482.

Halbach, P., Marchig, V. and Scherhag, C., 1980. Regional variations in Mn, Cu, and Co of ferromanganese nodules from a basin in the Southeast Pacific. Marine Geology, 38: M1-M9.

Halbach, P., Scherhag, C., Hebisch, U. and Marchig, V., 1981. Geochemical and mineralogical control of different genetic types of deep-sea nodules from Pacific Ocean. Mineral Deposita, 16: 59-84.

Halbach, P. and Puteanus, D., 1984. The influence of the carbonate dissolution rate on the growth and composition of Co-rich ferromanganese crusts from the central Pacific seamount areas. Earth and Planetary Science Letters, 68: 73-87.

Halbach, P., Friedrich, G. and von Stackelberg, U. (eds), 1988. The Manganese Nodule Belt of the Pacific Ocean geological environment, nodule formation, and mining aspects. Enke Verlag, Stuttgart, 254 pp.

Halbach, P., Kriete, C., Prause, B. and Puteanus, D., 1989. Mechanism to explain the platinum concentration in ferromanganese seamount crusts. Chemical Geology, 76: 95-106.

Han, X., Jin, X., Yang, S., Fietzke, J. and Eisenhauer, A., 2003. Rhythmic growth of Pacific ferromanganese nodules and their Milankovitch climatic origin. Earth and Planetary Science Letters, 211: 143-157

Harada, K. and Nishida. S. 1976. Biostratigraphy of some marine manganese nodules. Nature, 260: 770-771.

Harada, K. and Nishida. S. 1976. Biochronology of some Pacific manganese nodules and their growth histories. Colloques Internationaux du C.N.R.S. N° 289 – La Genèse des Nodules de Manganèse, pp. 211-216.

Hastings, D. and Emerson, M., 1986. Oxidation of manganese by spores of a marine bacillus: Kinetics and thermodynamic considerations. Geochimica et Cosmochimica Acta, 50: 1819-1824.

Hayes, S.P., 1979. Benthic currents observations at DOMES sites A, B, and C in the tropical North Pacific Ocean. In: Bischoff, J.L. and Piper, D.Z. (eds), Marine geology and oceanography of the Pacific manganese nodule province. Plenum Press, NY, pp. 83-112.

Hayes, S.P., 1988. Benthic currents in the deep ocean. In: Halbach, P., Friedrich, G. and von Stackelberg, U. (eds), The manganese nodule belt of the Pacific Ocean Geological environment, nodule formation, and mining aspects. Enke Verlag, Stuttgart, pp. 99-102.

Heath, G.R., 1981. Ferromanganese nodules of the deep sea. Economic Geology, 75: 736-76

Hein, J.R., 2004. Cobalt-rich ferromanganese crusts: Global distribution, composition, origin and research activities. In: Minerals other than polymetallic nodules of the international seabed area Proceedings of a workshop held on 26-30 June 2000 in Kingston, Jamaica Volume 1, pp.188-272 (http//:www.isa.org.jm)

Hein, J.R., Morgenson, L.A., Clague, D.A. and Koski, R.A., 1987. Cobalt-rich ferromanganese crusts from the exclusive economic zone of the United States and nodules from the oceanic Pacific. In: Scholl, D.W., Grantz, A. and Vedder, J.G. (eds), Geology and resource potential of the continental margin of western North America and the adjacent oceans-Beaufort Sea to Baja California. Circum-Pacific Council for Energy and Mineral Recources, Earth Science Series, Houston, Texas, pp. 753-771.

Hein, J.R., Schwab, W.C. and Davis, A.S., 1988. Cobalt- and platinum-rich ferromanganese crusts and associated substrate rocks from the Marshall Islands. Marine Geology, 78: 255-283.

Hein, J.R., Yeh, H.-W., Gunn, S.H., Sliter, W.H., Benninger,

L.M. and Wang, C.-H., 1993. Two major episodes of Cenozoic phosphogenesis recorded in equatorial Pacific seamount deposits. Paleoceanography, 8: 292-311.

Hein, J.R., Gibbs, A.E., Clague, D.A. and Torresan, M., 1996. Hydrothermal mineralization along submarine rift zones, Hawaii. Marine Georesources and Geotechnology, 14: 177-203.

Hein, J.R., Koschinsky, A., Bau, M., Manheim, F.T., Kang, J.-K., and Roberts, L., 2000. Cobalt-rich ferromanganese crusts in the Pacific. In: Cronan, D.S. (ed), Handbook of marine minerals. CRC Press, Boca Raton, Florida, pp. 239-279.

Hem, J.D., Roberson, C.E. and Lind, C.J., 1985. Thermodynamic stability of CoOOH and its coprecipitation with manganese. Geochim. et Cosmochim. Acta, 49: 801-810.

Heye, D., 1978. Changes in the growth rate of manganese nodules from the Central Pacific in the area of a seamount as shown by the Io nethod. Marine Geology, 28: M59-M65.

Hlawatsch, S., 1999. Mn-Fe-Akkumulate als Indikator für Schad- und Nährstoffflüsse in der westlichen Ostsee. Geomar Report 85, Geomar Research Center, Kiel (with English abstract)

Hlawatsch, S., Neumann, T., van der Berg, C.M.G., Kersten, M., Harff, J., and Suess, E., 2002. Fast-growing shallow-water ferro-manganese nodules from the western Baltic Sea: origin and modes of trace element incorporation. Marine Geology, 182: 373-387.

Hodge, V.F., Stallard, M., Koide, M. and Goldberg, E.D., 1985. Platinum and platinum anomaly in the marine environment. Earth and Planetary Science Letters, 72: 158-162.

Hogg, N.G., 2001. Quantification of the deep circulation. In: Siedler, G., Church, J. and Gould, J.J. (eds), Ocean circulation & climate observation and modeling of the global ocean. Academic Press, San Diego, pp. 259-270.

Horn, D.R. (ed.), 1972. Ferromanganese deposits of the ocean floor. National Science Foundation, Washington, D. C., 293 pp.

Huh, C-A. and Ku, T.-L. 1984. Radiochemical observations on manganese nodules from three sedimentary environments in the North Pacific. Geochimica et Cosmochimica Acta, 48: 951-963.

Ito, T., Usui, A., Kajiwara, Y., Nakano, T., 1998. Strontium isotopic composition and paleoceanographic implcation of fossil manganese nodules in DSDP/ODP cores, Leg 1-126. Geochimica et Cosmochimica Acta, 62: 1545-1554.

Jauhari, P. and Pattan, J.N., 2000. Ferromanganese nodules from the Central Indian Ocean Basin. In: Cronan, D.S. (ed.), Handbook of marine minerals. CRC Press, Boca Raton, Florida, pp. 171-195.

Jeong, K.S., Kang, J.K. and Chough, S.K., 1994. Sedimentary processes and manganese nodule formation in the Korea Deep Ocean Study (KODOS) area, western part of Clarion-Clipperton fracture zones, northeast equatorial Pacific. Marine Geology, 122: 125-150.

Jeong, K.S., Kang, J.K., Lee, K.Y. Jung, H.S., Chi, S.B. and Ahn, S.J., 1996. Formation and distribution of manganese nodule deposits in the western margin of Clarion-Clipperton fracture zones, northeast equatorial Pacific. Geo-Marine Letters, 16: 123-131.

Jeong, K.S., Jung, H.S., Kang, J.K., Morgan, C.L. and Hein,

J.R., 2000. Formation of ferromanganese crusts on northwest intertropical Pacific seamounts: electron photomicrography and microprobe chemistry. Marine Geology, 162: 541-559.

Johnson, C.E. and Glasby, G.P., 1969. Mössbauer Effect determination of particle size in microcrystalline iron-manganese nodules. Nature, 222: 376-377.

Johnson, D.A., 1972. Eastward-flowing bottom currents along the Clipperton Fracture Zone. Deep-Sea Research, 19: 253-257

Johnson, K.S., Coale, K.H., Berelson, W.M. and Gordon, R.M., 1996. On the formation of the manganese maximum in the oxygen minimum. Geochimica et Cosmochimica Acta, 60: 1291-1299.

Johnson, K.S., Gordon, R.M. and Coale, K.H., 1997. What controls dissolved iron concentrations in the world ocean? Marine Chemistry, 57: 137-161.

Johnston, J.H. and Glasby, G.P., 1978. The secondary iron oxidehydroxide mineralogy of some deep sea and fossil manganese nodules: A Mössbauer and X-ray study. Geochemical Journal, 12: 153-164.

Johnston, J.H. and Glasby, G.P., 1982. A Mössbauer spectroscopic and X-ray diffraction study of the iron mineralogy of some sediments from the Southwest Pacific Basin. Marine Chemistry, 11: 437-448.

Jung, H.-S. and Lee, C.-P., 1999. Growth of diagenetic ferromanganese nodules in an oxic deep-sea environment, northeast Pacific. Marine Geology, 157: 127-144.

Kadko, D. and Burckle, L.H., 1980. Manganese growth rates determined by fossil diatom dating. Nature, 287: 725-726.

Kasten, S., Glasby, G.P., Schulz, H., Friedrich, G. and Andreev, S.I., 1998. Rare earth elements in manganese nodules from the South Atlantic Ocean as indicators of oceanic bottom water flow. Marine Geology, 146: 33-52.

Knie, K., Korschinek, G., Faestermann, T., Dorfi, E.A., Rugel, G and Wallner, A., 2004. [60]Fe anomaly in a deep-sea manganese crust and implications for a nearby supernova source. Physical Review Letters, 93: 171103.

Knoop, P.A., Owen, R.M. and Morgan, C.L., 1998. Regional variability in ferromanganese nodule composition: north-eastern tropical Pacific Ocean. Marine Geology, 147: 1-12.

Köning, I., Drodt, M., Suess, E. and Trautwein, A.X., 1997. Iron reduction through the tan-green color transition in deep-sea sediments. Geochimica et Cosmochimmica Acta, 61: 1679-1683.

Kontar, E.A. and Sokov, A.V., 1994. A benthic storm in the northeastern tropical Pacific over fields of manganese nodules. Deep-Sea Research I, 41: 1069-1089.

Koschinsky, A. and Halbach, P., 1995. Sequential leaching of marine ferromanganese precipitates: Genetic implications. Geochimica et Cosmochimica Acta, 59: 5113-5132.

Koschinsky, A. and Hein, J.R., 2003. Uptake of elements from seawater by ferromanganese crusts: solid-phase associations and seawater speciation. Marine Geology, 198: 331-351.

Koschinsky, A., Halbach, P., Hein, J.R. and Mangini, A., 1996. Ferromanganese crusts as indicators for paleoceanographic events in the NE Atlantic. Geologische Rundschau, 85: 567-576.

Koschinsky, A., Stascheit, A.-M., Bau, M. and Halbach, P., 1997. Effects of phosphatization on the geochemical and mineralogical composition of marine ferromanganese crusts. Geochimica et Cosmochimica Acta, 61: 4079-4094.

Kotlinski, R., 1995. InterOceanMetal Joint Organization: Archievements and Challenges. Proceedings of the ISOPE (The International Society of Offshore and Polar Engineers)-Ocean Mining Symposium, Tsukuba, Japan, November 21-22, 5-7.

Krishnaswamy, S., 1976. Authigenic transition elements in Pacific pelagic clays. Geochimica et Cosmochimica Acta, 40: 425-435.

Ku, T.-L. and Glasby, G.P., 1972. Radiometric evidence for the rapid growth rates of shallow-water, continental margin manganese nodules. Geochimica et Cosmochimica Acta, 36: 699-703.

Ku, T.-.L., 1977. Rates of accretion. In: Glasby, G.P. (ed.) Marine manganese deposits. Elsevier, Amsterdam, pp. 249-267.

Kuma, K., Usui, A., Paplawsky, W., Gedulin, B. and Arrhenius, G., 1994. Crystal structures of synthetic 7 Å and 10 Å manganates substituted by mono - and divalent cations. Mineralogical Magazine, 58: 425-447.

Kunzendorf, H., Glasby, G.P., Stoffers, P. and Plüger, W.L., 1993. The distribution of rare earth and minor elements in manganese nodules, micronodules and sediments along an east-west transect in the southern Pacific. Lithos, 30: 45-56.

Kusukabe, M. and Ku, T.L., 1984. Incorporation of Be isotopes and other trace metals into marine ferromanganese deposits. Geochimica et Cosmochimica Acta, 48: 2187-2193.

Lalou, C., Brichet, E., Jehanno, C. and Perez-Leclaire, H., 1983. Hydrothermal manganese oxide deposits from Galapagos mounds, DSDP Leg 70, hole 509B and „Alvin" dives 729 and 721. Earth and Planetary Science Letters, 63:63-75.

Lavasseur, S., Frank, M., Hein, J.R. and Halliday, A.N., 2004. The global variation in the iron isotope composition of marine hydrogenetic ferromanganese deposits: impliations for seawater chemistry? Earth and Planetary Science Letters, 224: 91-105.

Lavelle, J.W., Cowen, J.P. and Massoth, G.J., 1992. A model for the deposition of hydrothermal manganese near mid-ocean ridge crests. Journal of Geophysical Research, 97: 7413-7427.

Lee, D.-C., Halliday, A.N., Hein, J.R., Burton, K.W., Christensen, J.N. and Günther, D., 1999. Hafnium isotope stratigraphy of ferromanganese crusts. Science, 285: 1052-1054.

Lei, G., 1996. Crystal structure and metal uptake capacity of 10 Å-manganates: An overview. Marine Geology, 133: 103-112.

Lenoble, J.-P., 2004. A comparison of the possible economic returns from mining deep-sea polymetallic nodules, sea floor massive sulphides and cobalt-rich crusts. Proceedings of a workshop held on 26-30 June 2000 in Kingston, Jamaica Volume 1, pp. 424-465 (http//: www.isa.org.jp)

Liebetrau, V., Eisenhauer, A., Gussone, N., Wörner, G., Hansen, B.T., Leipe, T., 2002. Ra/Ba growth rates and U-Th-Ba systematics of Baltic Fe-Mn crusts. Geochimica et Cosmochimica Acta, 66 : 73-83.

Lilley, M.D., Feely, R.A. and Trefry, J.H., 1995. Chemical and biochemical transformations in hydrothermal plumes. In: Humphris, S.E. Zierenberg, R.A., Mullineaux, L.S. and Thompson, R.E. (eds), Seafloor Hydrothermal systems: Physical, chemical, biological, and geological interactions.

American Geophysical Union Geophysical Monogrograph, 91: 369-391.

Ling, H.F., Burton, K.W., O'Nions, R.K., Kamber, B.S., von Blanckenberg, F., Gibb, A.J. and Hein, J.R., 1997. Evolution of Nd and Pb isotopes in central Pacific seawater from ferromanganese crusts. Earth and Planetary Science Letters, 146: 1-12.

Lonsdale, P., 1976. Abyssal circulation of the southeastern Pacific and some geological implications. Journal of Geophysical Research, 81: 1163-1176.

Lonsdale, P., 1981. Drifts and ponds of reworked pelagic sediment in part of the Southwest Pacific. Marine Geology, 43: 153-193.

Lupton, J., 1998. Hydrothermal helium plumes in the Pacific Ocean. Journal of Geophysical Research, 103: 15,835-15,868.

Lyle, M., 1983. The brown-green color transition in marine sediments: A marker of the Fe(III)-Fe(II) redox boundary. Limnology and Oceanography 28: 1026-1033.

Lyle, M., Owen, R.M. and Leinen, M., 1986. History of hydrothermal sedimentation at the East Pacific Rise, 19°S. In: Initial Reports of the Deep Sea Drilling Project. U.S. Government Printing Office, Washington, D.C., 92: 585-596.

McCave, I.N., 1988. Biological pumping upwards of the coarse fraction of deep-sea sediments. Journal of Sedimentary Petrology, 58: 148-158.

Macdonald, K.C. and Hollister, C.D., 1973. Near-bottom thermocline in the Samoan Passage, west equatorial Pacific. Nature, 241: 461-462.

McMurtry, G.M., von der Haar, D.L., Eisenhauer, A., Mahoney, J.J. and Yeh, H.W., 1994. Cenozoic accumu-lation history of a Pacific ferromanganese crust. Earth and Planetary Science Letters, 125: 105-118.

Manceau, A., Llorca, S. and Calas, G., 1987. Crystal chemistry of cobalt and nickel in lithiophorite and asbolane from New Caledonia. Geochimica et Cosmochimica Acta, 51: 105-113.

Manceau, A., Gorshkov, A.I. and Drits, V.A., 1992a. Structural chemistry of Mn, Fe, Co, and Ni in manganese hydrous oxides: Part I. Information from EXAFS spectroscopy. American Mineralogist, 77: 1133-1143.

Manceau, A., Gorshkov, A.I. and Drits, V.A., 1992b. Structural chemistry of Mn, Fe, Co, and Ni in manganese hydrous oxides: Part II. Information from EXAFS spectroscopy and electron and X-ray diffraction. American Mineralogist, 77: 1144-1157.

Manceau, A., Drits, V.A., Silvester, E., Bartoli, C. and Lanson, B., 1997. Structural mechanism of Co^{2+} oxida-tion by the phyllomanganate buserite. American Mine-ralogist, 82: 1150-1175.

Manceau, A., Lanson, B. and Drits, V.A., 2002. Structure of heavy metal sorbed birnessite. Part III. Results from powder and polarized extended X-ray absorption fine structure spectroscopy. Geochimica et Cosmochimica Acta, 66: 2639-2663.

Mandernack, K.W., Post, J. and Tebo, B.M., 1995. Manganese mineral formation by bacterial spores of the marine Bacillus, strain SG-1: Evidence for the direct oxidation of Mn(II) to Mn(IV). Geochimica et

Cosmochimica Acta, 59: 4393-4408.

Mangini, A., 1988. Growth rates of manganese nodules and crusts. In: Halbach, P., Friedrich, G. and von Stackelberg, U. (eds), The Manganese Nodule Belt of the Pacific Ocean Geological environment, nodule formation, and mining aspects. Enke Verlag, Stuttgart, pp. 142-151.

Mangini, A., Dominik, J., Muller, P.J and Stoffers, P., 1982. Pacific deep circulation: A velocity increase at the end of the interglacial stage 5? Deep-Sea Research, 29A: 1517-1530.

Mangini, A., Halbach, P., Puteanus, D. and Segl, M., 1987. Chemistry and growth history of Central Pacific Mn - crusts and their economic importance. In: Teleki, P.G., Dobson, M.R., Moore, J.R. and von Stackelberg, U. (eds), Marine minerals Advances in research and resource assessment. D. Reidel, Dordrecht, pp. 205-220.

Mangini, A., Eisenhauer, A. and Walter, P., 1990a. Response of manganese in the ocean to climate change in the Quaternary. Paleoceanography, 5: 811-821.

Mangini, A., Segl, M., Glasby, G.P, Stoffers, P. and Plüger, W.L., 1990b. Element accumulation rates in and growth histories of manganese nodules from the Southwestern Pacific Basin. Marine Geology, 94: 97-107.

Manheim, F.T., 1986. Marine cobalt resources. Science, 232: 600-608.

Manheim, F.T. and Lane-Bostwick, C.M., 1988. Cobalt in ferromanganese crusts as a monitor of hydrothermal discharge on the Pacific sea floor. Nature, 335: 59-62.

Mantyla, A.W., 1975. On the potential temperature in the abyssal Pacific Ocean. Journal of Marine Research, 33: 341-354.

Marchig, V., 2000. Hydrothermal activity on the southern, ultrafast-spreading segment of the East Pacific Rise. In: Cronan, D.S. (ed.), Handbook of marine minerals. CRC Press, Boca Raton, Florida, pp. 309-325..

Marchig, V. and Erzinger, J., 1986. Chemical composition of Pacific sediments near 20°S: changes with increasing distance from the East Pacific Rise. In: Initial Reports of the Deep Sea Drilling Project. U.S. Government Printing Office, Washington, D.C., pp. 371-381.

Marcus, M.A., Manceau, A. and Kersten, M., 2004. Mn, Fe, Zn and As speciation in a fast growing ferromanganese marine nodule. Geochimica et Cosmochimica Acta, 68: 3125-3136.

Martin, J.H., Knauer, G.A. and Broenkow, W.W., 1985. VERTEX: the lateral transport of manganese in the northeast Pacific. Deep-Sea Research, 32: 1405-1427.

Medford, R.D., 1969. Marine mining in Britain. Mining Magazine, 121(5): 369-381 and 121(6): 474-480.

Mellin, T.A. and Lei, G., 1993. Stabilization of 10 Å-manganates by interlayer cations and hydrothermal treatment: Implications for the mineralogy of marine manganese concretions. Marine Geology, 115: 67-83.

Mero, J.L., 1965. Mineral resources of the sea. Elsevier, Amsterdam, 312 pp.

Mero, J.L., 1977. Economic aspects of nodule mining. In: Glasby, G.P. (ed.), 1977. Marine manganese deposits. Elsevier, Amsterdam, pp. 327-355.

Meylan, M.A., Glasby, G.P., McDougall, J.C. and Kumbalek, S.C., 1982. Lithology, colour, mineralogy, and geochemistry of marine sediments from the Southwestern Pacific and Samoan Basin. New Zealand Journal of Geology and Geophysics, 25: 437-458.

Meylan, M.A., Glasby, G.P., Hill, P.J., McKelvey, B.C., Walter, P. and Stoffers, P., 1990. Manganese crusts and nodules from the Manihiki Plateau and adjacent areas: Results of HMNZS Tui cruises. Marine Mining, 9: 43-72.

Michard, G. and Albarède, F. 1986.The REE content of some hydrothermal fluids. Chemical Geology, 55: 51-60.

Miller, S. and Cronan, D.S., 1994. Element supply to surface sediments and interrelationships with nodules along the Aitutaki-Jarvis Transect, South Pacific. Journal of the Geological Society of London, 151: 403-412.

Moore, J.G. and Clague, D.A., 2004. Hawaiian submarine manganese-iron oxide crusts — A dating tool? GSA Bulletin, 116: 337-347.

Morgan, C.L., 2000. Resource estimates of the Clarion - Clipperton manganese nodule deposits. In: Cronan, D.S. (ed.), Handbook of marine minerals. CRC Press, Boca Raton, Florida, pp. 145-170.

Morgan, J.J., 2005. Kinetics of reaction between oxygen and Mn(II) species in aqueous solutions. Geochimica et Cosmochimica Acta, 69: 35-48.

Mukhopadhyay, R., Iyer, S.D. and Ghosh, A.K., 2002. The Indian Ocean Nodule Field: petrotectonic evolution and ferromanganese deposits. Earth-Science Reviews, 60: 67-130.

Müller, P.J., Hartmann, M. and Suess, E., 1988. The chemical environment of pelagic sediments. In: Halbach, P., Friedrich, G. and von Stackelberg, U. (eds), The manganese nodule belt of the Pacific Ocean Geological environment, nodule formation, and mining aspects. Enke Verlag, Stuttgart, pp. 70-99.

Murray, J.W. and Brewer, P.G., 1977. Mechanism of removal of manganese, iron and other trace metals from seawater. In: Glasby, G.P. (ed), Marine manganese deposits. Elsevier, Amsterdam, pp. 291-325.

Murray, J.W., Balistrieri, L.S. and Paul, B., 1984. The oxidation state of manganese in marine sediments and ferromanganese nodules. Geochimica et Cosmochimica Acta, 48: 1237-1247.

Nemoto, K. and Kroenke, L.W., 1981. Marine geology of the Hess Rise 1. Bathymetry, surface sediment distribution and environment of deposition. Journal of Geophysical Research, 86: 10734-10752.

Nicholson, K., Hein, J.R., Bühn, B. and Dasgupta S. (eds), 1997. Manganese mineralization: Geochemistry and mineralogy of terrestrial and marine deposits. Geological Society Special Publication, 119: 370 pp.

Norris, R.D. and Röhling, U., 1999. Carbon cycling and chronology of climate warming at the Palaeocene/Eocene transition. Nature, 401: 775-778.

Nozaki, Y., 1997. A fresh look at element distribution in the North Pacific. EOS Transactions of the American Geophysical Union, 78: 221.

O'Nions, R.K., Frank, M., von Blanckenburg, F. and Ling, H.F., 1998. Secular variation of Nd and Pb isotopes in ferromanganese crusts from the Atlantic, Indian and Pacific Oceans. Earth and Planetary Science Letters, 155: 15-28.

Overnell, J., 2002. Manganese and iron profiles during early diagenesis in Loch Etive, Scotland. Application of two diagenetic models. Estuarine, Coastal and Shelf Science, 54: 33-44.

Overnell, J., Harvey, S.M. and Parkes, R.J., 1996. A biogeochemical comparison of sea loch sediments: Manganese and iron contents, sulphate reduction rates and oxygen uptake rates. Oceanologica Acta, 19: 41-55.

Overnell, J., Brand, T., Bourgeois, W. and Statham, P.J., 2002. Manganese dynamics in the water column of the upper basin of Loch Etive, a Scottish fjord. Estuarine, Coastal and Shelf Science, 55: 481-492.

Piper, D.Z. and Williamson, M.E., 1977. Composition of Pacific Ocean ferromanganese nodules. Marine Geology., 23: 285-303.

Piper, D.Z. and Blueford, J.R., 1982. Distribution, mineralogy, and texture of manganese nodules and their relation to sedimentation at the DOMES Site A in the equatorial North Pacific. Deep-Sea Research, 29A: 927-952.

Piper, D.Z., Basler, J.R. and Bischoff, J.L., 1984. Oxidation state of marine manganese nodules. Geochimica et Cosmochimica Acta, 48: 2347-2355.

Piper, D.Z., Swint, T.R., Sullivan, L.G. and McCoy, F.W., 1985. Manganese nodules, seafloor sediment, and sedimentation rates in the Circum-Pacific region. Circum-Pacific Council for Energy and Mineral Resources Circum-Pacific Map Project. American Association of Petroleum Geologists, Tulsa, Oklahoma.

Piper, D.Z., Swint-Iki, T.R. and McCoy, F.W., 1987. Distribution of ferromanganese nodules in the Pacific Ocean. Chemie der Erde, 46: 171-184.

Plüger, W.L., Friedrich, G. and Stoffers, P., 1985. Environmental controls of the formation of deep-sea ferromanganese concretions. Monograph Series on Mineral Deposits, 25: 31-52.

Plunkert, P.A. and Jones, T.S, 1999. Metal prices in the United States through 1998. USGS. 181 pp. (http://minerals.usgs.gov/minerals/pubs/metal_prices)

Post, J.E., 1999. Manganese oxide minerals: crystal structures and environmental significance. Proceedings of the National Academy of Science of the USA, 96: 3447-3454.

Pulyaeva, I., 1997. Stratification of ferromanganese crusts on the Magellan seamounts. Proceedings of the 30th International Geological Congress, Vol. 13: 111-128.

Puteanus, D., and Halbach, P., 1988. Correlation of Co concentration and growth rate-a model for age determination of ferromanganese crusts. Chemical Geology, 69: 71-85.

Raab, W.J. and Meylan, M.A., 1977. Morphology. In: Glasby, G.P. (ed.), Marine manganese deposits. Elsevier, Amsterdam, pp. 109-146.

Rehkämper, M., Frank, M., Hein, J.R. and Halliday, A.N., 2004. Cenozoic marine geochemistry of thallium deduced from isotopic studies of ferromanganese crusts and palagic sediments. Earth and Planetary Science Letters, 219: 77-91.

Reid, J.L. and Lonsdale, P.F., 1974. On the flow of water through the Samoan Passage. Journal of Physical Oceanography, 4: 58-73.

Reyss, J.L., Lemaitre, N., Ku, T.L., Marchig, V., Southon, J.R., Nelson, D.E. and Vogel, J.S., 1985. Growth of manganese nodule from the Peru Basin: A radiochemical anomaly. Geochimica et Cosmochimica Acta, 49: 2401-2408.

Roemmich, D., Huatala, S. and Rudnick, D.L., 1996. Northward abyssal transport through the Samoan Passage and adjacent regions. Journal of Geophysical Research, 101: 14,039-14,055.

Roy, S., 1981. Manganese deposits. Academic Press, London, 458 pp.

Rudnick, D.L., 1997. Direct velocity measurements in the Samoan Passage. Journal of Geophysical Research, 102: 3293-3303.

Sanderson, B., 1985. How bioturbation supports manganese nodules at the sediment-water interface. Deep-Sea Research, 32A: 1281-1285.

Sawlan, J.J. and Murray, J.W., 1983. Trace metal remobilization in the interstitial waters of red clay and hemipelagic marine sediments. Earth and Planetary Science Letters, 64: 213-230.

Schlosser, P., Bullister, J.L., Fine, R., Jenkins, W.J., Key, R., Roether, W. and Smethie, W.M., 2001. Transformation and age of water masses. In: Siedler, G., Church, J. and Gould, J.J. (eds), Ocean circulation & climate observation and modeling of the global ocean. Academic Press, San Diego, pp. 431-452 + Plate 5.8.17 (P. 428)

Schmitz, W., Mangini, A., Stoffers, P., Glasby, G.P. and Plüger, W.L., 1986. Sediment accumulation rates in the southwestern Pacific Basin and Aitutaki Passage. Marine Geology, 73: 181-190.

Schmitz, W.J., 1995. On the interbasin-scale thermohaline circulation. Reviews of Geophysics, 33: 151-173.

Scholten, J.C., Scott, S.D., Garbe-Schönberg, D., Fietze, J., Blanz, T and Kennedy, C.B., 2004. Hydrothermal manganese crusts from the Pitcairn region. In: Hekinian, R., Stoffers, P. and Cheminée, J.-L., (eds), Oceanic hotspots intraplate submarine magmatism and tectonism. Springer-Verlag, Berlin, pp.375-405.

Segl, M., Mangini, A., Bonani, G., Hofmann, H.J., Nessi, M., Suter, M., Wölfli, W., Friedrich, G., Plüger, W.L., Wiechowski, A. and Beer, J., 1984. 10Be-dating of a manganese crust from the central North Pacific Ocean and implications for ocean palaeocirculation. Nature, 309: 540-543.

Segl, M., Mangini, A., Beer, J., Bonani, G., Suter, M. and Wölfli, W., 1989. Growth rate variations of manganese nodules and crusts induced by paleoceanographic events. Paleoceanography, 4: 511-530.

Shaw, T.J., Gieskes, J.M. and Jahnke, R.A., 1990. Early diagenesis in differing depositional environments: The response of transition metals in pore waters. Geochimica et Cosmochimica Acta, 54: 1233-1246.

Shevenell, A.E., Kennett, J.P. and Lea, D.W., 2004. Middle Miocene Southern Ocean cooling and Antarctic cryosphere expansion. Science, 305: 1766-1770.

Shilov, V.V., 2004. Stratigraphy of Upper Cenozoic deposits in the Clarion-Clipperton Fracture Zone (Pacific Ocean). Unpublished Candidate thesis, VSEGEI (All-Russian Institute of Geology), St. Petersburg. 158 pp. (In Russian)

Silvester, E., Manceau, A., Drits, V.A., 1997. Structure of synthetic monoclinic Na-rich birnessite: II. Results from chemical studies and EXAFS spectroscopy. American Mineralogist, 82: 962-978.

Skornyakova, N.S. and Murdmaa, I.O., 1992. Local variation in distribution and composition of ferromanganese nodules in the Clarion - Clipperton Nodule Province. Marine Geology, 103: 381-405.

Sorem, R.K. and Fewkes, R.H., 1979. Manganese nodule research data and methods of investigation. IFI/ Plenum Press, NY, 723 pp.

Stoffers, P., Glasby, G.P., Thijssen, T., Shrivastava, P.C. and Melguen, M., 1981. The geochemistry of co - existing manganese nodules, micronodules, sediments and pore waters from five areas in the equatorial and South - West Pacific. Chemie der Erde, 40: 273-297.

Stoffers, P., Glasby, G.P. and Frenzel, G., 1984. Comparison of the characteristics of manganese micronodules from the equatorial and south - west Pacific. TMPM Tschermaks Mineralogische und Petrographische Mitteilungen, 33: 1-23.

Stoffers, P., Schmitz, W., Glasby, G.P., Plüger, W.L. and Walter, P., 1985. Mineralogy and geochemistry of sediments in the Southwestern Pacific Basin: Tahiti - East Pacific Rise - New Zealand. New Zealand Journal of Geology and Geophysics, 28: 513-530.

Stüben, D., Glasby, G.P., Eckhardt, J.-D., Berner, Z., Mountain, B.W. and Usui, A., 1999. Enrichments of platinum-group elements in hydrogenous, diagenetic and hydrothermal marine manganese and iron deposits. Exploration and Mining Geology, 8: 233-250.

Sverjensky, D.A. 1984. Equilibrium redox equilibria in aqueous solution. Earth and Planetary Science Letters, 67: 70-78.

Szefer, P., Glasby, G.P., Kunzendorf, H., Görlich, E.A., Latka, K., Ikuta, K., and Ali, A.A., 1998. Distribution of rare earth and other elements, and the mineralogy of the iron oxyhydroxide phase in marine ferromanganese concretions from within Slupsk Furrow southern Baltic Sea, off Poland. Applied Geochemistry, 13: 305-312.

Taft, B.A., Hayes, S.P., Friederich, G.E, Codispoti, A, 1991. Flow of abyssal water into the Samoan Passage. Deep-Sea Research, 38, Supplement 1: S103-S128.

Tebo, B.M., Ghiorse, W.C., van Waasbergen, L.G., Siering, P.L. and Caspi, R., 1997. Bacterially-mediated mineral formation: Insights into manganese (II) oxidation from molecular genetic and biochemical studies. Reviews in Mineralogy, 35: 225-266.

Tebo, B.M., Bargar, J.R., Clement, B.G., Dick, G.J., Murray, K.J., Parker, D., Verity, R. and Webb, S.M., 2004. Biogenic manganese oxides: Properties and mechanisms of formation. Annual Review of Earth and Panetary Sciences, 32: 287-328.

Teleki, P.G., Dopson, M.R., Moore, J.R. and von Stackelberg, U. (eds), 1987. Marine minerals Advances in research and resource assessment. D. Reidel, Dordrecht, 588 pp.

Thamdrup, B. and Canfield, D.E., 1996. Pathways of carbon oxidation in continental margin sediments off central Chile. Limnology and Oceanography, 41: 1629-1650.

Thiel, H. (ed.), 2001. Environmental impact studies for the mining of polymetallic nodules from the deep sea. Deep-Sea Research II, 48: 3427-3882.

Thijssen, T., Glasby, G.P., Friedrich, G., Stoffers, P. and Sioulas, A., 1985. Manganese nodules in the central Peru Basin. Chemie der Erde, 44: 1-46.

Thomson, J., Higgs, N.C., Hydes, D.J., Wilson, T.R.S. and Sorensen, J., 1987. Geochemical oxidation fronts in NE Atlantic distal turbidites and their effects in the sedimentary record. In: Weaver, P.P.E. and Thompson, J. (eds), Geology and geochemistry of abyssal plains. Geological Society Special Publication, 31: 167-177.

Turekian, K.K. and Bacon, M.P., 2004. Geochronometry of marine deposits. In: Elderfield, H. (ed.), Treatise on geochemistry vol. 6 The oceans and marine geochemistry. Elsevier, Amsterdam, pp. 321-341,

Turner, S. and Buseck, P.R., 1981. Todorokites: A new family of naturally occuring manganese oxides. Science, 212: 1024-1027.

Usui, A., 1979. Nickel and copper accumulation as essential elements in 10Å manganite of deep-sea manganese nodules. Nature, 279: 411-413.

Usui, A., 1992. Hydrothermal manganese minerals in Leg 126 cores. Proceedings of Ocean Drilling Program Scentific Results, Volume 126: 113-123.

Usui, A. and Nishimura, A., 1992. Submersible observations of hydrothermal manganese deposits on the Kaikata Seamount, Izu-Ogasawara (Bonin) Arc. Marine Geology, 106: 203-216

Usui, A. and Ito, T., 1994. Fossil manganese deposits buried within DSDP/ODP cores, Leg 1-126. Marine Geology, 119: 111-136.

Usui, A. and Mita, N., 1995. Geochemistry and mineralogy of a modern buserite deposit from a hot spring in Hokkaido, Japan. Clays and Clay Minerals, 43: 116-127.

Usui, A. and Terashima, S., 1997. Deposition of hydrogenetic and hydrothermal manganese minerals in the Ogasawara (Bonin) arc area, northwest Pacific. Marine Georecources and Geotechnology, 15: 127-154.

Usui, A. and Glasby, G.P., 1998. Submarine hydrothermal manganese deposits in the Izu-Bonin-Mariana arc: An overview. The Island Arc, 7: 422-431.

Usui, A., Yuasa, M., Yokota, S., Nohara, M. and Nishimura, A., 1986. Submarine hydrothermal manganese deposits from the Ogasawara (Bonin) Arc, off the Japanese islands. Marine Geology, 73: 311-322.

Usui, A., Mellin, T.A., Nohara, M. and Yuasa, M., 1989. Structural stability of marine 10 Å manganates from the Ogasawara (Bonin) Arc: Implications for low-temperature hydrothermal activity. Marine Geology, 86: 41-56.

Usui, A., Nishimura, A. and Mita, N., 1993. Composition and growth history of surficial and buried manganese nodules in the Penrhyn Basin, Southwest Pacific. MarineGeology, 114: 133-153.

Usui, A., Bau, M. and Yamazaki, T., 1997. Manganese microchimneys buried in Central Pacific Pacific pelagic sediments: evidence of intraplate water circulation. Marine Geology, 141: 269-285.

van Andel , Tj.H., Heath, G.R. and Moore, T.C., 1975. Cenozoic history and paleoceanography of the central equatorial Pacific Ocean. Geological Society of America Memoir, 143: 134 pp.

van de Fliert, T., Frank, M., Halliday, A.N., Hein, J.R., Hattendorf, B., Günther, D. and Kubik, P.W., 2003. Lead isotopes in North Pacific deep - water implications for past changes in input sources and circulation patterns. Earth and Planetary Science Letters, 209: 149-164.

van de Fliert, T., Frank, M., Halliday, A.N., Hattendorf, B.,

Günther, D. and Kubik, P.W., 2004a. Tracing the history of submarine hydrothermal inputs and the significance of hydrothermal hafnium for the seawater budget—a combined Pb-Hf-Nd approach. Earth and Planetary Science Letters, 222: 259-273.

van de Fliert, T., Frank, M., Halliday, A.N., Hein, J.R., Hattendorf, B., Günther, D. and Kubik, P.W., 2004b. Deep and bottom water export from the Southern Ocean to the Pacific over the past 38 million years. Paleoceanograpy, 19: PA1020

van de Fliert, T., Frank, M., Lee, D.-C., Halliday, A.N., Reynolds, B.C. and Hein, J.R., 2004c. New constraints on the sources and behavior of neodymium and hafnium in seawater from Pacifc Ocean ferromanganese crusts. Geochimica et Cosmochimica Acta, 68: 3827-3843.

Varentsov, I.M. and Grasselly, Gy. (eds), 1980. Geology and geochemistry of manganese. Hungarian Academy of Sciences, Budapest, 3 volumes.

Varentsov, I.M., Drits, V.A., Gorshkov, A.I., Sivtsov, A.V. and Sakharov, B.A., 1991a. Mn-Fe oxyhydroxide crusts from Krylov Seamount (Eastern Atlantic): Mineralogy, geochemistry and genesis. Marine Geology, 96: 53-70.

Varentsov, I.M., Drits, V.A. and Gorshkov, A.I., 1991b. Rare earth element indicators of Mn-Fe oxyhydroxide crust formation on Krylov Seamount, Eastern Atlantic. Marine Geology, 96: 71-84.

von Blanckenburg, F. O'Nions, R.K. and Hein, J.R., 1996a. Distribution and source of pre-anthropogenic lead isotopes in deep ocean water from Mn-Fe crusts. Geochimica et Cosmochimica Acta, 60: 4957-4963.

von Blanckenburg, F., O'Nions, R.K., Belshaw, N.S., Gibb, A. and Hein, J.R., 1996b. Global distribution of beryllium isotopes in deep ocean water as derived from Fe-Mn crusts. Earth and Planetary Science Letters, 141: 213-226.

von der Haar, D.L., Mahoney, J.J. and McMurtry, G.M., 1995. An evaluation of strontium isotopic dating of ferromanganese oxides in a marine hydrogenous crust. Geochimica et Cosmochimica Acta, 59: 4267-4277.

von Langen, P., Johnson, K.S., Coale, K.H. and Elrod, V.A., 1997. Oxidation kinetics of manganese (II) in seawater in nanomolar concentrations. Geochimica et Cosmochimica Acta, 61: 4945-4954.

von Stackelberg, U., 1997. Growth history of manganese nodules and crusts of the Peru Basin. In: Nicholson, K., Hein, J.R., Bühn, B. and Dasgupta S. (eds), Manganese mineralization: Geochemistry and mineralogy of terrestrial and marine deposits. Geological Society Special Publication, 119: 153-176.

von Stackelberg, U., 2000. Manganese nodules of the Peru Basin. In: Cronan, D.S. (ed.), Handbook of marine minerals. CRC Press, Boca Raton, Florida, pp. 197-238.

von Stackelberg, U., Kunzendorf, H., Marchig, V. and Gwozdz, R., 1984. Growth history of a large manganese crust from the Equatorial North Pacific Nodule Belt. Geologishes Jahrbuch, 75A: 213-235.

von Stackelberg, U. and Marchig, V., 1987. Manganese nodules from the equatorial North Pacific Ocean. Geologishes Jahrbuch, 87D: 123-227.

von Stackelberg, U. and Beiersdorf, H., 1991. The formation of manganese nodules between the Clarion and Clipperton fracture zones southeast of Hawaii. Marine Geology, 98: 411-423.

Wallace, H.E., Thomson, J., Wilson, T.R.S., Weaver, P.P.E., Higgs, N.C., Hydes, D.J., 1988. Active diagenetic formation of metal-rich layers in N.E. Atlantic sediments. Geochimica et Cosmochimica Acta, 52: 1557-1569.

Wallner, C., Faestermann, T., Gerstmann, U., Knie, K., Korschinek, G., Lierse, C. and Rugel, G., 2003. Supernova produced and anthropogenic [244]Pu in deep sea manganese encrustations. New Astronomy Reviews, 48: 145-150.

Waychunas, G.A., 1991. Crystal chemistry of oxides and oxyhydroxides. Reviews in Mineralogy, 25: 11-68.

Wiltshire, J.C., 2000. Innovations in marine ferromanganese oxide tailings disposal. In: Cronan, D.S. (ed.), Handbook of marine minerals. CRC Press, Boca Raton, Florida, pp. 281-305.

Wiltshire, J.C., Wen, X.Y and Yao, D., 1999. Ferromanganese crusts near Johnson Island: Geochemistry, stratigraphy and economic potential. Marine Georesources and Geotechnology, 17: 257-270.

Xu, D., 1997. Paleo-ocean events and mineralization in the Pacific. Proceedings of the 30th International Geological Congress, Vol. 13: 129-144.

Yeats, P.A. and Strain, P.M., 1990. The oxidation rate of manganese in seawater: Rate constants based on field data. Estuarine Coastal and Shelf Science, 31: 11-24.

Zachos, J., Pagani, M., Sloan, L., Thomas, E., and Billups, K., 2001. Trends, rhythms, and aberrations in global climate 65 Ma to present. Science, 292: 686-693.

12 Quantification and Regionalization of Benthic Reflux

Matthias Zabel and Christian Hensen

12.1 Introduction

The observations made in the preceding chapters have demonstrated that the upper sediment layers usually contain the highest biochemical reaction rates within marine deposits.

The decomposition of organic matter represents the major motive force of the benthic system and its processes which are mostly catalyzed by microbial activity. How important is the boundary between free water mass and the pore water system? In this chapter, we will be concerned with finding answers to this question.

It is well established by a large number of long-term studies that most part of the primary organic substance is remineralized in the upper water zones after the organisms have died (cf. Chapter 4 and Section 5.5). Estimates on the marine carbon cycle, which were made for pelagic and hemipelagic domains of the ocean, demonstrate this process (Fig. 12.1). According to Berger et al. (1989), on average only about 1% of the organic matter reaches the sediment surface in the open ocean. Approximately 97% of this amount is decomposed in early diagenetic reactions so that only 0.03% of the organic carbon produced in the photic zone will ultimately become buried. Later on we will more thoroughly discuss the differences between pelagic and coastal regions of the ocean where the burial rates are estimated to be up to 30 times higher. Although the information contained in Figure 12.1 might vary depending on the applied method (e.g. de Baar and Suess 1993), in most cases relationships and dimensions usually remain quite consistent. This approximate balance may also apply to most nutrient cycles in the ocean.

Regarding the water body simplistically as an indivisible entity, without concerning to its nume-rous internal cycles, only the boundary conditions of the system will ultimately determine its potential changes in the course of time. Under such an assumption, the following boundary systems will have to be considered: 1) the

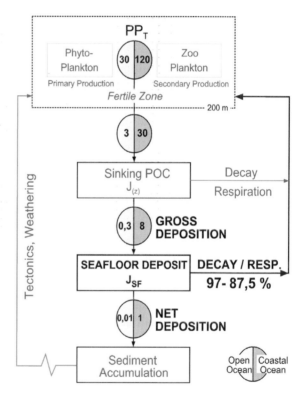

Fig. 12.1 Schematic representation of the particulate, organic carbon cycle in the ocean according to Berger et al. (1989). The numerical values have the dimension of gC m^{-2} yr^{-1}. Differences between the open ocean and coastal areas are shown. Of the 1% primary production (PP$_T$) that reaches the ocean floor, only 3% are embedded in the open ocean for a period of geological time. In contrast, 97% are decomposed by microbial activity and returned to the water column in the form of dissolved constituents (cf. Fig. 6.6).

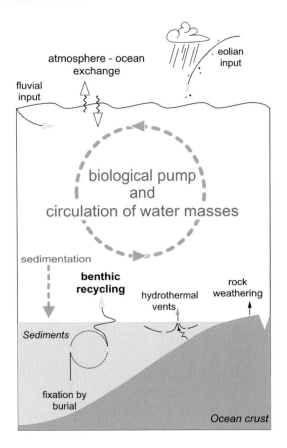

Fig. 12.2 Simplified representation of the imports and exports pertaining to oceanic cycles.

atmosphere, 2) the continents and 3) the ocean floor (Fig. 12.2).

Of all ocean boundaries the contact made with the atmosphere is surely of greatest significance with regard to medium- and short-term alterations in global cycles. Accordingly, one of the most fundamental issues in the discussion of climate change is concerned with the exchange processes of dissolved and free gases across this boundary (cf. Chapter 10). Considering the huge area of the seafloor where carbonate-rich sediments dominate (cf. Chapter 1), the marine deposits offer an enormous sink, which in the long run, may compensate for the input of greenhouse gases from the atmosphere (cf. Chapter 9).

The fluvial input of dissolved and suspended material to the ocean presents a boundary condition, which can reliably be estimated (cf. Fig. 1.2). Apart from eolian particle transport, which may be of great importance in areas adjacent to the great desert regions, this mode of transport represents the major pathway of element input into the marine system. In this regard, Chester (2000) gives an extensive overview especially of processes with terrestrial origin. In comparison, the local inputs originating from a weathering of the oceanic crust, or from hydrothermal activity (cf. Chapters 13 and 14), are quantitatively far less significant on a global scale.

The role of the benthic interface is not to be defined exclusively by one transport direction. On the one hand, deposition and burial of marine sediments remove elements from the marine cycles over geological time periods. Yet, as the example of the carbon cycle shown above already demonstrated, an immense proportion of accumulated particles are subject to dissolution or microbial decomposition in the course of early diagenesis (cf. Chapter 9). Marine sediments therefore also act as a 'secondary' source of remineralized dissolved components. The coexistence of these two fundamental, but oppositely directed mass movements constitutes one of the most essential phenomena at the seafloor. Next to studies on particle fluxes through the water column and element-specific accumulation rates, the quantification of benthic flux rates across the sediment/water boundary represents the third pre-condition for obtaining a complete balance of the marine material cycles.

The object of this chapter is to illuminate the existing approaches, which are used for the quantification and interpretation of benthic material fluxes. The emphasis is put on the regional and global interactions with other parameters by introducing different concepts for the investigation of spatial distribution patterns. The comparison of results from geostatistical methods and the opportunities given by modern geographical information systems (GIS) will be placed in the center of the observations.

12.2   Fundamental Considerations and Assessment Criteria for Benthic Flux Rates

Locally conducted investigations are the base of all spatial descriptions. Depending on the subject under study, the quality of regional and global observations, in principle, improves in accordance with the quantity of single studies and the type of their spatial distribution. However, scientific pro-

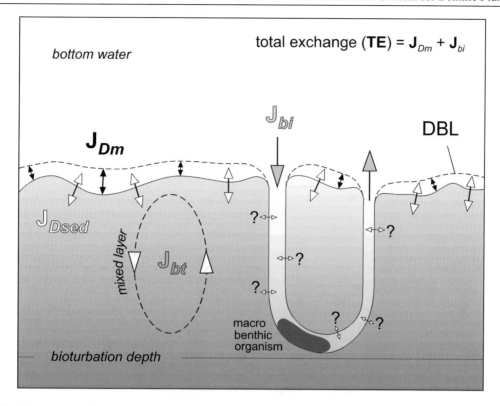

Fig. 12.3 Transport mechanisms of dissolved substances in the sediment/pore water system. The diffusive transport is substance-specific and is solely based on the distribution of material concentrations. Its calculation is performed on the basis of concentration gradients in pore water (J_{Dsed}) and/or within the diffusive boundary layer (J_{Dm}). The morphology of the sediment surface is usually not considered. To determine total exchanges (TE), transport processes induced by macrobenthic activity must be additionally included. Whilst the unspecific transport by means of bioturbation (J_{bt}) can mostly be neglected, bioirrigation (J_{bi}) assumes great importance especially in densely populated and nutrient-rich sediments. A rarely noticed boundary exchange takes place across the walls of housing or feeding habitats which are densely populated by microbes.

gress also implies the improvement and often diversification of the applied methods, which may lead to a limited comparability of the collected data. This is emphasized by a comprehensive summary of the technical equipment and the analytical methods currently employed for geochemical studies at the benthic boundary in Chapter 3. As the knowledge of detailed results obtained from applying these methods is an imperative criterion for selecting literature references on benthic flux rates, a brief compilation of the essential aspects related to this subject will follow.

Irrespective of the differences between *in situ* and *ex situ* measurements (see below and Chapter 6), there are generally two ways to determine the transport rates of dissolved substances across the water/sediment boundary: determination of the concentration differences along the depth profiles (gradient approach) and measurement of the time-dependent concentration changes of dissolved species within a closed

reservoir (chamber experiments; Reimers et al. 2001; cf. Chapters 3 and 6).

The fundamental difference between both approaches is caused by the different transport mechanisms involved (Fig. 12.3). Whereas only a diffusion-controlled transport of material can be measured by applying microelectrodes in the diffusion boundary layer (DBL) and in the pore water (cf. Section 5.2.3), enrichment and/or depletion in the overlying bottom water deliver rates of total exchange.

12.2.1 Depth Resolution of Concentration Profiles

Let us first have a look at depth profiles of concentrations and how they are interpreted. Since the DBL represents the transition zone between the sediment/pore water system and the turbulent movements within water column above, all diffusive exchange must pass this layer of laminar flow.

Since biogeochemical reactions within the DBL are negligible relative to the surface sediments, measurements of concentration gradients across this zone provide an accurate estimate of diffusive benthic exchange. As has already been explained in the Chapters 3 and 6, the necessary application of microsensor technology in deep-sea environments is only feasible for analyzing very few parameters (O_2, CO_2, Ca, H_2S and pH). All other profiling measurements are consequently based on the conventional methods of pore water extraction, or are conducted by applying ion exchange resins in so-called gel peepers. The

sampling methods mentioned differ in their underlying principles of measurement and in their power of depth resolution, which is of enormous significance concerning the rates of diffusive exchange. As the concentration changes in the pore water fraction situated immediately below the DBL are linear only in exceptional cases, the distance between each single reported concentration value markedly influences the determination of the exchange rate (cf. Chapter 3). An increased sampling density can therefore easily lead to changes of a factor of 2. Usually this implies more pronounced gradients, as for example shown in Figure 12.4, i.e. higher rates of transport. The problem of having to de-cide upon which result should be used for regio-nalization is likely to occur, particularly when older and recent studies are compared.

A careful examination of the database is always necessary if the flux rates were calculated from pore water profiles of variant resolution. A commonly used method to minimize such discrepancies consists of the application of mathematical fit-functions (e.g. Sayles et al. 1996). As long as the sampling procedure includes at least all essential features of an anticipated concentration curve (zones of release and fixation), the application of non-linear functions will permit a 'theoretical' depth resolution of any desired precision and thus a theoretical flux calculation in close proximity to the boundary.

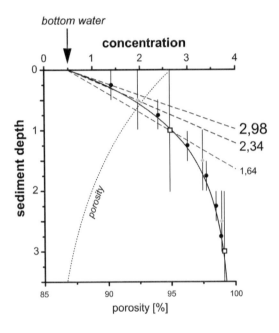

Fig. 12.4   Effects of the depth resolution in pore water concentration profiles on calculating the rates of diffusive transport. Three samples drawn from surface sediments are shown to possess different resolutions (intervals: 0.5 cm - dots, 1.0 cm diamonds, 2.0 cm - squares). All values are sufficient to plot the idealized concentration profile within the bounds of analytical error, yet very different flux rates are calculated in dependence on the depth resolution values. In the demonstrated example, the smallest sample distance indicates the highest diffusion (2.98 mmol cm^{-2}yr^{-1}). As soon as the vertical distance between single values increases, or, when the sediment segments under study grows in thickness, the calculated export across the sediment-water boundary diminishes (2.34 – 1.64 mmol cm^{-2}yr^{-1}). In our example, this error which is due to the coarse depth resolution can be reduced by applying a mathematical Fit-function. A truncation of 0.05 cm yields a flux rate of 2.84 mmol cm^{-2}yr^{-1}. (The indicated values were calculated under the assumption of the presented porosity profile according to Fick's first law of diffusion - see Chapter 3. A diffusion coefficient of 1 cm^{-2}yr^{-1} was assumed. Adaptation to the resolution interval of 2.0 cm was accomplished by using a simple exponential equation).

12.2.2   Diffusive versus Total Solute Exchange

Diffusion-controlled exchange represents only one more or less large proportion of the total transport activity between the sediment and the bottom water (Fig. 12.3). In recent years, numerous studies have shown that the negligence of macrobenthic activity – dominantly driving bioirrigation processes in the surface sediments – would lead to a significant underestimation of flux rates particularly in high productivity areas and marginal seas. These effects of non-local transport are best investigated for oxygen (e.g. Archer and Devol 1992; Glud et al. 1994; Wenzhöfer and Glud 2002; cf. Chapter 6). Except for coarse-grained sediments, the non-diffusive transport in marine sediments is primarily controlled by the activity of macrobenthic organisms (e.g. Glud et al. 1994; cf. Fig. 6.1). Compiling results from several studies in different areas, Figure 12.5

gives an impression of the range of relationships between diffusive and total flux. Obviously, the significance of molecular diffusion decreases with increasing respiration rate and thus, increasing input of organic carbon. The more degradable organic matter is supplied, the larger becomes the population of macrobenthic organisms and the higher is the demand for oxygen. However, because the major part of the world ocean is oligotrophic Jahnke (2001) argues that the activity of macrobenthic organisms is probably of minor importance for the total exchange of solutes at the seafloor. This conclusion is supported by benthic chamber experiments with soluble tracers (Sayles and Martin 1995) and investigations of the spatial distribution of DOU and TOU fluxes in the South Atlantic (Wenzhöfer and Glud 2002).

Nevertheless, because measurements of total exchange are still sparse, most studies dealing with the budgeting of benthic flux rates are restricted to observations below the 1,000m isobath to minimize the effect of non-local transport (Jahnke 1996; Zabel et al. 1998; Hensen et al. 1998; Seiter et al. 2005). This depth is frequently described as something like a 'boundary' below which the influence of macrobenthic organisms strongly diminishes and diffusive fluxes are approximately identical to the total transport rates (cf. Section 6.4). However, strictly speaking, this assumption is merely an arbitrary simplification.

12.2.3 *In situ* versus *ex situ*

In order to assess the comparability of literature values, it is relevant to distinguish measurements and sample withdrawals, which were conducted directly at the ocean floor (*in situ*) from those conducted on board of research vessels or in laboratories (*ex situ*). In Section 6.3.2 it was shown in detail that identical conditions of measuring the highly reactive elements oxygen and nitrogen (in the form of NO_2, NO_3, and NH_4) are of decisive importance for examining stoichiometric balances during early diagenesis. Indeed, the same is true for regional differences of single parameters. Depending on the sensitivity to extreme changes of pressure and temperature, the analyzed concentrations might differ strongly (e.g. Jahnke et al. 1982; Glud et al. 1994; cf. Fig. 6.16). This applies to both profiling measurements and to incubation experiments, implying an important, yet hardly quantifiable, effect on the measurement of benthic transport rates.

12.2.4 Time-Dependent Variances and Spatial Variations in the Micro-Environment

Estimates of the significance of seasonal and micro-environmental variations of the benthic

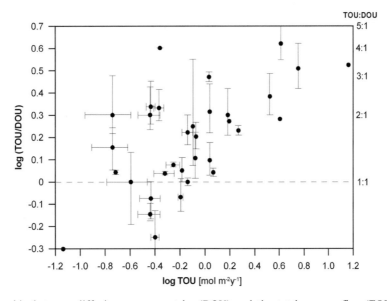

Fig. 12.5 Relationship between diffusive oxygen uptake (DOU) and the total oxygen flux (TOU). The compilation of several data sets (Reimers 1987; Reimers et al. 1992; Glud et al. 1994; Hales and Emerson 1997; Weber et al. 2001; Wenzhöfer and Glud 2002) shows a significant positive correlation between total respiration rates and the portion of nonmolecular transport processes (cf. Jahnke 2001).

system are very difficult to perform. Nevertheless, information on both factors is very important for assessing the comparability of results obtained from single case studies. For instance, the reproduction of measurements carried out at 'identical' locations, but performed in successive years and in different seasons, in fact exhibit a significant variation of concentrations within the near-surface zone (e.g. Zabel et al. 1998). Two reasons for this phenomenon are presently being discussed: a) seasonal and interannual variations of the particle flux to the ocean floor and b) differences in relatively small natural environments.

Temporal variations of export production are well known from long-term sediment trap experiments (e.g. Wefer and Fischer 1993) and can nowadays also be deduced by satellite monitoring. However, important effects of sediment pulses on biogeochemical reactions at the sediment surface are primarily confined to eutrophic areas, predominately along continental margins (Sayles et al. 1994). Such rapid deposition events of organic-rich aggregates known as 'marine snow' induce a very strong increase of benthic reaction rates, especially as far as the oxygen consumption is concerned

(Witte et al. 2003a/b, Bühring et al. 2005; Aspetsberger et al. subm. a/b; cf. Fig. 6.21). In contrast, interannual variation of benthic flux rates in the pelagic zone of the deep-sea has been estimated to be rather low (Smith et al. 1992).

Apart from the variances in particulate input, there are also small-scale micro-environmental variations at the ocean floor, which primarily depend on the morphology of the sediment surface and the intensity of the bottom water current (see Fig. 3.22; compare with Chapters 3 and 6). Time-lapse recordings of the ocean floor revealed that depending on the relief of the sediment surface marked sedimentation events are recorded to varying extent (e.g. Lampitt 1996). Hence, the so-called 'fluffy' layer is not a homogenous layer on top of the sediment, instead, deposits of phytodetritus accumulate in little hollows. It is evident that, in such a case, significantly different rates are determined, depending on the precise spot of sediment recovery or micro-electrode measurement.

In addition to these heterogeneities, pore water profiles can also be significantly affected by chemical variations on even smaller scale, which

Table 12.1 Factors for assessing the comparability of differently indicated benthic flux rates.

methods*	ex situ					in situ				
	conventional pore water sampling	whole core squeezer	gelpeeper DET / DGT	microelectrodes / -optodes	incubation experiments	harpoon sampler	whole core squeezer	gelpeeper DET / DGT	microelectrodes / -optodes	incubation experiments
main characteristics in this context										
flux calculation via gradient	yes	yes	yes	yes	**direct (!)**	yes	yes	yes	yes	**direct (!)**
depth resolution	some mm	mm	µm	µm	/	cm	mm	µm	µm	/
concentration	average	average	relative	"real"	"average"	average	average	relative	"real"	"average"
parameter set	most	most	many	some	"all"	most	most	many	some	"all"
transport mechan.	diffusive	diffusive	diffusive	diffusive	(total)	diffusive	diffusive	diffusive	diffusive	**total !!**
effects on flux calculations										
reliability**	+	++	++	+++	+++	--	++	++	+++	+++
bioturbation is neglected	yes	yes	yes	yes	(yes)	yes	yes	yes	yes	no
artefacts during sed. recovery	yes	yes	yes	yes	(yes)	no	no	no	no	no

*   for the detailed descriptions see Chapter 3
** this critical examination excludes mathematical data fits or subsequent computer modeling

are related to the formation of sub- and anoxic microenvironments. Burial of larger particles stemming from organisms and fecal pellets may create hotspots of microbial activity within the sediment. Locally restricted deficits of oxidants, particularly of oxygen and nitrate, are the consequence. In this context Jørgensen (1977) reported reducing microniches of about 50-200µm in diameter, where sulfate reduction may take place in a basically oxidized environment. The retardation of diffusion is likely caused by the formation of organic coatings or precipitation processes narrowing the pore space. Developing a simple mathematical model, Jahnke (1985) could already show that processes in microenvironments can affect pore water profiles in a different way. Applying a three-dimensional model to results of high resolution DGT profiles (see Section 3.5) Harper et al. (1999) concluded that the common one-dimensional view of the system may underestimate relevant process rates by a factor of at least 3.

For the sake of completeness, advective effusions also need to be mentioned here (see Chapters 13 and 14). Although their significance for global balances of materials (e.g. CO_2, see Chapter 5) has been hardly investigated yet, the hot vents and cold seeps, usually locally confined, seem to be rather unimportant for the interaction described here. Nevertheless, they may be relevant when looking on specific element cycles (e.g. Li, B or Ba).

In summary, knowledge of the applied methods and the local conditions is indispensable for the compilation of a database which should be as homogenous as possible. However, a clearly defined objective is required for characterizing patterns of regional and global distributions in order to successfully avoid illicit comparisons. As we will see, to merely resign to data produced under identical conditions is by no means a realistic conclusion. Good judgment is always of ultimate importance. Table 12.1 gives an overview on the important aspects of this section.

12.3 The Interpretation of Patterns of Regionally Distributed Data

One important objective in characterizing regional differences of early diagenetic processes is to obtain information on the rate-determining control parameters and the benthic material fluxes. It has

been stated before that, for various reasons, there can be no global database covering all biogeochemical and sedimentary environments. It is therefore only possible to design global balances of geochemical material cycles by means of deterministic approaches. It is thus reasonable to focus on those control parameters with the highest data density available, which are for example primary production or the C_{org} content of surface sediments (see Section 12.5).

The interrelations between the individual areas of the oceans have already been outlined at the beginning of this chapter. Owing to an intense research of the last decades, many principles of interaction and modes of possible interference among single processes could be demonstrated (e.g. a high rate of primary production yields large amounts of organic substance in the sediment, which intensifies the growth of benthic populations; cf. Section 12.2.2). According to the principle of actuality, any statement on environmental changes in Earth's history, as well as any prognoses on global change in the future, must be based on a thorough understanding of the present situation. However, empirical relations mostly apply only for a limited range of locations. In the following, the most important control parameters will be presented, along with a brief comment on the potential relevance they have for the benthic interface.

12.3.1 Input and Accumulation of Organic Material

The amount of nutrients recycled back to the ocean by benthic processes is closely related to the availability of organic substances, which can be degraded by microorganisms (cf. Chapters 3-10). There are two fundamental strategies to make use of this correlation as a control parameter for the benthic biogeochemical system and as a tool for the estimation of benthic material fluxes.

An indirect way to estimate the amount of organic substance in the benthic system uses the close coupling between the biological phytoplankton reproduction in the photic zone with the C_{org} rain to the ocean bottom. The overall advantage of this approach is that satellite data are available in high resolution in space and time, respectively (e.g. Antoine et al. 1996; Behrenfeld and Falkowski 1997; http://www.Oceancolor.gsfc.nasa.gov/SeaWiFS/). Figure 12.6

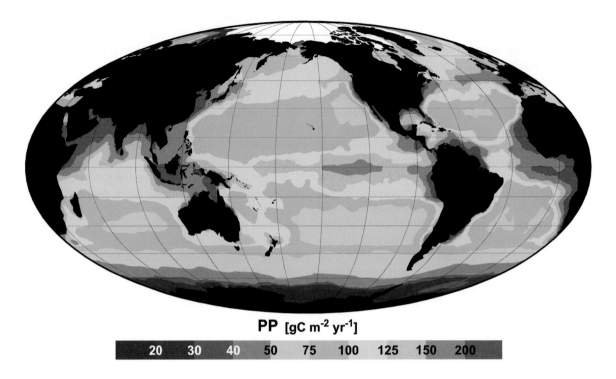

PP **[gC m⁻² yr⁻¹]**

| 20 | 30 | 40 | 50 | 75 | 100 | 125 | 150 | 200 |

Fig. 12.6 Map representing the mean annual primary production [gCm⁻²yr⁻¹]. Among other parameters, the distribution of pigment concentration, surface-water temperature and surface irradiance were used for its construction. The representation is the result of model calculations (redrawn after Antoine et al. 1996).

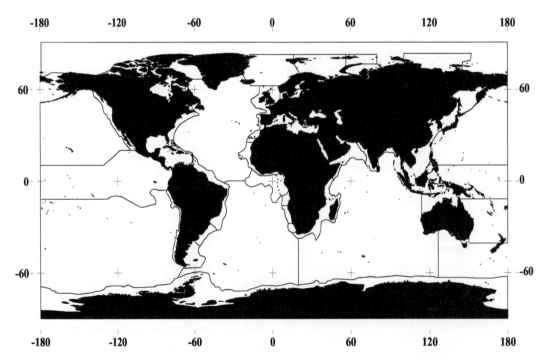

Fig. 12.7 Subdivision of the world's oceans into 33 TOC-based regional zones. The geographical boundaries of the single regions were defined by applying combined qualitative and quantitative-geostatistical analyses of the spatial dependencies of the data points. In principle, each benthic province is characterized by a specific combination of oceanographic and sedimentary characteristics (after Seiter et al. 2004).

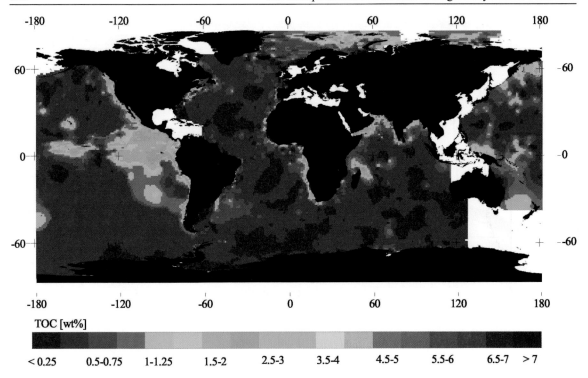

Fig. 12.8 Global distribution pattern of the organic carbon content (TOC) in surface sediments (<5 cm sediment depth) (with permission from Elsevier reprinted from Seiter et al. 2004).

shows the global distribution of the mean annual primary production of phytoplankton as simulated by Antoine et al. (1996). Based on the knowledge on how oceanic circulation and other physical parameters like radiation and wind mixing affect the production of phytoplankton, four major zones can be distinguished: 1) the polar zones 2) the west wind zone 3) the trade wind zone, and 4) the coastal zone (Sathyendranat et al. 1995). According to Longhurst et al. (1995) this rough division can be subdivided into 57 biogeochemical provinces that display different dynamics regarding the processes of primary production. The deposition flux of organic carbon may then be derived from application of simple relationships (cf. Fig. 6.3), which can on average provide useful results. But, as described below, such calculations may be critical especially along continental margins.

The second way to estimate the availability of organic substance in the benthic realm is to directly determine the organic carbon concentration in surface sediments (TOC) or to calculate the organic carbon burial rate. The simplistic approach to describe the dependence between the C_{org}-content and benthic flux rates has proven

successful in specific regions (cf. Figs. 6.23b and 6.25b). However, thousands of TOC measurements exist worldwide, but problems related to the internal comparability of the data may arise here due to sampling artifacts or the different analytical methods applied. After all, organic substances are not identically prone to the same degree of microbial decomposition (cf. Chapter 5). Since the analytical effort is enormous to classify the organic carbon either according to its origin (terrigenous or marine, e.g. by means of $\delta^{13}C$ – measurements, see Chapter 10), or its labile and refractory proportions (e.g. Dauwe and Middelburg 1998; Dauwe et al. 1999; Schubert et al. 2005), most C_{org} amounts are reported as unspecified total values. Nevertheless, Seiter et al. (2004) were able to calculate the global distribution of TOC in surface sediments by defining 33 regional benthic provinces, which consider regional oceanographic and sedimentary conditions controlling the deposition of organic material. (Figs. 12.7 and 12.8).

Comparing the results of both approaches outlined above it becomes obvious that the latter, in general, provides the more precise and realistic results, which is mainly caused by the complexity

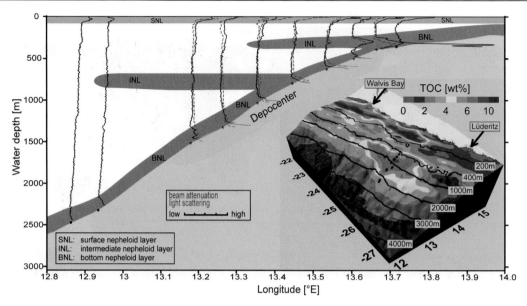

Fig. 12.9 Occurrence of nepheloid layers (based on beam attenuation and light scattering profiles) across the continental margin off Namibia at 25.5°S. As a result of the lateral particle transport, a TOC-rich depot center has develloped on the mid-slope (compiled from Inthorn et al. subm. a and b).

of sedimentation processes. On the one hand, the export production cannot be estimated directly based on remote sensing data, because there is a lack of essential empirical data. On the other hand, results obtained from sediment trap experiments have revealed that particulate matter is mostly not transported exclusively vertical to the ocean floor. Thus the pre-condition of a locally bound and vertical projection of biochemical processes in surface waters to the accumulation of biogenic particles on the ocean floor is not absolutely fulfilled. Other unknown factors are introduced by the variable sedimentation velocities of different particles and aggregates and by the varying sampling-efficiencies of sediment traps. Furthermore, the time particles spend in the water column effectively influences the rates of dissolution and microbial decomposition reactions in the course of the sedimentation process.

Despite these limitations, empirically determined transfer functions are frequently applied, which predominately describe the relation between water depth and particle flux (e.g. Suess 1980; Pace et al. 1987; Antia et al. 2001). Thus, the flux of particulate organic carbon to ocean floor is applied as the rate limiting control parameter (Schlüter et al. 2000; Wenzhöfer and Glud 2002). Although less precise the use of transfer functions is also very helpful for the use in coupled benthic-pelagic models (cf. Fig. 12.20). Some of these

mathematical equations are described in Chapter 6 (Fig. 6.3). Bishop (1989) compared the most commonly used functions with results from sediment trap studies.

12.3.2 Vertical versus Lateral Input of Particulate Matter

Ocean currents may cause a variable lateral drift of settling particles on their way to the ocean floor. This effect is particularly emphasized along continental margins, yet results from sediment traps have shown that this is also a typical phenomenon in open ocean settings. Frequently, sediment traps deployed closer to the sea floor receive higher particle fluxes than those some hundreds or thousands of meters above. Similar discrepancies are obtained when particle flux rates from sediment trap experiments are compared with benthic reaction rates (e.g. Wenzhöfer and Glud 2002; Seiter et al. 2005).

Chapter 3 contains some examples on how the amount of decomposed organic material can be very easily calculated from pore water profiles or the depletion of the corresponding electron acceptors (O_2, NO_3, etc.). If the C_{org} burial rate is estimated from the depth profile of the C_{org} content one can simply add the decomposition and preservation fluxes to obtain the original flux to the sediment surface. A number of studies have proven, however, that a comparison of the particle fluxes – either determined in the water column or calculated on the basis of the primary production – with benthic remineralization data do not produce a conclusive picture (e.g. Walsh 1991; Rowe et al. 1994; Hensen et al. 2000). Increasing current velocities may lead to erosion and re-suspension of older sediments and thus, provide an additional transport component. This so-called *winnowing effect* is of significance for material balances of most shelf regions. Depending on the velocity of the bottom currents, the accumulation of the smaller or lighter components (like organic material) can be reduced or completely prevented. The effect of resuspension processes result in a rise in the number of particles in the water layer near the bottom, the *nepheloid boundary layer* (Fig. 12.9). Its thickness might range from several tens to hundreds of meters. Where the currents

subside the particles previously kept in suspension will deposit. This accumulation process is referred to as *focussing* and occurs preferentially at mid slope depths. Frequently described, organic-rich *depot centers* develop (Fig. 12.9). The experimental results shown in Figure 12.10 were obtained from a study on the central Californian continental margin. Illustrating the clear discrepancy between sediment trap based flux estimations and the benthic response on the organic carbon input, they give strong indication for a lateral particle transport from the shelf and the upper slope region down into the deep-sea.

Although the flux relationships and resulting distribution patterns focus on outlined so far specific regions, it is evident that they are essential in order to provide reliable interpretations and estimates on benthic biogeochemical processes.

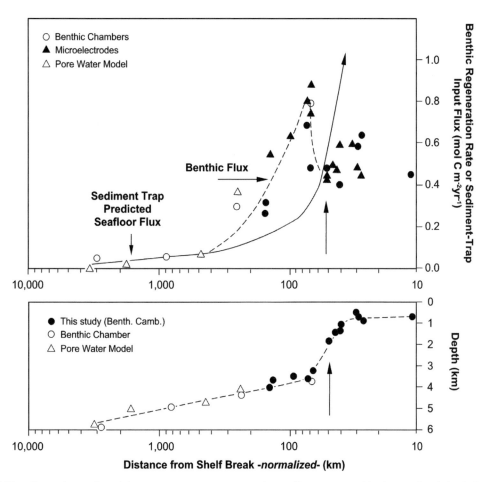

Fig. 12.10   Comparison of particle transport measurements using sediment traps with the results derived from *in situ*-measurements at benthic boundaries (after Jahnke et al. 1990). The lateral offset of maximal values of organic material import, as predicted by trap studies, and the highest O_2-depletion rates (converted in terms of organic substance decomposition) indicates a C_{org} - transport close to the ocean bottom, which is directed from the shelf and upper slope region to the deep-sea margins.

439

12.3.3   Composition of the Sediment

Chapter 1 presented a detailed introduction into the distribution and diversity of marine sediments. Here, we will highlight a few aspects on how the sediment composition, might affect the exchange of dissolved components moving across the benthic interface.

The problems that occur from applying biogenic sediment components as control parameters of benthic flux rates is demonstrated by the marine silicon cycle. The dispersal and the

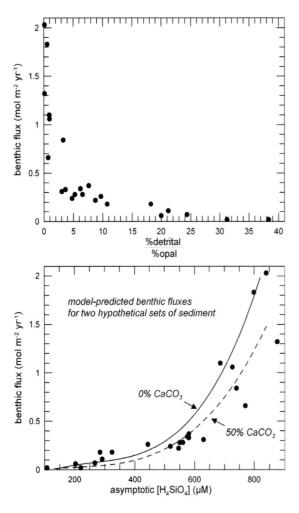

Fig. 12.11   The ratio between the detrital and the opal content of sediments may be the major controlling factor for the dissolution capacity of opal and therefore for the asymptotic increase of silicic acid pore water concentrations with sediment depth and the release of silicic acid to the bottom water. a) benthic silicic acid flux as a function of the detrital to opal mass ratio in surface sediments; b) model-predicted fluxes for two hypothetical sets of sediment. The TOC is assumed constant with 3 wt%. The data originate from different sources (modified after Dixit and Van Cappellen 2003).

proportion of biogenic opal in sediments depends on three factors: a) the rain rate of biogenic opal, b) the accumulation of other particles, and c) the degree of preservation in the sediment. At first sight the significance of these three factors appears to be trivial, but all in all there are some complex interactions. The rain rate of biogenic silicate as well as the degree of opal preservation in the sediment are both a function of specific dissolution rates. These may decrease in the course from production in the surface waters to burial in the sediments by some orders of magnitude due to alteration of the opal skeletons and changes of the environmental conditions (Van Cappellen et al. 2002). However, it may occur that the amount of biogenic opal in the sediment is below the limit of analytical detection, despite a significant production and export of siliceous plankton, which is caused by effective dissolution before final deposition. From a thermodynamic point of view, the dissolution of 'pure' opal is mainly a function of temperature, degree of undersaturation and surface area. However, pure phases almost never occur in nature and thus, simple approaches proved to be inappropriate in order to describe the dissolution processes. It is known from numerous studies that dissolution kinetics of biogenic opal are seriously affected by the concentration of absorbed or incorporated trace elements (e.g. van Bennekom et al. 1991; Dixit et al. 2001). On the one hand the presence and the amount of trace metals is a primary signal, but it also depends on the availability and the concentration of adequate lithogenic elements and may be subject to strong regional diversification. Van Cappellen and Qui (1997) and later Dixit et al. (2001) and Dixit and Van Cappellen (2003) were able to describe the relation between the primarily asymptotical increase of silicon concentrations in pore water, the silicon-specific ratio of lithogenic components, and the content of opal (Fig. 12.11). Since no simple correlation between the concentration of silicon in pore water and opal in the sediment could be found, it is not possible to estimate the extent of the benthic reflux by simply knowing the opal flux to the sediment. Furthermore, the diagenetic alteration of diatom shells in sediments seems to be accompanied with a change in the surface chemical structure of the frustules, which consequently results in a progressive loss of the reactivity of biogenic silica (Dixit and Van Cappellen 2002). Although the dissolution of diatom frustules itself is a thermo-

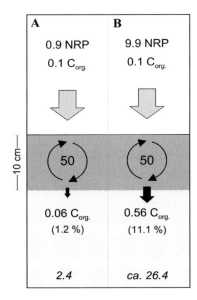
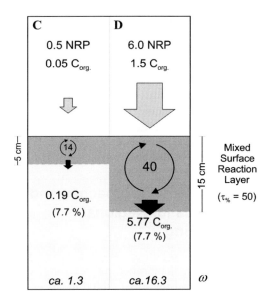

Fig. 12.12 Simplified representation of the burial efficiency of reactive substances (e.g. C_{org}) dependent on the accumulation rate of non-reactive components (NRP, esp. carbonate and lithogenic particles), modified after Jahnke (1996). The following assumptions are fundamental to the simulations shown: 1) The sedimenting particles have a density (δ) of 2.5 g cm^{-3}, 2) the surface-near reaction layer is subject to constant bioturbation (homogenization); 3) the porosity (ϕ) is constantly 85%. Flux rates are reported in the unit mg cm^{-2} yr^{-1}, whereas the resulting sedimentation rates (ω) are in cm kyr^{-1}. The decomposition rate of organic carbon compounds (τ) is reported as 50%/50 yrs. The juxtaposition of scenarios A and B demonstrates the dependence of the C_{org} burial rate on the rate of NRP accumulation. In case of identical decomposition rates, an increase of the NRP/C_{org} ratio (10/1-100/1) will result in a more efficient burial of C_{org} (1.2%-11.1%). The scenarios C and D show the relevant principle differences between the open ocean (C) and the coastal ocean (D). Compared to Figure 12.1, the selected C_{org} values are equivalent to rain rates of 50, or respectively, 225 gC m^{-2} yr^{-1} of primary production. The assumptions demonstrate that identical burial rates can prevail in both systems due to variable boundary conditions (decomposition efficiency, thickness of the mixed layer; cf. Fig. 6.6)

dynamic process caused by the undersaturation of pore fluids with respect to biogenic silica, the alteration process is probably closely linked to bacterial activity (Bidle and Azam 1999). Some recent studies have provided evidence that bacterial ectoprotease action on marine diatom detritus strongly accelerates silica dissolution rates by removing the organic coatings that normally protect frustules from direct exposure to the unsaturated pore and seawater (Bidle and Azam 2001; Bidle et al. 2002, 2003). The systematic investigation of pore water profiles of oxygen and silica revealed this inherent linkage of oxic respiration and opal dissolution. Holstein and Hensen (subm.) could derive a simple empirical formulation relating the Si flux to the oxygen flux with

$$F_{Si} = 0.54 \cdot F_{O_2} \qquad (12.1)$$

where the calculated Si-Flux reaches a confidence level of approximately 70%. This approach has

successfully been applied by Seiter et al. (subm.) who estimated the total Si reflux and thus, the minimum opal rain rate to the sea floor for the southern Atlantic Ocean. The combination of such simple empirical equations may provide a powerful tool in order to estimate total dissolution or material fluxes.

Dilution by quasi-non-reactive terrestrial particles, mainly clay minerals, feldspars and quartz provides another example emphasizing the significance of the sediment composition on benthic material fluxes. This process is equivalent to the effect of the accumulation rate of all biogeochemical non-reactive particles (NRP). More simply, a high accumulation rate, especially of carbonate (foraminifera, coccolithophorids etc.) and or lithogenic particles leads to an increasing burial efficiency or preservation of all reactive components (C_{org}, opal etc.). In case of constant inputs, for instance of organic substance, the time these substances spend in the highly reactive surface layer decreases with an increase in the rain rate of non-reactive compounds. A reversal of the effect, the reduction of the burial rate due to decreasing input of non-reactive phases, may be induced by physicochemical dissolution of carbonate below the CCD (see Chapter 9). Figure 12.12 demonstrates both effects schematically as simplified balances. The dissolution of carbonate related to the degradation of organic material is neglected in this case.

12.4 Conceptual Approaches and Methods for Regional Balancing

As pointed out in the preceding sections, the establishment of basin-wide or global balances of benthic material cycles always requires extrapolation procedures, because all available data sets are limited in size and usually consist of inhomogeneously distributed point data. Various conceptual approaches are available to approach this goal. In this section, fundamental aspects to the most frequently used methods will be briefly presented.

12.4.1 Statistical Key Parameters and Regression Analysis

The probably easiest method of regional balancing is based on the calculation of statistical key parameters, such as the mean value or the median value. In this context, the key parameters are always related to geographically defined regions. Therefore, the regions must be defined prior to statistically evaluating the presumed representative data. The definition of the region is not the result of the evaluation process. The knowledge of the regional distribution in each region is consequently an important pre-condition to the application of the procedure. The sufficient fulfillment of these conditions permits balancing performance on a global level (e.g. Nelson et al. 1995).

The options and conclusions to be drawn from regression analysis are essentially more diverse. Data sets of two or more parameters are compared with each other (cf. Figs. 12.11 or 12.18a/b). Such mathematical descriptions of relations between the single measured values and potential control parameters enable the statistical extrapolation of isolated measured values across the area (see Sections 12.5.1 and 12.5.2). Furthermore, they can

be employed for the characterization and demarcation of provinces, i.e. region-dependent validity assessment for the correlations to be investigated (cf. Section 12.3.1).

12.4.2 Variograms and Kriging

Fundamental publications related to this complex geostatistical procedure originate from Krige (1951) and Matheron (1963). In the broadest sense, geostatistics deal with the spatial variability of location-dependent variables. The central question is simply: are single and spatially dispersed values capable of giving a representation of coherent structures, i.e. do the results of local measurements show a spatial correlation? A solution to this fundamental problem can be found by means of *variogram analysis*. In contrast to regression analysis, only the relationships of one parameter are examined (e.g. values of benthic O_2 depletion or TOC content). Information on eventual control parameters, such as the biodegradable amount of organic substance, as mentioned in the example above, are not required. As long as the single values are interrelated, their mathematical description by the various kriging methods will serve to provide the desired spatial interpolation. Hence, it is evident that the proper application of kriging methods requires a rigorous pre-analysis by variogram analysis. Below, the essential theoretical basics are briefly outlined. A more advanced background to this field of study can be obtained by specific literature (e.g. Journel and Huijbregts 1978; Davis 1986; Akin and Siemes 1988; Schlüter 1996; Wackernagel 1996).

The construction of iso-linear maps implicates that the individual values obtained from local measurements are related to each other. These interrelations are usually subject to directional changes, i.e. closely neighbored values are more similar than distant ones. Variogram analysis is designed to identify such 'trend structures' showing decreasing similarity with increasing distance. In particular, variograms include vector functions designed to determine the one-half, medium, and squared differences between measurements conducted at two discrete locations (variances). The calculation of these variances will consequently consider the distance between the points. Thus, distance clusters need to be defined to permit an allocation (Eq. 12.2) of each variogram value (as a sum function - $\gamma_{(h)}$) to a number of point pairs ($n_{(h)}$) defined by the chosen distance

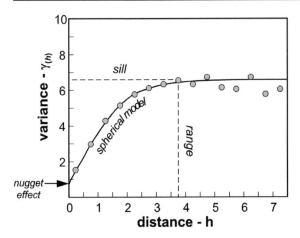

Fig. 12.13 Schematic representation of a variogram. The mean distances of point-pairs (h) and the corresponding variances of their measured values (γ(h)) are plotted. The relation between individual values diminishes with increasing distance, i.e. the γ(h)-values increase. However, the structural dependence between the point-pairs is only valid up to a specific distance (range). From this point the variances tend to scatter around a certain value (sill), which represents the total variance of all values. The nugget effect, or the apparent mismatch of the variogram to go through the origin, indicates for a regionalized variable that it is highly variable over distances less than the sampling/cluster interval. A spherical model was adapted to the idealized data.

criterion (h). In the example shown in Figure 12.13, a cluster size of 0.5 (e.g. degree or km) was chosen, i.e. differences between measured values with distances of up to 0.5 are plotted in the first variogram, should the distance lie between 2 and 2.5, the difference value will be entered in the fifth variance calculation.

$$\gamma_{(h)} = \frac{1}{2n_{(h)}} \cdot \sum_{i=1}^{n} \left(z_{(x_i)} - z_{(x_i + h)} \right)^2 \qquad (12.2)$$

$z_{(x_i)}$ and $z_{(x_i+h)}$: known values at the locations x_i and $x_i + h$

Mostly, data are related up to a certain distance. The variogram values usually start scattering at greater distances around a sill value, which is equivalent to the total variance of the parameter studied (transitive variogram type). The distance at the sill indicates the range of structural dependence. Frequently the ranges of a variogram vary depending on the chosen direction (geometrical or spatial anisotropy).

Different types of mathematical model functions (linear, spherical, exponential, etc.) are

available to describe the inherent relationships. These functions build the basis of the subsequent regionalization (or kriging) procedure. There are various geostatistical software packages designed for performing variogram analyses (share ware: e.g. GEO-EAS (Englund and Sparks 1988); commercial: e.g. VARIOWIN (Pannatier 1996), SURFER (Golden Software)).

In the kriging process selected data points are used for spatial weighting, which satisfy the outcome of the variogram analysis. Spatial estimates or interpolation of data is performed in such a manner that the estimated variances become minimal. Equation 12.3 demonstrates the formulation of ordinary kriging.

$$z^*_{(x)} = \sum_{i=1}^{n} \lambda_i \cdot z_{(x_i)} \qquad (12.3)$$

$z^*_{(x)}$: estimated value for a real, but unknown value $z_{(x)}$

$z_{(x_i)}$: known values at the locations x_i

λ_i : weighting factor dependent on distance and eventually on direction

A comprehensive system of equations needs to be solved in order to solve the optimization problem, making the sum of weighting factors turn to zero, i.e. not allowing distortion/strain to occur in the course of the estimation procedure. The limitation on applying this powerful regionalization procedure consists in the geographical distribution, or density, of the existing data. Apart from the ordinary kriging method there are a number of complex extensions (universal kriging, co-kriging, external-drift-kriging), which allow the use of additional information on parameter interactions for interpolation.

12.4.3 Geographical Information Systems (GIS)

Like demonstrated before, a reliable regional or global balancing of benthic processes requires information on a great variety of potential control parameters. Therefore it is not surprising that conventional calculation programs often cannot cope with the permanently growing data sets. During the last years geographical information systems (GIS) have been established as a very powerful

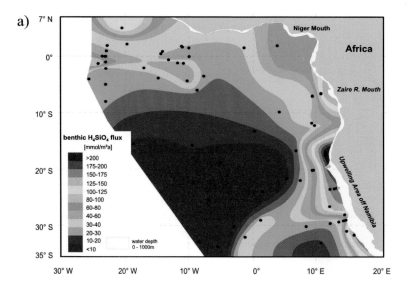

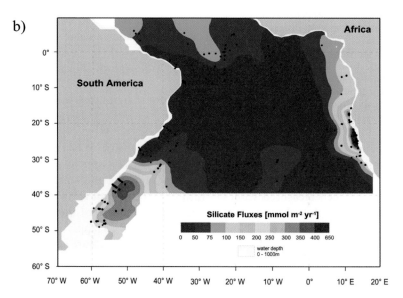

Fig. 12.14 Distribution pattern of rates of benthic silicon release in the South Atlantic. a) Corrected result of the kriging procedure on the basis of 76 measurement points in the eastern part of the South Atlantic (redrawn after Zabel et al. 1998). b) results of regionalization not corrected for, valid for the entire Atlantic ocean on the basis of 180 single measurements (redrawn after Hensen et al. 1998). The results of both procedures in general reflect the distribution pattern of surface-water productivity. Single significant deviations (western Argentinean Basin, eastern equatorial region) can be explained as due to various influences exerted on the particle flux down to the seafloor.

tool for handling and processing nearly unlimited amount of different types of data. Based on the background of a more or less internal data bank management system (DBMS), their main advantage is the possibility of comparison and combination of different information levels to determine structural relationships between all types of data. As a possible result, new structures with formerly unknown characteristics can be developed.

Apart from locally confined data, such as the benthic fluxes at definite locations, most GIS programs process grid data or vector data. One example for grid data are the already mentioned remote sensing data of pigment distribution in the surface water, or the ice cover of the ocean which is currently recorded by the Coastal Zone Color Scanner of the NASA or the earth-orbiting NOAA-satelites (cf. Section 12.3.1). These data are available as a grid of discrete image points and cover extensive surface areas of the ocean. Other examples are data sets, which contain global information on topographic height or bathymetric depth (e.g. http://www.ngdc.noaa.gov/mgg/global/seltopo.html)

Contrary to computer-aided programs used in map construction, GIS enables a linkage between various mapping levels and levels of information. For instance, set operations are performed to blend and calculate intersections, or enable sectional map coupling procedures. A bathymetric map can be, for instance, combined with the distributive pattern of benthic particle fluxes. As a result, the transport across the sediment/water interface is balanced for specific regions, such as the deep-sea, the continental slope, and the shelf area. As an example, Figure 12.19 described in Section 12.5.2 could not have been produced without a GIS program.

12.5   Applications

Following the theoretical explanations of the preceding sections, the different conceptual approaches of regionalization and balancing will now be demonstrated on several practical examples.

12.5.1   Balancing the Diffusion Controlled Flux of Benthic Silicate in the South Atlantic - Applications of Kriging

Apart from the benthic oxygen respiration (cf. next section), there are no globally valid functions of correlation between transfer rates of other pore water solutes and their possible control parameters. Ultimately, all processes of early diagenesis are in fact affected by the supply of organic substance, although other control parameters are also of relevance for the reflux rates of most nutrients (e.g. C:N:P - ratio of the remineralized material, cf. Section 6.2). The complexity of the benthic system demands a treatment of spatial differentiation. Like the biogeochemical provinces applied for calculating primary production or the TOC-based benthic provinces mentioned before, correlative dependencies between benthic flux rates and rate-limiting control parameters only possess regionally restricted validity. However, a greater number of reliable analytical data exist for a

limited number of parameters. In some oceanic regions, the available database, as well as its internal geographical resolution, already permit the application of the kriging method. The following example of the regional distribution of benthic silicon fluxes in the South Atlantic will make both aspects evident.

On the basis of comparative pore water measurements carried out at 76 locations in the eastern part of the South Atlantic, Zabel et al. (1998) demonstrated the distribution pattern of rates of benthic silicate release. As explained in Section 12.4.2 this extrapolation method always averages the measured values. From this it follows that the estimated result of regionalization may strongly differ from the real values in regions showing great differences between neighboring locations (especially in regions of the intensively studied continental slope). This effect depends on the resolution of the chosen grid (clustering) and can require a subsequent manual adaptation of the kriging results to the database. Figure 12.14a shows the corrected result of kriging-regionalization.

To regionalize the entire South Atlantic, Hensen et al. (1998) used an extended database, which contained 180 single measurements. Contrary to the more detailed map of the eastern parts, the distribution pattern obtained was not manually corrected subsequently (Fig. 12.14b). In spite of minor differences in the overall pattern, the results obtained from both procedures reflect the distribution of surface-water activity.

There are, however, differences in two areas: a) in the eastern equatorial Atlantic and b) the western part of the Argentine Basin. Despite a high level of production, the benthic silicon release is quite low in the Guinea and the northern part of the Angola Basin. Although high productivity prevails in the western part of the Argentine Basin, export production is not supposed to supply sufficient amounts of biogenic opal to match the high flux rates observed in this area. These specific deviations would never have been detected by simply applying empirical functions. Nevertheless it is a fundamental task to explain such regional deviations from normal ones.

As already described in Section 12.3.3, the solubility rate of biogenic opal depends on a large number of processes and environmental conditions. The quality of the various opal phases and their alteration over time are crucial in this regard (Archer et al. 1993; Van Cappellen et al.

2002). The Rivers Niger and Zaire, two of the largest streams worldwide discharge into the northern Angola Basin where they release tremendous amounts of dissolved and particulate substance (compare with Fig. 1.2). As outlined above (Sect. 12.3.3) the solubility of opal decreases when the proportion of aluminiferous lithogenic detritus increases (e.g. Van Cappellen and Qui 1997; Dixit et al. 2001). The high opal content in the sediments of river estuaries should therefore result from an effective preservation of amorphous silicates. With greater distance from equatorial upwelling, it should also be considered that remineralization processes in the water column may be more intense as described by generalizing transfer functions (cf. Fig. 6.3), which leads to a reduction of specific accumulation rates. Hence, both increased dissolution prior to deposition, and a reduced dissolution after deposition are probably the main reasons for the low benthic silicon release.

The sedimentary environment of the Argentine Basin is largely controlled by powerful current systems and intense gravitational mass transport (Ledbetter and Klaus 1987). Sinking particles, or particles already deposited, are subjected to a lateral drift over wide passages, or become resuspended (compare with Section 12.3.2). This

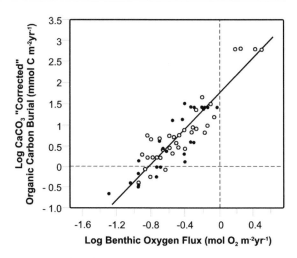

Fig. 12.16 Correlation between C_{org}-burial rates and benthic oxygen consumption. To compensate for deviant determinations made in regions with high and low $C_{org}/CaCO_3$ ratios. $CaCO_3$-burial rates are subtracted from the rates of total accumulation, thus producing *new* C_{org} burial rates (COB) by multiplying the concentrations of C_{org} with the accumulation rate corrected for $CaCO_3$. Open symbols represent Atlantic sites; solid symbols Pacific sites (after Jahnke 1996; cf. Fig. 6.6).

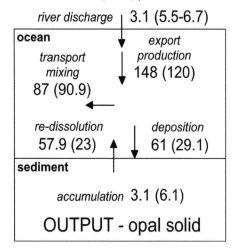

Fig. 12.15 Simplified structure and flux rates of the marine silicon cycle as resulting from the application of a prognostic, coupled water column-sediment, global biogeochemical ocean general circulation model. For comparison, flux values given in brackets base on field observations and were calculated by Tréguer et al. 1995 (after Heinze et al. 2003).

results in a relatively low accumulation of biogenic material on the shelf. Particularly light organic compounds are subject to winnowing and accumulate further downslope (*focusing*). Thus, high productivity is still reflected in this area, however, with a strong horizontal offset towards the deep sea. This effect is additionally enhanced by the import of altered biogenic opal from the Southern Ocean, transported northwards by the Antarctic Bottom Water (AABW).

What are the conclusions, to be drawn from this kind of regionalization? Figure 12.14b shows that approximately 80% of the total release occurs in the predominately oligotrophic open ocean. The balance for the South Atlantic (>1000m water depth) revealed a silicon recycling rate of $2.1 \cdot 10^{12}$ moles Si yr^{-1}. Assuming that the area under study specifies a representative part of the world's oceans, the global release rate may be extrapolated up to a value of $1.95 \cdot 10^{13}$ moles Si yr^{-1} (Hensen et al. 1998; Tab 12.2). These rough approximations probably underestimate the real flux rate, since opal-rich sediments from the Southern Ocean and the equatorial Pacific are not realistically taken into account (cf. Section 12.5.3). In all cases, these rough balances demonstrate that the benthic reflux of silicon outweighs the import by rivers and streams by a factor of 3 ($7.3 \cdot 10^{12}$ moles Si yr^{-1}, Wollast and Mackenzie

Table 12.2 Estimated benthic fluxes from deep-sea sediments (>1000m water depth) in 10^{12} mol yr^{-1}.

Parameter	Area	Flux	Reference
Oxygen	North Atlantic (60°N - 80°N)	0,33	after Schlüter et al. (2000)[1] and Sauter et al. (2001)[1]
	Atlantic (60°N - 60°S)	14,8	Jahnke (1996)
		22,1	Christensen (2000)
		23,8	Wenzhöfer and Glud (2002)
	Global	54	Jahnke (1996)
		80	Christensen (2000)
		98	after Wenzhöfer and Glud (2002)
		54	after Seiter et al. (2005)[2]
Silicate	South Atlantic (10°N - 50°S)	2,1	Hensen et al. (1998)
	Antarctic/Atlantic (40°S - 80°S)	4,28	Schlüter et al. (1998)
	Global	19,5	Hensen et al. (1998)
		23	Tréguer et al. (1995)
		83	Ridgwell et al. (2002)
		58	Heinze et al. (2003)
Phosphate	South Atlantic (10°N - 50°S)	0,035	Hensen et al. (1998)
	Global	0,31	Hensen et al. (1998)
Nitrate	South Atlantic (10°N - 50°S)	1,55	Hensen et al. (1998)
	Global	14,9	Hensen et al. (1998)

[1] Water depths > 500 m; [2] only DOU

1983). The balances confirm the results of other studies on the marine silicon cycle (e.g. Tréguer et al. 1995). However, more recent results by Heinze et al. (2003) using a coupled biogeochemical water column-sediment model, which is driven by a general circulation model, indicate that these fluxes may be underestimated by up to a factor of 3, implying a generally increased Si turnover in the ocean as suggested before (Fig. 12.15). However, using the potential coupling of the organic carbon mineralization and the dissolution rate of biogenic opal mentioned before (cf. Section 12.3.3) Seiter et al. (subm.) conclude that the calculations of Hensen et al. (1998) could be underestimations and therefore may support the model results.

12.5.2 Global Distribution of Benthic Oxygen Depletion Rates - An Example of Regression Analysis

The immense importance of dissolved free oxygen for the microbial decomposition of organic substance has already been discussed at length in the Chapters 3, 5 and 6. Due to its chemical properties which make it one of the most effective

energy sources and a reactive oxidizing agent, dissolved oxygen consumed in the uppermost layers of the sediment. Since the benthic concentrations of oxygen are additionally affected by nearly all processes and factors mentioned in Section 12.2, the precise determination of benthic flux rates still proves to be difficult. The number of reliable and comparable measurements is consequently very low. Moreover, most field studies are limited to selected oceanic areas. The low number of analyses and the extreme heterogeneity of the geographical distribution exclude an application of kriging procedures for regionalization as well as the construction of contour plots with the aid of GIS methods. Jahnke (1996) therefore chose regression analysis for the interpolation and extrapolation for a set of data consisting of 68 single measurements.

The close relation to bioavailable organic matter was used as the rate-limiting control parameter for benthic oxygen consumption. Jahnke (1996) determined an empirical function based on the correlation between the oxygen flux across the sediment/water interface to the concentrations of C_{org} and $CaCO_3$, and the overall

447

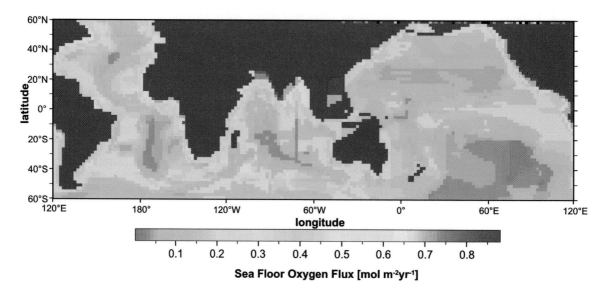

Fig. 12.17 Distribution pattern of benthic oxygen consumption (redrawn after Jahnke 1996).

sediment accumulation rate by relating oxygen depletion to the burial rate of organic carbon compounds (Fig. 12.16; Section 12.3.3). The accumulation rate was estimated based on different data sets from carbonate-rich (e.g. the Atlantic Ocean) and carbonate-poor regions (e.g. the Pacific Ocean). The calculation of oxygen fluxes by this simple method merely produced a maximum deviation from real values by a factor of 2.

The result obtained from regression between single measurements and the control parameters was then used for extrapolation. Global oxygen consumption could be reliably estimated for the first time on a 2° grid encompassing many regions where not a single data point exists (Fig. 12.17) The general pattern follows very much the one of primary production, which is of course not surprising since primary production is closely related to the available amount of benthic organic matter. Despite all inaccuracies the regionalization process made huge progress in terms of global element balances. Jahnke (1996) calculated a rate of global deep-sea consumption of dissolved oxygen as high as $1.2 \cdot 10^{14}$ moles yr^{-1}. Assuming that microorganisms have consumed the oxygen completely in the course of C_{org}-degradation and by applying the estimated burial rates of C_{org}, a flux rate of particulate organic carbon (POC) was estimated to be $3.3 \cdot 10^{13}$ moles yr^{-1} in the deep-sea. This value is equivalent to 45% of the POC-flux over the 1000 m depth horizon ($7.2 \cdot 10^{13}$ moles C yr^{-1}).

Using a somewhat different approach also Seiter et al. (2005) calculated the global POC-flux to the sea floor. In contrast to Jahnke (1996), the diffusive benthic oxygen uptake (DOU) was correlated with TOC and the oxygen concentration in bottom waters. The significant effect of the availability of oxygen for the decay of organic matter becomes obvious when comparing a region with oxygen-depleted bottom waters (~50-70 mmol m^{-3}) with those of normal oxygen levels (mostly close to saturation level; Fig. 12.18). Whereas under oxygen-limited conditions highest TOC contents correspond to lowest oxygen fluxes (Fig. 12.18a, cf. Fig. 6.23), there are mostly no differences between the two methods, if the oxygen concentration is high (Figure 12.18b). Based on a previous study of Cai and Reimers (1995), modified multiple regression analysis was performed for 11 benthic provinces, which represent different sedimentary and geochemical regimes (cf. Section 6.5.2). The extrapolation of the resulting, specific empirical fit functions led to a global map of DOU (Fig. 12.19). A similar transformation of DOU to the POC-flux like the one described above results in an accumulation flux of ~0.5 Gt $C_{org.}$ yr^{-1}. Slightly higher by a factor of about 1.3, this number corresponds well with the balance made by Jahnke (1996). However, the method of investigating the POC-flux in the deep sea applied by Seiter et al. (2005) offers a clear advantage compared to estimates based on sediment trap results. Certainly, the general distri-

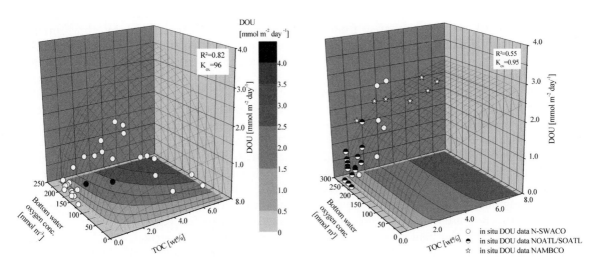

Fig. 12.18 Relation between the benthic oxygen respiration, TOC and the oxygen concentration in bottom waters (after Seiter et al. 2005). a) A significant influence of oxygen-depleted bottom water on the benthic respiration rates and the TOC contents for water depths below 1000 m could be observed for the northeast Pacific region when O_2 drops below 50–70 mmol m³. As to be expected, the highest TOC contents correspond to lowest concentration values of bottom-water oxygen. Above 125 mmol m³, the organic carbon content decreases with increasing bottom water oxygen concentration. b) If oxygen concentrations at the sediment-water interface are maintained by bottom water mixing, there is no effect of this parameter on the overall benthic fluxes. Nevertheless, scattering of data might result from the seasonal and interannual variations in bottom water oxygen content and the local variations of the sedimentation rates of inorganic components, which are likely to dilute the input of organic carbon.

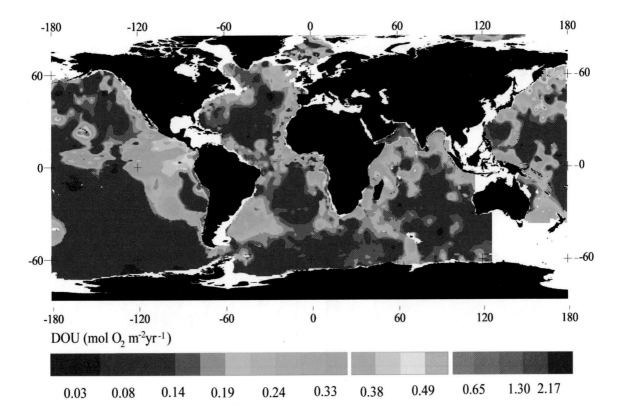

Fig. 12.19 Global distribution pattern of the diffusive oxygen uptake (DOU) at the sea floor (1° x 1°; >1000m water depth; after Seiter et al. 2005).

bution patterns are similar, but in detail significant discrepancies exist, which likely arise from the negligence of lateral transport processes.

12.5.3   Use of Numerical Models to Investigate Benthic-Pelagic Coupling

The rapid development of field studies improving our knowledge on benthic systems over the past decade is strongly coupled to the growing understanding that benthic mineralization has a considerable effect on total ocean geochemistry. Since the regionalization approaches outlined above are difficult to use in order to forecast how and to which extent the benthic reflux of e.g. dissolved carbon and nutrients will affect ocean productivity or atmospheric CO_2-levels, there are increasing efforts to construct coupled models, which are capable of predicting dynamic changes of the environmental conditions. Up to now these

models are far more limited in their regional resolution (mainly caused by their internal complexity and required calculation time) than data-based regionalization approaches. However, towards a better understanding of the interaction of global biogeochemical cycles coupled benthic-pelagic models are of outstanding significance in order to address many important themes as the CO_2 storage capacity of the ocean and marine sediments and long-term productivity changes and climate feedbacks. Since this is another vast field of study we will only develop a brief overview with respect to the most important acquisitions of the past few years.

Stand-alone numerical models to simulate various diagenetic processes in marine sediments have been referred to in various chapters of this book (i.e. Van Cappellen and Wang 1996; Boudreau 1996; Pfeifer et al. 2002; Luff and Wallmann 2003). Some of these models have been

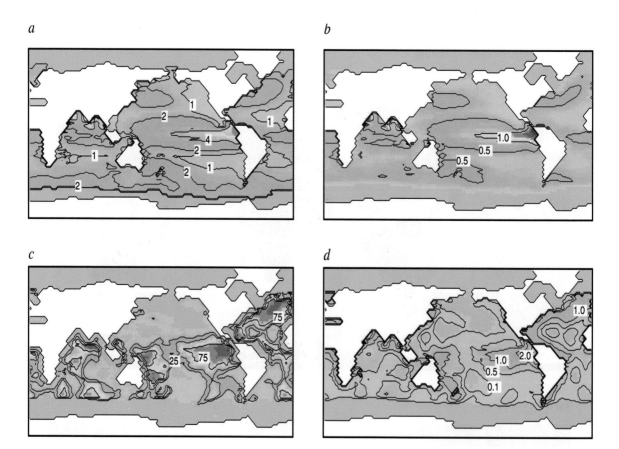

Fig. 12.20   Results of a steady-state simulation with a coupled model for ocean circulation, water chemistry and sediment diagenesis. Major control parameters and forcings comprise a large-scale geostrophic flow field, primary productivity controlled by nutrient advection, export production and sediment accumulation, as well as CO_3^{2-} input by weathering and CO_2-exchange with the atmosphere. a) $C_{org.}$ Export production (mol m^{-2} yr^{-1}), b) $CaCO_3$ export production (both mol m^{-2} yr^{-1}), c) wt% $CaCO_3$, d) $CaCO_3$ mass accumulation rate (g cm^{-2} kyr^{-1}) (from Archer et al. 1998).

tested with respect to their capability in order to adequately respond to dynamic forcing by organic matter supply (Soetaert et al. 1996), to calculate the global benthic oxygen demand (Christensen 2000) or the complete oxic/anoxic cycle in a global gridded domain in order to gain a better understanding of control parameters and kinetic rate constants (Archer et al. 2002). Benthic models coupled to the external forcing of more or less complex pelagic models exist in great variety. In general, the right balance between complexity and calculation time has to be found. Soetaert et al. (2000) tested various approaches of physical, benthic-pelagic models with respect to oxygen, carbon, and nutrient cycling and concluded that at least dynamic sediment models without depth integration are required to produce useful results. In particular, the definition of source and sink terms or the definition of imposed fluxes at the sediment/water interface is inappropriate. However, a certain number of models with a layered benthic module exist that have been tested in order to explore various overarching themes related to global change.

The steady-state model output of a coupled model simulating ocean circulation, biogeo-chemically-controlled surface ocean productivity, export of POC and $CaCO_3$, and sediment diagenesis is shown in Figure 12.20 (Archer et al. 1998). Without going into detail on the background of the study, this was chosen as an arbitrary example in order to show some potentials and limitations of these kind of approaches. The general patterns of export production (a,b) and sediment accumulation (c,d) follow very much the known data distributions, so that we can conclude a general agreement between observation and model prediction. Overall, the graphs show that it is possible to calculate parameter distributions of different compartments representing a consistent overall system. However, it is evident that the spatial resolution of these model is very low and regional characteristics can only be considered and hence, investigated to a limited extent. Up to now, the major advantage of these models is to investigate long-term trends controlled by a number of internal and external forcings.

As outlined in Chapter 6, phosphate is strongly linked to ocean productivity and its availability on geological time scales is largely controlled by the oxidation state of sediments and bottom water (Van Cappellen and Ingall 1994). According to their model, global-scale events of

anoxia/dysoxia would lead to a more efficient recycling of phosphorous and thus, to an enhancement of ocean productivity and organic carbon export. The increased POC accumulation, in turn, would further cause the depletion of oxygen in the bottom water and thus, further enhance the recycling of phosphate at the seafloor and lead into a positive feedback loop. This, general mechanism could be proven by Wallmann (2003) who invented a more sophisticated model to simulate the marine cycles of POC, oxygen, nitrate, and phosphorus. Two additional aspects deserve to be mentioned in this regard: (1) Significant enhancement of productivity requires sufficient supply of nitrate which needs to be provided by N_2-fixation. (2) Productivity of the glacial ocean could have been enhanced by the falling sea level as the marine regression favors an increased POC export to the deep ocean.

Heinze et al. (1999) developed a 10-layer sediment module coupled to a global-ocean circulation model (GCM) specifically designed to study the interactions of Si and C cycles and capable to handle important processes like the ocean-atmosphere exchange of CO_2, biogenic particle export production, as well as sediment accumulation and pore water geochemistry. The central outcome of the study is that one critical and very sensitive parameter in order to obtain reasonable fits to various geochemical data in ocean, atmosphere and sediment is the export production of POC and the corresponding rain ratio ($Si/C_{CaCO_3}/C_{Corg}$). This is basically the same problem, which complicates the prediction of benthic mineralization processes by empirically derived export functions (see above). Another very important factor is the solubility of opal and thus, the release of silicic acid from the sediments. A simple test of halving or doubling the solubility in the pore water leads to a significant increase or decrease of atmospheric pCO_2-levels (by either enhancing or outcompeting $CaCO_3$ production). This emphasizes the inherent coupling of Si and C cycles and the capability of coupled models to produce predictions, which would be possible to deduce from isolated analyses of the specific environmental compartments. The strong potential of drawing down atmospheric pCO_2-levels by significantly increasing Si-concentrations in the ocean could be approved by subsequent studies (Archer et al. 2000; Ridgwell 2002). Additionally, a higher turnover rate of Si in the ocean (Heinze et al. 2003; see Section 12.3.3) might support this idea. However, if glacial/interglacial pCO_2-fluctuations can really be ascribed to the Si-turnover remains questionable since the major unresolved factor in the jigsaw is the potential driving force behind the required perturbations of the Si input to the system. In this regard, the implementation of new perceptions concerning Si-dissolution and control parameters needs to be tackled by future generations of coupled models.

12.6   Concluding Remarks

The examples on regionalization of benthic fluxes have emphasized the value of such studies for solving a variety of problems related to establishing global element balances of the ocean. They have also shown that 1) many aspects related to the selection and compilation of the database need to be taken into account prior to the performance regional studies, and 2) that the interpretation of distribution patterns requires the knowledge of regional oceanographic and geochemical conditions (e.g. sediment composition). Beyond this, regionalization can be used as a very useful tool to prove the plausibility of prevailing concepts on global biogeochemical cycles, which are investigated with complex coupled ocean models. Thus, hypotheses on trans-regional to global biogeochemical interactions can be verified. More and more attention is given to the creation and maintenance of databases in order to gather as many results as possible obtained from many successful, national and international research programs. Apart from the balance of essential single components, and as follows from the current publications on the distribution of primary production, it is the characterization of the benthic biogeochemical provinces that can be now formulated as the main objective. The enormous quantities of the single benthic flux estimates selected in Table 12.2 may

give a concluding impression of the great importance of benthic processes and flux rates within the ocean cycles.

Acknowledgements

This is contribution No 0329 of the Research Center Ocean Margins (RCOM), which is financed by the Deutsche Forschungsgemeinschaft (DFG).

12.7   Problems

Problem 1

Use the simple model approach from Figure 12.12 and the general values given ($\delta = 85\%$, $\varphi = 2,5$ g cm^{-3}, $\tau_{1/2} = 500$ yr) to calculate the degree of organic carbon preservation when accumulation rates are 1 and 0,1 (mg cm^{-2} yr^{-1}) for NRP and $C_{org.}$, respectively, and a reactive mixed layer of 10 cm thickness. Please consider that the degradation of organic matter has an effect on the sedimentation rate (ω).

Problem 2

What is the main reason that rain rate measurements with sediment traps should not be used to estimate benthic biogeochemical processes and especially their distribution patterns?

Problem 3

Why any regionalization with the kriging method without previous application of variogram analysis is – strictly speaking – inadmissible?

Problem 4

Discuss advantages and disadvantages of estimating benthic fluxes by large-scale numerical models.

References

Akin, H. and Siemes, H., (eds), 1988. Praktische Geostatistik – Eine Einführung für den Bergbau und die Geowissenschaften.- Springer Verlag, Berlin, Heidelberg, pp. 304.

Antia, N.A., Koeve, W., Fischer, G., Blanz, T., Schulz-Bull, D., Scholten, J., Neuer, S., Kremling, K., Kuss, J., Peinert, R., Hebbeln, D., Bathmann, U., Conte, M., Fehner, U. and Zeitschel, B., 2001. Basin-wide particulate carbon flux in Atlantic Ocean: Regional export patterns and potential for atmospheric CO_2 sequestration.- Global Biogeochemical Cycles, 15, 845-862.

Antoine, D. André, J.-M. and Morel, A., 1996. Oceanic primary production - 2. Estimation at global scale from satellite (coastal zone color scanner) chlorophyll.- Global Biogeochemical Cycles, 10(1), 57-69.

Archer, D.E. and Devol, A., 1992. Benthic oxygen fluxes on the Washington shelf and slope: a comparison of in situ microelectrode and chamber flux measurements.- Limnology and Oceanography, 37, 614-629.

Archer, D., Lyle, M., Rogers, K. and Froelich, P., 1993. What controls opal preservation in tropical deep-sea sediments?- Paleoceanography, 8, 7-21.

Archer, D., Kheshgi, H. and Maier-Reimer, E., 1998. Dynamics of fossil fuel CO_2 neutralization by marine $CaCO_3$. Global Biogeochemical Cycles, 12, 259-276.

Archer, D., Winguth, A., Lea, D. and Mahowald, N., 2000. What caused the glacial/interglacial atmospheric pCO_2 cycles? Rev. Geophys., 38, 159-189.

Archer, D.E., Morford, J.L. and Emerson, S.R., 2002. A model of suboxic sedimentary diagenesis suitable for automatic tuning and gridded global domain.- Global Biogeochemical Cycles, 16, 1017, 10.1029/2000GB001288

Aspetsberger, F., Zabel, M., Ferdelman, T., Struck, U., Mackensen, A., Ahke, A. and Witte, U., subm. a. Instantaneous benthic response to varying organic matter quality: In situ experiments in the Benguela Upwelling System.- subm. to Limnology and Oceanography.

Aspetsberger, F., Zabel, M., Ferdelman, T., Ahke, A. and Witte, U., subm. b. Microbial response to changing organic matter quality: bacterial productivity and results from medium-term labeling experiments.- subm. to Biogeosciences.

Behrenfeld, M.J. and Falkowski, P.G., 1997. Photosynthetic rates derived from satellite-based chlorophyll concentration.- Limnology and Oceanography, 42, 1-20.

Berger, W.H., Smetacek, V. and Wefer, G., 1989. Ocean productivity and paleoproductivity - an overview.- In: Berger, W.H., Smetacek, V. and Wefer, G. (eds.) Productivity of the ocean: present and past.- Dahlem Konferenzen, John Wiley, Chichester, 1-34.

Bidle, K.D. and Azam, F., 1999. Accelerated dissolution of diatom silica by natural marine bacterial assemblages.- Nature, 397, 508-512.

Bidle, K.D. and Azam, F., 2001. Bacterial control of silicon regeneration from diatom detritus: Significance of bacterial ectohydrolases and species identity.- Limnology and Oceanography, 46, 1606-1623.

Bidle, K.D., Manganelli, M. and Azam, F., 2002. Regulation of oceanic silicon and carbon preservation by temperature control on bacterial activity.- Science, 298, 1980-1984.

Bidle, K.D., Brzezinski, M.A., Long, R.A., Jones, J.L. and Azam, F., 2003. Diminished efficiency in the oceanic pump caused by bacteria-mediated silica dissolution.- Limnology and Oceanography, 48, 1855-1868.

Bishop, J.K.B., 1989. Regional extremes in particulate matter composition and flux: Effects on the Chemistry of the ocean interior.- In: Berger, W.H., Smetacek, V. and Wefer, G. (eds) Productivity of the ocean: present and past.- Dahlem Konferenzen, John Wiley, Chichester, 117-137.

Boudreau, B.P., 1996. A method-of-lines-code for carbon and nutrient diagenesis in aquatic sediments.- Computers and Geosciences, 22, 479-496.

Bühring, S., Moodley, L., Lampadariou, N., Tselepides, A. and Witte, U., 2005. Microbial response patterns modified by food availability: in situ pulse chase experiments in the deep Cretan Sea (Eastern Mediterranean).- Limnology and Oceanography, in press.

Cai, W.J. and Reimers, C.E., 1995. Benthic oxygen flux, bottom water oxygen concentration and core top organic carbon content in the deep northeast Pacific Ocean.- Deep-Sea Research I, 42, 1681-1699.

Chester, R.C., 2000. Marine Geochemistry.- Blackwell Science Ltd, Oxford, 2nd ed., pp. 506.

Christensen, J. P., 2000. A relationship between deep-sea benthic oxygen demand and oceanic primary productivity. Oceanologica Acta, 23, 65-82.

Davis, J.C. (ed.) 1986. Statistics and data analysis in geology.- Wiley & Sons, NY, 2nd ed., pp. 646.

Dauwe, B. and Middelburg, J.J., 1998. Amino acids and hexosamines as indicators of organic matter degradation state in North Sea sediments.- Limnology and Oceanography, 43, 782-798.

Dauwe, B., Middelburg, J.J., Hershey, A.E. and Heip, C.H.R., 1999. Linking diagenetic alteration of amino acids and bulk organic matter reactivity.- Limnology and Oceanography, 44, 1809-1814.

de Baar, H.J.W. and Suess, E., 1993. Ocean carbon cycle and climate change - An introduction to the interdisciplinary Union Symposium.- Global and Planetary Change, 8, VII-XI.

Dixit, S., Van Cappellen, P. and Van Bennekom, A.J., 2001. Processes controlling solubility of biogenic silica and pore water build-up of silicic acid in marine sediments.- Marine Chemistry, 73, 333-352.

Dixit, S. and Van Cappellen, P., 2002. Surface chemistry and reactivity of biogenic silica.- Geochimica et Cosmochimica Acta, 66, 2559-2568.

Dixit, S. and Van Cappellen, P., 2003. Predicting benthic fluxes of silicic acid from deep-sea sediments.- Journal of Geophysical Research, 18, C10, 3334, 10.1029/2002JC001309.

Glud, R.N., Gundersen, J.K., Jørgensen, B.B., Revsbech, N.P. and Schulz, H.D., 1994. Diffusive and total uptake of deep-sea sediments in the eastern South Atlantic Ocean: in situ and laboratory measurements.- Deep-Sea Research, 41, 1767-1788.

Englund, E. and Sparks, A., 1991. GEO-EAS 1.2.1 - Geostatistical environmental assessment software - User's guide.- US-EPA Report #600/8-91/008, EPA-EMSL, Las Vegas, Nevada.

Hales, B. and Emerson, S., 1997. Calcite dissolution in sediments of the Ceara Rise: in situ measurements of pore water O2, pH, and CO2(aq).- Geochimica et Cosmochimica Acta, 61, 501-514.

Harper, M.P., Davison, W. and Tych, W., 1999. One dimensional views of three dimensional sediments.- Environmental Science & Technology, 33, 2611-2616.

Heinze, C., Maier-Reimer, E., Winguth, A.M.E. and Archer, D., 1999. A global oceanic sediment model for long-term climate studies.- Global Biogeochemical Cycles, 13, 221-250.

Heinze, C., Hupe, A., Maier-Reimer, E., Dittert, N. and Ragueneau, O., 2003. Sensitivity of the marine biospheric Si cycle for biogeochemical parameter variations.- Global Biogeochemical Cycles, 17, 1086, doi:10.1029/2002GB001943.

Hensen, C., Landenberger, H., Zabel, M. and Schulz, H.D., 1998. Quantification of diffusive benthic fluxes of nitrate, phosphate and silicate in the Southern Atlantic Ocean.- Global Biogeochemical Cycles, 12, 193-210.

Hensen, C., Zabel, M. and Schulz, H.D., 2000. A comparison of benthic nutrient fluxes from deep-sea sediments off Namibia and Argentina.- Deep-Sea Research II, 47, 2029-2050.

Holstein, J.M. and Hensen, C., subm. Microbial control of benthic biogenic silica dissolution.- subm. to Marine Geology.

Inthorn, M., Mohrholz, V. and Zabel, M., subm. a. Nepheloid layer distribution in the Benguela upwelling area offshore Namibia.- subm. to Deep Sea Research I.

Inthorn, M., Wagner, T., Scheeder, G. and Zabel, M., subm. b. Lateral transport controls distribution, quality and burial of organic matter along continental slopes in high-productivity areas.- subm. to Geology.

Jahnke, R.A., Heggie, D., Emerson, S.R. and Grundmanis, V., 1982. Pore waters of the Central Pacific Ocean: Nurient results.- Earth Planetary Science Letters, 61, 233-256.

Jahnke, R.A., 1985. A model of microenvironments in deep-sea sediments: Formation and effects on porewater profiles.- Limnology and Oceanography, 30, 956-965.

Jahnke, R.A., Reimers, C.E. and Craven, D.B., 1990. Intensification of recycling of organic matter at the sea floor near ocean margins.- Nature, 348, 50-54.

Jahnke, R.A., 1996. The global ocean flux of particulate organic carbon: Areal distribution and magnitude.- Global Biogeochemical Cycles, 10, 71-88.

Jahnke, R.A., 2001. Constraining organic matter cycling with benthic fluxes.- In: Boudreau, B.P. & Jørgensen B.B. (eds.) The benthic boundary layer.- Oxford Univ. Press, 302-319.

Jørgensen, B.B., 1977. Bacterial sulfate reduction within reduced microniches of oxidized marine sediments.- Mar. Biol., 41, 7-17.

Journel, A.G. and Huijbregts, C., (eds.) 1978. Mining Geostatistics.- Academic press, London, pp. 600.

Krige, D.G., 1951. A statistical approach to some basic mine valuation problems on the Witwatersrand.- Journal Chem. Metall. Min. Soc. South Africa, 52, 119-139.

Lampitt, R.S., 1996. Snow falls in the open ocean.- In: Summerhayes, C.P. & Thorpe, S.A. (eds.) Oceanography - An illustrated guide.- Manson Publ., Southampton Oceanogr. Centre, 96-112.

Ledbetter, M.T. and Klaus, A., 1987. Influence of bottom currents on sediment texture and sea-floor morphology in the Argentine Basin.- In: Weaver, P.P.E. & Thomson, J. (eds.) Geology and Geochemistry of abyssal plains, Geological Society Special Publications, 31, 23-31.

Longhurst, A., Sathyendranath, S., Platt, T. and Caverhill, C., 1995. An estimate of global primary production in the ocean from satellite radiometer data.- Journal of Plankton Research, 17, 1245-1271.

Luff, R. and Wallmann, K., 2003. Fluid flow, methane fluxes, carbonate precipitation and biogeochemical turnover in gas hydrate-bearing sediments at Hydrate Ridge, Cascadia Margin: Numerical modeling and mass balances.- Geochimica et Cosmochimica Acta, 67, 3403–3421.

Matheron, G., 1963. Principles of geostatistics.- Economic Geology, 58, 1246-1266.

Nelson, D.M., Tréguer, P., Brzezinski, M.A., Leynaert, A. and Québuiner, B., 1995. Production and dissolution of biogenic silica in the ocean: Revised global estimates, comparison with regional data and relationship to benthic sedimentation.- Global Biogeochemical Cycles, 9, 359-372.

Pannatier, Y., 1996. VARIOWIN: Software for spatial data analysis in 2D.- in Statistics and Computing, 91 S., Springer Verlag, Berlin, Heidelberg, NY.

Pace, M.L., Knauer, G.A., Karl, D.M. and Martin J.H., 1987. Primary production, new production and vertical flux in the eastern Pacific.- Nature, 325, 803-804.

Pfeifer, K., Hensen, C., Adler, M., Wenzhöfer, F., Weber, B. and Schulz, H.D., 2002. Modeling of subsurface calcite dissolution, including the respiration and reoxidation processes of marine sediments in the region of equatorial upwelling off Gabon.- Geochimica et Cosmochimica Acta, 66, 4247-4259.

Reimers, C.E., 1987. An in-situ microprofiling instrument for measuring interfacial pore water gradients: methods and oxygen profiles from the north Pacific Ocean.- Deep-Sea Research, 34, 2019-2035.

Reimers, C.E., Jahnke, R.A. and McCorkle, D.C., 1992. Carbon fluxes and burial rates over the central slope and rise off central California with implications for the global carbon cycle.- Global Biogeochemical Cycles, 6, 199-224.

Reimers, C.E., Jahnke R.A., and Thomsen, L., 2001. In situ sampling in the benthic boundary layer.- In: Boudreau, B.P. & Jørgensen, B.B. (eds.) The benthic boundary layer.- Oxford Univ. Press, 245-268.

Ridgwell, A.J., Watson, A.J. and Archer, D.E., 2002.

Modeling the response of the oceanic Si inventory to perturbation and consequences for atmospheric CO_2.- Global Biogeochemical Cycles, 16, 1071, doi:10.1029/2002GB001877.

Rowe, G.T., Boland, G.S., Phoel, W.C., Anderson, R.F. and Biscaye, P.E., 1994. Deep-sea floor respiration as an indication of lateral input of biogenic detritus from continental margins.- Deep-Sea Research I, 41, 657-668.

Sathyendranath, S., Longhurst, A., Caverhill, C.M. and Platt, T., 1995. Regionally and seasonally differentiated primary production in the North Atlantic.- Deep-Sea Research I, 42, 1773-1802.

Sauter, E.J., Schlüter, M. and Suess, E., 2001. Organic carbon flux and remineralization in surface sediments from the northern North Atlantic derived from porewater oxygen microprofiles.- Deep-Sea Research I, 48, 529-553.

Sayles, F.L., Martin, W.R. and Deuser, W.G., 1994. Response of benthic oxygen demand to particulate organic carbon supply in the deep sea near Bermuda.- Nature, 371, 686-689.

Sayles, F.L. and Martin, W.R., 1995. In situ tracer studies of solute transport across the sediment-water interface at the Bermuda Time Series site.- Deep-Sea Research I, 42, 31-52.

Sayles, F.L., Deuser, W.G., Goudreau, J.E., Dickinson, W.H., Jickells, T.D. and King, P., 1996. The benthic cycle of biogenic opal at the Bermuda Atlantic Time Series site.- Deep-Sea Research I, 43, 383-409.

Schlüter, M. (ed.) 1996. Einführung in geostatistische Verfahren und deren Programmierung.- Enke Verlag, Stuttgart, pp. 326.

Schlüter, M., Rutgers van der Loeff, M.M., Holby, O. and Kuhn, G., 1998. Silica cycles in surface sediments of the South Atlantic.- Deep-Sea Research I, 45, 1085-1109.

Schlüter, M., Sauter, E.J., Schäfer, A. and Ritzau W., 2000. Spatial budget of organic carbon flux to the seafloor of the northern North Atlantic (60N–80N).- Global Biogeochemical Cycles, 14, 329-340.

Schubert, C.J., Niggemann, J., Klockgether, G. and Ferdelman, T.G., 2005. Chlorin Index: A new parameter for organic matter freshness in sediments.- Geochemistry Geophysics Geosystems 6, Art. No. Q03005 MAR 8.

Seiter, K., Hensen, C., Schröter, J and Zabel, M., 2004. Organic carbon content in surface sediments – defining regional provinces. Deep-Sea Research I, 51, 2001-2026.

Seiter, K., Hensen, C. and Zabel, M., 2005. Benthic carbon mineralization on a global scale.- Global Biogeochemical Cycles, 19, GB1010, doi:10.1029/2004GB002225.

Seiter, K., Holstein, J.M., Hensen, C. and Zabel, M., subm. The benthic silica release and its implication for the estimation of the non-lithogenic particle fluxes to the sea floor.- in review at GeoMarine Letters.

Smith, K.L.jr., Baldwin, R.J. and Williams, P.M., 1992. Reconciling particulate organic carbon flux and sediment community oxygen consumption in the

deep North Pacific.- Nature, 359, 313-316.

Soetaert, K., Herman, P.M.J. and Middelburg, J.J., 1996. Dynamic response of deep-sea sediments to seasonal variations: A model.- Limnology and Oceanography, 41, 1651-1668.

Soetaert, K., Middelburg, J.J., Herman, P.M.J. and Buis, K., 2000. On the coupling of benthic and pelagic biogeochemical models.- Earth-Science Reviews, 51, 173-201.

Suess, E., 1980. Particulate organic carbon flux in the oceans: Surface productivity and oxygen utilization.- Nature, 288, 260-263.

Tréguer, P., Nelson, D.M., van Bennekom, A.J., DeMaster, D.J., Leynaert, A. and Quéguiner, B., 1995. The silica balance in the world ocean: A re-estimate.- Science, 268, 375-379.

Van Bennekom, A.J., Buma, A.G.J. and Nolting, R.F., 1991. Dissolved aluminium in the Weddell-Scotia Confluence effect of Al on the dissolution kinetics of biogenic silica.- Marine Chemistry, 35, 423-434.

Van Cappellen, P. and Ingall, E.D., 1994. Benthic phosphorus regeneration, net primary production, and ocean anoxia: A model of the coupled marine biogeochemical cycles of carbon and phosphorus.- Paleoceanography, 9, 677-692.

Van Cappellen, P. and Wang, Y., 1996. Cycling of Iron and Manganese in Surface Sediments: A General Theory for the Coupled Transport and Reaction of Carbon, Oxygen, Nitrogen, Sulfur, Iron, and Manganese.- American Journal of Science, 296, 197-243.

Van Cappellen, P. and Qui, L., 1997. Biogenic silica dissolution in sediments of the Southern Ocean. I. Solubility.- Deep-Sea Research II, 44, 1109-1128.

Van Cappellen, P., Dixit, S. and Van Beusekom, J., 2002. Biogenic silica dissolution in the oceans: Reconciling experimental and field-based dissolution rates.- Global Biogeochemical Cycles, 16, 23 NO. 4, 1075, doi:10.1029/2001GB001431.

Wackernagel, H. (ed.) 1996. Multivariate Geostatistics.- Springer Verlag, Berlin, Heidelberg, NY, pp. 256.

Wallmann, K., 2003. Feedbacks between oceanic redox states and marine productivity: A model perspective focused on benthic phosphorus cycling.- Gobal Biogeochemical Cycles, 17, NO. 3, 1084, doi:10.1029/2002GB001968.

Walsh, J.J., 1991. Importance of continental margins in the marine biogeochemical cycling of carbon and nitrogen.- Nature, 350, 53-55.

Weber, A, Riess, W., Wenzhöfer, F and Jørgensen, B.B., 2001. Sulfate reduction in Black Sea sediments: in situ and laboratory radiotracer measurements from the shelf to 2000m depth.- Deep-Sea Research I, 48, 2073-2096.

Wefer, G. and Fischer, G., 1993. Seasonal pattern of vertical particle flux in the equatorial and coastal upwelling areas of the eastern Atlantic.- Deep-Sea Research I, 40, 1613-1645.

Wenzhöfer, F. and Glud, R.N., 2002. Benthic carbon mineralization in the Atlantic: a synthesis based on in situ data from the last decade.- Deep-Sea Research I, 49, 1255-1279.

Witte, U., Aberle, N., Wenzhöfer F., 2003a. Rapid response of a deep-sea benthic community to POM enrichmen: an in situ experimental study.- Marine Ecology Progress Series, 251, 27-36.

Witte, U., Wenzhöfer, F., Sommer, S., Boetius, A., Heinz, P., Aberle, N., Sand, M., Cremer, A., Abraham, W.-R., Jørgensen, B.B. and Pfannkuche, O., 2003b. In situ experimental evidence of the fate of a phytodetritus pulse at the abyssal sea floor.- Nature, 424, 763-766.

Wollast, R. and Mackenzie, F.T., 1983. The global cycle of silica.- In: Aston, S.R. (ed.) Silicon Geochemistry and Biogeochemistry, Academic Press, London, 39-76.

Zabel, M., Dahmke, A. and Schulz, H.D., 1998. Regional distribution of diffusive phosphate and silicate fluxes through the sediment water interface - The eastern South Atlantic.- Deep-Sea Research I, 45, 277-300.

13 Input from the Deep: Hot Vents and Cold Seeps

Peter M. Herzig and Mark D. Hannington

The discovery of black smokers, massive sulfides and vent biota at the crest of the East Pacific Rise at 21°N in 1979 (Francheteau et al. 1979; Spiess et al. 1980) confirmed that the formation of new oceanic crust through seafloor spreading is intimately associated with hydrothermal venting and the formation of metallic mineral deposits at the seafloor. The 350°C hydrothermal fluids discharging from the black smoker chimneys at this site at a water depth of about 2,600 m continuously precipitate metal sulfides in response to mixing of the high-temperature hydrothermal fluids with ambient seawater. Seawater which penetrates deeply into the oceanic crust at seafloor spreading centers is being converted to a hydrothermal fluid with low pH, low Eh, and high temperature during water-rock interaction above a high-level magma chamber. This fluid is capable of leaching large amounts of metals and other elements from the rocks. Metal sulfides, including pyrite, sphalerite, and chalcopyrite which are precipitated from the hydrothermal fluids, gradually accumulate at and just below the seafloor where they can form large sulfide deposits. The TAG hydrothermal mound at the Mid-Atlantic Ridge at 26°N, for example, has a diameter of about 200 m and a height above seafloor of about 50 m. Drilling during Leg 158 of the Ocean Drilling Program indicated that the deposit contains 2.7 Mt of massive sulfide, containing 2 wt.% Cu, and 1.2 Mt of stockwork mineralization extending 125 m below the seafloor and containing 1 wt.% Cu (Hannington et al. 1998). The deposit is capped by a large black smoker complex with as many as 100 black smoker vents at a temperature of more than 360°C.

Seafloor hydrothermal activity at mid-ocean ridges and back-arc spreading centers has a major impact on the chemistry of the oceans (Edmond et al. 1979a, 1982) and has been responsible for

extensive alteration of modern and ancient oceanic crust (Alt 1995). It has been estimated that 25-30% of the earth's total heat flux is transfered from the lithosphere to the hydrosphere by the circulation of seawater through oceanic spreading centers (Lowell 1991; Stein and Stein 1994). Early estimates of the total discharge of hydrothermal vents at oceanic ridges based on heat flow data were on the order of 5×10^6 L/s. More recent estimates based on geochemical mass balances and geophysical measurements are of the same order of magnitude (Elderfield and Schultz 1996). The calculated fluxes require that the entire volume of the world's oceans is circulated through thermally active seafloor rift zones every 5-11 Ma (Wolery and Sleep 1976; Morton and Sleep 1985). If convective heat flux from the flanks of the ridges is included in this calculation, the cycling time for the world's oceans through the ridges is less than 1 million years. Neglecting any component of diffuse flow, the estimated flux of high-temperature fluids would require at least one black smoker with a mass flux of approxi-mately 1 kg/s and an estimated power of 1.5 megawatts (Converse et al. 1984) for every 50 meters of ridge crest (55,000 km in total). Off-axis diffuse flow accounts for as much as 70-80% of the total heat loss at oceanic ridges (Wheat and Mottl 1994; Stein and Stein 1994) and thus represents an important component of seafloor hydrothermal activity, although the actual chemical fluxes through the ridge flanks have not yet been documented. According to Wheat and Mottl (1994) and Ginster et al. (1994), up to 90% of the heat in the axial rift zones may be removed as a result of diffuse discharge resulting from subseafloor mixing of high-temperature hydro-thermal fluids with cold seawater. Although the discharge of hydrothermal fluids through ridge-crests is small by comparison

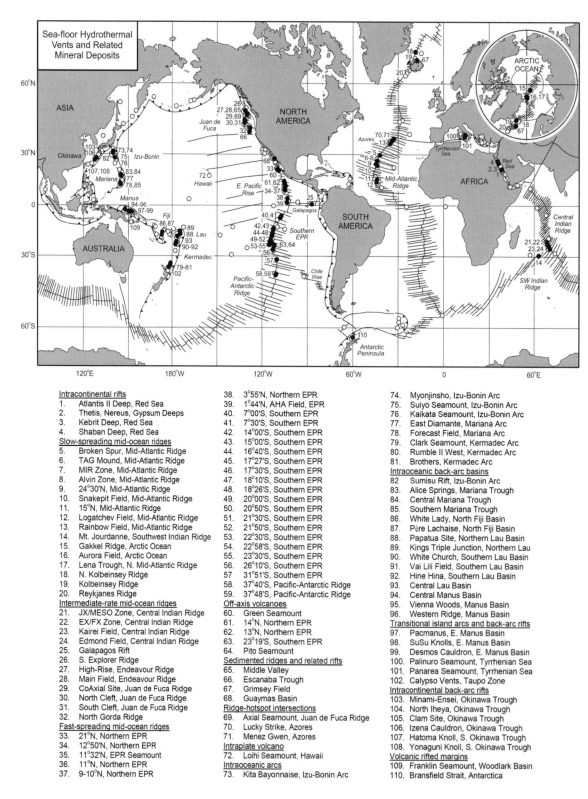

Fig. 13.1 Distribution of sea-floor hydrothermal vents and related mineral deposits. Numbers refer to high-temperature hydrothermal vents and related polymetallic sulfide deposits (closed circles). Other hydrothermal deposits and low-temperature vent sites, including Fe-Mn crusts and metalliferous sediments, are indicated by open circles (from Hannington et al., 2005). Major spreading ridges and subduction zones are indicated.

Intracontinental rifts
1. Atlantis II Deep, Red Sea
2. Thetis, Nereus, Gypsum Deeps
3. Kebrit Deep, Red Sea
4. Shaban Deep, Red Sea

Slow-spreading mid-ocean ridges
5. Broken Spur, Mid-Atlantic Ridge
6. TAG Mound, Mid-Atlantic Ridge
7. MIR Zone, Mid-Atlantic Ridge
8. Alvin Zone, Mid-Atlantic Ridge
9. 24°30'N, Mid-Atlantic Ridge
10. Snakepit Field, Mid-Atlantic Ridge
11. 15°N, Mid-Atlantic Ridge
12. Logatchev Field, Mid-Atlantic Ridge
13. Rainbow Field, Mid-Atlantic Ridge
14. Mt. Jourdanne, Southwest Indian Ridge
15. Gakkel Ridge, Arctic Ocean
16. Aurora Field, Arctic Ocean
17. Lena Trough, N. Mid-Atlantic Ridge
18. N. Kolbeinsey Ridge
19. Kolbeinsey Ridge
20. Reykjanes Ridge

Intermediate-rate mid-ocean ridges
21. JX/MESO Zone, Central Indian Ridge
22. EX/FX Zone, Central Indian Ridge
23. Kairei Field, Central Indian Ridge
24. Edmond Field, Central Indian Ridge
25. Galapagos Rift
26. S. Explorer Ridge
27. High-Rise, Endeavour Ridge
28. Main Field, Endeavour Ridge
29. CoAxial Site, Juan de Fuca Ridge
30. North Cleft, Juan de Fuca Ridge
31. South Cleft, Juan de Fuca Ridge
32. North Gorda Ridge

Fast-spreading mid-ocean ridges
33. 21°N, Northern EPR
34. 12°50'N, Northern EPR
35. 11°32'N, EPR Seamount
36. 11°N, Northern EPR
37. 9-10°N, Northern EPR

38. 3°55'N, Northern EPR
39. 1°44'N, AHA Field, EPR
40. 7°00'S, Southern EPR
41. 7°30'S, Southern EPR
42. 14°00'S, Southern EPR
43. 15°00'S, Southern EPR
44. 16°40'S, Southern EPR
45. 17°27'S, Southern EPR
46. 17°30'S, Southern EPR
47. 18°10'S, Southern EPR
48. 18°26'S, Southern EPR
49. 20°00'S, Southern EPR
50. 20°50'S, Southern EPR
51. 21°30'S, Southern EPR
52. 21°50'S, Southern EPR
53. 22°30'S, Southern EPR
54. 22°58'S, Southern EPR
55. 23°00'S, Southern EPR
56. 26°10'S, Southern EPR
57. 31°51'S, Southern EPR
58. 37°40'S, Pacific-Antarctic Ridge
59. 37°48'S, Pacific-Antarctic Ridge

Off-axis volcanoes
60. Green Seamount
61. 14°N, Northern EPR
62. 13°N, Northern EPR
63. 23°19'S, Southern EPR
64. Pito Seamount

Sedimented ridges and related rifts
65. Middle Valley
66. Escanaba Trough
67. Grimsey Field
68. Guaymas Basin

Ridge-hotspot intersections
69. Axial Seamount, Juan de Fuca Ridge
70. Lucky Strike, Azores
71. Menez Gwen, Azores

Intraplate volcano
72. Loihi Seamount, Hawaii

Intraoceanic arcs
73. Kita Bayonnaise, Izu-Bonin Arc

74. Myonjinsho, Izu-Bonin Arc
75. Suiyo Seamount, Izu-Bonin Arc
76. Kaikata Seamount, Izu-Bonin Arc
77. East Diamante, Mariana Arc
78. Forecast Field, Mariana Arc
79. Clark Seamount, Kermadec Arc
80. Rumble II West, Kermadec Arc
81. Brothers, Kermadec Arc

Intraoceanic back-arc basins
82. Sumisu Rift, Izu-Bonin Arc
83. Alice Springs, Mariana Trough
84. Central Mariana Trough
85. Southern Mariana Trough
86. White Lady, North Fiji Basin
87. Père Lachaise, North Fiji Basin
88. Papatua Site, Northern Lau Basin
89. Kings Triple Junction, Northern Lau
90. White Church, Southern Lau Basin
91. Vai Lili Field, Southern Lau Basin
92. Hine Hina, Southern Lau Basin
93. Central Lau Basin
94. Central Manus Basin
95. Vienna Woods, Manus Basin
96. Western Ridge, Manus Basin

Transitional island arcs and back-arc rifts
97. Pacmanus, E. Manus Basin
98. SuSu Knolls, E. Manus Basin
99. Desmos Cauldron, E. Manus Basin
100. Palinuro Seamount, Tyrrhenian Sea
101. Panarea Seamount, Tyrrhenian Sea
102. Calypso Vents, Taupo Zone

Intracontinental back-arc rifts
103. Minami-Ensei, Okinawa Trough
104. North Iheya, Okinawa Trough
105. Clam Site, Okinawa Trough
106. Izena Cauldron, Okinawa Trough
107. Hatoma Knoll, S. Okinawa Trough
108. Yonaguni Knoll, S. Okinawa Trough

Volcanic rifted margins
109. Franklin Seamount, Woodlark Basin
110. Bransfield Strait, Antarctica

with the fluxes from rivers, seafloor hydrothermal systems have a large impact on the geochemical budgets of certain elements in the oceans that are highly concentrated in vent fluids (Elderfield and Schultz 1996).

This chapter examines a variety of seafloor hydrothermal systems, which are dominated by high-temperature black- and white smoker discharge (250-400°C) and the formation of polymetallic massive sulfide deposits. These systems are currently far better known than the widespread diffuse low-temperature discharge on the ridge flanks. However, the significance of diffuse discharge indicates that the chemical fluxes to the ocean cannot be simply calculated from the composition and venting rates of high-temperature hydrothermal fluids at the ridge crests (cf., Alt 1995).

The history of discovery of seafloor hydrothermal systems has been reviewed by Rona (1988) and Rona and Scott (1993). After the initial discovery of hot metalliferous brines in the Red Sea (Miller et al. 1966), low-temperature venting and associated biological communities were located at the Galapagos hot springs (Corliss et al. 1979) and, eventually, the first black smokers and polymetallic sulfide deposits were found on the East Pacific Rise at 21°N (Francheteau et al. 1979; Spiess et al. 1980). This initiated an intensive investigation of the mid-ocean ridge systems in the Pacific, Atlantic and Indian Ocean, which

resulted in the deleniation of numerous new sites of hydrothermal activity. In 1986, the first inactive hydrothermal sites were found at the active back-arc spreading center of the Manus Basin in the Southwest Pacific (Both et al. 1986). Subsequently, active hydrothermal systems and associated sulfide deposits were reported from the Marianas back-arc (Craig et al. 1987; Kastner et al. 1987), the North Fiji back-arc (Auzende et al. 1989), the Okinawa Trough (Halbach et al. 1989), and the Lau back-arc (Fouquet et al. 1991). Today, more than 100 sites of high-temperature hydrothermal venting and related mineral deposits are known on the modern seafloor (Fig. 13.1; Hannington et al. 2005). The history of discovery of these deposits is given in reviews by Rona (1988) and Rona and Scott (1993).

13.1 Hydrothermal Convection and Generation of Hydrothermal Fluids at Mid-Ocean Ridges

At mid-ocean ridges, seawater penetrates deeply into layers 2 and 3 of the newly formed oceanic crust along cracks and fissures, which form in response to thermal contraction and seismic events in zones of active seafloor spreading (Fig. 13.2). The seawater circulating through the oceanic crust at seafloor spreading centers is converted into a metal-bearing hydrothermal fluid in a

Table 13. 1 Conversion of seawater to a hydrothermal fluid through water/rock interaction above a high-level magma chamber.

temperature:	2°C	→	> 400°C	magma chamber (1200°C)
pH (acidity):	7.8	→	< 4	$H_2O \rightarrow OH^- + H^+$
				$2OH^- + Mg^{2+} \rightarrow Mg(OH)_2$ (> 350°C: Ca instead of Mg)
				$Mg(OH)_2$ fixed in smectite (<200°C) chlorite (>200°C)
				excess H^+ = pH decrease
E_H (redox state):	+	→	-	$Fe^{2+} \rightarrow Fe^{3+}$ (in basalt)
				seawater $SO_4^{2-} \rightarrow S^{2-}$ (H_2S) at > 250°C
				note: a significant amount of the reduced S^{2-} in the hydrothermal fluid results from leaching of sulfide inclusions in the basalt

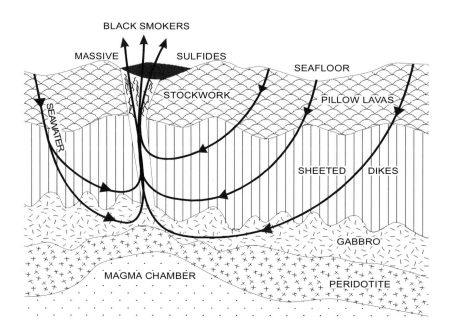

Fig. 13.2 Model showing a seawater hydrothermal convection system above a subaxial magma chamber at an oceanic spreading center. Radius of a typical convection cell is about 3-5 km. Depth of the magma chamber usually varies between 1.5 and 3.5 km (see text for details).

reaction zone situated close to the top of the subaxial magma chamber. The major physical and chemical changes to the seawater include increasing temperature, decreasing pH, and decreasing E_H (Table 13.1).

Temperatures higher than 400°C are reached as seawater is circulated close to the frozen top of the subaxial magma chamber (Cann and Strens 1982; Lowell et al. 1995). High-resolution seismic reflection studies have indicated that some of these magma reservoirs may occur only 1.5-3.5 km below the seafloor (Detrick et al. 1987; Collier and Sinha 1990, 1992; Fig. 13.3). The heat from the magma chamber drives the hydrothermal convection system and gives rise to black smokers at the seafloor. The crustal residence time of seawater in the convection system has been constrained to be 3 years or less (Kadko and Moore, 1988). Data from water/rock interaction experiments indicate that, with increasing temperatures, the Mg^{2+} dissolved in seawater (ca. 1280 ppm or 52 mmol/kg) combines with OH-groups (which originate from the dissociation of seawater at higher temperatures) to form $Mg(OH)_2$. The Mg^{2+} is incorporated in secondary minerals such as smectite (<200°C) and chlorite

(>200°C) (Hajash 1975; Seyfried and Mottl 1982; Seyfried et al. 1988; Alt 1995). The removal of OH-groups by $Mg(OH)_2$ creates an excess of H^+ ions and is the principal acid-generating reaction responsible for the drop in pH from seawater values (pH 7.8 at 2°C) to values as low as pH 3 (lower pH values are observed in some back-arc settings where magmatic gases are present in the hydrothermal fluids: e.g., Fouquet et al. 1993a). Further exchange of H^+ for Ca^{2+} and K^+ in the rock, releases these elements into the hydrothermal fluid and initially balances the removal of Mg^{2+} from seawater. However, at the high temperatures in the reaction zone, the formation of epidote (Ca fixation) contributes further to the acidity of the hydrothermal fluid. These reactions take place at water/rock ratios of less than five and commonly close to one (Von Damm 1995). In most cases, the removal of Mg^{2+} from seawater is quantitative. Therefore, the concentrations of major elements in mixtures of seawater and hydrothermal fluid can be extrapolated back to a common high-temperature "end-member" along mixing lines projected to Mg = 0 (see below). Seawater sulfate (SO_4^{2-}) also is removed, mainly by precipitation of anhydrite and partly by reduction to H_2S. Although reduction of seawater SO_4^{2-} contributes some H_2S to the hydrothermal fluid, but most of the reduced S is derived from the rock. The fluids become strongly reduced by reaction of igneous pyrrhotite to secondary pyrite and oxidation of Fe^{2+} to Fe^{3+} in the basalt (Alt 1995).

This highly corrosive fluid is now capable of leaching elements such as Li, K, Rb, Ca, Ba, the transition metals Fe, Mn, Cu, Zn, together with Au, Ag and some Si from the oceanic crust (Mottl 1983; Table 13.2). Metals and S are leached mainly from immiscible sulfide droplets trapped in the crystallized magma (e.g., Keays 1987; Shanks 2001). The metals are transported as chloride complexes at high temperatures and, in some cases, as bisulfide complexes (in particular Au) at lower temperatures.

Due to its buoyancy at high temperatures, the hydrothermal fluid rises rapidly from the deep-seated reaction zone to the seafloor along major faults and fractures within the rift valley or close to the flanks of the rift. In particular the inter-sections of faults running parallel and perpen-dicular to the ridge axis are the loci of high-velocity discharge, black smokers and massive sulfide mounds. The sulfide precipitation within

Table 13.2 Temperature dependence of water/rock in-teraction in the oceanic crust.

Temperature	Basalt	Seawater / Hydrothermal Fluid
< 150°C	K, Rb, Li, B, ^{18}O	←
> 70°C	Mg	←
> 350°C	Ca	←
>> 150°C	→	Fe, Mn, Zn, Cu, Ba, Si, Ca, Rb, Li, K, B, H^+, ^{18}O, S
all T	H_2O, CO_2, Na	←

the upflow zone (stockwork) and at the seafloor (massive sulfides) is mainly a consequence of mixing of high-temperature (ca. 350°C), metal-rich hydrothermal fluids with cold (about 2°C), oxygen-bearing seawater.

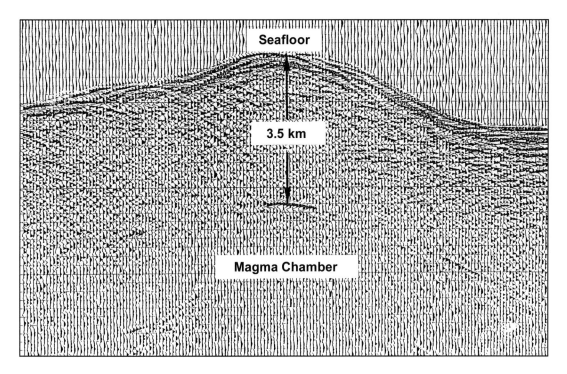

Fig. 13.3 Seismic profile showing the top of a high-level magma chamber at a depth of 3.5 km below the seafloor at a spreading center in the southwest Pacific (after Collier and Sinha 1992; provided by M.C. Sinha).

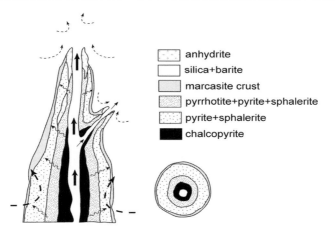

Fig. 13.4 Cross section of a typical high-temperature (350°C) black smoker chimney (from Haymon, 1983).

13.2 Onset of Hydrothermal Activity

In order for high-temperature hydrothermal fluids to discharge, they must first displace a large volume of cold seawater that occupies fractures and pore spaces within the permeable upper crust. The initial flushing of the system may be very rapid, especially on medium- and fast-spreading ridges that are undergoing extension at rates of tens of centimeters per year. Where there are frequent intrusions of magma close to the seafloor and fissure-fed eruptions along the axial rift, seafloor hydrothermal activity may begin with the sudden release of a large volume of hydrothermal fluid forming a "megaplume" in the overlying water column. Observations at hydrothermal fields on the Juan de Fuca Ridge (Embley et al. 1993; Baker 1995) and the East Pacific Rise (Haymon et al. 1993) suggest that the initiation of hydrothermal activity is directly linked to discrete volcanic eruptions and that megaplumes are triggered by dike emplacement (Embley and Chadwick 1994; Baker 1995). The intrusion of the dikes close to the seafloor and major eruptions of lava are coincident with the displacement of large volumes of hydrothermal fluid. Shortly after these eruptions, widespread diffuse flow of low-temperature (≤ 100°C) fluids begins through fractures in the fresh lavas and between new pillows (Butterfield and Massoth 1994). Within a period of about 5-10 years, a low-temperature vent field may be sealed by hydrothermal precipitates, allowing sub-seafloor temperatures to rise and fluid discharge to become focussed into deeper fractures. On a fast-spreading ridge such as the East Pacific Rise, the

cycle of dike injection, eruption, and hydrothermal discharge may repeat itself with each new eruption, perhaps as often as every 3-5 years (Haymon et al. 1993).

Diffuse venting typically occurs throughout the life of a hydrothermal system. It may be the earliest form of discharge in a new hydrothermal field (see above) but commonly also occurs at the margins of existing high-temperature upflow where rising hydrothermal fluids mix with cold seawater. Diffuse venting also typically dominates the last stages of activity in a waning hydrothermal system as high-temperature upflow collapses around a cooling subvolcanic intrusion. Periods of diffuse flow can sustain large biological communities, but are not generally associated with extensive sulfide mineralization because the low temperatures of the fluids (<10°C to 50°C) do not allow transport of significant concentrations of dissolved metals. The mineral precipitates associated with diffuse venting typically consist of amorphous Fe-oxyhydroxides, Mn-oxides, and silica.

13.3 Growth of Black Smokers and Massive Sulfide Mounds

Black smoker activity begins when the hydrothermal fluids contain enough metals and sulfur to cause precipitation of sulfide particles during mixing at the vent orifice. In order to carry these metals in solution, fluids arriving at the seafloor are usually hotter than 300°C. At many black smoker chimneys measurements of vent temperatures in the range of 350°-400°C are common.

The sequence of mineral precipitation which attends mixing of these high-temperature fluids with cold seawater has been modelled extensively (Janecky and Seyfried 1984; Bowers et al. 1985; Tivey 1995; Tivey et al. 1995). Most descriptive models of the growth of black smoker chimneys involve an early asssemblage of chalcopyrite, pyrrhotite, and anhydrite at high temperatures, followed by pyrite and sphalerite at lower temperatures (Haymon and Kastner 1981; Haymon 1983; Goldfarb et al. 1983; Oudin 1983; Graham et al. 1988; Zierenberg et al. 1984; Lafitte et al. 1985; Fouquet et al. 1993b). The black smoke emitted from the chimneys consists mainly of Fe-sulfide and Fe-oxide particles, together with amorphous silica. At lower-temperature "white and grey smokers" (<300°C), the smoke consists of amor-

phous silica and/or barite and anhydrite with lesser amounts of sulfide.

The characteristic feature that distinguishes black smoker vents from white smokers is their central chalcopyrite-lined orifice. In the models of Haymon (1983) and Goldfarb et al. (1983), anhydrite is precipitated around a black smoker vent at the leading edge of chimney growth, where hot hydrothermal fluids first encounter cold seawater (Fig. 13.4). The anhydrite is precipitated from Ca^{2+} in the vent fluids and SO_4^{2-} in ambient seawater, although some anhydrite in the outer walls of the chimney or in the interior of hydrothermal mounds (cf., Humphris et al. 1995) may also be formed simply by the conductive heating of seawater above 150°C. The anhydrite which forms the initial wall of the chimney is gradually replaced by high-

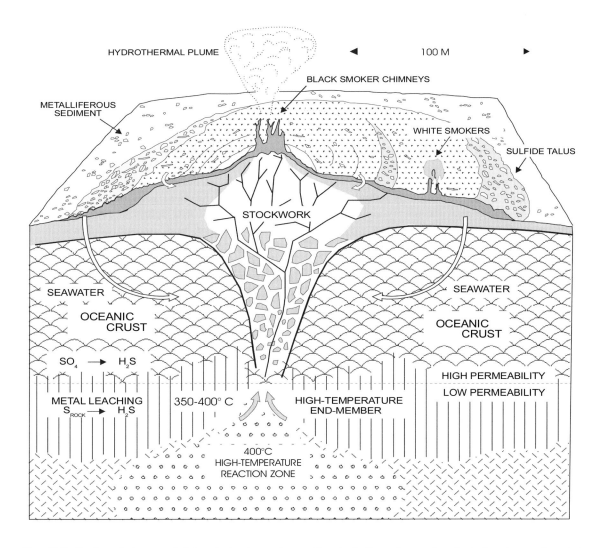

Fig. 13.5 Surface features and internal structure of an active hydrothermal mound and stockwork complex at an oceanic spreading center.

463

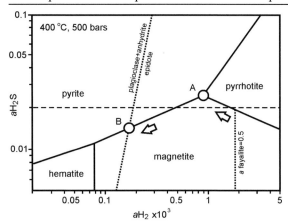

Fig. 13.6 The highest-temperature end-member fluids in black smokers have compositions that are close to equilibrium with pyrite-pyrrhotite-magnetite (A). Somewhat more oxidized fluids may result from equilibrium with an assemblage of epidote-plagioclase-epidote-quartz-magnetite-anhydrite-pyrite (PEQMAP buffer; B). As a result of this buffer assemblage, the end-member hydrothermal fluids have a narrow range of aH_2S and aH_2 (modified after Seyfried et al. 1999).

temperature Cu-Fe-sulfides as the structure grows upward and outward. Because of its retrograde solubility, anhydrite is not well preserved in older chimney complexes and eventually dissolves at ambient temperatures and seafloor pressures (Haymon and Kastner 1981). As a result, many black smokers that are cemented by anhydrite are inherently unstable and ultimately collapse to become part of a growing sulfide mound.

Most chimneys have growth rates that are rapid in comparison to the half-life of ^{210}Pb (22.3 years), and radioisotope ages of several days or less for precipitates of some active vents are consistent with growth rates observed from submersibles (5-10 cm per day: Hekinian et al. 1983; Johnson and Tunnicliffe 1985). Larger vent complexes commonly have measured ages on the order of decades (Koski et al. 1994), but the data for entire vent fields may span several thousands of years (Lalou et al. 1993).

At typical black smoker vents, a very large proportion (at least 90%) of the metals and sulfur carried in solution are lost to a hydrothermal plume in the overlying water column rather than deposited as chimneys. The metals are precipitated as sulfide particles in the plume above the black smokers and are rapidly oxidized and dispersed over distances of several kilometers from the vent (Feely et al. 1987, 1994a,b; Mottl and McConachy 1990). Due to oxidation and dissolu-

tion, these particles also release some of the metals back into seawater (Feely et al. 1987; Metz and Trefry 1993). Particle settling models indicate that only a small fraction of the metals is likely to accumulate as plume fallout in the immediate vicinity of the vents (Feely et al. 1987) and observations of particulate Fe dispersal confirm that most of the metals produced at a vent site are carried away by buoyant plumes (Baker et al. 1985; Feely et al. 1994a,b).

Black smokers usually grow on hydrothermal mounds that are large enough to be thermally and chemically insulated from the surrounding seawater (Fig. 13.5). The sulfide mounds also serve to trap rising hydrothermal fluids and impede the loss of metals and sulfur normally caused by direct venting into the hydrothermal plume. The largest sulfide deposits are often composite bodies which appear to have evolved from several smaller hydrothermal mounds (e.g., Embley et al. 1988). Many of the original chimney structures are overgrown and eventually incorporated in the larger mound, destroying primary textural and mineralogical relationships by hydrothermal replacement. Large sulfide mounds are also constructed from the accumulation of sulfide debris produced by collapsing chimneys, and most large deposits are littered with the debris of older sulfide structures. In many places, new chimneys can be seen growing on top of the sulfide talus, and this debris is eventually overgrown, cemented, and incorporated within the mound (Rona et al. 1993; Hannington et al. 1995). At the same time, high-temperature fluids circulating or trapped beneath the deposit precipitate new sulfide minerals in fractures and open spaces. Chimneys that are perched on the outer surface of an active mound apparently tap these high-temperature fluids but account for only a small part of the total mass of the deposit.

The common presence of high-temperature chimneys at the tops of the deposits and lower-temperature chimneys on their flanks suggests that most large mounds are internally zoned, similar to many ancient massive sulfide deposits on land (cf., Franklin et al. 1981). The most common arrangement of mineral assemblages is a high-temperature Cu-rich core and a cooler, Zn-rich outer margin. Mineralogical zonation within a deposit is principally a result of hydrothermal reworking, whereby minerals that are soluble at low-temperature such as sphalerite are dissolved by later, higher-temperature fluids and redistributed to the outer margins of the deposit. Sulfide samples that have been recovered from the interiors of hydrothermal mounds show signs of extensive hydrothermal recrystallization and annealing, especially when compared to the delicate, fine-grained sulfides found in surface precipitates (Hannington et al. 1998; Petersen et al. 2000). As a result of the extensive hydrothermal reworking in large sulfide mounds, it must be considered that hydrothermal fluids arriving at the seafloor are also likely to have been substantially modified by interaction with pre-existing hydro-

thermal precipitates in their path (e.g., Janecky and Shanks 1988), and caution should be used when interpreting measurements made at the surface of a large mound to infer processes related solely to the direct venting of a high-temperature end-member fluid.

13.4   Physical and Chemical Characteristics of Hydrothermal Vent Fluids

Most black smoker fluids are strongly buffered close to equilibrium with pyrite-pyrrhotite-magnetite, although the proximity of the fluids to this buffer assemblage is not necessarily a reflection of the state of saturation of the minerals in solution (Janecky and Seyfried 1984; Bowers et al. 1985; Tivey et al. 1995). Because of the high concentrations of reduced components such as ferrous iron, H_2 and H_2S, the fluids do not deviate significantly from the pyrite-pyrrhotite redox buffer, even after substantial mixing and cooling, and the common occurrence of both pyrite and pyrrhotite in many sulfide chimneys reflects conditions close to pyrite-pyrrhotite equilibrium throughout their venting history. Janecky and Seyfried (1984) note that mixing does not significantly alter the strong f_{O2} buffering of the fluids until relatively large amounts of seawater have mixed with the end-member solutions, a characteristic which reflects the abundance of reducing agents in the vent fluids and the ineffective buffer capacity of seawater.

Because the vent fluid compositions are determined largely by fluid-rock interactions that take place in the source region, the equilibrium mineral assemblage that is precipitated during mixing will depend to a large extent on the chemistry of the source rocks. Fluids that are buffered to lower f_{O2} and f_{S2} values produce a pyrite- and pyrrhotite-dominated assemblage; fluids at higher f_{O2} and f_{S2} values will precipitate only pyrite. Whereas the buffer assemblage in most volcanic rocks is close to pyrite-pyrrhotite-magnetite, fluids that have reacted extensively with organic-rich sediments or ultramafic rocks may have significantly more reduced compositions. As a result, seafloor spreading in areas of high sedimentation near the continental margins (e.g., Guaymas Basin) has given rise to a class of deposits that are mineralogically quite different

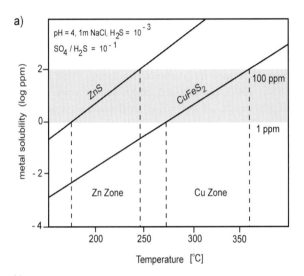

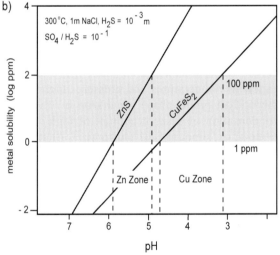

Fig. 13.7 Solubilities of chalcopyrite and sphalerite as a function of temperature and pH (modified after Large et al., 1989). (a) A typical end-member fluid with concentrations of Cu and Zn between 1 and 100 ppm (shaded area) on cooling will precipitate a zone of chalcopyrite first (between 360 and 275°C) followed by a zone of sphalerite (between 250 and 175°C). (b) Similar zonation can result from an increase in pH.

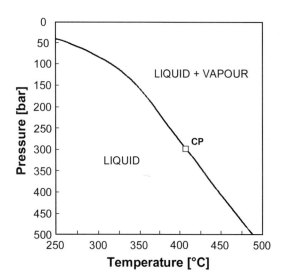

Fig. 13.8 Pressure-temperature curve for seawater (CP = critical point for seawater). Note that 100 bar equal 1,000 m water depth (after Bischoff and Rosenbauer 1984).

from those of bare-ridge massive sulfides. In these deposits, reaction of heated seawater with basaltic rocks controls the initial composition of the hydrothermal fluid, but interaction of the fluid with sediments in the upflow dramatically modifies the fluid chemistry. Chemical buffering of the fluids by sediments results in generally higher pH, and reactions with organic matter in the sediments result in significantly lower f_{O_2} than in bare-ridge systems (Von Damm et al. 1985a; Bowers et al. 1985). Pyrrhotite tends to be the dominant Fe-sulfide phase, and other minerals typically formed at low f_{O_2} may also be common (e.g. Koski et al. 1988; Zierenberg et al. 1993).

The metal zonation within black smoker chimneys and hydrothermal mounds reflects the precipitation of sulfide minerals according to their respective solubilities at different temperatures and pH (e.g., Fig. 13.7). For the most part, metals such as Cu and Zn are carried in solution as aqueous chloride complexes with stabilities that are enhanced at high temperatures and low pH (Bourcier and Barnes 1987; Crerar and Barnes 1976; Hemley et al. 1992). During mixing with and cooling by seawater, decreasing temperature and increasing pH cause the sequential deposition of chalcopyrite and sphalerite. The Cu-rich cores of black smokers generally reflect saturation with respect to chalcopyrite at high temperatures, whereas the outer Zn-rich zones reflect saturation

with sphalerite at lower temperatures (e.g., caused by mixing with seawater that penetrates through the chimney walls). Although the pH of a hydrothermal fluid will increase dramatically as a result of mixing, during conductive cooling the pH changes in the fluid are moderated by the production of acid associated with sulfide precipitation (e.g., $FeCl_2 + 2H_2S + 1/2O_2 = FeS_2 + 2Cl^- + 2H^+ + H_2O$), and the lower pH of the cooled fluids may inhibit the precipitation of certain minerals (Janecky and Seyfried 1984).

An important constraint on the P-T path for some seafloor hydrothermal fluids is the two-phase curve for seawater (Fig. 13.8; Bischoff and Rosenbauer 1984; Bischoff and Pitzer 1985). Variations in the salinities of some vent fluids indicate that phase separation is occurring in many seafloor hydrothermal systems (Von Damm 1988, 1990; Butterfield 2000; see below). At the depth range of the most mid-ocean ridge vent sites (2,500-3,000 m), the two-phase boundary of seawater occurs at temperatures between 385°C and 405°C (Fig. 13.8). Because temperatures of vent fluids at these depths are typically well below the two-phase, they are unlikely to undergo phase separation. However, other vent fluids have been observed that are clearly within the two-phase region (e.g., 420°C and 220 bars at Endeavour Ridge: Delaney et al. 1984), and similar conditions may be common in the high-temperature reaction zones or upflow conduits of some systems. At pressures and temperatures below the critical point for seawater (407°C and 298 bars), fluids which intersect the two-phase curve will separate a small amount of low-salinity, vapor. At temperatures and pressures higher than the critical point, phase separation involves the condensation of a small amount of high salinity brine (i.e., supercritical phase separation). The existence of such high-salinity fluids at depth has implications for the development of metal-rich brines. Bischoff and Rosenbauer (1987) noted that during supercritical phase separation, both the acidity and the concentration of heavy metals increase in the chloride-rich phase, and the solubilities of metals as aqueous chloride complexes in these fluids may be several orders of magnitude greater than in fluids of ordinary seawater composition.

At shallow water depths, subcritical boiling may have a major impact on seafloor mineralization. At 350°C, a 21°N-type fluid will intersect the two-phase curve for seawater when the

Table. 13.3   Chemical composition of hydrothermal fluids from selected hydrothermal fields at mid-ocean ridges and back-arc spreading centers in comparison to seawater.

	Tmax (°C)	pH (25°C)	Cl (ppt)	H₂S (ppm)	Na (ppm)	K (ppm)	Ca (ppm)	Ba (ppm)	Sr (ppm)	Fe (ppm)	Mn (ppm)	Zn (ppm)	Cu (ppm)	Si (ppm)	Reference
MARK (MAR 23°N)	350	3,9	19,8	201	11,725	931	421	-	5	122	27	3	1.0	514	Von Damm (1995)
TAG (MAR 26°N)	366	3,8	22,5	119	12,805	669	1,235	-	9	313	37	3	10.0	583	Edmond et al. (1995)
Lucky Strike (MAR 37°N)	332	-	19,3	-	-	-	-	-	-	35	21	-	-	-	Von Damm (1995)
EPR (11°N)	347	3,7	24,3	416	13,265	1,287	1,411	-	12	361	51	-	-	579	Von Damm (1995)
EPR (13°N)	(380)	3,3	26,9	279	13,702	1,165	2,204	-	16	603	160	-	-	618	Von Damm (1995)
EPR (21°N)	355	3,8	20,5	286	11,725	1,009	834	2,2	9	136	55	-	-	548	Von Damm (1995)
Guyamas Basin	315	5,9	22,6	204	11,794	1,924	1,663	7,4	22	10	13	-	-	388	Von Damm (1995)
Escanaba Trough	217	5,4	23,7	51	12,874	1,580	1,339	-	18	1	1	-	-	194	Von Damm (1995)
Middle Valley (JFR)	276	5,5	20,5	102	9,150	731	3,247	2,1	23	1	4	0,1	0,1	298	Von Damm (1995)
South Cleft (JFR)	285	3,2	31,8	(102)	15,196	1,459	3,395	-	20	575	143	24	1.0	640	Von Damm (1995)
North Cleft (JFR)	327	3.0	31.0	124	15,679	1,580	2,922	-	20	165	65	16	0,5	559	Von Damm (1995)
Axial Seamount (JFR)	328	3,5	22,1	242	11,472	1,048	1,876	3,6	17	59	63	7	0,6	424	Von Damm (1995)
Endeavour (JFR)	370	4,4	16,2	167	8,230	1,004	1,447	-	12	54	14	2	1,3	455	Von Damm (1995)
Lau Basin (SW-Pacific)	334	2.0	28.0	-	13,564	3,089	1,655	> 5.4	2	140	390	196	2,2	393	Fouquet et al. (1993a)
Seawater	2	7,8	19,4	-	10,759	399	413	0,02	8	$6 \cdot 10^{-5}$	$5 \cdot 10^{-5}$	$7 \cdot 10^{-4}$	$5 \cdot 10^{-4}$	5	

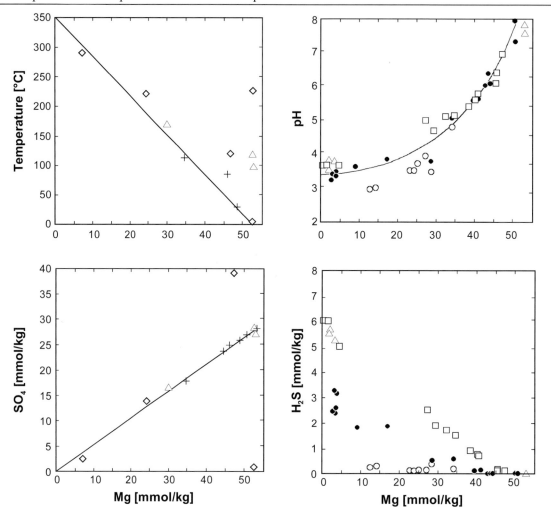

Fig. 13.9 Temperature, pH, SO$_4$ and H$_2$S versus Mg. Symbols indicate measurements of different black and white smoker fluids from the East Pacific Rise 21°N, the TAG site at 26°N Mid-Atlantic Ridge, and the Snakepit hydrothermal field at 23°N Mid-Atlantic Ridge (after Von Damm et al. 1983 and Edmond et al. 1995).

hydrostatic pressure is equivalent to about 1,600 m of water depth (160 bars or 16 Mpa; Fig. 13.8). If the water depth is less than 1,600 m, the fluid will boil beneath the seafloor and may separate a vapor-rich phase. At these lower pressures, the density difference between the vapor and liquid is large, which facilitates the separation of a low-salinity, gas-rich phase. Such fluids were first documented at vents in the caldera of Axial Seamount (1,540 m water depth and a fluid temperature of 349°C) where low salinity and high gas contents have been measured (Massoth et al. 1989; Butterfield et al. 1990). Similar boiling is now recognized in a large number of shallow submarine hydrothermal systems (German and Von Damm 2004; Hannington et al. 2005).

13.5   The Chemical Composition of Hydrothermal Vent Fluids and Precipitates

Since the discovery of high-temperature hydrothermal vents at the East Pacific Rise 21°N in 1979, hydrothermal fluids have been sampled at numerous sites at mid-ocean ridges and back-arc spreading centers. As noted above, in most high-temperature vent fluids, both Mg and SO$_4$ show a negative correlation with temperature, and an extrapolation to zero Mg and zero SO$_4$ intersects the temperature axis at a point corresponding to the end-member temperature (Fig. 13.9). Controls on the major element compositions of these fluids are summarized in Von Damm (1990) and Von Damm (1995), and some examples are given in Table 13.3. Time series measurements on the order of a decade at a number of sites have indicated that the chemical composition of vent fluids at individual sites does not show significant temporal variablity and is essentially „steady state" once the system has stabilized after a new volcanic event (Von Damm 1995). Phase separation is responsible for much of the chemical variability observed (cf., Von Damm 1995), but major differences are also recognized between fluids originating in different volcanic and tectonic settings with different host rock compositions (Hannington et al. 2005; see below). In some areas, contributions from a degassing magma are also inferred (Lupton et al. 1991; de Ronde 1995).

The salinities of vent fluids have been shown to range from values of about 30% (176 mM/kg; Von Damm and Bischoff 1987) to 200% (1090 mM/kg; Massoth et al. 1989) seawater values (546 mM/kg). These variations are important because Cl is the major complexing anion in the hydrothermal fluids. The observed salinities cannot be accounted for by hydration of the oceanic crust (Cathles 1983) or by precipitation and dissolution of Cl-bearing mineral phases (Edmond et al. 1979b; Seyfried et al. 1986) but are interpreted to be a result of phase separation at the top of the magma chamber followed by mixing of the brines and more dilute

hydrothermal fluids during ascent to the seafloor (cf. Butterfield 2000).

In mid-ocean ridge hydrothermal fluids, ^{3}He is significantly enriched over air-saturated seawater, with ^{3}He/^{4}He ratios between about 7 and 9. These ratios represent typical mantle values and correspond to ^{3}He originating from MORB magma that has degassed into the hydrothermal system (Lupton 1983; Baker and Lupton 1990). Lupton and Craig (1981) have demonstrated that ^{3}He behaves extremely conservatively in hydrothermal plumes and can be traced in the water column over distances up to 2000 km from the point of origin (Fig. 13.10). In unsedimented areas, CH_4 also is commonly enriched in the fluids, as a product of inorganic reactions in the hydrothermal system and also from magmatic degassing. Methane and total dissolvable Mn (TDM) have been found to be enriched about 10^6-fold over ambient seawater in high-temperature vent fluids (e.g., Fig. 13.11; Von Damm et al. 1985b). In diluted buoyant hydrothermal plumes, these values are still 100-fold enriched relative to seawater (Klinkhammer et al. 1986; Charlou et al. 1991) which makes CH_4 and Mn valuable tracers when prospecting hydrothermal vent areas.

Concentrations of trace elements such as Ag, As, Cd, Co, Se, and Au have been measured in some vent fluids (Von Damm 1990; Campbell et al. 1988a,b; Fouquet et al. 1993a; Trefry et al. 1994;

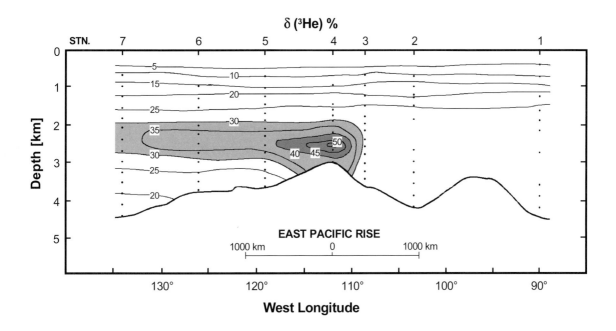

δ (^{3}He) %

Fig 13.10 Helium plume at the East Pacific Rise (after Lupton and Craig 1981).

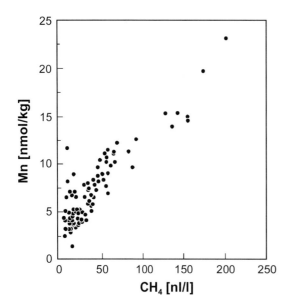

Fig. 13.11 Positive correlation of Mn and CH_4 in a hydrothermal plume (after Herzig and Plüger 1988).

Edmond et al. 1995; Table 13.3) and certain metals such as Cu, Co, Mo, and Se appear to show a strong positive temperature-concentration relationship which likely accounts for observed enrichments of these elements in the sulfides from high-

temperature chimneys (Hannington et al. 1991). Other elements such as Zn, Ag, Cd, Pb, and Sb appear to be signficantly enriched in lower-temperature fluids (Trefry et al. 1994). However, these data remain scarce, reflecting the difficulties of reliably measuring trace metal concentrations in the vent fluids.

The typical REE pattern of end-member hydrothermal fluids exhibits a strong enrichment of LREE and a pronounced positive Eu anomaly (Michard et al. 1983; Campbell et al. 1988c; Michard 1989; Fig. 13.12), in contrast to the source rocks (MORB) which have a nearly flat, LREE depleted spectrum, and to seawater which is characterized by a strong negative Ce anomaly. The overall pattern is strongly influenced by leaching of plagioclase from MORB under hydrothermal conditions.

The chemical compositions of volcanic and sedimentary rocks with which the end-member hydrothermal fluids react have a major impact on the metal supply and nature of subseafloor alteration. The bulk compositions of the deposits generally reflect the rock types from which the metals were leached (Doe 1994; Herzig and Hannington 1995). In mid-ocean ridge environments, the dissolution of sulfides and the destruction of ferromagnesian minerals in mid-ocean ridge

Table 13.4 Bulk chemical composition of seafloor polymetallic sulfides from mid-ocean ridges and back-arc spreading centers (after Herzig and Hannington 1995).

Element	Mid-Ocean Ridges [1]	Back-Arc Ridges [2]
n	890	317
Fe (wt%)	23.6	13.3
Cu	4.3	5.1
Zn	11.7	15.1
Pb	0.2	1.2
As	0.03	0.1
Sb	0.01	0.01
Ba	1.7	13.0
Ag (ppm)	143	195
Au	1.2	2.9

[1] Explorer Ridge, Endeavour Ridge, Axial Seamount, Cleft Segment, East Pacific Rise, Galapagos Rift, TAG, Snake Pit, Mid-Atlantic Ridge 24.5°N
[2] Mariana Trough, Manus Basin, North Fiji Basin, Lau Basin

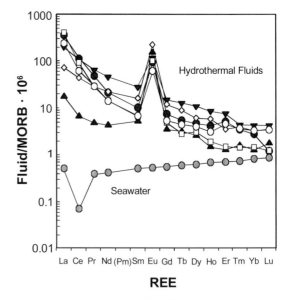

Fig. 13.12 Rare earth element (REE) concentrations in hydrothermal fluids versus Atlantic seawater at 2,500 m (Elderfield and Greaves 1982) normalized to MORB (Sun 1980). Symbols indicate measurements of different vent fluids from the Lucky Strike hydrothermal field at 37°17'N Mid-Atlantic Ridge (after Klinkhammer et al. 1995).

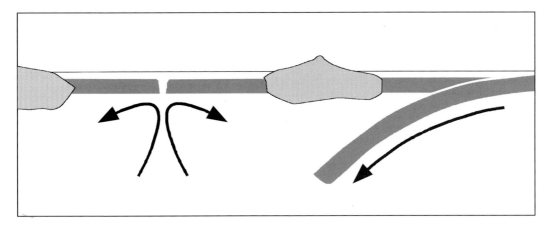

Fig. 13.13 Schematic diagram showing the geotectonic setting of a seafloor back-arc spreading center, showing the relationship to the active volcanic front of the arc and the subducting slab.

basalt (MORB) are the major sources of Cu, Fe, Zn, Au, and S in the hydrothermal system. In contrast, elements such as Pb and Ba are derived mainly from the destruction of feldspars. Feldspars are particularly abundant in felsic volcanic rocks, but less so in basalt, so elements such as K, Ba and Pb are enriched in fluids from environments with a high proportion of flesic rocks (e.g., volcanic arcs; see below). Even in mid-ocean ridge settings, the bulk compositions of the deposits appear to be quite sensitive to the chemistry of the source rocks (Table 13.4). For example, the high Ba contents of some mid-ocean ridge deposits may be related to the enrichment of incompatible elements in some lava suites due to an enriched melt source or greater degrees of fractionation (Perfit et al. 1983; Delaney et al. 1981; Clague et al. 1981).

Hydrothermal fluids at immature back-arc rifts, which occur behind island-arcs in areas where oceanic crust is being subducted (Fig. 13.13), clearly reflect the compositional differences of the source rocks (fractionated, calc-alkaline suites versus MORB). The fluids at immature back-arcs are typically enriched in Zn, Pb, As and Ba (and depleted in Fe) compared to mid-ocean ridge fluids, which is related to the higher concentrations of these elements in the associated lavas (Herzig and Hannington 1995; Table 13.5).

Evidence for direct contributions of magmatic fluid and gases to seafloor hydrothermal systems also has been noted in back-arc vent fluids and precipitates (Sakai et al. 1990; Gamo et al. 1997; Herzig et al. 1998). This is of major importance, as magmatic fluids and gases can be responsible for a significant input of metals into the hydrothermal

system (Hedenquist and Lowenstern 1994). Magmatic components, however, are extremely difficult to identify as they are usually masked by the large amount of seawater in the circulation system.

The recent discovery of active hydrothermal vents in the summit calderas of andesitic volcanoes along a number of submarine volcanic arcs (e.g., Izu-Bonin arc, Mariana arc, Tonga-Kermadec arc; Fig. 13.1) highlights the possible importance of

Table. 13.5 Chemical composition of hydrothermal fluids at mid-ocean ridges (EPR, East Pacific Rise 21°N) and back-arc rifts (Valu Fa Ridge, Lau Basin) relative to seawater (sw).

	EPR 21°N [1]	Lau Basin [2]	seawater
T (°C)	350	334	2
pH	3.6	2	7.8
salinity	sw	1.5 · sw	
Fe (ppm)	80	140	$6 \cdot 10^{-5}$
Mn (ppm)	49	390	$5 \cdot 10^{-5}$
Cu (ppm)	1.4	2.2	$5 \cdot 10^{-4}$
Zn (ppm)	5.5	196	$7 \cdot 10^{-4}$
Ba (ppm)	1.4	5.4	$2 \cdot 10^{-2}$
Pb (ppb)	54	808	$2 \cdot 10^{-3}$
As (ppb)	17	450	1.7

[1] Von Damm et al. (1985a); [2] Fouquet et al. (1993a)

direct contributions magmatic volatiles to the hydrothermal fluids in this setting. Widespread sulfur-rich fumaroles and low-pH vent fluids, similar to magmatic-hydrothermal systems in subaerial arc volcanoes, have been documented at a number of locations and reflect magmatic degassing of high-level magma chambers beneath the summits of the volcanoes (Tsunogai et al., 1994; de Ronde et al., 2003). A number of other aspects of this setting also result in dramatically different deposit types compared to those found on the mid-ocean ridges and in deeper back-arc basins. In particular, the shallow water depths (commonly <1,000 m) result in widespread boiling and generally lower temperatures of hydrothermal venting, with potentially significant sub-seafloor stockwork mineralization. However, the nature of the hydrothermal activity in these volcanoes is not yet well characterized. Some clearly involves high-temperature hydrothermal circulation of seawater and black smoker venting, but many hydrothermal plumes above the calderas appear to be related mainly to passive degassing of the volcano rather than high-temperature hydrothermal activity (Massoth et al., 2003). Nevertheless, considering the composition of the arc magmas, the flux of volatiles from such systems is likely much greater than that from the mid-ocean ridges.

Deposits on sediment-covered mid-ocean ridges commonly have lower Cu and Zn contents and higher Pb contents than bare-ridge sulfides, and some deposits such as in the Guaymas Basin contain abundant carbonate. The low metal contents of deposits in the Guaymas Basin are a consequence of the higher pHs of the fluids that arise from chemical buffering by the carbonate in the sediments, and the high CO_2 in the fluids is derived directly from the sediments themselves (Bowers et al. 1985; Von Damm et al. 1985a). A high ammonium content reflects the thermocatalytic cracking of buried organic matter (e.g., immature planktonic carbon; cf., Von Damm et al. 1985a). The fluids in this environment are also strongly reduced (in the pyrrhotite stability field) as a consequence of reaction with organic matter in the sediments (e.g., $C_{org} + 2H_2O = CO_2 + 2H_2$), and metal deficient relative to a volcanic-hosted hydrothermal system as a result of sulfide precipitation within the sediments. However, interaction of the hydrothermal fluids with sediments also may result in enrichments in certain trace elements derived from the sediments (e.g., Pb, Sn, As, Sb, Bi, Se: Koski et al. 1988; Zierenberg et al. 1993). The higher Pb

contents, in particular, reflect the destruction of feldpars from continentally-derived turbidites, a process supported by Pb isotope studies (LeHuray et al. 1988).

13.6   Characteristics of Cold Seep Fluids at Subduction Zones

In addition to hydrothermal activity at divergent plate boundaries, low-temperature fluid venting at both passive and convergent plate margins is an important global process. Cold seeps have now been documented at numerous sites along the circum-Pacific subduction zones and elsewhere on continental shelf, and it is estimated that this type of oceanic venting is also of major significance for the chemical budget of seawater. The most significant fluxes are from fluids that are expelled from thick organic-rich marine sediments trapped in accretionary wedges along the subduction trenches (e.g., Nankai Trench, Oregon and Alaskan margins). Pore fluids in the sediment account for as much as 50-70% of their volume, and this fluid is literally squeezed from the sediment through diffuse flow at the toe of the accretionary wedge or by focussed flow along major fault structures in the accreted sediments.

In order to estimate the mass flux from cold seeps, flow rates have to be known, and these have been difficult to measure because of the large areas of diffuse flow involved. Fluid flow rates determined from simple advection-diffusion modelling of temperature and chemical profiles are often unreliable if applied to sites which are densely populated with bottom macrofauna. Calculations by Wallmann et al. (1997), based on a biogeo-chemical approach using oxygen flux and vent fluid analyses and taking into account the high pumping rates of bivalves, have arrived at a mean value of 5.5 +/- 0.7 L m^{-2} d^{-1}. Von Huene et al. (1998) used sediment porosity reduction to calculate a fluid flow of 0.02 L m^{-2} d^{-1}. Suess et al. (1998) calculated an average rate of 0.006 L m^{-2} d^{-1} based on the occurrence of vent biota at the seafloor. Although these estimates span several orders of magnitude, the fluid flux at subduction zones is thought to be sufficient to cycle the volume of water in the oceans every 500 Ma. The total flux may be lower than that of hydrothermal circulation at mid-ocean ridges, but the fluids expelled from subduction zones are a significant player in the carbon cycle of the oceans as a result of the recycling of organic matter in marine sediments.

Methane plumes have been found to be a characteristic feature of cold seep areas. The concentrations of dissolved CH_4 in some areas are lower than in hydrothermal vent areas at mid-ocean ridges, but in other areas they exceed ridge values by more than an order of magnitude and have been successfully used as a tracer for detecting cold seep venting. The methane and carbon dioxide that are expelled with the pore waters are derived mainly from the breakdown of organic matter in the sediments. Methane together with H_2S support abundant chemosynthetic bacteria, pogonophorans, vestimentifera, and bivalves which differ from those that inhabit high-temperature vent sites on the mid-ocean ridges (Suess et al. 1998). As the temperatures of venting are low compared to hydrothermal vents at the mid-ocean ridges, metals are not significantly mobilized.

Typical precipitates for areas of subduction venting are carbonate and barite crusts and carbonate-cemented sediment (Dia et al. 1993; Suess et al. 1998). The formation of carbonate is related to the microbial oxidation of methane which causes the precipitation of $CaCO_3$ from pore fluids (Wallmann et al. 1997). The source for barium is thought to be the high concentration of biogenic barite buried in sediments at high productivity areas which is mobilized from the reduced sediment column due to sulfate depletion. Barite in the cold seep areas forms as a result of mixing between Ba-rich fluids upwelling from the sediment with seawater sulfate (Ritger et al. 1987; Torres et al. 1996).

Ice-like gas hydrates (clathrates) also have been found at numerous sites along the convergent plate margins (e.g., Suess et al. 1997). Methane hydrates are stable in solid form only in a narrow temperature-pressure window (Dickens and Quinby-Hunt 1994; Fig. 14.3 in chapter 14). In theory, $1 m^3$ of methane hydrate can contain up to $164 m^3$ of methane gas at standard conditions (Kvenvolden 1993). It has been estimated that the amount of carbon in gas hydrates considerably exceeds the total of carbon occurring in all known oil, gas and coal deposits worldwide (Kvenvolden and McMenamin 1980; Kvenvolden 1988). This raises the possibility that gas hydrates may be a future energy source of global importance.

If disturbed by a thermal anomaly or pressure change, large volumes of gas hydrates can break down into water, methane and carbon-dioxide. The dissolution of metastable gas hydrates is a natural consequence of tectonic uplift of accretionary prisms at plate margins and is seen to be partly responsible for the extensive methane plumes reported from these areas (Suess et al. 1997). Because of the limited stability of gas hydrates (low temperature and high pressure), they are of major relevance for the budget of greenhouse gases (Suess et al. 1997). The destabilization and dissolution of gas hydrates due to environmental changes (increase in ocean bottom water temperature or change of sea level) could liberate enormous amounts of methane to the water column and eventually to the atmosphere where they could potentially accelerate greenhouse warming and have an effect on future global climate (Leggett 1990).

13.7   Problems

Problem 1

A fluid sample collected at a mid-ocean ridge hydrothermal vent has a temperature of 250°C and Mg concentration of 15 mmol/kg (Fig. 13.9). If the fluid contains 100 ppm Fe at 250°C, what is the likely concentration of Fe in the end-member fluid at 350°C?

Problem 2

A high-temperature black smoker has a calculated end-member Fe concentration of 150 ppm (e.g., Table 13.3). If the fluid is venting from the chimney at a rate of 1 kg/s, how much Fe is being discharged into the ocean by this vent every year? If the global flux of black smoker fluids along the midocean ridges is equivalent to approximately one vent for every 50 m of ridge-crest, how much Fe would be discharged along the ridges every year?

Problem 3

A 100 m diameter sulfide deposit on the mid-ocean ridge has been drilled to a depth of 25 m. The metal concentrations of the sulfides recovered from the core are 2 wt.% Cu, 5 wt.% Zn, and 15 wt.% Fe, and the samples have a bulk density of $3 g/cm^3$. What is the total metal content of the deposit? If the deposit formed from a black smoker complex containing 100 vents of the type described in question 2, how long would it have taken to deposit this much metal, assuming a depositional efficiency of 1%?

Problem 4

The average depth of the summit calderas along a submarine volcanic arc in the western Pacific Ocean is about 1000 m. What is the pressure at this depth? What is the maximum temperature of a vent fluid discharging at 1000 m, assuming a composition similar to that of mid-ocean ridge vents? What is the maximum temperature of vents at 500 m depth?

Problem 5

Write a balanced chemical reaction (or reactions) to represent the breakdown of organic matter in sediment to form carbon dioxide and methane in a cold seep environment. If the average temperature of the sediments where this reaction is taking place is 5°C, what is the minimum depth at which you would expect to find solid gas hydrates forming from the methane? What would be the depth if the bottom temperature was 1°C?

References

Alt, J.C., 1995. Subseafloor processes in mid - ocean ridge hydrothermal systems. Seafloor Hydrothermal Systems: Physical, Chemical, Biological and Geological Interactions, AGU Geophysical Monograph, 91: 85-114 pp.

Auzende, J.M., Urabe, T., Deplus, C., Eissen, J.P., Grimaud, D., Huchon, P., Ishibashi, J., Joshima, M., Lagabrielle, Y., Mevel, C., Naka, J., Ruellan, E., Tanaka, T. and Tanahashi, M., 1989. Le cadre geologique d'un site hydrothermal actif. la campagne Starmer 1 du submersible nautile dans le bassin Nord Fidjien. C. R. Acad. Sci. Paris, 309: 1787-1795.

Baker, E.T., Lavelle, J.W. and Massoth, G.J., 1985. Hydrothermal particle plumes over the southern Juan de Fuca Ridge. Nature, 316: 342-344.

Baker, E.T. and Lupton, J.E., 1990. Changes in submarine hydrothermal 3He/ heat ratios as an indicator of magmatic/ tectonic activity. Nature, 346: 556-558.

Baker, E.T., 1995. Characteristics of hydrothermal discharge following a magmatic intrusion. Hydrothermal Vents and Processes, In: Parson, L.M., Walker, C.L. and Dixon, D.R. (eds) Geological Society Special Puplication, 87: 65-76 pp.

Bischoff, J.L. and Rosenbauer, R.J., 1984. The critical point and two-phase boundery of seawater, 200°C - 500°C. Earth and Planetary Science Letters, 68: 172-180.

Bischoff, J.L. and Pitzer, K.S., 1985. Phase relation and adiabats in boiling seafloor geothermal systems. Earth and Planetary Science Letters, 75: 327-338.

Bischoff, J.L. and Rosenbauer, R.J., 1987. Phase separation in seafloor geothermal systems: An experimental study of the effects on metal transport. American Journal of Science, 287: 953-978.

Both, R.A., Crook, K., Taylor, B., Brogan, S., Chapell, B., Frankel, E., Liu, L., Sinton, J. and Tiffin, D., 1986. Hydrothermal chimneys and associated fauna in the Manus back - arc basin, Papua New Guinea. American Geophysical Union Translations, 67: 489-490.

Bourcier, W.L. and Barnes, H.L., 1987. Ore solution chemistry VII. Stabilities of chloride and bisulfide complexes of zinc to 350°C. Economic Geology, 82: 1839-1863.

Bowers, T.S., von Damm, K.L. and Edmond, J.M., 1985. Chemical Evolution of mid - ocean ridge hot springs. Geochimica et Cosmochimica Acta, 49: 2239-2252.

Butterfield, D.A., Massoth, G.J., McDuff, R.E., Lupton, J.E. and Lilley, M.D., 1990. Geochemistry of hydrothermal fluids from Axial seamount hydrothermal emissions study vent field, Juan de Fuca Ridge: Subseafloor boiling and subsequent fluid rock and interaction. Journal of Geophysical Research, 95: 12895-12921.

Butterfield, D.A. and Massoth, G.J., 1994. Geochemistry of north Cleft segment vent fluids: Temporal changes in chlorinity and their possible relation to recent volcanism. Journal of Geophysical Research, 99: 4951-4968.

Butterfield, D.A., 2000. Deep ocean hydrothermal vents. In: Sigurdsson, H., Houghton, B.F., McNutt,

S.R., Rymer, H., Stix, J., and Ballard, R.D. (eds) Encyclopedia of Volcanoes, Academic Press, San Diego: 857-875.

Campbell, A.C., Bowers, T.S., Measures, C.I., Falkner, K.K., Khadem, M. and Edmond, J.M., 1988a. A time - series of vent fluid composition from 21°N, EPR (1979, 1981, 1985), and the Guaymas Basin, Gulf of California (1982, 1985). Journal of Geophysical Research, 93: 4537-4549.

Campbell, A.C., German, C., Palmer, M.R. and Edmond, J.M., 1988b. Preliminary report on the chemistry of hydrothermal fluids from the Escanaba Trough. American Geophysical Union Transactions, 69: 1271.

Campbell, A.C., Palmer, M.R.,Klinkhammer, G.P., Bowers, T.S., Edmond, J.M., Lawrence, J.R., Casey, J.F., Thompson, G., Humphris, S., Rona, P. and Karson, J.A. 1988c. Chemistry of hot springs of the Mid–Atlantic Ridge. Nature, 335: 514-519.

Cann, J.R. and Strens, M.R., 1982. Black smokers fuelled by freezing magma. Nature, 298: 147-149.

Cathless, L.M., 1983. An analysis of the hydrothermal system responsible for massive sulfide deposition in the Hokuroku Basin of Japan. Economic Geology, Monograph, 5: 439-487.

Charlou, J.L., Bougalt, H., Appriu, P., Nelsen, T. and Rona, P., 1991. Different TDM/CH4 hydrothermal plume signatures: TAG site at 26°N and serpentinized ultrabasic diapier at 15°05'N on the Mid-Atlantic Ridge. Geochimica et Cosmochimica Acta, 55: 3209-3222.

Clague, D.A., Frey, F.A., Thompson, G. and Rindge, S., 1981. Minor and trace element geochemistry of volcanic rocks dredged from the Galapagos spreading center: Role of crystal fractionation and mantle heterogeneity. Journal of Geophysical Research, 86: 9469-9482.

Collier, J. and Sinha, N., 1990. Seismic images of a magma chamber beneath the Lau Basin back - arc spreading center. Nature, 346: 646-648.

Collier, J.S., and Sinha, M.C., 1992. Seismic mapping of a magma chamber beneath the Valu Fa Ridge, Lau Basin. Journal of Geophysical Research, B97: 14,031-14,053.

Converse, D.R., Holland, H.D. and Edmond, J.M., 1984. Flow rates in the axial hot springs of the East Pacific Rise (21°N) : implications for the heat budget and the formation of massiv sulfide deposits. Earth and Planetary Sience Letters, 69: 159-175.

Corliss, J.B., Dymond, J., Gordon, L.I., Edmond, J.M., von Herzen, R.P., Ballard, R.D. Green, K., Williams, D., Bainbridge, A., Crane, K. and van Andel, T.H., 1997. Submarine thermal springs on the Galapagos Rift. Science, 203: 1073-1083.

Craig, H., Horibe, Y., Farley, K.A., Welhan, J.A., Kim, K.R. and Hey, R.N., 1987. Hydrothermal vents in the Mariana Trough: Results of the first Alvin dives. American Geophysical Union Transactions, 68: 1531.

Crerar, D. and Barnes, H.L., 1976. Ore solution chemistry V. Solubilities of chalcopyrite and chalcocite assemblages in hydrothermal solutions at 200°C to 350°C. Economic Geology, 71: 772-794.

de Ronde, C.E.J., 1995. Fluid chemistry and isotopic characteristics of seafloor hydrothermal systems and associated VMS deposits: potential for magmatic contributions. In: Thompson, JFH (ed) Magmas, Fluids, and Ore Deposits. Mineralogical Association of Canada Short Course Notes, 23: 479-509.

de Ronde, C.E.J., Massoth, G.J., Baker, E.T., and Lupton, J.E., 2003. Submarine hydrothermal venting related to volcanic arcs. In: Simmons, S.F., and Graham, I.J. (eds) Giggenbach Memorial Volume: Volcanic, Geothermal and Ore-Forming Fluids: Rulers and Witnesses of Processes within the Earth, Society of Economic Geologists, Special Publication, 10: 91-109.

Delaney, J.R., Johnson, H.P. and Karsten, J.L., 1981. The Juan de Fuca Ridge-hot spot-propagating rift system. Journal of geophysical Research, 86: 11747-11750.

Delaney, J.R., McDuff, R.E. and Lupton, J.E., 1984. Hydrothermal fluid temperatures of 400°C of the Endeacour Segment, northern Juan de Fuca Ridge. American Geophysical Union Transactions, 65: 973.

Detrick, R.S.P., Buhl, E., Vera, J., Mutter, J., Orcutt, J., Madsen, J. and Brocher, T., 1987. Multi-channel seismic imaging of a crustal magma chamber along the East Pacific Rise. Nature, 326: 35-41.

Dia, A.N., Aquilina, L., Boulègue, J., Bourgois, J., Suess, E. and Torres, M., 1993. Origin of fluids and related barite deposits and vent sites along the Peru convergent margin. Geology, 21: 1099-1102.

Dickens, G.R. and Quinby - Hunt, M.S., 1994. Methane hydrate stability in seawater. Geophysical Research Letters, 21: 2115-2118.

Doe, B.R., 1994. Zinc, copper, and lead in mid - ocean ridge basalts and the source rock control on Zn/Pb in ocean-ridge hydrothermal deposits. Geochimica et Cosmo-chimica Acta, 58: 2215-2223.

Edmond, J.M., Measures, C.I., McDuff, R.E., Chan, L.H., Collier, R., Grant, B., Gordon, L.I. and Corliss, J.B., 1979a. Ridge crest hydrothermal activity and the balance of the major and minor elements in the ocean: The Galapagos data. Earth and Planetary Science Letters, 46: 1-18.

Edmond, J.M., Measures, C.I., Mangum, B., Grant, B., Sclater, F.R., Collier, R., Hudson, A., Gordon, L.I. and Corliss, J.B., 1979b. On the formation of metal-rich deposits at ridge crests. Earth and Planetary Science Letters, 46: 19-30.

Edmond, J.M., von Damm, K.L., McDuff, R.E. and Measures, C.I., 1982. Chemistry of hot springs on the East Pacific Rise and their effluent dispersal. Nature, 297: 187 - 191.

Edmond, J.M., Campbell, A.C., Palmer, M.R., Klinkhammer, G.P., German, C.R., Edmonds, H.N., Elderfield, H., Thompson, G. and Rona, P., 1995. Time series studies of vent fluids from the TAG and MARK sites (1986, 1990) Mid - Atlantic Ridge: a new solution chemistry model and a mechanism for Cu/Zn zonation in massiv sulphide orebodies. In: Parson, L.M., Walker, C.L. and Dixon, D.R. (eds). Hydrothermal Vents and Processes, Geological Society Special Publication, 87: 77-86.

Elderfield, H. and Greaves, M.J., 1982. The rare earth elements in seawater. Nature, 296: 214-219.

Elderfield, H., and Schultz, A., 1996. Mid-ocean ridge hydrothermal fluxes and the chemical composition of the ocean. Annual Review of Earth and Planetary Science, 24: 191-224.

Embley, R.W., Jonasson, I.R., Perfit, M.R., Franklin, J.M., Tivey, M.A., Malahoff, A., Smith, M.F. and Francis, T.J.G., 1988. Submersible investigation of an extinct hydrothermal system on the Galapagos Ridge: Sulfide mounds, stockwork zone, and differentiated lavas. Canadian Mineralogist, 26: 517-539.

Embley, R.W., Chadwick, W.W., Jonasson, I.R., Petersen, S., Butterfield, D., Tunnicliffe, V. and Juniper, K., 1993. Geologic inference from a response to the first remotely detected eruption on the mid - ocean ridge: Coaxial Segment, Juan de Fuca Ridge. American Geophysical Union Transactions, 74: 619.

Embley, R.W. and Chadwick, W.W.(jr.), 1994. Volcanic and hydrothermal processes associated with a recent phase of seafloor spreading at the northern Cleft segment: Juan de Fuca Ridge. Journal of Geophysical Research, 99:4735-4740.

Feely, R.A., Lewison, M., Massoth, G.J., Robert-Galdo, G., Lavelle, J.W., Byrne, R.H., von Damm, K.L. and Curl, H.C.(jr.), 1987. Composition and dissolution of black smoker particulates from active vents on the Juan de Fuca Ridge. Journal of Geophysical Research, 92:11347-11363.

Feely, R.A., Massoth, G.J., Trefry, J.H., Baker, E.T., Paulson, A.J. and Lebon, G.T., 1994. Composition and sedimentation of hydrothermal plume particles from North Cleft segment, Juan de Fuca Ridge. Journal of Geophysical Research, 99: 4985-5006.

Feely, R.A., Gedron, J.F., Baker, E.T. and Lebon, G.T., 1994. Hydrothermal plumes along the East Pacific Rise, 8°40' to 11°50'N: Particle distribution and composition. Earth and Planetary Science Letters, 128: 19-36.

Fouquet, Y., von Stackelberg, U., Charlou, J.L., Donval, J.L., Erzinger, J., Foucher, J.P., Herzig, P.M., Mühe, R., Soakai, S., Wiedicke, M. and Whitechurch, H., 1991. Hydrothermal activity and metallogenesis in the Lau back - arc basin. Nature, 349: 778-781.

Fouquet, Y., von Stackelberg, U., Charlou, J.L., Erzinger, J., Herzig, P.M., Mühe, R. and Wiedicke, M., 1993a. Metallogenesis in back-arc environment: the Lau Basin example. Economic Geology, 88: 2154-2181.

Fouquet, Y., Auclair, P., Cambon, P. and Etoubleau, J., 1993b. Geological setting, mineralogical, and geochemical investigation on sulfide deposits near 13°N on the East Pacific Rise. Marine Geology, 84: 145-178.

Francheteau, J., Needham, H.D., Choukroune, P., Juteau, T., Seguret, M., Ballard, R.D., Fox, P.J., Normark, W., Carranza, A., Cordoba, D., Guerrero, J., Rangin, C., Bougault, H., Cambon, P. and Hekinian, R., 1979. Massiv deep-sea sulphide ore deposits discovered on the East Pacific Rise. Nature, 277: 145-178.

Franklin, J.M., Lydon, J.W. and Sangster, D.F., 1981. Volcanic - associated massiv sulfide deposits. Economic Geology, 75: 485-627.

Gamo, T., Okamura, K., Charlou, J.L., Urabe, T., Auzende, J.M., Ishibashi, J., Shitashima, K., Chiba, H., ManusFlux Shipboard Scientific Party, 1997. Acidic and sulfate - rich hydrothermal fluids from the Manus back-arc basin, Papua New Guinea. Geology, 25: 139-142.

German, C.R., and Von Damm, K.L., 2004, Hydrothermal processes. In: Holland, H.D., and Turekian, K.K. (eds) Treatise on Geochemistry Volume 6: The Oceans and Marine Geochemistry, Chapter 6.07, Elsevier: 181-222.

Ginster, U., Mottl, M.J. and von Herzen, R.P., 1994. Heat flux from black smokers on the Endeavour and Cleft segments, Juan de Fuca Ridge. Journal of Geophysical Research, 99: 4937-4950.

Goldfarb, M.S., Converse, D.R., Holland, H.D. and Edmond, J.M., 1983. The genesis of hot spring deposits on the East Pacific Rise, 21°N. Economic Geology, Monograph, 5: 184-197.

Graham, U.M., Bluth, G.J. and Ohmoto, H., 1988. Sulfide-sulfate chimneys on the East Pacific Rise, 11° and 13°N latitudes. Part 1: Mineralogy and paragenesis. Canadian Mineralogist, 26: 487-504.

Hajash, A., 1975. Hydrothermal processes along Mid-Ocean Ridges: an experimental investigation. Contributions in Mineralogy and Petrology, 53: 205-226.

Halbach, P., Nakamura, K., Wahsner, M., Lange, J., Sakai, H., Käselitz, L., Hansen, R.D., Yamano, M., Post, J., Prause, B., Seifert, R., Michaelis, W., Teichmann, F., Kinoshita, M., Märten, A., Ishibashi, J., Czerwinski, S. and Blum, N., 1989. Probable modern analogue of Kuroko - type massiv sulphide deposits in the Okinawa Trough back-arc basin. Nature, 338: 496-499.

Hannington, M.D., Herzig, P.M., Scott, S.D., Thompson, G. and Rona, P.A., 1991. Comparative mineralogy and geochemistry of gold-bearing sulfide deposits on the mid-ocean ridges. Marine Geology, 101: 217-248.

Hannington, M.D., Petersen, S., Jonasson, I.R. and Franklin, J.M., 1994. Hydrothermal activity and associated mineral deposits on the seafloor. Geological Survey of Canada Open File Report, 2915C: Map 1:35,000,000 and CD-ROM.

Hannington, M.D., Jonasson, I.R., Herzig, P.M. and Petersen, S., 1995. Physical, chemical processes of seafloor mineralization at mid - ocean ridges. In: Humphris, S.E. et al. (eds) Seafloor Hydrothermal Systems. Physical, Chemical, Biological and Geological Interactions, AGU Geophysical Monograph, 91: 115-157.

Hannington, M.D., Galley, A.G., Herzig, P.M., and Petersen, S., 1998. A comparison of the TAG Mound and stockwork complex with Cyprus-type massive sulfide deposits. In: Humprhis, S.E., Herzig, P.M., and Miller, D.J. (eds) Proceedings of the Ocean Drilling Program, Scientific Results, 158: 389-415.

Hannington, M.D., de Ronde, C.E.J., and Petersen, S., 2005. Sea-floor tectonics and submarine hydrothermal systems. 100th Anniversary Volume of Economic Geology (in press).

Haymon, R.M. and Kastner, M., 1981. Hot spring deposits on the East Pacific Rise at 21°N: Preliminary description of mineralogy and genesis. Earth and Planetary Science Letters, 53: 363-381.

Haymon, R.M., 1983. Growth history of hydrothermal black smoker chimneys. Nature, 301: 695-698.

Haymon, R.M.; Fornari, D.J., von Damm, K.L., Lilley, M.D., Perfit, M.R., Edmond, J.M., Shanks, W.C. III, Lutz, R.A., Grebmeier, J.M., Carbotte, S., Wright, D., McLaughlin, E., Smith, M., Beedle, N. and Olson, E., 1993. Volcanic eruption of the mid - ocean ridge along the East Pacific Rise crest at 9°45-52'N: Direct submersible observations of seafloor phenomena associated with an eruption event in April, 1991. Earth and Planetary Science Letters, 119: 85-101.

Hedenquist, J.W. and Lowenstern, J.B., 1994. The role of magmas in the formation of hydrothermal ore deposits. Nature, 370: 519-526.

Hekinian, R., Francheteau, J., Renard, V., Ballard, R.D., Choukroune, P., Cheminée, J.L., Albaréde, F., Minster, J.F., Charlou,J.L., Marty, J.C. and Boulégue, J., 1983. Intense hydrothermal activity at the rise axis of the East Pacific Rise near 13°N: Submersible witnesses the growth of sulfide chimney. Marine Geology Research, 6: 1-14.

Hemley, J.J., Cygan, G.L., Fein, J.B., Robinson, G.R. and D'Angelo, W.M., 1992. Hydrothermal ore - forming processes in the light of studies in the rock-buffered systems: I. Iron-copper-zinc-lead sulfide solubility relations. Economic Geology, 87: 1-22.

Herzig, P.M. and Plüger, W.L., 1988. Exploration for hydrothermal mineralisation near the Rodriguez Triple Junction, Indian Ocean. Canadian Mineralogist, 26: 721-736.

Herzig, P.M. and Hannington, M.D., 1995. Polymetallic massive sulfides at the modern seafloor: A review. Ore Geology Reviews, 10: 95-115.

Herzig, P.M., Hannington, M.D. and Arribas, A.(jr.)., 1998. Sulfur isotopic composition of hydrothermal precipitates from the Lau back-arc: implications for magmatic contributions to seafloor hydrothermal systems. Mineralium Deposita, 33: 226-237.

Humphris, S.H., Herzig, P.M., Miller, D.J.,Alt, J.C., Becker, K., Brown, D., Brügmann, G., Chiba, H., Fouquet, Y., Gemmell, J.B., Guerin, G., Hannington, M.D., Holm, N.G., Honnorez, J.J., Itturino, G.J., Knott, R., Ludwig, R., Nakamura, K., Petersen, S., Reysenbach, A.L., Rona, P.A., Smith, S., Sturz, A.A., Tivey, M.K. and Zhao, X., 1995. The internal structure of an active sea-floor massive sulfide deposit. Nature, 377: 713-716.

Janecky, D.R. and Seyfried, W.E.(jr.), 1984. Formation of massiv sulfide deposits on ocean ridge crests: Incremental reaction models for mixing between hydrothermal solutions and seawater. Geochimica et Cosmochmica Acta, 48: 2723-2738.

Janecky, D.R. and Shanks, W.C.III, 1988. Computational modelling of chemical and isotopic reaction processes in seafloor hydrothermal systems: chimneys, massive sulfides, and subjacent alteration zones. Canadian Mineralogist, 26: 805-825.

Johnson, H.P. and Tunnicliffe, V., 1985. Time series measurements of hydrothermal activity on the northern Juan de Fuca Ridge. Geophysical Research Letters, 12: 685-688.

Kadko, D. and Moore, W., 1988. Radiochemical constraints on the crustal residence time of submarine hydrothermal fluids: Endeavour Ridge. Geochimica et Cosmochimica Acta, 52: 659-668.

Kastner, M., Craig, H. and Sturz, A., 1987. Hydrothermal deposition in the Mariana Trough: Preliminary mineralogical investigations. American Geophysical Union Transactions, 68: 1531.

Katz, D.L., Cornell, D., Kobayashi, R., Poettmann, F.H., Vary, J.A., Elenblass, J.R. and Weinaug, C.F., 1959. Handbook of Natural Gas Engineering, 802. McGraw-Hill, NY, 802 pp.

Keays, R.R., 1987. Principles of mobilization (dissolution) of metals in mafic and ultramafic rocks - The role of immiscible magmatic sulphides in the generation of hydrothermal gold and volcanogenic massiv sulphide deposits. Ore Geology Reviews, 2: 47-63.

Klinkhammer, G.P., Elderfield, H., Greaves, M., Rona, P.A. and Nelson, T., 1986. Manganese geochemistry near high-temperature vents in the Mid-Atlantic Ridge rift valley. Earth and Planetary Science Letters, 80: 230-240.

Klinkhammer, G.P., Chin, C.S., Wilson, C. and German, C.R., 1995. Venting from the Mid-Atlantic Ridge at 37°17'N: the Lucky Strike hydrothermal site. In. Parson, L.M., Walker, C.L. and Dixon, D.R. (eds) Hydrothermal vents and processes. Geological Society Special Publication, 87: 87-96.

Koski, R.A., Shanks, W.C.I., Bohrson, W.A. and Oscarson, R.L., 1988. The composition of massiv sulfide deposits from the sediment-covered floor of Escanaba Trough, Gorda Ridge: implications for depositional processes. Canadian Mineralogist, 26: 655-673.

Koski, R.A., Jonasson, I.R., Kadko, D.C., Smith, V.K. and Wong, F.L., 1994. Compositions, growth mechanisms, and temporal relations of hydrothermal sulfide-sulfate-silica chimneys at the northern Cleft segment, Juan de Fuca Ridge. Journal of Geophysical Research, 99: 4813-4832.

Kvenvolden, K.A. and McMenamin, 1980. Hydrate of natural gas. A review of their geological occurrence. U. S. Geological Survey Circular, 825: 11.

Kvenvolden, K.A., 1988. Methane hydrate - a major reservoir of carbon in the shallow geosphere. Chemical Geology, 71: 41-51.

Kvenvolden, K.A., 1993. Gas hydrates - Geological Perspective and global change. Reviews of Geophysics, 31: 173-187.

Lafitte, M., Maury, R., Perseil, E.A. and Boulegue, J., 1985. Morphological and analytical study of hydrothermal sulfides from 21° north East Pacific Rise. Earth and Planetary Science Letters, 73: 53-64.

Lalou, C., Reyss, J.L., Brichet, E., Arnold, M., Thompson, G., Fouquet, Y. and Rona, P.A., 1993. New age data for Mid-Atlantic Ridge hydrothermal sites: TAG and Snakepit chronology. Journal of Geophysical Research, 98: 9705-9713.

Large, R.R., D.L. Huston, P.J. McGoldrick, P.A. Ruxton,

and G. McArthur, 1989. Gold distribution and genesis in Australian volcanogenic massive sulphide deposits and their significance for gold transport models. Economic Geology Monograph 6: 520-536.

Leggett, J., 1990. The nature of the greenhouse threat. In: Leggett, J. (ed) Global Warming, The Greenpeace Report. Oxford University Press, NY: 14-43.

LeHuray, A.P., Church, S.E., Koski, R.A. and Bouse, R.M., 1988. Pb isotopes in sulfides from mid-ocean ridge hydrothermal sites. Geology, 16: 362-365.

Lowell, R.P., 1991. Modeling continental and submarine hydrothermal systems. Reviews in Geophysics, 29: 457-476.

Lowell, R.P., Rona, P.A. and von Herzen, R.P., 1995. Seafloor hydrothermal systems. Journal of Geophysical Research, 100: 327-352.

Lupton, J.E. and Craig, H., 1981. A major helium-3 source at 15°S on the East Pacific Rise. Science, 214: 13-18.

Lupton, J.E., 1983. Fluxes of helium-3 and heat from submarine hydrothermal systems; Guaymas Basin versus 21°N EPR. American Geophysical Union Transactions, 64: 723.

Lupton, P., Lilley, M., Olson, E. and von Damm, K.L., 1991. Gas chemistry of vent fluids from 9°-10°N on the East Pacific Rise. American Geophysical Union Transactions, 72: 481.

Massoth, G.J., Butterfield, D., Lupton, J.E., McDuff, R.E., Lilley, M.D. and Jonasson, I.R., 1989. Submarine venting of phase-separated hydrothermal fluids at Axial Volcano, Juan de Fuca Ridge. Nature, 340: 702-7ß5.

Massoth, G.J., de Ronde, C.E.J., Lupton, J.E., Feely, R.A., Baker, E.T., Lebon, G.T., and Maenner, S.M., 2003. Chemically rich and diverse submarine hydrothermal plumes of the southern Kermadec volcanic arc (New Zealand). In: Larter, R. and Leat, P. (eds) Intra-Oceanic Subduction Systems: Tectonic and Magmatic Processes, Geological Society of London Special Publication 219: 119-139.

Metz, S. and Trefry, J.H., 1993. Field and laboratory studies of metal uptake and release by hydrothermal precipitates. Journal of Geophysical Research, 98: 9661-9666.

Michard, A., Albarede, F., Michard, G., Minster, J.F. and Charlou, J.L., 1983. Rare-earth elements and uranium in high-temperature solutions from East Pacific Rise hydrothermal vent field (13°N). Nature, 303: 795-797.

Michard, A., 1989. Rare earth element systematics in hydrothermal fluids. Geochimica et Cosmochimica Acta, 53: 745-750.

Miller, A.R., Densmore, C.D., Degens, E.T., Hathaway, F.C., Manheim, F.T., McFarlen, P.F., Pocklington, H. and Jokela, A., 1966. Hot brines and recent iron deposits in deeps of the Red Sea. Geochimica et Cosmochimica Acta, 30: 341-359.

Morton, J.L. and Sleep, N.H., 1985. A mid - ocean ridge thermal model: Constraints on the volume of axial heat flux. Journal of Geophysical Research, 90: 11345-11353.

Mottl, M.J., 1983. Metabasalts, axial hot - springs, and the structure of hydrothermal systems at mid-ocean ridges. Geological Society of American Bulletin, 94: 161-180.

Mottl, M.J. and McConachy, T.F., 1990. Chemical processes in buoyant hydrothermal plumes on the East Pacific Rise near 21°N. Geochimica et Cosmochimica Acta, 54: 1911-1927.

Oudin, E., 1983. Hydrothermal sulfide deposits of the East Pacific Rise (21°N) part I: descriptive mineralogy. Marine Mining, 4: 39-72.

Perfit, M.R., Fornari, D.J., Malahoff, A. and Embley, R.W., 1983. Geochemical studies of abyssal lavas recovered by DSRV Alvin from eastern Galapagos Rift, Inca Transform, and Equador Rift, 3. Trace Element abundances and pertogenesis. Journal of Geophysical Research, 88: 10551-10572.

Petersen, S., Herzig, P. M., and Hannington, M. D., 2000. Third dimension of a presently forming VMS deposit: TAG hydrothermal mound, Mid-Atlantic Ridge, 26°N. Mineralium Deposita, 35: 233-259.

Ritger, S., Carson, B. and Suess, E., 1987. Methane - derived authigenic carbonates formed by subduction-induced pore-water expulsion along the Oregon/Washington margin. Geological Society of American Bulletin, 98: 147-156.

Rona, P.A., 1988. Hydrothermal mineralization at ocean ridges. Canadian Mineralogist, 26: 431-465.

Rona, P.A. and Scott, S.D., 1993. A special issue on sea - floor hydrothermal mineralization: new perspectives. Canadian Mineralogist, 88: 1935-1976.

Rona, P.A., Hannington, M.D., Raman, C.V., Thompson, G., Tivey, M.K., Humphris, S.E., Lalou, C. and Petersen, S., 1993. Active and relict sea-floor hydrothermal mineralization at the TAG Hydrothermal Field, Mid-Atlantic Ridge. Economic Geology, 88: 1989-2017.

Sakai, H., Gamo, T., Kim, E-S., Tsutsumi, M., Tanaka, T., Ishibashi, J., Wakita, H., Yamano, M., and Oomori, T., 1990. Venting of carbon dioxide-rich fluid and hydrate formation in mid-Okinawa trough backarc basin. Science, 248: 1093-1096.

Seyfried, W.E.(jr.) and Mottl, M.J., 1982. Hydrothermal alteration of basalt by seawater under seawater-dominated conditions. Geochimica et Cosmochimica Acta, 46: 985-1002.

Seyfried, W.E.(jr.), Berndt, M.E. and Janecky, D.R., 1986. Chloride depletions and enrichments in seafloor hydrothermal fluids: Constraints from experimental basalt alteration studies. Geochimica et Cosmochimica Acta, 50: 469-475.

Seyfried, W.E.(jr.), Berndt, M.E. and Seewald, J.S., 1988. Hydrothermal alteration processes at mid - ocean ridges: constraints from diabase alteration experiments, hot - spring fluids and composition of the oceanic crust. Canadian Mineralogist, 26: 787 - 804.

Seyfried, W.E.(jr.), Ding, K., Berndt, M.E., and Chen, X., 1999. Experimental and theoretical controls on the composition of mid-ocean ridge hydrothermal fluids. In: Barrie, C.T., and Hannington, M.D. (eds) Volcanic-associated Massive Sulfide Deposits: Processes and Examples in Modern and Ancient Settings: Reviews in Economic Geology, 8: 181-200.

Shanks, W.C., III, 2001. Stable isotopes in seafloor

hydrothermal systems. In: Valley, J.W., and Cole, D.R., (ed) Stable Isotope Geochemistry, Mineralogical Society of American, Reviews in Mineralogy, 43: 469-526.

Spiess, F.N., Macdonald, K.C., Atwater, T., Ballard, R., Carranza, A., Cordoba, D., Cox, C., Diaz Gracia, V.M., Francheteau, J., Guerro, J., Hawkins, J.W., Haymon, R., Hessler, R., Juteau, T., Kastner, M., Larson, R., Luyendyk, B., Macdougall, J.D., Miller, S., Normark, W., Orcutt, J. and Rangin, C., 1980. East Pacific Rise. Hot springs and geophysical experiments. Science, 207: 1421-1433.

Stein, C.A. and Stein, S., 1994. Constraints on hydrothermal heat flux through the ocean lithosphere from global heat flow. Journal of Geophysical Research, 99: 3081-3095.

Stein, C.A., Stein, S. and Pelayo, A.M., 1995. Heat flow and hydrothermal circulation. In: Humphris, S.E. et al. (eds) Seafloor Hydrothermal Systems: Physical, Chemical, Biological and Geological Interactions, AGU Geophysical Monograph, 91: 425-445.

Suess, E. Bohrmann, G., Greinert, J., Linke, P., Lammers, S., Zuleger, E., Wallmann, K., Sahling, H., Dählmann, A., Rickert, D. and von Mirbach, N., 1997. Methanhydratfund von der FS Sonne vor der Westküste Nordamerikas. Geowissenschaften, 15: 194-199.

Suess, E., Bohrmann, G., von Huene, R., Linke, P., Wallmann, K., Lammers, S. and Sahling, H., 1998. Fluid venting in the eastern Aleutian subduction zone. Journal of Geophysical Research, 103: 2597-2614.

Sun, S.S., 1980. Lead isotopic study of young volcanic rocks from mid-ocean ridges, ocean islands arcs. Philosophical Transactions of the Royal Society of London, 297: 409-445.

Tivey, M.K., 1995. Modeling chimney growth and associated fluid flow at seafloor hydrothermal vent sites. In: Humhris, S.E. et al. (eds) Seafloor Hydrothermal Systems: Physical, Chemical, Biological and Geological Interactions, AGU Geological Monograph, 91: 158-177.

Tivey, M.K., Humphris, S.E., Thompson, G., Hannington, M.D. and Rona, P.A., 1995. Deducing patterns of fluid flow and mixing within the active TAG mound using mineralogical and chemical data. Journal of Geophysical Research, 100: 2527-2556.

Torres, M.E., Bohrmann, G. and Suess, E., 1996. Authigenic barites and fluxes of barium associated with fluid seeps in the Peru subduction zone. Earth and Planetary Science Letters, 144: 469-481.

Trefry, J.H., Butterfield, D.B., Metz, S., Massoth, G.J., Trocine, R.P. and Feely, R.A., 1994. Trace metals in hydrothermal solutions from cleft segment on the southern Juan de Fuca Ridge. Journal of Geophysical Research, 99: 4925-4935.

Tsunogai, U., Ishibashi, J., Wakita, H., Gamo, T., Watanabe, K., Kajimura, T., Kanayama, S., and Sakai, H., 1994. Peculiar features of Suiyo Seamount hydrothermal fluids, Izu-Bonin arc: Differences from subaerial volcanism. Earth and Planetary Science Letters, 126: 289-301.

Von Damm, K.L., Grant, B. and Edmond, J.M., 1983. Preliminary report on the chemistry of hydrothermal

solutions at 21° North, East Pacific Rise. In: Rona, P.A., Bostrom, K., Laubier, L. and Smith, L. (eds) Hydrothermal Processes at Seafloor Spreading Centers. Plenum Press: 369-390 pp.

Von Damm, K.L., Edmond, J.M., Grant, B. and Measures, C.I., 1985a. Chemistry of submarine hydrothermal solutions at 21°N, East Pacific Rise. Geochimica et Cosmochimica Acta, 49: 2197-2220.

Von Damm, K.L., Edmond, J.M., Measures, C.I. and Grant, B., 1985b. Chemistry of submarine hydrothermal solutions at Guayamas Basin, Gulf of California. Geochimica et Cosmochimica Acta, 49: 2221-2237.

Von Damm, K.L. and Bischoff, J.L., 1987. Chemistry of hydrothermal solutions from the southern Juan de Fuca Ridge. Journal of Geophysical Research, 92: 11334-11346.

Von Damm, K.L., 1988. Systematics of and postulated controls on submarine hydrothermal solution chemistry. Journal of Geophysical Research, 93: 4551-4935.

Von Damm, K.L., 1990. Seafloor hydrothermal activity: Black smoker chemistry and chimneys. Annual Reviews in Earth and Planetary Science, 18: 173-204.

Von Damm, K.L., 1995. Controls on the chemistry and temporal variability of seafloor hydrothermal fluids. In: Humhris, S.E. et al. (eds) Seafloor Hydrothermal Systems: Physical, Chemical, Biological and Geological Interactions, AGU Geological Monograph, 91: 222-247.

Von Huene, R., Klaeschen, D., Gutscher, M. and Frühn, J., 1998. Mass and fluid flux during accretion at the Alaskan margin. Geological Society of America Bulletin, 110: 468-482.

Wallmann, K., Linke, P., Suess, E., Bohrmann, G., Sahling, H., Schlüter, M., Dählmann, A., Lammers, S., Greinert, J. and von Mirbach, N., 1997. Quantifying fluid flow, solute mixing, and biogeochemical turnover at cold vents of the eastern Aleutian subduction zone. Geochimica et Cosmochimica Acta, 61: 5209-5219.

Wheat, C.G. and Mottl, M.J., 1994. Hydrothermal circulation, Juan de Fuca Ridge eastern flank: Factors controlling basement water composition. Journal of Geophysical Research, 99: 3067-3080.

Wolery, T.J. and Sleep, N.H., 1976. Hydrothermal circulation and geochemical flux at mid-ocean ridges. Journal of Geology, 84: 249-275.

Zierenberg, R.A., Shanks, W.C.I. and Bischoff, J.L., 1984. Massiv sulfide deposits at 21°, East Pacific Rise: Chemical composition, stable isotopes, and phase equilibria. Geological Society of America Bulletin, 95: 922-929.

Zierenberg, R.A., Koski, R.A., Morton, J.L., Bouse, R.M. and Shanks, W.C.III., 1993. Genesis of massive sulfide deposits on a sediment-covered spreading center, Escanaba Trough, southern Gorda Ridge. Economic Geology, 88: 2069-2098.

14 Gas Hydrates in Marine Sediments

GERHARD BOHRMANN AND MARTA E. TORRES

14.1 Introduction

Gas hydrates are naturally occurring ice-like crystalline compounds in which gases are trapped within a lattice of water molecules. The presence of gas hydrates is controlled by temperature, pressure and the availability of appropriate gases and water. The first discovery of gas hydrate goes back to 1810, with the pioneering synthesis of chlorine hydrate by Sir Humphrey Davy (Davy 1811). In the 1930s crystalline substances were observed to form spontaneously within natural gas pipelines in permafrost regions, and these deposits, which were clogging the pipelines, were identified as being hydrates of mixed hydrocarbon gases (Hammerschmidt 1934). The recognition that natural gas hydrates can block gas transmission lines, led the hydrocarbon industry to invest in efforts aimed at understanding gas hydrates, and thus begins the modern research in this subject.

Russian scientists (Vasil'ev et al. 1970) were the first to recognize that methane in natural systems could form gas hydrate deposits wherever the pressure and temperature conditions were favourable. These ideas were followed by discovery of gas hydrate, first in the permafrost regions of Russia (Makogon et al. 1971) and Canada's MacKenzie Delta (Bily and Dick 1974), and subsequently in sediments of the Caspian Sea and Black Sea (Yefremova and Zhizhchenko 1974). Interest in these deposits prompted the development of geophysical prospecting tools, which were used to predict the occurrence of gas hydrate in sediments of the Blake Ridge, of the western Atlantic Ocean (Stoll et al. 1971) and elsewhere (Shipley et al. 1979). In the early 1980s, hydrate was recovered from sediments of the Middle America Trench offshore Mexico by the Deep Sea Drilling project (Shipley and Didyk 1982). Since then, deep sea drilling has recovered hydrate from subsurface sediments along the Pacific and Atlantic continental slopes (Kvenvolden 1993). In addition, hydrate has been recovered from many near-surface environments along continental margins worldwide (Mazurenko and Soloviev 2003).

The number of hydrate publications, scientific sessions and workshops dedicated to gas hydrate research has increased substantially during the last 10-15 years, reflecting the development of a broad national and international hydrate research effort in this field. The interest in gas hydrates emerges from the awareness that these deposits may play significant roles in global and regional processes with societal and economic significance. A global hydrate assessment, although still uncertain, suggests that methane hydrates might represent an important future energy resource (Kvenvolden 1998; Collet 2002). In addition, other important hydrate questions that have attracted attention include: 1) Is there a feedback between methane hydrate stability and climate? 2) What is the role of methane hydrate in the carbon cycle? and 3) How much does gas hydrate contribute to seafloor stability on continental slopes?

The purpose of this chapter is to summarize some of the fundamentals of our current understanding of gas hydrate in marine sediments, its interactions with the environment, and recent findings from ongoing research programs that illustrate key aspects of gas hydrate dynamics. We start with general information on the structure and composition of gas hydrates and address their presence and distribution in the marine sediments based on their thermodynamic stability and environmental conditions. Because here we emphasize topics that are relevant to the scope of this textbook, we review the sources and migration mechanism of gases needed to stabilize the hydrate structure; the chemical and isotopic anomalies associated with hydrate formation; and the interaction of hydrates with fluid flow along continental margins.

14.2   Hydrate Crystal Chemistry and Stability of Gas Hydrates

14.2.1   Cages and Three Crystal Structures

Gas hydrates are non-stoichiometric, solid compounds similar to ice crystals (Sloan 1998). In these compounds, also called clathrates (latin clatratus for cage), water molecules form cage-like structures in which low molecular weight gases are enclosed as guest molecules (Fig. 14.1). The gas molecules interact with water molecules through van der Waals (nonpolar) forces. Since no bonding exists between the guest and host molecules, the guest molecules are free to rotate inside the cages, and this rotation can be measured by spectroscopic techniques (e.g. Gutt et al. 1999). Gas hydrate can contain different types of gas molecules in separate cages, depending on the gas composition in the environment of formation. Methane is the main gas in naturally occurring gas hydrates; however H_2S, CO_2 and, less frequently, other hydrocarbons, can also be found within the hydrate structure.

To date gas hydrates have been found to occur in three different crystal structures (Sloan 1998). Structures I and II both crystallize within a cubic system, whereas the third structure (also denominated H) crystallizes within a hexagonal system, analogous to water-ice (Fig. 14.1; Table 14.1). The structure of gas hydrate can be seen as a packing of polyhedral cages. Five types of hydrate cages are known, from which the simplest polyhedron is formed by twelve five-sided polygons (5^{12}) known as pentagonal dodecahedra. This cage is the smallest one that occurs in all three clathrate crystals (Fig. 14.1; Table 14.1). Larger diameter cages can be formed by adding two, four or eight hexagonal faces, and these are denoted as $5^{12}6^2$ in structure I, $5^{12}6^4$ in structure II, and $5^{12}6^8$ in structure H (Table 14.1). In addition, structure H has a medium-sized cavity with square, pentagonal and hexagonal faces ($4^35^66^3$). Figure 1 depicts the five cavities of all three structures that are known to occur naturally.

Structure I is most frequently observed. Its unit cell consists of 8 cages: 2 small (5^{12}) and 6 large cavities ($5^{12}6^2$). Inside each cavity resides a maximum of 1 guest molecule, such that 8 guests molecules are associated with 46 water molecules in structure I ($2[5^{12}]$ $6[5^{12}6^2]$ $46H_2O$). A unit cell of structure II consists of 24 cages, i.e. 16 small cavities (5^{12}) and 8 large ones ($5^{12}6^4$), which account for 136 water molecules ($16[5^{12}]$ $8[5^{12}6^4]$ $136H_2O$). Structure H forms a more complicated crystal composed of 3 small (5^{12}), 2 medium-sized ($4^35^66^3$) and 1 exceptionally large ($5^{12}6^8$) cavity associated with 34 water molecules ($3[5^{12}]$ $2[5^{12}6^4]$ $1[4^35^66^3]$ $34H_2O$).

When all hydrate cages are filled, the three crystal types have similar concentrations of 85 mol% water and 15 mol% guest molecules. Structure I hydrate with CH_4 and C_2H_6 has minimum (stoichiometric) hydration numbers of 5.75 and 7.67, respectively. Only large cavities in the Structure II hydrate are occupied with C_3H_8 (and i-C_4H_{10}), and such hydrates have a hydration number of 17 (e.g. Sloan 1998). However, hydration

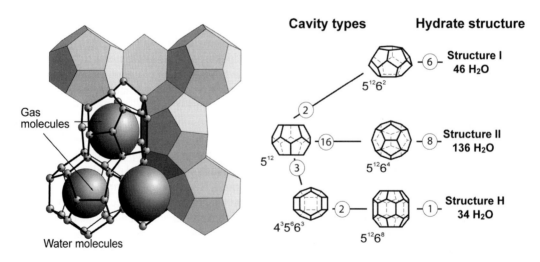

Fig. 14.1 Left: Gas hydrate of type structure I; small spheres are water molecules forming cages; large spheres are gas molecules. Right: Cage types and the number of individual cages forming the three common hydrate crystal structures. The circled numbers denote the numbers of the cages used to form the hydrate structure.

Table 14.1   Summary of some characteristics from the three crystal hydrate structures (from Sloan 1998).
* = Estimates of structure H cavities from geometric models.

Hydrate crystal structure	I		II		H		
Symmetry	Cubic		Cubic		Hexagonal		
Cell constant (Å)	12.03		17.31		a = 12.26; c = 10.17		
Cavity	Small	Large	Small	Large	Small	Medium	Large
Description of cavity	5^{12}	$5^{12}6^2$	5^{12}	$5^{12}6^4$	5^{12}	$4^35^66^3$	$5^{12}6^8$
Number of cavity / cell unit	2	6	16	8	3	2	1
ø cavity radius (Å)	3.8	4.33	3.91	4.73	3.9*	4.06*	5.71*
Coordination number	20	24	20	28	20	20	36
n H_2O/unit cell	46		136		34		

numbers of naturally occurring gas hydrates are highly variable, and are generally depleted in gas relative to its stoichiometric value. Samples from the Middle America trench off Guatemala and from the Green Canyon area of the northern Gulf of Mexico have hydration numbers of 5.91 and 8.2, respectively (Handa 1990). Matsumoto et al. (2000) reports a hydration number of 6.2 for hydrate from the Blake Ridge.

14.2.2   Guest Molecules

Gas molecules in sufficient amount are a prerequisite to stabilize the hydrate structures. In principle, the occupied hydrate cage is a function of the size ratio of the guest molecule to the host cavity. Figure 14.1 illustrates the guest/cavity size ratio for hydrates formed of a single guest component in either structure I or structure II (Sloan 1998). Molecules smaller than 3.5 Å will not stabilize hydrates and those larger than 7.5 Å are too large to fit in the cavities of structures I and II. Some molecules are too large to fit the smallest cage of each structure (e.g. C_2H_6 fits in $5^{12}6^2$ of structure I), whereas other molecules such as CH_4 and N_2 are small enough to enter both cavities (denoted as either 5^{12} and $5^{12}6^4$ in structure I). At pressures greater than 0.5 kbar two N_2 molecules can be accommodated in the $5^{12}6^4$ cage (Kuhs et al. 1996). The largest molecules determine which structure will form. Because propane and i-butane are present in many thermogenic natural gases, they will cause structure II to form. In such cases methane will occur in both cages of structure II and ethane will enter only the $5^{12}6^4$ cage of structure II.

Table 14.2 shows the size ratio of several gas molecules within each of the four cavities of structures I and II. A ratio of molecule to cage size of approximately 0.9 is necessary for stability of a hydrate composed of a single gas. When the size ratio exceeds unity, the gas will not fit within the cage structure and hydrate

will not form. When the ratio is significantly less than 0.9 the molecule cannot lend significant stability to the cage (Sloan 1998).

Structure I, which is by far the most commonly found in marine deposits, contains small guest molecules with diameters ranging from 4 to 5.5 Å.

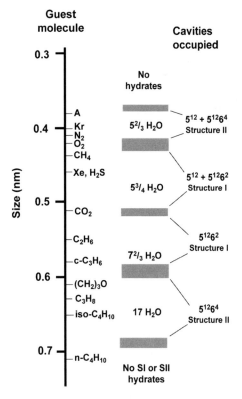

Fig. 14.2 Guest molecules versus hydrate cage size range (from Sloan 1998). Left line shows the size of typical hydrate-forming guest molecules. The number of water molecules in gas hydrates shown, corresponds to single guest gas occupants listed on the left. The related type of structures formed are listed on the left. As an example, methane has a typical hydration number of $5^3/_4$ and occupies both cages of structure I.

Table 14.2 Ratios of molecular diameters (obtained from von Stackelberg and Müller 1954) to hydrate cavity diameters for various gases, including those commonly found in natural gas hydrate (from Sloan 1998). [F] = indicates the cavity occupied by a single guest.

Molecule	Guest diameter Å	Structure I		Structure II	
		5^{12}	$5^{12}6^4$	5^{12}	$5^{12}6^4$
N_2	4.10	0.804	0.700	0.817 [F]	0.616 [F]
CH_4	4.36	0.855 [F]	0.744 [F]	0.868	0.652
H_2S	4.58	0.898	0.782	0.912	0.687
CO_2	5.12	1.00	0.834	1.02	0.769
C_2H_6	5.50	1.08	0.939 [F]	1.10	0.826
C_3H_8	6.28	1.23	1.07	1.25	0.943 [F]
$i\text{-}C_4H_{10}$	6.50	1.27	1.11	1.29	0.976 [F]
$n\text{-}C_4H_{10}$	7.10	1.39	1.21	1.41	1.07

Structure I cages can therefore enclose gas molecules that occur naturally in marine sediments and are smaller in diameter than propane, such as CH_4, CO_2 or H_2S. The natural occurrence of this crystal structure depends on the presence of biogenic gas in sufficient amounts, as commonly found in sediments of the ocean floor underlying areas of high biologic productivity. Cubic structure II generally occurs with guest molecules ranging between 6-7 Å. Hence, it contains natural mixtures of gases with molecules bigger than ethane and smaller than pentane, and it is therefore usually confined to areas where a thermogenic gas is present in the sediment. Hexagonal structure H may be present in either environment, but only with mixtures of both small and very large (8-9 Å) molecules, such as methylcyclohexane.

The smallest guest molecules that form hydrate structure II have diameters smaller than 4 Å, (e.g. Ar, Kr, O_2 and N_2). Such nitrogen and oxygen clathrates are known as air clathrates, and have been observed in ice-cores from Antarctica and Greenland below ~ 1100 m. In these samples, individual air bubbles have reacted with polar ice under high pressure to form clathrates (Shoji and Langway 1982).

14.2.3   Stability and Phase Boundaries of Gas Hydrates

The presence of gas hydrates is controlled by several factors, among which, temperature, pressure, ionic strength of the water and gas composition and abundance are key parameters (Sloan 1998). The pressure/temperature conditions required for pure methane hydrate stability are illustrated in Fig. 14.3. In this case, methane hydrate is stable at temperatures higher than 15°C only at high pressures (> 10 MPa). At lower pressures, the stability of methane hydrate requires colder temperatures (e.g. for P < 6 MPa; T < 10°C). In the pressure/temperature field, the phase boundary is determined by the gas composition and also by the ionic strength of the water. The presence of CO_2, H_2S, ethane and/or propane will have the effect of shifting the stability curve to a higher temperature at a given pressure, increasing the stability of methane hydrate. The presence of dissolved ions in the pore fluids, on the other hand, inhibits the stability of hydrate. There is a −1.1°C offset in dissociation temperature of methane hydrate in 33% NaCl, relative to that of hydrate formation in pure water (e.g. Dickens and Quinby-Hunt 1994). Thus, an increase in salinity of the fluids from which the hydrate is forming shifts the phase boundary to the left (Fig. 14.3).

Accurate and precise prediction of the P/T conditions for natural gas hydrate stability is a field of active research, and numerous methods for predicting methane hydrate stability can be found in the literature (summarized in Sloan 1998). Dickens and Quinby-Hunt (1994) estimate the P/T conditions for hydrate stability by interpolating experimentally determined dissociation data. Since their experiments were conducted in both seawater and freshwater matrices, their results are useful in evaluating the effects of pore fluid salinity. Other methods to estimate the stability conditions are based on minimizing the Gibbs Free Energy of the system. The most commonly used of these computer-based methods is the Sloan (1998) PC-DOS program CSMHYD, which allows for stability estimates at varying salinities. Simpler calculation methods are also available for the rapid estimation of hydrate formation conditions (Carroll 2003).

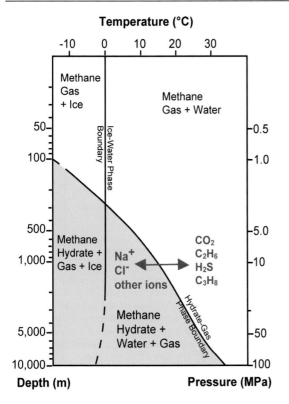

Temperature (°C)

Fig. 14.3 Phase diagram showing the boundary between methane hydrate (in yellow) and free methane gas (white) for a pure methane/H₂O system. Addition of ions shifts the boundary to the left, decreasing the P/T stability field. The presence of gases like carbon dioxide, hydrogen sulphide or other high-molecular hydrocarbons shifts the curve to the right, thus increasing the P/T field in which methane hydrate is stable (after Kvenvolden 1998).

14.3 Hydrate Occurrence in the Oceanic Environment

14.3.1 Gas Hydrate Stability Zone in Marine Sediments

Gas hydrates form wherever appropriate physical conditions exist and concentrations of low molecular weight gases, mostly methane, exceed saturation. The P/T factors for the presence of methane hydrates (Fig. 14.3) are present in marine sediments as shown by the phase boundary in Fig. 14.4. The dashed line shows a typical temperature profile through the water column in the Atlantic Ocean. Near surface temperatures are too warm and pressures too low for methane hydrate to be stable. Below the major thermocline there is change in the temperature gradient, and the temperature profile intersects the phase boundary at ~450 m water depth, which defines the upper limit for methane hydrate stability in that part of the ocean. If methane is sufficiently abundant, methane hydrate would form. However, since the density of hydrate is around 0.913 g cm⁻³ (Sloan 1998) any crystalline hydrate that may form in the water column (e.g. at sites of methane discharge) will rise due to its relative buoyancy and it will dissociate when it reaches depths above its stability field. However, if methane hydrate forms within the sediment pore space, it will be bound in place. If water temperatures are colder the upper limit for methane hydrate is shallower. This limit of

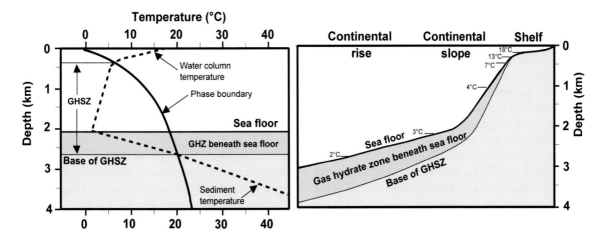

Fig. 14.4 Left: Stability field of pure methane hydrate at normal seawater salinity, as defined by temperature and pressure expressed as water depth. Intersections of the temperature profiles (stippled lines) with the phase boundary (heavy line) define the area of the gas hydrate stability zone (GHSZ). Right: Inferred thickness of the gas hydrate zone in sediments at a schematic continental margin assuming a typical geothermal gradient of 28°C km⁻¹. Typical bottom water temperatures are marked, and range from 18°C on shallow shelf regions to 2°C at the bottom of the continental rise (after Kvenvolden and McMenamin 1980).

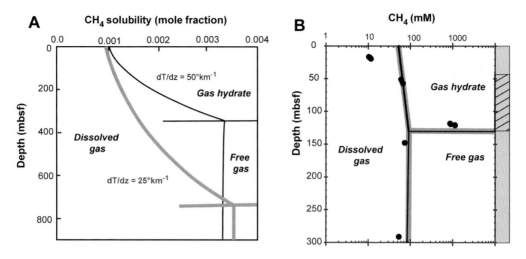

Fig. 14.5 A. Methane solubility as a function of depth in the sediment (mbsf = meters below seafloor) based on thermodynamic functions and assuming two different geothermal gradients of 50° km⁻¹ (black line) and 25° km⁻¹ (gray line). It illustrates the effect of temperature changes on the vertical gradient of methane solubility and on the depth of the GHSZ, which is defined by the discontinuity in the slope of the gas solubility curves and demarked by horizontal lines. In this example the pressure is assumed to be hydrostatic. The water depth is assumed to be 2000 m, and bottom water temperature used is 2.5° C (from Zatsepina and Buffet 1997). **B.** Approximate phase boundaries where dissolved gas, gas hydrate and free gas are predicted for Site 1245, drilled on 880 meters of water depth during ODP Leg 204 offshore Oregon. Uncertainties (~30%) in the position of these boundaries result from variations in subsurface thermal gradient, gas composition and pore fluid salinity. The closed circles represent methane concentration in sediments recovered at in situ pressure, revealing that there is not enough methane in the upper 45 meters to support hydrate formation, thus defining the gas hydrate occurring zone (GHOZ) as the interval between 45 and 135 mbsf. These inferences are consistent with observations of hydrate in the sediment, as indicated by the shaded region in the column to the right (from Tréhu et al. 2003).

hydrate occurrence in sediments is deeper in closed ocean basins where the temperature of the bottom water is higher. For example, within the Black Sea where the bottom water temperature is 9°C, the upper limit of hydrate stability is around 700 m water depth (Bohrmann et al. 2003). In contrast, in the polar oceans, gas hydrate can be stable in 300 m of water (Kvenvolden 1998).

The local geothermal gradient in marine settings determines the temperature profile below the sea-floor (Fig. 14.4 dashed line in the sediment sequence). As temperature in the sediments increases with depth, the sediment temperature will eventually get high enough to cross the phase boundary, such that gas hydrate will no longer be stable beneath this depth. Other factors like the gas composition and salt content of the pore water influence the precise location of the lower boundary of the gas hydrate stability zone (GHSZ). Thus, the base of the GHSZ is itself a phase boundary. Since the geothermal gradient is often quite uniform across broad regions beneath the seafloor, the thickness of the GHSZ is quite constant for a given water depth. However, a change in water depth will influence the thickness of the hydrate stability zone (Fig. 14.4). Due to the P/T conditions for hydrate

stability, the thickness of the GHSZ can reach 800 to 1000 m below seafloor in deep water areas, and the base of the GHSZ will shoal up as water depth decreases (Fig. 14.4).

Even though P/T conditions in most of the ocean floor lie within the hydrate stability field, no such deposits are found in the abyssal plain because there is not enough gas in these sediments to stabilize the hydrate structure. This fact illustrates the third fundamental requirement for gas hydrate formation. In addition to moderately high pressures and low temperatures, gas hydrates will only form if the mass fraction of methane exceeds its solubility. Methane solubility itself is a function of pressure and temperature. At depths within the GHSZ, the equilibrium concentration in the presence of hydrate decreases almost exponentially towards the seafloor (Fig. 14.5A). At greater depths, the equilibrium is defined between aqueous solution and free gas. In Figure 14.5B this relationship is shown for sediments recovered from the flanks of Hydrate Ridge (ODP Site 1245), in the Cascadia margin. Here the entire sediment column above 134 meters lies within the GHSZ; however, sediments above 40 meters do not have enough methane to support hydrate formation (Tréhu et al.

2003). Consistent with inferences based on methane concentration measured on cores collected at in situ pressures, gas hydrate in these sediments is only present between 45 and 134 meters, in what is known as the gas hydrate occurrence zone (GHOZ).

14.3.2   Seismic Evidence for Gas Hydrates

The first indications of methane hydrate in marine sediments were based on the observation of a seismic reflection called "bottom-simulating- reflector" or BSR, because it approximately mimics the sea-floor (Shipley et al. 1979). The BSR cuts across reflections of stratigraphic origin, making it readily apparent in marine seismic records (Fig. 14.6). This reflection occurs approximately at the depth where the base of the gas hydrate stability zone is predicted based on thermodynamic equilibria (e.g. Tucholke et al. 1977; Shipley et al. 1979; Hyndman and Spence 1992). Because of the temperature dependence of hydrate stability, the depth of the BSR provides a means of mapping the thermal gradient and heat flux in the overlying sediment (e.g. Davis et al. 1990). The negative polarity of this reflection indicates that it results from a decrease in acoustic impedance (defined as a product of density and seismic velocities) with depth. Fig. 14.6 illustrates the presence of a BSR on Blake Ridge, which demarks the impedance contrast between gas hydrate-cemented sediments above the BSR and the sequence below it, where free gas is present.

Although some details of the seismic reflection properties are not yet fully understood, it appears that the strength and the characteristics of the BSR is determined by the presence of free gas below the gas hydrate zone (Paull et al. 1996). The presence of free gas represents a very large change in seismic velocity, and therefore produces a very strong and sharply defined reflection. Theoretical models (Xu and Ruppel 1999) and synthetic studies (Wood and Ruppel 2000) indicate that the BSR is not a necessary condition for the presence of hydrate, as it only occurs when there is free gas beneath the distinct gas hydrate phase boundary. If there is no free gas below a deposit of gas hydrate, there will be no BSR. Indeed, sediments containing gas hydrate have been recovered from areas where there is no BSR (Mathews and von Huene 1985).

Models on gas hydrate concentration based on analyses of the BSR properties depend on a number of poorly constrained parameters, and thus these geophysical estimates need to be calibrated against direct measurement of hydrate abundance. The *Ocean Drilling Program* has sampled various BSR horizons on the continental slopes around the Pacific Rim (e.g. Peru, Chile, Costa Rica, Oregon/Washington, Japan) and on the passive US Atlantic margin (Blake Ridge) with the aim of understanding gas hydrate characteristics, distribution and concentration in continental margin settings. These efforts, in particular recovery of samples under in situ pressure and calibration of

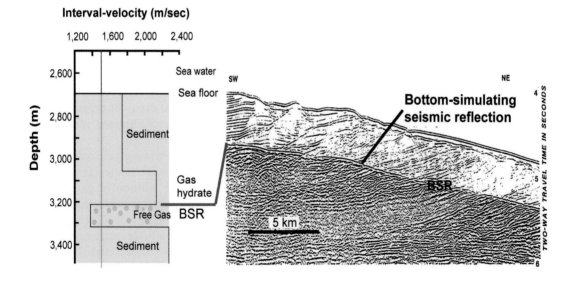

Fig. 14.6 Seismic record from Blake Ridge (Shipley et al. 1979), showing a distinct reflection, known as bottom simulating reflection (BSR), which indicates the presence of methane hydrate within sediments (right). Below the BSR there are strong reflections caused by free gas in the pores. A seismic velocity model (left) shows the strong contrast of velocity across the BSR.

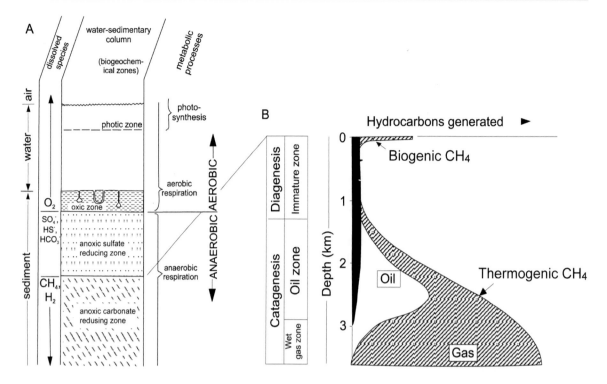

Fig. 14.7   A: An idealized cross section of a marine organic rich sedimentary environment, showing biogeochemical zones in ecological succession (from Claypool and Kaplan 1974). **B:** Hydrocarbon generation by diagenesis and catagenesis processes as a function of depth (from Tissot and Welte 1992).

various proxies for hydrate abundance (e.g. Tréhu et al. 2004a) will continuously improve our imperfect knowledge of the distribution of gas hydrate in the seafloor.

14.3.3   Generation of Gases for Hydrate Formation

In their classic work on the origin and distribution of methane in marine sediments, Claypool and Kaplan (1974) place biogenic methane generation within the ecological succession of microbial ecosystems in the marine sedimentary environment (Fig. 14.7A). These zones are characterized by successively less efficient modes of respiratory metabolism, which are interlinked by microbially-induced environmental changes: one microbe's metabolic waste serves as substrate for another organism. Details of the microbiological pathways during early organic matter diagenesis are covered in chapter 4. Here we focus on the generation of methane needed for gas hydrate formation.

Biogenic methane is produced as an end product of the metabolism of a diverse group of

obligate anaerobic archaea (killed by even traces of oxygen), generally known as methanogens. These organisms can live in a wide range of temperature, salinity and pH, but are limited in the substrates they can utilize for growth. The most important substrates for bacterial methanogenesis are acetate (acetoclastic methanogenesis) and $H_2 : CO_2$ (carbonate reduction). A detailed description of the pathways involved in methanogenesis from the bacterial decay of organic matter in marine and freshwater sediments is given by Wellsbury et al. (2000).

Deep ocean sites containing gas hydrate have been analyzed to determine bacterial numbers, activity rates, cultural metabolic groups and estimates of biodiversity using molecular genetic analyses (Reed et al. 2002; Colwell et al. 2004). Bacterial population usually decreases in number with increasing depth (Wellsbury et al. 2000), but significant bacterial counts and activities have been measured within and beneath the GHSZ in Blake Ridge and Hydrate Ridge sediments (Wellsbury et al. 2000; Colwell et al. 2004).

Deeper in the sediment, thermal alteration of organic matter generates methane and higher order hydrocarbons by catagenesis (Fig. 14.7B). Catagenesis occurs within the temperature range of 50° to 200° C, and gases (methane to butane) are produced at rates that are proportional to temperature.

At typical oceanic geothermal gradients of 20° to 50° C km⁻¹, sediment depths larger than 1km are required to produce significant amounts of gas by thermochemical action. Because thermogenic gas generation occurs at temperatures significantly deeper than those found within the GHSZ, high concentration of thermogenic gases within the GHSZ generally indicates the existence of a hydrocarbon migration pathway.

Biogenic and thermogenic gases can usually be distinguished on the basis of chemical and isotopic composition. Biological gas is dominantly composed of methane, which is depleted in ^{13}C relative to thermogenic methane (Whiticar 1999), as shown in Figure 14.8A. Methane derived from $H_2 : CO_2$ is in general more depleted than that derived from acetate. The hydrogen isotope signature may also provide information on the metabolic pathways, as acetate fermentation yields methane with a δD value lower than –250 ‰, whereas carbonate fermentation leads to δD values ranging from –150 to –250 ‰ (Whiticar et al. 1986).

Isotopic discrimination, nevertheless, should be used with caution. Various environmental factors such as substrate limitation and temperature may obscure the $\delta^{13}C$ distinction between thermogenic and biogenic sources. In addition,

laboratory experiments have shown that during acetate methanogenesis some of the methyl hydrogen atoms can exchange with water, affecting the δD of the methane produced (de Graaf et al. 1996).

The ratio of methane (C_1) to heavier hydrocarbons, usually expressed as the sum of ethane and propane ($C_2 + C_3$), also provides information on the methane source. Biogenic gas consists predominantly of methane and typically has values for the $[C_1 / (C_2 + C_3)]$ ratio that are greater than 10^3. In thermogenic gas, this ratio is usually less than 100 (Bernard et al. 1976). Although recent studies indicate that bacterial activity can indeed generate higher level hydro-carbons (C_2 to C_4), they do not occur at high enough concentration. In Figure 5B, the hydro-carbon ratio is plotted against the methane isotopic composition, showing the thermogenic versus biogenic gas fields.

Elemental and isotopic analyses of hydrate samples from a variety of settings show that microbial activity is the dominant methanogenic pathway in marine sedimentary environments, such as Blake Ridge (Dickens et al. 1997), Hydrate Ridge (Suess et al. 2001), Nankai Trough (Takahasi et al. 2001), Congo-Angola basin (Charlou et al. 2004) and the Sea of Okhotsk (Ginsburg et al. 1993). Hydrates with thermogenic methane have been recovered from the Gulf of Mexico (Brooks et al. 1984) and the Caspian Sea (Ginsburg et al. 1992).

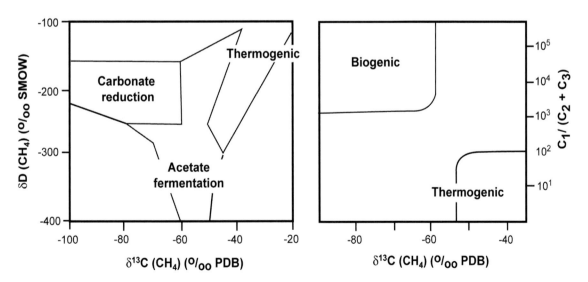

Fig. 14.8 Discrimination of biogenic and thermogenic methane sources based on **A**. The carbon and hydrogen isotopic composition of the methane (after Schoell 1988), and **B**. The ratio of methane (C_1) to higher hydrocarbons ($C_2 + C_3$) plotted against the carbon isotopic composition of methane (from Claypool and Kvenvolden 1983).

14.3.4   Methane Transport and Hydrate Formation

If methane, either from biogenic or thermogenic sources, is present in high enough concentration to stabilize the hydrate structure at thermodynamically favourable conditions (Figs. 14.3 and 14.4), it will combine with water to form hydrate. For methane hydrate to occur, the rain rate of carbon to the seafloor must be high enough to supply the required methane via degradation of organic matter in the sediment. Hydrate stability requires gas concentration in the hydrate at least two orders of magnitude greater than gas solubility in the liquid phase. Thus, methane generation and transport processes are key factors for constraining global hydrate inventories.

Hydrate Formation by in situ Biogenic Methane Generation and Transport in Advecting Fluids

The amount of biogenic methane is essentially controlled by both the availability and reactivity of organic matter in the upper hundreds of meters of the sedimentary sequence. Davie and Buffett (2001, 2003) demonstrated the critical need for quantitative models of biogenic methane production to describe the distribution of gas hydrate in the top few hundred meters of sediment. Key parameters are rates of sedimentation, quality and quantity of the organic matter and biological activity rates. They show that hydrate accumulation from in situ production in sediment with a TOC of 1.5%, will be less than 7% of the pore volume

If in situ production of biogenic methane is not adequate to support observed accumulations within the GHSZ, then additional methane must migrate from below. Paull et al. (1994), proposed a mechanism to concentrate methane via recycling at the base of the GHSZ in the Blake Ridge hydrate-bearing province. Progressive burial and subsidence through geologic time shifts the base of the GHSZ upward, so that deep-seated hydrate decomposes. As hydrate dissociates, the methane solubility is surpassed, and free gas permeates fissures in the overlying hydrate stability layer, enhancing gas hydrate contents via precipitation of the "recycled" methane.

Davie and Buffett (2001) also show that both in situ methane production and transport in upward migrating saturated fluids are needed to explain the dissolved chloride profiles observed in Blake Ridge sediment. Similarly, Hensen and Wallmann (2005) show that, although organic carbon degradation in the upper sediments of the Costa Rica margin can account for 0.4 to 1.1 % of hydrate content of the sediment, it alone cannot explain the hydrate distribution in this region. Furthermore they show that fluid flow may increase the total amount of hydrate that can be formed from the organic reservoir in this margin by more than 50%.

Fluid flow can scavenge methane from a broad region, thus it is expected that active margins with pervasive fluid transport would have higher abundance of gas hydrate. Nevertheless, even the small rates of fluid flow in passive margins, play a controlling role on the accumulation of gas hydrate (Egeberg and Dickens 1999). In fact, using a mechanistic model for the distribution of hydrate in marine sediment, Buffet and Archer (2004) conclude that the global inventory of gas hydrate is particularly sensitive to both, methane generation from organic matter and the rate of fluid flow.

Methane Transport in the Gas Phase

Most disseminated hydrate in marine sediment is thought to occupy less than 8% of the pore space of sediments integrated over the GHSZ. However, there are regions where massive hydrate is known to form near or at the seafloor. These shallow hydrate deposits are usually associated with areas of fluid venting and gas ebullition (Mazurenko and Soloviev 2003). Geochemical modelling of the shallow hydrate at the summit of southern Hydrate Ridge demonstrates the need for methane transport in the gas phase. Because of the low solubility of methane in water, advection of methane-saturated water is not enough to sustain the rapid hydrate growth in this system (Torres et al. 2004). In general, methane hydrate will only form large concentrated deposits where gas flow is present.

Methane concentration increases with depth in the sediment due to a combination of processes including microbial generation, methane recycling at the base of the GHSZ, and thermochemical generation at depth. When methane concentration in the pore water exceeds saturation, methane gas will exolve. However, the difficulty of nucleating bubbles of small size in fine-grained porous media can lead to significant supersaturations. Clennell et al. (2000) provide a comprehensive review of processes involved in movement of methane in marine sediment, with an emphasis on how porosity, pore size distribution and permeability of the sediment control the rates and mode of transport of gas. They discuss issues associated with capillary theory, multi-phase flow, invasion percolation, catenary transport, and flow in faults and fractures, which are important to fully understand gas hydrate formation dynamics in marine systems, but are beyond the scope of this book. Here we provide a simplified overview of two mechanisms that are important in gas transport, but refer the reader to Clennell et al. (2000) for a more quantitative treatment of these processes.

Gas Migration Induced by Diapirism

In regions of known diapirism, when overpressure builds up due to gravitational or fluid loading, gas migration can occur via faults and fractures. Here a combination of overpressure and the buoyancy of expanding gas drive the flow. If the flow reaches the surface, a mud volcano will form. The high advective rates transport methane bearing fluids, and relatively high temperatures and sometimes enhanced salinities preclude gas hydrate formation during methane migration to the seafloor. These characteristics are typical in regions such as the Gulf of Mexico (Ruppel et al. 2005), where overpressured fracture zones that surround moving salt diapirs provide active conduits for vertical migration from deep reservoirs to shallow subsurface (e.g. Sassen et al. 1994). Carbon elemental and isotopic analyses demonstrate high input of thermogenic methane in this region (Sassen et al. 1994). Other examples exist in the Eastern Mediterranean (De Lange and Brumsack 1998) and the Black Sea (Bohrmann et al. 2003). Another region of shallow hydrate formation associated with mud volcanism is the Håkon Mosby mud volcano in the Norwegian-Greenland Sea. Mud flow in the volcano is thought to be driven by the rise of lower density pre-glacial biogenic silica oozes buried beneath higher density glacial marine sediments. The methane in the hydrate here has a mixture of thermogenic and biogenic sources (Lein et al. 1999). A temperature model (Fig. 14.9) has been used to show how these fast-rising hot fluids serve as a methane transport mechanism to the seafloor, where hydrate content ranges from 10-20% to 0% by weight (Ginsburg et al. 1999).

Because fluid migration in diapir systems bring large amounts of gas to very shallow sub-bottom depths, gas hydrate formation in these soft sediments can create its own space by deforming the surrounding matrix (Bohrmann et al. 1998, Clennell et al. 1999; Torres et al. 2004). Thus, gas hydrate associated with rapid transport along faults and fractures are generally more localized and massive than biogenic deposits commonly found dispersed within the sediment.

Gas Pressure Driven Flow

Another driving force for methane transport is the generation of critical pressures in the gas phase (Flemings et al. 2003; Tréhu et al. 2004). Interconnection of gas-filled pores below the GHSZ transmits hydrostatic pressures from greater depths because of the low density of the gas phase. The excess (non-hydrostatic) pressure at the top of the gas layer may be sufficient to

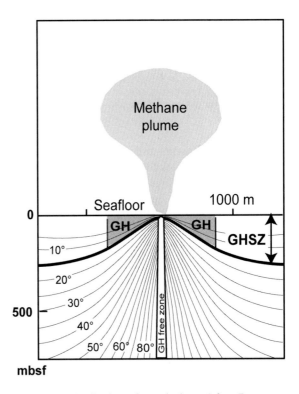

Fig. 14.9 Distribution of gas hydrate (after Egorov et al. 1999) superimposed on a schematic vertical model of the temperature field (after Ginsburg et al. 1999) in the Håkon Mosby Mud Volcano. The gas hydrate stability zone (GHSZ, shown by bold lines) is determined by pressure and temperature conditions; the zone of gas hydrate (GH) accumulation depends on both the thermal gradient and the flux rate of methane.

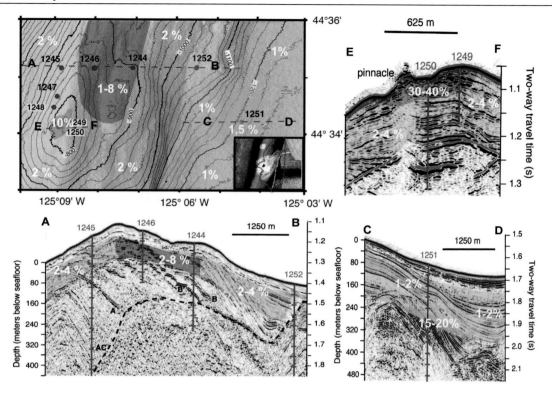

Fig. 14.10 Average gas hydrate concentrations in sediments from southern Hydrate Ridge deduced by drilling during ODP Leg 204 using a multi proxy approach (from Tréhu et al. 2004a). Upper left: Bathymetric map of the region studied during ODP Leg 204 and the lateral extend of zones of different gas hydrate content, estimated by averaging the data from the sea floor to the BSR. Location of seismic profiles and the drill sites are also shown. Gas hydrate concentrations as percentage of pore space shown in white bold numbers were estimated as average concentrations in the GHSZ.

fracture the sediments and drive gas towards the seafloor. This process has been postulated for both passive (Flemings et al. 2003) and active (Tréhu et al. 2004b) regions, where the volume fraction of gas is $\geq 10\%$ (Flemings et al. 2003).

In their analyses of past and future state of the hydrate reservoir, Buffett and Archer (2004b), suggest that if elevated gas pressures do occur as a transient response to warming, a rapid release of methane may be triggered by the development of critical pressures in the gas phase. Critical gas pressure below the base of the gas hydrate stability zone can trigger vertical migration of free gas to the seafloor.

14.3.5 Gas Hydrate Accumulation in Sediments and Fabric of Natural Gas Hydrates

Drilling of marine sediment cores as well as seafloor sampling by research vessels confirmed the presence of gas hydrate in sediments defined by the stability field as described in section 14.3.1. Conventional research vessels are only able to sample shallow sediments close to the seafloor, thus drilling campaigns are needed to investigate the distribution of gas hydrates deeper within the stability field. Because gas hydrates decompose rapidly when removed from the high-pressure, deep-water environments in which they form, the in situ distribution of gas hydrate must be estimated using various proxy techniques, each of which may have different sensitivity and spatial resolution (Tréhu et al. 2004a). Leg 164 of the Ocean Drilling Program (ODP) drilled several sites on the Blake Ridge, in the first dedicated academic effort to investigate naturally occurring gas hydrates in marine sediments (Paull et al. 1996). Estimates made using diverse gas-hydrate proxies revealed that gas hydrate occupies ~ 1% to 10% of the pore space in the sediment interval from 200 to ~450 mbsf. The hydrate occurs dispersed within the pore-space of fine-grained sediments or within fractures and faults (Paull et al. 1996). The distribution of fine grained gas hydrate within the lithologically uniform drift sediments of the

Blake Ridge was surprisingly heterogeneous and could not be explained in detail, except for the observation of two weakly defined zones where higher hydrate concentrations may indeed be caused by small differences in lithology.

Researchers involved in ODP Leg 204 generated the first high-resolution data set on the three-dimensional distribution of gas hydrate within Hydrate Ridge, in the Cascadia subduction zone (Tréhu et al. 2003). Several gas hydrate proxies were combined, and thus, the problem of spatial under-sampling inherent in methods traditionally used for estimating the gas hydrate was overcome (Tréhu et al. 2004a). The average gas hydrate content of sediments within the gas hydrate stability zone was estimated to be 1-2% of the pore space. Patchy zones of locally higher concentrations on the ridge flanks occur below ~ 40 mbsf, are structurally and stratigraphically controlled and occupy up to 20% of the pore space (Fig. 14.10). In contrast to this overall hydrate distribution, a high average gas hydrate content of 30-40% of pore space was found on the upper 30-40 mbsf at the ridge summit. Cores containing hydrate in massive chunks, lenses, plates and nodules, where recovered from an area where there is persistent and vigorous venting of methane gas (Heeschen et al. 2003).

A variety of gas hydrate samples were recovered from the southern summit of Hydrate Ridge by deploying a TV-guided grab on visible hydrate outcrops (Suess et al. 1999, 2001). Due to a self-preservation effect (Yakushev and Istomin 1992), massive hydrate shows little indication of decomposition, and samples from the inner part of the TV-grab appeared to be relatively pristine (Fig. 14.11). Scanning electron microscopic work revealed that only in very porous samples there was water-ice formation (Fig. 14.11B; Kuhs et al. 2004). On a macroscopic scale, pure white gas hydrate occurs in layers or joints several millimeters to centimeters thick. The layers are generally oriented parallel to the bedding planes and in some cases very massive hydrates of up to 10 cm in thickness have been observed (Fig. 14.11). Gas hydrate either fills large pore space in fractures or joints, or it creates its own space by fracturing or pushing apart the sediment framework during growth, most often along bedding planes. The result of such an active crystal growth is that the original sediment fabric is disturbed and mud clasts are formed. In many cases internal brecciation of the sediment was observed in which the angular edges of the clasts often fit with the edges of neighboring clasts (Fig. 14.11A).

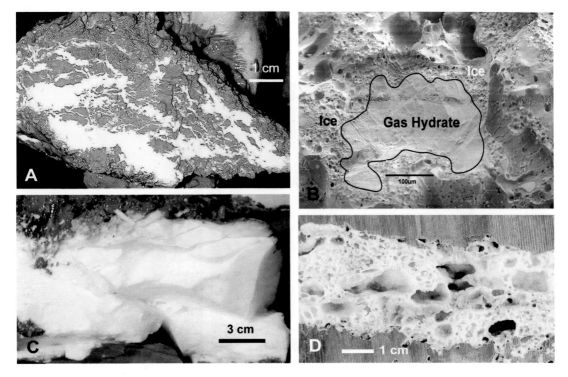

Fig. 14.11 Hydrate fabrics typical for shallow gas hydrate specimens (A, C and D): sediment-hydrate interlayering (A), pure dense hydrate layer (C), and highly porous bubble-shaped framework (D) B: Field-electron scanning micrograph of hydrate surrounded by bubble-shaped ice.

The internal fabric of pure gas hydrate has a peculiar structure with pores that result from rising methane gas. Such pores occur in variable sizes, and in some specimens very large pores of up to 3-4 cm in diameter can be observed (Fig.14.11D). The fabric is similar to that of gas hydrates experimentally formed on the sea-floor (Brewer et al. 1997). There are several lines of evidence that support migration of methane gas from a reservoir located beneath the GHSZ, which either turns into macroscopic porous gas hydrates or escapes at the seafloor. A variety of mechanisms are currently under investigation to determine how free gas pass through the gas hydrate stability zone. Gas may migrate through fractures or along tensional faults, in which all water is trapped in the gas hydrates, or gas hydrate formation may be inhibited by capillary forces or by localized high salinity zones. The free gas stream may move upwards very fast up to an area where conditions are favorable to form gas hydrates. Hydrate formation may plug up the migration conduits, and as high gas pressure builds up, the gas may be rerouted into soft sediment layers. The dynamic processes that interact with a complicated plumping system may be responsible for the large variety of gas hydrate and sediment fabrics observed.

Macroscopic hydrate fabrics deeper within the stability zone are very different from the near-surface deposits because at depth hydrate formation is constrained by the pore space in which hydrate precipitates. Abegg et al. (submitted) have investigated whole-round sediment samples from hydrate intervals, which were frozen in liquid nitrogen immediately after recovery. Nearly 60 frozen hydrate samples, covering a wide depth range of the gas hydrate occurrence zone (GHOZ) of southern Hydrate Ridge, were investigated by X-ray computerized tomography (CT). All sub-surface hydrate samples appear as veins or veinlets with dipping angels of more than 30° up to vertical dipping. Such hydrates are clearly precipitates filling tectonic fractures and/or faults deeper in the sediments (Fig. 14.12), where the geo-mechanical properties of the sediment preclude massive hydrate formation. These structures are in clear contrast to those of the gas hydrate that outcrops at the seafloor (Fig. 14.11).

14.4   Pore Water Anomalies Associated with Gas Hydrate Formation and Decomposition

Gas hydrate formation involves the removal of water molecules from the surrounding pore water, as they are sequestered in the clathrate lattice. Removal of water, with the exclusion of the dissolved ions, leads to changes in the concentration of salts in the pore water. Because chloride is an abundant and usually conservative ion in pore waters of shallow marine sediment, changes in dissolved chloride content are

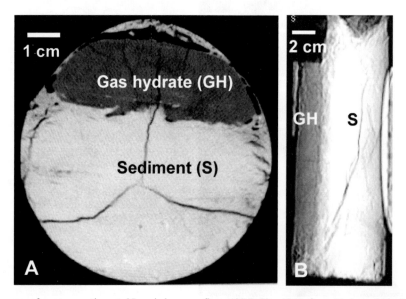

Fig. 14.12 CT-images of a core section at 87 m below sea-floor (ODP Site 1248 from Hydrate Ridge) showing that gas hydrate is filling a vertical fracture (low density is displayed in dark and high density is shown by lighter colour). A: CT-slice through the core B: CT-overview of the core section documenting the dipping of the hydrate-filled fracture parallel to the core (from Abegg et al. subm.).

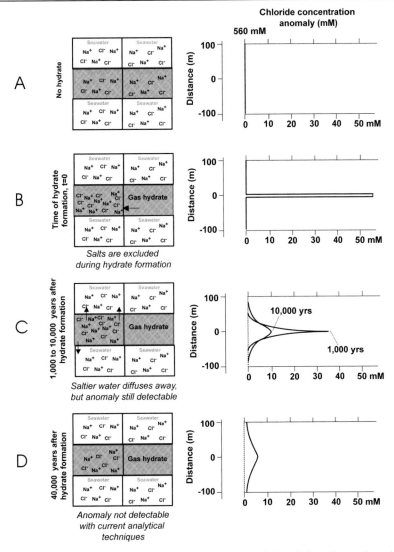

Fig. 14.13 Cartoon illustrating how gas hydrate formation increases the salinity of the adjacent interstitial pore fluid, and subsequent dissipation of the chloride anomaly via diffusion over time. A. Shows system before hydrate formation, sodium and chloride ions homogeneously distributed in the pore fluid. B. When gas hydrate forms, ions are excluded from the crystal lattice, and the pore fluids become saltier at the foci of hydrate formation. Right panel illustrates a 56 mM anomaly created by formation of gas hydrate that occupies 9% of the pore space. C. Over time the excess ions diffuse away, as illustrated by the diffusional decay model showing dissolved chloride profiles at 1,000 and 10,000 years. D. After 100,000 years, the chloride anomaly is smaller than that which can be detected with current analytical techniques. The 1-dimensional model assumes that the half width of the concentration spike to be 5 meters, a sediment porosity of 50% and the free solution diffusion coefficient for the chloride ion of 1.86×10^{-5} cm^2s^{-1} at 25 °C (modified from Ussler and Paull 2001).

commonly used to monitor formation and decomposition of gas hydrate deposits. In addition, formation of the hydrate lattice results in preferential uptake of the heavy oxygen and hydrogen atoms in the solid phase, with consequent depletion in the pore water. These two pore water parameters: dissolved chloride and the isotopic composition of the water itself, have been widely used to identify and quantify hydrate distribution and the dynamic processes involved in formation and destabilization of these deposits.

14.4.1   Gas Hydrate and Chloride Anomalies

The "Ion Exclusion" Effect

It has long been recognized that the formation and decomposition of gas hydrate lead to changes in dissolved chloride concentration of marine pore fluids (e.g. Hesse and Harrison 1981). Gas hydrates, like normal ice, exclude salts from the crystal structure, thus increasing the salinity of the surrounding water. The change in interstitial ion concentration resulting from this "ion exclusion" mechanism is proportional to the amount of gas hydrate that is formed. Ussler and Paul (2001) use a simple cartoon to represent the effect on pore water salinity at the foci of gas hydrate formation (Fig. 14.13). They further modelled the diffusive attenuation of the chloride anomaly over time, and showed that a positive anomaly of 56 mM (created by formation of hydrate that occupies ~9% of the pore space) will not be detected with current analytical methods after about 40,000 years (Fig. 14.13).

Various numerical models have shown that generating gas hydrate to a concentration of ~10% of the pore space, in both passive and active settings (e.g. Blake Ridge, Hydrate Ridge; Nimblett and Ruppel 2003) probably required formation times of at least 10^3, and perhaps as much 10^6 years. Therefore, if chloride behaved conservatively, the pore water in contact with these deposits should have a chloride concentration similar to seawater. This is, however, not commonly the case. Indeed, fluids with chloride concentration significantly lower than seawater have been sampled from most convergent margins and such "freshening" has been attributed to gas hydrate dissociation and dehydration of hydrous minerals at depth (e.g. Gieskes et al. 1990; Kastner et al. 1991). The issue of background chloride concentration, and an example of a chloride ano-maly created by natural gas hydrate dissociation is described in the following sections.

In situ chloride concentration in pore fluids of hydrate-bearing sediments also show enrichments relative to seawater in some natural systems. These occur when the geological setting supports formation of brines, or when gas hydrate forms so rapidly that the resulting excess ions do not have sufficient time to diffuse away. These scenarios are also discussed below.

Estimating Gas Hydrate Abundance Using Dissolved Chloride Data

Because gas hydrate is not stable at the temperature and pressure conditions that exist at the sea surface, most estimates of the in situ distribution and concentration of gas hydrate rely on a variety of proxies. Perhaps the most widely used of these proxies is based on the accurate measurement of dissolved chloride in the pore fluids. During core recovery, gas hydrate

dissociates, resulting in dilution of the chloride concentration by addition of water sequestered in the gas hydrate lattice prior to core recovery. The negative chloride anomalies relative to in situ chloride concentrations are proportional to the amount of gas hydrate in a sediment sample. Uncertainties in the estimates of gas hydrate abundance using the dissolved chloride proxy arise from a paucity of information on (1) the in situ dissolved chloride values, (2) the chloride content potentially trapped within the pores of the gas hydrates, and (3) the spatial sampling resolution.

There is to date no reliable data on the amount of Cl^- sequestered by the hydrate cage because the physical separation of the water released by natural hydrate dissociation from pore water contamination can be very difficult. Suess et al. (2001) suggest that there may be residual chloride trapped within the hydrate pore space. Nevertheless, since this number is small and very poorly defined, most estimates of hydrate abundance in marine sediments assume that hydrate formation excludes all dissolved ions.

If the amount of chloride ions trapped in the hydrate structure is assumed to be negligible, the measured chloride concentration after hydrate dissociation can be related to the hydrate abundance by the following equation (see Ussler and Paull 2001 for derivation).

$$Cl_s^-/Cl_o^- = 1-[V_h/(w-V_h(w-1))] \qquad (1)$$

where Cl_s^- is the chloride concentration in the sample (i.e. after hydrate decomposition), Cl_o^- is the pore water concentration in situ (prior to decomposition), V_h is the volume fraction of hydrate filling pore space, and w represents the occupancy-density characteristics of the gas hydrate formed, as calculated from:

$$w = \rho_w M_h/(\rho_h M_w m_w) \qquad (2)$$

Here, ρ_w and ρ_h are the densities of fresh water and gas hydrate respectively, and m_w is the number of moles of fresh water contained in 1 mole of gas hydrate. M_w and M_h represent the molecular weights of water and gas hydrate, respectively. The value of M_h depends on the degree of occupancy of the hydrate structure. When the structure is fully occupied, 1 mole of gas hydrate contains 5.9 moles of water, its density is 910 kg m^{-3} and its molecular weight is 122.2 g mol^{-1} (Ussler and Paull 2001).

The use of the chloride proxy is predicated on the assumption that the background chloride concentration is known and that the rate of hydrate formation is slow enough that high chloride anomalies resulting from salt exclusion during hydrate formation have been removed by diffusion and advection. A recurrent issue in these studies is the need for a robust estimate of the background chloride values (Cl_o in equation 3) against which the anomalous discrete excursions can be calculated.

Ussler and Paull (2001) nicely illustrate the effect that selecting various values for Cl_o has on the estimate of gas hydrate concentration, by comparing two approaches for estimating pore water baselines (Fig. 14.14). Using examples from a passive (Site 997, Blake Ridge) and active margin (Site 889, northern Cascadia margin), they clearly illustrate that simply assuming seawater dissolved chloride values as a baseline against which to measure the degree of dilution, in most cases will give wrong results. The smoothed baseline approach produces an estimate of gas hydrate more consistent with other independent hydrate proxies, and with the phase change that

occurs at the bottom of the GHSZ. To construct robust estimates of the hydrate abundance based on the dissolved chloride proxy, it is important to understand the processes that affect the in situ chloride distribution at each location.

The low chloride values measured below the BSR at sites drilled on the Blake Ridge (passive margin setting) have been attributed to long-term hydrate melting below the gas hydrate stability zone. A more pronounced freshening is observed in pore waters from active margin settings, as observed at Sites 889 in the northern Cascadia margin (Fig. 14.14), and in the Middle America trench at Sites 497, 498 (Harrison and Curiale 1982) and 568 (Hesse et al. 1985). It was unclear, though, if this deep freshening effect was due to gas hydrate processes or to other reactions independent of hydrate formation (Ussler and Paull 2001).

Data generated by drilling along an east-west transect in the southern Hydrate Ridge region

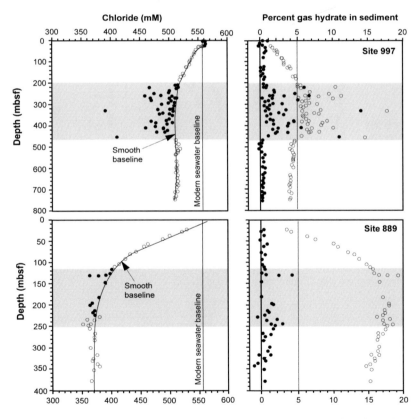

Fig. 14.14 Comparison of estimates of hydrate concentration based on two approaches for estimating the background chloride concentration (Cl⁻$_o$ in equation 3). Upper panel uses data from a passive margin (Site 997, Blake Ridge) and bottom panel shows data collected at an active margin (Site 889, northern Cascadia margin). In both cases, the shaded area denotes the region where gas hydrate is believed to be present, and the BSR denotes the geophysical reflector that indicates the bottom of the gas hydrate stability zone. The use of a modern seawater baseline predicts much larger amounts of gas hydrate, and suggests the presence of gas hydrate below the GHSZ (from Ussler and Paull 2001).

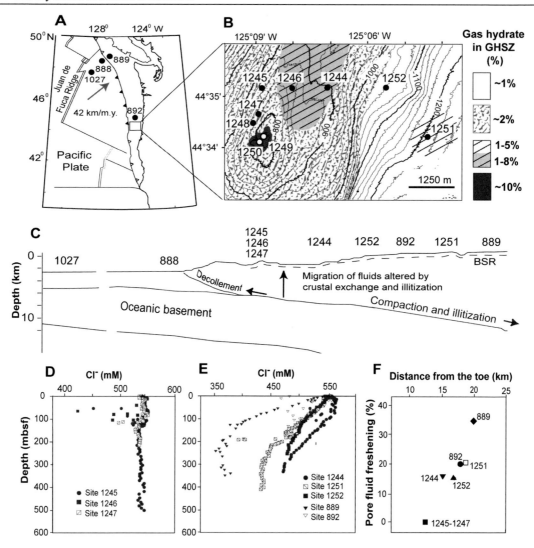

Fig. 14.15 Chloride freshening due to progressive illitization along the Cascadia margin accretionary margin. A. Tectonic setting. B. Details of sites drilled during ODP Leg 204, showing the gas hydrate distribution. C. Location of the sites relative to a schematic transect arcward from the incoming plate, the relative site locations are not to scale. D. Dissolved chloride at sites drilled less than 10 km away from the toe of the prism, showing no significant freshening at depth. Gas hydrate is apparent in discrete anomalies in the GHSZ. E. Freshening of deep fluids from sites drilled at various distances from the prism toe. F. Increase in pore fluid freshening of mélange samples with distance from the prism toe, consistent with progressive illitization as mélange sequences are exposed to higher temperatures over longer time periods (Figure modified from Torres et al. 2004).

has recently shown the separate effects of clay dehydration reactions and gas hydrate dissociation on the dissolved Cl⁻ distribution. These data provide geochemical evidence to evaluate the baseline question, and provide an example of a system where hydrate is present and background chloride contents do not deviate significantly from seawater values (Torres et al. 2004). As shown in Fig. 14.15, Sites 1244 and 1245 both have very similar gas hydrate contents, averaging 2-4 % within the gas hydrate stability zone, and

concentrated in patchy zones that contain up to 20 % hydrate (Tréhu et al. 2004). These two sites, however, have highly different chloride baselines (Fig. 14.15). In addition, there is very little gas hydrate presence at Site 1252, as evidenced by various proxy measurements, including chloride data (Tréhu et al. 2004a), even though the trend to low chloride values is well defined at this site.

The observed freshening with depth and distance from the prism toe is consistent with enhanced conversion of smectite to illite, driven by increase in temperature and age of accreted sediments (Fig. 14.15). Whereas discrete negative anomalies within the GHSZ are indeed the result of gas hydrate dissociation during core recovery, the smooth decrease with depth is independent of gas hydrate processes, and instead reflects the degree of illitization at depth. These results indicate that the smooth decrease with depth, commonly observed at sites drilled over accreted mélanges, is not directly related to gas hydrate abundance. Instead, chloride anomalies associated with gas hydrate should be calculated from discrete excursions to negative values against a background defined by the envelope of the measurements. In order to confidently define the dissolved chloride background concentration, care should be taken in obtaining enough resolution of the pore fluid sampling.

The presence of negative "spikes" in the chloride distribution suggests that the distribution of gas hydrate in marine sediments is highly heterogeneous. Whereas some observations reveal association of hydrate with coarse, high porosity horizons (Clennell et al. 1999), the factors controlling distribution of gas hydrate are not fully understood. Nevertheless, the question remains as to whether the patchy distribution of these deposits can be adequately mapped with pore water analyses. Limitations on how much pore water can be extracted from a section of the core, how many core sections can be dedicated to these analyses, and the time needed for each measurement, usually only allow for sparse measurements of the pore water composition.

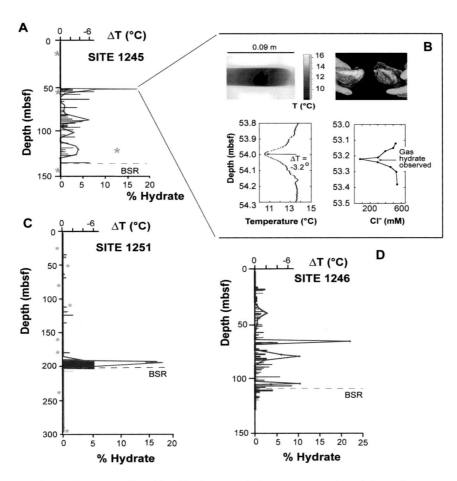

Fig. 14.16 Comparison of ΔT anomalies (blue lines) to gas hydrate content estimated from discrete anomalies in the dissolved chloride distribution (red lines), and given as percent occupancy of the pore space, for 3 sites drilled during ODP Leg 204. Green lines denote estimates based on data from pressure core barrel deployments. Horizontal (dashed) lines denote the depth of seismic reflectors corresponding to the bottom of the GHSZ (BSR). Location of the sites is shown in Figure 11. Insert B shows the temperature profile derived from an infrared image in the vicinity of a 2 cm-hydrate layer recovered from Site 1245, and the corresponding chloride concentration in closely-spaced pore water samples. The apparent offset in depth between the two graphs is due to the removal of core as gas expansion voids between the time when the IR data was collected and the pore water samples were taken (modified from Tréhu et al. 2004a).

Typical deep-sea drilling sampling resolution is on the order of one sample every 3 to 10 meters.

Other proxies have been developed to produce a more continuous, high resolution record of hydrate distribution in marine sediments. Among these, the use of an infrared (IR) camera to map cold spots in the core resulting from the endothermic decomposition of gas hydrate has proven to be highly effective (e.g. Weinberg et al. 2005). However the absolute value of the temperature measured by the IR camera depends on many different variables, including time of day, core depth and coring technique (Tréhu et al. 2003), and plots of temperature along the core are very noisy. A simple way of parametizing and displaying the IR temperature data is to define ΔT as the temperature anomaly relative to the local background. High-resolution measurements of chloride anomalies in core sections previously imaged with an IR camera can provide a calibration function needed to correlate the temperature anomaly with hydrate content, as shown in Figure 14.16B. These data, not only provide a means of calibrating the temperature anomalies, but also illustrates how a discrete hydrate layer can be easily missed with coarse sampling resolution. Samples collected from a 2-cm-thick hydrate layer and as much as 5 cm away from it show significant anomalies in the chloride content, whereas samples collected at distances >10 cm from the hydrate layer do not show any deviation from the background chloride values (Fig. 14.15).

Whereas the dissolved chloride measurements alone may not fully constrain the gas hydrate distribution, the good correlation between dissolved chloride and temperature anomalies shown in Figure 14.16 (Tréhu et al. 2004a), gives support to the use of a combined ΔT-ΔCl approach to best define an heterogeneous hydrate distribution. An understanding of the spatial variability in gas hydrate distribution may provide valuable insights into the possible response of these deposits to tectonic and environmental change.

Gas Hydrate Destabilization via Natural Processes

If environmental changes induce gas hydrate dissociation, the negative anomaly associated with water release would be attenuated over time by diffusion processes. The mathematical treatment of the signal attenuation is analogous to that described above for hydrate formation. An example of the chloride attenuation from natural dissociation processes on Hydrate Ridge is described by Bangs et al. (2005). These authors explain the presence of a double BSR in the seismic records as a remnant of a BSR_s that probably formed during the last glacial maximum

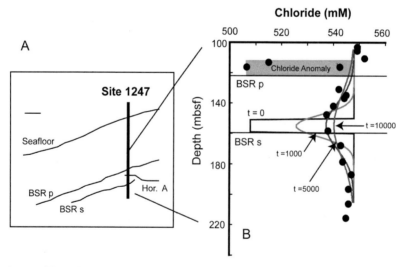

Fig. 14.17 A. Diagram illustrating the double BSR observed in seismic data in the vicinity of Site 1247. B. Chloride concentration in pore waters from site 1247, compared with expected values derived from a diffusive attenuation model following gas hydrate dissociation. The assumed hydrate content at time zero has a width of 10 m and a magnitude comparable to the anomaly observed just above the present BSR (BSRp). The data suggest that the hydrate dissociation occurred 5000 yrs ago. The authors postulate that pressure and temperature changes in the period of 8000 to 4000 years ago, led to a shift in the depth of the hydrate stability zone, creating the double BSR (modified from Bangs et al. 2005).

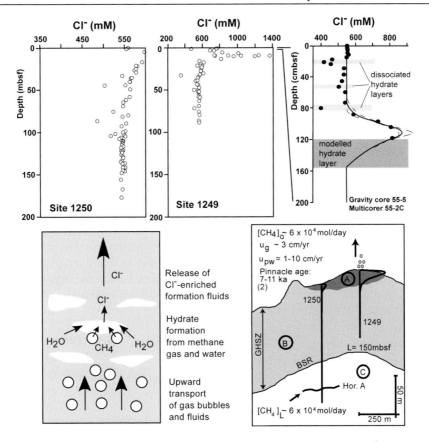

Fig. 14.18 Upper panel illustrates dissolved chloride concentration in pore waters collected from the summit of Hydrate Ridge during ODP leg 204 (Sites 1249, 1250, from Torres et al. 2004) and from a gravity core recovered from this area during RV SONNE expedition SO-143 (Haeckel et al. 2004). These data (panels A-C) indicate that hydrate is forming at very fast rates, so as to maintain the extremely high chloride values. Furthermore, to sustain the rapid formation rates, Torres et al. (2004) and Haeckel et al. (2004) show that methane must be supplied in the gas phase, as illustrated by the cartoon in panel. Methane solubility in seawater is too low for aqueous transport to deliver sufficient methane to form the observed hydrate deposits. D. Mass balance calculations based on a simple box model (E) indicate that the massive deposits recovered from the Hydrate Ridge summit probably formed in a period of the order of 100's to 1000's of years, highlighting the dynamic nature of these near-surface deposits (modified from Torres et al. 2004 and Haeckel et al. 2004).

(18,000 years ago). An increase in temperature of 3.3 °C that followed the last glaciation will shift of the hydrate stability by ~140 meters, which coupled with a concomitant sea-level rise of ~120 meters, results in an approximate net shift of the hydrate stability horizon of ~20 meters (Bangs et al. 2005). Among other evidence, the authors use a diffusion driven attenuation of the freshening signal induced by hydrate dissociation when the GHSZ shifted to a shallower depth. They show that the observed dissolved chloride distribution at a site drilled through the double BSR is consistent with a BSR shift that occurred 4,000 to 8,000 years ago (Fig. 14.17). This time frame, when analyzed in the context of thermal propagation lag in the sediment section and potential lag due to latent heat needed to dissociate hydrate, is

consistent with P/T changes in the water column that occurred at the end of the LGM (Bangs et al. 2005). A shift of the depth of hydrate stability associated with post-glaciation P/T changes, has also been suggested by others for Northern Cascadia (Westbrook et al. 1994), southwestern Japan (Foucher et al. 2002) and the Norwegian margin (Mienert et al. 1998).

Pore Water Brines

In the large body of gas-hydrate bearing locations drilled to date, the dissolved chloride show lower than seawater values (see reviews by Ussler and Paull 2001; Hesse 2003). However, there are examples of gas hydrate bearing sites in which the dissolved chloride in the pore fluids is higher than that of seawater. Most commonly these brines are associated with regions where the presence of old evaporites (e.g. Milano Dome, ODP Site 970 in the eastern Mediterranean, DeLange and Brumsack 1998), or salt-diapir intrusions (e.g. Blake Ridge Diapir ODP Site 996, Egeberg and Dickens 1999; mud volcanoes in the Northern Gulf of Mexico, Ruppel et al. 2005) leads to the enhanced chloride content. In addition to these settings, high dissolved chloride concentration associated with hydration reactions in the vicinity of an active spreading ridge was reported from ODP Sites 859 and 860 in the accretionary wedge at the Chile Triple Junction (Froelich et al. 1995).

In contrast to these regions in which brines are produced by geological processes, at the Hydrate Ridge summit, the high chloride brines observed (Haeckel et al. 2004; Torres et al. 2004) are generated by the rapid formation of gas hydrate deposits near the seafloor (Fig. 14.18). A one-dimensional transport-reaction model was used to simulate this chloride enrichment and place constrains on the mechanisms and time frames necessary to produce the observed concomitant massive hydrate deposition at the ridge summit. The models of Torres et al. (2004) and Haeckel et al. (2004) demonstrate the need for the presence of a fluid-gas mixture through the GHSZ, since the observed chloride enrichment cannot be generated exclusively from the transport of methane dissolved in the pore fluids. These massive hydrate deposits are forming very rapidly, and the continuous supply of methane gas maintains the pore water brines and the shallow gas hydrate deposits in contact with the methane-poor bottom seawater.

14.4.2 Gas Hydrate and Water Isotope Anomalies

The water sequestered in the hydrate lattice is preferentially enriched in ^{18}O and deuterium (D), thus the isotopic composition of the water in the pore spaces collected from gas hydrate bearing sediment can provide additional information on the abundance and the characteristics of these deposits. Pore fluid samples that had been modified by hydrate decomposition upon core recovery during ODP Legs 146 (Kastner et al. 1998), and 164 (Matsumoto and Borowski 2000) provided the first field data to derive the oxygen isotope fractionation factor for in situ hydrate formation. A more comprehensive sampling

protocol was subsequently conducted during Leg 204 (Tomaru et al. submitted). These calculations are based on the percent variation of Cl^- relative to background (ΔCl^-):

$$\Delta Cl^- = (1 - f) \times 100 \qquad (3)$$

where f is a fraction of formation water in sampled water given by:

$$f = \frac{Cl^-{}_s}{Cl^-{}_0} \qquad (4)$$

$Cl^-{}_s$ and $Cl^-{}_0$ are the Cl^- concentrations of sampled and formation water (i.e., in situ interstitial water) determined as background, respectively. The fractionation factors for oxygen (α_O) and hydrogen (α_H) can be determined from equilibrium equation, such that:

$$\Delta\delta = \delta_{GH} - \delta_0 = 1000 \cdot \ln\alpha \cdot (1 - f) \qquad (5)$$

where α_{GH} and α_0 are is $\delta^{18}O$ or δD values for gas hydrate and formation (background) water.

Figure 14.19 illustrates how the fractionation of ^{18}O to ^{16}O and H to D between pore water, and water derived from hydrate dissociation is related to the fractionation under in situ conditions, assuming a closed system. The average values of α_O and α_H from Leg 204 samples with negative ΔCl^- are calculated to be 1.0025 and 1.022. These fractionation factors agree with previously estimated α_O values from Leg 146 (Kastner et al. 1998) and Leg 164 (Matsumoto and Borowski 2000), and with the extrapolated α_H value from Leg 112 (Kvenvolden and Kastner 1990).

Figure 14.19 illustrates the fractionation factors for in situ hydrate formation that correspond to experimentally obtained values for oxygen (α_o: 1.0023 to 1.0032) and for hydrogen (α_H: 1.014 to 1.022) (Maekawa 2004). Analyses of pore water samples from a pore water brine sampled during Leg 204 reveal the oxygen and hydrogen isotopic fractionation during hydrate formation in natural systems. There are special challenges in fully constraining these values, since the dissolved chloride data from these brines reflects a mixture of the in situ fluids, with an unknown amount of fresh water added by hydrate dissociation during sample recovery. Nevertheless, Tomaru et al. (submitted) show that the isotopic fractionation in these massive deposits departs significantly from experimental data. More research is needed to fully understand these deviations.

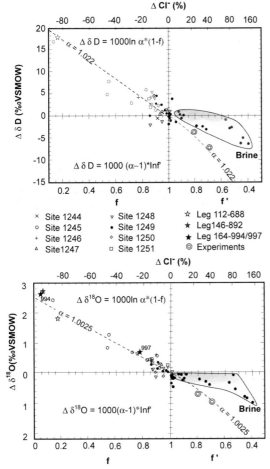

Fig. 14.19 Isotopic fractionation between water in the pore fluid and water in the hydrate lattice as a function of chloride anomalies (ΔCl). Hydrate dissociation causes chloride dilution and ^{18}O, D enrichment. The fractionation factors $\alpha_0 = 1.0025$ and $\alpha_H = 1.022$ are based on data from low-chloride pore waters recovered from Hydrate Ridge during ODP Leg 204. They are in agreement with previous estimates from Legs 146 and 164, as well as with experimentally determined values during hydrate formation shown by open circles. Samples collected from pore water brines deviate considerably from expected values (from Tomaru et al., submitted).

14.5   Gas Hydrate Carbonate Formation and Anaerobic Oxidation of Methane

14.5.1   Petrographic Characteristics of Clathrites

Authigenic carbonates are common features at seafloor seepage sites where fluids enriched in methane or oversaturated in bicarbonate escape from seafloor. Various investigators have described a particularly

large variety of carbonates from the Cascadia margin (Kulm et al. 1986; Ritger et al. 1987; Sample and Reid 1998; Greinert et al. 2001). Detailed petrographic, mineralogical and isotopic work was performed on a wide collection of samples that document several petrographically distinct lithologies. Carbonates occurr in boulder fields or in massive autochthonous chemoherm complexes (Teichert et al. 2005a). Other carbonates were sampled in direct contact with hydrates and in others, a direct relationship to gas hydrates was recognized (Bohrmann et al. 1998; Teichert et al. 2004). There are two main lithologies: a breccia composed of micrite-cemented monomict clasts, and pure aragonite of various appearances.

The breccia show angular clasts composed of the same fine-grained material as the terrigenous soft sediment on the seafloor, and submicrometer anhedral Mg-calcite crystals have been observed in the intergranular pore space between the terrigenous components (Bohrmann et al. 1998; Greinert et al. 2001). Although the grain-supported texture (Figs. 14.20A and 14.20B) shows up to 20-30% pore space, the clasts do not appear to have been transported over longer distances. The breccia is thought to form by the collapse of the clasts when gas hydrate in the sediment dissociates, followed by cementation with Mg-calcite and aragonite.

The second obvious carbonate lithology is composed of aragonite precipitates, that appear either as pure isopachous fringe cements (Fig. 14.20B) or as yellow layers of remarkable purity (Figs. 14.20A, 14.20C, and 14.20E). Pieces of isolated yellow aragonite layers have often been found associated with gas hydrates. Such layers have variable thicknesses of 1 to 3 cm, occur often in pieces of 10 to 20 cm in diameter and reveal truncated edges. The continuous aragonite layers grow directly within pure gas hydrate layers parallel to stratification and are therefore free of terrigenous sediment impurities. In several cases the aragonite precipitates have been directly recovered from within pure gas-hydrate layers (Greinert et al. 2001). The precipitates often exhibit a shape that partially images the inner surface morphology of the gas hydrate bubble fabric (Fig. 14.20E).

Such gas hydrate carbonates are also called clathrites and form archives in which geochemical processes of clathrate and clathrite formation is well documented (Teichert et al. 2005b). Their carbon isotope values range from -40‰ to -54‰ PDB, identifying methane as the dominant carbon source (Fig. 14.21). Bohrmann et al. (1998) analyzed mixtures of Mg-calcite and aragonite and showed that their oxygen isotopic composi-

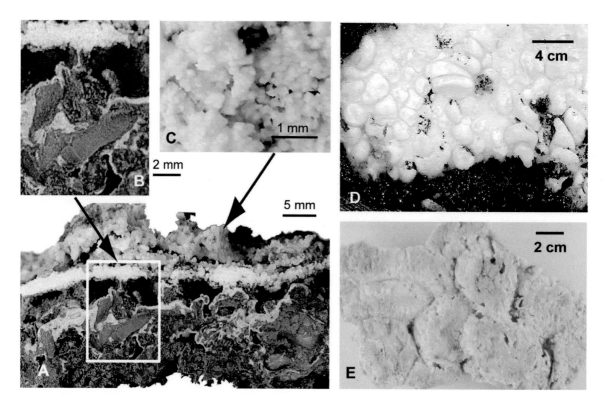

Fig. 14.20 (A) Vertical section trough an authigenic carbonate layer, showing a continuous aragonite (light) layer and fringe cements around Mg-calcite-cemented clasts. (B) Detail of the breccia. (C) Botryoidal features from the surface of the pure aragonite layer. (D) Bubble fabric of a pure methane hydrate layer and (E) corresponding aragonite layer imaging the bubble structure.

tion varies as a function of mineralogy (Fig. 14.21). The $\delta^{18}O$ value of the aragonite end-member (+3.68‰ PDB) is lower than the $\delta^{18}O$ of Mg-calcite (+4.86‰ PDB). By using appropriate isotope fractionation equations for each mineral, Bohrmann et al. (1998) calculated the oxygen isotopic composition of the pore water from which the carbonates precipitated. They found that the aragonite incorporates the isotopic composition of standard mean ocean water (SMOW) under recent seafloor conditions, when gas hydrates are also forming. In contrast, Mg-calcite most likely precipitated in response to destabilization of gas hydrates, because the pore water from which Mg-calcite precipitated is enriched in ^{18}O relative to SMOW. Similar associations have since been documented for authigenic carbonate recovered from the Gulf of Mexico (Formolo et al. 2004), further establishing that these minerals are valuable records of gas hydrate formation and destabilization through geologic time.

14.5.2   Carbonate Precipitation through Microbial Activity

Methane from gas hydrates greatly stimulates the entire ecosystem at cold seeps. (Suess et al. 2001, Sahling et al. 2002). On the basis of quantitative analyses of pore water sulfate and methane profiles, corroborated by isotopic mass balance models, geochemists postulated the anaerobic oxidation of methane (AOM) via sulfate reduction, as a dominant microbial process at cold seeps. (Suess and Whiticar 1989; Borowski et al. 1999), However, the AOM remained controversial for several years because the microbes responsible for this reaction proved to be very elusive. Only recently was a microbial consortium of methanotrophic archaea and sulfate-reducing bacteria identified on gas hydrate-bearing samples from Hydrate Ridge (Boetius et al. 2000). This interesting discovery was followed by similar findings on cold seeps and hydrate deposits in the Eel river basin (Orphan et al. 2004) and the Gulf of Mexico (Joye et al. 2004). These consortia consists of an inner sphere

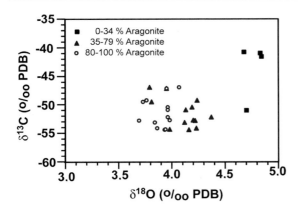

Fig. 14.21 Carbon and oxygen isotope values from gas hydrate carbonates of southern Hydrate Ridge. The carbonates are mixtures between Mg-calcite and aragonite; note the variation in oxygen isotope values with changing aragonite content (Bohrmann et al. 1998).

containing about 100 archaeal cells surrounded by about 200 cells of sulfate reducing bacteria (Fig. 14.22), and it operates via two possible separate reactions.

The archaea oxidize methane:

$$CH_4 + 2H_2O \rightarrow CO_2 + 4H_2 \qquad (6)$$

And sulfate reducing bacteria may act in two ways, indicated by reactions (7) and (8)

$$SO_4^{2-} + 4H_2 + H^+ \rightarrow HS^- + 4H_2O \qquad (7)$$

$$SO_4^{2-} + CH_3COOH + 2H^+ \rightarrow 2CO_2 + H_2S + 4H_2O \qquad (8)$$

Both reaction pathways are under discussion and it is not totally clear whether hydrogen is directly consumed (equation 7) or acetate is used (equation 8), though scavenging of H_2 will enhance the effectiveness of reaction (6). The net reaction can be summarized in the following equation,

$$CH_4 + SO_4^{2-} \rightarrow HCO_3^- + HS^- + H_2O \qquad (9)$$

The metabolic coupling involved in AOM, produces sulfide and dissolved inorganic carbon. Both methane and sulfate needed for AOM, are available in large amounts where methane vents are present at the seafloor. In the case of Hydrate Ridge, gas hydrates provide an almost inexhaustible supply of methane and the ocean water constitutes a large sulfate reservoir. Here the anaerobic methane oxidation rate is large because of the conti-nuous supply of methane from deeper sediments.

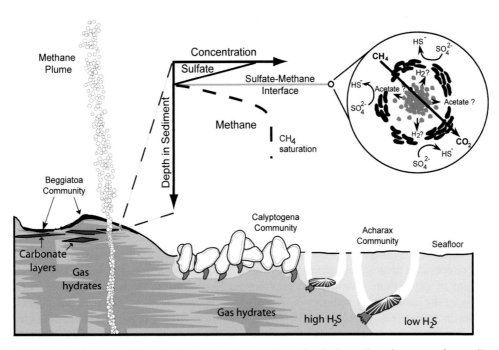

Fig. 14.22 Schematic illustration of gas hydrate deposits and biogeochemical reactions in near-surface sediments on southern Hydrate Ridge. High gradients in pore water sulfate and methane are typical of methane hydrate-rich environment close to sulfate-rich seawater. At the sulfate-methane interface (also named sulphate-methane transition in earlier chapters of the book) a microbial consortium of methanothrophic archaea and sulfate-reducing bacteria (Boetius et al. 2000) perform anaerobic oxidation of methane (AOM) leading to carbonate precipitation. AOM rates influence hydrogen sulfide fluxes and gradients, which are reflected on the seafloor by the distribution of vent communities around active gas seeps and gas hydrate exposures (Sahling et al. 2002).

The formation of hydrogen sulfide constitutes an energy source for chemoautotrophic organisms living on the seafloor. The colonization of the seeps depends on the local H₂S-gradient generated by AOM (Barry and Kochevar 1998; Sahling et al. 2002). The sulfide-oxidizing bacterium *Beggiatoa*, is usually found forming mats in areas with very high sulfide flux. *Calyptogena* clams, typically colonize areas with lower sulfide concentrations and surround the *Beggiatoa* mats. *Acharax* clams live in burrows within the sediment and are restricted to environments of very low sulfide concentration (Fig. 14.22). In addition to sulfide production, AOM increases carbonate alkalinity, which drives pervasive carbonate precipitation. The high concentration of bicarbonate as respiration product (equation 9), the presence of microbial surfaces, and the exudation of organic polymers that can bind calcium ions are all factors that support active carbonate precipitation (Iversen and Jørgensen 1985). Near-surface deposits of porous gas hydrate (Fig. 14.11) are ideal sites for AOM because sulfate can migrate through the porous space to the inner parts of the hydrates, where the microbial consortia can thrive. The aragonite precipitates observed within the sponge-like bubble structure of gas hydrates are evidence for such microbial processes. In addition, biomarker analyses of those layers show extremely high amounts of components (e.g. isoprenoids crocetane and pentamethylicosane) typical of those produced by methane-consuming and sulfate-reducing microorganisms (Elvert et al. 2001).

14.6   Concluding Remarks

Although the existence of gas hydrates has now been known for decades, our understanding of their potential impact on slope stability, the biosphere, carbon cycling, and climate change is still in its infancy. Laboratory and field studies at gas-hydrate-bearing sites, including several drilling expeditions in the past decade, have provided critical background data on the conditions of gas hydrate stability, and provide overall view of the composition and distribution of gas hydrates in nature (e.g. Dickens 2003). These results have sparked the development of models relating hydrate dynamics to tectonic and slope stability, and the possible impact of this system on global climate (Dickens 2003; Davie and Buffett 2001; Sloan 1998; Clennell et al. 1999).

Although the total amount of carbon trapped in gas hydrate is poorly constrained, existing knowledge suggest that these deposits may constitute a significant carbon reservoir, but a quantitative evaluation of its resource potential depends on reliable global and national inventories, and a better understanding of the geologic factors that lead to highly concentrated hydrate deposits.

Methane is a powerful greenhouse gas with a Greenhouse Warming Potential (GWP) 23 times that of CO_2 on a per-molecule basis. Sudden release of methane from gas hydrate therefore has the potential to affect global climate, and current hypotheses attribute past climate variations to methane release from gas hydrates in response to ocean warming and/or sea level change (Paull et al. 1991; Kennett et al. 2002; Dickens et al. 1995; Haq 1998). However, these hypotheses have yet to be confirmed and more research is needed to evaluate hydrate response to environmental change; the fate of steady fluxes of methane from hydrate reservoirs to the seabed, ocean surface and the atmosphere; and the radiative forcing of methane on climate change.

The impact of gas hydrate on seafloor stability is important for evaluating the safety of offshore structures as well as for understanding its role in rapid release of methane, which may affect climate change. Since gas hydrate encases large volumes of methane, when destabilized, these deposits may transform the host sediment into a gassy, water rich fluid. However, any buildup of overpressure from excess gas will depend on the balance between hydrate dissociation and pressure dissipation through possible permeability barriers. Freshening of the pore water may trigger slope instabilities through a possible „quick clay" behavior, which in turns would depend on the clay mineralogy of the sediment. Although massive landslide triggered by gas hydrate destabilization has not been directly observed, various investigators have shown that vast stretches of the oceanic margins where there is evidence for major large-scale slides and slumps coincide with deep water gas hydrate horizons (Mienert et al. 1998; Nisbet and Piper 1998; Paull et al. 2000). There are still gaps in our understanding of the mechanisms through which decaying hydrate may affect slope stability, on the triggering mechanism for gas hydrate decay, and on the environmental response to slope failure, in particular the possible generation of tsunamis (Driscoll et al. 2000). There are ongoing efforts to understand these phenomena and to develop predictive models, for example, in the region of the Storrega slide, off the coast of Norway (Bouriak et al. 2000; Bryn et al. 2003).

A full understanding of the complex interrelationships associated with the presence of gas hydrate in nature requires comprehensive inter-disciplinary studies that inlude laboratory experiments, numerical modelling and field observations. Because the factors that influence gas hydrate stability and the processes that occur as a consequence of gas hydrate formation are highly dynamic, these interrelationships can only be understood through time-series monitoring of complementary parameters over space and time though the installation of seafloor observatories. Efforts to establish such observatories are underway at a few key gas hydrate locations. These and other ongoing studies may well provide key answers to our current challenges of evaluating the role of these deposits in the global energy resources of the future, and on the global carbon cycle, climate change, and perhaps biotic evolution through our planet's geologic history.

Acknowledgements

This research used data provided by the Ocean Drilling Program (ODP), which is sponsored by the US National Science Foundation and participating countries under management of Joint Oceanographic Institutions (JOI), Inc. Funding for this research was provided by the U.S. Science Support Program (USSP) grant F001557, WCNURP grant PF806880 and by NSF grant OCE-9731157. The Federal Ministry of Education and Research (BMBF, Berlin) supported the studies by grant 03G0604A (collaborative project METRO). This is publication GEOTECH-194 of the program GEOTECHNOLOGIEN of the BMBF and DFG and publication No 0336 of the Research Center Ocean Margins of the University of Bremen.

14.7 Problems

Problem 1

The thermodynamically defined gas hydrate zone in sediments of the deep sea is much thicker than in sediments underlying shallow water, such as those in the upper continental slope (see Fig. 14.4). Does this mean that gas hydrate concentrations are higher in the deeper ocean than on the upper slope?

Problem 2

Discuss different methane sources in sediments. Is there a difference in hydrate formation whether the gas is biogenic or thermogenic in origin?

Problem 3

A sample collected from 90 meters below seafloor on Hydrate Ridge has an in situ methane concentration of 300 mM. Based on the phase boundary diagram shown in Fig.14.5B, do you expect hydrate to be present in this sample? If a sample with the same methane concentration was recovered from 250 mbsf, would there be hydrate in it? Why/why not. Discuss also a methane concentration of 10 mM in samples from 90 and 250 mbsf using Fig 14.5 B.

Problem 4

A water sample recovered from a gas-bearing region has a chloride concentration of 530 mM and sulfate concentration of 26.5 mM. Do you expect methane hydrate to be present?

Problem 5

Pore water samples from a hydrate-bearing core were shown to have a dissolved chloride concentration of 507 mM. If the background concentration at this site is known to be 550 mM, calculate the percent of the pore space that is occupied by gas hydrate, assuming full occupancy of the hydrate structure.

Problem 6

How could you explain the formation of shallow brines are formed in marine pore water, where there is no association with evaporites?

References

Abegg, F., Bohrmann, G., and Kuhs, W.F., subm., Data Report: Shapes and structures of gas hydrates from Hydrate Ridge imaged by CT-imaging of samples drilled during ODP Leg 204.

Bangs, N.L.B., Musgrave, R.J., and Tréhu, A.M., 2005. Upward shifts in the southern Hydrate Ridge gas hydrate stability zone following postglacial warming, offshore Oregon. J. Geophys. Res., 110:10.1029/2004JB003293.

Barry, J.P., Kochevar, R.E., 1998. A tale of two clams: Differing chemosynthetic life styles among vesico-myids in Monterey Bay cold seeps. Cahiers de Biologie Marine 39 (3-4): 329-331.

Bernard, B.B., Brooks, J.M., and Sackett, W.M. 1976. Natural gas seepage in the Gulf of Mexico. Earth and Planetary Science Letters 31(1): 48-54.

Bily, C., and Dick, J.W.L., 1974. Naturally occurring gas hydrates in the Mackenzie Delta. Bulletin of Canadian Petroleum Geology, 22: 320-352.

Boetius, A., K. Ravenschlag, K., Schubert, C.J., Rickert, D., Widdel, F., Gieseke, A., Amann, R., Jørgensen, B.B., Witte, U., and Pfannkuche, O., 2000. A marine microbial consortium apparently mediating anaerobic oxidation of methane. Nature, 407: 623-626.

Bohrmann, G., Greinert, J., Suess, E., and Torres, M., 1998. Authigenic carbonates from the Cascadia subduction zone and their relation to gas hydrate stability. Geology 26: 647-650.

Bohrmann, G., Ivanov, M.K., Foucher, J.P., Spiess, V., Bialas, J., Greinert, J., Weinrebe, W., Abegg, F., Aloisi, G., Artemov, Y., Blinova, V., Drews, M., Heidersdorf, F., Krabbenhöft, A., Klaucke, I., Krastel, S., Leder, T., Polikarpov, I., Saburova, M., Schmale, O., Seifert, R., Volkonskaya, A., and Zillmer, M., 2003. Mud volcanoes and gas hydrates in the Black Sea: new data from Dvurechenskii and Odessa mud volcanoes. Geo-Marine Letters, 23: 239-249.

Borowski, W.S., Paull, C.K., and Ussler, W., III, 1999. Global and local variations of interstitial sulfate gradients in deep-water, continental margin sediments: Sensitivity to underlying methane and gas hydrates. Marine Geology, 159: 131-154.

Bouriak, S., Vanneste, M., and Saoutkine, A., 2000. Inferred gas hydrate and clay diapirs near the Storegga Slide on the southern edge of the Vøring Plateau, offshore Norway. Marine Geology, 163:125-148.

Brewer, P.G., Franklin, M.J., Friedrich, G., Kvenvolden, K.A., Orange, D., McFarlane, J., and Kirkwood, W., 1997. Deep-ocean field test of methane hydrate formation from a remotely operated vehicle. Geology, 25: 407-410.

Brooks, J.M., II, M.C.K., Fay, R.R., and McDonald, T.J., 1984. Thermogenic gas hydrates in the Gulf of Mexico. Science, 223: 696-698.

Bryn, P., Solheim, A., Berg, K., Lien, K., Forsberg, C.F., Haflidason, H., and Ottensen, D., 2003. The Storegga Slide complex: repeated large scale sliding in response to climatic cyclicity, in Locat, J., and Mienert, J.,

(eds.), Submarine mass movements and their consequences. Advances in natural and technical hazards research, 19: 215-222.

Buffet, B., and Archer, D., 2004. Global inventory of methane clathrate: sensitivity to changes in the deep ocean: Earth and Planetary Science Letters, 227: 185-199.

Carroll, J., 2003. Natural gas hydrates - a guide for engineers: Burlington, Ma, Elsevier Science, 270 p.

Charlou, J.L., Donval, J.P., Fouquet, Y., Ondreas, H., Cochonat, P., Levaché, D., Poirier, Y., Jean-Baptiste, P., Fourré, E., Chazallon, B., and Party, T.Z.L.S., 2004. Physical and chemical characterization of gas hydrates and associated methane plumes on the Congo-Angola Basin. Chemical Geology: 205: 405-425.

Claypool, G.E., and Kaplan, I.R., 1974. The origin and distribution of methane in marine sediments, in Kaplan, I.R., ed., Natural gases in marine sediments, pp. 99-139.

Claypool, G.W., and Kvenvolden, K.A., 1983. Methane and other hydrocarbon gases in marine sediments: Ann. Rev. Earth Planetary Science, 11: 299-327.

Clennell, M.B., Hovland, M., Booth, J.S., Henry, P., and Winters, W.J., 1999. Formation of natural gas hydrates in marine sediments, 1, Conceptual model of gas hydrate growth conditioned by host sediment properties: Journal of Geophysical Research, 104: 22985-23003.

Clennell, M.B., Judd, A., and Hovland, M., 2000. Movement and accumulation of methane in marine sediments: Relation to gas hydrate systems, in Max, M.D., ed., Natural Gas Hydrate in Oceanic and Permafrost Environments: Dordrecht, Kluwer Academic Publ., pp. 105-122.

Collett, T., 2002. Energy resource potential of natural gas hydrates: AAPG Bulletin, 86: 1971-1992.

Colwell F., Matsumoto R., Reed D., 2004. A review of the gas hydrates, geology, and biology of the Nankai Trough. Chemical Geology, 205 (3/4): 391-404.

Davie M.K. and Buffett B.A., 2001. A numerical model for the formation of gas hydrate below the seafloor Journal of Geophysical Research, 106B (1): 497-514.

Davie, M.K., and Buffett, B.A., 2003. Sources of methane for marine gas hydrate: inferences from a comparison of observations and numerical models. Earth and Planetary Science Letters, 206: 51-63.

Davis, E.E., Hyndman, R.D., and Villinger, H., 1990. Rates of fluid expulsion across the Northern Cascadia accretionary prism: Constraints from new heat flow and multichannel seismic reflection data. Journal of Geophysical Research, 95: 8869-8889.

Davy, H., 1811. On a combination of oxymuriatic gas and oxygen gas: Philosophical Transactions of the Royal Society, 155 pp.

De Graaf, W., Wellsbury, P., Parkes, R.J., and Cappenberg, T.E., 1996. Comparison of acetate turnover in methanogenic and sulfate-reducing sediments by radio- and stable-isotope labeling and specific inhibitors: evidence for isotopic exchange. Appl. Env. Microbiol., 62: 772-777.

De Lange, G.J., and Brumsack, H.-J., 1998. The occurrence of gas hydrates in Eastern Mediterranean mud dome structures as indicated by pore-water composition, in Henriet, J.P., and Mienert, J., eds., Gas Hydrates: Relevance to World Margin Stability and Climate Change, Special Publications, 137: London, Geological Society, pp. 167-175.

Dickens, G.R., and Quinby-Hunt, M.S., 1994. Methane hydrate stability in seawater: Geophysical Research Letters, 21: 2115-2118.

Dickens, G.R., O'Neil, J.R., Rea, D.K., and Owen, R.M., 1995. Dissociation of oceanic methane hydrate as a cause of the carbon isotope excursion at the end of the Paleocene. Paleoceanography, 10: 965-971.

Dickens, G.R., Castillo, M.M., and Walker, J.C.G., 1997. A blast of gas in the latest Paleocene: Simulating first-order effects of massive dissociation of oceanic methane hydrate. Geology, 25: 259-263.

Dickens, G.R., 2003. Rethinking the global carbon cycle with a large, dynamic and microbially mediated gas hydrate capacitor. Earth and Planetary Science Letters, 213: 169-183.

Driscoll, N.W., Weissel, J.K., and Goff, J.A., 2000. Potential for large-scale submarine slope failure and tsunami generation along the U.S. mid-Atlantic coast. Geology, 28: 407-410.

Egeberg, P.K., and Dickens, G.R., 1999. Thermodynamic and pore water halogen constraints on gas hydrate distribution at ODP Site 997 (Blake Ridge). Chemical Geology, 153: 53-79.

Egorov, A.V., Crane, K., Vogt, P.R., and Rozhkov, A.N., 1999. Gas hydrates that outcrop on the sea floor: stability models. Geo-Marine Letters, 19: 89-96.

Elvert, M., Greinert, J., and Suess, E., 2001. Carbon isotopes of biomarkers derived from methane-oxidizing microbes at Hydrate Ridge, Cascadia Convergent Margin, in Paull, C., ed., Natural gas hydrates: Occurrence, distribution, and detection: Geophysical Monograph 124, American Geophysical Union, pp. 115-129.

Flemings, P., Liu, C.-S., and Winters, W.J., 2003. Critical pressure and multiphase flow in Blake Ridge gas hydrates. Geology, 31: 1057-1060.

Formolo, M.J., Lyons, T.W., Zhang, C., Kelley, C.A., Sassen, R., Horita, J., and Cole, D.R., 2004. Quantifying carbon sources in the formation of authigenic carbonates at gas hydrate sites in the Gulf of Mexico. Chemical Geology, 205: 253-264.

Foucher, J.-P., Nouzé, H., Henry, P., 2002. Observation and tentative interpretation of a double BSR on the Nankai slope. Marine Geology 187 (1-2): 161-175.

Froelich, P.N., Kvenvolden, K.A., Torres, M.E., Waseda, A., Didyk, B.M., and Lorenson, T.D., 1995. Geochemical evidence for gas hydrate in sediment near the Chile Triple Junction. In Lewis, S.D., Behrmann, J.H., Musgrave, R.J., and Cande, S.C. (Eds.), Proc. ODP, Sci. Results, 141: College Station, TX (Ocean Drilling Program), pp. 279-286.

Gieskes, J.M., Blanc, G., Vrolijk, P., Elderfield, H., and Barnes, R., 1990. Interstitial water chemistry—major constituents. In Moore, J.C., Mascle, A., et al., Proc. ODP, Sci. Results, 110: College Station, TX (Ocean Drilling Program), pp. 155-178.

Ginsburg, G.D., Guseynov, R.A., Dadashev, A.A., Telepnev, E.V., Askeri-Nasirov, P.Y., Yesikov, A.A., Mal'tseva, V.I., Mashiroav, Y.G., and Shabayeva, I.Y., 1992. Gas hydrates of the southern Caspian: International Geology Review, 43: 765-782.

Ginsburg, G.D., Soloviev, V.A., Cranston, R.E., Lorenson, T.D., and Kvenvolden, K.A., 1993. Gas hydrates from the continental slope, offshore Sakhalin Island, Okhotsk Sea. Geo-Marine Letters, 13: 41-48.

Ginsburg, G., Milkov, A.V., Soloviev, V.A., Egorov, A.V., Cherkashev, G.A., Vogt, P.R., Crane, K., Lorenson, T.D., and Khutorskoy, M.D., 1999. Gas hydrate accumulation at the Håkon Mosby Mud Volcano. Geo-Marine Letters, 19: 57-67.

Greinert, J., Bohrmann, G., and Suess, E., 2001. Gas hydrate-associated carbonates and methane-venting at Hydrate Ridge: Classification, distribution, and origin of authigenic lithologies, in Paull, C. and Dillon W.P. ed., Natural gas hydrates: Occurrence, distribution, and detection: Geophysical Monograph 124: 87-98, American Geophysical Union, pp. 99-113.

Gutt, C., Asmussen, B., Press, W., Merkl, C., Casalta, H., Greinert, J., Bohrmann, G., Tse, J., and Hüller, A., 1999. Quantum rotations in natural methane-clathrates from the Pacific sea-floor. Europhysics Letters, 48: 269-275.

Haeckel, M., Suess, E., Wallmann, K., and Rickert, D., 2004. Rising methane gas bubbles form massive hydrate layers at the seafloor: Geochimica et Cosmochimica Acta, 68: 4335-4345.

Hammerschmidt, E.G., 1934. Formation of gas hydrates in natural gas transmission lines. Industrial and Engineering Chemistry, 26: 851 pp.

Handa, Y.P 1990. Effect of hydrostatic pressure and salinity on the stability of gas hydrates. J Phys. Chem., 94: 2651-2657.

Haq, B.U., 1998. Natural gas hydrates: searching for the long-term climatic and slope-stability records, in Henriet, J.P., and Mienert, J., eds., Gas Hydrates: Relevance to World Margin Stability and Climate Change, Volume 137: Special Publications: London, Geological Society, pp. 303-318.

Harrison, W.E., and Curiale, J.A., 1982. Gas hydrates in sediments of holes 497 and 498A, Deep Sea Drilling Project Leg 67, in Aubouin, J., and von Huene, R., eds., Initial Reports of the Deep Sea Drilling Project, Volume 67, U.S.Governmental Printing Office, pp. 591-594.

Heeschen, K., Tréhu, A., Collier, R.W., Suess, E., and Rehder, G., 2003. Distribution and height of methane bubble plumes on the Cascadia Margin characterized by acoustic imaging: Geophysical Research Letters, 30: 10.1029/2003GL016974.

Hensen, C., and Wallmann, K., 2005, Methane formation at Costa Rica continental margin - constraints for gas hydrate inventories and cross-décollement fluid flow. Earth and Planetary Science Letters: 236: 41-60.

Hesse, R., and Harrison, W.E., 1981. Gas hydrates (clathrates) causing pore-water freshening and oxygen istope fractionation in deep-water sedimentary sections of terrigenous continental margins: Earth and Planetary Science Letters, 55: 453-462.

Hesse, R., Lebel, J., and Gieskes, J.M., 1985. Intersitial water chemistry of gas-hydrate bearing sections on the Middle America trench slope, DSDP Leg 84, *in* von Huene, R., and Aubouin, J., eds., Inital Reports of DSDP, Volume 84: Washington, U.S. Governmental Printing Office, p. 727-737.

Hesse, R., 2003. Pore water anomalies of submarine gas-hydrate zones as tool to assess hydrate abundance and distribution in the subsurface - what have we learned in the past decade. Earth-Science Reviews, 61: 149-179.

Hyndman, R.D., and Spence, G.D., 1992. A seismic study of methane hydrate marine bottom-simulating reflectors. Journal Geophys. Res, (97): 6683-6698.

Iversen, N., Jørgensen, B.B. 1985. Anaerobic methane oxidation rates at the sulfate-methane transition in marine sediments from Kattegat and Skagerrak (Denmark). Limnology & Oceanography 30 (5): 944-955.

Joye, S.B., Boetius, A., Orcutt, B.N., Montoya, J.P., Schulz, H.N., Erickson, M.J., and Lugo, S.K. 2004. The anaerobic oxidation of methane and sulfate reduction in sediments from Gulf of Mexico cold seeps. Chemical Geology 205: 219-238

Kastner, M., Elderfield, H., and Martin, J.B., 1991. Fluids in convergent margins: what do we know about their composition, origin, role in diagenesis and importance for oceanic chemical fluxes: Phil. Trans. R. Soc. London A, 335: 243-259.

Kastner, M., Kvenvolden, K.A., and Lorenson, T.D., 1998. Chemistry, isotopic composition, and origin of a methane-hydrogen sulfide hydrate at the Cascadia subduction zone. Earth and Planetary Science Letters, 156: 173-183.

Kennett, J., Cannariato, K., Hendy, I., and Behl, R., 2002. Methane hydrates in Quaternary climate change: the clathrate gun hypothesis, AGU books board, 216 p.

Kuhs, W.F., Bauer, F.C., Hausmann, R., Ahsbahs, H., Dorwarth, R., and Hölzer, K., 1996. Single crystal diffraction with X-rays and neutrons. High quality at high pressure? High Press. Res., 14: 341-352.

Kuhs, W.F., Genov, G.Y., Goreshnik, E., Zeller, A., Techmer, K., and Bohrmann, G. 2004. The impact of porous microstructures of gas hydrates on their macroscopic properties. International Journal of Offshore and Polar Engineering, 14: 305-309.

Kulm, L.D., Suess, E., Moore, J.C., Carson, B., Lewis, B.T., Ritger, S.D., Kadko, D.C., Thornburg, T.M., Embley, R.W., Rugh, W.D., Massoth, G.J., Langseth, M.G., Cochrane, G.R., and Scamann, R.L. 1986, Oregon subduction zone: venting, fauna, and carbonates. Science, 231: 561-566.

Kvenvolden, K.A., and McMenamin, M.A., 1980. Hydrates of natural gas: A review of their geological occurrence. US Geological Survey Circular, 825, 0-11.

Kvenvolden, K.A., 1988, Methane hydrate - a major reservoir of carbon in the shallow geosphere? Chemical Geology, 71: 41-51.

Kvenvolden, K.A., 1998. A primer on the geological

occurence of gas hydrate, *in* Henriet, J.P., and Mienert, J., eds., Gas Hydrates: Relevance to World Margin Stability and Climate Change, Volume 137: Special Publications: London, Geological Society, pp. 9-30.

Kvenvolden, K.A., and Kastner, M., 1990. Gas hydrates of the Peruvian outer continental margin, *in* Suess, E., and von Huene, R., (eds.), Proc. ODP., Scien. Results, Volume 112: College Station, pp. 517-526.

Kvenvolden, K.A., 1993. A primer on gas hydrates. In: Howell. D.G. et al. (eds.) The future of energy gases, U.S. geological survey professional paper, 1570: pp. 279-291.

Lein, A., Vogt, P., Crane, K., Egorov, A., and Ivanov, M., 1999. Chemical and isotopic evidence for the nature of the fluid in CH_4-containing sediments of the Håkon Mosby Mud Volcano. Geo-Marine Letters, 19: 76-83.

Maekawa, T., 2004. Experimental study on isotopic fractionation in water during gas hydrate formation: Geochemical Journal, 38:129-138.

Makogon, Y.F., Trebin, F.A., Trofimuk, A.A., Tsarev, V.P., and Cherskiy, N.V. 1971. Detection of a pool of natural gas in a solid (hydrated gas) state: Doklady p. 59-66. Akademii Nauk SSSR.

Mathews, M.A., and von Huene, R., 1985. Site 570 methane hydrate zone, *in* Orlofsky, S., ed., DSDP Initerum Reports, Volume 84, U.S. Gov. Printing Office, pp. 773-790.

Matsumoto, R., and Borowski, W., 2000. Gas hydrate estimates from newly determined oxygen isotopic fractionation and $\delta^{18}O$ anomalies of interstitial waters: Leg 164, Blake Ridge, *in* Paull, C., Matsumoto, R., Wallace, P.J., and Dillon, W.P., (eds.), Proc. ODP, Sci. Results, Volume 164: College Station, TX (Ocean Drilling Program), 196: pp. 203-206.

Mazurenko, L.L., and Soloviev, V.A., 2003. Worldwide distribution of deep-water fluid venting and potential occurrences of gas hydrate accumulations. Geo-Marine Letters, 23: 162-176.

Mienert, J., Posewang, J., and Baumann, M., 1998. Gas hydrates along the northeastern Atlantic margin: possible hydrate-bound margin instabilities and possible release of methane, *in* Henriet, J.-P., and Mienert, J., eds., Gas Hydrates: Relevance to World Margin Stability and Climate Change, Volume 137: London, Geological Society, pp. 275-291.

Nimblett J. ; Ruppel C., 2003. Permeability evolution during the formation of gas hydrates in marine sediments. Journal of Geophysical Research, 108 (B9) pp. EPM 2-1, Cite ID 2420, DOI 10.1029/2001JB001650.

Nisbet, E.G., and Piper, D.J.W., 1998. Giant submarine landslides. Nature, 392: 329-330.

Orphan, V.J., Ussler III, W., Naehr, T.H., House, C.H., Hinrichs, K.-U., and Paull, C.K. 2004. Geological, geochemical, and microbiological heterogeneity of the seafloor around methane vents in the Eel River Basin, offshore California. Chemical Geology 205: 265-289.

Paull, C.K., Ussler III, W., and Dillon, W.P., 1991. Is the

extent of glaciation limited by marine gas-hydrates: Geophysical Research Letters, 18: 432-434.

Paull, C.K., Ussler III, W., and Borowski, W., 1994. Sources of biogenic methane to form marine gas hydrates - In situ production or upward migration?: Annals of the New York Academy of Sciences, 715: 392-409.

Paull, C.K., Matsumoto, R., and Wallace, P.J., 1996. Proceedings of the ODP, Initial Reports: College Station, TX, 623 pp.

Paull, C.K., Matsumoto, R., Wallace, P.J., and Dillon, W.P., 2000, Proceedings of the Ocean Drilling Program, Scientifc Results: College Station TX, 459 pp.

Reed, D.W., Fujita, Y., Delwiche, M.E., Blackwelder, D.B., Sheridan, P.P., Uchida, T., and Colwell, F.S. 2002. Microbial communities from methane-hydrate bearing deep marine sediments in a forearc basin. Applied Environmental Microbiology, 20002, 3759-3770.

Ritger, S., Carson, B., and Suess, E., 1987. Methane-derived authigenic carbonates formed by subduction-induced pore-water expulsion along the Oregon/ Washington margin: Geological Society of America Bulletin, 98: 147-156.

Ruppel, C., Dickens, G.R., Castellini, D.G., Gilhooly, W., and Lizarralde, D., 2005. Heat and salt inhibition of gas hydrate formation in the northern Gulf of Mexico: Geophysical Research Letters, 32: L04605, doi:10.1029/2004GL021909.

Sahling, H., Rickert, D., Lee, R.W., Linke, P., and Suess, E., 2002. Macrofaunal community structure and sulfide flux at gas hydrate deposits from the Cascadia convergent margin: Marine Ecology Progress Series, 231: 121-138.

Sample, J., and Reid, M.R., 1998, Contrasting hydro-geological regimes along strike-slip and thrust faults in the Oregon convergent margin: Evidence from the chemistry of syntectonic carbonate cements and veins: GSA Bulletin, 110: 48-59.

Sassen, R., Mac Donald, I.R., Requejo, A.G., Guinasso Jr, N.L., Kennicutt II, M.C., Sweet, S.T., and Brooks, J.M., 1994. Organic geochemistry of sediments from chemosynthetic communities, Gulf of Mexico slope: Geo-Marine Letters, 14: 110-119.

Schoell, M., 1988. Multiple origins of methane in the Earth. Chemical Geology 71 (1-3): 1-10.

Shipley, T.H., Houston, M.H., Buffler, R.T., Shaub, F.J., McMillen, K.J., Ladd, J.W., and Worzel, J.L., 1979. Seismic evidence for widespread possible gas hydrate horizons on continetal slopes and rises. Am. Assoc. Petrol. Geol. Bull., 63: 2204-2213.

Shipley, T.H., and Didyk, B.M., 1982. Occurrence of methane hydrates offshore southern Mexico, in Watkins, J.S., and Moore, J.C., eds., Initial Reports of the Deep Sea Drilling Project, Volume 66, U.S. Government Printing Office, pp. 547-555.

Shoji, H., and Langway, C.C., 1982. Air hydrate inclusions in fresh ice core. Nature, 298:548-550.

Sloan, E.D.j., 1998. Physical/chemical properties of gas hydrates and application to world margin stability and climatic change, in Henriet, J.P., and Mienert, J., (eds.), Gas Hydrates: Relevance to World Margin

Stability and Climate Change, Volume 137: Special Publications: London, Geological Society, pp. 31-50.

Stoll, R.D., Ewing, J.I., and Bryan, G.M., 1971. Anomalous wave velocities in sediments containing gas hydrates: Journal Geophys. Res, 76: 2090-2094.

Suess, E., and Whiticar, M.J., 1989. Methane-derived CO_2 in pore fluids expelled from the Oregon subduction zone. Palaeogeography, Palaeoclimatology, Palaeo-ecology, 71: 119-136.

Suess, E., Torres, M.E., Bohrmann, G., Collier, R.W., Greinert, J., Linke, L., Rehder, G., Tréhu, A., Wallmann, K., Winckler, G., and Zuleger, E., 1999. Gas hydrate destabilization: enhanced dewatering, benthic material turnover and large methane plumes at the Cascadia convergent margin: Earth and Planetary Science Letters, 170: 1-15.

Suess, E., Torres, M.E., Bohrmann, G., Collier, R.W., Rickert, D., Goldfinger, C., Linke, P., Heuser, A., Sahling, H., Heeschen, K., Jung, C., Nakamura, K., Greinert, J., Pfannkuche, O., Tréhu, A., Klinkhammer, G., Whiticar, M.C., Eisenhauer, A., Teichert, B., and Elvert, E., 2001. Sea floor methane hydrates at Hydrate Ridge, Cascadia Margin, in Paull, C. and Dillon W.P. ed., Natural gas hydrates: Occurrence, distribution, and detection: Geophysical Monograph 124: 87-98, American Geophysical Union.

Takahashi, H., Japan Petroleum Exploration Co., L., Yonezawa, T., Corporation, J.N.O., Takedomi, Y., and Ministry of Economy, T.A.I., 2001. Exploration for Natural Gas Hydrate in Nankai-Trough Wells Offshore Japan, Offshore Technology Conference, Volume 13040: Houston, Texas, OTC.

Teichert, B.M.A., Gussone, N., Eisenhauer, A., and Bohrmann, G., 2004. Clathrites - Archives of near-seafloor pore water evolutions (^{44}Ca, ^{13}C, ^{18}O) in seep environments. Geology, 33: 213-216.

Teichert, B.M.A., Bohrmann, G., and Suess, E., 2005a. Chemoherms on Hydrate Ridge: Unique microbially mediated carbonate build ups in cold seep settings. Palaeogeography, Palaeoclimatology, Palaeoecology, 227: 67-85.

Teichert, B.M.A., Bohrmann, G., Torres, M., and Eisenhauer, A., 2005b. Fluid sources, fluid pathways and diagenetic reactions across an accretionary prism revealed by Sr and B geochemistry. Earth and Planetary Science Letters, 239: 106-121.

Tissot, B., and Welte, H., 1992. Petroleum formation and occurrence. Springer-Verlag, 536 pp.

Tomaru, H., Matsumoto, R., Torres, M.E., and Borowski, W.S., subm., Geological and geochemical constraints on the isotopic composition of interstitial waters from the Hydrate Ridge region, Cascadia continental margin.

Torres, M.E., Wallmann, K., Tréhu, A.M., Bohrmann, G., Borowski, W.S., and Tomaru, H., 2004. Gas hydrate growth, methane transport, and chloride enrichment at the southern summit of Hydrate Ridge, Cascadia margin off Oregon. Earth and Planetary Science Letters, 226: 225-241.

Tréhu, A.M, Bohrmann, G., Rack, F.R., Torres, M.E., et al., 2003. Proc. ODP, Init. Repts., 204 [CD-ROM].

Available from: Ocean Drilling Program, Texas A&M University, College Station TX 77845-9547, USA.

Tréhu, A.M., Long, P.E., Torres, M.E., Bohrmann, G., R., R.F., Collett, T.S., Goldberg, D.S., Milkov, A.V., Riedel, M., Schultheiss, P., Bangs, N.L., Barr, S.R., Borowski, W.S., Claypool, G.E., Delwiche, M.E., Dickens, G.R., Gracia, E., Guerin, G., Holland, M., Johnson, J.E., Lee, Y.-J., Liu, C.-S., Su, X., Teichert, B., Tomaru, H., Vanneste, M., Watanabe, M., and Weinberger, J.L., 2004a. Three-dimensional distribution of gas hydrate beneath southern Hydrate Ridge: constraints from ODP Leg 204. Earth and Planetary Science Letters, 222: 845-862.

Tréhu, A.M., Flemings, P.B., Bangs, N.L., Chevallier, J., Gràcia, E., Johnson, J.E., Liu, C.-S., Liu, X., Riedel, M., and Torres, M.E., 2004b. Feeding methane vents and gas hydrate deposits at south Hydrate Ridge. Geophys. Res. Lett., 31:10.1029/2004GL021286.

Tucholke, B.E., Bryan, G.M., and Ewing, J.I., 1977. Gas-hydrate horizons detected in seismic-profiler data from the western North Atlantic. American Association of Petroleum Geologists Bulletin, 61: 698-707.

Ussler III, W., and Paull, C., 2001. Ion exclusion associated with marine gas hydrate deposits, in Paull, C., and Dillon, W.P., (eds.), Natural gas hydrates-occurrence, distribution, and detection, Geophysical Monograph 124: Washington, AGU, pp. 41-51.

Weinberger, J.L., Brown, K.M., and Long, P.E., 2005. Painting a picture of gas hydrate distribution with thermal images. Geophysical Research Letters, 32: doi: 10.1029/2004GL021437.

Westbrook, G.K., Carson, B., Musgrave, R.J., et al. 1994. Proceedings, initial reports, Ocean Drilling Program, Leg 146: Cascadia Margin, 611 pp.

Whiticar, M.J., 1999. Carbon and hydrogen isotope systematics of bacterial formation and oxidation of methane. Chemical Geology, 161: 291-314.

Whiticar, M.J., Faber, E., and Schoell, M., 1986. Biogenic methane formation in marine and freshwater environments. CO_2 reduction vs. Acetate fermentation - Isotope evidence. Geochimica et Cosmochimica Acta, 50: 693-709.

Vasil'ev, V.G., Makogon, Y.F., Trebin, F.A., A.A., T., and Cherskiy, N.V., 1970. The property of natural gases to occur in the Earth crust in a solid state and to form gas hydrate deposits. Otkrytiya v SSR, pp. 15-17.

von Stackelberg, M., and Müller, H.R., 1954. Feste Gashydrate II: Struktur und Raumchemie. Z. Elektrochem., 58: 25-39.

Wellsbury, P., Goodman, K., Cragg, B.A., and Parkes, R.J., 2000. The geomicrobiology of deep marine sediments from Blake Ridge containing methane hydrate (Sites 994, 995 and 997), in Paull, C., Matsumoto, R., Wallace, P.J., and Dillon, W.P., eds., Proceeding of ODP, Volume 164: 379-391.

Wood, B.J., and Ruppel, C., 2000. Seismic and thermal investigations of the Blake Ridge gas hydrate area: a synthesis, in Paull, C., Matsumoto, R., Wallace, P.J., and Dillon, W.P., (eds.), Proceedings of ODP, Scientific Results, Volume 164: 253-264.

Xu, W., and Ruppel, C., 1999. Predicting the occurrence, distribution, and evolution of methane gas hydrate in porous marine sediments. Journal of Geophysical Research, 104: 5081-5095.

Yakushev, V.S., and Istomin, V.A., 1992. Gas-hydrate self-preservation effect, in Maeno, N., and Hondoh, T., (eds.) Sapporo, Hokkaido University Press, pp. 136-140.

Yefremova, A.G., and Zhizchenko, B.P., 1974. Occurrence of crystal hydrates of gases in the sediments of modern marine basins. Doklady Akademii Nauk SSSR, 214: 1,179-1,181.

Zatsepina, O.Y., and Buffett, B.A., 1997. Phase equilibrium of gas hydrate: implications for the formation of hydrate in the deep sea-floor. Geophysical Research Letters, 24: 1567-1570.

15 Conceptual Models and Computer Models

Horst D. Schulz

Upon recording *processes* of nature quantitatively, the term model is closely related to the term system. A *system* is a segment derived from nature with either real or, at least, imagined boundaries. Within these boundaries, there are processes which are to be analyzed. Outside, there is the *environment* exerting an influence on the course of the procedural events which are internal to the system by means of the *boundary conditions*. A *conceptual model* contains principle statements, mostly translatable quantitatively, with regard to the processes in a system and the influence of prevalent boundary conditions. If the systems to be reproduced are especially complex, any significant realization of the conceptual model is often only possible by applying computer models.

Depending on the field of interest and the nature of the task, the processes in a model will be either somewhat more physical/hydraulic, geochemical, biogeochemical, or biological. The systems and processes studied might then expand in size over the great oceans and their extensive stream patterns down to small segments of deep ocean floor measuring only few cubic centimeters, including the communities of microorganisms living therein. Consequently, there are no predefined dimensions or structures which are valid for the systems and their models, instead, these are exclusively determined by the understanding of the respective processes of interest and their quantitative translatability. Since geochemical and biogeochemical reactions occur relatively fast, and are thus bound to spaces of smaller dimensions, the models dealt with in this chapter will be preferentially concerned with dimensions encompassing few centimeters and meters of sediment, to some hundred meters of water column.

In the field of applied geological sciences, models frequently assume a primarily prognostic character. For instance, after accomplishing an adequate calibration, an hydraulic model applied

to groundwater, or a model of solute transport in groundwater, is primarily supposed to predict the system's reaction as precisely as possible under the given boundary conditions. Such models will find very practical and often well paid applications. In marine geochemistry such a prognostic function of models occurs rather seldom; here, the objective of obtaining a quantitative concept of a system predominates in most cases, which is 'function' of a particular segment of nature. Computer models can then check the plausibility of model concepts of complex systems, can draw attention to the areas not sufficiently understood, and might help discover parameters with which the system can be described reasonably well and effectively.

In the following, zero-dimensional geochemical reaction models will be distinguished from one-dimensional models, in which transport processes are coupled by various means to geochemical and biogeochemical reactions. The subsections will then introduce approaches to very different models, including models that have advanced to very different stages of development.

15.1 Geochemical Models

15.1.1 Structure of Geochemical Models

Zero-dimensional geochemical models are concerned with the contents of solutions of an aquatic subsystem (e.g. pore water, ocean water, precipitation, groundwater), the equilibria between the various dissolved species as well as their adjoining gaseous and solid phases. A computer model by the name of WATEQ (derived from water-equilibria) was published for the first time by Truesdell and Jones (1974). Unfortunately, this was done in the computer language PL1 which has become completely forgotten by now. Afterwards a FORTRAN version had followed (WATEQF) released by Plummer et al. (1976) as well as another revised version (WATEQ4F) by Ball and Nordstrom (1991).

All these programs share the common feature that they start out from water analyses which were designed as comprehensive as possible, along with conceiving the measured values as cumulative parameters (sums of all aquatic species), only to subsequently allocate them to the various aquatic species. This was done by applying iteration calculus that accounts for the equilibria related to the various aquatic species.

For instance, the measured value of calcium in an ocean water sample containing 10.6 mmoles/l is allocated to 9.51 mmol/l Ca^{2+}, 1.08 mmol/l $CaSO_4^0$, 0.04 mmol/l $CaHCO_3^+$, and to a series of low

Table 15.1 This ocean water analysis carried out by Nordstrom et al. (1979) has been frequently used to test and compare various geochemical model programs.

Seawater from Nordstrom et al. (1979)

pH	=	8.22
pe	=	8.451
Ionic strength	=	6.750E-1
Temperature (°C)	=	25.0

Elements	Molality
Alkalinity	2.406E-3
Ca	1.066E-2
Cl	5.657E-1
Fe	3.711E-8
K	1.058E-2
Mg	5.507E-2
Mn	3.773E-9
N (-III)	1.724E-6
N (V)	4.847E-6
Na	4.854E-1
O (0)	3.746E-4
S (VI)	2.926E-2
Si	7.382E-5

concentrated other calcium species in solution. The analysis of sulfate yielding 29.26 mmol/l is allocated accordingly to 14.67 mmol/l SO_4^{2-}, 7.30 mmol/l $NaSO_4^-$, 1.08 mmol/l $CaSO_4^0$, 0.16 mmol/l KSO_4^-, and, likewise, to a number of low concentrated sulfate species. The procedure is similarly applied to all analytical values. Table 15.2 summarizes the species allocations for the ocean water analysis previously shown in Table 15.1.

Based on the activities of the non-complex aquatic species (e.g. $[Ca^{2+}]$ or $[SO_4^{2-}]$), the next step consists in calculating the saturation indices (SI) of each compound mineral:

$$SI = \log\left(\frac{IAP}{K_{sp}}\right) \qquad (15.1)$$

In this equation IAP denotes the Ion Activity Product (in the example of gypsum or anhydrite these would be ($[Ca^{2+}]\cdot[SO_4^{2-}]$). K_{SP} is the solubility product constant of the respective mineral. A saturation index SI = 0 describes the condition in which the solution of the corresponding mineral is just saturated, SI > 0 describes the condition of supersaturation of the solution, SI < 0 its undersaturation. The activity [A] of a substance is calculated according to the equation:

$$[A] = (A)\cdot\gamma_A \qquad (15.2)$$

Here, (A) is the concentration of a substance measured in moles/l and γ_A is the activity coefficient calculated as a function of the overall ionic strength in the solution. For infinitesimal diluted solutions, the activity coefficients assume the value of 1; hence, activity equals concentration. In ocean water, the various monovalent ions, or ion pairs, display activity coefficients somewhere around 0.75; whereas the various divalent ions, or ion pairs, display values around 0.2 (cf. Table 15.2, last column).

After transformation, the equation 15.1 to calculate the saturation index of the mineral gypsum reads:

$$SI_{gypsum} = \log\left(\frac{[Ca^{2+}]\cdot[SO_4^{2-}]}{K_{sp,gypsum}}\right) \qquad (15.3)$$

If we now insert the solubility product constant for gypsum $K_{sp,gypsum} = 2.63\cdot10^{-5}$ in the equation and the exemplary values mentioned above for ocean water (activity coefficients of divalent ions ≈ 0.2; concentration of Ca^{2+} = 9.15/1000 mol/l;

Table 15.2 For an analysis of Table 15.1, this allocation of the aquatic species was calculated with the geochemical model program PHREEQC.

Species	Molality	Activity	Molality log	Activity log	Gamma log
OH$^-$	2.18E-06	1.63E-06	-5.661	-5.788	-0.127
H$^+$	7.99E-09	6.03E-09	-8.098	-8.220	-0.122
H$_2$O	5.55E+01	9.81E-01	-0.009	-0.009	0.000
C (IV)	2.181E-03				
HCO$_3^-$	1.52E-03	1.03E-03	-2.819	-2.989	-0.171
MgHCO$_3^+$	2.20E-04	1.64E-04	-3.659	-3.785	-0.127
NaHCO$_3$	1.67E-04	1.95E-04	-3.778	-3.710	0.068
MgCO$_3$	8.90E-05	1.04E-04	-4.050	-3.983	0.068
NaCO$_3^-$	6.73E-05	5.03E-05	-4.172	-4.299	-0.127
CaHCO$_3^+$	4.16E-05	3.10E-05	-4.381	-4.508	-0.127
CO$_3^{-2}$	3.84E-05	7.98E-06	-4.415	-5.098	-0.683
CaCO$_3$	2.72E-05	3.17E-05	-4.566	-4.499	0.068
CO$_2$	1.21E-05	1.42E-05	-4.916	-4.849	0.068
MnCO$_3$	3.18E-10	3.71E-10	-9.498	-9.430	0.068
MnHCO$_3^+$	7.17E-11	5.36E-11	-10.144	-10.271	-0.127
FeCO$_3$	1.96E-20	2.29E-20	-19.709	-19.641	0.068
FeHCO$_3^+$	1.64E-20	1.22E-20	-19.785	-19.912	-0.127
Ca	1.066E-02				
Ca^{+2}	9.51E-03	2.37E-03	-2.022	-2.625	-0.603
CaSO$_4$	1.08E-03	1.26E-03	-2.967	-2.900	0.068
CaHCO$_3^+$	4.16E-05	3.10E-05	-4.381	-4.508	-0.127
CaCO$_3$	2.72E-05	3.17E-05	-4.566	-4.499	0.068
CaOH$^+$	8.59E-08	6.41E-08	-7.066	-7.193	-0.127
CaHSO$_4^+$	5.96E-11	4.45E-11	-10.225	-10.352	-0.127
Cl	5.656E-01				
Cl$^-$	5.66E-01	3.52E-01	-0.247	-0.453	-0.206
MnCl$^+$	1.13E-09	8.41E-10	-8.948	-9.075	-0.127
MnCl$_2$	1.11E-10	1.29E-10	-9.956	-9.888	0.068
MnCl$_3^-$	1.68E-11	1.26E-11	-10.774	-10.901	-0.127
FeCl^{+2}	9.58E-19	2.97E-19	-18.019	-18.527	-0.508
FeCl$_2^+$	6.27E-19	4.68E-19	-18.203	-18.330	-0.127
FeCl$^+$	7.78E-20	5.81E-20	-19.109	-19.236	-0.127
FeCl$_3$	1.41E-20	1.65E-20	-19.850	-19.783	0.068
Fe (II)	5.550E-19				
Fe^{+2}	3.85E-19	1.19E-19	-18.415	-18.923	-0.508
FeCl$^+$	7.78E-20	5.81E-20	-19.109	-19.236	-0.127
FeSO$_4$	4.84E-20	5.65E-20	-19.315	-19.248	0.068
FeCO$_3$	1.96E-20	2.29E-20	-19.709	-19.641	0.068
FeHCO$_3^+$	1.64E-20	1.22E-20	-19.785	-19.912	-0.127
FeOH$^+$	8.23E-21	6.15E-21	-20.084	-20.211	-0.127
Fe (III)	3.711E-08				
Fe(OH)$_3$	2.84E-08	3.32E-08	-7.547	-7.479	0.068
Fe(OH)$_4^-$	6.60E-09	4.92E-09	-8.181	-8.308	-0.127
Fe(OH)$_2^+$	2.12E-09	1.58E-09	-8.674	-8.801	-0.127
FeOH^{+2}	9.46E-14	2.94E-14	-13.024	-13.532	-0.508
FeSO$_4^+$	1.09E-18	8.16E-19	-17.962	-18.089	-0.127
FeCl^{+2}	9.58E-19	2.97E-19	-18.019	-18.527	-0.508
FeCl$_2^+$	6.27E-19	4.68E-19	-18.203	-18.330	-0.127
Fe^{+3}	3.88E-19	2.80E-20	-18.411	-19.554	-1.143
Fe(SO$_4$)$_2^-$	6.36E-20	4.75E-20	-19.196	-19.323	-0.127
FeCl$_3$	1.41E-20	1.65E-20	-19.850	-19.783	0.068
K	1.058E-02				
K$^+$	1.04E-02	6.49E-03	-1.982	-2.188	-0.206
KSO$_4^-$	1.64E-04	1.22E-04	-3.786	-3.913	-0.127

	Species	Molality	Activity	Molality log	Activity log	Gamma log	
Mg		5.507E-02					
	Mg^{+2}	4.75E-02	1.37E-02	-1.324	-1.864	-0.541	
	$MgSO_4$	7.30E-03	8.53E-03	-2.137	-2.069	0.068	
	$MgHCO_3^+$	2.20E-04	1.64E-04	-3.659	-3.785	-0.127	
	$MgCO_3$	8.90E-05	1.04E-04	-4.050	-3.983	0.068	
	$MgOH^+$	1.08E-05	8.07E-06	-4.966	-5.093	-0.127	
Mn (II)		3.773E-09					
	Mn^{+2}	1.89E-09	5.86E-10	-8.724	-9.232	-0.508	
	$MnCl^+$	1.13E-09	8.41E-10	-8.948	-9.075	-0.127	
	$MnCO_3$	3.18E-10	3.71E-10	-9.498	-9.430	0.068	
	$MnSO_4$	2.38E-10	2.77E-10	-9.624	-9.557	0.068	
	$MnCl_2$	1.11E-10	1.29E-10	-9.956	-9.888	0.068	
	$MnHCO_3^+$	7.17E-11	5.36E-11	-10.144	-10.271	-0.127	
	$MnCl_3^-$	1.68E-11	1.26E-11	-10.774	-10.901	-0.127	
	$MnOH^+$	3.29E-12	2.45E-12	-11.483	-11.610	-0.127	
	$Mn(OH)_3^-$	5.36E-20	4.00E-20	-19.271	-19.398	-0.127	
	$Mn(NO_3)_2$	2.62E-20	3.06E-20	-19.582	-19.515	0.068	
Mn (III)		7.108E-26					
	Mn^{+3}	7.11E-26	5.12E-27	-25.148	-26.291	-1.143	
Mn (VI)		2.322E-28					
	MnO_4^{-2}	2.32E-28	7.21E-29	-27.634	-28.142	-0.508	
Mn (VII)		1.127E-29					
	MnO_4^-	1.13E-29	8.41E-30	-28.948	-29.075	-0.127	
N (-III)		1.724E-06					
	NH_4^+	1.58E-06	1.18E-06	-5.802	-5.929	-0.127	
	NH_3	9.36E-08	1.09E-07	-7.029	-6.961	0.068	
	$NH_4SO_4^-$	5.40E-08	4.03E-08	-7.267	-7.394	-0.127	
N (V)		4.847E-06					
	NO_3^-	4.85E-06	3.62E-06	-5.315	-5.441	-0.127	
	$Mn(NO_3)_2$	2.62E-20	3.06E-20	-19.582	-19.515	0.068	
Na		4,854E-01					
	Na^+	4.79E-01	3.38E-01	-0.320	-0.471	-0.151	
	$NaSO_4^-$	6.05E-03	4.51E-03	-2.219	-2.346	-0.127	
	$NaHCO_3$	1.67E-04	1.95E-04	-3.778	-3.710	0.068	
	$NaCO_3^-$	6.73E-05	5.03E-05	-4.172	-4.299	-0.127	
O (0)		3.746E-04					
	O_2	1.87E-04	2.19E-04	-3.728	-3.660	0.068	
S (VI)		2.926E-02					
	SO_4^{-2}	1.47E-02	2.66E-03	-1.833	-2.575	-0.741	
	$MgSO_4$	7.30E-03	8.53E-03	-2.137	-2.069	0.068	
	$NaSO_4^-$	6.05E-03	4.51E-03	-2.219	-2.346	-0.127	
	$CaSO_4$	1.08E-03	1.26E-03	-2.967	-2.900	0.068	
	KSO_4^-	1.64E-04	1.22E-04	-3.786	-3.913	-0.127	
	$NH_4SO_4^-$	5.40E-08	4.03E-08	-7.267	-7.394	-0.127	
	HSO_4^-	2.09E-09	1.56E-09	-8.680	-8.807	-0.127	
	$MnSO_4$	2.38E-10	2.77E-10	-9.624	-9.557	0.068	
	$CaHSO_4^+$	5.96E-11	4.45E-11	-10.225	-10.352	-0.127	
	$FeSO_4^+$	1.09E-18	8.16E-19	-17.962	-18.089	-0.127	
	$Fe(SO_4)_2^-$	6.36E-20	4.75E-20	-19.196	-19.323	-0.127	
	$FeSO_4$	4.84E-20	5.65E-20	-19.315	-19.248	0.068	
	$FeHSO_4^{+2}$	4.24E-26	1.32E-26	-25.373	-25.881	-0.508	
	$FeHSO_4^+$	3.00E-27	2.24E-27	-26.523	-26.650	-0.127	
Si		7.382E-05					
	H_4SiO_4	7.11E-05	8.31E-05	-4.148	-4.081	0.068	
	$H_3SiO_4^-$	2.72E-06	2.03E-06	-5.565	-5.692	-0.127	
	$H_2SiO_4^{-2}$	7.39E-11	2.29E-11	-10.131	-10.639	-0.508	

Table 15.2   continued.

concentration of SO_4^{2-} = 14.67/1000 mol/l), yields the following equation:

$$SI_{gypsum,\ seawater} =$$

$$\log\left\{\left(0.2\cdot\frac{9.15}{1000}\right)\cdot\left(0.2\cdot\frac{14.67}{1000}\right)\Big/2.63\cdot10^{-5}\right\} = -0.69$$

(15.4)

Matching to Tables 15.1 and 15.2 the saturation indices are demonstrated in Table 15.3, likewise calculated by applying the PHREEQC model (Parkhurst 1995; Parkhurst and Appelo 1999). Here, the majority of the saturation indices have been calculated on the basis of mineral phases, however, there are also gaseous phases denoted as '$CO_2(g)$', '$H_2(g)$', '$NH_3(g)$', '$O_2(g)$'. The saturation index stands for the common logarithm of the respective partial pressure. The number of minerals and gases listed in Table 15.3 depends on the fact that calculations can only be performed after all the ions involved have been analyzed, and furthermore, provided that the database file of the program contains the appropriate thermodynamic data.

It needs to be stated clearly that the conditions of saturation thus calculated in the process of water analyses are determined *exclusively* on the basis of the prevalent analytical data and the employed data available on equilibria. A mineral shown to be supersaturated must not immediately, or at a later instance, be precipitated from this solution, but *can* be formed, for example, when other conditions are fulfilled, e.g. such standing in relation to the reaction kinetics. At the same time, a mineral found to be *under*saturated does not have to become dissolved immediately or at a later time - after all, it is possible that this mineral will never come in contact with the solution. The result of such a model calculation should just be understood as the statement that certain minerals can be either dissolved or precipitated. It goes without saying that mostly such minerals are of particular interest that have a calculated saturation index close to zero, because this circumstance often refers to set equilibria and hence to corresponding reactions.

A huge number of such geochemical models are available. Apart from the models belonging to the WATEQ family which are only of interest to science history nowadays, the model SOLMINEQ (Kharaka et al. 1988), although designed for

waters in crude oil fields, is essential because it contains a special pressure and temperature corrective for the applied constants which might as well be important in marine environments. Other models make use of a temperature corrective only. Unfortunately, the SOLMINEQ model has not been developed any further nor improved since its first release, so that only a first, somewhat faulty version is now available.

Beside the model EQ 3/6 (Wolery 1993), the model PHREEQE (Parkhurst et al. 1980) and its recent successor PHREEQC (Parkhurst 1995; Parkhurst and Appelo 1999) are geochemical models in the true sense of the word. With these, one cannot just calculate any kind of saturation indices on the basis of a previously conducted analysis, afterwards to be interpreted more or less well by giving some meaning to them, instead, almost any process can be simulated almost without any limitation, after pre-selecting boundary conditions (e.g. exchanges with the gaseous phase and dissolution/precipitation of minerals), or specific processes (e.g. decomposition of organic matter with corresponding amounts of nitrogen and phosphorous, mixing of waters containing different solutes).

The only limitation is one's own knowledge as to the process to be modeled. The program PHREEQC can be obtained from the internet as public domain software, including an elaborate, very informative description and many examples, at:

*http://wwwbrr.cr.usgs.gov/projects/
GWC_coupled/phreeqc/index.html*

The examples presented in the following and in Section 9.2 have been calculated exclusively with the program PHREEQC.

15.1.2 Application Examples of Geochemical Modeling

Determination of Saturation Indices in Pore water, Precipitation of Minerals

Saturation indices were determined with the PHREEQC model for the core which pore water concentration profiles were previously shown in Chapter 3, Figure 3.1. Here, our attention was drawn to the supersaturation of several manganese minerals in various depths, and thus in differing redox environments. Figure 15.1 shows the depth profiles for the saturation indices of the minerals

Table 15.3 According to the species distribution shown in Table 15.2, the PHREEQC model (Parkhurst 1995) was also applied to calculate the saturation indices. Here, 'CO$_2$ (g)', 'H$_2$(g)', 'NH$_3$(g)', 'O$_2$(g)' do not stand for mineral phases, but for gaseous phases in which the saturation index is obtained as the logarithm of the respective partial pressure.

Phase	SI	log IAP	log KT	
Anhydrite	-0.84	-5.20	-4.36	CaSO$_4$
Aragonite	0.61	-7.72	-8.34	CaCO$_3$
Artinite	-2.03	7.57	9.60	MgCO$_3$:Mg(OH)$_2$:3H$_2$O
Birnessite	4.81	5.37	0.56	MnO$_2$
Bixbyite	-2.68	47.73	50.41	Mn$_2$O$_3$
Brucite	-2.28	14.56	16.84	Mg(OH)$_2$
Calcite	0.76	-7.72	-8.48	CaCO$_3$
Chalcedony	-0.51	-4.06	-3.55	SiO$_2$
Chrysotile	3.36	35.56	32.20	Mg$_3$Si$_2$O$_5$(OH)$_4$
Clinoenstatite	-0.84	10.50	11.34	MgSiO$_3$
CO$_2$ (g)	-3.38	-21.53	-18.15	CO$_2$
Cristobalite	-0.48	-4.06	-3.59	SiO$_2$
Diopside	0.35	20.25	19.89	CaMgSi$_2$O$_6$
Dolomite	2.40	-14.69	-17.09	CaMg(CO$_3$)$_2$
Epsomite	-2.36	-4.50	-2.14	MgSO$_4$:7H$_2$O
Fe(OH)$_{2.7}$Cl$_{0.3}$	5.52	-6.02	-11.54	Fe(OH)$_{2.7}$Cl$_{0.3}$
Fe(OH)$_3$ (a)	0.19	-3.42	-3.61	Fe(OH)$_3$
Fe$_3$(OH)$_8$	-12.56	-9.34	3.22	Fe$_3$(OH)$_8$
Forsterite	-3.24	25.07	28.31	Mg$_2$SiO$_4$
Goethite	6.09	-3.41	-9.50	FeOOH
Greenalite	-36.43	-15.62	20.81	Fe$_3$Si$_2$O$_5$(OH)$_4$
Gypsum	-0.64	-5.22	-4.58	CaSO$_4$:2H$_2$O
H$_2$ (g)	-41.22	1.82	43.04	H$_2$
Halite	-2.51	-0.92	1.58	NaCl
Hausmannite	1.78	19.77	17.99	Mn$_3$O$_4$
Hematite	14.20	-6.81	-21.01	Fe$_2$O$_3$
Huntite	1.36	-28.61	-29.97	CaMg$_3$(CO$_3$)$_4$
Hydromagnesite	-4.56	-13.33	-8.76	Mg$_5$(CO$_3$)$_4$(OH)$_2$:4H$_2$O
Jarosite (ss)	-9.88	-43.38	-33.50	(K$_{0.77}$Na$_{0.03}$H$_{0.2}$)Fe$_3$(SO$_4$)$_2$(OH)$_6$
Magadiite	-6.44	-20.74	-14.30	NaSi$_7$O$_{13}$(OH)$_3$:3H$_2$O
Maghemite	3.80	-6.81	-10.61	Fe$_2$O$_3$
Magnesite	1.07	-6.96	-8.03	MgCO$_3$
Magnetite	3.96	-9.30	-13.26	Fe$_3$O$_4$
Manganite	2.46	6.28	3.82	MnOOH
Melanterite	-19.35	-21.56	-2.21	FeSO$_4$:7H$_2$O
Mirabilite	-2.49	-3.60	-1.11	Na$_2$SO$_4$:10H$_2$O
Mn$_2$(SO$_4$)$_3$	-54.60	-9.29	45.31	Mn$_2$(SO$_4$)$_3$
MnCl$_2$:4H$_2$O	-12.88	-10.17	2.71	MnCl$_2$:4H$_2$O
MnSO$_4$	-14.48	-11.81	2.67	MnSO$_4$
Nahcolite	-2.91	-13.79	-10.88	NaHCO$_3$

Table 15.3 continued

Phase	SI	log IAP	log KT	
Natron	-4.81	-6.12	-1.31	$Na_2CO_3:10H_2O$
Nesquehonite	-1.37	-6.99	-5.62	$MgCO_3:3H_2O$
NH_3 (g)	-8.73	2.29	11.02	NH_3
Nsutite	5.85	5.37	-0.48	MnO_2
O_2 (g)	-0.70	-3.66	-2.96	O_2
Portlandite	-9.00	13.80	22.80	$Ca(OH)_2$
Pyrochroite	-8.01	7.19	15.20	$Mn(OH)_2$
Pyrolusite	7.03	5.37	-1.66	MnO_2
Quartz	-0.08	-4.06	-3.98	SiO_2
Rhodochrosite	-3.20	-14.33	-11.13	$MnCO_3$
Sepiolite	1.15	16.91	15.76	$Mg_2Si_3O_{7.5}OH:3H_2O$
Siderite	-13.13	-24.02	-10.89	$FeCO_3$
SiO_2 (a)	-1.35	-4.06	-2.71	SiO_2
Talc	6.04	27.44	21.40	$Mg_3Si_4O_{10}(OH)_2$
Thenardite	-3.34	-3.52	-0.18	Na_2SO_4
Thermonatrite	-6.17	-6.05	0.12	$Na_2CO_3:H_2O$
Tremolite	11.36	67.93	56.57	$Ca_2Mg_5Si_8O_{22}(OH)_2$

manganite (MnOOH), birnessite (MnO_2), manganese sulfide (MnS), and rhodochrosite ($MnCO_3$).

The precipitation of manganese oxides, and manganese hydroxides, is very obvious at a depth of about 0.1 m below the sediment surface. For the minerals manganite and birnessite, exemplifying a group of similar minerals, the saturation indices lie around zero. In deeper core regions these and all other oxides and hydroxides are in a state of undersaturation.

The slight but very constant supersaturation (saturation index about +0.5) found for the mineral rhodochrosite over the whole depth range between 0.2 and 12 m below the sediment surface is quite interesting as well. This can be understood as an indication for a new formation of the mineral in this particular range of depth. Below 13 m rhodochrosite and MnS are both *under*saturated, obviously absent in the solid phase, and therefore cannot have controlled the low manganese concentrations still prevalent in greater depths by means of mineral equilibria.

Calculation of Predominance Field Diagrams

The graphical representations of species distributions and/or mineral saturation as a function of E_H-value and pH-value, which are commonly typified as predominance field diagrams, or sometimes even not very correctly as stability field diagrams, can be done very accurately with the aid of geochemical model programs. The diagrams related to manganese, iron and copper shown in Chapter 11 (Figures 11.2, 11.4, and 11.16) as well as the Figure 15.2 referring to arsenic, have all been made by applying the model program PHREEQC.

The procedure consisted in scanning the whole E_H/pH range in narrow intervals with several thousand PHREEQC-calculations and a given specific configuration of concentrations (in this particular case an ocean water analysis at the ocean bottom, at the site of manganese nodule formation). Thus, each special case is individually calculated and adjusted in agreement to the appropriate temperature, the correct ionic strength (and hence the correct activity coefficient) and accounting for all essential aquatic species and all the minerals eventually present.

Such diagrams as these are not only much more closely adapted to the respective geochemical environment, but also contain essentially more information than the otherwise customary diagrams in which boundaries mostly result in a simplified

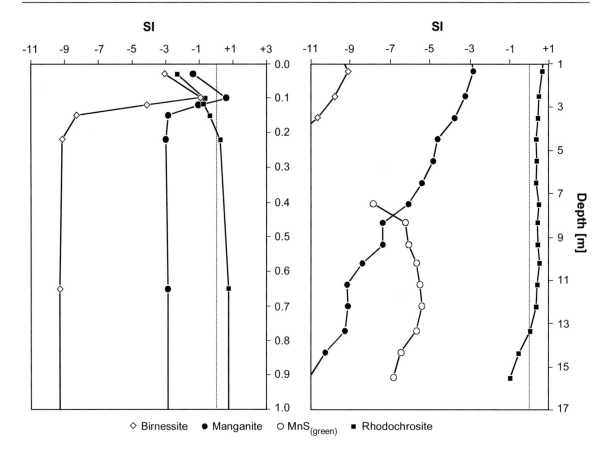

Fig. 15.1 Depth profiles of the saturation indices related to several manganese minerals in sediments approximately 4,000 m below the water surface, off the Congo Fan. The diagram depicted on the left-hand side covers the region close to the sediment surface; whereas the diagram on the right reflects the range down to the core's ultimate depth in another depth scale. Concentration profiles of the pore water pertaining to this core are shown in Figure 3.1; the concentration profile of manganese is shown in more detail in Figure 3.8, its diffusive fluxes are discussed more thoroughly in Section 3.2.3.

fashion from the given specific conditions. On the other hand, this information cannot any longer be included reasonably into one diagram only. Here, one diagram is used exclusively for the pre-dominance field of aquatic species (left diagram), whereas a second is used to visualize the sa-turation ranges of the various minerals (right diagram). This makes sense inasmuch as, wherever minerals become supersaturated, there will nevertheless be an according distribution of the aquatic species that cannot simply be omitted in preference of the minerals. The diagrams shown on the right-hand side reflect the satura-tion ranges. Here, it is reasonable to depict over-lapping ranges of various minerals, because which mineral might precipitate - and which one will not - is not determined on account of these statements or the degree of supersaturation.

Adjustment of Equilibria to Minerals and to the Gaseous Phase

Calcite is supersaturated in seawater which is described with regard to its solutes in the Tables 15.1 to 15.3. For calcite, the degree of supersa-turation is expressed by the saturation index of 0.76, as shown in Table 15.3. Moreover, the partial pressures values of the gaseous phases pCO_2 and pO_2 (log pCO_2 = -3.38; log pO_2 = -0.70) are indi-cated which differ only slightly from the respec-tive atmospheric values. Now it would make sense to ask what composition of this seawater is to be expected, if the equilibria were adjusted to the atmospheric CO_2 and O_2-partial pressure, and, for example, to the mineral calcite that eventually might need to be precipitated. Table 15.4 shows the result of such a calculation with regard to the essential properties of the solution. Naturally, the

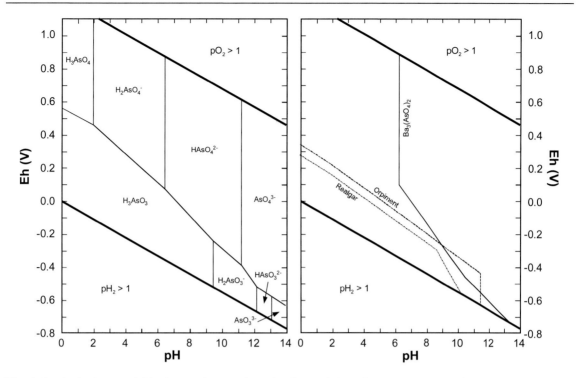

Fig. 15.2 Predominance Field Diagram for arsenic under the condition of manganese nodule formation at the bottom of the deep sea (after Glasby and Schulz 1999). The calculations in this diagram were performed with the model PHREEQC. The partial diagram on the left only represents the predominant ranges of the aquatic species. The diagram on the right covers the ranges of several minerals which are, under the given conditions, supersaturated (SI > 0).

program PHREEQC will then provide further detailed information on the distribution of the aquatic species and all the saturation indices according to the Tables 15.2 and 15.3, which presentation, however, was omitted in this context.

Decomposition of Organic Matter by Oxygen and Nitrate

If certain boundary conditions are maintained, the application of the model program PHREEQC leads to options of particular interest, whenever simulations of one, or several simultaneously running reactions are carried out. The oxidation of organic matter by dissolved oxygen - and after it is consumed, by nitrate - represents a very important reaction in seawater and especially in the sediment. In the model calculation presented in Figure 15.3 the following boundary conditions were selected: The seawater under study, or marine pore water, should be in a state of equilibrium with the mineral calcite at the beginning of the reaction and with a gaseous phase containing 21 % O_2 and 0.034 % CO_2. The

composition of this water sample is summarized in Table 15.4. Organic substance containing amounts of nitrogen and phosphorus according to the

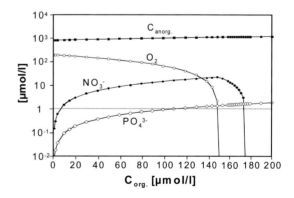

Fig. 15.3 Geochemical model calculation using the program PHREEQC. To an oxic seawater with calcite in equilibrium (cf. Table 15.4) an organic substance is gradually added automatically leading to a redox reaction. The system should continue to be open to calcite equilibrium, but sealed from the gaseous phase environment. First, free oxygen, and subsequently nitrate, will be consumed. Afterwards, the system shifts to the anoxic state (cf. continuance of reactions in Fig. 15.4).

Seawater from Nordstrom et al. (1979)

Phase	Si	log IAP	log KT
Calcite	0.00	-8.48	-8.47
CO_2 (g)	-3.47	-21.62	-18.15
O_2 (g)	-0.68	82.44	83.12
pH	=	7.902	Charge balance
pe	=	12.713	Adjusted to redox equilibrium
Ionic strength	=	6.734E-1	
Temperature (°C)	=	25.0	

Elements	Molality	
C	8.139E-4	Equilibrium with Calcite and pCO_2
Ca	9.876E-3	Equilibrium with Calcite
Cl	5.657E-1	not changed
Fe	3.711E-8	not changed
K	1.058E-2	not changed
Mg	5.507E-2	not changed
Mn	3.773E-9	not changed
N	6.571E-6	Adjusted to redox equilibrium
Na	4.854E-1	not changed
O(0)	3.951E-4	Equilibrium with pCO_2
S	2.93E-02	not changed
Si	7.382E-5	not changed

Table 15.4 In the seawater analyses shown in Tables 15.1 to 15.3, an equilibrium adaptation to the mineral calcite (SI = 0) and to the atmosphere with log pCO_2 = -3.47 (equivalent to pCO_2 = 0.00034) and log pO_2 = -0.68 (equivalent to pO_2 = 0.21) was calculated. This model calculation was performed with the program PHREEQC (Pankhurst 1995, Parkhurst and Appelo 1999).

Redfield ratio (Redfield 1958) of C:N:P = 106:16:1 (cf. Section 3.2.5) is then gradually added to this solution. Simplified, such a theoretical organic substance is composed of $(CH_2O)_{106}(NH_3)_{16}(H_3PO_4)_1$ or, divided by 106: $(CH_2O)_{1.0}(NH_3)_{0.15094}(H3PO4)_{0.009434}$. The horizontal axis in Figure 15.3 records the substance addition in terms of µmoles per liter. If the situation in the sediment is to be described, an equilibrium of calcite should exist as well. The system under study should not be open to the gaseous phase any more, instead, these influences should now be determined by the consumption of oxygen within the system and by the concomitant release of CO_2.

In the model, the organic substance added to the reaction automatically triggers a sequence of redox reactions which becomes evident due to the concentrations involved which are demonstrated in Figure 15.3. First, the concentration of oxygen declines, and the concentration of nitrate and phosphorus increases according to each of their proportions in the original organic substance. As soon as oxygen is consumed, nitrogen will be utilized for further oxidation of the continually added organic substance. Subsequently, the system shifts over to an anoxic state.

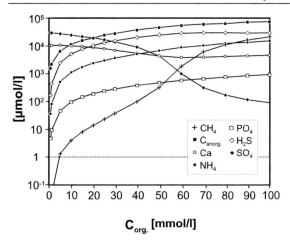

Fig. 15.4 Geochemical model calculation using the program PHREEQC. In an anoxic system (state at the end of the model calculation from Fig. 15.3), the gradual addition of organic matter to the redox reaction is continued, whereby the system is kept open for calcite equilibrium and sealed from the gaseous phase. Initially, the dissolved sulfate will be consumed, in the course of which low amounts (logarithmic scale) of methane will emerge. Only after the sulfate concentration has become sufficiently low, will the generation of methane display its distinct increase.

Decomposition of Organic Substance under Anoxic Conditions

An anoxic system, as obtained from the previous reaction sequence, will be capable of further on-going reaction, if the supply of organic matter is continued and the reaction is held under the same boundary conditions. The result consists in the continuation of the processes shown in Figure 15.3. The outcome of this experiment is presented in Figure 15.4. Here as well, the horizontal axis is calibrated with regard to the added amounts of organic matter in terms of mmoles per liter.

As expected, the sulfate concentration in solution initially decreases at a swift rate and, concomitantly, low amounts of methane already emerge. Only after the concentration of sulfate falls under a certain level, will methane be released in substantial amounts. The permanently maintained calcite equilibrium will result, due to precipitation, in the depression of the concentration of dissolved calcium ions (compare Schulz et al. 1994).

The sequence of reactions, as described and modeled above, has naturally been well established and frequently documented ever since the publications made by Froelich et al. (1979). However, it should not be overlooked that these processes were not pre-determined for the modeling procedure, instead, they rather 'automatically' evolved from the fundamental geochemical data supporting the model as well as from the pre-set, although very generalized, starting and boundary conditions. During the entire model calculation procedure, all quantitatively important species, the complex aquatic species, as well as the activity coefficients related to these solutions, were invariably taken into account. The dissolution/precipitation of calcite, a realistic and also quite well known reaction, was employed for an adjustment of the mineral phase equilibrium. Such a system can be made more complex, almost at will, by specifically operating with additional mineral phases (e.g. iron minerals or manganese minerals) that dissolve eventually, provided they are present and the condition of undersaturation prevails, or are allowed to precipitate in the case of supersaturation. However, the simulation of such mineral equilibria requires a very profound knowledge of the specific reactions of mineral dissolution or precipitation in the system under study.

15.2   Analytical Solutions for Diffusion and Early Diagenetic Reactions

Analytical solutions for Fick's Second Law of Diffusion were already discussed in Chapter 3, Section 3.2.4, without taking diagenetic reaction into account. With a diffusion coefficient D_{sed}, which describes the diffusion inside the pore cavity of sediments, Fick's Second Law of Diffusion is stated as:

$$\frac{\partial C}{\partial t} = D_{sed} \cdot \frac{\partial^2 C}{\partial x^2} \qquad (15.5)$$

This partial differential equation has different solutions for particular configurations and boundary conditions. In Section 3.2.4 the following solution was presented:

$$C_{x,t} = C_0 + \left(C_{bw} - C_0\right) \cdot erfc\left(\frac{x}{2 \cdot \sqrt{\left(D_{sed} \cdot t\right)}}\right)$$

$$(15.6)$$

The solution of the equation yields the concentration $C_{x,t}$ at a specific point in time t and a specific depth x below the sediment surface. Here, t=0 is a point in time at which the concentration C_0 prevails in the profile's entire pore water compartment. At all later points in time the bottom water displayed a constant value C_{bw}. An approximation for the complementary error function (erfc) according to Kinzelbach (1986) can be found in Section 3.2.4, where examples are given that employ this analytical solution of Fick's Second Law.

Advection and very simple reactions can also be included in Fick's Second Law of Diffusion (e.g. Boudreau 1997):

$$\frac{\partial C}{\partial t} = D_{sed} \cdot \frac{\partial^2 C}{\partial x^2} - v \cdot \frac{\partial C}{\partial x} - k \cdot C + R_{(x)} \qquad (15.7)$$

Here, v denotes the mean velocity of advection, and k is a rate constant of a reaction with first order kinetics. The last term in the equation $R_{(x)}$ is an unspecified source or sink related term which is determined by its dependence on the depth coordinate x. Instead of $R_{(x)}$, one might occasionally find the expression (ΣR_i) which emphasizes that actually the sum of different rates originating from various diagenetic processes should be considered (e.g. Berner 1980). Such reactions, still rather easy to cope with in mathematics, frequently consist of adsorption and desorption, as well as radioactive decay (first-order reaction kinetics). Sometimes even solubility and precipitation reactions, albeit the illicit simplification, are concealed among these processes of sorption, and sometimes even reactions of microbial decomposition are treated as first order kinetics.

These, or smaller variations of the equation, are occasionally referred to as 'general diagenetic equation', thus stating a rather simplistic comprehension of (bio)geochemistry in favor of a more mathematically minded approach. A comprehensive and elaborate treatment of the analytical solutions of the equation is presented by Boudreau (1997) who, in particular, considers its multiple variations and boundary conditions. Currently, this important book represents the 'state of the art' for the diagenesis researcher who is more interested in mathematics. For the more practical diagenesis scientist thinking in terms of (bio)geochemistry, this presentation remains somewhat unsatisfactory, despite the fact that it gives various inspirations and contributes to a vast number of important compilations (e.g. as to the state of knowledge on diffusion coefficients in the sediment). Ultimately, the (bio)geochemical appli-

cations of such a solution are in fact rather limited and illicitly restrict the complexity obtained by experimental measurement in the natural system.

In simpler stationary cases the author would always prefer solutions obtained by applying the 'Press-F9-method' (compare Section 15.3.1), simple explicit numerical two-step models (compare Section 15.3.2), and numerical solutions, particularly for non-stationary cases or in cases with complex reactions, by using models like STEADYSED1 or CoTReM (compare Section 15.3.3).

Van Cappellen and Gaillard (1996) gave indications that the activity coefficients in marine pore water, and the various complex species as well, need to be included as a very important part of the solution (also compare Sections 15.1.1 and 15.1.2). It has to be assumed that, for instance, the decomposition of sulfate via respective complex equilibria also affects the activity of other dissolved ions. Furthermore, it has to be considered that charged complexes and ions, transported by means of diffusion at different rates (differing diffusion coefficients !), must lead to a separation of the distributed charges in the course of transportation. The importance this has for the processes particularly taking place in the sediment is not yet sufficiently established. However, van Cappellen and Gaillard (1996) assume that such processes may probably be neglected in marine sediments as the charges there are essentially determined by sodium and chloride ions.

15.3   Numerical Solutions for Diagenetic Models

The previous section demonstrated that the analytical solutions of Fick's Second Law of Diffusion can only be applied to a very limited number of cases. Frequently, highly simplified boundary conditions have to be assumed which actually cannot be found in a natural environment. Additionally, the effort is almost always limited to merely one component dissolved in pore water, or to its behavior in space and/or time. Any complex relations between the numerous dissolved components, or even concentrations of complexes and ion pairs which become variable due to reaction sequences, or accordingly taking into account the difference between concentrations and activities, all are simply not regarded in the mathematical solutions of diagenetic processes. Nevertheless, these multiple component processes do not constitute a rare exception but represent the normal case, and must be considered the more one becomes concerned with measuring natural systems.

The solution to this problem consists in the application of numerical solutions when diagenetic processes are modeled. Such numerical solutions always divide the continuum of reaction space and reaction time into discrete cells and discrete time intervals. If one divides up the continuum of space and time to a sufficient degree into discrete cells and time steps (which is not the decisive problem with the possibilities given by today's computers), one will be able to apply much simpler and better manageable conditions within the corresponding cells, and with regard to the expansion of a time interval, so that, in their entirety, they still will describe a complex system. Thus, it is possible, for example, to apply the two-step-procedure (Schulz and Reardon 1983), in which the individual observation of physical transport (advection, dispersion, diffusion) or any geochemical multiple component reaction is made feasible within one interval of time.

15.3.1 Simple Models with Spreadsheet Software ('Press-F9-Method')

Structure of a Worksheet and Oxygen/Nitrate Modeling

Normal spreadsheet software (e.g. Microsoft - Excel®) present tools especially suited for modeling stationary and not too complex systems. The well-understood decomposition of organic substance by dissolved oxygen, and the concomitant release of nitrate, will serve as an example. Nitrate is also utilized in the oxidation of organic substance in deeper sections of the diffusion controlled profile. Figure 15.5 demonstrates the result of such a model procedure, the details of which will be further discussed in the following.

Table 15.5 shows excerpts from the worksheet used for modeling the decomposition of organic substance mediated by oxygen and nitrate. In the upper frame of the worksheet there are - apart from the headline – single parameters and constants that will be used in the lower part of the worksheet; they will be discussed later in that particular context.

In the lower frames of Table 15.5 the actual worksheet is depicted in rows 11 to 61 that divide

up the one-dimensional profile into 50 lines, which represent 50 discrete cells. The first column (A) contains the depth below the sediment surface. In the uppermost cell (11) this depth is set to zero and represents the bottom water zone immediately on top of the sediment. Line 61 is already beyond the boundary of the modeled area and will be discussed later in the context of setting the boundary conditions. In cell A12 to A61 the depth is increased by the value dx = 0.001 m per cell, as pre-selected in cell A6. The cell A12 contains the Excel® notation: =A11+$A$6. This field may be copied all the way down to field A61. In the next two columns (B) and (C) we find default porosity values (n) or diffusion coefficients (D_{sed}) valid for all cells. However, should the porosity constants or diffusion coefficients be known for the particular depth profile, then they may be entered individually into the respective cells. The columns (D), (E) and (F) are connected to each other by loops so that an error message will be announced,

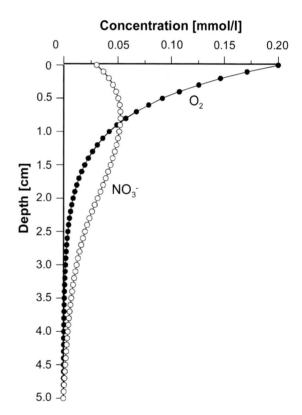

Fig. 15.5 Model of the decomposition of organic substance by dissolved oxygen and dissolved nitrate in a diffusion controlled pore water profile. Modeling was performed according to the 'Press-F9-method' with the spreadsheet software Excel®. Details pertaining to this model are explained in Table 15.5 and in the text.

Table 15.5 Worksheet (excerpts) from modeling the decomposition of organic substance by dissolved oxygen and dissolved nitrate in a diffusion controlled pore water profile. The model procedure was performed according to the 'Press-F9-method' with an Excel®. worksheet The columns are marked in the uppermost row with alphabetical characters, the rows are numbered in the first column. Details pertaining to the model are further explained in the text.

	A	B	C	D	E	F	G	H	I
1									
2	**Model oxidation C$_{org}$ with oxygen and nitrate**						switch	C:N	k(O$_2$)
3							0	3.0	7.9E-09
4					mmol l^{-1} = mol m^{-3}				
5	dx(m)		dt(sec)	C$_0$(O$_2$)	C$_0$(NO$_3^-$)	C$_b$(O$_2$)	C$_b$(NO$_3^-$)		k(NO$_3^-$)
6	0.001		1.0E+06	0.2	0.03	0.0	0.0		3.2E-09
7									
8									
9	Depth	ϕ	Ds	O$_2$-Flux	O$_2$-Red.	O$_2$	NO$_3^-$-Flux	NO$_3^-$-Red.	NO$_3^-$
10	m		m^2 sec^{-1}	mol m^{-2} s^{-1}		mmol l^{-1}	mol m^{-2} s^{-1}		mmol l^{-1}
11	0.000	0.60	5.4E-10	9.36E-09		2.00E-01	-2.11E-09		3.00E-02
12	0.001	0.60	5.4E-10	8.01E-09	1.35E-09	1.71E-01	-1.66E-09	0.00E+00	3.65E-02
13	0.002	0.60	5.4E-10	6.85E-09	1.16E-09	1.46E-01	-1.27E-09	0.00E+00	4.16E-02
14	0.003	0.60	5.4E-10	5.86E-09	9.90E-10	1.25E-01	-9.40E-10	0.00E+00	4.55E-02

continued

29	0.018	0.60	5.4E-10	5.65E-10	9.53E-11	1.21E-02	8.27E-10	0.00E+00	3.63E-02
30	0.019	0.60	5.4E-10	4.83E-10	8.16E-11	1.03E-02	8.55E-10	0.00E+00	3.37E-02
31	0.020	0.60	5.4E-10	4.13E-10	6.98E-11	8.83E-03	8.78E-10	0.00E+00	3.11E-02
32	0.021	0.60	5.4E-10	3.54E-10	5.97E-11	7.56E-03	8.07E-10	9.09E-11	2.84E-02
33	0.022	0.60	5.4E-10	3.03E-10	5.11E-11	6.46E-03	7.42E-10	8.29E-11	2.59E-02
34	0.023	0.60	5.4E-10	2.59E-10	4.37E-11	5.53E-03	6.81E-10	7.56E-11	2.36E-02

continued

55	0.044	0.60	5.4E-10	1.15E-11	1.40E-12	1.77E-04	1.30E-10	7.13E-12	2.23E-03
56	0.045	0.60	5.4E-10	1.04E-11	1.12E-12	1.41E-04	1.25E-10	5.84E-12	1.83E-03
57	0.046	0.60	5.4E-10	9.56E-12	8.62E-13	1.09E-04	1.20E-10	4.61E-12	1.44E-03
58	0.047	0.60	5.4E-10	8.93E-12	6.29E-13	7.96E-05	1.17E-10	3.42E-12	1.07E-03
59	0.048	0.60	5.4E-10	8.52E-12	4.11E-13	5.20E-05	1.15E-10	2.26E-12	7.07E-04
60	0.049	0.60	5.4E-10	8.32E-12	2.03E-13	2.57E-05	1.14E-10	1.13E-12	3.52E-04
61	0.050					0.00E+00			0.00E+00

if the appropriate Excel® setting ('Iterative Calculation') should not be selected. The diffusive O$_2$ flux is calculated by employing Fick's First Law of Diffusion (Equation 3.2 in Chapter 3), whereby the gradient obtained from the concentrations in column (F) is used. The concentrations in column (F) are calculated stepwise for each time interval from the preceding value, from the material flux in and out of the cell, and from the decomposition activity listed in column (D). Hence, the result in cell D11 reads in Excel® notation:

$$D11=B11*C11*((F11-F12)/\$A\$6)$$

D11 consequently contains the flux rate of material directed from the bottom water into the sediment. The cells D12 to D60 will be accordingly copied from this cell. Cell D61 remains empty.

The decomposition of organic substance in the uppermost sediment zone is written in the cell E12; cell E11 (bottom water) remains empty. In the example shown the decomposition of organic substance mediated by oxygen in a reaction following first order kinetics was still bound to the yet prevalent oxygen concentration. This does not mean to say that this must always be so. As for the model, any other kinetic, or, in general terms, any other value could be assigned to the respective cells. As for the first order kinetics example, the line E12 reads in Excel® notation as follows:

E12=$I$3*F12

Cells E13 to E60 are copied from this cell accordingly, cell E61 remains empty. Column (F) contains O_2-concentrations, whereby cell F11 contains the upper boundary condition via a constant bottom water concentration. Cell F61 contains the concentration at the lower boundary - which in this particular case is equivalent to 0.0. The cell F12 is calculated from cell F12 (hence from its own previous value) as well as from the cells D11, D12, and E12, each multiplied with the length of the time interval. Written in Excel® notation this reads:

F12=IF($G$3;$D$6;F12+(D11*$C$6)-(D12*$C$6)-(E12*$C$6))

(The function 'IF' refers to the English version of Excel®, e.g. for the German version this function would be 'WENN'.)

Cells F13 to F60 are copied from this cell accordingly. Cell F61 contains the lower boundary conditions (in this case 0.0) which will be further dealt with later. It appears of practical importance at this point of calculation that there is the option of stopping the calculation procedure by means of a switch function. If the switch cell G3 assumes zero value the calculation will be performed. If the cell contains the value 1, then all concentrations will be adjusted to the value present in cell D6 (concentration in bottom water). Designing the structure of the worksheet, the switch should be set to a value of 1, in order to avoid unnecessary error messages due to references to not yet properly filled cells. Additionally, it is easy to rebuild the worksheet whenever ill-chosen conditions pertaining to the iteration have led to extreme values and/or oscillations.

Column (G) contains nitrate concentrations analogous to column (D) used for oxygen. In this example, the same diffusion coefficients were used to calculate the fluxes. Here, different individually adjusted coefficients can certainly also be used. Again, cell G11 contains the flux rate directed from the bottom water to the sediment and reads written in Excel® notation:

G11=B11*C11*((I11-I12)/$A$6)

Cells G12 to G60 are copied from this cell accordingly, whereas cell G61 remains empty. Column (H) contains the decomposition mediated by nitrate which can be set voluntarily, as has been the case in oxygen mediated decomposition. In this example, a decomposition of 0.0 has been pre-determined for the cells H12 to H31, as well as the assumption of first-order kinetics in cells H32 to H60. Here, too, it is not intended to claim that this allocation must necessarily be as shown in the example. As for the model, any other type of kinetics would be possible as well as any values set otherwise . Hence, cell H32 reads in Excel® notation:

H32=$I$6*I32

Cells H33 to H60 are copied from this cell accordingly. Cell H61 remains empty.

Analogous to column (F), column (I) this time contains the calculated nitrate concentrations which again are connected with loops to the columns (G) and (H). However, in the case of nitrate, there is also some release into the pore water fraction due to the oxygen mediated decomposition of organic substances. Thus, each cell belonging to column (I) is calculated from its previous value, from the diffusive import from above, from the diffusive export downwards, from the import from the oxidation of organic material mediated by oxygen (here, selective C:N ratio), as well as from the decomposition mediated by nitrate. For cell I12, this reads in Excel® notation as follows:

I12=IF($G$3;$E$6;I12+(G11*$C$6)-(G12*$C$6)+((1/$H$3)*E12*$C$6)-(H12*$C$6))

(The function 'IF' refers to the English version of Excel®, e.g. for the German version this function would be 'WENN'.)
Cells I13 to I60 are copied from this cell accordingly. Cell I11 contains the bottom water concentration as a fixed boundary condition, whereas cell I61 assumes the value of 0.0 as the concentration at the lower boundary. In this column it is also recommended to link the calculation performance to the operational 'switch' of cell G3.

The model is now finished and the essential margin values compiled in the upper frame of Table 15.5 are added. The switch in G3 is then set to zero. Now the key F9 (re-calculation of the worksheet) is repeatedly pressed until the figures in the worksheet do not (significantly) change, i.e., until a (practically) steady state is reached. Depending on the speed of the computer and the spreadsheet software used, this can take some few minutes, so that one can hold the F9-key down during that time.

At this point it needs to be clearly stated that only the final iterative steady state of the model provides a correct result, because the diffusive flux is calculated according to Fick's First Law of Diffusion in the columns (D) or (G) which is only valid when stationary conditions are provided. Even if time intervals of specific length are employed in the process of iteration, it is not permitted, by no means at all, to draw conclusions on the time necessary for reaching the steady state, since the columns (D) and (G), and consequently (F) and (I) as well, contain false information relative to each of the respective non-steady states.

The length of the time intervals (cell C6) is only a mechanism used for iteration, whereby too short intervals will result in an unnecessarily long pressing of the key F9, whereas too large intervals would lead to escalating oscillations within the model (in our example after about dt=2.3E6). Only after the steady state is reached will this 'false way' of the iteration become 'consigned to oblivion', and the result will be correct in all columns. Non-steady states and intervals prior to the development of particular concentration patterns can only be managed by applying Fick's Second Law of Diffusion, either in form of analytical solutions (cf. Section 15.2) or in form of numerical solutions (cf. Sections 15.3.2 and 15.3.3).

The concentration-dependent boundary conditions assigned to the model's upper-boundary limits (unspecified but constant concentrations in the cells F11 and I11) are relatively unproblematic since they represent a constant concentration in the bottom water zone. This is an imperative prerequisite for assuming a steady state in pore waters from superficial sediments. The condition for the lower boundary of the profile is somewhat more problematical. In the above example, the concentrations of both oxygen and nitrate were set to zero. In the model, this is tolerable as long as the value 0.0, constituting the lower boundary

(within the limits) of the model, already results from the reactions, and as long as the set value 0.0 will not exert any practical influence on the lower parts of the profile. This 'zero concentration boundary condition' hence constitutes a 'zero flux boundary condition'. This also implies that the boundaries of the model must be extended downward until this state is sufficiently assumed. Selecting a zero concentration and, consequently a 'zero flux boundary condition', makes sense in this example, but it must not always be an adequate boundary condition. In the following example, we will consider the case in which a value is chosen which clearly deviates from 0.0 and thus indirectly provides a steady diffusive flux across the lower boundary.

Modeling the Reaction Sulfate/Methane

Niewöhner et al. (1998) describe profiles of pore water which were obtained from upwelling areas off the shores of Namibia. In these pore-water samples, sulfate reacted with methane in a ratio of 1:1. A similar example has already been introduced in Chapter 3 (Figure 3.31, also cf. Chapter 5), in context with problems occurring in methane analytics. In this example, the measured values will now be presented in Figure 15.6 together with the result of a model designed according to the 'Press-F9- method'.

In principle, the worksheet for the modeling of Figure 15.6 does not differ very much from the one described previously, so that it does not demand special description. The essential features of this model are:

The decomposition of methane and sulfate (cf. Chapter 8) occur for both reactants in specific depths at identical rates which were entered into the spreadsheet. The adjustment to the measured profiles was committed only by these (microbial) decomposition rates. The decomposition parameter is set to 0.0 in all other cells, so that a diffusion controlled transport occurs with a constant concentration gradient.

The concentration of the upper boundary condition for the model is again provided by the appropriate bottom water concentrations: sulfate = 28 mmoles/l; methane = 0.0 mmoles/l.

The concentration of the lower boundary condition for sulfate is 0.0 mmol/l, since it was decomposed to 0.0 mmol/l long before the lower boundary of the model had been reached. However, the concentration of the lower boundary

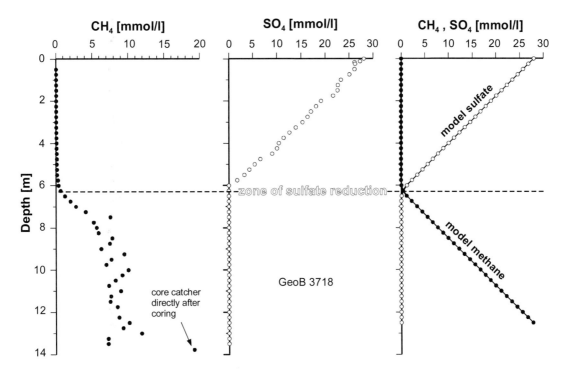

Fig. 15.6 Model results of the mutual decomposition of sulfate and methane as a 1:1-reaction in a diffusion controlled pore water profile. Modeling was performed according to the 'Press-F9-method' using the standard software Excel®. Details pertaining to the model and the calibration with data from a measured pore-water profile obtained from an upwelling area off Namibia (Niewöhner et al. 1998) are discussed in the text.

condition for methane has been concluded on the basis of pre-determining a steady decomposition in the same amount as for sulfate, and that this decomposition should continue in a specific depth until the value 0.0 is reached.

15.3.2 Two-Step Models with Explicit Numerical Solution of Fick's 2nd Law

In order to adequately account for the time course of processes which are mainly governed by diffusion, Fick's Second Law of Diffusion must be applied without exception (Equation 15.5). After all, Fick's First Law does not account for time at all. The analytical solutions described in Sections 3.2.4 and 15.2 and referring to Fick's Second Law may permit the calculation of a distribution of concentrations at a certain point in time, however, they apply to inert, non-reacting compounds. Only very specific and very simple reaction types, such as the completely reversible adsorption/ desorption of a compound or its degradation following first-order kinetics can be included in

analytical solutions. All reactions of higher complexity - and these encompass actually all the processes of early diagenesis - require reactions which are distinguished by the interdependencies between the various reactants and thus evade an analytical solution in which a distribution of concentrations has to be calculated in a single step, starting out from a given initial concentration.

Adequately modeling the combination of the physical transport by means of diffusion (including advection) on the one hand, and reactions which may be as complex as you like on the other, demands a subdivision of both the continuum of time and the transport pathway into discrete time intervals and discrete sections or cells, respectively. In each time step the physical transport of each cell belonging to the transport pathway under study is then calculated relative to each of the other cells, just as if the compound would not undergo reactions. Then the transport process is quasi brought to a halt and the concentrations in all cells will be changed just as if any imaginable (bio) geochemical reaction had in fact taken place. There is actually no type of reaction that cannot be applied for mathematical reasons in this regard. Then physical transport is calculated once again on the basis of the concentrations thus calculated, followed by a subsequent calculation of the reactions etc. This procedure is referred to as the "Two-Step Method" or "Operator Splitting". It will be discussed more thoroughly in the following. As much as we know, this method has been applied for the first time by Schulz and Readon (1983). Here, only the simplest form of such a modeling process, the so-called explicit solution, will be presented for introductory purposes. It rapidly illustrates the principle of the Two-Step Method and suffices completely to achieve adequate modeling results in most practical situations.

The explicit solution (e.g. Kinzelbach 1986; Sieger 1993; Landenberger 1998) calculates a concentration $C_{x,t+t}$ as the concentration at location x of the transport pathway with respect to the future time point (t+Δt), whereby Δt denotes the length of the time step:

$$C_{x,t+\Delta t} = C_{x,t} + \Delta t \left[\begin{array}{c} D_{sed}\left(\dfrac{C_{x+\Delta x,t} - 2 \cdot C_{x,t} + C_{x-\Delta x,t}}{\Delta x^2} \right) \\ -v_a\left(\dfrac{C_{x+\Delta x,t} - C_{x-\Delta x,t}}{2 \cdot \Delta x} \right) \end{array} \right]$$

(15.8)

Only concentrations at the known time point t are used on the right side of the equation sign for the same location x on the transport pathway, next to the two neighboring cells (x+ Δx) and (x-Δx), whereby Δx denotes the cellular length of the transport pathway. D_{sed} is the diffusion coefficient of the compound under study (Section 3.2.2). To a certain extent, the equation also permits the inclusion of advection with the flow velocity v_a. If v_a = 0.0, then only diffusion will be effective. However, in cases of greater flow velocities the much greater dispersion coefficient must be applied instead of the diffusion coefficient D_{sed}. Its dimension depends on the flow velocity. Such fluxes, including the adjusted dispersion coefficients, are preferably significant for the flow properties of groundwater in sandy aquifer (Schulz 2004).

Equation 15.8 hence permits the calculation of the distribution of concentrations in all cells of a transport pathway at the end of a given time period, starting out from any distribution of concentrations prevailing at the beginning. Some-

what problematical is merely the first (uppermost) cell and the last (lowest) cell of the given pathway, for each of these cells lack a neighboring cell either above or below it, a condition which Equation 15.8 demands to be fulfilled by all cells. This normally does not lead to a problem for the first, uppermost cell. It represents the bottom water above the sediment package and thus contains a constant or, at least, a time-specific concentration. The last, or lowest, cell presents more difficulty, as various margin conditions can be selected. Here, an example could consist in a constant or, at least, time-specific concentration as well. Another possibility would be an extension of the concentration gradient from the third-last cell over the second-last to the ultimately last cell. At least Equation 15.8 is only applied to the cells 2 to (n-1) in a series of n cells, and not to the cells 1 and n.

Each of the two variants applicable to determine the concentration in the last, lowest cell of the transport pathway have a flaw. A constant value for the concentration in this cell demands that, during model calculation, a concentration deviating essentially from this value by way of diffusion or reaction should never approach the lower margin, where the model calculation would be influenced by a constant value, without being capable of influencing this value itself. For example, if the constant value were 0.0, all higher concentrations appearing in the proximity would quasi become "swallowed". For this reason, the model territory always has to be defined large enough to ensure that this will never be the case. Such an effect does not appear with the same intensity in the second variant, which is used to determine the concentration at the lower margin by means of a constant gradient, but this solution also produces problems. The small errors which arise from the extrapolation of the gradient can add up to quite remarkable deviations in the course of numerous time steps. It is best to define the model territory large enough to avoid that the gradient near the lower margin will not essentially differ from 0.0 at any given concentration.

The explicit solution permits simple straightforward modeling, but has the disadvantage that the length of a time-step is very closely bound to the length of a cell and to the diffusion coefficient. Very unpleasant oscillations frequently appear which rapidly add up to values lying outside the permissible numerical range if the time-steps become too large. Equation 15.9 describes a maximum length a time-step may assume,

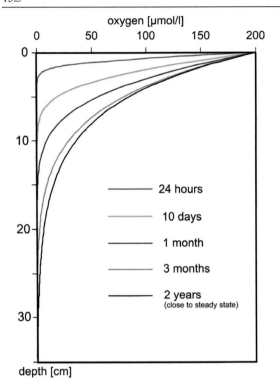

oxygen [µmol/l]

depth [cm]

Fig. 15.7 Calculation of the equilibrium between oxygen consumption occurring during the oxidation of sedimentary organic matter near the sediment surface and re-supply of oxygen by means of diffusion. Here, oxygen consumption is assumed to follow first-order kinetics. Application of Fick's Second Law in an explicit numeric solution permits a reliable calculation of the times required for the adjustment of a stationary condition.

whereby distinctly shorter time-steps may also be used without causing any problem.

$$\Delta t \leq \frac{\Delta x^2}{2 \cdot D_{sed}} \qquad 15.9$$

The explicit solution can be realized quite reasonably with an Excel® worksheet in the sense of the 'Press-F9-Method' (Section 15.3.1). However, in most cases a program written in a higher programming language will be applied, since this type of modeling is made *because* one desires to include more complex (bio)geochemical reactions. And sooner or later the limits of modeling are reached when an Excel® worksheet is used.

An easy to handle and rather generally designed computer model written in FORTRAN (EXPLICIT.FOR) can be downloaded on the internet at:

http://www.geochemie.uni-bremen.de/downloads/

along with brief user instructions written in English and data sets pertaining to several examples.

The following examples (Example 1 and 2) were calculated using this program. Subsequent examples (Examples 3 and 4) reveal how an explicit solution was realized with an Excel® worksheet in the sense of the 'Press-F9-Method' (Section 15.3.1). Downloads of these worksheets are also available at the above mentioned internet address.

Example 1: Non-steady state development of an oxygen profile

An oxygen concentration profile such as the one which has been measured in situ (Fig. 3.5) or modeled (Fig. 15.5) is mostly interpreted in terms of a stationary condition (steady state) resulting from the consumption of oxygen in the sediment and its resupply by means of diffusion. Its consumption in the sediment is due to the oxidation of organic matter and other redox processes, such as the re-oxidation of divalent manganese. But how long does it take until such a stationary condition is reached ?

Figure 15.7 shows an example for the development of a steady-state condition starting out from an extreme initial situation. The explicit solution to Equation 15.8 always requires an initial situation, a concentration profile from which the following profiles of the subsequent time-step are calculated. In this example, an oxygen concentration of 0.0 was chosen as the initial situation underlying the entire profile. In principle, it does not make any difference what kind of initial concentration profile is used to reach the steady-state concentration profile. The initial profile will be "lost", at the latest, after a steady state is reached. Only the time necessary to reach the steady state will differ. The more the initial situation already resembles the steady state, the shorter the time until this condition will be reached - and vice-versa.

In this example, the oxygen consumption in each cell was calculated by applying first-order kinetics, which means that, in this case, oxygen is depleted with a half-life of 50 days. This is only but one possibility imaginable, other rates or kinetics of degradation are also possible and can be applied in the model accordingly. The half-life of 50 days represents a sediment distinguished by a relatively poor turnover. Naturally, there is no consumption at the beginning when no oxygen is present in the entire profile. But any oxygen supplied from the bottom water into the sediment by means of diffusion will be immediately subject to the already mentioned consumption of oxygen displaying a half-life of 50 days.

If the sedimentary diffusion coefficient is given, the position of stationary condition will then be merely a function of the degradation rate, hence the half-life assumed in our example. Figure 15.7 illustrates that a concentration profile close to the stationary condition is reached within a time period as short as three months. This is the case even in our example in which the penetration depth of oxygen is at about 30 cm, and despite the fact that the initial concentration of 0.0 over the entire profile is far away from the stationary condition.

Example 2: Dating a sediment slide at the continental slope

While the previous example presents a purely theoretical case, Example 2 refers to a real situation (Zabel and Schulz 2001). The data shown in Figure 15.8 represent material from the cores sampled with a gravity corer at a water depth of approximately 4,000 m on the deep-sea fan off the estuary of the Congo

River (Locations GeoB1401 and GeoB4914). Further pore-water profiles from Location GeoB1401 are also shown in Fig. 3.1.

Particularly the sulfate profile in pore water (center diagram in Fig. 15.8) was difficult to interpret as a picture showing a steady-state condition, for one would expect a nearly constant gradient that reaches down to the sulfate-methane transition zone (e.g. Niewöhner et al. 1998). Figure 3.6 shows such a measured profile. The sulfate profile depicted in Figure 15.8., however, displays a distinct alteration in the gradient at a depth range between 8 m and approximately 10 m. This would normally be explained with a re-oxidation of sulfide yielding sulfate under stationary conditions. However, oxygen and nitrate are not available for the reaction, as they have already been exhausted in the uppermost centimeters of this very reactive sediment. A re-oxidation reaction of sulfide with trivalent iron originating from the solid phase of the sediment would be imaginable theoretically, but this possibility was ruled out because the trivalent iron content of the sediment and the low sedimentation rate (approximately 5 cm kyr^{-1}, as derived from the left diagram in Fig. 15.8) account for a flux of trivalent iron to the reaction site which would be markedly too low - by more than a power of magnitude.

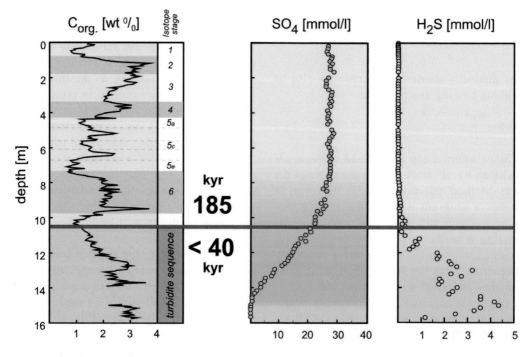

Fig. 15.8 Geochemical data obtained from a sediment core originating from a depth a 4,000 m off the Congo-River estuary (GeoB1401, GeoB4914) (Zabel and Schulz 2001). Left: C_{org} concentration and isotope stages. Center: Sulfate concentrations in pore water. Right: Sulfide concentrations in pore water. The red line crossing all profiles denotes a leap in time which is understood as an old surface, a basis of a sediment slide.

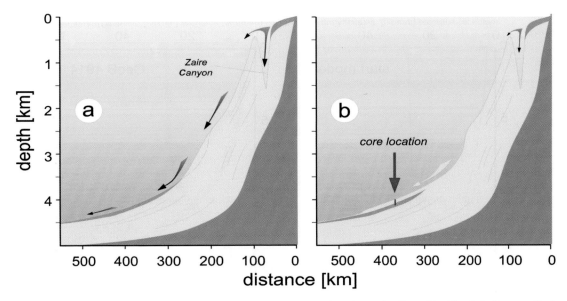

Fig. 15.9 Model concepts related to the development of the situation shown in Fig. 15.8. The left diagram displays the development of turbidite currents rich in organic matter at a water depth of approx. 4,000 m. The right diagram shows how the slide of an internally unperturbed sediment package covers these turbidite layers (Zabel and Schulz 2001).

A solution to this problem is only gained when this profile is interpreted as a non-state condition (Zabel and Schulz 2001), assuming that in recent geological time an approximately 10 m-thick sediment block slid over a very young series of turbidites which were rich in organic matter. This sediment slide must have proceeded so slowly that the sediment block remained unperturbed in its interior stratigraphy. Figure 15.9 shows the model concept which provided the basis in subsequent modeling of the pore water concentration profiles. The left diagram shows the situation before the slide when turbidite currents, rich in organic matter and originating from the overflow of the Zaire Canyon onto the deep-sea fan, had reached a depth of 4,000 meters. The right diagram shows the situation after the slide, including the sampling site of the sediment core examined.

But when did this slide occur? As a steady-state condition has quite obviously not been reached yet, a time-correct model according to Fick's Law and the Two-Step Method must be used to answer the question.

Figure 15.10 demonstrates in its left side the initial situation, hence the situation immediately after the slide in terms of sulfate and methane concentration profiles and alkalinity. Underneath the slide, i.e at the former sediment surface, a steady-state condition with a constant gradient reaching down to the sulfate-methane transition zone was then anticipated. We are familiar with such profiles from highly reactive sediments, for example, those obtained from the Congo Deep-Sea Fan. Unperturbed gradients displaying markedly less biogeochemical turnovers were applied to the sediment package lying on top of the sedimentary slide mass as the initial situation. We also know about such profiles in sediment cores originating from locations somewhat higher on this continental slope.

The right diagram in Figure 15.10 shows the modeling of a non-steady state condition about 300 years after the slide. The mutual degradation of sulfate and methane at the sulfate-methane transition zone and the release of bicarbonate (represented by alkalinity) were stoichiometrically applied as the involved geochemical processes according to the extent of methane oxidation. The model calculation was terminated after achieving a good visual adjustment of the calculated sulfate profile to the measured data. It clearly shows that this modeling procedure is able to explain entirely the measured concentration profiles of sulfate, methane, alkalinity and ammonia (not shown). Since the model was run using the correct time and the correct diffusion coefficients, anticipating that the slide incidence had taken place at about 300 years before the measurement of the concentration profiles may also be assumed to be correct. The potential error of the visual adjustment of the

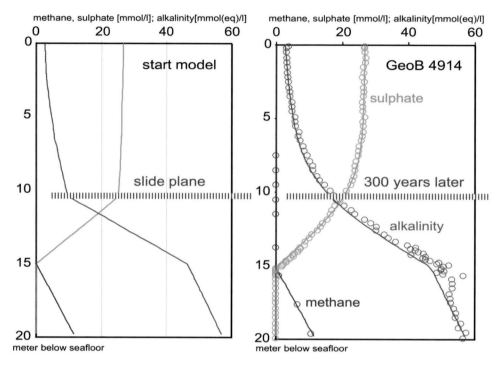

Fig. 15.10 Model calculation of sulfate in pore water, alkalinity and methane in an explicit numeric solution of Fick's Second Law, accounting for the reaction between sulfate and methane and the alkalinty which is affected thereby. The adjustment to the measured profiles depends on the time passed since the slide occurrence and thus permits the reconstruction of the time of its occurrence which took place about 300 years ago.

sulfate profile for the time span modeled is estimated to be somewhere between 5 to 10 % of this value.

Example 3: Reconstruction of the date of a sediment avalanche at the continental slope

Example 3 also illustrates the modeling of a sulfate profile really measured, including a sulfate-methane transition zone in the deep part of the profile. However, in this case the profile displayed such an unusual course that its interpretation as a steady-state condition did not appear justified by any means. As a consequence, this profile rested among many other measurement data for almost ten years. Only after we understood that non-steady-state conditions in marine sediment profiles - particularly in the continental slope region - are much more frequent than previously assumed were we able to understand this profile as well (Hensen et al. 2003).

The rather steep continental slope off the coast of Uruguay is distinguished by frequent irregular sedimentary transport events. Not only turbidity currents but also slides of relatively

unperturbed sediment packages occur here, such as sediment avalanches. In case of sediment avalanches the originally stratified sediment layer is mostly destroyed and dissolved into single blocks. However, the original sediment is not dissolved all the way down to each single grain as is the case with turbidite currents. Figure 15.11 shows a section of the same profile above the continental slope, where the examined core was sampled with a gravity corer (GeoB2809, Bleil et al. 1994). The sediment avalanche which had been deposited at this location can be clearly recognized in the lower part of this sediment acoustic profile as it lies, practically without displaying any structure, on top of the otherwise layered sediments. Once again, the non-steady state progression of the sulfate profile in the pore water would provide us with information on the age of the avalanche.

Figure 15.12 shows the result of a non-steady-state model run which was performed by applying the explicit solution according to the Two-Step Method, this time with an Excel® worksheet according to the ‚Press-F9-Method'. This worksheet can also be downloaded from the above-

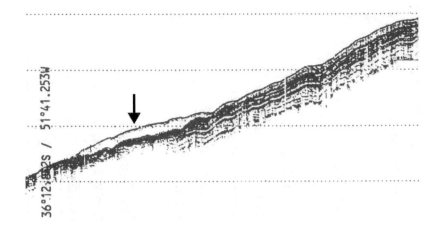

Fig. 15.11 Sediment-acoustic profile (PARASOUND) of the continental slope off the coast of Uruguay (Bleil and participants 1994). A sediment avalanche is recognizable below deposited as a compact sediment. The sediment core GeoB2809 also originated from this profile, including the pore water profiles shown in Fig. 15.12. (see also Hensen et al. 2003).

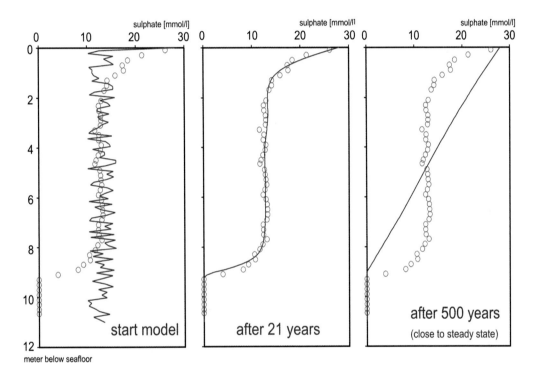

Fig. 15.12 Model calculation of the sulfate concentration profile in the deposition of a sediment avalanche on the continental slope off the coast of Uruguay (see also. Fig. 15.11). Red circles denote the measured sulfate concentrations. Blue lines indicate the model calculation at various time points after the avalanche: Left: Immediately after the occurrence of the avalanche. Center: approx. 21 years later. Right: approx. 500 years later.

mentioned website. The measured sulfate concentrations are marked by red circles in all partial diagrams of Figure 15.12. The left diagram shows in blue color the initial situation once again immediately after the deposition of the sediment avalanche. In this case, it was assumed that the concentration profile before the incidence of the avalanche decreased linearly from the concentration in ocean water (28 mmol l^{-1}) at the sediment surface down to a concentration of 0.0 in the region at the sulfate-methane transition zone. The profile was then perturbed by the avalanche, yielding irregular concentrations over the entire profile which scattered around the average of the ocean water values and the 0.0 value. This distribution of initial concentrations was calculated as a somewhat randomized 50 % ocean water concentration.

Starting out from this profile of initial concentrations, we then only introduced the time-correct diffusion in the model calculation, along with a correct diffusion coefficient, and the mutual degradation of sulfate and methane at the sulfate-methane transition zone.

First, the minor, randomly occurring differences between adjoining points were 'smoothed out' in the course of model calculation. At the same time, the upper part of the profile adapted more and more to the ocean water transition, while the lower part adapted to the low concentration of the sulfate-methane transition zone. After somewhat more than 20 years, an almost perfect adjustment to the measured profile was achieved without any other fitting technique (center diagram in Fig. 15.12). This core was obtained in 1994, and the sulfate profile was measured in the same year. The sediment avalanche consequently took place in the early 70s of the last century. The right diagram in Figure 15.12 shows the same profile model as it would look like in 500 years after the event. A new, almost stationary condition will be reached after this time, and only a faint suggestion of the S-shaped curve typical of the non-steady state condition will still remain recognizable.

Example 4: Formation of phosphorite in shelf sediments

It has been recently demonstrated in shelf sediments of the highly productive upwelling area off the coast of Namibia that certain sulfur bacteria release very high concentrations of phosphate into the pore water near the sediment surface. This produces a stratified precipitation of phosphorite minerals (Schulz and Schulz 2005). The questions to be answered by modeling are as follows:

- At which phosphate concentration must we expect the precipitation of which minerals ?
- The concentrations in pore water at a depth of only one centimeter below the sediment surface are considered as stationary. What release rates applying to phosphate must we postulate for the bacteria if it is known from experiments that the precipitation kinetics of phosphate minerals are in a certain range, and if the measured concentration profile in pore water is to be simulated correctly?
- How much time does it take for sulfur bacteria to release these phosphate concentrations till the amounts of phosphate found in the solid phase of the sediment accumulated ?

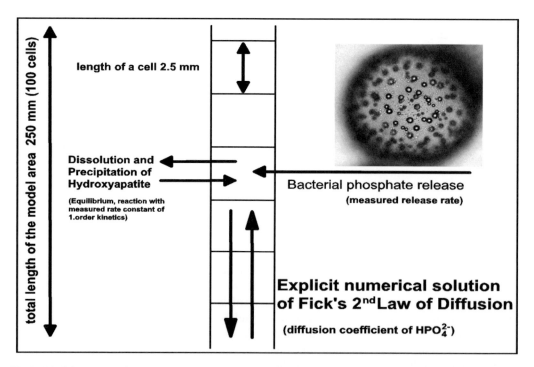

Fig. 15.13 Model concept about the development of phosphorite layer in an anoxic sediment on the shelf off the coast of Namibia (Schulz and Schulz 2005). Phosphate released by bacteria (*Thiomargarita*) reaches a stationary condition in pore water which is determined by release rates, precipitation and diffusion-dependent transport.

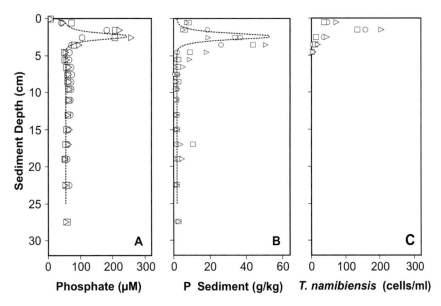

Fig. 15.14 Measured values and model calculation related to the model concept shown in Fig. 15.13. Open circles, triangles and squares denote the measured values of three parallel cores. The broken line curves represent the model calculation. Left: phosphate concentrations in pore water. Center: Sedimentary contents of phosphorus. Right: Bacterial cell counts in the sediment (after Schulz and Schulz 2005).

- How fast or slow is the phosphate mineral layer embedded in the sediment re-dissolved if phosphate concentrations increasing in the process of dissolution return to the ocean water by diffusion ?

Figure 15.13 shows the principle structure of the model which is supposed to give an answer to these questions. Here, realization was done in accordance with the explicit solution and the Two-Step Method using an Excel® worksheet, which can also be downloaded from the above mentioned website, along with other relevant information.

The model will record the upper 25 centimeters of the sediment if there are 100 cells each measuring a length of merely 2.5 mm. This accounts for a sufficiently good depth resolution in the interesting range of reactions, and it reaches deep enough so that the sensitive lower margin of the model is by no means influenced by the processes running inside the model. Apart from the time-correct diffusion of dissolved phosphate, a rate-controlled input (bacterial release of phosphate into the pore water fraction) and a rate-controlled output (precipitation of a phosphate mineral) are accounted for. Simultaneously all precipitations and dissolutions had to be recorded so that the total of precipitated phosphate could be compared with the sedimentary contents found in nature.

The input was only given for cells in which the presence of these large sulfur bacteria of the genus *Thiomargarita* that are even easily visible with the naked eye has been known to exist at the respective sediment depths. Precipitation of the mineral hydroxyapatite $[Ca_5OH(PO_4)_3]$ was conducted with first-order kinetics as soon as the saturation concentration (40 µmol l^{-1} phosphate) was exceeded. This concentration resulted from model calculations performed with the program PHREEQC (Section 15.1.2) on this mineral which is characteristic for such conditions and whose presence in this sediment appeared to be likely on account of x-ray diffractometry (XRD).

The results of measurements made with natural sediments are compiled in Figure 15.14, together with the results of the model calculations. Both pore water concentrations and the contents in the solid phase of the sediment were absolutely correctly simulated considering the range of available measurement accuracy and reproducibility.

The above questions concerning modeling can thus be answered, and the answers concomitantly describe the circumstances of the development of marine phosphorite deposits by sulfur bacteria:

- The specific situation of pore water concentrations at this location yielded a phosphate concentration of 40 µmol l^{-1} which is required in order to obtain saturation in the pore water fraction relative to the mineral hydroxyapatite which is known as a primary precipitate of phosphorite deposits.

- Phosphate release rates of sulfur bacteria ranging from 0.024 to 0.08 pmol l^{-1} cell^{-1} had to be postulated. In case of higher release rates, calcium which is required for the precipitation of hydroxyapatite [Ca$_5$OH(PO$_4$)$_3$] could not have been supplied fast enough by means of diffusion. In case of a lower release rate, diffusion would have caused a distinctly broader spreading of the concentration peak in the pore water fraction than the peak measured. A phosphate release rate of 0.011 to 0.028 pmol l^{-1} cell^{-1} for the bacteria was found in laboratory experiments. This range is very close to the postulated values, both ranges even overlap to certain extent. That the bacteria did not produce the highest release rates of the model may be explained with the fact that they simply did not appreciate their state of 'captivity' as much as their life in the wild. A precipitation rate applicable to hydroxyapatite emerged almost incidentally, which was also verified in laboratory experiments.

- Anticipating the fastest release rate of 0.08 pmol l^{-1} cell^{-1} it will take only three months until the contents measured in the sediment are reached. In case of the lowest rate of 0.024 pmol l^{-1} cell^{-1} it will take 14 months. Both values are plausible also against the background of sedimentation rates on the shelf.

- The phosphorite layers are quite well protected after being embedded in the sediment, for concentrations produced by the dissolution of these layers need a very long time to return to the sediment surface by diffusion. Even after 100 modeled years, the phosphate contents in the sediment will be only slightly lower than they were at the time of their development. And 100 years on the shelf is plenty of time during which everything from further and deeper deposition to erosion of the sediment can happen.

15.3.3 Two-Step Models for Combined Complex Transport/Reaction Processes

The main problem encountered in modeling combinations of advective/dispersive and/or diffusive transport, on the one hand, and any (bio)geochemical reaction, on the other hand, is that at least two totally different processes interact simultaneously on the same object:

Physical transport including diffusion, advection and dispersion, when reactions are absent, is one part of the problem. In their entirety, these processes are quantitatively well understood in model concepts. They are also applicable, without raising principle problems, in analytical and numerical solutions to the general partial differential equations of material transport.

Geochemical and (bio)geochemical processes which are independent of transport processes present the other part of the problem. Here as well, there is a number of far advanced model concepts. The models belonging to the PHREEQE or PHREEQC - type may be referred to in this respect (cf. Section 15.1.1 and 15.1.2). Other fields are object of the various chapters in this book.

First, attempts were made to develop models that could accomplish the coupling of both process groups in one single, either analytical or numerical procedure, although these often bore strong limitations for the application to geochemical reactions. Attempts now more often foresee solutions in two steps. This group of methods is referred to either as "operator splitting", or as the "two-step method". Boudreau (1997) mentions this procedure only briefly at the close of his book and describes it as 'an apparently crazy idea that works rather well in practice'. The first versions of such models, which couple physical transport to geochemical reactions in groundwater are already more than 20 years old (e.g. Schulz and Reardon 1983). Application examples for the 'two-step method' in modeling diagenetic processes in marine sediments over the recent years were published by Hamer and Sieger (1994), Van Cappellen and Wang (1995), Wang and Van Cappellen (1996), Hensen et al. (1997), Sieger (1993), Landenberger et al. (1997), Landenberger (1998), Adler et al. (2000a), Adler et al. (2001), Wenzhöfer et al. (2001), Pfeifer et al. (2002), Haese et al. (2003), Hensen et al. (2003).

The model CoTReM (**Co**lumn **T**ransport and **Re**action **M**odel) (Adler et al. 2000b) is primarily designed for its application to complex early diagenetic reactions in marine sediments. In the one-dimensional transport part of the model (as a continuance of the model DISPER by Flühler and Jury 1983) the model's territory is divided up into a variable number of REVs (representative elementary volumes). These REVs may differ in

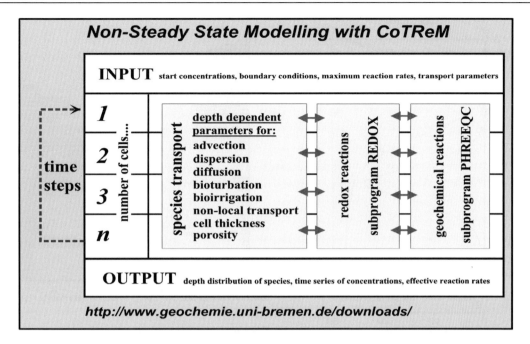

Fig. 15.15 Functional principle of the CoTReM model in the simulation of early diagenetic processes in sediments. The time series of (bio)geochemical reactions and material transport are alternately simulated in a two-step procedure. PHREEQC is used as a subroutine for geochemical reactions. CoTReM, including instructions and exemplary data, can be downloaded from the internet site indicated.

length, porosity, and in their dispersion and diffusion properties.

The time to be modeled is subdivided into a number of time steps. First, a numerical solution is used to solve the partial differential equation of transport (including adsorption and desorption) with respect to all substances under study. Afterwards, the geochemical processes in each REV are modeled independent of transport processes. For this step, CoTReM runs a subroutine called REDOX, as well as the geochemical equilibrium and reaction model PHREEQC. The next time step is processed by starting again with the physical transport, followed by the geochemical reactions etc., until the pre-determined operating time is completed. The structural principle of the model CoTReM is outlined in Figure 15.15.

The latest version of CoTReM together with a program description and data files of some examples is available at the internet web page, mentioned above.

An example which was published by Pfeifer et al. (2002) illustrates very clearly the functions provided by CoTReM. The authors processed a sediment core obtained from the equatorial upwelling area off the coast of West Africa (Location GeoB4906, water depth 1251 m). The left

diagram in Figure 15.16 shows the concentration profiles of oxygen, nitrate, manganese, iron, and sulfate in the upper six centimeters of the core, along with the results of the model calculation made with CoTReM. The concentrations of ammonia and calcium, the pH and pCO_2 were also measured and included into the calculation (not shown). The following processes had to be included into the modeling procedure:

- Oxidation of organic matter by oxygen
- Oxidation of organic matter by nitrate
- Oxidation organic matter by iron oxide or iron hydroxide
- Oxidation of organic matter by sulfate
- Re-oxidation of dissolved divalent manganese by oxygen
- Re-oxidation of dissolved divalent iron by nitrate
- Re-oxidation of dissolved divalent iron by manganese oxide
- Dissolution/precipitation of iron sulfide and calcite

The adjustment of all reactions to the measured profiles was made by the depth-dependent turnover rates or reaction kinetics of the indivi-

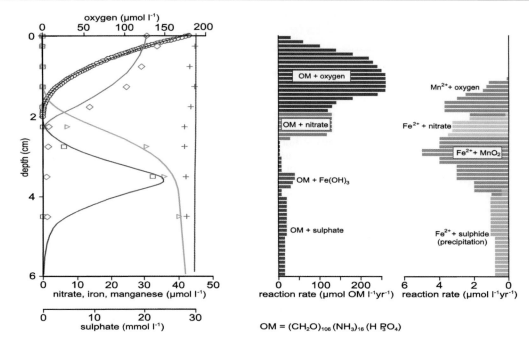

Fig. 15.16 Measurements and results of a model calculation performed with CoTReM applying to the upper centimeters of a sediment core from equatorial West Africa (GeoB4906, 1251 m water depth) (after Pfeifer et al. 2002). The symbols in the left diagram represent measured values of pore-water concentrations (circles: oxygen, diamonds: nitrate, triangles: manganese, squares: iron, crosses: sulfate). The unbroken lines denote the concentrations calculated by the model. The right diagram illustrates the assumed (bio)geochemical reactions which take place in various depths and display dissimilar turnover rates. (OM: Organic matter).

dual processes. In this regard CoTReM permits modeling to start with the maximum turnover rates, which the program subsequently reduces to realistic values. As some of the reactions interact with each other, the adjustment of the model to the concentration profiles of the single parameters often involves a laborious process of step-by-step procedure that requires a repeated subsequent correction of the depth-related turnover rates. However, there is no other way which could be better and especially easier to understand the complex interacting system of early diagenesis in the sediment. The right diagram in Figure 15.16 shows the depth distribution of the turnover rates, which were ultimately necessary for adjustment. Hence they also provide the basis for a quantitative understanding of all processes that are relevant to this sediment.

An also very versatile program designed for modeling the essential processes in early diagenesis originates from Wang and Van Capellen (1996), or Van Cappellen and Wang (1995). This model STEADYSED1 can be obtained from the authors as 'public domain software' and

enables the operator to change and adjust all essential sets of data to a particular problem. Yet, one can also select and apply far more unreasonable combinations than reasonable ones - which is in the nature of the model considering the complexity the model offers. The agreeable 'learning by playing' which is typical for so many simple models, turns out to be not so easy after all.

The model uses the following physical data: the sedimentation rate for solid phase advection, temperature for correcting diffusion, porosity and the derived formation factor. A description of bioturbation is obtained by utilizing a mixture-related coefficient, comparable to the diffusion coefficient. An indication of depth informs how far below the sediment surface bioturbation will be effective.

Constant concentrations are assumed for the following components in bottom water: O_2, NO_3^-, SO_4^{2-}, Mn^{2+}, Fe^{2+}, NH_4^+, salinity, alkalinity and pH-value. As the model STEADYSED1 exclusively tolerates steady state situations, this assumption naturally requires constant concentrations in bottom water as a prerequisite.

As for the solid phase of sediment, the input of iron and manganese oxides, the amount of organic substance, and the C:N:P ratio of the organic substance are used. The diagenetic decomposition of the organic substance mediated by the electron acceptors oxygen, nitrate, manga-

nese and iron oxides, sulfate, all the way to methane fermentation, is simulated according to the reactions published by Froelich et al. (1979). However, the user may also select the C:N:P ratio of the organic substance. The re-oxidation of iron sulfides, and manganese sulfides, or of methane is

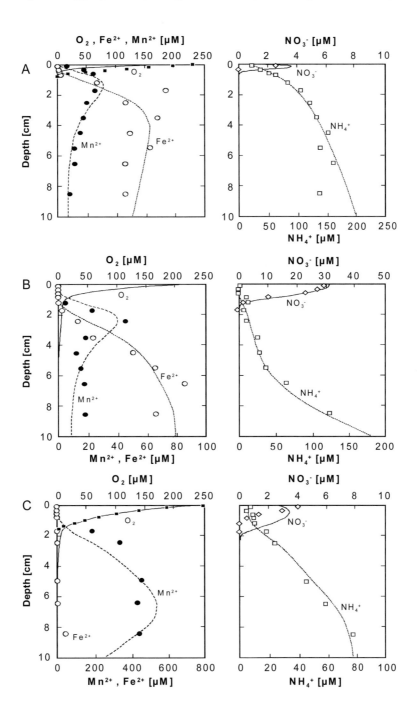

Fig. 15.17 Concentration profiles in pore water of three examples from marine sediments. Data represented by symbols were taken from Canfield et al. (1993 a,b), whereas the curves demonstrate the respective simulations carried out with the model STEADYSED1 by Wang and Van Cappellen (1996).

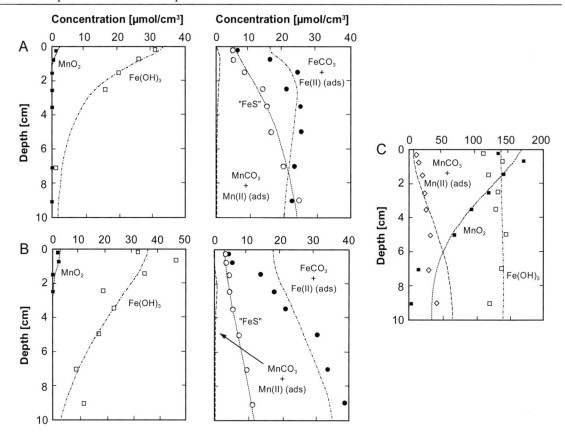

Fig. 15.18 Concentration profiles of iron and manganese solid phases in sediments derived from three marine examples. Data represented by symbols were taken from Canfield et al. (1993 a,b), whereas the curves demonstrate the respective simulations carried out with the model STEADYSED1 by Wang and Van Cappellen (1996).

preferably performed as 'secondary redox reactions'. Moreover, the adsorption of ammonium, and the precipitation/dissolution of iron carbonate, manganese carbonate, and iron sulfide can be included. For all these reactions, the analyst is able and *required* to select the adequate rates.

This is where the general problems in applying such models occur: with all the parameters and boundary conditions that need to be set, there will be so many 'buttons' and control options ultimately allowing for almost any kind of modeling of the measured data. Only if so many examples are available as Wang and Van Cappellen (1996) found in the data of Canfield et al. (1993a, b) based on measurements, will it be possible to come to substantial and reasonable results. Figure 15.17 shows concentration profiles for O_2, NO_3^-, NH_4^+, Mn^{2+}, Fe^{2+} measured in the pore water of three cores. Figure 15.18 demonstrates the solid phase concentration profiles of Fe- and Mn participating in the reactions.

If one compares the two model programs CoTReM by Adler et al. (2000b), on the one hand, and the model STEADYSED1 published by Wang and Van Cappellen (1996) on the other, the following aspects need to be emphasized:

- Both models have been made freely available by the respective authors as public domain software and hence can be used by everyone.

- Both models are not easy to operate and require at least some background knowledge in modeling procedures and early diagenetic reactions.

- Both models simulate the complex situation of combining diffusive, advective and bioturbational transports in pore water and in the solid phase. They contain the various reactions of early diagenesis and require a large number of reliable measurements. Otherwise, too many degrees of freedom would remain.

- Both models contain the possibility to simulate bioirrigation; STEADYSED1 generally treats bioirrigation and bioturbation with much more thoroughness and reliability.

- STEADYSED1 is quite limited as far as its chemical processes are concerned (absence of real activities in the pore water, and hence only apparent equilibrium constants, complex species are not included). However, it contains all the essential processes currently known. CoTReM is much more flexible owing to the utilization of PHREEQC (Parkhurst 1995) as a subroutine, yet it requires proportionally more information.

- The model CoTReM works with Fick's Second Law of Diffusion and thus permits the calculation of any possible, especially non-steady state situations. STEADYSED1 can only be applied to calculate steady state situations which accordingly demand the existence of steady state boundary conditions.

15.4 Bioturbation and Bioirrigation in Combined Models

Bioturbation is normally modeled by applying a bio-diffusion coefficient to the pore water fraction *and* the solid phase. This means that a bio-diffusion in the solid phase will be simulated with the same model concept which is applied in the form of Fick's laws to the dissolved constituents in pore water. Under the assumption of a given steady state (e.g. STEADYSED1), this requires that a sufficient length of time is studied which allows the macroorganisms contributing to bioturbation to display activity all over and as many times as appears necessary. Only then will each part of the solid phase be turned over often enough in the statistical balance, hence providing a workable model concept. As for cases of non-steady state and short observation periods the model concept actually cannot be considered as valid.

Figure 15.19 shows, in one exemplary model calculation carried out with STEADYSED1, the conversion rates of iron and manganese in the upper sediment zone and their exchange with the supernatant bottom water layer. It is evident that the amounts of iron which ultima-

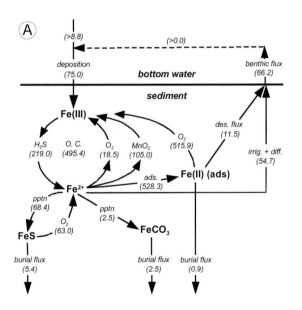
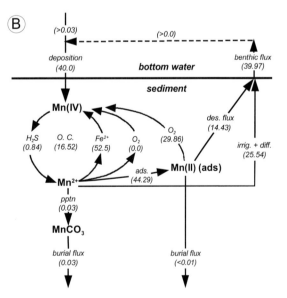

Fig. 15.19 Iron and manganese fluxes in superficial marine sediments and bottom water, as obtained from calculations using the model STEADYSED1 by Wang and Van Cappellen (1996). Here, it is significant that bioturbation and bioirrigation were included in the task.

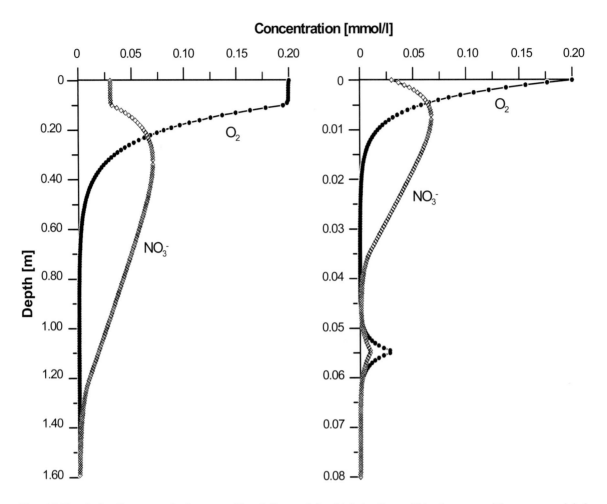

Fig. 15.20   Left: Oxygen and nitrate profiles influenced by bioirrigation within the upper 10 cm as modeled with an Excel® spreadsheet according to the 'Press F9 method'. The simulation was performed on the basis of anticipating a partial coupling of pore water to the concentrations prevalent in bottom water in the upper 10 cm of the sediment. Right: Here, a quantity of bottom water was continually added in a specific depth. The result is related to the measurement shown in Figure 3.24 after Glud et al. (1994).

tely become deposited in the sediment (burial flux) as FeS, $FeCO_3$, or in an adsorbed state, are essentially smaller than the amounts undergoing conversions in the upper parts of the sediment and which have therefore undergone redox processes several times before. The lower part of Figure 15.19 demonstrates the corresponding turnovers of manganese with again quite specific ratios between the burial flux, conversion rates in the upper sediment zone, and exchange with the bottom water.

The simulation of bioirrigation is much more difficult, a circumstance which pertains to its nature, since the process, endowed with great inhomogeneity and strong dependence on the organism under study, combines bottom water with pore water at the particular depths concerned. How this combination becomes effective in any one particular case, and to which degree it also influences the pore water not involved directly, naturally depends on the macroorganisms presently responsible for the effect. Thus, the process actually eludes a generalized solution and should be treated individually for each special case. To this end, workable models are not available in sufficient number, neither are there any adequately applicable modeling techniques at hand.

In Figure 15.20, and by using the 'Press-F9-method' described in Section 15.3.1, two steady states displaying bioirrigation have been calculated. The case shown on the left-hand side anticipates an upper sediment zone continually influenced by bioirrigation, whereas the case shown on the right merely assumes surface coupling in a specific depth - e.g., by the effect of a particular worm-hole. The result may be compared with the measurements conducted by Glud et al. (1994) and shown in Figure 3.32. Just as Wang and Van Cappellen (1994) have described it in one application of the model STEADYSED1, the situation here is that bioirrigation conveys nitrate into the sediment, whereas, at the same time, molecular diffusion transports nitrate from the sediment back into the bottom water. This results from the fact that both processes are coupled to the bottom water layer in different depths of the sediment. In the case of diffusion, these are the superficial nitrate-rich zones, whereas in the case of bioirrigation, the deeper situated zones are effective, which contain only small amounts of nitrate.

Ultimately, all these agreements between the measured profile and the resultant model are not really significant since they merely represent a direct application of the model concept of bioirrigation, i.e., the pore water input conducted from the bottom water at a certain depth in the sediment. Yet, these models are only interesting inasmuch as the non-diffusive transport between sediment and bottom water results from them. These are exemplified for iron and manganese in Figure 15.19, calculated with a similar model concept using the program STEADYSED1. It should be kept in mind that these calculations are only workable to the point as they comply to a rather simplified model concept and to a stationary situation anticipated not only in the 'Press F9 method' but in STEADYSED1 as well. Only by conducting measurements of the material fluxes *in-situ* at very dissimilar locations, and after repeated comparisons with the adequate models have been made, will the previous model concepts and the calculated fluxes ever become verified in the future.

Acknowledgements

This is contribution No 0337 of the Research Center Ocean Margins (RCOM) which is financed by the Deutsche Forschungsgemeinschaft (DFG) at Bremen University, Germany.

15.5 Problems

Problem 1

Table 15.1 shows an analysis of ocean water made by Nordstrom et al. (1979). A condition of oversaturation of the minerals calcite (SI = 0.76) and dolomite (SI = 2.4) at a measured pH value of 8.22 has been calculated for this analysis with the model PHREEQC (see also Table 15.3). What pH is to be expected if we calculate the saturation state of both minerals in this ocean water sample with PHREEQC under the condition of $\log pCO_2 = -3.47$?

Problem 2

At which pCO_2 would the mineral calcite in the ocean water sample in Table 15.1 (Nordstrom et al., 1979) be saturated if a pH value of 8.22 had been measured ? How could this pCO_2 which is markedly below the value of the atmospheric be explained ?

Problem 3

What penetration depth of oxygen would you expect in a sediment, if the removal of oxygen can be described in terms of a half-life of 5 days? Sediments have a temperature of 10°C. Their porosity is at 70 percent. For the diffusion coefficients use Tables 3.1 and 3.2. The bottom water shall have an oxygen concentration of 200 µmol l^{-1}. Use an oxygen concentration less than 1 µmol l^{-1} as "zero".

Problem 4

A different sulfate concentration gradient is measured above the sulfate-methane transition zone than underneath the methane zone. How could this be explained, despite the fact that sulfate and methane react with each other in a ratio of 1:1 ?

References

Adler M., Hensen C., Kasten S. and Schulz H.D. (2000a) Computer simulation of deep sulfate reduction in sediments of the Amazon Fan. Internat. Journal of Earth Sciences, 88, pp. 641-654

Adler M., Hensen C. and Schulz H.D. (2000b) CoTReM – Column Transport and Reaction Model, Version 2.3, User Guide. http://www.geochemie.uni-bremen.de/downloads/cotrem/index.htm, 59 pp.

Adler M., Hensen C., Wenzhöfer F., Pfeifer K. and Schulz H.D. (2001) Modeling of calcite dissolution by oxic respiration in supralysoclinal deep-sea sediments. Marine Geology, 177, pp. 167-189

Ball J.W. & Nordstrom D.K. (1991) WATEQ4F - User´s manual with revised thermodynamic data base and test cases for calculating speciation of major, trace and redox elements in natural waters. US Geol. Surv., Open-File Report 90-129, 185 pp

Berner R.A. (1980) Early Diagenesis: A Theoretical Approach. Princeton Series in Geochemistry, Princeton University Press, 241 pp

Bleil U. and participants (1994) Report and preliminary results of *Meteor* cruise M29/2 Montevideo-Rio de Janeiro, 15.7.-08.08.1994, Vol. 59, Fachbereich Geowissenschaften, Universität Bremen, 153 pp.

Boudreau B.P. (1997) Diagenetic Models and Their Implementation: Modelling Transport and Reactions in Aquatic Sediments. Springer-Verlag, Berlin, Heidelberg, New York, 414 pp.

Canfield D.E., Thamdrup B. and Hansen J.W. (1993a) The anaerobic degradation of organic matter in Danish coastal sediments: Iron reduction, manganese reduction, and sulfate reduction. Geochim. Cosmochim. Acta 57, pp. 3867-3883

Canfield D.E., Jørgensen B.B., Fossing H., Glud R., Gundersen J., Ramsing N.B., Thamdrup B., Hansen J.W., Nielsen L.P., Hall P.O.J. (1993b) Pathways of organic carbon oxidation in three continental margin sediments. Mar. Geol. 113, pp. 27-40

Froelich P.N., Klinkhammer G.P., Bender M.L., Luedtke N.A., Heath G.R., Cullen D., Dauphin P., Hammond D., Hartman B. (1979) Early oxidation of organic matter in pelagic sediments of eastern equatorial Atlantic: suboxic diagenesis. Geochim. Cosmochim. Acta, 43, pp. 1075-1090

Flühler H. and Jury W.A. (1983) Estimating solute transport using nonlinear, rate dependent, two-site-adsorption models. Microfiche, Eidg. Anst. forstl. Versuchswesen, 245, Zürich, 48 pp.

Garrels R.M. (1960) Mineral Equilibria at Low Temperature and Pressure. New York (Harper), 254 pp.

Garrels R.M. and Christ Ch.L. (1965) Solutions, Minerals and Equilibria. New York-Evanston-London (Harper & Row)-Tokyo (Weatherhill), 450 pp.

Glasby G.P. and Schulz H.D. (1999) E_H, pH diagrams for Mn, Fe, Co, Ni, Cu and As under seawater conditions: Application of two new types of E_H, pH diagrams to the study of specific problems in marine geochemistry. Aquatic Geochemistry, 5, pp. 227-248

Glud R.H., Gundersen J.K., Jørgensen B.B., Revsbech N.P., Schulz H.D. (1994) Diffusive and total oxygen uptake of deep-sea sediments in the eastern South Atlantic Ocean: in situ and laboratory measurements. Deep-Sea Res., 41, pp. 1767-1788

Haese R.R., Meile C., Van Cappellen P., De Lange G.J. (2003) Carbon geochemistry of cold seeps: Methane fluxes and transformation in sediments from Kazan mud volcano, eastern Mediterranean Sea. Earth and Planetary Science Letters, 212, pp. 361-375

Hamer K. und Sieger R. (1994) Anwendung des Modells CoTAM zur Simulation von Stofftransport und geochemischen Reaktionen. Verlag Ernst & Sohn, Berlin, 186 pp.

Hensen C., Landenberger H., Zabel M., Gundersen J.K., Glud R.N. and Schulz H.D. (1997) Simulation of early diagenetic processes in continental slope sediments off Southwest Africa: The computer model CoTAM tested. Marine Geology, 144, pp. 191-210

Hensen C., Zabel M., Pfeifer K., Schwenk T., Kasten S., Riedinger N., Schulz H.D. and Boetius, A. (2003) Control of sulfate pore-water profiles by sedimentary events and the significance of anaerobic oxidation of methane for the burial of sulfur in marine sediments. Geochimica and Cosmochimica Acta, 67, pp. 2631-2647

Kharaka Y.K., Gunter W.D., Aggarwal P.K., Perkins E.H. and DeBraal J.D. (1988) SOLMINEQ88: a computer program for geochemical Modeling of water-rock-interactions. US Geol. Surv., Water-Resources Investigations Report 88-4227, 207 pp.

Kinzelbach W. (1986) Groundwater Modelling - An Introduction with Sample Programs in BASIC. Elsevier, Amsterdam - Oxford - New York - Tokyo, 333 pp.

Landenberger H., Hensen C., Zabel M. und Schulz H.D. (1997) Softwareentwicklung zur computergestützten Simulation frühdiagenetischer Prozesse in marinen Sedimenten. Ztschr. dt. geol. Ges., 148, pp. 447-455

Landenberger H. (1998) CoTReM, ein Multi-Komponenten Transport- und Reaktions-Modell. Berichte, Fachbereich Geowissenschaften, Univ. Bremen, 110, 142 pp.

Niewöhner C., Hensen C., Kasten S., Zabel M., Schulz H.D. (1998) Deep sulfate reduction completely mediated by anaerobic methane oxidation in the upwelling area off Namibia. Geochim. Cosmochim. Acta, 62, pp. 455-464

Nordstrom D.K., Plummer L.N., Wigley T.M.L., Wolery T.J., Ball J.W., Jenne E.A., Basset R.L., Crerar D.A., Florence T.M., Fritz B., Hoffman M., Holdren G.R., Jr., Lafon G.M., Mattigod S.V., McDuff R.E., Morel F., Reddy M.M., Sposito G. and Thrailkill J. (1979) A comparison of computerized chemical models for equilibrium calculations in aqueous systems: in Chemical Modeling in aqueous systems, speciation, sorption, solubility, and kinetics. Jenne, E.A., ed. Series 93, American Chemical Society, pp. 857-892

Parkhurst D.L. (1995) User´s guide to PHREEQC: a computer model for speciation, reaction-path, advective-transport, and inverse geochemical calculations. US Geol. Surv., Water-Resources Investigations Report 95-4227, 143 pp.

Parkhurst D.L. and Appelo C.A.J (1999) User's guide to PHREEQC (Version 2): A computer program for speciation, batch-reaction, one-dimensional transport, and inverse geochemical calculations. US Geol. Surv., Water-Resources Investigations Report 99-4259, 312 pp.

Parkhurst D.L., Thorstenson D.C. and Plummer L.N. (1980) PHREEQE - a computer program for geochemical calculations. US Geol. Surv., Water-Resources Invest. 80-96, 219 pp.

Pfeifer K., Hensen C., Adler M., Wenzhöfer F., Weber B. and Schulz H.D. (2002) Modeling of subsurface calcite dissolution, including the respiration and reoxidation processes of marine sediments in the region of equatorial upwelling off Gabon. Geochimica and Cosmochimica Acta, 66, pp. 4247-4259

Plummer L.N., Jones B.F. and Truesdell A.H. (1976) WATEQF - a fortran IV version of WATEQ, a computer program fpr calculating chemical equilibrium of natural waters. US Geol. Surv., Water-Resources Invest. 76-13, 615 pp.

Redfield, A.C., 1958. The biological control of chemical factors in the environment. Am. Sci., 46: 206 226.

Schulz H.D. (2004) Auswertung von Markierungsversuchen. In Käss W. (Ed.) Geohydrologische Markierungstechnik, Lehrbuch der Hydrogeologie Bd.9, S.347-356, Borntraeger-Verlag, Berlin, Stuttgart.

Schulz H.N. and Schulz H.D. (2005) Large Sulfur Bacteria and the Formation of Phosphorite. Science, 307, pp. 416-418

Schulz H.D. and Reardon E.J. (1983) A combined mixing cell/analytical model to describe two-dimensional reactive solute transport for unidirectional groundwater flow. Water Resour. Res, 19, pp. 493-502

Schulz H.D., Dahmke A., Schinzel U., Wallmann K., Zabel M. (1994) Early diagenetic processes, fluxes, and reaction rates in sediments of the South Atlantic. Geochim. Cosmochim. Acta, 59, pp. 2041-2060

Sieger R. (1993) Modellierung des Stofftransports in porösen Medien unter Ankopplung kinetisch gesteuerter Sorptions- und Redoxprozesse sowie thermodynamischer Gleichgewichte. Berichte, Fachbereich Geowissenschaften, Univ. Bremen, 110, 142 pp.

Truesdell A.H. and Jones B.F. (1974) WATEQ, a computer program for calculating chemical equilibria on natural waters.- Jour. Research US Geol. Surv., Washington D.C., 2, pp. 233-248

Van Cappellen P. and Wang Y. (1995) STEADYSED1: A Steady-State Reaction-Transport Model for C, N, S, O, Fe and Mn in Surface Sediments. Version 1.0 User's Manual, Georgia Inst. Technol., 40 pp.

Van Cappellen P., Wang Y. (1996) Cycling of iron and manganese in surface sediments: a general theory for the coupled transport and reaction of carbon, oxygen, nitrogen, sulfur, iron, and manganese. Amer. Journal of Science, 296, pp. 197-243

Van Cappellen P. and Gaillard J.-F. (1996) Biogeochemical Dynamics in Aquatic Sediments. In: Lichtner P.C., Steefel C.I. and Oelkers E.H. (eds.) Reviews in Mineralogy, Vol. 34: Reactive Transport in Porous Media, The Mineralogical Society of America, Washington DC., pp. 335-376

Wang Y. and Van Cappellen P. (1996) A multicomponent reactive transport model of early diagenesis: Application to redox cycling in coastal marine sediments. Geochim. Cosmochim. Acta, 60, pp. 2993-3014

Wenzhöfer F., Adler M., Kohls O., Hensen C., Strotmann B., Boehme S. and Schulz H.D. (2001) Calcite dissolution driven by benthic mineralization in the deep-sea: In situ measurements of Ca^{2+}, pH, pCO_2 and O_2. Geochimica et Cosmochimica Acta, 65, pp. 2677-2690

Wolery T.J. (1993) EQ 3/6, A Software Package for Geochemical Modeling of Aqueous Systems. Lawrence Livermore National Laboratory, California, 247 pp.

Zabel M. and Schulz H.D. (2001) Importance of submarine landslides for non steady state conditions in pore water systems – lower Zaire (Congo) deep-sea fan. Marine Geology, 176, pp. 87-99

Answers to Problems

Chapter 1

Answer 1

Deep-sea or red clay covers the deep ocean basins below CCD. Calcareous sediments cover topographic highs like ocean ridges above CCD. Siliceous sediments occur in the Southern Ocean around Antarctica, along the equator in the Indian and Pacific oceans. Terrigenous sediments drape the continental margins.

Answer 2

Because of chemical tropical weathering, Kaolinite is considered the "clay mineral of low latitudes". As a product of continental weathering Illite shows high concentration at mid-latitudes of the northern oceans surrounded by great land masses. Chlorite as a product of physical weathering is essentially inversely related to kaolinite and referred to as the "high latitude mineral". Smectite showing a very variable distribution in the oceans is generally considered as an indicator of a "volcanic regime".

Answer 3

Large quantities of sediment are provided by rivers and directly from land and deposited on the shelf and/or directly beyond the shelf edge on the continental slope. The youngest crust occurs along the mid-Atlantic ridge and increases with distance from the ridge to the margin where more time is available to accumulate greater sediment thicknesses.

Answer 4

Basically, sedimentation rate decreases with increasing distance from the sediment source. Since for terrigenous sediments fluvial input is the major source sedimentation rate decreases with distance from river mouth and/or coast. The lowest values are reached in the oceanic basins far offshore where red clay accumulates extremely slowly. Sedimentation rate for biogenic sediments – calcareous as well siliceous – depends on high biogenic production, on carbonate compensation depth CCD) and dilution by terrigenous sediments is low.

Answer 5

Deep sea sediments are generally best described by *composition* according to particle origin, e.g. non-biogenic (terrigenic, volcanic) versus biogenic and can further be differentiated according to *mineralogy* (carbonate or opal) and/or *texture* (mud, ooze, clay).

With respect to relative distance from shore or continents sediments are called "pelagic sediments" consisting of biogenic oozes (carbonate and/or siliceous) and only very little admixtures of extremely fine-grained land derived material. They are distinguished from "hemipelagic sediments" typically deposited along continental margins and containing a substantial portion of land or shallow water derived components.

Chapter 2

Answer 1

Bulk parameters are based on a volume-oriented model which only depends on the total amount of pore fluid and sediment grains. Acoustic/elastic parameters like P- and S-wave velocity and attenuation are based on a micro-structure oriented model which considers both the amount *and* distribution of pore fluid and sediment grains.

Answer 2

A direct method is the analysis by weight and volume, indirect methods are gamma ray attenuation and electrical resistivity measurements.

Analysis by weight and volume primarily determines the wet bulk density, gamma ray attenuation the wet bulk density (by assumption of a processing porosity), and electrical resitivity the porosity. Porosity values can be converted to density values (and vice versa) by using equation 2.3.

Answer 3

Calcareous and siliceous sediment differ significantly by their sediment grain density. It amounts to about 2.7 g cm^{-3} for calcareous ooze and to about 2.1 g cm^{-3} for siliceous ooze, so that a downcore grain density model has to be developed from the carbonate (and opal) content, in order to calculate wet bulk densities from porosities.

Answer 4

The (mean) grain size mainly influences the ultrasonic wave propagation. P-wave velocity and attenuation increases with increasing (mean) grain sizes.

Answer 5

The strength of a reflection horizon is determined by the reflection coefficient $R = (I_2 - I_1) / (I_2 + I_1)$, which depends on the acoustic impedance contrast of the two adjacent layers. The acoustic impedance (I) depends on the P-wave velocity (v_P) and wet bulk density (ρ) and is defined as the product of both parameters, $I = v_P \rho$.

Chapter 3

Answer 1

Measurement in fresh sediment: *oxygen, pH-value, sulfide*. As soon as possible in squeezed or centrifuged pore water: *alkalinity, ammonia, total iron*. After preservation within some weeks: *manganese, nitrate, potassium*. Without preservation even after long time: *chloride, sodium, sulfate*.

Answer 2

Diffusion coefficient in seawater D^{SW} for SO_4^{2-} at 5°C is 5.72E-10 m^2 s^{-1}. Tortuosity for a sediment with a porosity of $\phi = 0.7$ is $\theta^2 = 1.71$. Diffusion

coefficient for SO_4^{2-} in pore water under these conditions is $D_{sed} = 3.3E-10$ m^2 s^{-1}. The gradient in the depth interval between 0 and 6 m is 28 mol m^{-3} per 6 m = 4.7 mol m^{-3} m^{-1}. The downward flux of SO_4^{2-} is $J_{sed} = 0.7 \cdot 3.3E-10 \cdot 4.7$ mol m^{-2} s^{-1} = 1.1E-9 mol m^{-2} s^{-1} or 34 mmol m^{-2} a^{-1}.

Answer 3

Diffusion coefficient in seawater D^{SW} for O_2 at 5°C is 1.23E-9 m^2 s^{-1}. Tortuosity for a sediment with a porosity of $\phi = 0.7$ is $\theta^2 = 1.71$. Diffusion coefficient for O_2 in pore water under these conditions is $D_{sed} = 7.2E-10$ m^2 s^{-1}. The steepest gradient close to the sediment surface (depth interval between 0 and ca.1 mm) is 28 mol m^{-3} m^{-1}. The downward flux of O_2 is $J_{sed} = 0.7 \cdot 7.2E-10 \cdot 28$ mol m^{-2} s^{-1} = 1.4E-8 mol m^{-2} s^{-1} or 0.45 mol m^{-2} a^{-1}. With an O_2/DIC-ratio of 138/106 (cf. Eq. 3.10 and Fig. 3.11) this rate is equivalent to 0.34 mol DIC m^{-2} a^{-1} or 4.1 g C m^{-2} a^{-1}.

Answer 4

A glove box with an inert gas atmosphere is necessary for all steps from sampling to analyzing anoxic pore water and/or sediment for constituents which would be oxidized in contact with the normal atmosphere. Such constituents are e.g. Fe^{2+}, Mn^{2+}, NH_4^+, S^{2-}. Only especially pure and expensive nitrogen does not contain interfering oxygen concentration. Normal argon usually contains sufficiently low oxygen concentration and is much cheaper.

Answer 5

Elements of mainly terrestrial origin are: Ti, Al, Fe, K, Rb, Mg. A mainly marine input is characteristic for Ca, Sr, (Ba). The elements Mn, S, (Fe) reach the sediment solid phase mainly by diagenetic processes.

Chapter 4

Answer 1

Enrichment of organic matter in marine sediments is mainly favored by the combined effects of low oxygen concentrations, either already in the water column or at shallow depth in the pore water system of the sediments, and high sedimentation rates, particularly in areas of high primary bioproductivity. The stagnant basin and the productivity model are used to illustrate favorable conditions for organic matter preservation. In addition, high sinking velocities, adsorption or organic matter to mineral matter (clay) and small grain sizes of the deposits, restricting exchange with the oxygen-containing ocean water after sedimentation, contribute to organic matter preservation.

Answer 2

Molecular oxygen and seawater sulfate are the main electron acceptors which mediate organic matter metabolization during early diagenesis by aerobic and anaerobic bacteria, respectively. Nitrate, iron(III), and manganese(IV), often in that order of significance, are additional electron acceptors in organic matter respiration. Ultimately, methanogenesis uses carbon as electron acceptor at greater sediment depth, when sulfate is depleted.

Answer 3

Not all parts of the biomass of decayed organisms have the same chance of being preserved in the fossil record. Water-soluble organic compounds or organic biopolymers easily hydrolyzed into small water-soluble monomers are rapidly recycled already in the water column or early during diagenesis. Examples are proteins and carbohydrates and their monomeric forms, sugars and amino acids, respectively. Hydrophobic lipids have a significantly higher preservation potential, which increases with increasing molecular weight. Reaction of organic matter with reduced sulfur inorganic sulfur species (e.g. hydrogen sulfide) formed during bacterial sulfate reduction, further enhances the preservation potential by the formation of high-molecular-weight organic material.

Answer 4

Biological markers are organic molecules which preserve their carbon skeleton after the decay of organisms and through the processes of sedimentation, diagenesis and eventually catagenesis in a way that a correlation with a (group of) precursor organism(s) is still possible. This relationship may reveal important information of the environmental conditions at the time of sediment deposition. Cyclic lipids like steroids and penta-cyclic triterpenoids have proven particularly useful as biological markers or molecular fossils.

Answer 5

A key question to be answered by organic geochemists is that of the marine or terrigenous origin of organic matter in sediments and the extent of mixing of the two components, respectively. A number of bulk and molecular organic geochemical parameters are available to address this question, including elemental analysis, carbon stable isotope analysis, organic petrography and specific biological markers for marine and terrestrial organic matter. Sea surface temperatures can be reconstructed by the analysis of the degree of unsaturation of long-chain alkenones from haptophytes preserved in sediments ($U^{K'}_{37}$ index) or by the determination of the number of cyclo-pentane rings in fossilized Crenarchaeotal membrane lipids (TEX_{86} index). Information on the sedimentary methane cycle can be obtained by the combined analysis of specific membrane lipids of the organisms involved in this process and by the determination of the carbon stable isotope signal of lipids from methanotrophic microorganisms. Subtle changes in molecular structure and their sterical orientation reveals information of the progress of diagenesis and catagenesis.

Chapter 5

Answer 1

A simplified calculation of diffusion distance, $L = (2Dt)^{0.5}$ (Eq. 5.7), is made here for chloride: The diffusion coefficient for Cl^- at deep-sea temperature (4°C) is $D = 11 \cdot 10^{-10}$ m^2 s^{-1} (see Chapter 3, Table 3.1); the time is 10,000 years = $3.2 \cdot 10^{11}$ s; thus, $L = (2 \cdot 11 \cdot 10^{-10} \cdot 3.2 \cdot 10^{11})^{0.5} = 26$ m. So, the lowered salinity of the world ocean since the end of the last glaciation can be traced in the pore water profile of chloride down to about 26 m below seafloor.

Answer 2

The following processes are energetically feasible: 2, 3, and 5. Process 2 has, however, not been detected in marine sediments and no microorganisms are known that can carry out the process. Process 3 proceeds chemically whereas process 5 is biologically catalyzed.

Answer 3

Thioploca are chemolithotrophs. If they assimilate CO_2 as the sole carbon source they are chemolithoautotrophs. If they assimilate only acetate they are chemolithoheterotrophs.

Answer 4

Concentrations of dissolved organic compounds are very low in the open ocean. Small cells are relatively more efficient than large cells in taking up these compounds and can therefore have higher biomass-specific metabolic rate.

Answer 5

The organic carbon of the plankton cell must first be buried to a sediment horizon below the penetration depth of sulfate. Here, the particulate organic material must be degraded by hydrolytic exoenzymes. The breakdown products are then fermented by bacteria in several steps to acetate, hydrogen and CO_2, which are ultimately converted to methane by methanogenic archaea.

Answer 6

High sensitivity in such a radiotracer experiment is achieved with a large amount of radioactivity and a long incubation time (both, of course, within certain limits). The efficient separation of the radioactive methane from the sample at the end of the experiment and the sensitive measurement of its radioactivity are also important. The sample size is not important as long as all produced radioactive methane is measured.

Chapter 6

Answer 1

Oxygen is the main electron acceptor during microbial degradation processes of organic substance. The energy yield for organisms when using oxygen is higher than for all other constituents. Products of the other reactions in the sequence of remineralization (e.g. iron or sulfate

reduction) are reduced species, which are partly re-oxidized by or under participation of oxygen. Under this point of view the total benthic consumption of oxygen is composed by direct oxygen assimilation and those secondary sinks. Just because re-oxidation is only one, but in most cases the dominant pathway for reduced species, this simplification is not correct. Of course, this approach of estimation can not be used under reduced benthic condition, which can prevail for example in more or less closed basins (e.g. the Black Sea).

Answer 2

Carbon limited means that not enough reactive organic carbon is available in order to pursue mineralization processes. If there is organic carbon available in excess all potential oxidants will be used up quickly. These two extremes are met on the deep sea plains and in shallow marine environments along the continental margins, respectively. All types of transitional environments are found in-between.

Answer 3

Denitrification is the only biological process that produces free nitrogen. It removes fixed nitrogen compounds and, therefore, exerts a negative feedback on eutrophication making it a crucial process for the preservation of life on earth.

Answer 4

Under anoxic conditions iron oxides are reduced and cannot act as a major carrier of phosphate in the sediments. Hence, phosphate, which is released from organic material in the course of mineralization processes, will mostly diffuse into the bottom water (and further enhance biological productivity), thus, decreasing the C_{org}/P_{org} ratio of the sediments.

Answer 5

The anaerobic oxidation of ammonia can be an important process in the absence of oxygen, if nitrite as an alternative electronacceptor is available. Nitrite does usually not accumulate in high concentrations but may be continuously delivered by bacteria using nitrate as electron acceptor for the oxidation of organic carbon. Thus, anammox can be expected in anaerobic environments, which are rich in ammonia, organic material and nitrate.

Chapter 7

Answer 1

a) 146,880 tons of total Fe; 64,260 tons of highly reactive Fe.
b) 25,704 tons of highly reactive Fe remains in the estuary and the near-coastal zone.
c) atmospheric input.

Answer 2

a) specific surface area; degree of crystallinity; and the standard free energy, ΔG^0 (or respective solubility constant, K)
b) Natural sediments comprise a wide range of iron oxyhydroxides with different dissolution behavior under the given extraction conditions. The extraction result is the fraction of iron which dissolved during the incubation; this fraction may result from the dissolution of various phases which contributed with different quantities.

Answer 3

Bioturbation is an effective transport mechanism to replenish freshly precipitated, highly reactive iron oxyhydroxide from the sediment surface to the zone of iron reduction. Vice versa, reduced iron phases, e.g. FeS, is transported to the oxic / suboxic zone, where it becomes oxidized.

Answer 4

Flux and composition of reactive iron input to surface sediments; bioturbation; competitive iron reduction pathways, e.g. by sulfide.

Answer 5

a) $$DOP = \frac{Fe - pyrite}{Fe - pyrite + Fe - oxide};$$

DOP is the degree of pyritization, a measure for the fraction of reactive iron being bound to pyrite.
b) Fe-pyrite is determined with the Cr(II)/HCl extraction; Fe-oxide is determined with the dithionite-extraction.

Chapter 8

Answer 1

No, it does not really matter. The end products may be similar by the two pathways. This makes it difficult to distinguish them, either from geochemical profiles or from experiments.

Answer 2

The H_2S is oxidized chemically by MnO_2 to elemental sulfur, S^0. The S^0 is then disproportionated by bacteria to H_2S and SO_4^{2-} in the ratio 3:1. The new H_2S formed may again be oxidized by MnO_2 to S^0, and the cycle is repeated until the H_2S is completely oxidized to SO_4^{2-} (Eq. 8.21).

Answer 3

By a relatively thin oxic and suboxic zone, e.g. in organic-rich sediments or in oxygen-deficient environments, relatively more organic material is buried down into the sulfate zone. Bioturbation enhances the burial of relatively fresh organic material into the sulfate reduction zone. At the same time, however, bioturbation and bioirrigation deepen the oxic-suboxic zones and thereby push down the zone of sulfate reduction.

Answer 4

Sulfate reduction is generally most intensive just below the oxic and suboxic zones, yet advective processes such as bioirrigation in the upper part of the sediment counteract a decrease in sulfate concentration. Anaerobic oxidation of methane, in contrast, provides a deep and focused sink for downward diffusing sulfate that tends to control the depth of sulfate penetration, independent of the near-surface activity.

Answer 5

The lock-in of a diagenetic front may be caused by non-steady state diagenesis due to a shift in, e.g. the rate of sediment deposition, the type of sediment or its organic content, the oxygen concentration of the bottom water, or the methane flux from below. As a result, the diagenetic front may be stationary within a sediment horizon for a prolonged period.

Chapter 9

Answer 1

The $CaCO_3$ production on continental shelves is about $24.5 \cdot 10^{12}$ moles per year. The carbonate production of the continental slope areas is only about $6 \cdot 10^{12}$ moles per year, whereas the production of the pelagic ocean is about $60-90 \cdot 10^{12}$ moles per year (cf. Tab. 9.5).

Answer 2

$CaCO_3$ production comprises the total amount produced in the surface waters. Accumulation is the - much smaller - portion which becomes buried in the deeper sediment. The difference between both comprises the dissolution in the water column and in the upper parts of the young sediment, from where dissolved inorganic carbon (DIC) still can be transferred to the bottom water. Thus the sediment below the calcite compensation depth (CCD) is characterized by no calcite accumulation. Below the CCD calcite particles may reach the seafloor and may even be found in the upper few decimeters of the young sediment, but they will be dissolved according to processes of early diagenesis.

Answer 3

At high latitudes the calcite compensation depth (CCD) is at a water depth of about 3000 m, whereas in low latitudes the CCD is at a depth of about 5000 m. The undersaturation with respect to calcite of the deep ocean is caused by cold water, which sinks to the seafloor at high latitudes and flows as deep water current towards the low latitudes. The closer the deep current comes to the low latitudes, the more calcite from the production in surface waters is dissolved in the upper parts of the deep current. Thus the CCD of the deep water current is continuously lowered on the way from high to low latitudes.

Answer 4

The predominant carbonate species in normal sea-water is bicarbonate (HCO_3^-), close to 70 % of the dissolved inorganic carbon (DIC). About 10 % of the total calcium in normal sea-water exists as $CaSO_4$ complex. This complex is rather insignifi-

cant in anoxic pore-water if the sulfate is consumed by anaerobic methane oxidation (AOM).

Answer 5

Oxidation of C_{org} results in a production of dissolved inorganic carbon (DIC) and thus in an undersaturation of the pore water with respect to calcite. If calcite is present in the sediment solid phase, it will be dissolved until saturation is reached. Due to these reactions the concentration of DIC in pore-water is usually higher than in ocean bottom water. This gradient causes a diffusive flux from the sediment to the bottom water. Upwelling areas are characterized by very high C_{org} content of the sediment and high fluxes, deep-sea sediments by low C_{org} content and low fluxes.

Chapter 10

Answer 1

$0.02 \cdot -45$ ‰ (ice caps) $+ 0.98 \cdot 0.0$ ‰ (modern ocean) $= -0.9$ ‰.

Answer 2

$38,000 \, Gt \cdot 0.3$ ‰ $+$ (modern ocean)
    $x \, Gt \cdot -25$ ‰ $=$ (terrestrial carbon)
$(38,000 + x) \, Gt \cdot 0.0$ ‰   (glacial ocean)
Therefore: 456 Gt, which is about 20% of the today's terrestrial carbon pool. Other explanations would include the change in ocean pH (LGM: +0.2) and the resulting influence on $\delta^{13}C$ values in foraminifers (Spero et al. 1997) or a change in the ratio of inorganic to organic carbon contributed to the sediments (Derry and France-Lanord 1996).

Answer 3

Using equations (10.9) and (10.13), the effect of the b-value on pCO$_2$ would range between 192 and 384 µatm, which underlines the importance of physiological processes for paleo-PCO$_2$ determinations.

Answer 4

$\delta^{15}N = 3$‰ $\Rightarrow f = 0.50$

higher nitrate availability during the upwelling season, less nitrate utilization.

$\delta^{15}N = 8‰ => f = 0.07$

quick depletion of nitrate, low productivity.

Answer 5

The glacial pH was about 0.2 units higher. Hence, the carbonate ion concentration was higher and the atmospheric pCO2 was lower during the last glacial, in accordance with ice core observations.

Chapter 11

Answer 1

To be meaningful, an E_H, pH diagram must reflect the environmental conditions of the system under consideration. The E_H, pH diagram of Mn presented in Fig. 11.2 was calculated for the following conditions pertaining to deep seawater from the Angola Basin; temperature 2°C, density (1.023 g cm^{-3}), Ca 427 ppm, Mg 1,284 ppm, Na 10,760 ppm, K 390 ppm, Cl 19,350 ppm, Mn 0.012 ppb and Fe 0.021 ppb (Glasby and Schulz 1999). These data give a realistic value for the concentration of Mn in deep seawater and permit calculation of the degree of complexation of dissolved Mn species with anions in solution. As such, this diagram reflects as accurately as possible the environmental conditions in which deep-sea manganese nodules form in this region. By contrast, the E_H-pH diagram of Mn taken from Garrels and Christ (1965) is plotted for a temperature of 25°C at a pressure of 1 atmosphere in pure water in which the activity of dissolved Mn is taken to be is 10^{-6}. This, in no way, reflects the conditions under which deep-sea manganese nodules form. An additional problem is that there are no standard free energy data for the fine-grained manganese oxide minerals such as 10 Å manganate and δMnO$_2$ which make up manganese nodules. This is a major limitation in using these diagrams to infer the conditions of formation of marine manganese deposits.

Answer 2

See answer to Problem 1. Again, there are no standard free energy data for the fine-grained iron oxyhydroxide minerals such as akagenéite (β-FeOOH) which make up marine manganese deposits.

Answer 3

a) Diffusion coefficient in seawater D^{SW} for Mn^{2+} at 10°C is 4.46E-10 m^2 s^{-1}. Tortuosity for a sediment with a porosity of $\phi = 0.84$ is $\theta^2 = 1.45$. Diffusion coefficient for Mn^{2+} in pore water under these conditions is D$_{sed}$ = 3.3E-10 m^2 s^{-1}. The steepest gradient in the depth interval between 0.75 and 1.5 cm is 18.85 mol m^{-3} m^{-1}. The upward flux of Mn^{2+} is J$_{sed}$ = 0.84 · 3.3E-10 · 18.85 mol m^{-2} s^{-1} = 5.2E-9 mol m^{-2} s^{-1} or 0.45 mmol m^{-2} d^{-1}.

b) The highest concentrations of Mn^{2+} in pore water were measured at between 4 and 6 cm below the sediment surface. This is therefore the depth at which Mn^{2+} is released.

c) Yes. The gradient between 0.25 and 0.75 cm under sediment surface is 10.54 mol m^{-3} m^{-1} which is only 56 % of the steepest gradient.

Answer 4

The NASC-normalized REE patterns for hydrogenous manganese deposits generally display a positive Ce anomaly which is a function of the oxygen content of the ambient water mass. The Ce/La ratio of these deposits can therefore be used as a redox indicator (Glasby et al. 1987, Kunzendorf et al. 1993, Kasten et al. 1998, Szefer et al. 1998). The NASC-normalized REE patterns for submarine hydrothermal manganese deposits, on the other hand, sometimes display a negative Ce anomaly and a positive Eu anomaly. In principle, submarine hydrothermal manganese crusts should display a negative Ce anomaly typical of submarine hydrothermal fluids and seawater (de Baar et al. 1985). In practice, many hydrothermal crusts contain an additional hydrogenous component characterized by a positive Ce anomaly (Scholten et al 2004). From the observed Ce anomaly, it is possible to calculate the relative contribution of these two components in the crust (Fleet 1983, Scholten et al. 2004, Fig. 12.14). However, this calculation ignores the possible presence of volcaniclastic material which frequently acts as a cementing agent in the crust and which can significantly increase the REE content of the crust (Scholten et al. 2004). A positive Eu anomaly, on the other hand, indicates that the hydrothermal fluid has exceeded 250°C at some stage during its evolution. At this temperature, Eu^{2+} becomes the stable form of Eu in hydrothermal fluids resulting in elevated concentrations of Eu relative to the other REE in submarine hydrothermal manganese crusts (Sverjensky 1984, Michard and Albarède 1986). Because the REE are preferentially adsorbed on Fe oxyhydroxide minerals, it follows that the ΣREE contents of marine manganese deposits are a function of the Fe content of the deposit.

For the data presented here, the Ce/La ratios of the samples lie in the sequence SW Pacific nodule (5.7) > Co-rich Mn crust (4.5) > C-C F.Z. (3.5) > Peru Basin (2.0). The SW Pacific nodule therefore formed under the most oxygenated conditions and the Peru Basin under the least oxygenated conditions. This reflects the strength of the ocean bottom currents and the oxygen content of the bottom waters at these sites. The Vani Mn deposit is a fossil stratiform hydrothermal deposit from Milos island, Greece and the submarine hydrothermal crust both display negligible Ce anomalies which is surprising in view of the fact that they are clearly submarine hydrothermal deposits which would be expected to have pronounce negative Ce anomalies (Fleet 1983). However, the Baltic Sea concretion displays a slight negative Ce anomaly reflecting the less oxidizing conditions prevalent during their formation (Szefer et al. 1998). The Eu anomaly, on the other hand, is marked for the Vani Mn deposit, minor for the submarine hydrothermal crust and slight for the Baltic Sea concretion. These data indicate that the Vani Mn deposit formed from hydrothermal fluids which had been well above 250°C at some stage and the submarine during their evolution. The slightly positive Eu anomaly for the Baltic Sea concretion probably reflects the incorporation of plagioclase into the concretion during its growth.

Answer 5

A growth rate of 0.8 mm Ma^{-1} of Co-rich Mn crusts makes them of the slowest growing minerals on Earth. This low growth rate for hydrogenous deep-sea manganese deposits is a reflection of the extremely low concentration on Mn in deep-sea water (~0.01 ppb, Fig. 11.1) and the slow rate of removal of Mn from seawater. Growth rates of manganese nodules formed on a red clay substrate have been estimated to be 1-2 mm Ma^{-1}, of nodules formed on a siliceous ooze substrate 3-8 mm Ma^{-1} and on nodules formed on a hemipelagic clay substrate 20-50 mm Ma^{-1} (Huh and Ku

1984). Growth rates of nodules formed as a result of oxic (siliceous ooze) and suboxic (hemipelagic clay) diagenesis are higher than those of hydrogenous nodules (red clay) because of an additional contribution from Mn derived remobilization of Mn from the sediment pore water (Sawlan and Murray 1983, Dymond et al. 1984, Müller et al 1988). This frequently results in higher growth rates on the underside of the nodules than on the upper surface (Finney et al. 1984). By contrast, very much higher growth rates are encountered for ferromanganese concretions from the Baltic Sea and from submarine hydrothermal manganese crusts. In the case of ferromanganese concretions from Mecklenburg Bay, this reflects the enhanced migration of Mn from below the redoxocline to shallower, more oxidizing environments during the spring and summer in this part of the Baltic (Liebetrau et al. 2002). In the case of submarine hydrothermal manganese crusts, reducing hydrothermal fluids migrate upwards through the volcanic sands to form a manganese-cemented hardpan within the upper several cm of the sea floor (Usui and Nishimura 1992). Dense submetallic layers of pure manganate minerals can then develop beneath this hardpan to form the submarine hydrothermal manganese crust which acts as a trap to ascending hydrothermal fluids. Several authors have proposed that these crusts grow downwards on the underside of the crusts. This conclusion is supported by the younger ages of the underside of submarine hydrothermal manganese crusts from the Izu-Bonin Arc (0-3,000 years) compared to those from the upper surfaces (40,000-50,000 years; Usui and Nishimura 1992). The rapid formation of these crusts is facilitated by the high concentrations of Mn in submarine hydro-thermal fluids which range from 0.5 to 390 ppm (German and Von Damm 2004).

Chapter 12

Answer 1

ω = 2.7 cm kyr^{-1}, burial rate $C_{org.}$ = 0.0125 mg $cm^{-2} yr^{-1}$, degree of organic carbon preservation = 12.5 %.

Answer 2

Sediment traps cannot resolve the influence of lateral particle transport processes. At best, these values give some indication of the mass accumulation at the sea floor.

Answer 3

Without the justification from variogram analysis that a statistically verifiable regional structure exists within the database, any result of regionalization by kriging will be extremely questionable. The pure kriging method is only a weighting process. The necessary boundary conditions, like the range of a structure, can only be determined by geostatistical methods.

Answer 4

Coupled benthic pelagic models are useful to apply, because they enable a better understanding of the interaction of global biogeochemical cycles. For example, the impact of variations of mineralization processes in the sediment can be investigated straightforward in terms of potential changes in ocean water chemistry and hence, climate feedbacks. On the other hand, their degree of complexity and spatial resolution are rather limited to date and hence, predictions are not very precise when compared to data-based regionalization approaches.

Chapter 13

Answer 1

A 350°C end-member fluid would have a Mg concentratrion of 0 mmol kg^{-1}, and seawater has a Mg concentration of 52 mmol kg^{-1} (Fig. 13.9). Therefore, the 250°C fluid, with a Mg concentration of 15 mmol kg^{-1}, contains about 71% of the high-temperature end-member and about 29% of mixed seawater. If the 250°C fluid contains 100 ppm Fe, then the high-temperature end-member must contain about 140 ppm Fe (0.71 x 140 ppm = 100 ppm).

Answer 2

A concentration of 150 ppm Fe is equivalent to 150 mg of Fe per kilogram of fluid. A black smoker with a flux of 1 kg s^{-1} is discharging $3.2 \cdot 10^7$ kg of hydrothermal fluid every year containing $4.8 \cdot 10^9$ mg of Fe (150 mg kg^{-1} Fe $\cdot 3.2 \cdot 10^7$ kg = $4.8 \cdot 10^9$ mg

Fe or 4.8 t Fe). One black smoker every 50 m along the 55,000 km of mid-ocean ridge is equivalent to 1,100,000 black smoker vents. This number of vents along the world's mid-ocean ridges could deliver more than 5 million tonnes of Fe to the seafloor every year. (Note: this calculation neglects any component of diffuse flow along the ridges. Less than 10% of the fluids vented at the ridge axis are discharged at black smoker temperatures. Therefore the total number of black smoker vents is likely much smaller. For a more complete analysis of metal fluxes at ridge crests, see Elderfield and Schultz 1996).

Answer 3

A sulfide deposit with a diameter of 100 m and a thickness of 25 m will contain a total volume of sulfide equivalent to 196,250 m^3. For a bulk density of 3 g/cm^3, the total amount of sulfide present is 588,750 tonnes. The deposit therefore contains 11,775 tonnes of Cu metal (2 wt.%), 29,437 tonnes of Zn (5 wt.%), and 88,312 tonnes of Fe (15 wt.%). A vent complex with 100 black smokers (each discharging 4.8 tonnes Fe per year) and a depositional efficiency of 1% could deposit this much Fe in 18,400 years [88,312 tonnes Fe per (4.8 tonnes/year · 100 vents · 1/100 efficiency) = 18,400 years].

Answer 4

The pressure at 1000 m depth is 100 bars (10 m = 1 bar). The maximum temperature of venting is determined by the pressure on the two-phase curve for seawater (Fig. 13.8). At 1000 m or 100 bars, a hydrothermal fluid venting at the seafloor cannot be hotter than about 310°C. At 500 m depth or 50 bars, the maximum temperature will be about 270°C.

Answer 5

Many reactions can be written, but the majority of methane and carbon dioxide are produced in marine sediments by the oxidation of buried organic matter, following reactions of the type: $C_{org} + 2H_2O = CO_2 + 2H_2$ and $CO_2 + 4H_2 = CH_4 + 2H_2O$. At 5°C, solid methane hydrates will be stable at a pressure of about 45 bars (45 atm, Fig. 13.14), which is equivalent to a depth of 450 m. At 1°C, the hydrate will form at about 300 m depth.

Chapter 14

Answer 1

No! Defined by its stability field the potential volume for gas hydrate is much larger in the deep sea than on the upper slope; however, the availability of sufficient methane is a critical requirement for gas hydrate formation. Due to high plankton productivity continental slopes have, in general, higher organic carbon concentrations than deep sea areas. Higher amounts of methane and gas hydrate are therefore restricted to continental margins.

Answer 2

In principle there is no difference which methane molecule is used to stabilize the clathrate structure. Hydrates formed from biogenic methane are thought to be more widespread in marine sediments, and these can be formed from methane gas produced in situ. In contrast, the presence of thermogenic methane requires gas migration form deeper sediment sources (e.g. in mud volcanoes). Thermogenic methane is also accompanied by the presence of higher hydrocarbons, which may result in the formation of gas hydrate structure II, whereas methane hydrate from biogenic methane is restricted to structure I.

Answer 3

A methane concentration of 300 mM is clearly above methane saturation and in a depth of 90 mbsf well in the stability for methane hydrate. It is obvious that the same concentration of methane in 250 mbsf is outside the stability field where it occurs as free methane gas. A methane concentration of 10 mM is clearly below the methane solubility at both 90 and 250 mbsf, which means that methane hydrate cannot be formed and the methane occurs as a dissolved phase in the pore water.

Answer 4

The chloride concentration is typical for an open marine environment, suggesting no dilution by gas hydrate dissociation. More important, the high sulfate concentration indicates that there is no methane in the sediment. Even if methane were migrating to this region, it would be consumed by anaerobic methane oxidation (AOM).

Answer 5

Around 10 % of the pore space are filled with gas hydrate.

Answer 6

Since chlorides and other ions from marine pore water cannot be incorporated into the clathrate structure, those constituents are enriched in residual pore fluids when hydrate is formed in marine sediments. When hydrate formation is faster than the rate of salt removal by diffusion and advection, the excess salt would accumulate in the pore water resulting in the formation of a brine.

Chapter 15

Answer 1

The pH-value is 7.6.

Answer 2

The $\log pCO_2$ is -4.1. The lower partial pressure of CO_2 in ocean surface water is usually caused by algal photosynthesis.

Answer 3

A steady state penetration depth is expected at about 12.5 cm.

Answer 4

Sulfate and methane have different diffusion coefficients. Methane diffusion is about 50 % faster. Different porosities at different depths of a sediment profile are possible.

Index